Bio#Futures

Emmanuel Koukios • Anna Sacio-Szymańska
Editors

Bio#Futures

Foreseeing and Exploring the Bioeconomy

Springer

Editors
Emmanuel Koukios
Research Group BIOTOPOS
Organic Technologies Laboratory
National Technical University of Athens
Athens, Greece

Anna Sacio-Szymańska
4CF Strategic Foresight Company
Warsaw, Poland

ISBN 978-3-030-64968-5 ISBN 978-3-030-64969-2 (eBook)
https://doi.org/10.1007/978-3-030-64969-2

This Springer imprint is published by the registered company Springer Nature Switzerland AG
The registered company address is: Gewerbestrasse 11, 6330 Cham, Switzerland

Foreword

Turning Ideas to Visions, and Visions to Actions

The world is facing tremendous challenges. This is nothing new. The Club of Rome already described some 48 years ago in its epochal report "*Limits to Growth?*" that this unbridled desire to increase production and consumption will create pollution and poverty beyond control. Whereas the Europe 2020 strategy recommended bio-economy as a key element for smart and green growth in Europe, it is clear that even the most ambitious programs that are currently designed are insufficient to turn the present crisis around.

We do not need to improve performance, nor do we have to double our efforts to reduce our impact. Instead, we have to bluntly shy away from the rhetoric and public declarations and pass on to action. It has been mentioned that all key innovations required to turn the present production and consumption model around were already invented before the Second World War. I agree. True innovations have been invented by Nature. Perhaps it is time to embrace a bioeconomy where we strive to be as intelligent as Nature.

This requires a fundamental shift in the business model. We cannot pursue the quest for evermore efficiency, through economies of scale and standardization, blended with a blind belief that free trade will make the difference for citizens around the world. We know that time has come to replace the fetish of efficiency with a strong commitment to resilience. Our management of natural resources has to embrace a simple principle: we cannot expect Nature to produce more but instead we must do more with what Nature produces.

This implies that we will first and foremost submit ourselves to the most basic laws governing matter and energy: the laws of physics. The basket of opportunities offered by the laws of physics should be used and depleted before envisioning any sort of involvement in chemical engineering or genetic manipulation. This innovative approach teaches us that forest regeneration relies on a profound understanding of the difference between the temperature of the soil and the rain, and that the management of resulting dew not only brings forests back but this can also provide the ability to tomato farms in becoming nothing less than net producers of drinking water. This is the kind of innovation we are looking for: moving from a situation of aquifer depletion for the production globally traded tomatoes to farming with locally grown tomatoes that generate more water than needed.

The core shift the book *Bio#Future*s highlights – which also happens to be my main interest – is the design of not only new technologies, rather the development and implementation of new business models that respond to the local needs of everyone, implying that we take care of all stakeholders, the communities on which we depend, and the ecosystems we value. If we are not prepared to question the core business model, which builds on core competences, then we are not getting anywhere close to Nature's time-proven promotion of life. Imagine: a business model that promotes life!

There are many cases that demonstrate this; the bioeconomy is not only able to offer a fresh look at how society could respond to all basic requirement for water, nutrition, health, housing, and energy, but also generate so much more value that it also secures jobs. Case studies are key to understand how this actually works. The communities of Las Gaviotas in Colombia, The Songhai Centre in Benin, Montfort Boys Town in Fiji, El Hierro in Spain, and Rumpan in Sweden have all demonstrated clearly that it is possible to convert an idea into a vision through the combination of what the best of science has to offer with the deployment of the best minds in business. Those visions were turned into reality precisely thanks to that.

The *Bio#Futures* book takes this to the level of the skills and tools required as well as the methodologies applied. However, we have to submit to the readers the need to apply precautionary principles. Any innovation can have unintended consequences that could never have been foreseen. The spraying of glyphosate applied to corn and sugar cane as biofuels is one unintended error that clearly affects the health of a whole population. Illustrations of that are not scarce: the use of palm oil has led to the destruction of the habitat of the orangutan. The demand for shiitake mushrooms grown on oaks has decimated oak forests in some countries.

So, in all our enthusiasm to move forward, we have to maintain a clear objective: the Common Good. And it is with this purpose in mind that I fully subscribe to this exercise of Prof. Emmanuel Koukios and Dr. Anna Sacio-Szymańska to put the best insights that are available today together in this book.

Prof. Gunter Pauli, Dr. h.c.

Contents

Introduction

Bio#Futures: An Overview

About 10,000 years ago, humans began
to domesticate plants and animals.
Now it's time to domesticate molecules.
– *Susan Lindquist*, MIT Professor (1949–2016)[1]

Scope and Objectives of the Bio#Futures *Mission*

In order to cope with an increasing global population, rapid depletion of vital resources, growing environmental pressures, and climate change, world economies and societies need to radically change their approach to production, consumption, processing, storage, recycling, and disposal of biological resources. Thus, the *Europe 2020* and other similar strategies have recommended *bioeconomy* as a key element for smart and green growth in Europe and other parts of the world. Advancements in bioeconomy research and innovation uptake will allow Europe and the world to improve their management of natural resources and open new and diversified markets for food and bio-based applications.

According to an increasing number of indications, a new mega-wave of socio-technical change, expected to peak after 2030, will include the emergence of bio-economy, that is, the whole spectrum of applications of biological sciences and technologies in all socio-economic areas and sectors, radically transforming our societies and the world as we know it. The resulting action task and mission will be to foresee, map, and assess the emergence of this great wave we call "*bio-tsunami*,"

[1] Comment by the late MIT Professor of Biology Susan Lindquist, then director of the Whitehead Institute for Biomedical Research, at a 2003 Conference in Crete on the subject of self-assembling peptides and proteins
https://news.mit.edu/2003/peptides-science (accessed on 27 September 2020).

thus preparing societies to harness its huge potential for health, work, prosperity, and quality of life, while avoiding getting crashed by its risks and side-effects. The expected research findings will also include the critical technological breakthroughs affecting the bio-tsunami's emergence and the socio-economic clusters describing its main pathways of social change.

Therefore, coming up with innovative roadmaps for implementing sustainable bioeconomy requires an innovative approach. The successful roadmaps will depend upon new agricultural practices, new industrial technologies, new business models, new social practices (such as sharing and circular economies), and new skill profiles. This complex task requires a sense of urgency to move forward timely and mobilize human and other key resources of this procedure, that is, a true Copernican Revolution. A *forward-looking* approach is therefore needed.

The existing literature contains analyses of some of the key elements of bioeconomy; they are usually considered independently from each other and typically approached from unilateral perspectives, technical, economic, or social. What is missing from such a fragmented and fast-depreciated knowledge capital is a *systems analysis*–based discussion, coupled with a long-term, futures-thinking orientation; this will ensure the required impact and sustainability of the new and emerging bioeconomy-based products, processes, services, and attitudes.

The Four Pillars of the **Bio#Futures** *Mission*

The aim of this book is to bring together analyses, case studies, and experiences from all parts of the world that describe effective foresight exercises and other forward-looking actions oriented to map and assess the emergence of bioeconomy. The resulting synthesis will be of value in initiating stakeholders' dialogues, building visions, and developing scenarios responding to global bioeconomy-induced systemic challenges. This will make possible a widening of the menu of potential bioeconomy-based actions, strategies, and policies to ensure that future generations are not deprived of means (be them technological or other), alternatives, and choices that would guarantee them an expected sustainable quality of life.

The primary audiences for this book include all professionals working in leadership and other decision-making positions in business, policy, and research, especially those who seek innovative yet feasibly implementable and financially viable approaches in tackling complex problems; other recipients include local and regional non-profit organizations and activists and also students, pupils and trainees who wish to positively influence the future of planet Earth. The so targeted *Bio#Futures* mission is built on the following four Action Pillars:

- *PILLAR I: Learning from Case Studies.* They will describe regional, national, or other stories and bioeconomy experiences, based on the use of foresight and other forward-oriented approaches. Potential impacts might include: (i) initiating stakeholders' dialogue and building visions and scenarios responding to global bioeconomy-induced systemic challenges; (ii) establishment of actions, strategies, or policies to satisfy long-term technological, environmental, economic, societal, etc., development needs related to bioeconomy uptake; and (iii) establishment of actions, strategies, or policies, assuring that the successful futures-oriented practices regarding bioeconomy will be sustained.

- *PILLAR II: Mapping the Bioeconomy Geography.* In their ongoing joint research[2], the present book's co-editors have identified the following 10 areas of bio-based emergence, determined by a Horizon Scanning exercise: (1) ageing, (2) health care and cure, (3) food and nutrition, (4) human re-engineering, (5) the Green Economy, (6) management and governance, (7) new risks and barriers, (8) research and innovation drivers, (9) skills and education, and (10) mystery and disrupting factors.
- *PILLAR III: Assessing the Bioeconomy Toolbox.* Composed of bio-resources; bio-process technologies; biochemical engineering – cultivating molecules, cells, and tissues; bioenergy and biofuels – shift away from fossil fuels; biomaterials; biochemicals and other bio-products; biochar and soil organic carbon; bio-wastes – symbiotic economy; bio-remediation; bio-medical applications – artificial organs; bio-agricultural aspects – organic, biological, and alterative farming; bio-pharmaceuticals; pre/pro-biotics; bio-informatics; bio-info-nano convergence; biorefineries; metabolic engineering; bio-systems; bio-communities; bio-skills; bio-education; bio-politics; and bio-ethical aspects.
- *PILLAR IV: Relevant Foresight Elements.* Methodologies that can be employed include scenarios, environmental scanning, expert panels, road-mapping, simulation gaming, brainstorming, storytelling, and experiential futures; visions and scenarios including drivers, challenges, milestones, obstacles, and key actors; futures policy issues including impact, security, economic growth, and (tech) innovation; futures strategy issues including new roles for research entities, enterprises, educational institutions, civil society organizations, and their impact on specific indicators (employment, finance); context-related issues including global challenges (SDGs), trends in research (RRI), and workforce.

Process and Product of the **Bio#Futures** *Mission*

Potential chapter authors were identified and invited by the book co-editors in late 2019 and early 2020. Following a period of discussions, the book structure has been finally put together. Drafting the book chapters took place in the period after April 2020, under the guidance of the book editors.

Due to the outbreak of the COVID-19 pandemic in the same period, the deadline for the submission of chapters by their authors had to be extended to September 2020. Despite the delays, this bio-based crisis offered the opportunity to chapter authors to reflect on the situation and even devote parts of their work (up to a whole chapter) to that.

[2] For more details, please see Chap. 3 of this volume.

The resulting volume consists of 27 chapters, presented in 8 parts. These are authored by 70 persons (60% female, 40% male) from 35 groups and institutions, from 17 countries in 4 continents (in alphabetical order): Belgium, Brazil, Denmark, Egypt, Finland, Czech Republic, Germany, Greece, India, Italy, Lithuania, Netherlands, Poland, Russia, Sweden, the United Kingdom, and the United States. In particular,

- *Part I* aims at setting the scene for the *Bio#Futures* mission, as defined above, with the help of three chapters addressing foresight topics and issues applied to bioeconomy.
- The three chapters in *Part II* focus on key aspects of the critical transformation of economies and societies to circular systems and the role of bioeconomy in such transition.
- *Part III* presents recent findings from the *BioEcoJust* project; its 3 chapters contribute to the development of an appropriate conceptual frame to evaluate *Bio#Futures* options.
- The 6 chapters grouped as *Part IV* cast light on two essential clusters of the emerging bioeconomy, that is, the agro-food and the healthcare one, as well as their vital interface.
- The three chapters in *Part V* address topics related to the role of bioeconomy in enhancing sustainability strategies and responding to complex environmental challenges.
- Biofuels and other innovative bioenergies are expected to play a key role in the transition from a fossil-carbon based economy, as shown by the three chapters in *Part VI*.
- The three chapters of *Part VII* propose alternative approaches to some key socio-cultural aspects/effects of the emerging bioeconomy: education, search for meaning, and ideology.
- The three chapters of the closing *Part VIII* elaborate on factors expected to play critical roles as driving forces and barriers, that is, global trends, responsible innovation, and bioethics.

Finally, we, the book editors would like to express our sincere thanks to the following, who have contributed to the success of the *Bio#Futures* project: the publisher's ever-helpful staff; the project heroes, authors of the 27 chapters, who worked for several months within a pandemic crisis; and Dr. Gunter Pauli, who accepted our invitation to write the book's foreword. The first editor is also grateful to the members of his Research Team *BIOTOPOS* for long and productive discussions and constant encouragement.[3]

Athens, Greece — Emmanuel Koukios
Warsaw, Poland — Anna Sacio-Szymańska

---

[3]Irene Daouti-Koukiou, Lazaros Karaoglanoglou, Rea-Clarissa Koukiou, Dimitrios Koullas, Nikolaos Kourakos, Sofia Papadaki.

Part I
Setting the Scene for Bio#Futures

Chapter 1
Bioeconomy as a Driver for the Upcoming Seventh K-Wave (2050–2100)

Markku Wilenius

Abstract This chapter sets out to project the future role of bioeconomy in our socio-economic structure in the light of the so-called Kondratieff long-wave theory. This framework presupposes 40–60 year-long economic cycles, and according to the author's interpretation of the theory, we are now embarking into the sixth wave (2010–2050), followed by the seventh wave 2050–2100. This chapter sets out to understand the development of the waves in the light of the changing relationship of humans with nature. The assumption is that as we move towards mid-century, our relationship with nature will turn into a more collaborative form, away from the extractive practice of current economies. This chapter explores four potential domains for bioeconomy development in this respect: (1) agriculture, turning to more regenerative and becoming a part of the solution to climate problem; (2) forestry, expanding its role as a source of new materials, medicine and well-being; (3) algae production, becoming an essential source of energy and new materials production; and (4) biomimicry, being extensively deployed as a design principle for emerging technologies. All this potential development will signify exponential growth of the impact of bioeconomy for our societies to come.

Keywords Long-wave theory · Seventh Kondratieff wave · Bioeconomy futures · Human relationship with nature

1 Introduction: The Corona Shock and the Fragility of the Planet

As I am writing this chapter, a coronavirus called COVID-19 is ravaging the world economic system, threatening to bring it down. This tiny biological entity is setting the public agenda and determining the actions of societies around the globe. What kind of biofuture does this virus mean for us? What does the virus outbreak tell us about our

---

M. Wilenius (✉)
Professor of Futures Studies and UNESCO Chair of Learning for Transformation and Planetary Futures, Finland Futures Research Centre, University of Turku, Turku, Finland
e-mail: markku.wilenius@utu.fi

E. Koukios, A. Sacio-Szymańska (eds.), *Bio#Futures*,
https://doi.org/10.1007/978-3-030-64969-2_1

society as a whole? How could societies prepare better for these kinds of unexpected events? What do these events imply in terms of our relationship with nature?

The COVID-19 pandemic has really brought home the utter fragility of our world and our planet. The economic and social havoc wreaked by the crisis goes far beyond anything our global system has experienced in times of peace. There are three specific reasons why this has happened. According to our analysis, the first reason has to do with the nature of our current capitalistic economic model. During the last 40 years, as Jacobs and Mazzucato and others have noted, our economic system in the western world has consistently given precedence to short-term profits over long-term benefits (Jabobs and Mazzucato 2016). There seems to be a strong bias to ramp up income inequality, as clearly shown by Thomas Piketty. In contrast to most of the twentieth century when the trickle-down effect still created a rather equal distribution of wealth among the socio-economic classes, there has been a sharp turnaround from the early 1980s as the uppermost quadrant of earners are now relatively much better off (Piketty 2014).

While social stratification has intensified over the last 40 years, it has also had a natural counterpart. Our natural habitats have become much more fragile, and the ecosystem has lost much of its resilience. The level of biodiversity, one of the ultimate indicators of planetary development, has declined dramatically. According to the WWF Living Planet Index, the population size of vertebrates[1] has declined by 60% from 1970 to 2014. Moreover, measurements show that our ecological footprint – a measure of our consumption of natural resources – has risen by a staggering 190% in the past 50 years.[2] Additionally, recent FAO calculations indicate that we have lost 420 million hectares of forest since 1990. The main cause of deforestation is large-scale commercial agriculture (the cultivation of soya beans and palm oil), accompanied by local subsistence agriculture.[3] Also, our most critical asset of all, the top soil of agricultural lands, is being rapidly depleted. According to FAO estimates, we can look forward, on average, to some 60 years of harvests before the top soil is degraded beyond use.[4] Without the top soil, the capacity of land to filter water, absorb carbon and ultimately produce healthy and nutritious food will be severely limited. All this is due to modern agriculture that relies on intensive tilling and pesticide and fertilizer use.

It is clear, then, that we have created a global economy that is not only socially unjust but is also causing a major ecosystem catastrophe that further amplifies the real pandemic threat to the world: runaway climate change. The Intergovernmental Panel on Climate Change's (IPCC) landmark 1.5 °C report stated convincingly that the current measures taken to avoid breaching the critical 1.5 °C temperature threshold are inadequate. On the other hand, it is known that once the planet warms up by

[1] https://s3.amazonaws.com/wwfassets/downloads/lpr2018_summary_report_spreads.pdf

[2] https://www.footprintnetwork.org/resources/data/

[3] http://www.fao.org/state-of-forests/2020/en/

[4] http://www.fao.org/news/story/en/item/357059/icode/

2°, the world system will enter a whole new stage, a path of extreme events from which there is no return.[5]

Our planetary ecosystem, it seems, is being stripped of its protective, regenerative capacity. But the same has happened to our social systems. Norms, regulations, redundancies, human-made and natural buffers, a whole host of checks and balances that used to be there to ensure the necessary rejuvenation of the system – all this has been brushed aside to give way to short-term profits. As sociologist Ulrich Beck famously stated in his analysis of risk society, these are the globe-wide impacts no one intended (Beck 1992). These unintentional consequences of human action have depleted the resilience of human-built systems, at the same time as we have destroyed the resilience of natural systems.

To oversimplify somewhat, societies are generally speaking run by their intentions. Are these intentions currently leading our systems towards greater resilience and towards building up a capacity to transform, the two critical qualities of any complex system? Let's consider this in the context of the World Economic Forum's Global Competitiveness Report,[6] which ranks the countries of the world on an index that is said to measure the capacity of countries to provide prosperity for their citizens.

In practice, the index consists of 103 indicators. These indicators, however, say very little about things like decarbonization, new energy availability, environmental health or resource security. In other words, the measure of competitiveness in the index effectively excludes consideration of resilience and the capacity of agility and transformation. It ignores the whole idea of long-term sustainability. Current responses to the COVID-19 crisis, for instance, have made clear the failure of the US public health care system to tackle and contain the coronavirus – and yet the United States ranks second in the global competitiveness index. Indeed as we move forward, the increasing complexities and uncertainties faced in future societies will give added weight and importance to two major requisites that human systemns are facing the capacity to be resilient and the capacity to transform.

2 The Long-Term View to the Future

In this chapter, we attempt to project the long-term future based on the K-wave theory of societal development (see Wilenius 2017). We apply a theory of long waves in order to try to understand the evolution of societies from one phase to another as they respond to unfolding challenges. We postulate that bioeconomy will have an increasingly prominent role as societies seek to bring greater well-being to their citizens while struggling to overcome the problems they have accumulated over time. The K-wave theory of societal change posits that the technologies, modes of organization and values that will define a wave are already burgeoning and identifiable from the very start of the previous wave. In other words, what we see now

[5] https://www.ipcc.ch/sr15/

[6] http://www3.weforum.org/docs/WEF_TheGlobalCompetitivenessReport2019.pdf

emerging, even if it still marginal, can potentially grow and evolve into something predominant much later – in our case, in the latter part of this century.

These waves are described and explained in more depth later in the chapter. In the meantime, let us briefly introduce the waves that are now unfolding. Since the birth of industrialization, we can count five waves that have occurred in cycles of 40–60 years. We have currently moved into the sixth wave.

2.1 Sixth Wave (2010–2050)

Having started around 2010, the sixth wave will see the focus of technology development shift towards resource efficiency, the building of a new energy system and a transition to a more digital and circular economy, showing the first real signs of a post-industrial model of material economy. Post-material values and systems thinking, key ideas in circular society, will slowly come to prevail in the cultural domain, marking a departure from the material values derived from previous phases. This shift is well manifested by long-term data from the World Values Survey.[7] In the sixth wave, both the growth rate of the traditional economy and resource use will be challenged as the convergence of world economies continues and large emerging economies decelerate (Guillemette and Turner 2018). Around 2050, climate emissions will overshoot the 1.5 °C limit, as anticipated in the IPCC's business-as-usual scenario (IPPC 2018). This will lead to increasingly dramatic weather phenomena, floods, fires and droughts, extending to areas previously unaffected by such extreme conditions. Furthermore, various irreversible second order changes will make climate patterns increasingly unpredictable.

2.2 Seventh Wave (2050–2100)

Artificial intelligence has become ubiquitous around the world, and together with advanced blockchain systems has helped to create created a new form of global interconnectedness and consciousness. A changing world is moving from climate shocks towards a rebalancing of resource use and resource renewal. Industrial heritage systems have been superseded by biological systems, resulting in a transformation from industrial thinking to full-scale circular economy. Non-renewables are increasingly substituted by biologically manufactured materials, and new resource bases for commercial harvesting have been developed. In the oceans, for example, kelp forests are now exploited for a variety of goods and services. The huge potential of algae to produce food, energy, medicines and a wide array of other products is being fully utilized. People are refining organic materials, and the entire planet

[7] http://www.worldvaluessurvey.org/wvs.jsp

has been turned into a garden that is carefully nurtured and harvested. The massive shocks caused by global warming and changed weather patterns have resulted in a global ban on the use of coal, oil and natural gas for energy production. The bioeconomy has become the new economic paradigm, coupled with advanced concepts of circulation and social equality. As global agricultural practices have shifted towards a more regenerative model and large habitats have been reforested, levels of carbon dioxide in the atmosphere have fallen dramatically. The manipulation of natural systems has pushed many species close to extinction, but massive revitalization efforts have been put in place. New forms of cultivating biomaterials are flourishing everywhere, and work is underway to restore the planet's soil and forests, inspired by the new ethos to save the planet and humans as part of the planet.

These condensed images of the future draw attention to the key dynamic of the waves: each phase of growth is fed by a need arising from the previous wave, but as attempts gather momentum to find solutions, so novel issues unfold that begin increasingly to pressure the system as the wave is nearing its end. This type of schematic approach provides a useful heuristic tool for anticipating very long-range futures. At the same time, it helps us break loose from the constraints and lock-ins that tend to tie our thinking to the present reality.

Let us now go back to exploring how human cultures have evolved over time in terms of their relationship with nature.

3 From Past to Future: The Development of the Human Relationship with Nature

In order to understand the role of organic life in our cultural systems, we really need to go back all the way to the switch from hunter-gatherer to agriculture. The cultivation of land marked a critical shift indeed in our relationship with nature. As agriculture came to dominate our relationship with nature, it laid the ground for a more structured approach to human collaboration. Compared to hunting, agriculture was a much more predictable and organized pattern of behaviour. It also created a very different kind of social structure: it made possible the creation of institutions such as the church and the army, all because our relationship with nature had changed (see also Laloux 2014).

As this change began to unfold some 10,000 years ago, something very different started to emerge as well. Suddenly, societies found themselves in a position where they were able to grow, even if that growth was still very modest compared to the explosion seen in the last century. Communication with the land through cultivation helped societies grow and expand, and early technologies such as ploughing and harrowing ensured a more stable harvest and paved the way to more intensive agriculture – and ultimately to industrial revolution.

The rise of agricultural society brought stability and organization. At the same time, it marked the beginning of the taming and harnessing of nature for human purposes. As steady seasonal crops provided for a regular food supply, the human relationship with nature became more instrumental. But farmers still had to deal with various fluctuations: in some years the weather could be extremely dry; in others there might be a surplus of tormenting rains. All the work in the fields was done by hand, with simple tools. It was imperative to take good care of the soil since no other fertilizers were available except the manure provided by domesticated animals.

The nature of this society that was being built was, according to the model created by futurist Pentti Malaska, a society of basic needs: it attended to the needs that are quintessential for human life on earth, i.e. food, shelter and clothing. Agriculture provided the food; stable settlements provided the shelter. Malaska's model of socio-economic evolution is well-described in this diagram (Malaska 1999):

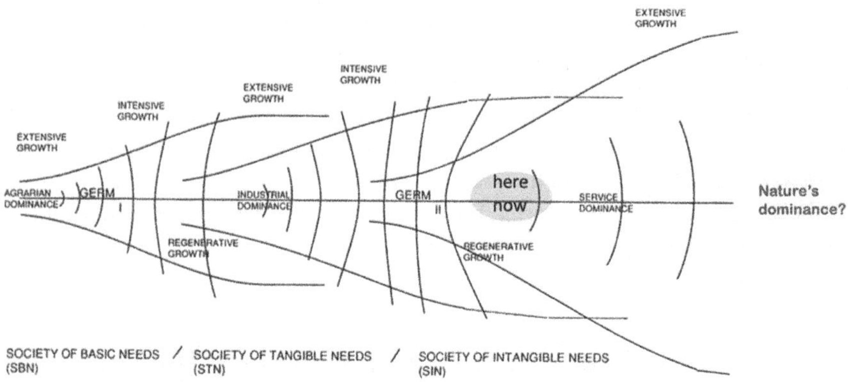

Original by Pentti Malaska (1999), further developed by Markku Wilenius

From its modest origins, the agrarian way of life evolved over the millennia by experiment and experience into the dominant form of culture and a source of livelihood to most people. As we approach the industrial turn, something happens that Malaska calls *autopoietic transformation*. That is, societies begin to transform and turn, in a self-organized manner, onto a totally new trajectory. The emergence of industrial society marked a step towards a redefined order of needs (Wilenius 2014). These new emerging needs can best be described as tangible, entailing everything from household wares to skyscrapers. Meeting these needs was supported by a growing range of technologies, all designed to expand the range of human activities. By the early nineteenth century, the march to modern society was truly underway.

The advent of industrial society heralded a very different relationship with nature. Natural resources, particularly everything that could be drilled and extracted from the earth, became the core engine of industrialization (Bardi 2014). The raw materials were needed to build factories in which products were produced for trading. The relationship with nature thus became very utilitarian. If the farmer used to prepare the land with hardly any external inputs, industrialization suddenly provided access to all manner of inputs. Essentially an extended human arm, technology was there to facilitate the extraction process. Financial markets emerged to make the necessary transactions possible.

Industrial development can be seen and understood as a sequence of phases where humans discover needs that have to be met in order to improve the quality of human life create a decent human life. Humans become increasingly creative in building new machinery and in inventing materials and processes that allow them to break from the way of life that is centred around food, shelter and clothing. Meeting multiple tangible needs becomes the lynchpin of human aspiration. Values become more material and whole new social classes are born. Natural resources and trade become a source of wealth accumulation.

At some point, however, Malaska says, industrial dominance begins to give way to service dominance. While industrial dominance brought in its wake its own logic (hierarchically run factories, mass production, etc.), the new service dominance logic is based on tailoring, assisted by new algorithm-based technologies such as artificial intelligence and the Internet of Things. All these new services are now leading to a society that functions in a fundamentally different way. The key motive in service dominance is efficient resource allocation. This is what all these new technologies are geared to: communicating and distributing data and information in a way that allows systems to adapt to take away the word "new" changing circumstances.

But there is more. Malaska's model posits that with services, we also move from tangibles to intangibles. This is a well-known phenomenon, yet still poorly understood (see Haskel and Westlake 2018). At the core of the phenomenon is a shift in investments towards software, design and branding, for example, at the same time as values become more immaterial. Whether in manufacturing or service production, intangibles have gained increasing prominence, and that in itself has given a whole new tone to our economies. It also seems to be leading to new consumption patterns: given the increasing amount of software at our disposal, our consumption of human relations has increased to unprecedented levels. According to US statistics, young people (16–24 years) today spend on average 3 hours on social media every single day.[8] Through social media, social relationships have become consumables that increasingly permeate our behavioural patterns.

In order to deepen our understanding of the transformation we are now witnessing, it is necessary to look at how industrialization itself has developed over time. There is no better way to do that than to examine the long-term waves of socio-economic development, the so-called Kondratieff waves. The notion of long-term waves was first introduced in the international debate by the Russian economist Nikolai Kondratieff at the beginning of the twentieth century (Kondratieff 1928/1984). The theory postulates that societies develop in cycles of 40–60 years. Each wave begins with a period of growth and ends with a period of decline and depression. In-between each wave there is always a period of crisis that stimulates the birth of a new cycle. Kondratieff himself was a traditional economist who relied

[8] https://review42.com/how-much-time-do-people-spend-on-social-media/

on data describing economic activities, but in his wake economist Joseph Schumpeter gave a somewhat broader interpretation to the theory and incorporated technological cycles into the waves. In this chapter, we offer a much more holistic interpretation of the cycle theory. However, economic data can still be used to illustrate the unfolding of cycles, as is done in the following graph which uses rolling 10-year yields of the Standard & Poor equity index coupled with the notion of technological shifts (Fig. 1.1).

Based on these 40–60 year patterns, it is possible to see how each wave has been driven by the creation of specific technologies and their applications, which in turn have instigated new social behaviours. With the onset of industrialization and the steam engine, the race was on for human invention. The use of steam brought a dramatic increase in labour productivity. The second wave, which brought the age of steel and railroads, contributed to expand the infrastructure and created the first mass transport system. In the third wave, new innovations around electricity use had a tremendous impact on industrial productivity. At the same time, new chemistry came along, paving the way to the production of paper from pulp, for instance. In the fourth wave from the 1930s onwards, automobiles burst onto the scene, accompanied by a sharp rise in the consumption of petrol and other petrochemicals. The last wave, emerging in the early 1970s after the oil crisis, turned out to be a triumphal march for digitalization and the expansion of various communication technologies. That era came to an end with the financial crisis that very nearly brought down the whole global financial system.

Each of these waves has signified brought together human intentions, igniting the creation of new technological fields and aspirations of economic growth with a view to providing material goods and promoting new cultural habits. Industrialization has moulded human life and created ever new ways of harnessing natural resources. All this has created the challenge we are now facing: how to stop the plundering of nature and how to establish a more balanced relationship and interaction with the natural world (Wijkman and Rockström 2011). Indeed, along with rapid industrialization, we have denied our planetary boundaries and brought ourselves to a situa-

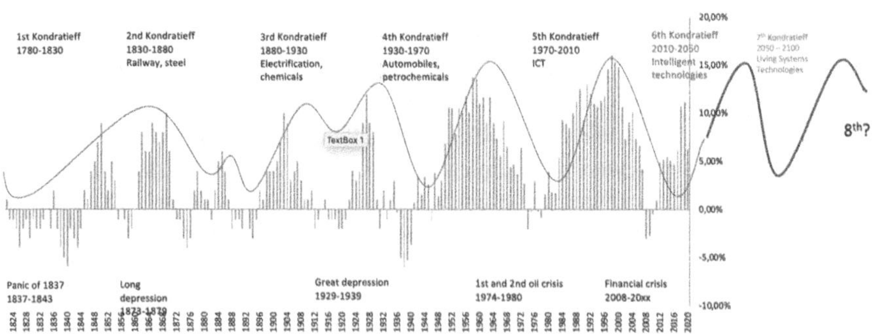

Fig. 1.1 Kondratieff waves. (Rolling 10-year return on the S&P 500 from Jan 1814 to March 2020 (% per year). Source Datastream, Bloomberg, Helsinki Capital partners (illustration), Markku Wilenius)

tion where many ecosystems are being depleted by the intensive extraction of natural resources.

From very early on, the human meaning of industrialization was progress. Its essence was the intention to create a decent human life and a comfortable everyday existence. This was not something that would happen overnight: in early industrialization, conditions in the rising urban developments were miserable. However, modern sewage systems and other new technologies boosted the development of modern urban systems, even though that development was often thwarted by contradictions and challenges. As political scientist Marshall Berman observes, modernization meant empowerment and destruction at one and the same time: it was in the nature of modernism to build cities and industries while tearing down everything representing the old. To be modern was to live in the constant maelstrom of disintegration and renewal in a world where "all that is solid melts into air" (see Berman 1981). As much as human experience has been shaped by modernization, it has also brought new ideas about what the future should look like. The development of industrialization was thus shaped by the human capacity for anticipation.

In the context of the K-waves, my argument is that the anticipatory power of the waves stems from the observation that the technologies adopted cleared the way for the appearance of other technologies. If technology is understood as an extension of human power, then movement in the course of industrialization has meant the process of taming ever new technologies for human good. The following matrix attempts to illustrate the point (Fig. 1.2):

4 The Sixth Wave Revisited

We are currently ploughing ahead at full steam in the sixth wave, which will take us up to the middle of this century (see Wilenius 2017). This wave is characterized by a deepening understanding of the huge inefficiencies and unintended consequences we have been building into our economies. This is particularly true with regard to our energy sources, infrastructure and consumption patterns, all of which suggest some alarming conclusions about our capacity to mitigate climate change. There is overwhelming evidence that this change has already started. Recently NASA reported that average global temperatures over the last 5 years have been the warmest on record, pushing the level of warming beyond 1.1 °C.[9] The 2015 Paris Agreement recommended that warming should not be allowed to exceed 1.5 °C, but signatory countries have been very slow to adopt the agreement as a part of their national policies.[10] The IPCC estimates that hundreds of millions of lives will be at stake if the world exceeds the 1.5 °C warming target and begins to move towards the

[9] https://climate.nasa.gov/news/2945/nasa-noaa-analyses-reveal-2019-second-warmest-year-on-record/

[10] https://www.nationalgeographic.com/science/2019/11/nations-miss-paris-targets-climate-driven-weather-events-cost-billions/

THE SUCCESSION OF DEVELOPMENT WAVES IN INDUSTRIAL SOCIETIES

K-Waves	1st wave	2nd wave	3rd wave	4th wave	5th wave	6th wave
Period	1780–1830	1830–1880	1880–1930	1930–1970	1970–2010	2010–2050
Drivers	Steam Machine	Railroad Steel	Electricity Chemicals	Automobiles, Petrochemicals	Digital communication technologies	Intelligent, resource efficient technologies
Prime field of application	Clothing industry and energy	Transport, infrastructure and cities	Utilities and mass-production	Personal mobility and freight transport	Personal computers and mobile phones	Materials and energy production and distribution
Human interest	New means for decent life	Reaching out and upwards	Building maintenance	Allowing for freedom	Creating new space	Integrating human, nature and technology

Fig. 1.2 Development waves in industrial societies

2 °C level. Beyond 2 °C, the research shows, we will begin to see large chunks of the Antarctic and Arctic ice melt into the sea – something that is in fact already happening.[11,12]

It is estimated that a staggering 70% of all energy production is lost to waste heat.[13] At the same time, we are still hugely dependent on fossil fuels, which account for some 80% of primary energy production.[14] The shift towards cleaner energy has been a painfully slow process, in no small part because of the 5 trillion dollars of subsidies that still go each year to the fossil fuel industry according to IMF estimate.[15] At the same time, it is estimated that under the current baseline scenario regarding the necessary steps to reduce emissions levels, the global economy could benefit up to 26 trillion dollars by 2030 for taking stringent measures.[16]

In other words, it seems that in the current sixth wave, we will inevitably have to focus on addressing the structural inefficiencies inherited from the previous period. The ambitions to achieve such efficiency gains have been there for a long time, yet

[11] https://phys.org/news/2020-02-oceans-antarctic-ice-sheet-collapse.html

[12] https://www.washingtonpost.com/climate-environment/rapid-arctic-meltdown-in-siberia-alarms-scientists/2020/07/03/4c1bd6a6-bbaa-11ea-bdaf-a129f921026f_story.html

[13] https://e360.yale.edu/features/waste-heat-innovators-turn-to-an-overlooked-renewable-resource

[14] https://www.iea.org/data-and-statistics/charts/global-primary-energy-electricity-generation-final-consumption-and-co2-emissions-by-fuel-2018

[15] https://www.imf.org/en/Publications/WP/Issues/2019/05/02/Global-Fossil-Fuel-Subsidies-Remain-Large-An-Update-Based-on-Country-Level-Estimates-46509

[16] See the report of the Global Commission on the Economy and Climate: https://newclimateeconomy.net/publications

the results so far have been quite disappointing (von Weizsäcker et al. 1998; von Weizsäcker 2009). Even though the price of renewable energy has come down to competitive levels and new energy investments have soared to around 300 billion dollars, we still have a very long way to go before we will see renewables actually replacing fossil fuel-based energy production.

However, productivity gains will certainly be pursued with every technological advance possible. This will involve not only an abundance of new technologies but also new emerging social movements as growing numbers of people, particularly in the younger generation, will seek to find ways to adjust their lifestyles according to the planetary boundaries.[17]

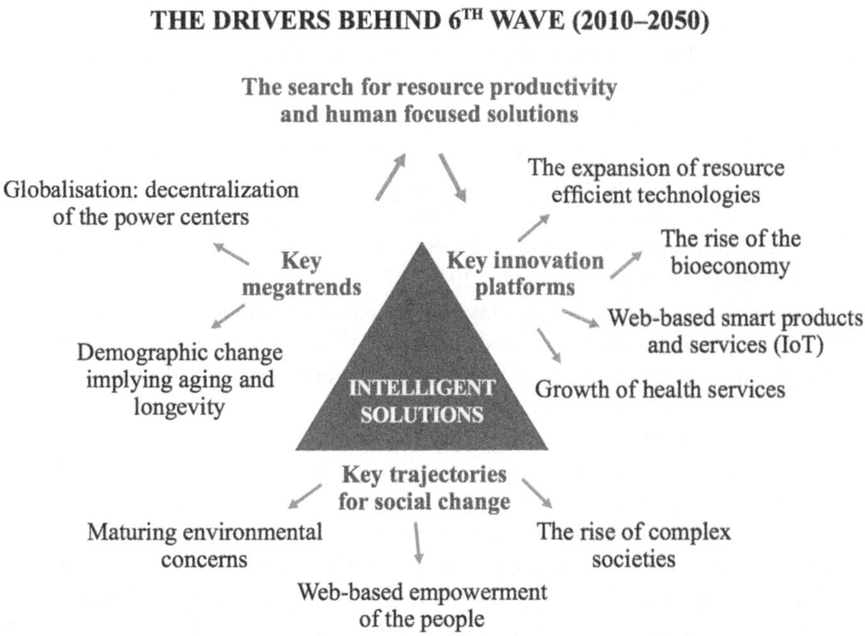

New resource-efficient technologies will come on stream as we move further along the sixth wave. Blockchain and virtual currency will increasingly be adopted by virtue of their capacity to enhance transparency and security in any transactional system and ultimately pave the way to a more neutral global currency. Artificial intelligence and machine learning are also rapidly developing and expanding, helping to speed up communication between machines and humans and to introduce learning in various technological systems. Their growth and expansion is expected to continue over the next decades, which will at once cause much concern about the control-invoking mechanisms embedded in these technologies.

[17] https://www.thenational.ae/uae/young-people-value-environment-over-money-according-to-global-survey-1.190473

Quicker and more immersive learning will be also greatly enhanced in the sixth wave. Augmented and virtual reality technologies will be widely deployed, given their capacity to improve learning through experience and transcend time and place. We will also see the continued spread of Internet of Things (IoT) technologies, enabling the integration and coordination of physical assets and digital systems. Big data technologies will also expand on the strength of their capacity to extract information from large volumes of data. Furthermore, 3D printing technologies will gain greater prominence, allowing for the production of three dimensional objects using digital data – and saving large amounts of materials.

We can also expect to see the growing interest in the secrets of human health materialize in new health technologies and preventive medicine. At the same time as personalized medicine becomes more readily available, new stem cell and nanoscale medicine will also be widely deployed. Digital technologies will be extensively used, and societies strapped by the high costs of running their public health systems will invest heavily in preventive health care in order to cut the costs of treatments for the various ill effects of cardiovascular disease. In the United States alone, cardiovascular diseases are the direct cause of one in three deaths, and the annual cost of treatment is estimated at 316 billion dollars.[18] Globally, according to some estimates, annual health-care costs are projected to rise from the current figure of 4.3 trillion to some 18 trillion dollars by 2040.[19]

In the sixth wave, the bioeconomy will be based on providing solutions to two major and very much interconnected challenges that face humanity as we move on to the next phase. The first set of bioeconomic solutions have to do with new bio-based materials that need to be developed in order to provide a replacement for non-renewable materials that have become increasingly scarce, such as critical and rare metals including lithium, copper, uranium, gold and so-called rare earth elements (REEs). The second set of solutions should arrive in the areas of land degradation, food scarcity and the western dietary bias. As Lester Brown has compellingly stated, providing adequate food supplies could become a weak link for our civilization in the same way as it was for many preceding civilizations, starting from Mesopotamia: many have ultimately perished because of the degradation of their cultivated agricultural land (Brown 2012). All these developments will mature as we move on to the seventh wave of development around the middle of the century.

[18] https://millionhearts.hhs.gov/learn-prevent/cost-consequences.html

[19] https://image.health.allianzcare-emails.com/lib/fe9b12747766047874/m/1/30a0836e-6ce7-4b9c-8b47-99a206299502.pdf

5 The Seventh Wave

As the sixth wave draws to a close by the middle of the century and the population on the planet reaches some 9 billion, the inability of societies to respond adequately and timely to the challenge of climate change will lead to a massive crisis. With ever-stronger storms, enduring droughts and rising sea levels due to the meltdown of large parts of the Antarctic ice bed, decision-makers will finally recognize that the runaway climate change must be stopped at any cost. This will bring in new global policies whose sole aim is to reverse climate change and elicit material and even spiritual revolution on earth as it is dawning upon us that we have persisted too long with our old model of industrial, material intensive development. The awareness of the earth system and the carbon cycle is about to increase radically.

Moreover, there will be wider acknowledgement of the long-term cycles of geological ages, and the present age, aptly named by aquatic ecologist Eugene Stoermer as the Anthropocene, is understood to postpone the next glacial period (Stager 2011). Humans of the twenty-first century have changed forever the deep future of earth, the next thousands or even millions of years. Whatever humans will be able to do in terms of regulating carbon emissions, there is no reversing the systematic changes that have already happened. However, the changes to the climate and indeed to the earth system will bring a huge boost to all efforts to rebalance our skewed relationship with nature. The focus in the latter part of the century will be on gaining a deeper understanding of living systems, and many technologies, both old and new, will be geared to learning from nature. It will be a period of nature-orientated technologies, in the spirit that futurist Pentti Malaska coined the term (Pouru et al. 2018).

So this is the backdrop for the seventh wave. Let us proceed now to look more closely at some of the key fields of operation. My projection is that the notion of bioeconomy will have at least four critical dimensions in the seventh wave:

1. A renaissance of agriculture, which will get underway as agricultural practices are geared towards regenerating the soil, including massive carbon sequestration.
2. Forests will become an extensive source of new materials, but at the same time their role as a promoter of human health and well-being will be fully realized.
3. Algae will be cultivated and used on a grand scale as a source of energy and new materials.
4. Biomimicry will be extensively deployed as a design principle for new technologies, bringing substantial gains in resource efficiency and enabling totally new technologies in all fields of life.

The following explores each of these dimensions in closer detail.

5.1 *Soil and Agriculture*

The potential of soil for the bioeconomy and its role in addressing climate change has largely been overlooked. The fight against climate change has been dominated by actions aimed at transferring to low carbon renewable energy by increasing resource efficiency and cutting emissions in many sectors of the economy, including buildings, transport and infrastructure. While great advances have been made on many of these fronts, the huge potential for using cultivated agricultural land as a carbon sink has not been considered a major option for carbon storage.

For some time now, it has been known that the agriculture sector as a whole is an important contributor to climate change, accounting for up to 30% of all human emissions (OECD 2015).[20] The potential of carbon sequestration in soil is well known, but as said indicated, the ideas of moving towards more regenerative agricultural practices has been largely dismissed as an overall policy goal (Zomer et al. 2017).

Recent research has shown that agricultural land has been a major anthropogenic source of carbon dioxide and contributed significantly to the atmospheric concentration of CO_2, which has risen from the pre-industrial level of 278 ppm to the current figure of 417 ppm.[21] Today, it is estimated that the soil carbon stock has been depleted by around 133 Pg (petagram, a unit of mass). Since the estimated stock of organic carbon at 2 metre's depth is 2047 Pg and the stock of inorganic carbon 1558 Pg, even minor changes in the organic carbon stock may have a substantial impact on atmospheric concentration levels. In fact, the calculations prove that the technical cumulative potential of carbon sequestration at 178 Pg in soil and 155 Pg in vegetation between 2020 and around 2100 and at the end of seventh wave could cause a massive drawdown of 178 ppm, bringing atmospheric CO_2 back to pre-industrial levels (Lal 2020). Moving organic carbon from the atmosphere to soil and vegetation, all achievable by human activity, could thus have hugely positive effects by creating climate-resilient soils and generating new, post-industrial agricultural practices throughout the world.

Given the current state of the food chain and indeed our food and agricultural policies, it is imperative now to launch a rebalancing act on a planetary scale.[22] Longitudinal studies have hinted that the only viable long-term solution is to combine modern scientific agriculture with selected aspects of small-scale traditional, regenerative agricultural practices. This will balance our biological need for food with the environmental impact of food production while building on those types of cultural, social and psychological needs that are deeply integrated into our societies (Cleveland 2014).

Global agricultural subsidies currently amount to 700 billion dollars a year. It is notable that out of this vast sum, only 1% is spent on pro-environmental purposes.

[20] https://www.oecd.org/agriculture/ministerial/background/notes/4_background_note.pdf

[21] https://www.co2.earth/

[22] https://thecarbonunderground.org/

According to a report from the Food and Land Use Coalition, we need to transform our food and land use systems within the next 10 years.[23] By reallocating our subsidies to support sustainable farming, we might be able in the next decade to develop a land use and food production regime that can help safeguard biodiversity, provide a healthy diet, improve food security, create more sustainable methods of agriculture and ensure a better income for farmers. In other words, the global farming community that is currently contributing to global environmental and social destruction could become a key provider of solutions to the problems of global warming and soil degradation. As agriculture covers some 30% of the planet's land surface, the shift from carbon emitter to carbon absorber would critically redefine the role of agriculture in the global economy bioeconomy.

5.2 Forest Sector

The seventh wave will bring a reassessment of the key role of forests in the global ecosystem. According to the Global Forest Watch, which is run by the University of Maryland, we are currently losing primary forest coverage at a rate of over ten million hectares a year. One-third of that destruction is happening in the tropical regions of the world where it is almost impossible to restore the native rainforest. Another source indicates that a forest area equivalent to the size of India will be lost by 2050 if the current rate of deforestation is allowed to continue.[24]

A recent foresight study using the Delphi process and consulting a large number of forest experts explored the future of the forest industry and the bioeconomy and concluded that this sector is highly dependent on the future agenda of international climate and energy policies (Hurmekoski et al. 2019). It was concluded that the forest sector need to invest heavily in developing and widening its concepts, products and services, while the focus for the bioeconomy should be on eco-system services. The assumption, endorsed by the vision of the experts interviewed, was that in the future, the forest sector will be an increasingly strong part of the bioeconomy overall and as such a major component of the sustainability pathway for 2050 and beyond (ibid.). Earlier studies have also suggested that European forests could easily double their carbon sequestration by introducing so-called climate-smart forestry. They suggest that the total mitigating impact of forests and the forest sector up to 2050 could be as high as 20% of total EU emissions and thus represent a significant part of the EU's mitigation strategy (Gert-Jan Nabuurs et al. 2017).

It seems quite probable, then, that forests and the forest industry will have an elevated role in the seventh wave economy, for a number of reasons. The first has to do with the fact that the transformation that started with digitalization around the turn of the millennium – which saw a reduced demand for paper and somewhat later

[23] https://www.foodandlandusecoalition.org/global-report/

[24] https://www.cgdev.org/media/future-forests

a sharp increase in the demand for board and pulp – will finally come to completion. As easily recyclable material from woody biomass will become the norm, substituting plastic-based and other non-renewable materials, the sustainability potential of wood fibre-based materials and products will be fully realized.

The second factor is that we will need a deeper knowledge of the role of forests in working towards a climate balance. The Global Climate Risks Index shows clearly that weather extremes are already affecting many parts of the world, particularly countries in tropical regions.[25] The trend is continuing to gather momentum, and the future outlook is for more extreme weather patterns and for more heat. According to the IPCC's 1.5 °C report, we will be seeing more unusually hot days and weather extremes in the future in many parts of the world, with tropical countries being the hardest hit (IPCC 2018). There is a growing body of evidence that land use strategies will become the hot topic in climate change negotiations in the coming years and decades, surpassing the role of energy, industry and transport. The IPCC Special Report on Land Use observes that there is major potential for carbon sequestration with reforestation and afforestation strategies (IPCC 2019).

The message to the forest sector is this: high-end knowledge about the forestry industry, whether in Finland, Canada or elsewhere, will continue to grow and flourish, partly because a massive reforestation of Earth has already started in the sixth wave. Back in 2006, the United Nations Environmental Program launched its One Billion Trees campaign,[26] which has now spread across the world.[27] In the seventh wave, we will probably see huge reforestation plans all around the globe in a bid to balance the carbon cycle by sequestrating carbon into the soil and trees. A study at the Technical University of Zürich (ETH) showed that there is space available on Earth for about 1 trillion trees that would cover 0.9 billion hectares of land, particularly in Russia, China, Canada and Australia. The combined effect of this forestation exercise would be a drawback of 2005 billion tonnes of carbon, the equivalent of two-thirds of the carbon released into atmosphere since the industrial revolution.[28] There is little doubt that as the climate keeps shifting towards ever more diverse weather patterns, these strategies will be applied in order to save human lives and to prevent infrastructure destruction.

The growing ambitions to save the Earth from runaway climate change will most likely also spur a renaissance of wood construction, which can be thought of as a method of carbon storage. Large timber buildings have been built and tested during the last 20 years, and their future looks very promising as wood-based building materials are stronger and safer than concrete and even steel.[29] As the technology

[25] https://germanwatch.org/sites/germanwatch.org/files/20-2-01e%20Global%20Climate%20Risk%20Index%202020_14.pdf

[26] https://www.unenvironment.org/resources/publication/plant-planet-billion-tree-campaign

[27] https://gulfnews.com/uae/environment/the-billion-tree-campaign-1.130096

[28] https://ethz.ch/en/news-and-events/eth-news/news/2019/07/how-trees-could-save-the-climate.html

[29] https://www.nature.com/news/the-wooden-skyscrapers-that-could-help-to-cool-the-planet-1.21992

continues to evolve and improve and as consumers increasingly turn to ecological options, there is good reason to expect that wooden buildings and structures will be endorsed in the seventh wave.

Furthermore, forest-based textiles and bioplastics have moved to the rapid development phase and are set to grow and expand in the future.[30] Innovations whereby plastic is made from wood, textiles are made from wood-based sources and large buildings are made from timber all open up interesting avenues for the transfer and circulation of consumption onto more sustainable paths (Hetemaki et al. 2017). Bioreactors and refineries are definitively the type of industrial activities that greatly diversifies the output of high-value chemical and pharmaceutical streams as well as biofuel as an energy source. The forest sector itself is a hub for multisector networks that offer diverse outputs ranging from climate regulation and water and soil protection to new schemes of eco-tourism and cultural services. In short, the forest represents a vital source of alternative materials and services for the future, and indeed there are substantial grounds to argue that the maintenance and restoration of forests will be one key element of planetary politics in the seventh wave.

5.3 *The Case of Algae*

Current UN estimates are that by 2050, the world population will number around 9.7 billion.[31] The world will need new sources of food in order to meet the growing demand. In fact, calculations show that food production will need to be stepped up by 70% in order to meet demand levels in 2050. Algae have a high protein content and are very nutritious. They can be cultivated on non-arable land, and they also grow in sea water, which means they do not compete for resources with other forms of food production. Recently, much research has been done in different parts of the world to understand the potential of algae in food production, materials development and the promotion of human health. Algae have substantial potential to play an important part in reinventing the global food system for the second half of the century.[32]

The reason why algae are attracting such huge interest in research and development has to do with the well-known planetary "hockey curve" problem.[33] The exponential growth of consumption and emissions is creating a very different kind of reality for the second half of the century, and in order to solve the growing problem, we need to reinvent the way we produce food. Algae are one of the most promising

[30] https://arstechnica.com/science/2020/01/are-bioplastics-all-hype-or-the-future-of-textiles/

[31] https://population.un.org/wpp/

[32] https://ethz.ch/en/the-eth-zurich/global/eth-global-news-events/2017/11/eth-meets-you-at-the-aaas-2018-in-austin-texas/AAASAlgae.html

[33] https://openresearch-repository.anu.edu.au/bitstream/1885/13126/3/1259855.full.pdf

avenues to explore, as are certain other biotechniques that will allow us to disconnect food production from traditional agriculture. In the future, we will be making food "from air", simply by adding microbes to the fermentation process. Already there is a food-producing company whose carbon footprint is one-hundredth that of meat production.[34]

For the seventh wave, the critical systems that need to dramatically improve their sustainability performance are food, agriculture and energy. The major issue for our society is how we are going to change our manufacturing and food production systems in order to increase resource efficiency while at the same time reducing emissions to a fraction of what they are today. According to FAO, climate change and warming will contribute to constraining the amount of arable land in many parts of the world.[35] The challenge of how to produce more food on less arable land is acute indeed. By 2050, it is simply imperative that we find new ways of producing food, fuel and some critical materials.

In all these critical sectors, emerging technologies in the area of algae production may become a major industry. In terms of water usage, for instance, algae are 800 times more efficient in producing protein than meat. Moreover, algae have enormous genetic diversity, which means that they have an extremely complex biochemistry and can thus serve as an almost endless source of materials (Fabris et al. 2020). There is a good chance that much of our food in the latter part of the century will come from ponds and vats, with algae serving as a source of food, energy and medicine. Moreover, algal biomass may contain large amounts of oils, and it has the great advantage of being able to convert almost all its energy from feedstock into a source of energy. Instead of using existing precious resources, algal biomass is produced by taking CO_2 from the atmosphere and small quantities of waste water. The technology is already there, ready to go. In other words, scientists seem to agree that algae feedstock is one of the major opportunities for a far more sustainable method of fuel production (Adeniyi et al. 2018). Since algae are plentiful and still a largely untapped opportunity, there is real potential for them to become a commercially attractive product for the global economy.

Algae can be classified into three main groups, i.e. microalgae, algae and macroalgae. Microalgae are single cell organisms that produce sugars using sunlight and carbon dioxide. They are essentially phytoplankton, which produce 50% of the world's oxygen. Macroalgae are multi-cellular tissues that are found in seaweeds, for instance. A good example of an algae product is spirulina, which has a wide range of positive health effects.

Microalgae are very rich in lipids and provide a raw material for biodiesel. They need very little land to grow and therefore do not compete with agricultural crops. It is a highly efficient source of oil compared to other similar sources. Algae are also a source of amino acids and omega 3. Furthermore, microalgae biomass can be used

[34] https://edition.cnn.com/2020/01/20/europe/solar-foods-solein-scn-intl-c2e/index.html

[35] http://www.fao.org/3/I9542EN/i9542en.pdf

to produce carbon-neutral plastics and foams. In terms of food security, algae have another important use in tying up phosphorus, a non-renewable source that is rapidly being depleted as there is only one country in the world, Morocco, with substantial amounts of this feedstock. As some studies suggest, in the process of using phosphate as a fertilizer in agricultural lands, up to 80% is wasted as run-off, causing eutrophication.[36]

In sum then, algae offer major potential in at least three distinct areas: in food security, in addressing global nutritional deficiencies and in producing inexpensive but high-quality protein to feed the growing population of the world. Up to 70% of the dry weight of algae is protein, so it offers a terrific alternative for the world which is now getting much of its protein intake in the form of meat. In algae, we have a non-soya-based vegan alternative that can be used as flour, oil and many other foodstuffs – and even as raw material for beer.[37] Secondly, in the world of chemicals, algae-based biofuels can become a significant replacement for fossil fuels. Thirdly, there are also major uses for algae in the field of medicine. Most synthetic vitamins are currently derived from petroleum. Algae could be a great source for all our vitamins and an important dietary supplement as they contain all the essential vitamins, minerals and amino acids that human beings need. In fact, we can eat the whole algae and have all the benefits of natural vitamins. In countries such as Japan and South Korea, where seaweed is traditionally an important part of the diet, people generally enjoy better health and have a longer life expectancy than people in Western countries. There is a strong evidence that dietary algae is a major explanatory factor (Wells et al. 2017). Astaxanthin is a carotenoid pigment that occurs in microalgae and in salmon, and it is known to be a powerful antioxidant with many health benefits, including a healthier skin, improved endurance, a healthier heart and reduced joint pain. It is even used as a cancer treatment.[38,39]

Algae could indeed well become a major cornerstone for our daily life in the latter part of this century, helping it transform into a bioeconomy (Ibid.). We know that they can be economically viable and that they are non-polluting. They conserve rather than squander energy. They also conserve other natural resources as we can use algae to produce the food, chemicals and energy we need.

Since algae have this real potential to become substantial trigger for the bioeconomy, there is much ongoing research and development around algae in different parts of the world, and it is expected that algae could indeed revolutionize the food we eat, the medicine we take and the energy we rely on (Khan et al. 2018). There is immense potential for algae to be a source of solutions critical for the seventh wave. Researchers say that algae factories of the future could produce "liquid, solid and gaseous biofuels [that] may become commercially available in the years

[36] https://www.responseable.eu/wp-content/uploads/key-story-eutrophication-0518.pdf

[37] https://www.williamsbrosbrew.com/beer/kelpie

[38] https://www.healthline.com/health/health-claims-astaxanthin

[39] https://www.clinicaleducation.org/resources/reviews/astaxanthin-the-key-to-a-new-you/

2020–2025" (Raslavicius et al. 2018, p643). They also conclude that "future climate change mitigation will rely on a synergistic combination of CO_2 capture and utilization technologies, with microalgal carbon capture and biomass production playing a significant role" (Raslavicius et al. 2018, p652).

As Fabris and colleagues have convincingly pointed out, the scaling up of algae use would help to tackle the hardest equation to solve in the world today: the growing needs for food, energy and materials cannot be met simply by stepping up the intensity of agriculture and fossil fuel extraction (Fabris et al. 2020). Microalgae have higher photosynthetic efficiency than plants and therefore greater potential to produce biomass that can be cultivated on non-arable land. Moreover, algae function as next-generation, cell-sized industrial plants, with the capacity to produce highly diverse products ranging from chemicals to skin creams.

In the seventh wave, this tremendous capacity of algae could be put to full use by drawing on advances in synthetic biology and IoT automation to develop algal manufacturing technology. As we need to move decisively from petroleum-based manufacturing and energy production, this new dominant form of economy will establish itself and flourish once production is ramped up some time in the 2050s, if not earlier.

6 Conclusions

Humans have a long history with nature, and hopefully a long future, too. In this introductory chapter to Bio#Futures book compilation I have used various frameworks to show how our relationship with nature has developed over time. As we move forward in this century, we will see how this relationship will intensify and assume new forms via two very different routes: through our crash course that has already destroyed many natural habitats and changed weather patterns, but also through innovation and development, bringing in more technology and science in an attempt to understand how we can cater for human needs without crossing planetary boundaries.

The seventh wave will be a defining time for the human race, a last chance to rehabilitate its relationship with nature. This is why I call this phase a wave of living systems technology. During this phase, our technology will increasingly begin to intelligently mimic the way in which nature operates. Inspired by the capacities of nature, a whole new set of technologies will be developed that will allow us to build a much more efficient and productive economy, while creating a much more human-friendly but at the same time nature-oriented way of life. As we move on in 2020s, it will become ever more apparent with more violent and extreme weather events that we are indeed currently on the crash-course with nature.

The real challenge is this: how do we learn to collaborate with nature instead of just extracting and using its resources? We need to and hopefully we will make biology work for us as a primary source of economic wealth and human well-being. In the case of algae, which currently produce half of the world's oxygen, we have

50,000 documented species, which means that they represent a huge genetic diversity. If we can learn how to put to use even a fraction of the intelligence and potential inherent in different algae species, leave alone all other intelligence in nature, then we would be well on our way to sustainably access much of the food, energy and medicine we need. Then, and only then, can we call our economy a true bioeconomy.

References

Adeniyi, O.A., Azimov, U. & Burluka, A. (2018). Algae biofuel: Current status and future applications. Renewable and Sustainable Energy Reviews, volume 90, July 2018, Pages 316–335

Bardi, U. (2014). Extracted. How the Quest for Mineral Wealth Is Plundering the Planet. Chelsea Green Publishing, London.

Beck, U. (1992). Risk society: Towards a new modernity. London; Newbury Park, Calif.: Sage Publications.

Berman, M. (1981). All that is Solid Melts into Air: the Experience of Modernity. New York: Simon & Schuster

Brown, L. R (2012). Outgrowing the Earth. The food security challenge in an age of falling water tables and rising temperatures. London, Earthscan.

Cleveland, D. (2014). Balancing on a planet. The future of Food and Agriculture. University of California Press.

Fabris, M., Abbriano, R.M., Perinice, M., Sutherland, D., Commault, C., Hall, C.C., Leen, L., McCauley, J.I., Unnikrishnan, K., Ray, P., Kahlke, T., Ralph, P.J. Emerging Technologies in Algal Biotechnology: Toward the Establishment of a Sustainable, Algae-Based Bioeconomy. Frontiers in Plant Science. Front. Plant Sci., 17 March 2020

Gert-Jan Nabuurs, G-J., Delacote, P., Ellison, D., Hanewinkel, M., Hetemäki, L., Lindner, M. (2017). By 2050 the Mitigation Effects of EU Forests Could Nearly Double through Climate Smart Forestry. Forests 2017, 8, 484

Guillemette, Y. and D. Turner (2018), "The Long View: Scenarios for the World Economy to 2060", OECD Economic Policy Papers, No. 22, OECD Publishing, Paris, https://doi.org/10.1787/b4f4e03e-en

Haskel, J. & Westlake, S. (2018). Capitalism Without Capital. The Rise of the Intangible Economy. Princeton University Press.

Hetemaki, L., Hanewinkel, M., Muys, B., Ollikainen, M., Palahí, Marc., Trasobares, A. (2017). Leading a way to European cricular bioeconomy strategy. Euroepan Forest Institute, From Science to Policy 5.

Hurmekoski, E., Lovrić, M., Lovrić, N., Hetemäki, L., Winkel G. (2019). Frontiers of the forest-based bioeconomy – A European Delphi study. Forest Policy and Economics 102 (2019) 86–99

IPCC. 2018. Summary for Policymakers. In: Global Warming of 1.5 °C. Special Report of the Intergovernmental Panel on Climate Change.

IPCC. 2019. Climate Change and Land: An IPCC Special Report on climate change, desertification, land degradation, sustainable land management, food security, and greenhouse gas fluxes in terrestrial 31 ecosystems.

Jabobs, M. & Mazzucato, M. (2016). Rethinking Capitalism: An Introduction. In: Jabobs, M. & Mazzucato, M. (eds.). Rethinking Capitalism: Economics and Policy for Sustainable and Inclusive Growth. Hoboken, N.J.: Wiley / Political Quarterly

Khan, M.I., Shin, J.H., Kim, J.D. The promising future of microalgae: current status, challenges, and optimization of a sustainable and renewable industry for biofuels, feed, and other products. Microbial Cell Factories. 2018;17(1):1–21

Kondratieff, N. (1928/1984). 'Long wave cycle' and 'Theses of N.D. Kondratieff's paper: Long cycles in economic conditions' in the long wave cycle. New York: Richardson & Snyder (trans: Daniels, Guy), 25–99 and 101–5, 137–8. In: F. Louçã & J. Reijnders (Eds.), The foundations of long wave theory. Models and methodology. Volume I. Cheltenham, UK: Edward Elgar

Lal, R. (2020). Managing soils for negative feedback to climate change and positive impact on food and nutritional security. Soil Science and Plant Nutrition, Volume 66, 2020 - Issue 1

Laloux, F. (2014). Reinventing Organisations. A guide to creating organisations. Brussels: Nelson Parker.

Malaska, P. (1999) A Conceptual Framework for the Autopoietic Transformation of Societies. FUTU-publication 5/99. Futures Research Centre. Turku School of Economics and Business Administration. Turku.

Piketty, T. Capital In the Twenty-First Century. Cambridge Massachusetts: The Belknap Press of Harvard University Press, 2014.

Pouru, L., Wilenius, M., Holstius, K., Heinonen, S. (2018). Pentti Malaska. A Visionary and Forerunner. Futures Series 9, The Finnish Society for Futures Studies.

Raslavicius, L., Striugas, N., Felneris, M., (2018) New insights into algae factories of the future. Renewable and Sustainable Energy Reviews. Vol. 81, 643–654.

Stager, C. (2011). Deep Future. The Next 100 000 Years of Life on Earth. Thomas Dunne Books, New York.

von Weizsacker E. U., Hargroves, C, Smith M.H., Desha, C., Stasionpoulos, P. (2009). Factor Five. Transforming the Global Economy through 80% improvements in Resource Productivity. London, Earthscan.

von Weizsacker E. U., Lovins, A.B., Lovins, L.H. (1998). Factor Four. Doubling wealth, halving resource use. London, Earthscan.

Wells, M.L., Potin, P., Craigie, J.S., Raven, J.A., Merchant, S.S., Helliwell, K.E., Smith, A.G., Camire, M.E., Brawley, S.H. Algae as nutritional and functional food sources: revisiting our understanding. Journal of Applied Phycology, 2017; 29(2): 949–982.

Wijkman, A. & Rockström J. (2011). Bankrupting Nature. Denying our planetary Boundaries. Earthscan, Routledge.

Wilenius, M. (2014). Society, Consciousness and Change. An Inquiry into Pentti Malaska's futures thinking. Futures Volume 61, September 2014, Pages 58–67

Wilenius, M. (2017) Patterns of the Future: Understanding the Next Wave of Global Change. World Scientific, London

Zomer et al. (2017) Global Sequestration Potential of Increased Organic Carbon in Cropland Soils. Nature, Scientific Reports 7,

Chapter 2
What Can *We* Do? Participatory Foresight for the Bioeconomy Transition

Simone Kimpeler, Ariane Voglhuber-Slavinsky, Bärbel Hüsing, and Elna Schirrmeister

Abstract Several future scenarios were developed and illustrate how differently the bioeconomy can be shaped. The contribution of individual consumption and lifestyles is highlighted. None of the scenarios describe a bioeconomy that was perceived exclusively positive by the involved people. In each of the elaborated scenario, there are also assumptions that were critically and controversially discussed. It became clear that the individual assessments differ considerably on how desirable the presented pictures of the future are. The bioeconomy is therefore not a solution to all problems, and not to the same extent for all people.

Why do we want a bioeconomy? The future scenarios show that a sustainable bioeconomy will require changes in many different subfields of society at the same time. Individual developments need to be critically examined, but also the interplay of different elements needs to be assessed in terms of their impact on specific aspects of sustainability.

The alternative futures of the bioeconomy assist to become aware of the effects on our everyday life and to initiate a discourse that, on the one hand, supports a conscious examination of critical points and, on the other hand, addresses the plurality of needs.

Keywords Foresight · Participatory future dialogues · Scenario development · Alternative futures · Bioeconomy · Sustainability · Storytelling

S. Kimpeler (✉) · A. Voglhuber-Slavinsky · B. Hüsing · E. Schirrmeister
Fraunhofer Institute for Systems and Innovation Research (ISI), Karlsruhe, Germany
e-mail: Simone.Kimpeler@isi.fraunhofer.de

E. Koukios, A. Sacio-Szymańska (eds.), *Bio#Futures*,
https://doi.org/10.1007/978-3-030-64969-2_2

1 Introduction

Climate change, mainly due to the extensive use of finite fossil resources, is one of the grand challenges which demands the transformation to a circular economy based on renewable resources. This contribution to climate protection and sustainability goals will have far-reaching impacts on our living environment. The transformation to a bio-based economy – a bioeconomy – results from the interplay of several factors: It is being decisively driven not only by technical innovations but also by social processes and depends on the interactions of and collaboration between stakeholders from society, technology, economy and ecology. The transformation will be profound. How these changes will look like is not clear yet and will be determined by the decisions we take today and in the upcoming years: How and what will we produce with which biogenic raw materials? Which products will we use in the future, and how will our consumer behaviour change? Is this bio-based economy more sustainable than our current economy? Are there differences in sustainability between different forms of bioeconomy? And in which bioeconomy do we want to live?

In the research project "BioKompass – Communication and Participation for the Societal Transformation towards the Bio-economy", funded by the German Federal Ministry of Education and Research (BMBF) from 2017 to 2020,[1] technology experts, together with pupils, students and citizens, developed alternative scenarios for the bioeconomy in the year 2040. These scenarios were subsequently used in public future dialogues on how life will change in a bioeconomy, in teaching courses at schools, and were core elements in a participatory exhibition on bioeconomy "Shaping our future – How do we want to live?[2]" at the Senckenberg Natural Museum in Frankfurt, Germany. The BioKompass future dialogues conducted so far have shown that sustainability is key for a successful bioeconomy transition – but underlying developments are often controversially assessed with regard to sustainability and desirability. It is therefore worth discussing alternative futures in order to become aware of the diverse and different impacts of the bioeconomy on everyday life – as well as the impact of consumer behaviour on the sustainability of the bioeconomy. None of the scenarios developed in the BioKompass project fulfil all the future wishes of those stakeholders involved – advantages in some areas go hand in hand with limitations in other areas – in particular with respect to sustainability goals. This is one of the main functions of participatory future dialogues in the bioeconomy transformation: to help participants in better understanding the cross-impacts, opportunities and risks of new technologies in interaction with demand and consumer behaviour.

---

[1] BioKompass has been coordinated by Fraunhofer Institute for Systems and Innovation Research ISI; partners were Senckenberg Natural Museum, Fraunhofer Institute for Chemical Technology ICT, Fraunhofer Institute for Computer Graphics Research IGD and Institute for Socio-ecological Research ISOE.

[2] Digital version of the exhibition: https://zukunftgestalten.senckenberg.de/

In the following, the goals and methodology of the BioKompass project will be outlined, and the four scenarios that have been developed will be presented.

2 Why Is a Transformation Towards a Bioeconomy Necessary?

A bioeconomy is based on three pillars: firstly, it uses and converts biogenic raw materials in industrial production processes instead of fossil feedstocks. These raw materials can originate from plants cultivated in agriculture and forestry, animals, bacteria, algae or from organic waste. Secondly, industrial production processes make use of the metabolic capabilities of living organism to convert the biogenic raw materials into products, e.g. in the fermentative production of bio-plastics, or the use of enzymes as biocatalysts. Thirdly, it offers totally new ways of satisfying human basic needs – e.g. by replacing livestock farming for meat production with all its negative impacts on climate, environment and health by providing highly nutritious proteins through culturing meat "animal-free" in bioreactors or by processing plant, algae or insect proteins into "meat-like" products. The bioeconomy comprises all economic sectors that produce, process or use plants, animals, microorganisms or organic waste. These are, for example, agriculture and forestry, plant and animal breeding, food industry, chemical and pharmaceutical industry, paper, leather and textile production. Moreover, the bioeconomy offers increasingly solutions for sectors like mechanical engineering, automotive engineering and the construction industry, and it is strongly driven by the use of information and communication technologies.

As a bioeconomy is based on renewable biogenic resources, it also requires arable land or innovative solutions to produce these raw materials. This bears the risk that industrial and energetic use of biogenic resources could be economically favoured over food production. However, it is an explicit goal of all bioeconomy efforts to give food production a priority. As a consequence, a bioeconomy will have to rely to a major extent on non-food biogenic resources, e.g. wood, algae or biogenic waste, and must rely on an industry structure that is based on a cascading, most efficient and high-value use of biogenic resources and the principle of circularity, in which material and nutrient cycles are closed, e.g. by extensive reuse and recycling.

3 The Approach of Scenario-Based, Participatory Future Dialogues

The BioKompass research project supports the societal transformation process towards a bioeconomy through appropriate information and communication formats, with the aim that experts as well as citizens and young people can develop a better understanding of a bio-based economy. Therefore, the BioKompass project aims at stimulating stakeholder participation in the transformation towards a bioeconomy by combining various participatory Foresight methods and formats. Foresight in this case supports the thinking in alternatives about a future development that is uncertain regarding its concrete outcomes and impacts. The following methods were applied: first a stakeholder mapping to identify relevant actors, today and in the future, in different roles in the innovation system. Then the research team carried out a future dialogue with more than 60 citizens and experts. The aim was to discuss with the participants how bioeconomy could affect everyday life both from supply, demand and user perspectives, e.g. in consumption, housing, work, economy or mobility. This discussion formed the basis for jointly deriving foci for the following scenario development process. In this expert-based scenario process key factors were identified, and for each key factor 3–4 future projections were elaborated, which were then combined into six alternative scenarios. In a final future dialogue at the museum, the story-telling method was used to co-create narratives for each scenario together with citizens, students, pupils and experts from research and industry. The narratives are easy to understand for laypersons and provide a link to everyday life settings of various target groups. The scenarios and the results of these future dialogues, e.g. the co-created narratives, formed the basis for subsequent dissemination via different media channels and for participatory measures: Firstly, an interactive exhibition in the Senckenberg Nature Museum was designed, using narratives from three scenarios to illustrate the impacts and challenges of the bioeconomy for a broader audience, accompanied by a virtual tour on their website and an App with educational content to be used in the museum. The exhibition now provides artefacts, statistics as well as explanations around three alternative scenarios of the bioeconomy in order to show how differently the bioeconomy can be understood, designed and developed and above all, how individual consumption patterns and lifestyles can influence it and vice versa will be influenced by it. Secondly, the scenarios were used for an educational programme with pupils aged 15–17. Since none of the future scenarios describe an ideal type of bioeconomy, there are several assumptions about future projections in each scenario that stimulate critical reflection and discussion. In addition, all participatory formats and communication measures were evaluated in an accompanying research process to better understand the effects of the different participatory formats on the understanding of such complex future developments as the bioeconomy transformation.

Prerequisite for a successful communication strategy is a well-defined target group and content of interest for the target group. The target group here is rather broad. Especially for the exhibition in the Senckenberg Nature Museum, it is defined as citizens interested in natural history-, biology- and biodiversity-related topics as well as other museum visitors such as parents and others with children, pupils and students. The content is built on alternative scenarios of a future bio-based economy, which – and this is key – have been developed by the target group itself together with experts. For each scenario, the participants themselves created everyday stories, each illustrating life in the bioeconomy under different conditions and in different regions of Germany. When using this so-called storylines during the project, in, for example, the exhibition or in a school context, the co-creation continues, as all visitors, pupils or participants contribute own opinions, thoughts and ideas to these pictures of the future. In the upcoming years the scenarios are intended to be used in dialogues and other discursive events at the museum or in science festivals with people from science, business and politics.

4 The BioKompass Scenarios

The six future scenarios that were developed in the BioKompass project vary in many aspects. Two of the six scenarios have a negative environmental impact, e.g. lead to loss of biodiversity. Since positive environmental impacts are the key legitimation for the transformation of the economy towards a bioeconomy, these two scenarios have not been further used in the BioKompass project formats, and will not be analysed in detail here. We focus here on only two selected key factors, namely, consumer behaviour and use of agricultural land. They represent the demand and supply side in each scenario and have direct influence on how much bio-based feedstock is available for industrial and energetic use in a bioeconomy, to which extent consumption can be bio-based, and to which extent negative impacts on food production, climate and environment and biodiversity can be expected.

The four scenarios without a negative environmental impact can be distinguished by their positioning with regard to these factors: Three of the four scenarios are associated with a change in consumer behaviour towards less or more sustainable consumption; only one represents a variation of the bioeconomy in which consumer behaviour remains unchanged and sustainability challenges of consumption get solved by technical means rather than social innovation. With respect to land use, two scenarios feature a reduction in agricultural land use in Germany, whereas in one there is no change in land area, while in the fourth one there is even an expansion. This is the result of varying degrees of autarky versus imports of biogenic resources, and of different agricultural practices, ranging from highly intensified conventional agriculture via organic agriculture with increased yields to permaculture-like practices. In each world, the effects on biodiversity are different, but overall they succeed in reducing environmental impacts, e.g. by financial incentives for sustainable agricultural practices, by importing biogenic resources (and thus either creating negative environmental impacts abroad or by making sustainable agricultural practices mandatory for imports) or by substantially expanding national parks.

To give an impression of the scenario narratives that have been developed, two are described below. A description of the key features of each scenario is followed by a story about a day in a person's life in such a future world. For the others see the projects website https://museumfrankfurt.senckenberg.de/en/biokompass/.

4.1 *Scenario 1: Rising High with High-Tech Bioeconomy*

By the year 2040, the German economy has fundamentally changed in the direction of a high-tech bioeconomy. Industry, politics and society have jointly initiated this change. Germany imports only little crude oil and petroleum-based products. Instead a great shift towards biomass as feedstock for various application areas has been undertaken.

The chemical industry uses biomass from European cultivation areas as raw material, but in order to meet the high demand for biomass, it is also procured globally and imported from all over the world. In addition, novel raw materials, e.g. based on wood, algae, grasses, organic waste and CO_2, are exploited. In Germany, agriculture and forestry are managed in a more sustainable way. In order to limit the negative environmental effects of intensive agriculture, modern, highly automated and digitalised production processes are used, in which fertilisers and pesticides are applied with great accuracy exactly meeting the plants' demand and which partly take place in multi-storey greenhouses. Farmers receive subsidies for the implementation of sustainable agricultural practices.

Biotechnological production processes are highly advanced: gene-optimised plants and waste materials are processed in fully integrated bio-refineries, making use of industry 4.0 concepts.

The recycling of products is highly developed and more important than the reuse of products and the extension of product life.

Germany is a lead market for products based on biomass or recycled raw materials.

Many people live in "Smart Eco-Homes", which were built using bio-based plastics and insulation materials. A more sustainable consumption is supported by "smart" products, which support resource-efficient use. Foods are similar to traditional products from 2020, but, for example, micro-algae replace butter and eggs in many foods. Therefore, products with a lower environmental footprint are available to all consumers.

Young people drive less cars and ride more bicycles than in 2020, but many do not want to give up their own vehicle – even if they are more expensive. At least they can justify their comfort by the argument that they are driving with alternative fuels and not with fossil fuels any more.

4.1.1 Future Story of Beate: High-Tech Savvy?, Environmentally Conscious and Self-Sufficient

"What a heat – and that already in the early morning", thinks Beate (50) and first takes a rainwater shower in her balcony cabin. She then uses the shower water for her potted flowers and the hanging vegetables on the kitchen window. Then she cleans the fermenter. With this home bioreactor, her latest purchase, she can now grow the microorganisms that are specifically adapted to herself, making individualised products possible. After initial experiments at home, she plans to use the bioreactor in her teaching classes soon. She is well known among her students for her experimenting with new technologies. The joy of experimenting can also be seen in her furnishings, a wild mixture of digital high-tech and wood as well as different bio-based materials.

It is important to Beate that everything is recyclable or compostable – and yet as durable as possible. A living room wall is clad with a thin algae reactor – similar to wallpaper – which Beate helped develop and programme to change colour depending on the amount of daylight. Now Beate is standing in the middle of the living room, looking anxiously at the flashing air sensors on the ceiling. The house technology is very prone to errors. Beate can often repair small things herself, but in this case she has to contact the facility management and describe her problem to an artificial intelligence algorithm via the hotline.

Then Beate has breakfast. Today, she eats – as an exception – wholemeal bread in traditional quality from the organic baker, which is a real (and expensive!) rarity. Here Beate ticks a little differently than her friends, who are of the opinion that genetic engineering is sustainable. After all, genetically modified plants, for example, also grow on barren soils, require little fertiliser and thus pollute the environment less. Although this makes sense in Beate's opinion, she still enjoys her traditionally baked wholemeal bread and fills it with homemade meat sausage: She has enriched the microorganisms grown in the bioreactor with soy protein and herbal aromas and has shaped this mixture with the 3D printer into a sausage. She enjoys the wide selection of food and the personal design possibilities. She can produce almost everything herself. Crockery and cutlery are made of easily degradable bioplastics; her coffee beans are grown in Germany, a cultivation based on top-quality coffee from Ecuador, vintage 2025.

While she goes to the garage and gets into her car, she decides to work through the complex topic in class with her students. She drives with biofuel, which has become very expensive due to the El Niño crop failures in Latin America the last year. That is why she wants to buy an electric car with built-in solar collectors in the near future.

4.1.2 Future Story of Barbara: Freelance Online Teacher and Forest Lover

Barbara (35) lives with her partner in a 2-room apartment in a fully digitalised passive house. The voice control of the building services and household appliances as well as a very good Internet connection were important decision criteria when buying the house. For cost reasons, the two women opted for an apartment in the country-side. The rural area is very barren and characterised by monocultures, and wind turbines and solar plants can be seen everywhere.

The distance to their employer, a school in the neighbouring town, is no problem, as most of the lessons take place online. Only for exams and special occasions teachers and students have to be physically present. Today a course will present project results. The task was to design bioeconomy start-ups that develop bio-based products such as new plant varieties, a bamboo grooming robot or a digital cat. This year, Barbara is teaching at a "drop-out" school, which was created as a counter-movement to the high-tech mainstream in the school system. The school was founded by a parents' initiative, who prefer to live in the countryside on a self-sufficient basis and want to teach their children alternative values in school.

Barbara is a bit dissatisfied with the curriculum. She has a different opinion on some of the topics taught. For example, the drop-out school would not work at all without high-tech, because without fast Internet there would be no online courses! Since she works as a freelance teacher, she looks for a new job for each school year. To do this, she uses a job platform based on AI to select the right school for her qualification profile. Barbara also uses algorithms to control her diet: She uses an app that adjusts her diet to her daily needs. In this way, she has her blood pressure and allergies well under control. The appropriate food is ordered via the app and delivered to the house in bioplastic packaging. Meat from the lab, algae jelly and fish soup, for example, are on the menu today. The sea has become the most important source of food.

Real forest is an attractive excursion destination, and forest reserves attract visitors with apps that combine virtual content with real forest experience. The app helps to identify animals or provides information about allergens in plants. Barbara also goes to the forest as often as possible.

4.1.3 Insights from the Future Dialogue

During the discussion, it became clear that in this scenario the definition of what is an "organic food" could change significantly. This is because in this future world, genetically modified plants and animals are accepted as bio-based food components, as long as they contribute to environmental sustainability.

The implications of globalizsation for the bioeconomy remain unclear in the scenario. The resource demand in this scenario is very high, and at the same time a lot of the biomass processed is imported, leading to a high dependence on international trade.

The scenario focusses on technological innovations. The ambivalence that such innovations are necessary, but most likely not sufficient and can have many trade-offs, is reflected by comments given by participants of a future dialogue:

Participants believe that "The scenario is desirable because of...":

- Improved climate
- Comfortable life
- High living standard
- A great deal of enthusiasm for technology
- Mankind accepts that there is no untouched nature
- No restrictions in consumption
- No postmodern compulsion for self-optimisation

Participants believe that "This scenario is not desirable because of...":

- Too much interference with nature
- Bioeconomy is primarily industrial biotechnology
- Technology friendliness as a prerequisite
- Technology does not provide the solution for all problems
- Growth credo is not questioned in this world
- Resource problems continue to exist, exacerbated by consumer behaviour
- Still much individual traffic
- The human becomes subordinate to machines
- Too individualistic, little personal contact with other people
- Loneliness (technology cannot replace personal relationships)
- State seems antisocial and liberalistic
- Global inequality is increasing
- Danger of "green-washing" without real sustainability

4.2 *Scenario 2: Bioeconomy Through Ecologically Conscious Lifestyle*

The majority of the population want to live more sustainably and are willing to accept restrictions. This is also reflected in legislation, which sanctions unsustainable products and practices by making them more expensive: e.g. meat is subject to a tax due to its resource-intensive production. Coffee-to-go cups and other packaging intensive drinks and food are no longer available. Living areas over 20 m^2 per household member are charged with an additional tax. Plastics are only permitted for certain applications. Petroleum-based plastic toys are prohibited. As a result, the use of non-bio-plastics has declined sharply, even bio-based plastics are used in a sensible way. Many products have become more expensive but also more durable.

The bioeconomy industry in Germany produces a wide range of products and has experienced strong growth, but this has not fully compensated for the losses in other economic sectors. Biogenic raw materials are mainly processed for the domestic market, which has a positive effect on the depth of value creation.

When it comes to food, people trust in regional products because the international certificates for sustainable production are not transparent and stringent enough.

Only little meat is produced and consumed. Waste avoidance is a major issue in industry, agriculture and also private consumption. Agricultural areas are set up wherever possible, but are managed in an integrated manner. Rapid technological progress in the use of wood, organic waste of different origin and CO_2 allows the production of a wide variety of bio-based materials, which was hardly conceivable in 2020.

Forestry and agriculture in Germany and Europe have not only become much more efficient but also more diversified and sustainable. This is based on an intelligent combination of approaches from permaculture and innovative automation.

The increased demand for biomass now requires a significant expansion of the areas under cultivation. In order to improve environmental protection and biodiversity, measures to maintain and increase biodiversity are therefore an integral part of agricultural and forest land use practices and are required by law.

4.2.1 Future Story of Oda: Environmentally Conscious and Thrifty by Conviction

"So this is how my daughter lives now," thinks her mother when Oda (23) shows her the photos of her new home. She is happy that everything went so well with her daughter's first own flat. After all, everything happened quite quickly: first of all the start of her working life with her first permanent job as a teacher. After that she looked for a flat and moved to the big city, to this 25 m^2 flat on the 2nd floor. The furniture takes some getting used to, thinks her mother – a mixture of old and new. The wardrobe from the state-owned manufacturer was very expensive, but has a lifetime guarantee. The rest of the furniture is self-made or from a flea market. "Look at your clothes", laughs Oda's mother, "everything is self-knitted or from the second-hand shop".

In the photo Oda's desk catches the eye: made by a carpenter as a present for her 18th birthday. Oda also still has the armchairs she made herself from pallets as a family activity, as well as the many woolen blankets, in order to be able to do without heating if possible. "Is there a supermarket nearby?", her mother continues. Oda explains that, as in many residential areas, there is a small shop for everyday necessities, where organic boxes are sold directly from the farmer. There Oda also receives milk and cheese from her "cow-sharing" – a project in which she shares the business risk of the farmer through a membership fee and solidarity-based sales prices for dairy products and meat. Oda thus buys almost exclusively food from the region. Imported fruits such as mangos and bananas are extremely expensive, but for climate protection reasons they are not even on Oda's menu. She doesn't eat canned pineapples, but knows many old apple varieties which are processed directly after harvesting and consumed all year round as puree, juice or in a dried version.

What she does not receive from "her" farmer, Oda obtains from the vertical farming supplier in the neighbouring district, thus implementing particularly short supply chains. There, for example, products such as fruit and salads are grown in an energy-saving way. Full of pride, Oda tells her mother that she keeps a beehive on the roof garden with her neighbours and wants to grow tomatoes on her balcony.

"Tell me about school" – Oda laughs. She has heard this sentence from her mother for 20 years. She then reports that she only teaches in projects and combines content from several subject areas. Important subjects are natural sciences, home economics including school garden and repair science. Next week she will take her project course on an excursion to the Ministry of Agriculture and to the Vertical Farm. Oda hopes that her students will also get inspiration for their career choice there.

"Will you be okay with your money?" Oda's mother wants to know. "Sure, Mum", Oda assures her, "My income is not only based on teaching, but I also work a few hours a week in the Share & Repair Lab. I can even afford to take a train ride to your home several times a year".

4.2.2 Future Story of Oskar: Successful Organic Farmer

Oskar is the proud owner of one of the few completely energy self-sufficient farms in his village. With the help of solar – and bioenergy and additional efficiency measures such as combined heat and power generation, he can cultivate his fields and greenhouses sustainably and economically. He offers a wide range of products for numerous customers – from vegetables and fruits to corn and cereals to CO_2 from biogas production, which is used in the chemical industry.

Oskar sees himself more as an entrepreneur than as a farmer who negotiates on an equal footing with his partners and customers. He is not only familiar with vegetable and cereal cultivation but also studied business administration with a focus on marketing and digital agriculture.

His primary goal is not only to maximise yields but also to make the best possible use of resources. Consumer confidence in the quality of his products is also very important to him.

For example, together with his neighbour, who specialises in the cultivation of medicinal plants, he introduced the weekly "Citizens' Country Day". People help with the work on the farm and thus learn how sustainable his business is. He also uses resource-conserving products for equipment and machinery, e.g. dandelion rubber instead of conventional rubber tires and recycled components. This has promoted both the appreciation of his organic farming and the demand for his organic products.

"Today", Oskar thinks, "I'm going to explain to people how important subsidies will continue to be for organic farming here in the future". Oskar sees himself as a modern organic farmer who not only uses innovative technologies but is also involved in sales. He is well aware of his strong negotiating position – and that this is an important part of modern organic farming.

4.2.3 Insights from the Future Dialogues

During the discussion the question arose whether there would be resistance to the state regulations. It was then assumed that these are socially accepted and legitimised. However, this would severely restrict people's freedom of action.

One participant remarked that the scenario could well be developed further with regard to various technologies and the industrial perspective. After all, all this could not be achieved without growth.

The scenario could also be understood as a reconciliation of technology with society. If there is a social consensus in favour of sustainability, government incentives for the circular economy make sense.

Participants believe that "The scenario is desirable because...":

- Focus on sufficiency and rethinking in society
- Sustainability is achieved through efficiency and innovation
- Environmentally friendly local production is promoted
- Chance of sustainable and sufficient food production
- Idea of a post-growth economy

Participants believe that "This scenario is not desirable because...":

- Production side of that scenario can be further elaborated
- This world is too narrow for me
- Social compulsion to self-renunciation does not correspond to my wish; high state regulation
- Smaller product range from abroad

5 Conclusion: No Bioeconomy Without Change in Consumption?

The diversity of the future scenarios developed illustrates how differently the bioeconomy can be shaped and how much individual consumption and lifestyles influence this or, conversely, are influenced by a change. None of the visions of the future describe a bioeconomy that was perceived exclusively positive by the involved people. In each scenario, there are also assumptions that were critically and controversially discussed by the participants of the future dialogue. During these dialogues, it became clear that the individual assessments differ considerably on how desirable the presented pictures of the future are. The bioeconomy is therefore not a solution to all problems, and not to the same extent for all people.

Why do we want a bioeconomy? The future scenarios show that a sustainable bioeconomy will require changes in many different subfields of society at the same time. Individual developments need to be critically examined, but also the interplay of different elements needs to be assessed in terms of their impact on specific aspects of sustainability.

During the first future dialogue at the beginning of the project, the ideas on the bioeconomy were still rather vague for many participants and allowed for extensive individual interpretation although the participants hardly discussed controversially. The elaborated scenarios then made the difference between today and future bioeconomy worlds clearer for the participants, and very controversial, but constructive discussions arose. This was fostered also by the "immersion" into the alternative futures via storytelling. Where some felt restricted, others saw great opportunities and vice versa. Often the same points were presented as both positive and negative arguments.

In summary, it can be said that it is worth taking a look at alternative futures of the bioeconomy in order to become aware of the effects on our everyday life and to initiate a discourse that, on the one hand, supports a conscious examination of critical points and, on the other hand, addresses the plurality of needs.

Chapter 3
The Emergence of Bioeconomy in the 6th Kondratiev Wave of Change: A Horizon Scanning-Based Approach

Emmanuel Koukios and Anna Sacio-Szymańska

Abstract This paper reports research results on the assessment of the emergence of bioeconomy as part of the mega-wave of socio-technical change that has taken off with the global financial crisis of ca. 2008 and is expected to peak around 2030. The appearance of bioeconomy-related phenomena on this wave is strongly related to the formation of clusters of opportunities, problems and other key factors. Nine such clusters were identified with the use of previous foresight exercises following the Horizon Scanning approach, with the help of a specially designed for the needs of this work Questionnaire, focusing on major expectations, worries and modes of action. A tenth, "mystery" cluster was found necessary to cover collectively the effects of disrupting factors. The various interactions of the identified clusters can explain the so far sporadic and sometimes inconsistent observations on the emergence patterns of bioeconomy as part of a Kondratiev Wave.

Keywords Bioeconomy · Foresight · Horizon Scanning · Kondratiev waves · Problems · Opportunities · Actions · Clusters

E. Koukios (✉)
Research Group BIOTOPOS, Organic Technologies Laboratory, National Technical University of Athens, Athens, Greece
e-mail: koukios@chemeng.ntua.gr

A. Sacio-Szymańska
4CF Strategic Foresight Company, Warsaw, Poland

E. Koukios, A. Sacio-Szymańska (eds.), *Bio#Futures*,
https://doi.org/10.1007/978-3-030-64969-2_3

1 Introduction

1.1 Defining Bioeconomy

The term "bioeconomy" includes all industrial and economic sectors that produce, manage and otherwise exploit biological resources, and related services, supply or consumer industries, including the following major examples (European Commission 2012; Cichocka et al. 2010):

- Agriculture, fisheries, forestry, aquaculture
- Agro-food, wood, fibre and other bio-industries
- Human and animal health, pharmaceuticals
- Biochemicals, biomaterials, "green" bio-chemistry
- Bioenergy, biofuels, other bio-products
- Bio-remediation, bio-waste management
- Bio/eco-systems management, rural development

1.2 A Lot at Stake

According to various indications (Koukios 2017), the new mega-wave of socio-technical change, expected to peak after 2030, will include the emergence of bio-economy, i.e. the whole spectrum of applications of biological sciences and technologies in all socio-economic areas and sectors, radically transforming our societies and the world as we know it.

The object of the research presented in this chapter is to foresee, map and assess the emergence of this great wave we call *BIO-TSUNAMI*, thus preparing societies to harness its huge potential for health, work, prosperity and quality of life, while avoiding getting crashed by its risks and side-effects.

Recent research activities by our group include the identification of critical technological breakthroughs affecting the bio-tsunami's emergence, and the socio-economic clusters describing its main pathways of social change (Koukios and Sacio-Szymańska 2018).

1.3 Risks and Opportunities

The emergence of bioeconomic applications is linked with a great number of strategic opportunities, and major risks and threats for Europe and the World (European Commission 2010; Koukios et al. 2005; Koukios 2015):

- The former includes a great innovation potential; environmental "greening"; mitigation of climatic change effects; substitution of fossil fuels; generation of employment; food quality and safety; growth of new SMEs and other businesses; policy coordination; regional development and smart specialization; and international cooperation.

- The latter list soil erosion and desertification; high-intensity farming; food imbalance (malnutrition vs. wastes); food chain and water supply risks; societal reactions to some biotechnologies; extreme pressure on resource; dominant culture of youth and beauty at all costs; centralized governance of systems; unauthorized use of genetic information; and bio-ethics.

1.4 Research Objective

The target of the work reported here is to initiate a foresight exercise on the emergence of bioeconomy, considered as a fast-approaching mega-wave of socio-technical transformation, following the Information and Communication (ICT) one, and defined as "bio-tsunami", which is built by a wide spectrum of converging elements, clusters and fields: biology, biotechnology, bio-engineering, new agriculture, novel foods, health, quality of life, cosmetics, bioenergy, environment, water, sustainability, education, knowledge management, design of smart applications and more. The bio-tsunami tends to "absorb" selective elements from the previous one, thus multiplying its power and enhancing its applications range, e.g. in bio-informatics.

2 Methodology

2.1 Riding the Kondratiev Wave of Change

The model of socio-technical change adopted in this work is that of Kondratiev's (K) long cycles, of 40–60 years each, starting since the industrial revolution. According to this concept, the K1 wave, dominated by textile and other industries, deployed between 1771 and 1829; K2, powered by steam and railroads, ran in 1829–1875; K3, driven by steel and heavy engineering, ran in 1875–1908; K4, dominated by petroleum, electricity, automobiles and mass production ran in 1908–1970; K5 was driven by information and telecommunication, in the period 1970–2010; and finally, K6 is the current wave (2010–2050?), which is to be explored for its bioeconomic elements (Nefiodow and Nefiodow 2014; Wilenius 2017).

In the first step of our approach, we reviewed the existing literature on the present Kondratiev Wave (K6) and identified wave features having any relevance with bioeconomy. In Table 3.1, we summarize the results of our literature survey of K6 elements relevant to bioeconomy, either directly or indirectly, i.e. through interaction

Table 3.1 Main literature suggestions for K6 elements

Health, especially Holistic Health Applications
Psycho-Social Health
Biotechnology
Biomedical Applications
Nano-Molecular Applications
Nuclear Energy Developments
Renewable Energy Breakthroughs
Hydrogen Energy
Robotics, Intelligent Machines
Artificial Intelligence
New Leadership and Management
ICT to support the above – Post-Info Applications
Converging Technologies (info-bio-nano, etc.)

Source: Nefiodow and Nefiodow (2014), Wilenius (2017), Mason (2015), Goldschmidt and Hilbert (2009), Smihula (2010), Wilenius (2014), Toffler (1980)

and convergence. From that list we can see that there is no clear pattern of bioeconomy emergence, with some authors even almost eliminating bioeconomy from their describing of K6 actions (Mason 2015; Goldschmidt and Hilbert 2009).

2.2 *Horizon Scanning Approach*

The method selected for the foresight of the bioeconomy emergence during K6 is the Horizon Scanning one (Horizon Scan Report 2007; OECD 2016). Specifically, we have assessed the available information on the problems and the opportunities as identified from previous Horizon Scans, having the same or a similar time horizon, in order to isolate the most relevant, directly or indirectly, to bioeconomy (OECD 2009a, b; Koukios 2014; Geels and Schot 2007).

In addition, we have used a tailored designed Questionnaire, shown in Table 3.2, in order to focus on the expectations, the worries and the preferred modes of action of experts and lay persons, including students and other youth. More than 100 responses were collected in the period of March–June 2017, from audiences in Poland, Greece and Italy.

The various types of parameters used in our analysis of the Horizon Scanning process are described by the following symbols:

A: FOCUS ON EXPECTATIONS according to Questionnaire
B: FOCUS ON WORRIES according to Questionnaire
C: FOCUS ON PROPOSED ACTIONS according to Questionnaire
P: PROBLEMS identified by Horizon Scanning
O: OPPORTUNITIES identified by Horizon Scanning

Following the identification of the most relevant to bioeconomy features, a number of Clusters will be put together, consisting of appropriate combinations of A, B, C, P and O.

Table 3.2 Questionnaire on major bioeconomic driving forces and barriers

The term "Bioeconomy" refers to the applications of biological sciences and technologies to Economy and Society. To help our project tentatively map such a complex landscape, please share with us your related major wishes and visions, as well as fears and threats.

A. EXPECTATIONS (*Please, select no more than 4 of the following*):

A1. Achievement of longevity ____________
A2. Fighting of serious illnesses ____________
A3. Quality foods – Fighting hunger and obesity ____________
A4. A new generation of cosmetics ____________
A5. Clean biofuels ____________
A6. New bio-materials for quality of life ____________
A7. Reproduction and regrowth of organs ____________
A8. Protection of environment and ecosystems ____________
A9. New employment possibilities ____________
A10. Regional development ____________
A11. New national development model ____________
A12. Other (please, explain) ____________

B. WORRIES (*Please, select no more than 4 of the following*):

B1. Hyper-intensive farming of plants/animals/others ____________
B2. Genetically modified foods ____________
B3. Cloning/genetic improvement of humans ____________
B4. A culture of beauty and youth at any cost ____________
B5. Biological weapons – Bio-risks ____________
B6. Domination of large multi-national companies ____________
B7. Control of bio-innovations through patents ____________
B8. Increase of unemployment due to new technologies____________
B9. Lack of protection of personal biological data ____________
B10. Bio-ethics and morality issues ____________
B11. Other (please, explain) ____________

C. PROPOSED ACTIONS (*Please, select no more than 3 of the following*):

C1. Research and innovation ____________
C2. New investment projects ____________
C3. New forms of project financing ____________
C4. Environmental management ____________
C5. Social tasks and initiatives ____________
C6. Education and training ____________
C7. Art and culture ____________
C8. Youth initiatives ____________
C9. Integrated policy framework ____________
C10. Citizens information/awareness ____________
C11. Other (please, explain) ____________

3 Results and Discussion

3.1 Identification of the Most Relevant to Bioeconomy Opportunities

Table 3.3 presents the results of our assessment of the list of 73 opportunities already identified by Horizon Scanning (HSc) from the point of view of our foresight study, i.e. regarding the emergence of bioeconomy (Wilenius 2014; OECD 2009b).

Table 3.3 List of very high bioeconomy relevance opportunities identified by HSc

O01	ATMOSPHERE: Climate change
O05	HYDROSPHERE: Minerals and gas hydrates from ocean floor
O08	BIOSPHERE: Inspiring life processes
O09	BIOSPHERE: Focused development of marine and other wild environments
O10	BIOSPHERE: (Re-) creating a living environment can contribute to more life
O11	BIOSPHERE: Intervention in humankind's own evolution
O17	FOOD – AGRICULTURE: Strategic function agriculture
O18	FOOD – AGRICULTURE: Modern craftsmanship
O19	FOOD – AGRICULTURE: Healthy, eating patterns-based, functional foods
O24	HEALTH CARE: Monitoring/developing vaccines for infectious diseases
O26	HEALTH CARE: Preventive medicine
O28	ENERGY PRODUCTION: Robust strategy for energy supply
O33	SCIENCE & TECHNOLOGY: Threats as leitmotif for innovation
O34	SCIENCE & TECHNOLOGY: Promising new fields of science
O35	SCIENCE & TECHNOLOGY: Converging technologies (CT)
O36	SCIENCE & TECHNOLOGY: CT for human cognition and communication
O37	SCIENCE & TECHNOLOGY: CT to improve health and physical capacities
O38	SCIENCE & TECHNOLOGY: CT to improve group and social processes
O39	SCIENCE & TECHNOLOGY: National security
O44	SCIENCE & TECHNOLOGY: Understand disease with developmental biology
O45	SCIENCE & TECHNOLOGY: Biological research into pathogens
O46	SCIENCE & TECHNOLOGY: Biomedicine and the prolongation of life
O49	EDUCATION: Educational system keeping up with global knowledge increase
O50	EDUCATION: Creativity in education
O53	SOCIAL, FAMILY, WORKING LIFE: Ageing as Silver Fleet
O54	SOCIAL, FAMILY, WORKING LIFE: Ageing actively
O59	NATIONAL ECONOMY: Employment at all education levels
O60	NATIONAL ECONOMY: Country branding
O61	NATIONAL ECONOMY: The right economic growth
O63	NATIONAL ECONOMY: Entertainment industry
O64	NATIONAL ECONOMY: Wellness
O65	NATIONAL ECONOMY: No use made of unique features
O67	NATIONAL ECONOMY: New technical products and services

The level of the bioeconomic relevance of each opportunity was expressed on a five-level scale (where + = Very Low, ++ = Low, +++ = Moderate, ++++ = High and +++++ = Very High). The 33 factors listed in Table 3.3 (45% of the total) are those ranked at the top level of the assessment. The numbers of the opportunities are those of the initial list in the literature (Horizon Scan Report 2007). The capital letters indicate the respective section of the opportunities list, whereas the identified opportunity is shown in lower case letters.

3.2 *Identification of the Most Relevant to Bioeconomy Problems*

Table 3.4 presents the results of our assessment of the list of 86 risks and problems already identified by Horizon Scanning (HSc), from the point of view of our foresight study, i.e. regarding the emergence of bioeconomy (Wilenius 2014; Toffler 1980; Horizon Scan Report 2007; OECD 2016, 2009a, b).

The level of the bioeconomic relevance of each problem was expressed on a five-level scale (where + = Very Low, ++ = Low, +++ = Moderate, ++++ = High and +++++ = Very High).

The 21 factors listed in Table 3.3 (25% of the total) are those ranked at the top level of the assessment. The numbers are those of the initial list (Horizon Scan Report 2007). For the use of capital and lower case letters, see above.

3.3 *Identification of the Most Popular Options for the Emergence of Bioeconomy*

Table 3.5 presents some results from the use of the Questionnaires. The spread of the responses was such that almost all options appear to be statistically significant. The relative emphasis of responses was given to opportunities, followed by actions to be taken and less so on risks and problems. The latter seem to be linked to unexpected/disrupting factors. Another interesting finding of the Questionnaires was their potential to express the attitude of the responding by their answer profile.

Table 3.4 List of very high bioeconomy relevance problems identified by HSc

P11	BIO-SPHERE: Loss of natural resources
P13	BIO-SPHERE: Infectious diseases constitute a permanent threat
P19	FOOD – AGRICULTURE: Hunger
P20	FOOD – AGRICULTURE: Obesity increases
P36	SAFETY AND EMERGENCY SYSTEMS: Crisis control
P37	SECURITY: No rational security policy
P41	BUSINESS OF S&T: Insufficient knowledge management
P42	BUSINESS OF S&T: The leading role of Europe is in danger
P43	BUSINESS OF S&T: Threat to intellectual property rights
P44	BUSINESS OF S&T: Decreasing confidence in science
P45	SCIENCE & TECHNOLOGY: New risks, ethical issues and social problems
P46	SCIENCE & TECHNOLOGY: Custom-made man
P47	SCIENCE & TECHNOLOGY: Developments give criminals opportunities
P48	SCIENCE & TECHNOLOGY: Robots ousting humans
P50	EDUCATION: Insufficient educational level of population
P51	EDUCATION: Educational system not attuned to the educational biography
P78	INTERNATIONAL SYSTEM: International agreements on tech developments
P79	TENSIONS: Shortage, a breeding ground for conflicts
P83	NATIONAL STATE: Lack of government, business and social partnerships
P84	NATIONAL STATE: Insufficient reflexive/strategic capacity of governments

Table 3.5 List of the ten most significant Questionnaire results

A1	Achievement of longevity
A2	Fighting of serious illnesses
A3	Quality foods – Fighting hunger and obesity
A8	Protection of environment and ecosystems
B3	Cloning/genetic improvement of humans
B10	Bio-ethics and morality issues
C1	Research and innovation
C6	Education and training
C9	Integrated policy framework
D	Other priorities indicated by survey participants and literature (see Table 3.6, below)

Symbols as defined in Table 3.2

Table 3.6 Possible disrupting trends (D) for bioeconomy in the period 2016–2018

(a) Disrupting socio-economic trends relevant to bioeconomy	
D1.	In the new brand experiences are more important than products
D2.	Shift from sharing economy to on-demand economy
D3.	Digital Darwinism of Companies towards social trends
D4.	Dynamic customers change brand dynamics away from legacy strategies
D5.	New jobs for tasks that computers need humans to complete – education as a constant
D6.	Marketing–innovation–technology–Human Resources synergies
D7.	Shift from R&D Departments to Innovation Centres
D8.	Investing in Culture 2.0!
D9.	Businesses operate under radical transparency to gain trust by customers
D10.	Schools pay students to learn how to become agents of expertise
(b) Disrupting technology trends relevant to bioeconomy	
D11.	Web transformed by mobile-first behaviour
D12.	The new Machine Age – Intelligent systems for all everyday life and work uses
D13.	From reactive to proactive medicine
D14.	Plugging tech into the Human Operating System (HOS) to create new "ecosystems"
D15.	Vertical uses of drones for care, delivery, exploration and more
D16.	3D printing applications disrupting supply chains in vertical industries
D17.	Autonomous, self-driving cars transforming transportation and infrastructures
D18.	Experiences get real virtual by immersive computing and next-gen imaging
D19.	In-home battery systems for eco-friendly power of households and cars
D20.	Smart fabrics and bluetooth tech in clothing for body and environment control

Source: Solis (2013)

3.4 Clustering of Opportunities, Problems and Actions

By combining the results of the previous two tables and those of the Questionnaire, we have been able to identify among the huge number of possible combinations (more than half a million) the following nine Clusters of socio-technical action (Geels and Schot 2007).

In parenthesis, we suggest the use of an appropriate for each Cluster symbolic figure taken from history, literature or mythology (Kereńyi 1951).

With asterisks (*) we indicate factors shared between Clusters.

CLUSTER #1: The Silver Society (Gilgamesh)

This cluster is formed by a rather small, but clear – and clearly opportunistic – presence on the horizon of the emerging bioeconomy.

A1 Achievement of longevity
O46 SCIENCE & TECHNOLOGY: Biomedicine and the prolongation of life
O53 SOCIAL, FAMILY, WORKING LIFE: Ageing as Silver Fleet
O54 SOCIAL, FAMILY, WORKING LIFE: Ageing actively

The story of Gilgamesh, the hero who wished eternal life to find out its limitations, could serve as an appropriate lesson to be learnt from the "silver" vision of longevity.

CLUSTER #2: The Cure and Care Economy (Aesculapius)

This cluster shows also a strongly opportunistic presence on the horizon of the emerging bioeconomy, with the exception of just one problem-limitation (P13) that could come as a warning: beware of the permanent threat of infectious diseases.

A2 Fighting of serious illnesses
P13 BIO-SPHERE: Infectious diseases constitute a permanent threat
O24 HEALTH CARE: Monitoring/developing vaccines for infectious diseases
O26 HEALTH CARE: Preventive medicine
O37 SCIENCE & TECHNOLOGY: CT to improve health and physical capacities
O44 SCIENCE & TECHNOLOGY: Understand disease with developmental biology
O45 SCIENCE & TECHNOLOGY: Biological research into pathogens

The case of Aesculapius – the founder of medicine and proponent of a holistic approach - could serve as the key lesson to accompany the emergence of a Cure-and-Care approach.

CLUSTER #3: The Horn of Affluence (Amalthea)

This cluster shows a more balanced emergence, as its foreseen opportunities appear to be of equal power to its scanned problems, in clear contrast to the Health Cluster.

A3 Quality foods – Fighting hunger and obesity
B1 Hyper-intensive farming of plants/animals/others
B2 Genetically modified foods
P19 FOOD – AGRICULTURE: Hunger
P20 FOOD – AGRICULTURE: Obesity increases
O17 FOOD – AGRICULTURE: Strategic function agriculture
O18 FOOD – AGRICULTURE: Modern craftsmanship
O19 FOOD – AGRICULTURE: Healthy, eating patterns-based, functional foods

The story of Amalthea, the sacred goat who secretly fed the hiding baby Zeus, offering him a Horn of Affluence full of the food of gods, could serve as the key symbol of the role of nutrition.

CLUSTER #4: The Garden of Mother Earth (Gaia)

This is the cluster of Green and Sustainable Development, full of opportunities, with just a touch of management and some pressure on resources. The list of features covers a variety of "Green" applications, with emphasis on energy.

A5 Clean biofuels
A6 New bio-materials for quality of life
A4 A new generation of cosmetics
A8 Protection of environment and ecosystems
C4 Environmental management
P11 BIO-SPHERE: Loss of natural resources
O01 ATMOSPHERE: Climate change
O05 HYDROSPHERE: Minerals and gas hydrates from ocean floor
O09 BIOSPHERE: Focused development of marine and other wild environments
O28 ENERGY PRODUCTION: Robust strategy for energy supply
O64 NATIONAL ECONOMY: Wellness
O67 NATIONAL ECONOMY: New technical products and services*

The case of Gaia represents a story of the Garden of Eden revisited, in which the green-ness could concern several elements of the emerging bioeconomy: resources, processes, products, services, social practices, etc.

CLUSTER #5: The Green Governance (Nemo)

This cluster is one of the richest in converging elements, including several Action and Management options. In addition, the cluster appears well balanced with respect to problems and opportunities – in contrast to some of the other clusters.

A11 New national development model
A10 Regional development
A9 New employment possibilities
B8 Increase of unemployment due to new technologies
C2 New investment projects
C3 New forms of project financing
C9 Integrated policy framework
P48 SCIENCE & TECHNOLOGY: Robots ousting humans
P78 INTERNATIONAL SYSTEM: International agreements on tech developments
P83 NATIONAL STATE: Lack of government, business and social partnerships
P84 NATIONAL STATE: Insufficient reflexive/strategic capacity of governments
O59 NATIONAL ECONOMY: Employment at all education levels
O60 NATIONAL ECONOMY: Country branding
O61 NATIONAL ECONOMY: The right economic growth
O63 NATIONAL ECONOMY: Entertainment industry
O65 NATIONAL ECONOMY: No use made of unique features
O67 NATIONAL ECONOMY: New technical products and services

Captain Nemo, the hero created by the pen of Jules Verne, would be the ideal leader to navigate the emerging bioeconomy in the rough, uncharted and uncertain waters of the twenty-first century.

CLUSTER #6: The Human Re-engineering (Pygmalion)

This cluster represents a case of almost balanced emergence, as opportunities are of the same power as problems. We also note that opportunities are mostly affected by health-care aspects, whereas problems by moral and ethical issues.

B3 Cloning/genetic improvement of humans
A7 Reproduction and regrowth of organs
B4 A culture of beauty and youth at any cost
P46 SCIENCE & TECHNOLOGY: Custom-made man
O08 BIOSPHERE: Inspiring life processes
O10 BIOSPHERE: (Re-) creating a living environment can contribute to more life
O11 BIOSPHERE: Intervention in humankind's own evolution

Pygmalion, a superb sculptor, created Galatea so perfect that he fell in love with his creation, with dire consequences. This story could teach Human (Re-) Engineers a great lesson.

CLUSTER #7: The New Risk Bio-Society (Sphinx)

This cluster is dominated by several types of problems and people's worries, whereas the opportunities are minimal. The multitude of such emerging bioeconomy features indicate that a new type of action territory will be spreading, whereas appropriate action plans are missing.

B5 Biological weapons – Bio-risks
B6 Domination of large multi-national companies
B7 Control of bio-innovations through patents
B9 Lack of protection of personal biological data
B10 Bio-ethics and morality issues
C5 Social tasks and initiatives
P36 SAFETY AND EMERGENCY SYSTEMS: Crisis control
P37 SECURITY: No rational security policy
P43 BUSINESS OF S&T: Threat to intellectual property rights
P45 SCIENCE & TECHNOLOGY: New risks, ethical issues and social problems*
P47 SCIENCE & TECHNOLOGY: Developments give criminals opportunities
P79 TENSIONS: Shortage, a breeding ground for conflicts
O33 SCIENCE & TECHNOLOGY: Threats as leitmotif for innovation*
O39 SCIENCE & TECHNOLOGY: National security

The new bioeconomic landscape of the emerging risks and worries can find its proper mythological symbol in the enigmatic smile and the deadly riddles of the Egyptian Sphinx.

CLUSTER # 8: Shaping the Future (Leonardo)

In this cluster, opportunities seem to be the main driving forces, but problems are also present and acting; with the whole equilibrium being regulated by a relevant action plan. Overall, this cluster represents forces that are shaping up the whole emerging bioeconomy.

C1 Research and innovation
P48 SCIENCE & TECHNOLOGY: Robots ousting humans*
P41 BUSINESS OF S&T: Insufficient knowledge management*
P42 BUSINESS OF S&T: The leading role of Europe is in danger*
O33 SCIENCE & TECHNOLOGY: Threats as leitmotif for innovation*
O34 SCIENCE & TECHNOLOGY: Promising new fields of science
O35 SCIENCE & TECHNOLOGY: Converging technologies (CT)
O36 SCIENCE & TECHNOLOGY: CT for human cognition and communication
O38 SCIENCE & TECHNOLOGY: CT to improve group and social processes

Proposing Leonardo's story as a symbolic element of the particular type of bioeconomy emergence was not because of Gioconda's enigmatic smile, but for Leonardo as an innovator, being well ahead of its time – a warning to colleagues aiming to shape the future by their research actions.

CLUSTER #9: For a Skilful Society (Piaget)

In contrast to research (Cluster #8), the main driving forces in the field of education in bioeconomy seem to be the problems, worries and risks. The second in power factors are those of action planning and management, with creativity and educational innovations playing a key role.

C6 Education and training
C7 Art and culture
C8 Youth initiatives
C10 Citizens information/awareness
P50 EDUCATION: Insufficient educational level of population
P51 EDUCATION: Educational system not attuned to the educational biography
P41 BUSINESS OF S&T: Insufficient knowledge management*
P42 BUSINESS OF S&T: The leading role of Europe is in danger*
P44 BUSINESS OF S&T: Decreasing confidence in science
O49 EDUCATION: Educational system keeping up with global knowledge increase
O50 EDUCATION: Creativity in education

The story of Jean Piaget has covered educational theory and practice in ways that have influenced, and keep affecting, educators and institutions. So, his story is useful as a shaping factor of people's minds, skills and lives, especially those involved in the emerging bioeconomy.

CLUSTER #10: "Mystery" Cluster – Action of Disrupting Forces (Agatha Christi)

The above-listed nine Clusters provide less than 50% of the explanatory power of the assessment; therefore there is a need to define a 10th Cluster, which can be justified as the result of disrupting forces. In Table 3.6, we present a list of possible disrupting trends in action for the current period of research.

The story we associate to this "mystery" cluster is the one of the most famous mystery story-teller, Agatha Christi; the basic element of her stories is the one of surprise and of an emerging unexpected outcome, which looks a posteriori normal and predictable with lots of hindsight.

3.5 Emerging Bioeconomy: The Role of Clusters

Using the Clusters identified as determining the dynamics of bioeconomy emergence, we can now revisit the list of factors presented in Table 3.1. Table 3.7 summarizes the effect of Clusters on the emergence elements, almost all of which can be expressed as the result of the interaction of a minimum of two (2) Clusters. Of particular importance are two of the nine clusters: Cluster #4 (Garden of Eden) and Cluster #8 (Shaping the Future), as well as their multiple combinations and interactions on energy and nano-molecular applications.

4 Concluding Remarks

The foresight of the emerging bioeconomy based on a Horizon Scanning approach has made possible the identification of the nine essential Clusters of opportunities, problems and stakeholder actions, as shown in Table 3.8.

Table 3.7 Emergence of bioeconomy in K6 explained by Clusters

Health, especially Holistic Health Applications **#1, 2**
Psycho-Social Health **#2, 7**
Biotechnology # **3, 4**
Biomedical Applications # **1, 2**
Nano-Molecular Applications # **4, 8**
Nuclear Energy Developments # **4, 7**
Renewable Energy Breakthroughs # **4, 8**
Hydrogen Energy **#4, 8**
Robotics, Intelligent Machines **#6, 9**
Artificial Intelligence **#6, 9**
New Leadership & Management **#5, 7**
ICT to support the above – Post-Info Applications **#8, 9**
Converging Technologies (info-bio-nano, etc.) **#8, 9**

Table 3.8 The essential Clusters of emerging bioeconomy

#1: *The Silver Society* - Ageing and Demographic Barriers
#2: *The Care and Cure Practice* – Towards Holistic Health
#3: *The Horn of Affluence* – Food for 10+ B People on Earth
#4: *The Garden of Mother Earth* – Green Biobased Economy
#5: *The Green Governance* – New Type of Managers and Leaders
#6: *The Human Re-Engineering* – Re-Designing Nature
#7: *The New Risk Bio-Society* – Re-Setting Boundaries
#8: *Shaping the Future* – Smart-to-Wise Research Priorities
#9: *For a Skilful Society* – Training Brains for the 10 Clusters
#10: *"Mystery" Game Changers* – Surprise Disrupting Factors

The products of interaction between those nine Clusters can explain the various empirical observations of emergence phenomena reported by various researchers, e.g. (Koukios et al. 2018; Timmis and Timmis 2016).

Of particular value is the possibility of this approach to explore the content of the ongoing Kondratiev Wave of socio-technical change (2010-?) with respect to its bioeconomy-relevant elements.

Acknowledgments The research for this chapter took place at the Innovation Strategies Department of the Institute for Sustainable Technologies – National Research Institute, in Radom, Poland.

This work was prepared in the frame of the research project titled "The New Tsunami of Socio-Technical Change: Foresight of Agro/Bio/Chemo/Eco/Cogno...Convergence (BIO-TSUNAMI)", supported by the POLONEZ 1 Programme of the National Science Centre (NCN) with contract no. UMO-2015/19/P/HS4/04103. This project has received funding from the European Union's Horizon 2020 Research and Innovation Programme under the Marie Skłodowska-Curie grant agreement No. 665778.

References

Cichocka, D. et al., European Union Research and Innovation Perspectives in Biotechnology, *Journal of Biotechnology*, Vol. **157**, 383–191, 2010.

European Commission, *Europeans and Biotechnology in 2010 - Winds of Change?,* Report of a Working Group, Brussels, October 2010.

European Commission, *Innovating for Sustainable Growth: A Bioeconomy for Europe*, COM (2012) 60 final, Brussels, 13.2.2012.

Geels, Frank W., and Johan Schot, "Typology of sociotechnical transition pathways", *Research Policy*, **36** (3), 2007.

Goldschmidt, A.J.W., and Hilbert, J., *Health Economy in Germany - Economical Field of the Future*. Wikom Publishing House, Wegscheid, 2009.

Horizon Scan Report, *Towards a future oriented policy and knowledge agenda*. Eds. R. In't Veld et al., Horizon Scan Publication, The Hague, 2007.

Kerenyi, Karl, *Greek Mythology*, James Hudson, London, January 1951.

Koukios, E., "The bio-tsunami project", In: Koukios, E. et al. (eds.), The emergence of bioeconomy - Opportunities and risks - A forward-looking study, ITeE, Radom (PL), 2017.

Koukios, E. et al., Final Report, Working Group "Biotechnologies", *Technology Foresight in Greece*, General Secretariat for Research and Technology, Athens, 2005 (in Greek).

Koukios, E. et al., "Targeting sustainable bioeconomy", *J. Cleaner Production*, Vol. **172**, 3931–3941 (2018).

Koukios, E., From biotechnologies to bioeconomy and biosociety, In: *Technology Foresight*, D. Agrafiotis (ed.), Ellinika Grammata, Athens, 2015 (in Greek).

Koukios E.G., "Technology management for the biobased economy: Mapping, dynamics and policies – The case of Greece", In: *International cases on innovation, knowledge and technology transfer*, Ed. D.M. Trzmielak and D.V. Gibson, Centre of Technology Transfer, University of Lodz, Lodz 2014.

Koukios, E.G., and Sacio-Szymańska, A., "Assessing the emergence of bioeconomy by the radical technology inquirer tool", European Journal of Futures Research, 6:23 (2018) (https://doi.org/10.1186/s40309-018-0152-x).

Mason, P., *PostCapitalism: A guide to our future*, Penguin Books, 2015

Nefiodow, L., and Nefiodow, S., *The Sixth Kondratieff - The New Long Wave of the Global Economy*, Charleston, 2014.

OECD, Directorate for Science, Technology and Innovation (DSTI), *An OECD scan of megatrends and technology trends in the context of future research policy*, Danish Agency for Science, Technology and Innovation, Copenhagen, 2016.

OECD, *The Bioeconomy to 2030,* Report, 2009a, http://biotech2030.ru/wp-content/uploads/docs/int/The%20Bioeconomy%20to%202030_OECD.pdf (accessed: 28 January 2019).

OECD, *The Bioeconomy to 2030,* Summary Report, 2009b, https://www.oecd.org/futures/long-termtechnologicalsocietalchallenges/42837897.pdf (accessed: 28 January 2019).

Smihula, D., "Waves of technological innovations and the end of the information revolution, Journal of Economics and International Finance", Vol. **2** (4), 58–67, 2010.

Solis, B., *What's the Future of Business? - Changing the Way Businesses Create Experiences*, Altimeter Group, San Francisco Bay Area, 2013.

Timmis K.N., and Timmis J.K, "A routine check-up at the digital medical centre". Crystal Ball Section, *Microbial Biotechnology J.*, 1–3, 16 October 2016.

Toffler, A., *The Third Wave*, New York, Bantam Books, 1980.

Wilenius, M., "Leadership in the sixth wave- Excursions into the new paradigm of Kondratieff cycle 2010-2050", *European J. of Futures Research*, Vol. **2,** no. 1, 2:36 (2014).

Wilenius, M., *Patterns of the future – Understanding the next wave of global change*, World Scientific Publishers, London, 2017.

Part II
Towards Circular Bioeconomy and Biosociety

Chapter 4
Strategies for a Rapid Transition to a Circular, Biobased Society

Lene Lange

Abstract Rapid transition towards circular, biobased economy includes upscaling current knowledge/technologies *and* at the same time investing in public/private collaboration for developing biobased technologies. What is ready to be done and where investment is urgently needed can be summarized as follows: Reduction of emissions through improved resource efficiency by upgrading waste; stopping loss of biodiversity by reducing pesticides and land-use; hereby strengthening industrial competitiveness, creating jobs of many skills and improve rural livelihood. An analysis is given for so far too slow implementation: Global climate agenda focuses on energy/transport, where CO_2 reduction is easily calculated, while indirect, but huge effects of improved resource efficiency are neglected. A strategy for rapid transition is outlined: unlocking full potential of the biomass, upgrading all components to the highest level. Upgrade all types of biomass – including industrial side streams and organic wastes. Communicate that biobased solutions address many societal challenges. Highlights are given of promising emerging bio-based technologies: negative emission technologies, circular textile industry technologies, BioAg, substituting pesticides, biological soil improvement and food and feed ingredients for improved health. Instruments and drivers are described, embracing that societal changes do not come from technology alone: Incentive structures, building markets; knowledge dissemination; transforming EU subsidies to drive implementation of greener solutions; international biobased collaboration to gain priority, including win/win strategy for growth economy collaborations; and dedicated bioeconomy alliance between Africa and Europe.

Keywords Circular biobased economy · Transition strategies · Valorizing industrial side streams · Upgrading organic wastes · Upscaling technologies · Public/private synergies · Emerging biobased technologies · Global climate change agenda · International cooperation · New alliance · Strengthening African biobased economy

---

L. Lange (✉)
BioEconomy, Research & Advisory, Karensgade 5, Valby, Copenhagen, Denmark

E. Koukios, A. Sacio-Szymańska (eds.), *Bio#Futures*,
https://doi.org/10.1007/978-3-030-64969-2_4

1 Introduction

The first step forward, implementing the biobased society, is to use the knowledge, knowhow and technologies we already have to upgrade the biological resources, producing value added products from what now goes wasted. Societal drivers for doing this are many. Improved use of the biological resources already produced in agriculture, forestry and aquaculture contributes to reducing emissions contributing to climate change mitigation; strengthening industrial competitiveness, hereby creating more jobs for many types of skills and competences; stimulating rural development by developing local biobased activities; as well as giving more room for biodiversity, by producing more animal feed as well as non-food products with no use of land and pesticides. However, in spite of the many value adding societal upsides, implementing new and more responsible use of the biological resources is still moving too slow. The primary three reasons for this are as follows: The climate agenda is still, also globally, focused on energy and transport, which can be easily calculated in CO_2 Excel sheets, while effects of more responsible consumption are much more complex to calculate, but not less important. The concepts of biology, plant biomass structures, enzymes and microbial conversion are not easily communicated in one-liners. Lastly, we need to attract much more public attention to our messages. As scientists we are often so enthusiastic about the new possibilities, just opened for new frontier research that we forget to tell enough about all the knowledge we already have, ready for use. This chapter aims to provide a background for empowering many more (scientists as well as informed laymen) to communicate the message about what can be done much better already now, hereby advancing responsible consumption of the local and global biological resources, along with communicating the next opportunities that are ready for implementing in only a few years.

2 Unlock the Full Potential of the Biomass by Upgrading All Components to the Highest Level

The first and foremost thing to do to forward a rapid transmission to a circular, biobased society is to recognize biomass as a precious and finite resource, which should be used primarily when upgraded to its full potentials. Exploiting only the energy content of the biomass, by, e.g. combustion of wood chips or conversion of the biomass to energy only, biofuel will in the future most probably not be seen as sustainable in a resource efficiency perspective; the long term most precious in the biomass is the structures, not the energy, as we in the future can get ample supply of truly emission-neutral, renewable green energy from wind, sun and geothermy. The following example illustrates how structures of the biomass can be used for making higher value products: It is possible to produce prebiotic, gut-health-promoting xylan oligoes (XOS) from the hemicellulose fraction of wheat straw

(Dotsenko et al. 2018). Thus, higher value feed and food ingredients can be recovered first from the wheat straw biomass, and the residual cellulose fibres can be used as basis for making cellulosic biofuel or maybe even used for producing higher value products, by using the fibres, upgraded to new types of textile fibres. Another example is to substitute the current practice of converting a full crop or biomass into biogas only, a common practice in many parts of EU where biogas prices are subsidized (as the only type of biobased product). New innovations allow for making much higher value instead of using the biogas directly as bioenergy. Biogas can be upgraded to higher value products, separating biogas into CO_2 and methane, using the methane as building block for a new technology platform, and producing biobased jet fuel by a GTL (Gas2Liquid or PtX) approach. In this process, naphtha will be produced as a side product in big volumes, constituting a large supply of new raw material for producing bioplastic (reference, H Wenzel, SDU). Furthermore, biogenic CO_2 (as well as atmospheric CO_2) can be hydrogenated by splitting water (using affordable surplus windmill electricity) into hydrogen and oxygen. Also the residual bacterial biomass from anaerobic digestion can be developed into valuable, precisely administrated, soil improvement products. The press pulp after production of plant oils can be used as basis for producing protein-rich food or animal feed; and the press pulp after juice production can be used for making textile fibres or higher value antioxidant-rich health-promoting products (e.g. from black berries). The new type of grass biorefineries can produce alternative, local protein to substitute for non-sustainable soy protein, imported from South America. The classical exploitation of sugar polymers from seaweed can be developed into modern seaweed-based blue biorefineries, in which an entire spectrum of components can be valorized. Not the least are the highly interesting, almost totally, unexploited unique marine, health-promoting products with documented anticancer effect, as well as prebiotic effect, or polymers with wound-healing potential, of relevance for the growing issue of diabetes-related not easily healing wounds. The examples are legio.

The first products in the new biomass conversion bioeconomy are built on the following process: using plant cell wall-degrading enzymes to break plant polymers completely down to short sugar oligoes or even to monomeric sugars. Such simple sugars were then used to grow microbes (bacteria or fungi), which then produced fuel (e.g. bioethanol by fungal yeast), organic acids from filamentous fungi or building blocks for chemicals and materials (primarily produced by bacteria). Notably, with the emerging technologies in sight, it is obvious that we right now are facing a paradigm shift in bioeconomy. Fuel, materials and chemicals will in the future be possible to make from atmospheric CO_2 by a carbon capture and use approach (see Sect. 6.1). With this in mind the next level of bioeconomy approach is expected primarily to follow a new biorefinery approach: to valorize the biomass through biorefining technologies where nature's complexity is kept intact and used as basis for production of higher value products for health, food and feed ingredients, as well as circular use of fibres (e.g. textiles) and polymeric building blocks. One of the first examples of such a biorefinery was the Green Biorefinery introduced above. Here it is the primary product. The protein content of the green grass is recovered by a simple screw press treatment, separating the grass into juice and a fibrous pulp. The protein is recovered from the juice by precipitation. From the hemicellulose of the residual side stream, the fibrous pulp, higher value animal feed additives can be made in the following way: By enzyme treatment, the hemicellulose fibres are broken down to short xylan oligoes (XOS), documented to have prebiotic effect, improving gut health of non-ruminant animals and people.

Also animal-derived complex molecules are upgraded in next level biorefineries. Excellent examples of this new trend are that all parts of the milk feedstock or all components of a potato feedstock can be upgraded to higher value such as food and feed ingredients (link to Arla and KMC to be inserted). Next in line could be full valorization of the press pulp after olive oil, sunflower oil or rape oil, full valorization of by-products from flourmills and full valorization of higher value products from innards of pigs, cattle and chicken as well as from keratinaceous by-products such as chicken feather, fish skeleton and fish skin, pig bristles, etc.

Thus, for a rapid transmission to a more biobased society, it is important to introduce a cascading, holistic integrated approach to the biomass conversion processes. Utilizing all components of the biomass to their full potential is called the cascading principle. It provides basis for more responsible consumption, contributing significantly to climate change mitigation and climate change adaptation, using the harvest fully instead of letting 30–50% be wasted as is currently done globally. In a future perspective of planet health as well as from an economic point of view, it makes good sense to use the raw material as efficiently as possible. In Denmark, we have several examples of full industrial upgrade of all side streams of a given feedstock, which have been developed and commercialized and only built on business-targeted investments. Return of investments was secured by delivering highly positive business opportunities, providing for expanding business and improved international competitiveness.

3 Upgrade All Types of Accessible and Sustainably Sourced Biomass – Not Just Crop Residues – But Also Industrial Side Streams and All Types of Organic Wastes

In the next era of biobased industries, we expect many new feedstocks to be included, processed by many new even more efficient and cost-conscious bio-processing regimes. Further, new products, with new user-friendly functionalities will be developed, addressing a broader spectrum of societal challenges (see Sect. 4 below). In order to unlock the full potential and optimized valorization of conversion of all these types of biomass, it is essential to distinguish and address separately the three major types of biomass feedstock: residues from primary production, industrial side streams and wastes (mixed, wet and dirty).

Residues from primary production: Forestry residues can be upgraded to a higher level than hitherto pursued. Bark holds a high content of antioxidants and oils, as well as nutritional components. New initiatives have been taken, e.g. in Africa to make upgraded use of bark Thorben G. Nielsen, Pesitho, pers.com. Historically, in the Nordic countries a food security tradition existed, when the harvest failed: bark was mixed into bread and porridge. This practice is even included in poetry in the text of one of the most famous Norwegian songs (quote in Norwegian: "Frosten tock vor grannes aker; vi får atter blande bark i brodet"; in English: "The frost took our neighbors harvest; we will again have to mix bark in our bread"). However, new technologies have taken valorization of forestry residues into new levels: converting wood paste into a substrate for fungal growth. By adding a source of nitrogen, e.g. seaweeds, it is possible to produce single cell proteins for animal feed, simply by growing the highly nutritious fungal yeast on the N-enriched wood paste. The resulting product is not only protein-rich; it is also rich in vitamins (B12). And the fungal cell wall materials are documented to have prebiotic, gut health-stimulating effect. The evidence for such claims are very solid: For more than a decade the fungal yeast biomass, left after production of human insulin, was used as animal feed for very successful large scale production of pigs. The pig farmers reported on healthy, fast-growing pigs.

In fishery a new challenge, which can also be seen as an opportunity, is emerging. EU has passed a new directive, banning dumping of by-catch at sea; the by-catch must be landed along with the primary catch product. This directive has the effect that a new fish-based protein- and oil-rich feedstock will become available for upgrading. Such upgrading may even reach high-priced gourmet level products, made from parts of hitherto unexploited fish species. Furthermore, harvesting of invasive species and species which by climate change has started occurring in ecologically unbalanced volumes can provide a new short cut for making higher value products from aquatic animals (e.g. protein-rich sea stars, chitosan-rich medusas, or (farmed?) invertebrates such as sea cucumber). Also for agricultural residues, a lot of additional steps can be taken to ensure improved bioresource efficiency. The following example illustrates the unexploited potential: Discarded, odd-sized and odd-shaped vegetables and fruits along with foliar parts of (also higher value organics) vegetables can be upgraded to new types of salted, dried or microbially fermented vegetables inspiring high gourmet grade in Japanese cuisine as well as leading to more, locally, produced nutritious animal feed.

4 Upcycling of the Biorefinery Concept to a New Level by Growing Heterotrophic Organisms on Residual Streams, Creating a New Protein-Rich Biomass

The available and sustainably accessible biomass cannot deliver to all the needs of a more biobased society. The bioresources are finite. Therefore, biobased products will be in higher demand than what we can deliver by even upgraded use of all sustainably sourced biomass components through optimized biorefining. This is due to the fact that biobased products will be needed for fulfilling many purposes, addressing an entire range of societal challenges. The most ambitious, among the current generation of biorefineries, applies a cascading principle to valorization of biomass. It produces several types of biobased products, all in all making optimized use of each of the components of the biomass. However, heterotrophic organisms like bacteria, fungi and insects can be grown also on biomass residues with a low level of proteins (if nitrogen of other sources are supplied). This enables us to get more protein out of the biomass than the protein present in the biomass itself, achieved by producing, e.g. single cell bacterial, fungal or insect protein. And it also opens for the opportunity to produce more essential and nutritional food and feed ingredients beyond proteins, such as lipids, vitamins, antioxidants, etc. in large scale. This is doable because the heterotrophic organisms (bacteria, fungi or insects) secrete enzymes for breakdown of the residual plant biomass-derived side streams. For insects it is the bacteria in the insect gut which break down plant cell wall materials.

Another sustainable approach to produce higher volumes of compounds found in nature is by making microbes, grown in a fermenter that produces the compound for us, with no requirement for land use, with minimal level of emissions and no harm done to the organisms, which produce such materials in nature. One example is to produce higher volumes of new types of antibiotics and other health-promoting compounds found in nature, e.g. from rare plants or extremophilic microorganisms. For doing this we need even more research on heterologous expression of both small peptides and complex secondary metabolites. Another fermenter-based fungal production of essential molecules is to produce different kinds of milk protein and different kinds of meat proteins. This represents a new approach to contributing to meeting the global demand for meat and milk without burdening the climate with emission of the highly potent greenhouse gas methane. Also production of specialized molecules, such as flavours or pigments, can be done in fungal or bacterial strains, optimized for microbial fermentation. All in all, growing microbes in fermenters is expected to be one of the most important approaches to providing for global supply of sustainably produced food and also new types of natural products (including new antibiotics) produced for the health of man and animals by pharmaceutical industries.

5 Develop Biobased Products, Addressing a Whole Spectrum of Societal Challenges

The first bioeconomy era was totally dominated by making biobased substitutes for fossil-based products such as fuels, chemicals and materials based on upgrading/refining crude oil to a wide spectrum of products. This focus was developed to address the most burning global issue: how to switch from fossil based to bio-based. This approach – for short described by one line, "from biomass to biofuel" – has been so well communicated that it shadowed for the many other approaches for valorizing biomass including the approaches, where much higher economic and societal value could be generated than just making low-priced bio-energy. Now the trend is to develop a wide spectrum of biobased products. Products which together are addressing many societal challenges, as well as contributing to meeting several of the UN Sustainable Development Goals: More responsible consumption, achieved by using all parts of the biomass improved health of man, animals and plants, achieved through smart and affordable technologies; upgrading biorefinery side streams for combating lifestyle diseases and improving gut health; clean and safe drinking water through cutting down use of pesticides, achieved through the development of new BioAg products, which strengthen resilience and robustness of the plants, instead of killing the intruders, diseases and pests; giving more room for biodiversity, by making smarter circular uses of cotton fibres and by producing animal feed from upgrading of side streams and by producing animal proteins (milk and meat protein) for the future in fermenters, viz. with no use of land; and not the least to improve social inclusiveness by creating jobs and stimulating rural and coastal development, just to mention a few of the very wide spectrum of advancements, which can be delivered by the next generation of microbial production, providing many new products for the biobased and bio-produced bioeconomy.

6 Invest in Emerging Biobased Technologies, Taking Leapfrog Steps Towards a Biobased Society

6.1 Negative Emission Technologies

Seen from a global resource-approach, the emissions now leading to the serious threat of climate change can also be seen simply as "misplaced resources". By capturing and upgrading such misplaced carbon resources, we can contribute to both reduction of climate change and resource efficiency. New negative emission technologies (NET) are going beyond the current focus on Carbon Capture and Storage. Captured carbon should of course preferably be used instead of just stored, making no value and with inherent risk of future leaks being harmful to climate. Biobased feedstocks, undergoing chemical and/or biological processing, can lead to new, highly needed NET innovations. Many such innovations can be embraced under the heading "Carbon Capture and Use, CCU". The last few years of research have shown that this can be done, making significant value from such emissions and hereby providing routes for producing useful products, which have a negative carbon footprint. An innovation, which is much needed to be recognized, up-scaled and commercialized, is the following case: biogenic biogas can be split into CO_2 and methane. Such biobased methane can subsequently (by Gas2Liquid or Power2X, PtX technologies) be used as building block to produce liquid e-fuels. This process is enabled by using electricity (low-priced and zero emission energy) from windmills, made affordable especially in the periods where the windmills produce surplus of electricity. Even more methane can be produced from biogas if also the CO_2 from the biogas is converted into methane (viz. hydrogenated by splitting water into hydrogen and oxygen). Such methane-based e-fuel is especially suitable as substitute for the currently used fossil-based jet fuel, bunker oil and diesel for heavy transport vehicles (flights, ships and trucks), where no other sustainable energy resource exists today (batteries will be too heavy with current battery technology). From this process, making higher value from biogas, another useful compound (Naphtha) is produced as a large volume side stream. Notably, Naphtha is a promising building block for making both bio-plastic and new types of biobased synthetic textile fibres (see Sect. 6.2 below).

6.2 Circular Technologies for a More Sustainable Textile Industry (Re-use Textiles and Fibres)

The paramount challenge: Textile industry is the most environmentally burdensome industry next only to the oil sector, not just by emissions but also by taking land from food production and biodiversity and being amongst the largest consumers of water and pesticides. Circularity in the textile sector is a necessity! Opportunity: The textile sector is aware of its burning platform and ready for change. Ambitious goals have over the last few years been expressed by a broad spectrum of players in the fashion and textile sector, indicating that access to more sustainably-produced fabrics and materials is a necessity for the sector to meet their sustaníability targets. Further, EU has taken drastic steps to ban combustion of textile (from 01.01.2025), hereby, de facto creating big volumes of used textiles ready for upgrade. The Danish Bioeconomy panel recommended the Danish government to give priority to developing upgrade technologies for discarded textiles. Technical options for textile circularity: (1) use and re-use cloths; (2) downgrading outsorted textiles to carpets, stuffing, etc.; (3) circular, upgraded use of discarded textiles as raw materials for new textiles; and (4) cleaning and re-using fibres of outsorted textiles (Fig. 4.1).

6.3 BioAg Products, Substituting for Pesticides Through Improved Plant Robustness

BioAg products are a highly interesting and promising part of the biobased era. Biobased BioAg products (microbes, microbial enzymes or metabolites) can be used for addressing essential challenges, such as strengthening the crop plants (adaptation to climate change), as substitutes for pesticides, to improve soil fertility, and contributing to global food security as well as – most importantly – leaving more room for biodiversity.

Fig. 4.1 Danish Marine Proteins: Sea stars, when appearing in invasive amounts can be processed into protein-rich animal feed

Pesticide use is a threat for drinking water quality and safety as well as for biodiversity and health of man. The new era of BioAg products has already started. A fungus (a species of the mould genus *Penicillium*) has been developed into an already commercialized product, "Jump Start". This type of fungus is known for having nutrient resource efficiency improvement effect by mobilizing otherwise for the plants inaccessible bound phosphorus from soil. A further effect of this microbial product is to give more robustness to newly germinated crop plants in case of spring draught. Other products, already available or being close to commercialization, have the effect of making the crop plants more robust to attacks of pests and diseases. Further, other microbially derived products can improve soil fertility by stimulating the soil microbiome, including a positive effect on the rhizosphere microbiome. The entire BioAg area is a very complex area as beneficial effects can be difficult to provide evidence for and therefore the mode of action being difficult to elucidate and document. The perspective for positive development of the BioAg product area is to be leveraged by improved understanding of microbiome composition and function, including the microbiome of the phyllosphere, rhizosphere or seed surface; all types of microbiomes are interacting with pests and diseases, being either soil-, wind- or seed-borne.

Finding substitutes for chemical pesticides is urgent. New very disturbing developments of fungal infections of (especially immuno-compromised) humans are taking place in growing frequency as we speak. Non-curable fungal infections are caused by fungal isolates (especially of species of the fungal mould *Aspergillus*) which have acquired resistance towards the rather few efficient fungal drugs we know of. Notably, the antibiotic resistance in the mould is not developed due to overuse of fungal drugs for treating humans. Research results point towards such resistant fungal strains that are developed as a response to frequent use of fungicides in rather high doses on agricultural crops. The world is in dire need for being able to reduce our use of pesticides, not the least fungicides. Examples have been recorded, where the patient survived leukaemia (or other types of cancer), but subsequently died from a non-curable fungal infection, invading the brain or the lungs.

6.4 New Biological, Soil Improving Products

Soil improvement has so far not been in focus for developing the biobased economy. However, selected side streams of next generation biomass conversion can through targeted processing be developed into optimized soil improving products (biochar and fibre-rich products). Notably, while we are getting closer to having a good overall grip on the carbon cycle on the planet, terrestrial as well as marine, we are slowly encountering that there might be new and unpleasant surprises in the nitrogen cycle. It has recently been shown that climate change may lead to a higher risk of nitrous oxide to be produced. Notably, laughter gas, N_2O, has approx. 300 times the greenhouse effect as compared to CO_2. More specifically, N_2O can be produced as an indirect effect of use of inorganic N-holding fertilizers. Another source of nitrous oxide is being produced by big container ships. It comes from N being present in the "dirty" bunker oil. N_2O is also produced in hitherto non-disclosed amounts by airplanes. Further, emission of nitrous oxide is also occurring, when low lying and wet agricultural soil is attempted cultivated.

Use of inorganic phosphorus as fertilizer constitutes another burning platform: inorganic P is getting close to being depleted. However, interesting new technologies have been made for recovering phosphorus by biological means: bacterial biomass from anaerobic digesters (AD for waste water treatment) are upconcentrating P in their bacterial cells. Such occurrence of phosphorus can be recovered from the AD bacterial biomass, hereby providing a new biogenic source of phosphorus for soil improvement, present in a form, which can be administrated precisely so that overdosing does not take place – as is the risk for both N and P from animal manure. A method for achieving circularity of nutrients by "Nutrient Capture and Use (NCU)" will most likely be the approach used already in the rather near future. Hereby two things are achieved in one go, rescuing resources, which are useful for improved soil fertility and being possible to administrate in correct doses; as well as reducing pollution of the nearby waters, by avoiding excess of run-off nutrients from agricultural fields; excess nutrients which could harm biodiversity. Nutrient Capture and Use (NCU) can be used in many parts of EU, where agricultural land is close to the sea as occurring, e.g. in the Baltic countries, Ireland, Sweden and Denmark. It is to be expected that such countries in the coming years will develop new Nutrient Capture and Use technologies. This could also be highly relevant for many other parts of the world, e.g. New Zealand, Bangladesh, African continent, China, Indonesia, the Philippines, etc. Overall, new types of biomass-derived or microbially derived soil improvement products are relevant for many parts of the globe, however especially a crucial need for tropical soils. Such soil types are under climate change challenged conditions under threat for being even faster depleted of soil nutrients. Sustainable and affordable, locally accessible soil improvement products are one of the single most important efforts for improving agricultural yields in Africa. Shortage of especially P but also N is one of the single most important reasons for lower yields in sub-Saharan African agriculture.

6.5 Climate-Friendly Food and Feed Ingredients for Improved Health of Animal and Man

A new generation of health-promoting, biobased products is emerging, different from, but supplementary to, pharma and drugs. Such new products are being developed based on increased knowledge and understanding of microbiomes: knowledge, not just about organismal composition but also insight in the functional role of the different groups of organisms in the microbiome and not the least, based on this gaining insight into how such composition and function can be modified by external interventions. The product candidates for such external interventions are gut health-promoting food ingredients and feed additives, stimulating the healthy part of the gut flora, hereby inhibiting the unhealthy part of the microbiome flora, to be used for one stomach as well as for ruminant animals and likewise, for new products for improved skin microbiome, as well as for improved, healthy microbiome composition and function in the respiratory track system. The next ambitious step to take (an emerging new innovation) is to develop biobased (probiotic or prebiotic) products, which can be documented to inhibit the replication of antibiotic resistance genes and a-biotic resistant microorganisms.

6.6 Biobased Public Health-Promoting Products

Valorization of marine products seems to have overall higher potential for health-promoting effects than the terrestrial products, e.g. for skin care, wound healing and protection against infections, microbiome boosters (in gut, skin and respiratory track system). A third biobased aspect, with strong upside for new health-related innovations to be gained, is to make new health-directed bioprospecting activities in the marine biosphere, to be applied for supporting both public health and the new generation of blue biorefineries. There is so much more to be found!

6.7 Improved Health of the Planet and Protection of Biodiversity by Biobased Products

Geoengineering efforts of relevance for the future health of our planet are currently taking place notably much below the public radar. The little, which is exposed or published, indicates a disturbing trend. Almost no biological expertise is involved. Therefore, possible risks to the biosphere are not fully taken into account. Further, on the other hand, the potential positive effect of new bio-relevant geoengineering mechanisms on the health of our planet is not included. Hopefully, action is taken to involve public researchers (here including biological expertise) also from Europe in the area of Geoengineering. 360-degree thinking and transparency are needed!

7 Instruments and Drivers for Speeding Up Transitions

7.1 Incentives for and building the market

(carbon tax, directives for responsible consumption, public procurement). Recently EU directives, aiming at improved circularity, improved resource efficiency and reduction of waste, have been decided upon in EU. To take full advantage of these new directives, upgrading technologies are needed to be ready for the new era of resource circularity. It is expected that there will be more such directives, taking EU towards more responsible use of natural resources also in the upcoming decades. However, no operational connection has been established between introducing new EU directives and ensuring that upgrading technologies will be among the priorities for the common EU research and innovation agenda. This dilemma has led to at least two burning platforms, where immediate follow-up R&D actions are needed: the directive for compulsory outsorting of textiles from waste (meaning that textile can no longer be burnt) and the directive on compulsory landing of fish by-catch (meaning that such fish can no longer be dumped at sea). Both of these examples of directives will lead to significant amounts of available and accessible feedstocks, with huge potential for valorization through circular upgrade. For the textile area, the challenge now is to asap develop new technologies for re-use of especially cotton fibres, hereby reducing the need for land, water and pesticides used for production of virgin cotton fibres; for making new types of fibres from leftover biomass from, e.g. the green biorefinery, or from leftover product from forestry and wood processing; and for using new alternative and local, sustainably produced plant fibres. The magnitude of making the textile sector more sustainable is put in perspective by the mere fact that textile industry globally is the most burdensome on global resources and climate, next only to the oil industry.

7.2 Transition through knowledge dissemination, public schools, life-long learning

To understand the basic concepts of a biobased society, you need to understand the basics of biology, microbes and plants. But more than that is needed. Also the microorganisms, forming microbiomes in your gut and on your skin, are of importance for nutrition, health and hygiene. However, biology does not have sufficient space and attention in the public school curricula to cover also such topics. This could be about to change! Corona epidemic brought hygiene into every people's home, work places and schools. Another sign of change was the message from IPCC, in fall 2019, that food practice (eating primarily meat based and wasting more than needed) is of significant importance for the climate. This was followed closely by alert warnings from experts and organizations that we lose biodiversity in a speed never seen before. Connecting excess in meat consumption with biodiversity loss is essential. Feed production takes up around 3/4 of the total arable land. Producing more feed from upgrade of residues and side streams can free land and space for biodiversity. Next came the message from industry that they need more bio-skilled people; and the message from the unions that their members need new skills and competences. Together this calls for concerted action, starting in public schools, to be given priority in higher education and offered broadly in life-long learning. The Danish Workers Union, "3F", took the initiative of making a folder communicating the basics of the "Biobased Society", written for their members, but used much more broadly by politicians and opinion makers and in schools. It is now translated into eight languages (Lange and Lindeblad 2016).

7.3 New, more resource-efficient and climate- and environment-friendly agriculture, forestry and fishery are an option within reach

Directives have been made in EU to introduce circularity as mentioned above, for plastic and textiles, and making landing of by-catch of fish compulsory. Next could be an obligation to adhere to resource efficiency like the already introduced industry regulation by Best Available Techniques (BAT, OECD) concepts. The BAT concept is already in operation and used as key policy tools for Pollution Prevention and Control in EU. The possible gain by introducing BAT concepts for improved resource efficiency can be illustrated by examples: Scientists at Wageningen Research Center, The Netherlands, and at Aarhus University, Denmark, have provided evidence for that changing from growing cereal (wheat and barley) to growing grass, to be converted in the green biorefinery into protein-rich animal feed, substituting for imported soy. This practice can lead to double yield of biomass per hectare as well as to reducing the excess run-off of nutrients with 50%, which again means reducing the pollution of the surrounding waters, currently threatening biodiversity. In fishery, in general only 50% of the total fish biomass is used for consumption. The other half is wasted. The cooperatively owned Codland Fish processing plant in Iceland has been able to increase this fraction to around 80 or even >90% being used for food purposes. Similarly, the Norwegian company Biomega has introduced technologies for upgrading of fish-cut-offs to higher value food ingredients, including fish proteins and fish oils (reference to NCM publication). For forestry, NGOs are arguing, evidence based and convincingly, for more untouched forest areas, while on the other side urge for phasing out fossil coal for electricity and heat has drawn forestry plantation practice in direction of producing wood chips for combustion, not just wood chips from forestry residues but downgrading entire trunks into wood chips. We need a new type of forestry: the climate- and biodiversity-friendly forest, from which can also be produced higher value products, upgrading wood biomass in much smaller volumes per hectare, but to much higher prices, providing also beauty for public recreational activities.

7.4 *EU subsidies of primary production in the future linked to more green practices?*

Subsidies especially for agriculture occupy a very large share of the total EU common budget. This harbours in itself a highly interesting opportunity. If for receiving EU subsidies it is required that you change agricultural practices, reduce GHG emissions, CO_2 and methane, cut down run-off of fertilizer to surrounding waterways, improve resource efficiency and/or protect environment and biodiversity, it can be the most potent driver for change in direction of a more sustainable Europe.

7.5 *Social inclusiveness, creating jobs and rural development*

Bioeconomy, new value chains and efficient upgrade of primary production are the strongest card on hand for stimulating rural development. Many societal and socio-economic challenges in our society are connected to the growing social and economic distance between urban and rural areas. This is of concern across the political spectrum and across the EU member states. The positive societal aspects of strengthening the biobased society have so far not been communicated broadly and in clear terms. We can and have to do better.

8 International Win/Win Collaboration, Contributing to a More Sustainable Planet

Opportunities for contributing more to both global sustainability and to EU competitiveness through international collaboration are almost not touched upon in recent high-level papers for Horizon Europe, HEU development. During most of Hor2020, two formal/informal drivers were prevalent for setting the scene for most of the international, third country collaboration activities in EU: A series of earmarked calls where international partners from specific countries were compulsory to include, in order for EU applicants to qualify and EU Commission (EC) bilateral initiatives based on opportunities identified during EC visits to such countries. However, during the last few years of Hor2020, interesting new international collaboration has started emerging, e.g. the portfolio of the EU Joint Undertaking, public private partnership, Bio-Based Industries (BBI) now includes a handful of interesting international win/win partnerships, reflecting growing interest in international RDTI investments in EU, creating hope for turning the trend prevalent for decades, where R&D investments in EU to a high degree led to business investments and establishment on other continents. Notably, new businesses are created based on advancements in research and technologies, funded by European RDTI investments. It is of utmost importance to continue and strengthen such development. Hopefully, the new EU common Research collaboration program, Horizon Europe, HEU, must include a strategically well-thought-through international RTDI strategy. To exemplify, two building blocks for such a strategy could be:

A dedicated bioeconomy alliance between Europe and Africa is timely and very important to have established as soon as possible, to work alongside with the already ongoing food-security collaboration programs between EU and Africa. If we do not expand from focusing only on food security, not on upgrading of residues, side streams and wastes, we risk in a few years to be in the situation that severe malnutrition and hunger is a reality in several parts of Africa – at the same time as 30–50% of all agricultural, forestry and aquatic production are still going wasted. We can help this not to happen by timely sharing knowledge and building partnerships in an alliance between Africa/EU for developing the African biobased bioeconomy.

Well-thought-through technology protection strategies and policies within the biorefinery area could pave the way for a highly interesting development: Noteworthy, a huge cumulative platform of biorefinery technologies are already published, in the form of scientific peer-reviewed papers, placed in the public domain. In the new biobased circular bioeconomy, common efforts could be invested into making such an open access technology platform more accessible for newcomers, including an overview of characterized commercialized enzymes and microbes and of open access microbial host systems (with no strings attached), developed for production of proteins in fermenters and of biorefinery technologies such as fermentation, product recovery and recirculation of processing water, etc. Such a platform of accessible knowledge, enzymes and microbes could serve several purposes: to function as an easy guideline to follow (for both academia and industries), when entering into international collaboration with, e.g. the BRICK countries, or highly industrialized countries like Canada, the USA and Australia making sure you are on safe ground in collaboration within the open access public domain. Further, such a platform would also be relevant for lowering the entrance threshold for new biobased EU companies to enter into the biorefinery/biobased area, hereby, paving the way for a more equal distribution of biobased industries in all regions of Europe.

9 Future Perspectives

Rapid transition to a more biobased society depends on decision-makers to see the societal value and opportunities in producing value adding products from the one third to half of all harvest, which now goes wasted, and on the establishment of integrated bioresource research collaboration throughout the entire knowledge- and value-chain, hereby opening for fast and efficient use of new knowledge and for combining known and novel technologies. For this to happen it is essential to relate the broad spectrum of biobased technologies to the wide range of societal needs: climate change mitigation and adaptation; feeding a growing population also under climate change challenged food production conditions (e.g. shortage of rain combined with high temperatures in Southern Europe and general climate change challenged conditions for agriculture in many parts of Africa); exploiting the new public health-relevant opportunities for making affordable gut-health-promoting food and feed ingredients from conversion of plants, fungi and algae and seaweed biomass; making building blocks for biobased materials and chemicals to substitute for fossil-based, through biorefinery and negative emission technologies (NET); making new, low-emission biomass derived soil improvement product; and making biosubstitutes for pesticides and building new knowledge for fighting the threat of growth in antibiotic resistance. Last but not the least, successful, just and timely transition of a sustainable, circular and biobased global society depends on that we stimulate international collaboration, promoting socially inclusive and responsible win/win knowledge sharing (Figs. 4.2, 4.3, and 4.4).

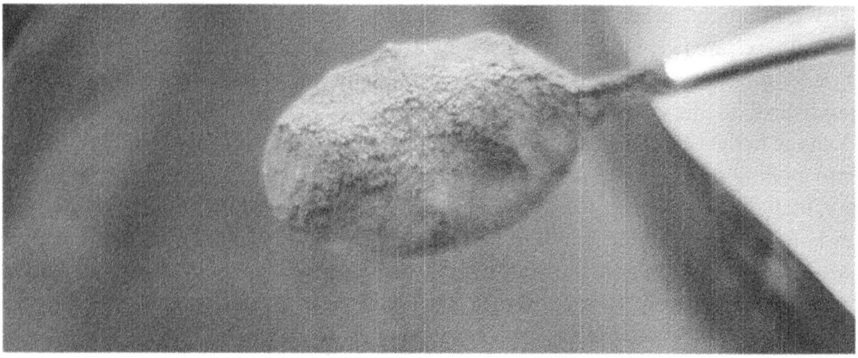

Fig. 4.2 Locally produced protein powder made from the juice of a screw press processing in the Green Grass Biorefinery. Production of the feedstock, perennial green grass or clover grass, as compared to growing, e.g. cereals, has an environmentally friendly profile: it can be cultivated without the use of pesticides, and it gives (under Danish conditions) twice as much biomass and only half the run-off of surplus nutrients per hectare. (Photo: Ida Marie Jensen, Aarhus University, Denmark)

Fig. 4.3 Seaweed, *Saccharina latissima*, cultivated in the deep and clean fjords of Faroe Island. The feedstock for making a cascading of products, including higher value gut-health-promoting feed and food products

Fig. 4.4 Small scale production of nutritious food in the form of oyster mushrooms, grown on the household's own coffee ground, otherwise wasted. (Photo by René Georg)

References

Dotsenko, G, Meyer, AS, Canibe, N, Thygesen, A, Krogsgaard-Nielsen, M and Lange. L (2018). Enzymatic production of wheat and ryegrass derived xylooligosaccharides and evaluation of their in vitro effect on pig gut microbiota. Biomass Conversion and Biorefinery (2018) 8: 497–507. https://doi.org/10.1007/s13399-017-0298-y

Lange, L and Lindeblad, J (2016). The fundamentals of the biobased society. Published by United Fereation of Danish Workers 3F, Copenhagen

Chapter 5
Circular Bioeconomy: A Path to Sustainable and Climate-Wise (Material) Economy?

Vafa Järnefelt, Anna Tenhunen, Laura Sokka, Pekka Tuominen, and Raija Lantto

Abstract Circular Bioeconomy – the symbiosis of bio- and circular economies is widely accepted as a solution for ensuring wise and frugal use of bioresources, and provide means to mitigate climate change. In this chapter, we apply circularity concept to scrutinize the use of bioresources and discuss their upcycling circularity potential during and at the end of life of the products and materials. Furthermore, the relationship, including potential synergies and conflicts, between circular bioeconomy and climate change mitigation is studied. The covered sectors include food, packaging, buildings and infrastructure and energy use of regenerated and produced bioresources.

Keywords Circular economy · Bioeconomy · Energy transition · Land use · Sustainability · Biomaterials · Bioplastics · Climate change mitigation · Carbon neutrality · Bioenergy

> Since the global biomass is not an unlimited resource, which applications are the most valuable and sustainable for its use? When biobased resources are used, do we make the economy more environmentally sustainable?

It is widely agreed that Circular Bioeconomy (CBE), the symbiosis of bio- and circular economies, should ensure wise and frugal use of the bioresources and provide means to fight against climate change and environmental degradation. Overall, the concept of circular economy (CE) has become a widely used term in the promotion of sustainable economic activities during the past years (e.g. Circular Economy Action plan (2015), Circular Economy package in (2018), European Commission (2021); Platform for accelerating the Circular Economy (PACE), PACE, WEF 2018). CBE is a part of CE that emphasizes the circulation of renewable resources and products.

V. Järnefelt (✉) · A. Tenhunen · L. Sokka · P. Tuominen · R. Lantto
VTT Technical Research Centre of Finland Ltd, Espoo, Finland
e-mail: Vafa.Jarnefelt@vtt.fi

E. Koukios, A. Sacio-Szymańska (eds.), *Bio#Futures*,
https://doi.org/10.1007/978-3-030-64969-2_5

From the environmental sustainability point of view, we are trying to solve two massive global challenges by using biomass: climate change with the reduction of greenhouse gas (GHG) emissions and transitioning our linear economy into sustainable circular operation model that is based on using bioresources.

Yet, the concept of circular economy is argued to be inherently vague and lacks scientific foundation and empirical research to support it (Kirchherr and van Santen 2019; Korhonen et al. 2018). If not evaluated properly, circular activities can backfire in increased emissions and loss of materials and therefore impact nature negatively (Carmona et al. 2017; Zink and Geyer 2017). Thus, circular activities are not inherently sustainable. Each process, material and product should be assessed to evaluate as for sustainability. Furthermore, it is vital to understand that biomass is not an unlimited resource. Even though technologies and different ways of use allow extending the resources available, there is a limit to how much biomass we can use sustainably. Furthermore, there are uses where we cannot replace biomass, most importantly as food and feed, while in other uses, such as energy, there are other alternatives. The crucial challenge to solve is how much biomass we have sustainably available after such vital and necessary uses, and what is the most efficient way to utilize it.

Biomaterials are typically categorized more sustainable than their non-renewable counterparts due to their regenerative nature. However, in many cases replacing a material source with more sustainable one alone is not enough from a lifecycle management perspective. If materials and products are made out of bioresources but are not designed for efficient circularity, their true potential for achieving sustainability is lost.

Our economic system is presently facing large sustainability challenges. On one side we have the Paris Agreement aiming at limiting global warming to well below 2.0 degrees Celsius, which means reducing the global CO_2 emissions by 30% between 2010 and 2030 and reaching net-zero emissions by 2050 (IPCC 2019). On the other side, we have a global linear economy that uses massive amount of virgin raw materials each year, and the extraction and production of these raw materials are fuelled mainly with fossil energy.

Transition from fossil energy resources to clean and renewable ones plays a vital role in reducing GHG emissions and mitigating climate change. In sustainable circular economy, products and materials should be re-circulated into use in a way, which does not degrade the value of the product or material and does not use excess energy or virgin materials in the re-circulation processes. Inevitably, complete avoidance of energy use is never possible. The outmost layers of circularity operations are often the most energy intensive. Thus, keeping products in longer use and avoiding unnecessary recycling contributes to climate change mitigation.

However, there is a limit to how many times materials and products can be re-circulated. Some materials preserve value better than others in recycling processes, even in advanced processing. This is not only due to design aspect but also due to the material type and properties. Design for circularity can extend the lifetime of the material but only to a certain point. Eventually all of the materials reach their very end-of-life, after which there is no longer value in re-circulating them back into use as materials due to severe downgrading and loss of critical properties. Or at this stage recycling processes may be unsustainable compared to value achieved, for instance, due to requirement of extensive energy or use of chemical in the process. Therefore, the most benefit may be gained by re-circulating material into energy. In some cases and for some material streams, energy recovery can be the best option also from the environmental sustainability point of view. As materials lose value in properties during use phases and reprocessing, the heat value might not be the same compared to virgin material. For example, when used for bioenergy, more energy might be gained from primary biomass. Some of the downgraded cast-off materials have less energy value at their end-of-life; despite this their use for energy is welcomed in replacing the primary biomass for energy.

In 2017, the global economy needed 100.6 Gt of materials for covering key societal needs, like nutrition, housing, consumables, mobility, health care, communication and services (Circle Economy 2020). According to forecasts of International Forecast Panel (IFP), the amount is estimated to grow between 170 and 184 Gt by 2050 (IFP 2017). In 2015, from 92.8 Gt input of the total resources entering the global economy, some 51.9 Gt remained unaccounted and were assumed scattered into the environment in form of emissions and unrecoverable wastes. Most of the extracted resources are lost already within the first year of production without any possibility to restore and re-circulate them back into economy. Thus, the challenge of adjusting our economic system within sustainable limit is enormous and requires ambitious mitigation actions across all economic sectors. During the past few years, several countries and regions have launched targets for carbon neutrality. For example, the EU is aiming at carbon neutrality by 2050, which is at the heart of the European Green Deal accepted in 2019. Among Nordic countries Finland has one of the most ambitious climate change mitigation strategies targeting carbon-neutraly by 2035 (Carbon-Neutral Finland 2035 2020).

In this chapter, we apply circularity concept to scrutinize the use of bioresources and discuss their upcycling circularity potential during and at the end of life (i.e. EOL phase) of the products and materials. Furthermore, the relationship, including potential synergies and conflicts, between circular bioeconomy and climate change mitigation is studied. The covered sectors include food, packaging, buildings and infrastructure and energy use of regenerated and produced bioresources. These biomass applications were chosen due to their volume, importance and relevance. The role of nutrition in use of biomass is self-evident, whereas packaging plays a vital role in making the food sector more frugal and sustainable and decreasing its contribution to climate change. In addition, demand for plastic packaging is growing tremendously due to increased consumption, increase of population and especially rise of middle class. Due to the intense increase in predicted plastic production, the emissions are also predicted to grow significantly. Infrastructure is one of the largest contributors to climate change, in which the use of bioresources is seen as a part solution for cutting down the emissions. Presently, biomass dominates the use of renewable energy (WBA 2019). We start the discussion with bioenergy since biomass use for energy is the most conflicting one within the circularity concept.

Although globally different strategies for the use of biomass pursue sustainability and green growth, they overlap and may be conflicting with each other depending on what applications for biomass are promoted. By exploring these sectors and applications where biomass is used, we put an emphasis on circularity and emissions and discuss critically if the replacement of unsustainable materials with bioresources makes the (material) economy environmentally more sustainable.

1 Conflicting Uses of Biomass and the Potential for Sustainable Circularity

The agriculture, forest and other land (AFOLU) is responsible for about 25% of the global net anthropogenic emissions (IPCC 2014). The emissions stem mainly from deforestation and agricultural emissions from soil and nutrient management and livestock. The AFOLU sector accounts for about a quarter ($\sim 10 - 12GtCO_2eq / yr$) of net anthropogenic GHG emissions (IPCC 2014). In order to get an understanding of how efficiently biomass works in climate change mitigation, it needs to be studied in relation to a reference situation where less biomass is harvested for energy (Heinonen et al. 2017). For agricultural biomass, the computational implications are relatively clear: the rotation is fast and the harvested biomass grows back quickly. Potential loss of carbon takes place either through land use change when new fields are cleared or through loss of carbon from soil. In the slow-growing forests, such as boreal forests, impacts of harvesting need to be calculated by comparing scenarios with different levels of extraction of forest biomass.

The assumption considering biomass a zero emission source is often based on the misunderstanding of the UNFCCC greenhouse gas emission inventories in which biogenic carbon emissions from energy production are considered zero (Wiloso et al. 2016). This is because the emissions from carbon stock changes are taken into account in the land use sector. Thus, they are not considered zero but are accounted for in land use instead of energy.

In addition to emissions from land-use changes, climate impacts of agricultural biomass production stem from production of fertilizers, energy and other inputs needed in growing, collecting and processing the biomass, and emissions from soils resulting from fertilizer application (e.g. N2O emissions). Furthermore, the role of indirect land use change (ILUC) in connection to bioenergy production has been widely recognized (e.g. Plevin et al. 2010; Wicke et al. 2012). ILUC refers to a change in land use elsewhere because direct land use change (LUC) causes either displaced production of agricultural food, feed or fibres in order to cover for the existing demand or because more land is brought into agricultural production due to price increases (Gerssen-Gondelach et al. 2017). Actions have been taken to mitigate ILUC as a result of bioenergy production from agricultural biomass such as the limitations on high ILUC risk biofuels in the EU Renewable Energy Directive, but its complete elimination is very difficult. Hence, the role of wood biomass in climate change mitigation has gained a lot of attention during the past years. Wood biomass is, when sustainably grown, a renewable resource and largely recyclable and reusable material. Only a small fraction of wood products cannot be re-used or recovered directly (e.g. hygiene paper or contaminated wood material). However, the role of forest biomass in climate change mitigation is more complicated (Fig. 5.1). In the forests, wood acts as carbon storage. It is crucial to manage and harvest the forests in a sustainable way. Furthermore, the present wood product mix typically consists mainly of short-lived products and energy. With this type of applications of wood, positive climate impacts cannot be obtained for at least decades (Seppälä et al. 2019b).

The idea of carbon neutrality of forest biomass use is based on the notion that in sustainable forestry, the extracted wood will eventually grow back and re-absorb the carbon that was released. As forests are slow-growing, in particular boreal forests, it takes from decades to even centuries for it to absorb the released CO_2 from the atmosphere. The so-called climate debt from the utilization of forest biomass for energy stems from exactly this: as biomass is taken from the forest, an unavoidable reduction in forest carbon stock is caused, compared to a situation where biomass is not taken. When biomass is removed from the forest, the amount of carbon stored in the forest decreases. Furthermore, had the forest not been harvested, it would have continued to absorb more carbon, thereby acting as a carbon sink – a phenomenon often referred to as foregone sequestration (e.g. Koponen and Soimakallio 2015) Fig. 5.2. In the longer term, more intensive harvesting leads to lower carbon stock on a landscape level than the scenario with less harvesting.

However the impact is not straightforward: the overall carbon balance of wood use is determined by such factors as the actual use of the harvested wood products,

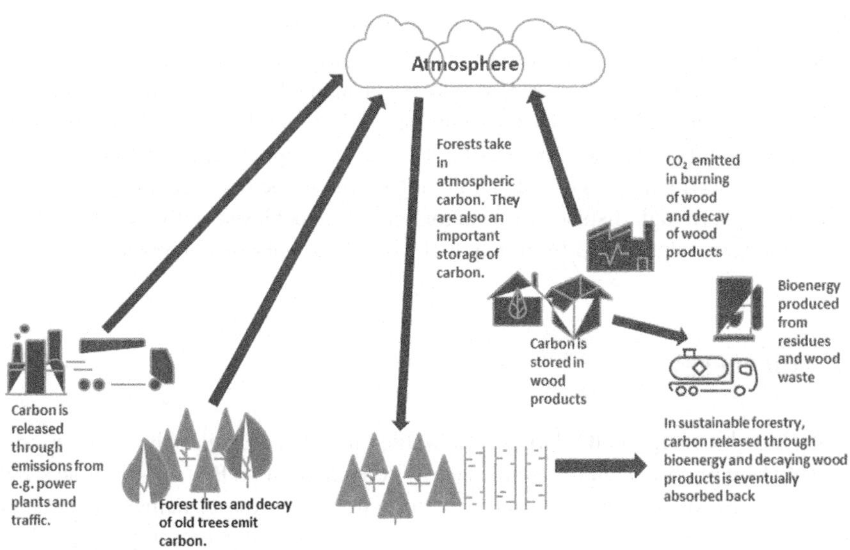

Fig. 5.1 Forests affect climate in many ways

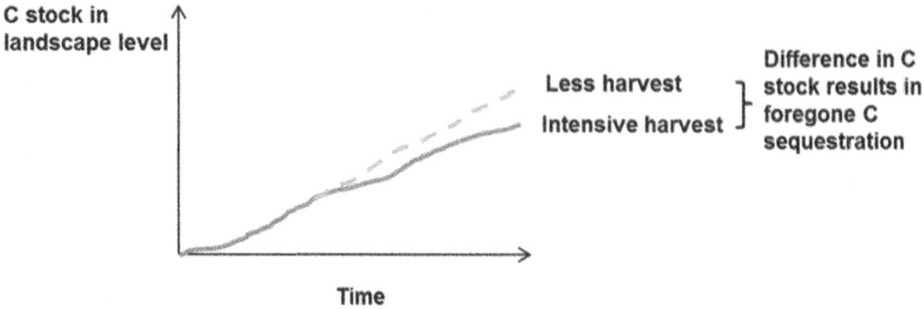

Fig. 5.2 Illustration of a comparison between intensive and less intensive harvest scenarios: carbon stock (C stock) increases in both scenarios but more in the "less harvest" scenario. (Koponen et al. 2015)

how long they remain in use and what kind of products they replace (substitute) in that use (Fig. 5.2). This means that carbon loss in forests could be compensated through production of long-lived products with high wood content, such as solid wood products used in wood buildings that could also substitute for emissions from other fossil-intensive materials such as steel, glass or concrete (Gustavsson et al. 2006; Pingoud et al. 2010). Displacement factors have been presented to express the efficiency of using biomass to reduce net greenhouse gas (GHG) emission through quantifying the amount of emission reduction gained per unit of wood use (Sathre and O'Connor 2010). Yet, these displacement factors typically do not consider the lost carbon sequestration in the forest.

In order to take into account the lost carbon sequestration in the forest, Seppälä et al. (2019a) have introduced "required displacement factors (RDF)", which depict the displacement factor needed in order to achieve zero $CO_{2eq.}$ emissions from increased forest utilization over time in comparison to a reference harvest scenario. Their results indicate that in Finland, which has large forested areas and important forest industry, compensating a 30% increase in annual harvest would necessitate an RDF of 2.4 tons carbon per each carbon ton in the wood-based products and fuels obtained from the increased harvest in 2017–2116 (Seppälä et al. 2019a). Most of the wood uses do not provide such high displacement factors. According to Kunttu (2020), high enough displacement factors could be achieved with only few products, e.g. textiles produced from pulp and wood-based plastic composites.

2 Bioenergy

Currently, bioenergy is the largest renewable energy source globally with an estimated share of 70% of all renewable energy consumption (WBA 2019). At the same time, the use of biomass for energy is in conflict with the principles of circular economy and in many cases also with climate change mitigation. Energy production is a one-off use of biomass, while there are other more sustainable uses of biomass that are longer lasting by binding carbon for longer periods of time and especially in replacing non-renewable materials that have high environmental impact. Since the amount of biomass available is limited by various other factors, such as other land uses, its use for energy production reduces opportunities for valuable non-energy uses. Regarding GHG emissions, the role of biomass is not straightforward because its impact on emissions is dependent on various land-use related phenomena such as climate debt and foregone sequestration that were discussed in the previous section. Direct emissions are comparable with those of fossil fuels, with CO_2 emission factors of approximately 110 g CO_2/MJ for solid biomass fuels, 70–80 g/MJ for biofuels and 50–60 g/MJ for biogas (Statistics Finland 2020). The overall GHG balance of biomass can vary widely depending on the type of biomass, the location it is grown (e.g. need of fertilization, growth rate, transportations needed) and how the fuel is processed (e.g. how much energy and other inputs are needed in the processing).

Other than GHG emissions from biomass are roughly comparable to the combustion of other chemically similar fuels. Pollutants such as nitrous oxides and particulate matter can be a local health and environmental problem, in some cases a serious one. Typically pollution-related health problems can be severe with small-scale combustion of fuels in densely populated areas. This is an issue in many developed countries where firewood still has a role in heating. For example, in Finland small scale burning of firewood causes approximately 200 excess deaths a year from a population of 5.5 million (Savolahti 2020). However, the problems are much more severe in the developing world, where in some localities, such as much of sub-Saharan Africa, the traditional use of biomass for heating and cooking is still the dominant form of energy use (WBA 2019). Moreover, the health impacts are often paired with other harmful effects such as heightened risk of fire and overuse of local resources. Some of these detrimental impacts will be alleviated with growing incomes and people gaining access to other energy carriers. However, this development will bring along other environmental and social burdens unless it is done relying on sustainable energy sources.

The world energy market still relies heavily on the use of fossil fuels, which is the main contributor to climate change globally. Even though the overall amount of renewables is increasing, its growth rate is not fast enough to meet the simultaneously growing global demand for energy. During 2016–2017 the primary net energy supply of fossil fuels had increased more than the supply of renewable energy sources that was lacking behind (0.7%). The increase of 1.5% in 2016–2017 in total primary energy supply was matched by coal, oil and natural gas. In 2017, combined (oil, oil products, coal and gas) fossil fuels still dominated the global energy mix with the share of 80%, and the renewables accounted for 17.7% in the final global energy consumption. The same year, among renewable energy sources, bioenergy accounted for 70% of the renewable energy consumption. The relative share of bioenergy has been declining slightly a fraction of a percentage point (0.5–1.0%) annually due to decreased use of traditional biomass sources that outpace the simultaneous growth of modern uses of bioenergy (WBA 2019). In absolute terms the use of bioenergy nevertheless is growing.

Most analysts are forecasting growth for the use of biofuels. For example, IEA forecasts a growth of 40% for total use of bioenergy globally by the year 2040 (IEA 2019), whereas Rogelj et al. (2018) foresee on average a tenfold increase in biomass use in electricity generation by the year 2050. The increases in these scenarios are in part driven by the fact that combined with CCS and CCU, bioenergy offers countries the unique ability to actually reach negative CO_2 emissions and contribute to international climate goals.

Globally, forestry is the largest contributor to the bioenergy mix. It accounts for more than 85% of all the biomass used for energy purposes, including forestry products like charcoal, fuelwood, pellets and wood chips. Recently the use of agriculturally produced feedstock for the production of liquid fuels such as ethanol and biodiesel has grown fast, roughly doubling in volume from 2008 to 2018 (REN21 (2019): Renewables 2019 Global Status Report). It now represents 8% of all bioenergy use and 3% of all energy for transport globally. The relatively high carbon footprint of farming however means that these fuels tend to be far from climate neutral, roughly halving the carbon emissions compared with similar crude oil-derived fuels (UK Department for Transport (2008): Carbon and Sustainability Reporting Within the Renewable Transport Fuel Obligation).

There are, however, still high-yield sources of bioenergy that are far underused. In terms of major energy content, IEA has identified biomass production on marginal lands (potential of 60–110 EJ), residues from agriculture (15–70 EJ), forest residues (30–150 EJ), dung (5–55 EJ) and organic wastes (5–50 EJ) as ones that do not interfere with existing agriculture and forest harvesting for other purposes (IEA 2007). Even though their potential is large, also in this case that same potential competes between application for energy and longer lasting material use.

Growing use of biomass for energy is difficult to reconcile with the concept of circularity. To resolve the conflicting demands for biomass in energy and other uses, the clearest path forward is to grow the share of other renewables such as solar, wind, hydroelectric, geothermal, ocean, hydrogen and nuclear. This can help to reduce the overall share of biomass, especially primary forest biomass that is being used for energy production. Bioenergy demand could be also partially met by certain products and materials that have reached their very end-of-life, and are more valuable and sustainable for energy application; although this kind of replacement requires massive optimization of resources and the volume of biomaterials reaching their very end-of-life may not be sufficient to meet the demand. In addition, there may be limits to this kind of optimization to match the local and temporal needs.

Despite this, it is evident that the shift to reducing primary use of biomass for energy will not happen overnight. Bioenergy has some benefits that make it ideal to certain roles at least for a transition period on the path to a more sustainable energy system. It is, for example, a promising option to balance variable energy sources, since it is a flexible and storable resource (Hakkarainen et al. 2019). In certain uses, such as heavy vehicles, ships and aircraft, it is difficult to replace liquid fuels, making biofuels often the most viable renewable replacement. Moreover, when biomass is used for energy production, carbon capture and utilization (CCU) technologies should be applied to re-circulate the emissions into raw materials, and further into new products and chemicals. Currently, the scale up and market uptake of CCU technologies is challenged by energy-intensive CCU processes. On-going research in the CCU field is aiming at finding sustainable and profitable solutions for CO_2 re-circulation technologies. When resolved, use of bioenergy combined with CCU can offer the triple benefit of renewable energy, CO_2 removal from atmosphere and renewable chemical products created from the flue gasses.

3 Nutrition

"Food is the single strongest lever to optimize human health and environmental sustainability on Earth. However, food is currently threatening both people and planet." These emphatic words begin the summary report written by experts of the EAT- Lancet Commission (EAT-Lancet Commission 2019). We live in the world of growing human population leading to increasing need for food and increasing challenges in the primary production of food raw materials, both threatening the food security. This dual problem calls for radical and systemic transformation in the global food system that can be supported by circular bioeconomic solutions, data technologies and digital platforms. The EAT-Lancet Commission's message is clear: Agriculture and fisheries must not only produce enough calories to feed a growing global population but must also support environmental sustainability (EAT-Lancet Commission 2019). This will be, self-evidently, realized firstly by shifting towards plant-based food, particularly in high and mid-income societies. In the second wave, healthy microbe biomass and specialty proteins, e.g. meat, milk or egg proteins, produced by microbial hosts in controlled environments may appear to enrich the plant-centred diet flavoured perhaps by means of augmented reality.

We continuously use more resources than what the planet is capable of producing. Overall global material extraction has multiplied tenfold since the beginning of the twentieth century. It is claimed that global socioeconomic material stocks rise 23-fold over the twentieth century and require half of annual resource use (Krausmann et al. 2017). The global extraction of biomass in 2017 was 24.6 billion tonnes, 21.3 billion tonnes of which was used for food processing (Circle Economy 2020). Food and feed together account for the majority of global biomass demand, crop production (sugar cane, maize, wheat, rice paddy, potatoes together) 5 billion tonnes, meat 340 million tonnes (Weindl et al. 2015), fisheries and aquaculture 170 billion tonnes (amounts approximated from FAOstat data from 2017 (FAO 2020a). It is estimated that about one third of food is lost or wasted throughout the agro-food chain, in the high- and middle-income societies mostly in the consumption stage and in the lower-income societies in the earlier stages of the food production value chain (FAO 2020b). This is not only wasting food raw materials and food itself, but also pure water, energy and soil, consequently increasing emissions to the environment.

Agriculture, fishery and aquaculture, including forestry, have significant impacts on water, soil, air quality and biodiversity. Increasing need for food and energy increases pressure to further expand extraction of both abiotic and biotic resources from the earth crust stretching the planetary boundaries (Steffen et al. 2015) to the utmost. Besides, agriculture is the largest user of the world's freshwater reservoirs (UNESCO 2020). This, rather chilling, situation calls for solutions that a circular bioeconomy can offer. These solutions must also be sustainable, which is not self-evident or straight forward, when the nexus of food, land, energy and water is to be mastered.

In bioeconomic operational and business models, supply and demand may conflict when it comes to offering to the societies healthy food, biobased materials and bio-energy, production of which necessitates water (potable in case of food), soil and arable land. A sustainable CBE maintains the value of the extracted natural resources by cascading the resource use (EEA 2018) and creating solutions by which resources are kept in value-sustaining use as long as possible and recycling them back into use at the end-of-life. In addition to food, the agro-food chain losses and waste have important implications for resource insufficiency in a broader sense causing ultimately poverty and limiting economic growth.

In the early stages of the agro-food chain (primary production and post-harvest), balanced land-use between crops cultivation and animal farming as well as wise use of agro-industrial side streams and residues is improving. A recent FAO study states that 86% of livestock feed (forages, crop residues and side streams) is not suitable for human consumption and edible grains account only for 13% of the global livestock dry matter intake (FAO 2020c). A decade ago about a half of the world's crop energy content was eaten directly by people, the rest being divided to livestock and energy or other industry purposes (Cassidy et al. 2013).The researchers also reported that crop energy content used for biofuel production increased fourfold between the years 2000 and 2010, from 1% to 4%, representing a net reduction of available food globally.

The transition to wasteless and loss-resistant food and agriculture will happen through harnessing broadly possibilities offered by data technologies, artificial intelligence, automation, novel food raw materials and production technologies as well as genuinely closing agro-food material, side stream and waste flows enabled e.g. in integrated production systems (Lantto et al. 2018).

Food production is predicted to move gradually from farming lands and natural waters to compact closed-loop systems that minimize the need for mineral fertilizers, feed for livestock, clean water and energy. The smart food production factories will be closely integrated to farming systems driven by advanced automation. In addition, coupling farming and food production systems with their process digital twins for constant and accurate assessment of the production and process situations, optimal efficiency and reliability can be achieved (Poutanen et al. 2017). It is also the way to manage the food production and supply chains more efficiently. Integrating platform economic solutions to food distribution will pave the way to reduced waste generation enabling shorter and more efficient food supply chains. This necessitates availability and sharing of accurate data, which is more easily said than done due to currently prevailing operational and business models that favour sub-optimization of operators in the chain.

A scalable and convertible closed-loop food factory (based on, e.g. recycled aquaculture, aquaponics or hydroponics systems) will enable resource efficiency and independence from regional and climatic boundaries. It will need data and sensor technologies, smart automation and robotics, water purification, chemical and food technologies as well as expertise in integrating, optimizing and making compatible with each other diverse production systems, such plant, insect and algae cultivation and aquaculture. Outfitting a close-loop food by a clean energy supply system (e.g. photovoltaic systems for electricity), anaerobic digestion of the factories' biowaste for energy carriers and recycled fertilizers and a collection system for rainwater and condensed atmospheric water, environmental sustainability and overall resource efficiency can be improved and independence from external factors minimized.

Sustainability of the first half of the agro-food chain (primary production, postharvesting and food processing) can be improved by the abovementioned integrated closed-loop solutions. The latter half of the chain is taken care of by the retail, and particularly consumers who play the most focal role to what direction the agro-food chains will eventually develop. By changing purchasing and consumption behaviour towards favouring non-meat food and accepting new protein and other nutrient sources, for example, insects or microbes, consumers can affect the whole agro-food chain consequently improving sustainable use of natural resources and sufficiency of food.

The bulk of the living matter is composed of elements and compounds found in the atmosphere. Instead of being restricted to selected locations, atmospheric resources are available everywhere. Sustainability of the future food system could be achieved by taking into use atmosphere and industrial exhaust gases as raw material resources. The atmosphere is a notable reservoir of CO_2 together with other essential elements and compounds such as nitrogen, oxygen and water. Comeback of the resource sufficiency will rely on technological innovations that will enable the use of agricultural and industrial effluents in addition to the atmospheric components as raw materials (Lantto et al. 2018; Lehtonen et al. 2019).

Green plants are naturally efficient in transforming CO_2 to sugars and other essential nutrients. The more we get our nutrients from plant sources the better is the food security globally. Shift towards plant-based diets is essentially important in the high- and middle-income societies. One of the consequences of the progressive climate change is the productivity loss in many agricultural regions in the world, but food needs to be produced in increasing amounts. Secondly, CO_2 should be captured and utilized also by other means than via photosynthesis of plants. Nature's immense synthesis power should be harnessed to support bringing back its resource sufficiency in order to secure sufficient amount of nutritious food for all. Several microorganisms exist that convert CO_2 and N_2 by nature to components the cells need for growth and multiplication. To our knowledge the food safety status of these microbes has not been studied or determined. Of course, ability to utilize both CO_2 and N_2 as sole carbon and nitrogen sources for growth can be engineered to an edible microbe. This possibility has to wait for permissive political decision-making and regulative environment before genetically engineered microbes could be used as food factories to produce all nutritive components a healthy human diet requires.

By utilizing the ability of microbes, whether natural or genetically modified, to fix gaseous nitrogen (N_2) and CO_2, food production could be decoupled from agriculture and aquaculture. The environmental impacts are minimized. Food production would be no longer dependent on any specific ambient temperature, humidity, light, soil type or region, and as such a food source can also be provided in locations that suffer from lack of arable land or soil due to the consequences of the climate change and unsustainable agricultural practices. Microbial cells as "factories" to produce animal proteins for food use are already seen potential from the commercial perspective.

Food production could take a giant leap forward by shifting towards plant- and microbe-based diets and taking into using gaseous components directly from air or human and industrial operations. Meat and meat-based food would be the loser in this battle. When feed protein is converted to meat protein, the most of it is lost to maintain the bodily functions and growth of the animal. An average of 3 kg of cereals are needed to produce 1 kg of meat at global level (FAO 2020c). Fresh meat and fish contain about 75% of water, while grains, nuts and peas a lot less. In addition, meat and fish products are very prone to spoilage in warm temperatures. These facts are crucial when packaging is considered for storage and transportation.

4 Plastic Packaging – The Villain – Decreases Food Sectors' GHG Emissions

Packaging is crucial, yet it is a controversial topic as the main material used for it – plastic – is accumulating as waste in the environment and especially oceans. The negative environmental impacts have set off trends to go packaging-free or even avoid plastics all together. However, plastics are not the bad guys – certainly there is a lot to fix in the current linear economy that is known for single-use plastics, but the answers lie in sustainable circular economy. To further tackle the contribution to climate change, it is immensely interesting to look at circular bioeconomy solutions.

If global food waste would be a country, it would be the third biggest GHG emitter – 30% of food produced globally is lost or wasted annually (FAO 2013). A study by Zero Waste Scotland calculated that in 2016, the carbon footprint of food waste from Scottish households was almost three times that of plastic waste collected (Zero Waste Scotland 2018). Food packaging is a crucial plastic application as it reduces food waste by preserving and protecting the food. Food waste is already a major concern because of its direct emission contribution to climate change, and it would increase even further without proper packaging. In Fig. 5.3, some examples are presented to highlight the impact of plastic packaging on the shelf life of different food products (FPA 2014).

Produce	Cucumber	Zucchini slices	Banana (distribution)	Cherries	Pear	Fish	Cheese	(Whole) Chicken
Shelf life: no packaging with plastic packaging	3 days 14 days	1–2 days 4–6 days	15 days 36 days	14 days 28 days	7–10 22–25	7 days 12 days	190 days 280 days	7 days 20 days
Difference	11 days	2–4 days	21 days	14 days	15 days	5 days	90 days	13 days

Fig. 5.3 Comparison of shelf life of different food products with and without plastic packaging. (Source: Tenhunen & Pöhler 2020)

The first task of the packaging is to serve the product, focused on preservation and protection, and these requirements for the packaging are based on the product. If the global diet would shift from mainly fresh meat and fish to dry grains, nuts and peas, the requirements for their safe preserving would be less intensive. A piece of meat today could typically be packaged into vacuum with multilayer and multimaterial nonrenewable virgin fossil-based plastic packaging – this type of packaging is challenging if even possible to recycle due to the fact that the laminated multilayers are not separable. Packaging of dry grains on the other hand could be made out of fibre-based packaging that is recyclable multiple times.

The packaged food products are expected to ensure long logistics chains from factories to intermediate storage and eventually shops; in many cases food is shipped from countries and continents to others. In addition to preservation properties, the packaging must be presentable and market the product accordingly. Today, we have a wide range of advanced packaging that are many times competing with each other; glass, wooden, paper and cardboard, composite, metals and fossil- and biobased plastics (Emblem & Emblem 2012).

The food production, logistics and sales of food have changed significantly as mass production of plastics began in the 1950s. Plastics have been designed for optimum product performance in a linear economy – they have been designed for single-use and low-cost production. In addition, plastics have great barrier properties, they are light-weight and easily mouldable into different shapes and sizes, which has made the mass production of plastics effortless. Linearity aspect of plastic products makes it challenging to switch to circular practices – the variety of materials is huge, and there is lack of infrastructure for collection and processing – furthermore, the markets for recycled polymers are still developing.

What further makes the shift to circularity challenging is the massive volumes of generated and used plastics. In 2018, global plastics production was almost 360 million tonnes and almost 62 million tonnes in Europe. Packaging represents the largest end-use markets in Europe – approximately 40% of all plastics produced are used for packaging (Plastics Europe 2018). The production of plastics is expected to increase even up to fourfold by 2050 (Fig. 5.4).

Developing societies and especially the rise of middle class is expected to have a massive impact in the increased predictions of future plastic productions. In the Western Europe and North America, average of annual plastic consumption per capita is around 100 kg, where in Asia it is only around 20 kg/capita/year. To

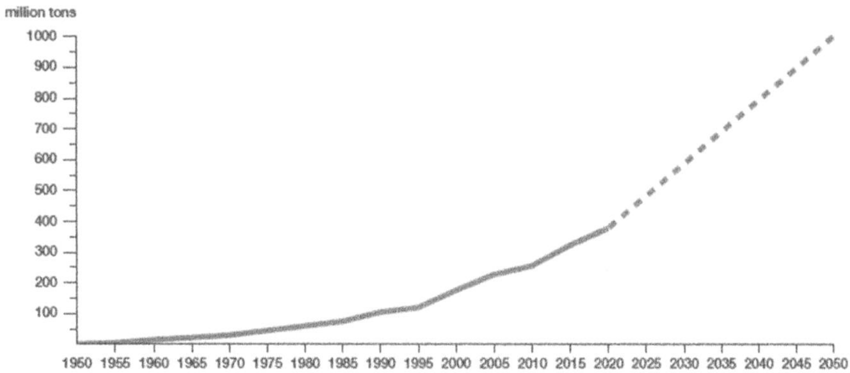

Fig. 5.4 Global plastic production volume development 1950–2050. (Source: Tenhunen & Pöhler 2020)

elaborate on the impact of developing societies and rise of middle class, in India, the overall plastics consumption has increased over 40 times since 1990 to 2017 from 0.4 Mt/year to 16.5 Mt/year (Babayemi et al. 2019).

Plastic packaging has positive impact on lowering food sectors emissions; nevertheless, plastic food packaging has an emission-intensive life cycle. As plastic packaging has a relatively short-term life cycle in a linear economy, the recycling of plastic food packaging is challenging and still not a well-established practice globally. Originally, linear economies were developed without a recycling capacity, resulting in our environment becoming a waste reservoir. Today's make-use-uncontrolled dispose economy of plastics has resulted in plastic waste leaking into nature and polluting the environment – humans, animals and living organisms on land and in the sea are susceptible to the effects of plastic waste in nature. Globally, 1.5–4% of global plastics production ends up in the ocean annually. The estimated release of microplastics in the environment in the EU is 75,000–300,000 tonnes each year (European Commission 2019). The economic viability of agriculture, fishery and other livelihoods is vulnerable due to the effects of plastic waste.

Furthermore, plastics contribute to climate change during several stages of their life cycle, e.g. during manufacturing, end-of-life management by incineration and even as pollutants in the environment as the waste begins to degrade. The current GHG emissions from plastics life cycles threaten our ability to meet global climate targets. Today, the GHG emissions from plastic production and incineration are more than 850 million metric tons, which is equal to the emissions from 189 (500 MW) coal power plants and very close to the total emissions of greenhouse gases in Germany in 2019 (858 Mt) (Clean Energy Wire 2020; Hamilton et al. 2019). With the current trajectory, by 2050 the cumulative emissions of the plastic life cycle will be over 56 gigatonnes CO_2eq, which is 10–13% of the global carbon budget calculated based on the 1.5 °C target (Hamilton et al. 2019). Recycled and renewable raw materials, such as biomass or carbon dioxide, will become even more important feedstock streams for future plastics in a circular plastics economy to further mitigate the impact to climate change.

Currently, polymers are mainly produced from fossil sources such as crude oil. Bioplastics represent approximately only 1% (2.1 Mt) of plastics produced. For biobased plastics, packaging is the largest field of application which accounted for more than 53% (1.14 Mt) in 2019. In 2019, material use took up 2% of the global agricultural area; from this biobased plastics took up approximately 0.016% (Fig. 5.5) (Bioplastics 2019). It has been calculated that if all world-wide fossil plastics (2015) would be produced from biomass, the demand for feedstock would be 5% of the total amount of biomass produced and harvested each year (Martien van den Oever, Karin Molenveld, Maarten van der Zee 2019). This does not take into account the use of side and waste streams. What would 5% mean in terms of taking up land from food and feed? Is it sustainable to utilize such agricultural land for material production? Instead, the priority should be in food production and in side and waste streams utilized for materials as well as utilizing wood biomass from forests. One of the challenges with food packaging is legislation, e.g. EU's current legislation allows mainly only the use of virgin material to be used as contact material for food applications (European Commission 2008). Recycled material, if collected as a mixture, contains contaminants like bio- and food waste, not applicable again as food packaging. An exception is PET bottles that can be reprocessed for bottles again due to separate collection systems and specific recycling process. Safety is something that cannot be compromised – for this reason it requires a lot from R&D&I to develop the means to process food packaging waste back into food packaging, to close the loop. Furthermore, the pressure is on to quickly develop reliable solutions. One interesting and promising solution is (thermo)chemical recycling, where the plastic waste material is recycled back to feedstock building blocks, monomers.

Plastics consists of polymers and additives and are typically named according to the polymers they contain. Polymers are processed and mixed with additives to help with processing or improve product performance such as plasticizers, flame

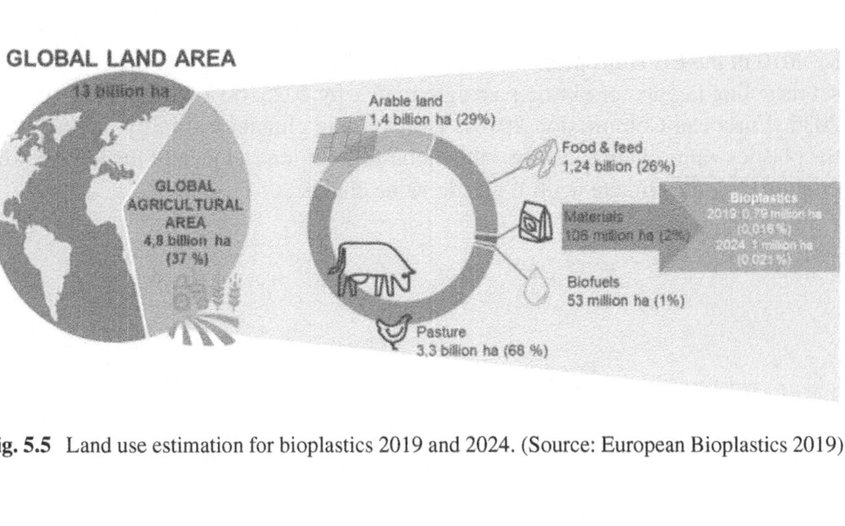

Fig. 5.5 Land use estimation for bioplastics 2019 and 2024. (Source: European Bioplastics 2019)

retardants, antioxidants, light and heat stabilizers, and pigments. Fossil-based plastics as a material, the polymers and a variety of additives, and food packaging applications are high performing and well optimized for the use phase – they have been developed and researched for a relatively long time compared to biobased plastic alternatives. In addition to biobased plastics, fibre-based solutions like carton, packaging papers, container and corrugated board are important packaging applications. The food packaging sector is one of the main sectors driving the growth of fibre-based packaging (Paper Industry World 2015). There is no doubt that the different packaging materials are competing with each other. Furthermore, we are seeing hybrid packaging materials where fibre-based and plastic materials are combined together. It is also important to note the difference in biobased materials and biodegradable materials – biobased materials do not mean that the material is biodegradable. Also, having biobased solution is not necessarily circular. The critical issue is how the material can be recycled back into materials without losing or minimizing value. Is it the best use of materials if they are biodegraded after one use? Or is it in this case, just a linear single-use biodegradable product that is only a bit better than its fossil-based non-biodegradable counterpart? For achieving the optimal solution, reuse and recyclability should be carefully assessed, and biobased and biodegradable products should be also designed for re-circulation.

The wide variety of available materials used in different packaging is making the recycling of these materials challenging. The challenge lies in getting high-value and quality recyclates from a very heterogeneous mixture of packaging waste. Smart sensor-based separation processes would assist in solving the challenge of heterogeneous packaging waste, but as the typical recycling infrastructure relies extensively on manual identification and separation, intense investments are needed in R&D and infra to support circularity. Another difficulty, experienced with especially biobased plastics, is that the volumes of biobased plastics are not sufficient enough to actually recycle them. Bioplastics are faced with the harsh reality that even though they are recyclable, they do not get recycled.

The shortcomings and implications of plastics have caused the regulatory framework to change and tighten quickly. For instance, EU has set ambitious policy measures and targets for all the packaging to be 100% recyclable, reusable or compostable by 2030 in the EU (European Commission 2019). Furthermore, in the EU there are set recycling targets for plastic packaging: 55% by 2025, 60% by 2030 and 65% by 2035 (European Commission 2018). The pressing climate neutrality targets effect the plastics value chains heavily – the biobased solutions are one of the key answers in unlocking the climate neutral plastic value chains.

5 Infrastructure: Wood in Long-Lived Products and Structures

Carbon loss in forests could be compensated through production of long-lived products with high wood content, such as solid wood products used in wood buildings that could also substitute for emissions from other fossil-intensive materials such as steel, glass or concrete. In 2017, of the total global material inputs of 100.6 billion tonnes, 48.0 billion tonnes were added to long-term stock, like machines and infrastructure (Circle Economy 2020). From the existing long-term stocks 17 billion tonnes of materials were removed the same year, making the remaining net addition 31 billion tonnes (Circle Economy 2020), resulting in the largest use of material resources. Growing economies and urbanization will even further increase the share of the raw materials used for infrastructure. Using wood as a construction material can potentially lead to large GHG benefits. In a Canadian study applying dynamic life cycle assessment, it was found that the climate impacts of wood product use ranged between −388 and −1264 kg CO_{2eq}/m^3 wood product in life spans of 50–100 years (Head et al. 2020).

In another study by Gustavsson et al. (Gustavsson et al. 2006), it was found that wood-framed buildings would result in 30–130 kg lower carbon emissions per m2 floor area compared to concrete buildings. They also found that use of wood as building material resulted in much lower emissions than its use as energy. Also Pingoud et al. (Pingoud et al. 2010) found in their study of wood use in Finland that highest climate benefits could likely be received with production of long-lived products that substituted fossil-fuel and energy-intensive materials.

However, in a review of different LCA/carbon footprint standards for building materials and products, Tellnes et al. (2017) found that none of them so far consider the effects of delayed emissions on global warming potential (GWP), an aspect that has been highlighted in emerging scientific methods (Helin et al. 2016). Lately, new methods have been introduced to include the temporal aspect into LCA of long-lived products. There are two groups of these methods: dynamic LCA (DLCA) and the bio-GWP approach (Breton et al. 2018). In DLCA, time-dependent characterization factors are calculated to assess the dynamic LCI for any given time horizon (Levasseur et al. 2013). In principal the method is applicable to any impact category, but it has mainly been used for global warming, based on the concept of radiative forcing.

Further challenge in compensating for carbon loss through long-lived products is that typically only part of the biomass is usable in products with long lifetime, which limits its contribution to climate change mitigation and circular bioeconomy. Thus, more research is needed about the potential of long-lived products in mitigating the climate change.

6 Final Words

Bioeconomy emphasizes the use of biogenic feedstocks, mostly forest and agricultural-based, and their components to end products that are mostly food, feed, energy, chemicals, textiles, packagings and paper. It can be considered to belong under the umbrella of circular economy in a sense that it focuses on using biogenic rather than fossil-based feedstocks to produce and manufacture products, but so far it has not emphasized reuse or recycling aspects of the biobased products well enough. Up to now, use of biomass has not always been neither circular nor sustainable. Thus, these two economy models alone are not the answer to sustainability. Use of biomass for each application discussed in this chapter requires careful calculations to avoid mistakes in the evaluation of sustainability of biomass use.

The use of bioresources is generally justified due to regeneration of needed resources compared to non-renewables, amount of embodied carbon, needed energy for recirculation processes and as carbon reservoirs. Food sector is the most crucial application of biomass which should be secured in the first place. Alongside this, packaging plays a prolonging and preserving role in food sector that has significant emissions and potential for decreasing emissions if biobased materials are used sustainably and designed for circularity. Globally, infrastructure has substantial role in emissions, and expanding urbanization is expected to grow the amount of emissions even further due to extensive amount of material input to long-lived stocks. Wood biomass has potential in contributing to GHG emissions positively, but its use and evaluation should be studied further.

Presently, biomass dominates the use of renewable energy (WBA 2019). Even though biomass is not a carbon-free energy source, it is renewable when sustainably used, and its role in the transition to an energy system based on carbon-free energy sources can be justified. On the other hand, use of valuable biomass, especially primary use for energy, fights against the principles of material circularity – cascading where raw material should be kept as material and used in a way which preserves value and enables re-circulation. In the future, bioenergy could be, for example, a promising option to balance variable energy sources, since it is a flexible and storable resource while the focus should be on increasing the share of other renewable energy sources.

References

Babayemi, J. O., Nnorom, I. C., Osibanjo, O., & Weber, R. (2019). Ensuring sustainability in plastics use in Africa: consumption, waste generation, and projections. *Environmental Sciences Europe*, *31*(1), 60. https://doi.org/10.1186/s12302-019-0254-5

Bioplastics, E. (2019). *Renewable feedstock*. https://www.european-bioplastics.org/bioplastics/feedstock/

Breton, C., Blanchet, P., Amor, B., Beauregard, R., & Chang, W.-S. (2018). Assessing the Climate Change Impacts of Biogenic Carbon in Buildings: A Critical Review of Two Main Dynamic Approaches. *Sustainability, 10*(6), 2020. https://doi.org/10.3390/su10062020

Carbon-Neutral Finland 2035. (2020). The Ministry of the Environment. Government's Climate Policy. https://www.ym.fi/en-US/The_environment/Climate_and_air/CarbonNeutral_Finland_2035

Carmona, L. G., Whiting, K., Carrasco, A., Sousa, T., & Domingos, T. (2017). Material Services with both eyes wide open. *Sustainability (Switzerland), 9*(9), 1–23. https://doi.org/10.3390/su9091508

Cassidy, E. S., West, P. C., Gerber, J. S., & Foley, J. A. (2013). Redefining agricultural yields: From tonnes to people nourished per hectare. *Environmental Research Letters, 8*(3). https://doi.org/10.1088/1748-9326/8/3/034015

Circle Economy. (2020). *Circularity Gap Report 2020*. 69. https://www.circularity-gap.world/

Clean Energy Wire. (2020). *Germany's greenhouse gas emissions and climate targets*. https://www.cleanenergywire.org/factsheets/germanys-greenhouse-gas-emissions-and-climate-targets

EAT-Lancet Commission. (2019). Healthy Diets From Planet; Food Planet Health. *Lancet*, 32.

EEA. (2018). The circular economy and the bioeconomy – Partners in sustainability. In *EEA Report 8/2018* (Issue 8). https://doi.org/10.2800/02937

Emblem, A. & Emblem, H. (2012). Packaging functions. In Packaging Technology (pp. 24–49). Elsevier. https://doi.org/10.1533/9780857095701.1.24

European Commission. (2008). *EUR-Lex – 32008R0282 – EN – EUR-Lex*. https://eur-lex.europa.eu/legal-content/EN/TXT/?uri=CELEX:32008R0282

European Commission. (2018). *Report from the Commission to the European Parliament, the Council, the European Economic and Social Committee and the Committee of the Regions on the implementation of EU waste legislation, including the early warning report for Member States at risk of. 2018*. http://ec.europa.eu/eurostat/web/waste/data/database,

European Commission. (2019). A circular economy for plastics – Insights from research and innovation to inform policy and funding decisions. In *European Commission*. https://doi.org/10.2777/269031

European Commission. (2021). https://ec.europa.eu/environment/circular-economy/first_circular_economy_action_plan.html

FAO. (2013). *Food wastage footprint – Impacts on natural resources*.http://www.fao.org/docrep/018/i3347e/

FAO. (2020a). *FAOSTAT*. Food and Agriculture Data. http://www.fao.org/faostat/en/#home

FAO. (2020b). *Food Loss and Food Waste*. Food Loss and Food Waste. http://www.fao.org/food-loss-and-food-waste/en/

FAO. (2020c). *More Fuel for the Food/Feed Debate*. More Fuel for the Food/Feed Debate. http://www.fao.org/ag/againfo/home/en/news_archive/2017_More_Fuel_for_the_Food_Feed.html

FPA. (2014). The Value of Flexible Packaging in Extending Shelf Life and Reducing Food Waste. In *Flexible Packaging Association – Report*.

Gerssen-Gondelach, S. J., Wicke, B., & Faaij, A. P. C. (2017). GHG emissions and other environmental impacts of indirect land use change mitigation. *GCB Bioenergy, 9*(4), 725–742. https://doi.org/10.1111/gcbb.12394

Gustavsson, L., Pingoud, K., & Sathre, R. (2006). Carbon Dioxide Balance of Wood Substitution: Comparing Concrete- and Wood-Framed Buildings. *Mitigation and Adaptation Strategies for Global Change, 11*(3), 667–691. https://doi.org/10.1007/s11027-006-7207-1

Hakkarainen, E., Hannula, I., & Vakkilainen, E. (2019). Bioenergy RES hybrids – assessment of status in Finland, Austria, Germany, and Denmark. *Biofuels, Bioproducts and Biorefining, 13*(6), 1402–1416. https://doi.org/10.1002/bbb.2019

Hamilton, L. A., Feit, S., Muffett, C., & Kelso, M. (2019). Plastic & Climate: The hidden costs of plastic planet. In *Center of International Environmental Law*. https://www.ciel.org/reports/plastic-health-the-hidden-costs-of-a-plastic-planet-may-2019/

Head, M., Levasseur, A., Beauregard, R., & Margni, M. (2020). Dynamic greenhouse gas life cycle inventory and impact profiles of wood used in Canadian buildings. *Building and Environment, 173*, 106751. https://doi.org/10.1016/J.BUILDENV.2020.106751

Heinonen, T., Pukkala, T., Mehtätalo, L., Asikainen, A., Kangas, J., & Peltola, H. (2017). Scenario analyses for the effects of harvesting intensity on development of forest resources, timber supply, carbon balance and biodiversity of Finnish forestry. *Forest Policy and Economics, 80*, 80–98. https://doi.org/10.1016/j.forpol.2017.03.011

Helin, T., Salminen, H., Hynynen, J., Soimakallio, S., Huuskonen, S., & Pingoud, K. (2016). Global warming potentials of stemwood used for energy and materials in Southern Finland: differentiation of impacts based on type of harvest and product lifetime. *GCB Bioenergy, 8*(2), 334–345. https://doi.org/10.1111/gcbb.12244

IEA. (2007). *Contribution of Bioenergy to the World ' s Future Energy.*

IEA. (2019). *World Energy Outlook 2019 – Analysis – IEA*. World Energy Outlook 2019 – Analysis – IEA. https://www.iea.org/reports/world-energy-outlook-2019

IFP. (2017). *ASSESSING GLOBAL RESOURCE USE. A systems approach to resource efficiency and pollution reduction.*https://www.resourcepanel.org/sites/default/files/documents/document/media/assessing_global_resource_use_amended_130318.pdf

IPCC. (2019). *Global Warming of 1.5°C. An IPCC Special Report on the impacts of global warming of 1.5°C above pre-industrial levels and related global greenhouse gas emission pathways, in the context of strengthening the global response to the threat of climate change,* (and T. W. (eds. . Masson-Delmotte, V., P. Zhai, H.-O. Pörtner, D. Roberts, J. Skea, P.R. Shukla, A. Pirani, W. Moufouma-Okia, C. Péan, R. Pidcock, S. Connors, J.B.R. Matthews, Y. Chen, X. Zhou, M.I. Gomis, E. Lonnoy, T. Maycock, M. Tignor (Ed.)). https://doi.org/10.1038/291285a0

IPCC, 2014. (2014). Summary for Policymakers. In R. Edenhofer, O., J. Pichs-Madruga, Y. Sokona, E. Farahani, S. Kadner, K. Seyboth, A. Adler, I. Baum, S. Brunner, P. Eickemeier, B. Kriemann, & T. Z. and J. C. M. (eds.). Savolainen, S. Schlömer, C. von Stechow (Eds.), *Climate Change 2014: Mitigation of Climate Change. Contribution of Working Group III to the Fifth Assessment Report of the Intergovernmental Panel on Climate Change*. Cambridge University Press, Cambridge, United Kingdom and New York, NY, USA.

Kirchherr, J., & van Santen, R. (2019). Research on the circular economy: A critique of the field. *Resources, Conservation and Recycling, 151*, 104480. https://doi.org/10.1016/J.RESCONREC.2019.104480

Koponen, K., & Soimakallio, S. (2015). Foregone carbon sequestration due to land occupation—the case of agro-bioenergy in Finland. *The International Journal of Life Cycle Assessment, 20*(11), 1544–1556. https://doi.org/10.1007/s11367-015-0956-x

Koponen, K., Sokka, L., Salminen, O., Sievänen, R., Pingoud, K., Ilvesniemi, H., Routa, J., Ikonen, T., Koljonen, T., Alakangas, E., Asikainen, A., & Sipilä, K. (2015). Sustainability of forest energy in Northern Europe. *Vtt.Fi*. http://www.vttresearch.com/impact/publications

Korhonen, J., Honkasalo, A., & Seppälä, J. (2018). Circular Economy: The Concept and its Limitations. *Ecological Economics, 143*, 37–46. https://doi.org/10.1016/J.ECOLECON.2017.06.041

Krausmann, F., Wiedenhofer, D., Lauk, C., Haas, W., Tanikawa, H., Fishman, T., Miatto, A., Schandl, H., & Haberl, H. (2017). Global socioeconomic material stocks rise 23-fold over the 20th century and require half of annual resource use. *Proceedings of the National Academy of Sciences of the United States of America, 114*(8), 1880–1885. https://doi.org/10.1073/pnas.1613773114

Kunttu, J. (2020). *Wood utilization scenarios and their sustainability impacts in Finland* [University of Eastern Finland]. https://doi.org/10.14214/df.292

Lantto, R, Järnefelt, V, Tähtinen, M, Jääskeläinen, A Laine-Ylijoki, J Oasmaa, A Sundqvist-Andberg, H Sözer, N. (2018). *Going beyond a circular economy. A vision of a sustainable economy in which material, value and information are integrated and circulate together.*

Lehtonen, Juha; Järnefelt, Vafa; Alakurtti, Sami; Arasto, Antti; Hannula, Ilkka; Harlin, Ali; Koljonen, Tiina; Lantto, Raija; Lienemann, Michael; Onarheim, Kristin; Pitkänen, Juha-Pekka;

Tähtinen, M. (2019). The Carbon Reuse Economy: Transforming CO2 from a pollutant into a resource. In *Report*. https://doi.org/10.32040/2019.978-951-38-8709-4

Levasseur, A., Lesage, P., Margni, M., & Samson, R. (2013). Biogenic Carbon and Temporary Storage Addressed with Dynamic Life Cycle Assessment. *Journal of Industrial Ecology*, *17*(1), 117–128. https://doi.org/10.1111/j.1530-9290.2012.00503.x

Martien van den Oever, Karin Molenveld, Maarten van der Zee, H. B. (2019). Biobased and biodegradable plastics. In *Bio-based and biodegradable plastics – Facts and Figures* (Vol. 4, Issue 1722). https://doi.org/10.1007/978-3-319-68255-6_103

Paper Industry World. (2015). *What is the role of fibre based packaging...?*https://www.paperindustryworld.com/what-is-the-role-of-fibre-based-packaging/

Pingoud, K., Pohjola, J., Valsta, L., & others. (2010). Assessing the integrated climatic impacts of forestry and wood products. *Silva Fennica*, *44*(1), 155–175.

Plastics Europe. (2018). *Plastics-the Facts 2018 An analysis of European plastics production, demand and waste data*.

Plevin, R. J., Jones, A. D., Torn, M. S., & Gibbs, H. K. (2010). Greenhouse Gas Emissions from Biofuels' Indirect Land Use Change Are Uncertain but May Be Much Greater than Previously Estimated. *Environmental Science & Technology*, *44*(21), 8015–8021. https://doi.org/10.1021/es101946t

Poutanen, K., Nordlund, E., Paasi, J., Vehmas, K., & Åkerman, M. (2017). Food economy 4.0. VTT's vision of an era of smart consumer-centric food production. *In VTT*. https://doi.org/10.2307/j.ctt20p56g8.8

Rogelj, J., D. Shindell, K. Jiang, S. Fifita, P. Forster, V. Ginzburg, C. Handa, H. Kheshgi, S. Kobayashi, E. Kriegler, L. Mundaca, R. Séférian, and M. V. V. (2018). Mitigation Pathways Compatible with 1.5°C in the Context of Sustainable Development. In Masson-Delmotte, S. C. V., P. Zhai, H.-O. Pörtner, D. Roberts, J. Skea, P.R. Shukla, A. Pirani, W. Moufouma-Okia, C. Péan, R. Pidcock, & and T. W. J.B.R. Matthews, Y. Chen, X. Zhou, M.I. Gomis, E. Lonnoy, T. Maycock, M. Tignor (Eds.), *Global Warming of 1.5°C. An IPCC Special Report on the impacts of global warming of 1.5°C above pre-industrial levels and related global greenhouse gas emission pathways, in the context of strengthening the global response to the threat of climate change,*. Intergovernmental Panel on Climate Change (IPCC).

Sathre, R., & O'Connor, J. (2010). Meta-analysis of greenhouse gas displacement factors of wood product substitution. *Environmental Science & Policy*, *13*(2), 104–114. https://doi.org/10.1016/J.ENVSCI.2009.12.005

Savolahti, M. (2020). Climate and health impacts of residential wood combustion in Finland. *Doctoral Dissertation*. https://aaltodoc2.org.aalto.fi/handle/123456789/43035

Seppälä, J., Heinonen, T., Pukkala, T., Kilpeläinen, A., Mattila, T., Myllyviita, T., Asikainen, A., & Peltola, H. (2019a). Effect of increased wood harvesting and utilization on required greenhouse gas displacement factors of wood-based products and fuels. *Journal of Environmental Management*, *247*, 580–587. https://doi.org/10.1016/J.JENVMAN.2019.06.031

Seppälä, J., Heinonen, T., Pukkala, T., Kilpeläinen, A., Mattila, T., Myllyviita, T., Asikainen, A., & Peltola, H. (2019b). Effect of increased wood harvesting and utilization on required greenhouse gas displacement factors of wood-based products and fuels. *Journal of Environmental Management*, *247*, 580–587. https://doi.org/10.1016/J.JENVMAN.2019.06.031

Statistics Finland. (2020). *Fuel classification 2020*.

Steffen, W., Richardson, K., Rockström, J., Cornell, S. E., Fetzer, I., Bennett, E. M., Biggs, R., Carpenter, S. R., de Vries, W., de Wit, C. A., Folke, C., Gerten, D., Heinke, J., Mace, G. M., Persson, L. M., Ramanathan, V., Reyers, B., & Sörlin, S. (2015). Planetary boundaries: Guiding human development on a changing planet. *Science*, *347*(6223), 1259855. https://doi.org/10.1126/science.1259855

Tellnes, L. G. F., Ganne-Chedeville, C., Dias, A., Dolezal, F., Hill, C., & Zea Escamilla, E. (2017). Comparative assessment for biogenic carbon accounting methods in carbon footprint of products: a review study for construction materials based on forest products. *IForest – Biogeosciences and Forestry*, *10*(5), 815. https://doi.org/10.3832/IFOR2386-010

Tenhunen, A., & Pöhler, H. (2020). *A Circular Economy of Plastics – A vision for redesigning plastics value chains.* https://cris.vtt.fi/en/publications/a-circular-economy-of-plastics-a-vision-of-redesigning-plastics-v/fingerprints/

UNESCO. (2020). *Fact2: Agricultural use | United Nations Educational, Scientific and Cultural Organization.* http://www.unesco.org/new/en/natural-sciences/environment/water/wwap/facts-and-figures/all-facts-wwdr3/fact2-agricultural-use/

WBA. (2019). *GLOBAL BIOENERGY STATISTICS 2019.* https://worldbioenergy.org/global-bioenergy-statistics

Weindl, I., Lotze-Campen, H., Popp, A., Müller, C., Havlík, P., Herrero, M., Schmitz, C., & Rolinski, S. (2015). Livestock in a changing climate: Production system transitions as an adaptation strategy for agriculture. *Environmental Research Letters, 10*(9). https://doi.org/10.1088/1748-9326/10/9/094021

Wicke, B., Verweij, P., van Meijl, H., van Vuuren, D. P., & Faaij, A. P. C. (2012). Indirect land use change: review of existing models and strategies for mitigation. *Biofuels, 3*(1), 87–100. https://doi.org/10.4155/bfs.11.154

Wiloso, E. I., Heijungs, R., Huppes, G., & Fang, K. (2016). Effect of biogenic carbon inventory on the life cycle assessment of bioenergy: challenges to the neutrality assumption. *Journal of Cleaner Production, 125*, 78–85. https://doi.org/10.1016/J.JCLEPRO.2016.03.096

Zero Waste Scotland. (2018). *The Carbon Footprint of Scotland's Waste* (Issue September). http://www.zerowastescotland.org.uk/content/scotland's-carbon-metric-impact

Zink, T., & Geyer, R. (2017). Circular Economy Rebound. *Journal of Industrial Ecology, 21*(3), 593–602. https://doi.org/10.1111/jiec.12545

Chapter 6
The Bioeconomy Perspectives in Transformation Towards a Circular Economy in Poland

Danuta Ciechanska, Joanna Kulczycka, Marta Kutyna-Bakalarska, Olga Janikowska, and Stanisław Bielecki

Abstract Development of the bioeconomy is one of the strategic tasks introduced in The Polish Circular Economy Roadmap and National Smart Specialization. The food sector is one of the most important and fastest growing branches of the Polish economy. 10.5% (1.7 million) of all employees employed in industry are involved in it. The increasing demand for eco-friendly products and packaging and at the same time focusing of the EU policies on carbon neutrality and ensuring resource and energy efficiency in a holistic approach creates new challenges and opportunities for food and beverage sector. Most of the companies start to analyse the economic and environmental impact from a value chain perspective to identify the environmental "hot spots" and value added from material supply to distribution from store supply to the customer and waste management. The paper analyses challenges and obstacles for turning bio-waste, residues and discards into valuable resources taking into account economic, environmental and social aspects based mainly on an example of Maspex (https://maspex.com/) which is one of the largest companies in the segment of food products in Central and Eastern Europe.

Keywords Bioeconomy · Circular economy · National specializations · Food and beverage sector

---

D. Ciechanska (✉)
Association Bioeconomy Cluster, Lodz, Poland

J. Kulczycka
AGH University of Sciences and Technology, Cracow, Poland

M. Kutyna-Bakalarska
Grupa Maspex Wadowice, Wadowice, Poland

O. Janikowska
Mineral and Energy Economy Research Institute of the Polish Academy of Sciences, Cracow, Poland

S. Bielecki
Lodz University of Technology, Faculty of Biotechnology and Food Sciences, Lodz, Poland

E. Koukios, A. Sacio-Szymańska (eds.), *Bio#Futures*,
https://doi.org/10.1007/978-3-030-64969-2_6

1 Introduction

A circular bioeconomy is an important part of the European sustainable development strategy. According to the European Commission, the bioeconomy comprises those parts of the economy that use renewable biological resources from land and sea to produce food, materials and energy.[1] Therefore bioeconomy should strengthen the connection between economy, society and the environment.[2] The updated bioeconomy strategy is part of the Commission's efforts to create a boost for employment, growth and investments in the EU. According to the strategy the main challenge for the future is to increase the sustainable use of renewable resources, to face both global and local challenges, such as climate change and sustainable development. The European Commission underlines that it is absolutely crucial to implement the systemic changes in the way products are being produced, consumed and utilized and brings attention to the fact that the development of the bioeconomy (a renewable segment of the circular economy) will be and should be a way of reaching innovative ways of providing food, clean water and energy in the future.

Implementation of the priorities by the European Bioeconomy Strategy is based on five strategic goals:

- Ensuring food and nutrition security due to the changing needs of consumers in the field of sustainable food production and consumption practices
- Managing natural resources in a sustainable manner, aimed at preventing soil
- degradation, restoring degraded ecosystems and ensuring their resilience to climate change and valorisation of natural resources and secondary raw materials
- Reducing dependence on non-renewable resources, which is key to achieving EU energy and climate policy goals
- Mitigation and adaptation to climate change, which is recognized as the global challenge of the present generation
- Strengthening European competitiveness and job creation by providing a framework for innovation and implementation and supporting the development of markets for biotechnology-based products.[3]

The diagnosis of the bioeconomy indicates that many regions in the Europe still have a low level of bioeconomic maturity, which means they cannot fully use their potential such as creating new jobs, use resource efficiently or plan a sustainable rural. Five main areas which leads to more effective implementation of the bioeconomy at both regional and national level are pointed out in the document Bioeconomy development in EU regions Mapping of EU Member States, and those are:

---

[1] https://ec.europa.eu/research/bioeconomy/index.cfm

[2] A sustainable Bioeconomy for Europe: strengthening the connection between economy, society and the environment. Updated Bioeconomy Strategy, 2018.

[3] A sustainable Bioeconomy for Europe: strengthening the connection between economy, society and the environment. Updated Bioeconomy Strategy, 2018.

- More effective strategic planning and bioeconomy management at national and regional level
- Support for value chains/circularity and in particular the involvement of SMEs
- Development of research and innovation in technology, knowledge transfer and new skills in the field of bioeconomy
- Coordinated financing and synergy between support instruments
- Increased social awareness and acceptance[4]

According to the report the progress in implementation of bioeconomy will have a significant impact on the planning of activities towards further development of the circular bioeconomy, as well as on the development of recommendations for the implementation of the BIO concept in the circular economy.

2 Strategy of Bioeconomy Development in Poland

2.1 Overview

The Polish Strategy for Responsible Development (SRD) is a recommendation for the future development of the country. Strategy follows the development vision contained in the 2030 Agenda. According to the document the circular economy should play an important role in achieving the goals of sustainable development in Poland, combining social, environmental and economic issues. Sustainable development requires fulfilment of some boundary conditions such as efficiency of raw material consumption, the use of renewable energy, waste management and reduction of greenhouse gas emissions. According to Bio-Based Industries Joint Undertaking (BBI JU), there is a strong correlation between three specific objectives of the Sustainable Development Goals (SCGs) and the bioeconomy:

sustainable economic growth based increasingly on knowledge, data and organizational excellence; socially sensitive and territorially sustainable development; and an effective state and institutions for growth and social and economic inclusion. Pursuant to the assumptions of the strategy, the main goal of all activities and undertakings envisaged in it is "to create conditions for the growth of income of Polish residents, while increasing cohesion in the social, economic, environmental and territorial dimensions[5]".

---

[4] Bioeconomy development in EU regions Mapping of EU Member States' Final Report 2017.

[5] The 2030 Agenda for Sustainable Development – Implementation in Poland.

2.2 Road Mapping

The Road Map Towards the Transition to Circular Economy in Poland is based on the Ellen MacArthur model of circular economy which assumes the existence of two cycles: biological (renewable raw materials) and technical (non-renewable raw materials). The aim of the Circular Economy Road Map, on the one hand, is to indicate horizontal actions which would affect the largest possible section of social and economic life. On the other hand, the Circular Economy Road Map prioritizes the areas in which development will allow for taking advantage of the opportunities facing Poland and at the same time will address the currently existing or expected threats. This document refers to five general areas: (1) *Sustainable industrial production* attention was drawn to the important role of industry in the Polish economy and new opportunities for its development; (2) *Sustainable consumption* the need to take action on this stage of the life cycle, so far often underestimated in the context of its contribution to the transition to CE, is justified; (3) *Bioeconomy* outlines the management of renewable raw materials (the biological cycle of CE), which seems to have an unexploited potential in Poland; (4) *New business models* discusses the opportunities for reorganizing functioning of various market participants based on the idea of CE; (5) the implementation, monitoring and financing of CE.[6]

According to the Polish Statistic Office, in 2017 the area of agricultural land in Poland amounted to 14,620 thousand ha (i.e. ca. 47% of the size of the country), which translates into a significant potential for the development of bioeconomy based on this source of biomass. Currently, apart from food production, biomass is most often used for energy purposes – mainly for direct combustion and, to a relatively small extent, for the production of liquid fuels. According to Polish Investment and Trade Agency, the food sector is one of the most important and fastest growing branches of the Polish economy. 10.5% (1.7 million) of all employees employed in industry are involved in it. Poland is the sixth largest market in Europe, with a capacity of 38.5 million inhabitants. Polish producers are characterized by high competitiveness both in the EU and in the world. 80% of all exports, in 2017, went to the EU's internal market, which after accession became one of the main driving forces for the sector with a potential of over 508 million consumers. Strengths of the Polish food sector are many years of tradition, high-quality product, competitive production and labour costs, qualified staff, solid educational base and R&D potential and well-developed network of suppliers.[7]

The CE Road Map focuses, on the one hand, on general actions aimed to create conditions for the development of bioeconomy in Poland and, on the other hand, on activities concerning the development of bioeconomy in selected areas, i.e. creating local value chains, in industry in general, and in the power industry in particular. Due to the cross-sectional and inter-sectoral nature of bioeconomy, there is no single ministry in Poland responsible for shaping this framework and defining its

[6] The Road Map Towards the Transition to Circular Economy, 2019.

[7] https://www.paih.gov.pl/sectors/food_processing

development directions at the central level. The implementation of individual actions provided in the CE Road Map will be carried out with competence of individual ministries. According to the document "development of bioeconomy contributes to the reduction of pressure on the natural environment, as well to creating new jobs, in rural areas in particular. The production of innovative materials and products within bioeconomy requires a continuous supply of quality biomass. Therefore, it is important to build local value chains in the areas around local biorefineries, which will be able to produce high quality bio-residue material in the quantities consistent with entrepreneurs' needs".[8]

Biomass is currently one of the most popular renewable energy sources in Poland. According to the assumptions biomass should be cascading, based primarily on its use for food production and as a raw material for the chemical, pharmaceutical, paper and building materials industries as well as for the production of organic fertilizers. Only residual biomass and waste from the final stages of recycling should be used for energy purposes, with priority given to the production of biofuels and biogas. Bioeconomy can provide a strong stimulus for increasing the innovation and competitiveness of entire industries; however, the use of biomass by industry in Poland is still not widespread. To support the development of the bioeconomy in the industrial sector, the principle of the cascading use of biomass is important, favouring the use of higher value-added technologies that allow the reuse and recycling of products and raw materials.

2.3 Bioeconomy as a Strategic Area of National Smart Specializations

According to the European Union innovation is a driving force for future economic growth and social development in Europe, and therefore better conditions for innovative processes should be created. This goal is to be achieved through the use of the concept of smart specialization. Smart specialization is a key element of EU efforts to support countries and regions in developing their own path of economic growth. The goal of smart specialization is to create new areas of economic activity and to increase the competitiveness of regions. The development of the European economy is to be based on knowledge and result from investments in the sphere of education, research and innovation. Smart specialization is a key element of the European Union's efforts to support this development model. It refers directly to the "Innovation Union" flagship project, which is the basic instrument for achieving the goals of the Europe 2020 strategy.[9]

In Poland, the bioeconomy plays an increasingly important role, being an important element of the National Smart Specializations (NSS). National specializations

---

[8] Road Map towards the Transition to Circular Economy, 2019, p. 21.

[9] Europe 2020 Flagship Initiative Innovation Union.

indicate preferences in providing support for the development of research, development and innovation (R & D & I) under the financial perspective for 2014–2020 which are specified in the government document entitled "National Smart Specializations". Issues related to bioeconomy are widely covered in two NSS: (1) innovative technologies, processes and products of the agriculture and food and forest-based sector and (2) biotechnological and chemical processes, bioproducts and products of specialist chemistry and environmental engineering.

The bioeconomy has become one of the economic priorities in the area of R & D & I that may contribute to the transformation of the national economy towards a circular economy by modernizing it, structural transformation, diversification of products and services and creating innovative socio-economic solutions, also supporting transformation towards a resource-efficient economy, including natural resources.

In the case of bioeconomy, a broad interdisciplinary profile of this key area of specialization has been outlined, which creates enormous opportunities for creating mutual cooperation links based on NSS for the development of integrated value chains. The bioeconomy is perceived as an area with a large potential of raw materials and technologies; all the more it seems necessary to strengthen the efforts to increase the technological readiness of bioeconomy solutions aimed at developing business initiatives.

In the Entrepreneurial Discovery Process, activities are carried out to continuously activate, update and absorb national specializations to implement strategic research and innovation development plans in the country. In the process of monitoring National Smart Specializations, periodic assessment of the activity of individual specializations is carried out, including data based on applications submitted and co-financed in operational programs. Based on this data, the NSS is periodically prioritized. In the area of bioeconomy for NSS 2 and NSS 3 in the first half 2019, there was quite a large variation in the number of submitted and co-financed projects, for NSS 2 mainly in the agri-food and forestry-wood areas and for NSS 3 in the development of bioprocesses and specialized chemistry. Thematic evaluation of the achievement of the national smart specialization goals carried out as part of the NSS monitoring provided information on the research and development potential of scientific units and its impact on the implementation of the NSS 19 goals. As part of the study, scientific units were assigned to individual NSS, and the strength of their links with specializations was assessed. NSS 2, being a specialization in the field of bioeconomy, came second to NSS 1 Health with the largest number of science institutions that reported membership in this specialization.[10]

As part of the evaluation of national smart specializations carried out by PARP, an assessment was made of the internationalization of national enterprises (Maciej Piotrowski, 2019). The aim of the study was to characterize the level of internationalization of Polish enterprises taking into account the specificity of individual NSS and non-NSS groups with significant internationalization potential and

[10] Ecorys Polska, Taylor Economics, 2019.

identification of activities, which should be implemented by the public sector to strengthen the internationalization potential of Polish enterprises, in particular under the NSS. The obtained data confirmed that NSS 2 is in the range with the highest potential for internationalization as well as in the range with the highest import intensity.

2.4 Bioeconomy: A Strategic Area of Regional Smart Specializations (RSS)

A condition for support in operational programs of activities implemented under the two thematic objectives of the European Regional Development Fund (ERDF) was the regions' identification of smart specialization. Its improvement should be based on strategy setting priorities and tools for using the opportunities and potentials of a given region and achieving competitive advantages. Therefore, each region, which is a beneficiary of cohesion policy in the 2014–2020 financial perspective, was obligated to prepare its own research and innovation strategy for smart specialization (the so-called RSS3 strategy) and identify in it the areas in which it wants to specialize and increase its competitiveness.

Bioeconomy is an area of smart specialization with high potential in the context of increasing the competitiveness of Polish regions. Its essence lies in the innovative use and management of renewable biological resources in order to generate new types of products and production techniques, while meeting the requirements of sustainable development. Bioeconomy is one of the most promising economic sectors in which Polish regions intend to specialize. The concept of bioeconomy has become one of the areas of strategic planning, especially in the sphere of innovation. The development of bioeconomy means the need to make internal changes in the sector, including strengthening the integration of the economy with science and the sphere of business with the social environment. This means that the bioeconomy should be considered comprehensively, from the theoretical point of view in micro, meso and macroeconomic terms, and from the side of economic policy in regional, national and European terms.

From a microeconomic perspective, bioeconomy includes the production of various products and services related to living organisms for food and utility purposes by farms, processing plants and other business units. From the meso-economic point of view, bioeconomy is a sector or sphere of production of these products and services and creation of local and regional systems for the production and consumption of products and services. In macroeconomic terms, attention is paid to economic structures and processes that are the basis for the sustainable use of biological, renewable production resources to produce healthy food, feed, materials, energy and other products while respecting the principles of food, health, energy and environmental protection.

13 Polish voivodships have declared the production of agricultural raw materials and their processing into food as specialization. The importance of agriculture and food processing sectors was not taken into account by three voivodships: Małopolska, Pomorskie and Śląskie. Specializations related to biomass production and processing as well as waste to energy conversion are not much less common. More than half of the voivodships have adopted activities in the sphere of environmental protection and biodiversity as well as in sectors of the industry processing biological raw materials into non-food products as areas of specialization within the bioeconomy. Sub-specialization covering knowledge gathering, scientific research, institutions as well as creating value chains and cooperation was indicated relatively least frequently.

3 Diagnosis of the Bioeconomy Potentiality in Poland

3.1 Potential Feedstock and Bioproducts Demands

According to the analysis of the Deloitte report entitled "Closed loop – open opportunities" (Report of Deloitte, 2018) in which development and growth prospects for the circular economy are presented, Poland is a prospective country for the development of the bioeconomy. According to the data, the food, feed and drink sectors account for the largest share in bioeconomy turnover in the EU and Poland, accounting for nearly half of the total turnover. In turn, the turnover of the bio-industry including the production of chemicals and chemical products, pharmaceutical products, plastics, paper, textiles, biofuels and bioenergy as well as the wood industry sector is worth EUR 600 billion.

The agricultural and forestry sectors are mainly responsible for the production of raw materials for processing industries using biomass. In Poland, this is a very rich base: the arable land in 2017 accounted for about 14.6 million ha, i.e. they occupy about 47% of the country's territory, by 10 pp. more than the average in the world.[11] Poland is also in the European lead when it comes to the area of forests – they grow on the area of 9.2 million ha, covering about 29.6% of the country's area.[12] Domestic extraction of biomass, i.e. biodegradable products, waste and residues from biological origin, in 2017, amounted to 191 million tonnes and was nearly 20% higher than in 2010.[13]

The structure of obtained biomass includes in % arable crops (except fodder), 34%; crop residues, fodder and grazing biomass, 53%; wood, 13%; and wild fish, plants and aquatic animals, 0.1%. In recent years, the turnover of Polish foreign trade in biomass and its products has been growing. Significant surpluses have been visible for many years, which have been gradually increasing over time. In

[11] Piotrowski S., Carus M., Carrez D., (BIC) European Bioeconomy in figures 2008–2015, 2018..

[12] Statistic Office, 2018.

[13] http://www.lasy.gov.pl/nasze-lasy/polskie-lasy

comparison with 2008, the tonnage of exported biomass in 2015 increased by 118%. In 2017, the value of agri-food sector exports reached a record EUR 27.3 billion, accounting for 13.4% of the value of Polish exports.[14]

An important source of raw materials for the bioeconomy is biomass resources from the biodegradable waste stream, which in the circular bioeconomy system should be used first in the production of new bioproducts and biomaterials (including those used in new bio-packaging) and then in the processes of transformation into biofuels and biogas. It is biomass coming from both agricultural activities, production and processing processes (pomace, sludge, decoctions, musts, fragments of raw materials, by-products, defective products, etc.), municipal waste (organic fraction) as well as wastewater treatment processes (primary and secondary).

In the cascade biorefinery concept, these processes are carried out in a fully integrated mode with bioenergy recovery to comprehensively and fully utilize the potential of biosurets. The same process assumptions may apply to the forestry/wood industry, which generates approx. 50% of biomass in the form of all types of forest waste associated with logging, all types of production residues related to wood processing and all types of wood waste obtained in recycling processes.

The implementation of a circular economy in Poland may be highly beneficial from a macroeconomic point of view. The performing of this type of economy would definitely have a positive impact on economic growth – "the estimated effect in the form of added value from saving 1% in material and energy costs for Polish GDP can be as much as PLN 19.5 billion[15]". The implementation of a circular economy allows the use of a number of solutions related to the reduction of material consumption, the amount of waste and its recovery as well as the increase in energy efficiency. The impact on GDP of the reduction of material and energy consumption by 1% in individual sectors of the economy was estimated. The analysis is a preliminary approximation of macroeconomic effects for Poland. The input-output model focuses on relationships and dependencies between various branches of the economy, thanks to which it is possible to examine how the activities of a given branch of the economy affect the development of others.

According to the Report of Deloitte, the reduction of material and energy consumption costs in the case of bioeconomy can be achieved, among others, by:

- Using production (including packaging) materials that are biodegradable and easy to recycle
- Transition to an agricultural system allowing soil regeneration and revitalizing the ecosystem in agricultural areas
- The use of new technologies for the recovery of nutrients and energy from waste (e.g. anaerobic digestion).[16]
- Sustainable waste management according to UE should be based on the following principles:

[14] Eurostat, 2017.

[15] https://www2.deloitte.com/pl/en/pages/zarzadzania-procesami-i-strategiczne/articles/innowacje/raport-zamkniety-obieg-otwarte-mozliwosci.html

[16] Report of Deloitte, 2018.

- Waste prevention
- Preparation for reuse
- Recycling
- Other recovery methods
- Storage

The goal of EU regarding recycling by 2030 is 65% for municipal waste and 75% for packaging waste, as well as limiting the amount of landfill for municipal waste generated by 2030, to a maximum of 10%.

What is especially important in Polish ecological policy, which is closely related to raw materials and energy policies, is the subject of analyses and discussions, but still in many aspects it needs to be updated and clarified. The amount of generated waste in Poland (excluding municipal waste) has been in the range of 110–130 million tonnes/year since 2000. In 2017, 126 million tonnes of waste was generated, of which municipal waste was 12.5 0 million tonnes (9.5%). The amount of waste generated annually remains at a similar level, with a constant increase in GDP, which may indicate positive trends in waste management. According to Statistics Poland recently, the largest amount of waste was generated in the Dolnośląskie and Śląskie voivodships. Lubelskie, Podkarpackie and Warmińsko-Mazurskie voivodships are those who generate the smallest amount of wastes. In Poland, on a regional scale, the amount of waste generated is in the range from about 190 to about 375 kg/inhabitant. In turn, the average amount of municipal waste generated per capita in the European Union in 2016 was 483 kg. The amount of municipal waste generated depends not only on the population density but also on consumption patterns determined by the standard of living of society. Between 2006 and 2017 from the total amount of municipal waste generated in the European Union, 29% was recycled, 27% was thermally neutralized, 26% was neutralized by landfilling, and 16% was composted.[17]

The methods of thermal processing of waste with energy recovery have been used in Poland since 2015. An important element conditioning the possibility of using waste for energy needs is the availability and transmission capacity of energy infrastructure, including heating. The highest values of the heat network density index were recorded in the Śląskie, Małopolskie and Łódzkie voivodships. The majority (approx. 77.2%) of heat produced is intended for the needs of heating residential buildings. The most heat for heating purposes was produced using solid fuel, followed by gas and liquid fuels. The share of waste as fuels used for heat production is marginal. There are eight waste incineration plants in Poland and one multifuel CHP plant using RDF fuel. Further RDF heating plants are under construction or at the planning stage. In 2016, the total amount of waste generated in the EU-28 in all sectors of the economy and households amounted to approx. 2.5 billion tonnes, while the share of individual countries in this amount varies. The highest levels of waste production were recorded for municipal services, households and manufacturing activities. According to Eurostat an overall increase (by around 10%) in the

[17] GUS Ochrona środowiska, 2018, Infrastruktura komunalna w 2017 r., GUS 2018, Infrastruktura komunalna w 2017 r., GUS 2018.

number of waste recycled and incinerated with energy recovery can be seen. There is a clear tendency to move away from landfilling municipal waste, mainly in favour of recycling and incineration, and to a lesser extent composting and other methods.

According to the United Nations, our production and consumption lead to large quantities of waste. "An important element on work on eco-cycles is therefore sustainable waste management. Articles that circulate in society contain large quantities of different materials. Many are energy-demanding to produce and contain substances that exist in limited quantities of different materials. Many are energy-demanding to produce and contain substances that exists in limited quantities. It is therefore necessary to manage common resources in long-term manner to achieve sustainable cycle". According to many scientists, current production of biofuels is rather unsustainable, so there is a strong need to improve current production methods. According to the framework, policy decisions concerning biofuels should take into account certain moral values. These include human rights and global justice, solidarity and the common good, sustainability and intergenerational justice. The five ethical principles and one ethical duty forming the core of the ethical framework are:

- Biofuels development should not be at the expense of people's essential human rights, including food, health and water.
- Biofuels should be environmentally sustainable.
- Biofuels should contribute to a net reduction of total greenhouse gas emissions.
- Biofuels should adhere to fair trade principles.
- The costs and benefits of biofuels should be distributed in an equitable way. It should not happen, for example, that the benefits occur in the developed world and the costs occur disproportionately in poor countries.

If these five principles are respected, depending on certain key considerations, such as absolute cost or whether there are even better alternatives, there is a duty to develop such biofuels.[18]

3.2 *Bio-Inspired Industrial Potentiality in Poland*

The analysis of the potential of companies interested in research in the bioeconomic area indicates the existing implementation potential in this area, which, however, requires quite intensive efforts to intensify activities related to the preparation of R&D potential that can be implemented in industrial practice. To this end, actions necessary to increase the technological readiness indexes of TRL solutions developed in scientific and research units are necessary, including investment activities in pilot installations for rescaling technological solutions with high financial risk, which include innovative solutions in the field of bioeconomy and industrial biotechnology. Good practices in the field of bioeconomy are implemented by some domestic companies (Table 6.1).

[18] https://www.ncbi.nlm.nih.gov/books/NBK196458/

Table 6.1 Examples of good practices in Poland

Company	Internet link
BIOTECHNIKA Biotech solutions for modern industry in Lodz is a leading company on the Polish market, a company implementing industrial biotechnology projects, including: Production of ethanol for any purpose (consumption, bioethanol, technical ethanol, pharmaceutical ethanol) Bioenergetics, including mainly classic agricultural biogas plants Industrial biogas plants adapted to the characteristics of a specific industry (for ethanol production plants, breweries, starch plants, dairies, sugar factories and basically for every industry using biomass in its various forms in the production process) Sewage treatment plants (mainly industrial) – often associated with the previously mentioned installations (ethanol production or industrial biogas plants), for industrial processing of agricultural products (production of starch, starch hydrolysates) Biotechnology processes of small-scale products ("fine chemicals")	*http://www.biotechnika.net/*
BOWIL Biotech Sp. z o.o. in Władysławowo is the first factory in the world biocellulose made in accordance with pharmaceutical GMP standards. The production of biocellulose, a natural biomaterial that is used in medicine, biotechnology, pharmacy, dentistry, cosmetics, food industry and many other fields, is the result of scientific and industrial cooperation with the Institute of Technical Biochemistry of the Lodz University of Technology	*https://www.bowil.pl/produkty/bioceluloza/*
NapiFeryn BioTech company was established in 2014 in Łódź, awarded in the competition "Strong in business 2016". The Economic Award of the Łódź Voivodeship in the start-up category has developed an innovative technology for the production of high-quality protein from the residue after pressing oil from rapeseed. Protein isolates will be used by oil mills and food producers as food ingredients. The discovery of NapiFeryn BioTech corresponds to future civilization challenges in the scope of the possibility of feeding a larger population without adverse environmental impact processes. The company's goal is to ensure that at least 5% of the world's rapeseed crops are processed using NapiFeryn BioTech technology to form natural functional proteins	*http://www.napiferyn.pl*
ORLEN, south company from the ORLEN Capital Group, is working under the research and development project of the INNOCHEM Program on an innovative method of biotechnological transformation of biorefinery by-products and plant-derived raw materials into lactic acid necessary for the agricultural, food, medical and pharmaceutical industries. The substance plays a key role in the production of the most popular, fully biodegradable polymer, polylactide (PLA), which is used, among others, for the production of biodegradable packaging	*https://www.orlen.pl*
FLUID SA, a company from Sędziszów, is the only company in the world that sells biochar. The raw materials used for its production are, among others, straw, energy willow, animal droppings, municipal waste or sewage sludge. As a result, biochar can be used in energy, agriculture, environmental protection and industry	*https://innpoland.pl/*

In the current conditions, the competitiveness of the economy is increasingly based on research, development and innovation (R & D & I) and the ability to absorb dynamically and participate in the creation and development of new technologies. To meet the abovementioned challenges, cooperation between stakeholders representing various environments, industries or technologies is necessary. The key to achieving this goal is, inter alia, clusters, which thanks to the naturally established cooperation of enterprises, research institutions, business environment institutions, non-governmental organizations and local authorities are referred to as the catalyst for innovative processes. Cluster initiatives dedicated/related to the bioeconomy area are presented in Table 6.2

3.3 Bio Value Chain Model: Case of Food and Beverage Sector in Poland

Grupa Maspex Sp. z o.o. Sp. K. is one of the largest companies of the food segment in Central and Eastern Europe. It owns leading brands in the food industry and offers primarily juices, nectars, drinks, mousses, fruit cocktails, energy drinks, pasta, cereal products, sauces, instant products (teas, coffee whiteners, cocoa, etc.) diet supplements, vitamins and sweets. The company offers products of the highest quality, which invariably enjoy great recognition of consumers and traders. Since 2016 the Maspex Group has been a founding member of the EIT Food consortium, co-financed by the European Institute of Technology, whose aim is to change the methods of production, distribution and consumption of food, taking into account the needs and consumer expectations and the organization of the entire value chains of the food sector. EIT Food is to stimulate innovation, develop talents and engage consumers in the agri-food sector as well as support entrepreneurship of start-ups. The Company's R&D activity focuses on developing process and product innovations in the food industry, and at the same time it has great potential in conducting development projects in the field of logistics and marketing.

As part of the project, R&D works were carried out regarding the possibilities of using side products of pomace from fruit and vegetable processing in finished products which is related to bio-based innovated projects. Maspex Group in a scientific and industrial consortium with the Institute of Biopolymers and Chemical Fibers, COBRO Packaging Research Institute and the University of Humanities and Sciences of Jan Długosza (Częstochowa. Poland) has developed a patent-protected method of producing biodegradable, thermoplastic and barrier films as well as biopolymer injection moulds from starch and protein raw material derived from the by-product production line of the Lubella company of the Maspex Wadowice Group company. The developed technology is in line with the development trends of bioeconomy in the meaning of circular economy, namely, it covers activities related to the material management of biosources wastes and by-products of the food industry.

Table 6.2 Cluster initiatives dedicated/related to the bioeconomy in Poland

Cluster name	Internet link
Association Bioeconomy Cluster, whose aim is to integrate and concentrate the scientific and economic environment operating in the field of bioeconomy with particular emphasis on its cross-sectoral impact, covering the basic pillars of cooperation, i.e. innovation, education, entrepreneurship development and social communication	http://klasterbio.pl/
Natureef Association, which implements goals through webinars on packaging trends, resource efficiency technologies and financing, joint research projects mainly in the field of biomaterials, business missions, conferences and workshops, forward-looking projects for companies and associations opening new spaces of cooperation, i.e. circular economy , Industry 4.0, urban agriculture	https://natureef.pl/
NUTRIBIOMED Cluster, Wrocław Technology Park S.A., whose main idea was to build Poland's strong position in the industry offering dietary supplements, nutraceuticals and biomedical preparations, as well as using native, natural raw materials and modern technology for their production	http://www.nutribiomed.pl/
Klaster Life Science, Kraków Foundation, whose aim was to develop an innovation ecosystem in the field of biotechnology and life science, including creating cooperation networks, supporting entrepreneurship and innovation as well as combining and developing resources and competences in the area of Life Science	https://lifescience.pl/
Waste Management and Recycling Cluster RECYCLING COOPERATION CENTER, not for profit system sp. z o.o., which creates Polish enterprises providing a full range of management services for most categories of waste throughout the country as well as in most EU countries and outside the EU	https://www.clustercollaboration.eu/cluster-organisations/waste-management-and-recycling-cluster

Starch and protein compositions can be used for the production of (bio) packaging and bio-clays for paper products (*https://maspex.com/; https://www.eitfood.eu/; project NCBiR nr PBS1/A5/22/2012*).

Maspex in the project "Development of innovative products based on processing pomace" comprehensively develops new effective methods for processing pomace, which are a side product of fruit and vegetable processing, and to develop recipes for innovative products based on carrot pomace or by adding pomace to traditional products. As a result of the project, new food products that meet the requirements of consumers are proposed, being a valuable source of vitamins and minerals, and – most importantly – a different perspective on the side products of vegetable and fruit processing. Fruit and vegetable waste from the processing industry is still a source of many valuable nutrients, because during the processing of the main raw material, many valuable ingredients remain in the pomace. There is still a lot of nutrients to be found in carrot pomace, including carotenoids, fibre, polyphenols, mineral salts, carbohydrates and protein. Carotenoids and polyphenols are primarily antioxidant. Dietary fibre has a positive effect on the human digestive system, intestinal peristalsis, carbohydrate and lipid metabolism, stimulates the growth of bacterial colon flora and plays an important role in the prevention of atherosclerosis and diabetes. Products developed on the basis of carrot pomace, or with its participation, will be enriched with valuable nutrients and will have a relatively low energy value. Especially pomace out of lactic acid fermentation will be very attractive for the consumers (including those not consuming animal products). Fermentation enriches vegetables with beneficial microorganisms and bioactive substances. Pickled vegetables are currently recommended by nutritionists and doctors because of their health promoting properties. Consumers are increasingly aware of the health problems associated with incorrect nutrition and are aware of the healthy potential of vegetables. They are also eager to look for low calories products due to the obesity risk (new products with added carrot fibre will have lower calorific value than traditional ones without such an addition). New products based on carrot pomace will broaden the range of vegetable products and will be able to satisfy consumer expectations and contribute to increased consumption of this food group. The research carried out as a result of the project will allow to determine the conditions for the production and use of biomass in the closed cycle and the conditions for the development of food use of agricultural products. Thanks to the implementation of the project, the Applicant acquired practical knowledge regarding the development of standards and norms conditioning sanitary safety of waste and side products of the agri-food industry, conditioning the implementation of the United Nations development goals for 2015–2030. As part of the pursuit of ensuring sustainable consumption and production patterns (objective 12), it is assumed to halve food waste at the stage of sales and consumption by 2030 and reduce losses at the stage of agriculture and processing.

Maspex in European consortium-industrial Group (Maspex; Roquette; University of Reading; Valio; VTT Technical Research Centre of Finland; Tymbark – MWS) realized the project "Novel concepts for creating dietary fibre-rich foods from side streams (ColloidFibre)". The daily dietary fibre (DF) intake in Europe is way below the recommended level, arising partly from difficulties in creating appealing DF-rich food products for the consumers. The challenges are most pronounced in high-moisture foods where insoluble DF often gives rise to an unpleasant coarse mouth-feel and selling of the insoluble particles to the bottom of the container. High viscosity and unpleasant flavour are additional factors hampering the applicability of DF in foods. A common practice for DF enrichment in high-moisture foods has been to use only highly modified soluble DF to ensure proper sensory quality. The most common soluble fibre ingredients, such as inulin or various oligosaccharides, typically have a relatively low molecular weight and tend to cause intestinal discomfort in consumers. However, for maximized health effects and improved intestinal comfort, both soluble and insoluble DF of relatively high molecular weight would be beneficial. The utilization of more natural, less degraded DF is also driven by a recent regulatory change in the USA, which does not anymore allow labelling of "isolated, synthetic" fibre as DF in food products. This activity aims at creating better opportunities for the consumers to increase their DF intake by developing DF ingredients with improved properties that enable their addition to high-moisture food products not traditionally rich in DF. Side streams from wheat, oats and pea processing industry will be used as DF sources. These nutritious side streams will be modified by novel processing concepts not yet widely adopted in the food industry. The target is to reach an optimum ratio of soluble/insoluble DF in terms of technological functionality and sensory properties, without causing excessive intestinal discomfort often associated with the soluble DF alternatives existing on the market. The innovativeness of the chosen solutions is maximized by bringing together actors from academia and the industry with multidisciplinary expertise in food technology, material science, lignocellulosic biomass processing, flavour chemistry, sensory perception, nutrition, consumer understanding and business creation. As the outcome, this approach will generate positive societal impact by enabling ingredient, process and product innovations which helps the food industry to introduce new types of appealing and easy-to-consume DF-rich products on the market. This will improve consumers' possibilities to increase their DF intake, even for those people not aware of lacking DF in their diet. This will on longer term have a positive impact on public health and decrease health-care costs. This activity will also help in utilizing food resources more efficiently by enabling the application of side streams produced in vast quantities as food source for humans instead of as animal feed.

The project "EIT Food Digital Marketplace for Side Streams" performed by European consortium of 15 partners (ACESUR; Colruyt; Döhler; DSM; Givaudan Nederland; Maspex; Nestle; PepsiCo; Puratos; Technische Universität München; University of Reading; Givaudan Switzerland; Tymbark – MWS; Vlevico; RethinkResource) is aimed to strengthen the EIT's activity in Circular Economy/ Resource Stewardship by the following advances:

- Comprehensive platform that accelerates side stream valorization
- Generation of easily accessible information on available side streams and secondary materials
- Enablement of large scale sourcing of secondary resources
- Product innovation through upcycling
- Open up new application fields within and outside the food sector
- A fee-based accessible marketplace

Maspex is additionally connected with investments related to the policy of efficient economy of sources, energy and water. The Cogeneration (CHP) system in fruit and vegetable processing plant in Tymbark, Poland, was launched. The purpose of developing this system was to reduce the total demand for electricity and technological heat of the juice production plant through the use of biogas generated in the wastewater treatment plant (WWTP) and natural gas as the main fuel for electricity, process steam, chilled water and heating water production. In order to achieve the designated goal, the CHP system, based on the aggregate cogeneration with an electricity capacity of 1.0 MWe, fed with bi-fuel (biogas and high-methane natural gas). The industrial symbiosis aspect is also realized by the Tymbark processing plant agreement for wastewater treatment received from Tymbark municipality and a local dairy plant OSM Limanowa which corresponds in about 15% of the total wastewater flow in the WWTP (Fig. 6.1).

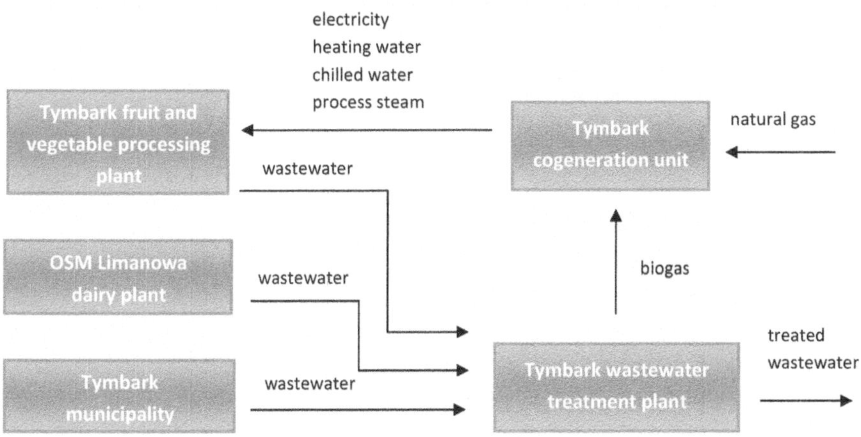

Fig. 6.1 The Cogeneration (CHP) system in fruit and vegetable processing plant in Tymbark. (On the base of Maspex own database)

4 Factors Conditioning Activities for the Development of Bioeconomy

The development of bioeconomy is inextricably linked to the use of an innovative approach, as well as the creation of new patterns of effective use of human capital. The potential inherent in human resources as well as financial and infrastructure capabilities should generate the development of new types of products and production techniques. It should also lead to the creation of an appropriate synergy of implemented policies, in particular scientific, scientific and technical, innovative, economic and social policies. Thanks to this impact, the economy should more effectively use both current and future resources in the production of basic raw materials, semi-finished and finished products in the food sector as well as industries and services processing or using biological resources. The key challenge for Poland in the macroeconomic dimension is to strengthen cooperation between science, entrepreneurs and public authorities, as part of the so-called triple helix, whose mission is to build an open and expansive economy. The positive impact on individual areas of the bioeconomy, the way it is shaped and implemented is ensured by the integration of the objectives of individual policies supporting finance, science, knowledge transfer and innovation. Less dispersion of development initiatives, concentration of public funds on science, research dissemination, knowledge transfer and innovation as well as activation and consolidation of extra-budgetary funds may become a key chain of sustainable activities to achieve positive effects conditioning the possibility of implementing bioeconomic innovations (Fig. 6.2). Creating a sustainable bioeconomy that fits into the idea of a circular economy requires coordinated efforts from public authorities and industry.[19]

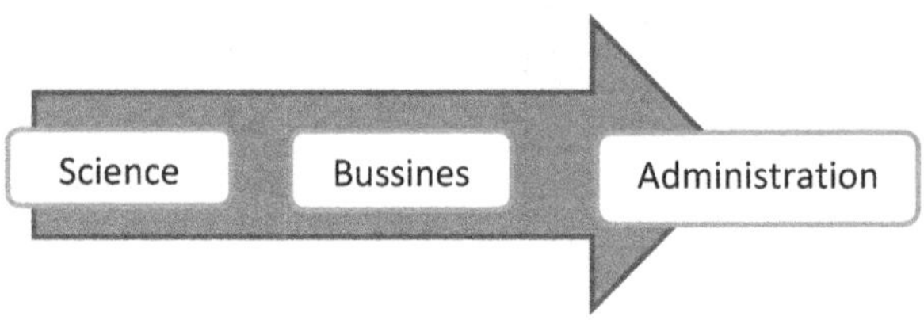

Fig. 6.2 Key structures conditioning the development of the bioeconomy

[19] Eugeniusz K. Chyłek, Monika Rzepecka „Biogospodarka – konkurencyjność i zrównoważone wykorzystanie zasobów", Polish Journal of Agronomy, 7, 3–13, 2011.

5 Recommendations for the Implementation of Bioeconomy in the Concept of Circular Economy

According to Wicki research and development until 2016 included in particular the following issues[20]:

- Developing processes for obtaining energy and chemicals with high added value from biomass from waste and vegetation using industrial biotechnological methods
- Obtaining new biomaterials and polymer composites with controlled biodegradability based on cellulose nanofibres and bio-nanocellulose
- Developing technologies for obtaining new biocatalysts and biocatalyst mimetics for the production of fuel and organic chemical compounds of significant industrial importance (from biomass)
- Developing biotechnological processes for the production of functional foods useful in the prevention and treatment of diet-related diseases
- Developing new ways to integrate fermentation and bioconversion processes with product separation, purification and dosing
- Development of biorefinery processes based on waste and renewable resources. While in the short term two further directions are envisaged: (1) strengthening innovation and competitiveness of the food industry and (2) developing technologies for the conversion of second generation biomass (residues from the food industry, household and municipal waste) into biofuels and raw industrial materials. The pressure on the food industry is well justified, given the importance of this sector in the Polish economy.

The EU's involvement in the transition from the linear economy model to the circular economy model has accelerated the development of bioplastics in Europe. The EU has begun to appreciate the benefits of bio-based materials, and the amended EU Waste Framework Directive and Packaging and Packaging Waste Directive introduced in 2018 are an incentive for Member States to expand their use of packaging materials and improve packaging market conditions for these products.

In Poland, one of the key sectors in terms of socio-economic impact inscribed in the bioeconomy is the agri-food sector. Its economic impact, both on the domestic and international market, makes it one of the most important sectors requiring special attention in the development of the latest technological trends. At the same time, it is one of the key sectors in the biological raw materials supply chain.[21]

---

[20] Wicki L., A. Wicka, 2016, Bio-Economy Sector In Poland And Its Importance In The Economy, Proceedings of the 2016 International Conference "Economic Science For Rural Development" No 41 Jelgava, LLU ESAF, 21–22 April 2016, pp. 219–219).

[21] Ciechańska D., 2019. Raport 'Identyfikacja dobrych praktyk, barier oraz kierunków rozwoju obszarów badawczych ułatwiających wdrażanie w Polsce w obszarze biogospodarki koncepcji GOZ', w ramach projektu "GOSPOSTRATEG – otoGOZ".

In recent years, the importance of the chemical industry in the development of bioeconomic initiatives has increased, especially in the field of advanced biorefineries, which use bio-chem processes to transform renewable resources into sustainable chemical products, biomaterials and fuels. These processes are of particular importance in the circular economy due to the maximization of the use of all valuable components of biomass raw materials, both secondary and primary as well as in the future perspective of municipal bio-waste.

Strategies for bioeconomic activities to improve the innovation and competitiveness of innovative sectors of the economy, in particular the packaging industry, focus on several important aspects that fit into the circular economy, i.e.:

- Development of biorefining technology in accordance with the cascade concept that guarantees effective material recovery of each, significant by-product and waste component in combination with the rational management of raw material residues for the production of biofuels and energy
- Development of technologies for rational and sustainable management of natural resources
- Development of technologies extending the life cycle of biomass-based products (including intelligent packaging technologies – functional, barrier, biodegradable) which will allow for efficient management of waste biomass stream from the consumption of goods

The development and implementation of support systems focusing on these activities will allow the use of huge resources at the disposal of the Polish bioeconomy and develop circular economy systems based on biomass raw materials, biotechnology and bio-recycling.[22]

The analysis of EU strategic documents, regional strategies, national smart specializations as well as the identification of good scientific and business practices have created the basis for prioritizing research issues in the field of bioeconomy, including some following technological issues[23]:

- Improvement of existing industrial, agricultural and transport technologies in the aspect of elimination of negative climate changes
- Production of high-quality biogas (>90% methane)
- Construction of cascading biorefineries – maximizing the use of primary and secondary biomass raw materials
- Bioplastics in the circular economy
- Organic recycling of plastics

[22] Ciechańska D., 2019. Raport 'Identyfikacja dobrych praktyk, barier oraz kierunków rozwoju obszarów badawczych ułatwiających wdrażanie w Polsce w obszarze biogospodarki koncepcji GOZ', w ramach projektu "GOSPOSTRATEG – otoGOZ".

[23] Ciechańska D., 2019.Stanisław Bielecki, Beata Gutarowska, in Raport 'Identyfikacja dobrych praktyk, barier oraz kierunków rozwoju obszarów badawczych ułatwiających wdrażanie w Polsce w obszarze biogospodarki koncepcji GOZ', w ramach projektu "GOSPOSTRATEG – otoGOZ".

- Sewage sludge management (for bioplastics)
- H_2 and chemical building blocks from biogas
- CO_2 as a raw material in the economy
- Naturally produced biopolymers and their functionalization
- Ecological biopreparations for agriculture
- Combined production of biomethane and biohydrogen from waste from the agri-food industry
- Construction of biological, molecular tools and systems (biocatalysts, bioreactors) with new functions, for new processes in the bioeconomy
- Joint management of organic waste (kitchen with green), new farming techniques and modern genetic modifications in agriculture for the preservation of food and bio-based products
- Utilization of plastic waste (complex, contaminated) and sewage sludge for coal production for restoration of degraded soils and water purification
- Water management and soil quality improvement and horizontal actions to improve the commercialization of research and horizontal activities to improve the process of research commercialization as follows:

  – Digitization of the bioeconomy
  – New legal regulations for the bioeconomy
  – Cross-sectoral initiatives – optimal raw material and product management in integrated value chains
  – Concentrators of bioeconomic activities – Hub BIO-GOZ
  – Entrepreneurship accelerators – start-up initiatives related to national specializations in bioeconomy

On the base of the diagnosis of bioeconomy status, some recommendations could be defined as effected on the implementation of circular bioeconomy in practice[24]:

1. Development of pilot installations for upscaling R&D solutions in the area of circular bioeconomy from TRL at levels 3–5 to TRL at levels 6–9
2. Implementation of strategic documents for the development of the bioeconomy in Poland – Road Mapping
3. Focus and prioritization of bioeconomy activities based on regional strategies – Bioregions cooperation platform
4. Rationalization of biowaste management in the country
5. Research internationalization – cooperation within European platforms
6. Strengthening inter-cluster cooperation building integrated value and supply chains for various sectors related to the bioeconomy

[24] Ciechańska D., 2019. Raport 'Identyfikacja dobrych praktyk, barier oraz kierunków rozwoju obszarów badawczych ułatwiających wdrażanie w Polsce w obszarze biogospodarki koncepcji GOZ', w ramach projektu "GOSPOSTRATEG – otoGOZ".

7. Debate with various stakeholder groups on the prospects for implementing the bioeconomy in Poland
8. Bio-based industry needs of skills and competencies identification
9. Bio-based education at all levels of education

Acknowledgements This paper is based on the original research prepared in the frame of the project oto-GOZ supported by National Center for Research and Development, Poland.

The authors are particularly grateful to MASPEX Company for their permission to use the Maspex own projects database.

References

The 2030 Agenda for Sustainable Development – Implementation in Poland

Bioeconomy development in EU regions Mapping of EU Member States'/regions' Research and Innovation plans & Strategies for Smart Specialisation (RIS3) on Bioeconomy, Final Report – February 2017

Bio-based Industries Consortium, European Bioeconomy in figures 2008–2015, 2018

Central Statistical Office, Land use and sown area in 2017 r., 2018; World Bank

COM (2015) 614 final (2015), Communication from the Commission to the European Parliament, The Council, The European Economic and Social Committee and the Committee of the Regions. Closing the loop – An EU action plan for the Circular Economy

Ciechańska D., 2019a. Raport 'Identyfikacja dobrych praktyk, barier oraz kierunków rozwoju obszarów badawczych ułatwiających wdrażanie w Polsce w obszarze biogospodarki koncepcji GOZ', w ramach projektu "GOSPOSTRATEG – otoGOZ"

Ciechańska D., 2019b. Stanisław Bielecki, Beata Gutarowska, in Raport 'Identyfikacja dobrych praktyk, barier oraz kierunków rozwoju obszarów badawczych ułatwiających wdrażanie w Polsce w obszarze biogospodarki koncepcji GOZ', w ramach projektu "GOSPOSTRATEG – otoGOZ"

Degórski, M., 2018. Gospodarka o obiegu zamkniętym. Circular economy – nowe podejście w rozumieniu relacji człowiek-środowisko. Studia KPZF, t. 183. ISSN 0079-3507

Eugeniusz K. Chyłek, Monika Rzepecka „Biogospodarka – konkurencyjność i zrównoważone wykorzystanie zasobów", Polish Journal of Agronomy, 7, 3-13, 2011

European Commission, 2016. Next steps for a sustainable European future – European action for sustainability. Communication from the Commission to the European Parliament, the Council, the European Economic and Social Committee and the Committee of the Regions COM(2016) 739 final, Strasbourg.

Ewaluacja potencjału badawczo-rozwojowego jednostek naukowych i jego wpływu na realizację celów KIS. 2019. Ecorys Polska, Taylor Economics spotkanie Grupy Konsultacyjnej ds. KIS, Sterdyń

Kirchherr, J., Reike, D., Hekkert, M., 2017. Conceptualizing the circular economy: An analysis of 114 definitions. Conservation and Recycling, 127, 221-232. https://doi.org/10.1016/j.resconrec.2017.09.005

Pardo, R., Schweitzer, J.P., 2018. A long-term strategy for a European circular economy – setting the course for success. Policy Paper produced for the Think 2030 project, Brussels.

Piotrowski, M., 2019.'Ocena internacjonalizacji krajowych przedsiębiorstw z obszaru specjalizacji KIS' spotkanie Grupy Konsultacyjnej ds. KIS, Sterdyń

Projet R2PI – Horyzont 2020

National Smart Specialisation version 6 (valid from January 1, 2020)

The Strategy for Responsible Development for the period up to 2020, Council of Ministries 2017

ROAD MAP towards the Transition to *Circular Economy*. Annex to Resolution No 136/2019 of the Council of Ministers of 10 September 2019. Bridging the circular economy and social enterprise: the Dutch Ministry of Defence and Biga Groep.
Soufani, K., et al., 2018. European Business Review.
Wicki L., A. Wicka, 2016, Bio-Economy Sector In Poland And Its Importance In The Economy, Proceedings of the 2016 International Conference "ECONOMIC SCIENCE FOR RURAL DEVELOPMENT" No 41 Jelgava, LLU ESAF, 21-22 April 2016, pp. 219-219)
'Zamknięty obieg – otwarte możliwości' Report of Deloitte, 2018

Internet Sources

(http://www.biotechnika.net/)
(https://www.bowil.pl/produkty/bioceluloza/)
(http://www.napiferyn.pl)
(https://www.orlen.pl)
(https://innpoland.pl/)
http://klasterbio.pl/
(https://natureef.pl/)
http://www.nutribiomed.pl/
https://lifescience.pl/
(https://www.clustercollaboration.eu/cluster-organisations/waste-management-and-recycling-cluster)
https://maspex.com
https://www.eitfood.eu/

Part III
BioEcoJust Themes and Approaches

Chapter 7
Open Biofutures: The Challenge of Maintaining Agency for Long-Term Futures

Amos Taylor and Nicolas A. Balcom Raleigh

Abstract The world is complex, and its developments are always uncertain. In this context, the bioeconomy represents a framework for innovating solutions which can enable a global transformation to a sustainable future. However, bioeconomy developments can also serve to repackage problematic or even unjust economic patterns from the past. This chapter proposes a heuristic of *open futures* and *closed futures* which can be used as lenses useful for ethically evaluating future imaginaries such as the bioeconomy. These lenses help actors imagine possible consequences of various developments on ecology or nature. Therefore, *open* and *closed biofutures* can serve as tools for engaging the ethicality of various development trajectories. These lenses also encourage actors to seek wider inclusion when considering who has agency and transformative agency in the bioeconomy conjecture. This chapter presents how this tool for thinking has been utilized in the Bioeconomy and Justice Project (BioEcoJust) to imagine long-range futures that represent particular complex challenges. As a research tool these lenses have enabled us to envisage a diverse range of futures and evaluate what is closing or opening about them. We conclude that the lenses of opened and closed biofutures can be used by innovators and decision-makers of all sectors for consideration of the ethicality of the work they choose to pursue today.

1 Introduction

In this chapter we propose a framing approach of 'open' vs. 'closed' futures and then demonstrate how this framing can be applied to illuminate ethical pathways in biofutures. We propose that impacts on agency of people in the future are a significant distinguishing factor between 'open' and closed' futures and a potential source

A. Taylor (✉) · N. A. Balcom Raleigh
Finland Futures Research Centre, Turku School of Economics – University of Turku, Turku, Finland
e-mail: amos.t.taylor@utu.fi; nabara@utu.fi

E. Koukios, A. Sacio-Szymańska (eds.), *Bio#Futures*,
https://doi.org/10.1007/978-3-030-64969-2_7

for motivation in taking action to develop the bioeconomy. Due to the difficulties of overcoming lock-ins of existing regimes of being, transformative agency is needed in attempts to open closed futures. We then discuss the implications of pursuing 'open futures' as a source of ethicality in biofutures, especially in considering long-range futures.

Today, from multiple sources, there is great pressure for humanity to undergo a transition in one way or another, due to the current challenges we face ranging from climate change to economic recovery. These forces suggest a shift from unsustainable to sustainable, from a high carbon economy to a low carbon economy, from high consumption to low consumption, from imbalance to balance with nature and from fossil-based single-use goods to renewable and circular based ones. The future imaginary of a widespread bioeconomy, or biofutures, has been proposed as one such mechanism to bring us to a more ecologically sound future (Bell et al. 2018; VTT 2018).

The bioeconomy represents a particular challenge for the field of futures studies when considered more widely and openly and for longer time periods. Likewise, the arena of the future bioeconomy concept is stretched when it is considered more holistically. The bioeconomy can be interpreted as an open, diverse and changeable concept that spans multiple converging sectors. It is itself complex and hard to define definitively, and its interpretations change over time. Each of these interpretations can replicate old meanings or produce new meanings beyond the mainstream industry-oriented interpretation. Often these meanings describe new societal and regionally specific formulations of bioeconomy ('Global Visions for the Bioeconomy' 2015; Taylor et al. 2019). The bioeconomy can be conceptually pinned down using the concept of sociotechnical imaginary from Science and Technology Studies (Jasanoff 2015; Jasanoff and Kim 2009). A futures studies perspective would go further and call the bioeconomy a *future imaginary* (cf. Clark 2011) that is being deployed by various actors to focus action of governments and industries to drive change and achieve specific objectives. In a sense, the task we are proposing here is to rigorously engage the variety of forms this future imaginary offers, in order to identify aspects that close or open what can emerge next. The aim of doing so is to illuminate just and ethical pathways for bioeconomy development.

2 Bioeconomy and Justice Approaches to Long-Range Futures

The Bioeconomy and Justice Project (BioEcoJust) aims to explore the long-term potential developments of the bioeconomy and their ethical implications. The project is funded as part of the Academy of Finland's BioFuture2025 program which includes many largely technical and business innovation-focused projects. Ours is given a mandate to explore the long-term potential developments of bioeconomy from a wide range of vantage points.

BioEcoJust combines futures research approaches from Finland Futures Research Centre at the University of Turku with Applied Ethics approaches from Aalto University. The goal is to identify justice-related issues arising from the development of the bioeconomy and ethical frameworks by which to address them. It is in this setting that we developed many sensemaking tools including the one presented in this chapter, open and closed biofutures (see, Taylor et al. 2019). This tool is based upon several perspectives concerning ethics from the field of futures studies.

3 A Future Studies Perspective

The futures field offers its own set of ethical perspectives concerning what is important to consider when exploring futures, and it gives insight concerning the interplay between future imaginaries and decision-making. We ask questions like: Who's desirable future is it? In what contexts? Who wins? Who loses? What impacts cascade to later generations? What assumptions are behind an *ethical stance*, and do they stand up to wider scrutiny when stated plainly? These kinds of questions can reveal key perspectives and situational framings regarding what should or should not be done today for the future.

Futures studies can be seen to have a clear emphasis on generating and exploring alternative and preferable futures rather than creating a singular official future. Eleonora Masini (1993) has called this aspect the 'third rule of futures studies', where futures are plural and that there is always an implied indirect colonization of the futures of others and thus require alternative options. Ilkka Niiniluoto describes this work as a tree of alternatives, where a futurist's role should be to construct alternative possible futures, assess the probability of alternative futures and evaluate the preferability or desirability of alternative futures (2001, 2017). With the general underlying virtuous aim of 'improving the freedom and welfare of humankind', he sees it as a design science which attempts to help the rational planning of the future (Niiniluoto 2001, 2017, 26). We can interpret this field as trying to actively affect positive change through action, utilizing science and theory, philosophy and critical discussion.

Anita Rubin (2017, 252) explains another crucial aspect of this, that there are contradictory images of the future that are built upon different worldviews – with their own interpretations and rationales. In this arena, rational decision-making and choices regarding the future that is dependent on, for example, cultural habits, language, values and beliefs are highly ambiguous and not necessarily rational at all. In this critical futures studies perspective, a person would take a step back to consider the complexity of a proposed future, not only to see what opportunities and options lie ahead but to comprehend their own values and reaction to it. By engaging these contradictions and viewpoints, it is possible to create futures that 'open up from new and different starting points' (Rubin 2017, 253).

What does this field contribute to this debate? Andersson who by critically looking at the history of futures studies arrives at a sobering assessment of the field as a whole; she poignantly describes its characteristics as (emphasis added):

> [...] how contemporary societies attempt to manage questions relating to the long term, and how societies produce visions, knowledge, and means of intervention aimed at future change and future control. It would seem that there are moments in which the future is future no more, but present—in other words, *when the future acquires a presence and requires urgent action.* (Andersson 2012, 1430)

The above conceptualizations regard the heritage of futures studies, exploring the very nature of how we explore and consider futures. An emerging conceptualization of what futures are and how people use them is recently ascending into focus – anticipation (e.g. Poli) and futures literacy (e.g. Miller 2018). Anticipation proposes that all life utilizes anticipatory systems (ala Robert Rosen 2012). Futures literacy refers to the capability to diversify how and why we *use the future* by deepening our understanding of anticipatory systems. These offer what can be called 'the bioeconomy project', new approaches to achieve reflexivity and ethical awareness in their work. It proposes that people can become skilled at noticing and questioning the futures they use to understand their options. It promotes challenging assumptions to critically assess futures, highlighting the distinct difference between anticipation *for* the future as well as anticipation *for* emergence.

An inherent feature of complex and adaptive systems is that it is difficult or impossible to foresee alternative paths in a paradigm shift (Kuhmonen 2017). Objectively engaging this type of system by seeing its closing or opening potentials suggests an approach to ethically engage a rapidly changing environment without fully itemizing its inner workings, but appreciate and evaluate it externally. Bussey has a similar use of the thematic categories of open and closed futures, concerning organizations and foresight, where closed futures 'correspond to a dominant pattern' of maintaining coherence and that this is why they 'risk all tomorrows for the stability of today'. On the other hand, 'Open futures by contrast are pluralistic, inclusive and participatory' (Bussey 2014). This represents for the BioEcoJust project a functioning ethical approach to engaging and making-sense of futures and comparatively also the consequences of the present. 'Openness to alternative futures is one of the defining factors of […]contemporary futures studies' (Minkkinen et al. 2019).

Open futures are future imaginaries which seek to identify the innovations we can pursue in the present that open options for future people (including ourselves depending on the time horizon) to thrive. When this mode of imagining is directed toward generating futures of humans in the ecosystems to which they belong, we have open biofutures and their opposite closed biofutures.

4 Closed Biofutures Versus Open Biofutures as Prompts for Imagining

In a sense, this period of the fossil fuel-dominant world and any scenario of climate change that includes average global temperatures above 1.5C are *closed futures*, where the future is all used up (see, e.g. Smith et al. 2018). Closed futures are often driven by what Sohail Inayatullah (Inayatullah 2008) calls 'used futures' – the imagined futures 'created by others in the past' which we continue to use in unquestioning and unexamined ways. There are many climate change impacts of decisions made according to these 'used futures'. Investments in speculated fossil resources have to recapture their investment while outdated fossil industries are locked-in for the long haul to receive a return on their investment for the next 40–50 years. They become 'closed futures' when the choices already made by previous generations are radically limiting the choices of those living in the future. When imagining what our descendants will need, a necessity will be life-sustaining ecosystems. However, closed futures instead give future people ecological systems that have reached their limits, biodiversity that is radically reduced and eroded agricultural land with topsoil that no longer sustains food production.

Closing the future reduces agency for those in the present by blinding them to a fuller range of options concerning which projects to apply their time and resources, and for future generations by removing opportunities to live differently than inscribed by people of the past. When seeing options through a closed futures lens, the problems become the most predominant features of the landscape, choices for action tend toward incremental manoeuvring around these problems, and transformative change is left unimagined. Actions of people acting from closed futures include postponing radical changes, manipulating public opinion and lobbying for the business as usual. When these actors produce 'new' futures, they tend to come up with yet more socio-technological imaginaries capable of perpetuating conventional industrial concerns and creating previously understood notions of value within the existing neo-liberal economic order.

During the Covid-19 global pandemic (in 2020), closed futures have also appeared. Decisions to prioritize conventional mundane futures, such as 'going to work', 'going to school' and 'taking summer holidays' over those futures where the problem is fully addressed before we move on, have cost many nations hundreds, thousands or even hundreds of thousands of lives. These COVID-19 deaths permanently close the options, for those of us who survive this pandemic, to enjoy our relationships with any family and friends who died from this disease. On the other hand, the moment of crisis feels like it is already closed or is closing. As one group of actors after another fail to prevent the spread of the disease, futures others imagined they would be enjoying now evaporate. Freedoms are minimized, movement is restricted, employment compromised, and financial limitations are felt. This situation is a suitable example of what we mean by closed futures – some futures can appear to be positive, but in fact carry consequences for others.

This example further introduces issues of human interrelations to our Earth's ecosystems. The source of the virus is argued to be a direct result of human encroachment onto wild nature systems (see, e.g. Everard et al. 2020; Hockings et al. 2020)[1]. In other words, it is the closing in on animal habitats that has enabled greater contact between the illnesses of the wild and human populations. Actions of others in the past, to increase built-up land and establish human habitats in previously wild ecosystems, have produced a closed present.

An *open future* on the other hand offers new modes of imagining the unexplored, uncertain and unthought futures (Sardar and Sweeney 2016) in addition to those that are desirable, possible or probable (Amara 1981; Niiniluoto 2001). To a large extent, 'normal times' are synthetic framings of what is happening in our world, while 'post-normal times' are always happening for some people somewhere and can be thought of as a persistent condition (Mayo 2020; Sardar and Sweeney 2016). While many widely discussed future imaginaries emphasize fixed endpoints demarked by clear indicators and goals, these emphases can limit our imaginations to 'what is' versus 'what is becoming' (De Roo 2018). Beyond focusing on fixed points in time when a sustainable future end state could be achieved, our imaginations about what could happen in the future benefit from remaining open and compatible with the fact that ideal end states never materialize as time continues regardless. For example, one can ask questions like what happens beyond the Paris Agreement? Or beyond the current UN Sustainability Development Goals? Questions such as these invite space for dynamic imagination involving stories in motion and future narratives that leave room for the yet to be expressed, named or considered.

Thinking beyond incremental sets of alternative futures which are necessarily bounded by the positionality, situation and worldviews of the actors imagining them, the practices we propose involve constant identification and critical exploration of closed and open futures while engaging agency and the interplay between diverse complex perspectives. Efforts to develop a global bioeconomy potentially provide both opportunities for opening future possibilities as well as radically closing them, or just a continuation of *used future* narratives. These three kinds of futures can be used as lenses through which we see dichotomies of bioeconomy choices: building human habitat vs. preserving non-human habitats, growing biomass for food vs. fuel, prioritizing ecological objectives vs. economic growth, etc. *Open* or *closed bio*-futures thus include in them, not only the relationship between resources, technology and society but also considerations of the rights of humans and non-humans to a future and general well-being of living nature, ecologies and biodiversity. The openness and closedness of this often-neglected consideration, of the priorities of *non-human life*, thus provides where bioeconomy-related future imaginaries can act as a 'a bridge to a better future' for a wider range of species (see

[1] https://www.theguardian.com/world/2020/jun/17/pandemics-destruction-nature-un-who-legislation-trade-green-recovery. Accessed 24 September 2020.

Matti Häyry[2]). A bioeconomy future imaginary which also considers impacts on the rest of the living nature changes our range of choices and sense of agency. They raise the question of what the next sequences of change in the development of bio-economies could be that would potentially follow such a transition, or what other desirable futures have we bypassed in the choosing or encouraging of one path over another. It is not satisfactory to take the visions proposed, the 'official futures' (see Dator 2009) of the bioeconomy at face value because doing so comes with a risk of advancing another closed future (e.g. neo-colonialism, or human-species-centrism). To break away from these 'official futures' requires exploring the potentiality landscape anew, especially when considering the long-term future where multiple generations, multiple new waves of economies and societies as well as multiple new trajectories and innovative paradigms are involved. A more holistic perspective is thus demanded when imagining open biofutures, and the opening vs. closing framing serves to incorporate follow-on effects.

5 Open Biofutures

The publishing of the *Limits to Growth* offers a first popular-facing introduction to the concept that economic growth as conventionally implemented entails systemic impacts on our planet's resources and ecosystems and there are limits which would be dangerous to cross (1972). It highlighted the need to question the effects of how we live and its implications on the planet and its occupants. It functioned as a warning and spurred ethical discussions that can be said to have profoundly influenced how we conceive of sustainable development today. While the authors of the report engaged in 'trying to know what could happen in the future', they ultimately built a set of futures imaginaries driven by simulations and models which people could use to make decisions in the present.

Today, there are many growth-oriented future-imaginaries which try to address the notion of ecological limits while holding onto assumptions that economic growth is good and necessary. Examples include the circular economy, EU green deal, green growth and the bioeconomy which are all launched to compel coordinated action that could enable societal transformation to a sustainable future, or at least one that does not destroy our planet's biosphere. Each of these future imaginaries offers certain virtues, policy regimes and solutions and yet also exposes biases and flaws, as well as deep assumptions about what can or should change and why.

The bioeconomy is fuzzy in its definition (Golembiewski et al. 2015), yet holds much promise to meet the challenge of producing a more ecologically symbiotic

[2] Professor Matti Häyry gave a presentation on this topic for the Finnish Academy, presenting a bridge to the future that would avoid the pitfalls of an unjust future bioeconomy. https://www.aka.fi/globalassets/32akatemiaohjelmat/biofuture2025/posterit-2019-lahti/hayry-bioecojust-poster-2.pdf

economic system. Purely seen as a diverse knowledge base, contributing to the next paradigm built upon the information age, the knowledge-based bioeconomy can be seen to articulate new valuable know-how and information and practices directly pertinent to future global sustainable challenges. Renewable or bio-energy alternatives, as well as endless varieties of bio-based chemicals and materials to replace fossil dependence, offer the chance for industries and consumption to switch to cleaner alternatives. Likewise, for example, forests offer biomass and carbon capture capabilities to provide the planet the basis to tackle climate change and rebuild the economy around it. And yet this paradigm shift is still in its infancy, and formulations of the bioeconomy do not yet cover the full spectrum of a fully emerged bio-based society.

However, there are also many criticisms of the current formulations of the bioeconomy. For example, a critical view of bioeconomy (Ahlqvist and Sirviö 2019) depicts the results of their Finnish analysis as a closed future:

> [...] there is little doubt that as a policy idea the bioeconomy is primarily designed to bring about a new round of capital accumulation [...] the bioeconomy clearly is an attempt (in the face of ecological exhaustion) to open up a new resource frontier of cheap nature to perpetuate the capital accumulation process.

They observe that the bioeconomy as a means to extract large biomass (in this case from forest) by the already established industries remains the main sentiment that overlooks the other new actors who might emerge and the opportunities of synthetic biology (ibid. 410). That is, the natural resources represent a means to accumulate wealth and replace the inputs of the fossil industry with them. This future socio-technological imaginary offers a colonization of certain interpretations of the future, in which nature is seen predominantly as an industrial resource and the subject of accumulation of capital.

They suggest rather a 'redefinition of the bioeconomy concept in such a way that enables environmental concerns to be articulated, concern about livelihoods and territorially balanced economic development to be voiced, as well as political pressure on economic praxis to be exerted' (ibid. 416). Clearly in their view the power relations asserted from top down have formed the singular narrative in which nation state or a region should define its economy and its connection to the global market. Ahlqvist and Sirvio tell how history has been inscribed in such a way to produce one specific economic relationship to nature, where the actors and markets are formulated toward that end. And that particular formulation of the bioeconomy defines its path toward the future. What is clear with this knowledge is that there are other paths open, some side-lined and some yet unidentified. Similarly, Birch examines from a neoliberal critique of how 'markets and natures are being imagined and constructed in the pursuit of the bio-economy', and where there is the need to 'identifying alternative bio-economies, reflecting different bio-economies that are not underpinned by market principles' (Birch 2016, 2017, 2019; Birch et al. 2010). This identifies a need to shift away from certain assumed perspectives where an industry perspective would be normative.

The diverse interpretations and evolution of the policy driven concept of bioeconomy suggest that it is absorbing new values and manifesting toward something novel. If you consider that it has at one time or another been referred to as the bio-based economy, the bioeconomy and the circular bioeconomy and now the circular sustainable bioeconomy (Hetemäki et al. 2017). You can imagine these as expanding rings on a tree, where time adds new sectors and interpretations upon the old, through necessity and changes in values. One notable interpretation by the Iceland Bioeconomy policy suggests a sphere with *freshwater, marine, human capital, wilderness, forests and farmland* with additional themes like *education, innovation, nature-based tourism, arts and crafts,* as well as *sustainable resources* amongst others occupying the inner rings (*backbone, service sector, secondary industries*, etc.) (see Mattis & Iceland Bioeconomy) (Smáradóttir et al. 2014). For this northern sector, multiple frames allow for a more holistic view of the complex relationships and the multiple emerging dimensions under one title. In addition, these are just policy developments that can be seen to be slow moving in comparison with what interpretations are happening in practice, where specific sectors are developing under its umbrella, defying current categorization. Many of them, like those working in synthetic biology ecosystems[3], *lignin* start-ups[4], novel carbon offsets[5], ocean forestry[6], nature tourism and those countless still to be discovered can be seen to be redefining the sector as a whole with further impacts and implications for society at large. Some of these niche areas could become adopted or converge with established industries or practices (like textiles industry, or biofuels); others could offer game-changing or creative destruction capabilities. The promise and vision of this new era is boundless in imagination.

However there have been clear warnings by researchers that the bioeconomy has been critically hijacked (Vivien et al. 2019) or not as sustainable as it is proposed, where sustainability has been used to greenwash dirty sectors that would lead to unsustainable or unjust futures (Ramcilovic-Suominen and Pülzl 2018). The risk is present in this ever-wider policy envelope where it acts like a Trojan Horse repackaging older or less desirable sectors and solutions under the greener paradigm transformation. This brings with it future problems that might perpetuate the old, or merely be a guise to reposition in a new market (Korhonen et al. 2020; Ramcilovic-Suominen and Pülzl 2018; Vivien et al. 2019). It must be assumed that there will always be problematic disingenuous factors that come along with the mix. For example, a circular economy does not automatically equate a good and just sustainable economy, just because it is more efficient, circular and a viable bio alternative to petroleum-based economy (Hetemäki et al. 2017). There is always the dangerous potential to enslave, colonize, restrict nature, dictate future land use, monopolize and affect equality all under a 'good' virtuous policy.

[3] See https://www.synbio.fi/ for example.

[4] See https://ligninclub.fi/ for example.

[5] See https://compensate.com/

[6] See https://www.nordicinnovation.org/news/growing-global-ocean-rainforest

A 'good' bioeconomy may serve the perceived 'greater good', as dictated by the decisions made toward how a future image 'should be'. In this vision, knowledge and even science may be guided toward results that match a minority belief and interest, or assumption about the future under the categories of relevance and strategic choices. A good example of this critical closing perspective is demonstrated by Andersson and Westholm (2019), who have indicated that the choices made toward certain dominant future images of the Nordic forest sector had the effect of closing futures, restricting certain research narratives and findings, to rather promote those that matched their desired official narrative.

Considering the above warnings, open biofutures would therefore entail the opposite of the above criticism. They would disrupt or end the perpetuation of colonialism, question the continuation of used futures which are no longer wanted and rebel against 'guiding hands' in science seeking to reinforce incumbent future imaginaries. Instead, open biofutures would prioritize the activation of sustainable and inclusive potentials that fully include justice for all living nature. Moreover, allowing space for considering new emerging pathways from diverse agents is suggestive of the unprecedented future and is conscious of the power of individuals' own agency to affect it.

6 Toward Transformative Agency

Agency can be seen as a central ethical concern when engaging closed and open futures. Can people and other living beings participate in change? Are we able to influence our future, and if so how and to what extent? Furthermore, do we have 'transformative agency'? Through this concept, agents are empowered in the transformation process, they affect their own destiny and also the futures of others, and they make meaningful and decisive contributions to the transformation of society. This locates future transformative discourse toward the individual level to engage larger systems.

Concerning change in a complex adaptive system, it can be difficult to identify who influences change in society. Does it happen top-down, through government, industry or an elite? Or does change happen bottom-up from the grassroots individual citizen level? In some perspectives the bioeconomy has located decision-making to be by high-level government and industry figures, resulting in the participation of citizens to be marginalized (Mustalahti 2018; Vainio et al. 2018). In this way industries are built on the basis of a future imaginary by the few upper echelons and not from within society. Another perspective could show that entrepreneurs as actors experiment and define new paths and make networks and communities at the grassroots level. These can be seen as individuals innovating new technologies, services or ecosystem services that open up new emerging pathways, although with uncertain outcomes. Yet another perspective would concern the agency and role of *others*, those non-human entities (e.g. other living species, as well as natural systems such as lakes, rivers, etc.) that are impacted by the future imaginaries humans invent and enact. In this manner, who's agency is prioritized becomes a large factor in ethically approaching complex adaptive systems.

As nature, ecosystems and ultimately the whole biosphere are added into society-industrial perspectives, agency and transformative agency take on a completely new meaning. Thus, the actors are broad in range within and affected by the bioeconomy, as it is assumed to be approaching global scale and potentially to become the next transformational shift for life on Earth from post-industrial to a *bio age* or *bio-society*. These diverse abundant agents then can be seen to exist simultaneously, and the whole is built upon the diverse interconnection of these perspectives. Milojević and Izgarjan suggest that alternative storytelling as a strong means to engage in futures offer agency, where they empower individuals to overcome trauma, or a closed state, toward opening up alternative narratives (Milojević and Izgarjan 2014). Imagining and including those 'outsiders' into new narratives becomes crucial for our collective journey. As climate change continues to dislocate populations (Sassen 2016) and to drive whole species extinctions, 'the earth is full of refugees, human and not, without refuge' (Haraway 2015).

We propose that *open and closed futures* can enable ethicality in transformative agency, defined in such a way that change occurs in a complex and adaptive system, that can be identified as a catalyst toward holistic and continuous ecologically just change, where agency is collaborative and social in nature, linking social-ecological systems and understood as involving a systemic change at all levels (Steward 2008, 2012; Westley et al. 2013).

7 Reconsidering Futures, Informing Action Toward Long-Term Ethical Development of the Bioeconomy

Where admittedly futures studies as a field is highly multidisciplinary, naturally drawing from multiple fields' method and theory bases, off the shelf methodologies do not do justice for the potential research needs of the complexity and un-anticipatable long range of the BioEcoJust project. To say anything useful about what could be just or unjust between 2018 and 2125, we developed these lenses of open and closed futures as a basis for ethical argumentation, as even considering positions toward the future that are highly visionary in nature has a fatal flaw in that they can form a specific closed future. For example, 'Cathedral Thinking'[7] suggests that some individual or group can create grand projects which take multiple generations to complete to benefit future generations, like architectural cathedrals for future generations to enjoy. Our challenge to this proposition is that they form a rather rigid framework about what future people will need or want and perpetuate values, systems and assumptions from the moments they are made. Such projects do

[7] https://cathedralthinking.com/

not question the need for such cathedrals in the future, hog resources and human attention while they are produced, block other developments and assume significant efforts to maintain.

A counter-perspective would be *moonshot* investment in development projects that potentially have positive open spillovers to society at large, created by public-led investments to form pluralistic platforms for development through strategic missions (see Mazzucato 2019).[8] These kinds of projects culminate massive research and development knowledge, taking advantage of the state of the art. Consider the Internet, for example, that can be seen to have vastly opened future potential for our society. However, that said, in recent years it has been observed the negative environmental cost of the digital age, that is supposed to be free from worldly material consumption, the Internet and its associated technologies are responsible for ever-increasing huge global energy use and extraction of precious materials that have put the planet in crisis. It can be noted that socially too, technologies initially that were seen as liberating can be seen to close our human capabilities to socialize, to imprison within a system of dopamine gratification through social media and video games. Indeed, like all technologies, Internet Communication Technology has tightly coupled itself to our minds and now shapes our very consciousness. As we imagine futures of the bioeconomy, we would benefit from utilizing this knowledge and question any grand solutions, cathedral thinking, especially those considered to be green, renewable and sustainable for the future. We've found that evaluating potentialities by viewing them through the lenses of open and closed futures supports such a questioning.

In practice this approach has coloured our research to engage complexity in a critical and open manner. We undertook several such activities. One approach that continuously informed on our project was *open horizon scanning*. That as a practice it was open in nature intended to seek emerging and novel issues, to widen the scope of possibility. These would often challenge a normative perspective to open up new dimensions, often these were ethical in nature, pushing the boundaries of what could be included in the developing topic or future imaginary. Commonly horizon scanning involves exploring the landscape of environmental changes through a 'comprehensive systematic examination of risk, uncertainty and emerging trends', in order to push thinking toward challenging assumptions (Rowe et al. 2017). We added the element of open participation and an expanded framing to our practice. These were also actively shared openly with colleagues and wider networks to form discussion and debate. Issues such as geoengineering, synthetic biology, CRISPR gene editing technology, etc. test our assumptions and offer contrasting perspectives (see BioEcoJust Open Horizon Scanning)[9]. Literature searches were also conducted to identify common directions within the field of classic and progressive bioeconomy, and also those that critically assessed or offered new potentials.

[8] https://ec.europa.eu/info/sites/info/files/research_and_innovation/contact/documents/ec_rtd_mazzucato-report-issue2_072019.pdf

[9] https://ffrc.wordpress.com/2018/09/24/bioecojust-open-horizon-scanning/

These processes formed a rather agnostic and objective approach to technology, policy and the economy. These resulted in generated observations on the complex relationship between the triangulation of humans, technology and nature. In a way we observed that each article or artefact expressed values about this relationship, and about its desired or undesirable future, for example, how technology should be used for humans toward nature. We observed about five general common worldviews that reoccurred in the literature and horizon material that we refer to as *bioworlds* (Balcom-raleigh et al. 2018; Taylor et al. 2019). Some articulated or implied perspectives on what was 'good' way for a bioeconomy, i.e. resources as a solution, inspiration from natures design, edit nature through technology, rejuvenate and restore nature and the equality of all nature (ibid.). Agency and transformative agency can be seen to be expressed differently in all of these world archetypes, where *what is good and bad* or what innovations to prioritize going forward toward a bioeconomy imaginary. What becomes interesting is to see the combinations of these perspectives, the discord and discourse between these perspectives and all worldviews present in some manner or other. Convergence and divergence then lead to entanglement and engagement of opening and closing futures. Our interest has been allowing settings in which we can observe potential future decision-making that have especially ethical dilemmas. Those dilemmas represent turning points in complex systems.

Continuing in this process of participatory research allows an exploration of these *ethical stances* through structured interviews, Delphi (see Chap. 11, this book), and through the gamification of workshops (see Chap. 10, this book). Ethical stances can be seen to be positions more noticed in situational impasse's, places where decisions are made based on ethical value-based assumptions, where there are also opposing views or challenges to be faced. In exploring long-range futures, understanding and looking out for these ethical stances, informed by the changing nature of the interaction of bioworlds, allows decision-making to be explored in such a way as to embrace uncertainty.

Encouraging exploration of these ethical stances we approach scenario making in such a way as to cultivate near impossible situations, which are crisis moments where there were no clear right or wrong answers of how to proceed and resolve the situation. These at first have been developed as *vignette* scenarios, which are short scenes that depict a certain critical situation. Furthermore, these future settings often explored seemingly extreme situations, although resembling current day issues or launching emerging technologies or practices to extreme ends. The scenarios in final form will be action oriented, to communicate to challenge holders who could continue to be aware of ethical stances, situations and the degree of openness of future imaginaries be it in policy or society, or technological form.

The basis for this approach is informed by the Futures Literacy Laboratory meta-design and approach (Miller 2018) that encourages reframing as an exercise in expanding the variety of ways people imagine futures. This encourages encounters with potential 'change in the conditions of change' (ibid.). This resonates with how agency and transformative agency can be achieved, to allow new narratives to be made. Examining this type of phenomena has informed our approach to disassociate ourselves from forecasting the future of current bioeconomy, toward an open futures perspective that accepts as fact the persistent and natural diversity in the dynamics of change. Causation between actors and situations then can contribute to causal changes to the whole.

Consider biodiversity as an example of a complex adaptive system; the recent warning by the IPBES[10] has highlighted the loss of biodiversity as a serious threat to the planet. The rapid loss of biodiversity effects the ecosystems, and humans need nature, closing and limiting species and further contributing to climate change (Isbell et al. 2015). The rapid decrease in global insect, bird and fish population threatens to push ecosystems to a breaking point that in turn affect food availability and quality of life for humans. This is referred to as a 'silent killer'[11] as it happens without most people noticing the massive loss until it is too late. This closing biofuture represents both a long-range future threat in which agency has already been stripped away long before it becomes a perceived threat, and it represents a depiction of the acute incompatibility of the current pathways in which human growth bound activity is at odds with global ecosystems. It also highlights the unseen value that it offers to humans to safeguard the planet and tackle climate change.

Therefore, an open biodiverse future would place diversity as a key issue, shifting away from scaled-up monocrops in agriculture, for example, or more generally identifying negative practices and processes that critically threaten species and create momentum to counter those. Ecosystems thus become collaborators with humankind. Considering also the emerging understanding about the vast underground networks of enzymes [12] that sprawl our planet under surface signals the scientifically yet unknown and unexplored aspects of our planet and the future roles that they could have. In this perspective humans are clearly understood to be but one of many species on this biodiverse planet.

8 Conclusion

Our chapter has proposed that maintaining agency can be seen to be a challenge when considering long-range biofutures. Both practically in terms of futures research considering its complexity and assumptions but also when considering all the critical observations that suggest the current future imaginaries relate to a closing bioeconomy. We have proposed open biofutures as a tool for reintroducing ethical and justice-related considerations to bioeconomy developments. This requires continually acting upon the present to provide future people and other beings with

[10] https://ipbes.net/news/media-release-biodiversity-nature%E2%80%99s-contributions-continue-%C2%A0dangerous-decline-scientists-warn

[11] https://www.theguardian.com/environment/2018/nov/03/stop-biodiversity-loss-or-we-could-face-our-own-extinction-warns-un

[12] https://www.theguardian.com/science/2018/dec/10/tread-softly-because-you-tread-on-23bn-tonnes-of-micro-organisms

open options within which they can thrive. An open biofuture is one that can be used in the present to identify actions and new ways of being which can provide choices to people of tomorrow.

We propose this analytical concept as an antidote to the commonly used bioeconomy futures which are often utilized by governments and industries to attract resources and innovate to compete within existing extractive economic systems. These economic systems place limits on what degree any new bioeconomic product or service can do to regenerate soil, restore wild places, reduce human consumption and land use and restore ocean ecosystems. Taking on this mode of working can enable well-intentioned innovators to go further with their impacts and take into mind ethical considerations along the way. Our criticisms sited in this article might be misinterpreted at first as anti-industry and technology; our emphasis on critique of these is in reaction what is perceived as a normative perspective that requires viable alternatives.

As our ethical stance, this approach enforces the idea that there are no predefined pathways to the future, and to imagine alternatives to dominant futures can be an essential reflective process in which the openness or closedness of the future imaginary is considered. This allows for a space in which critical awareness can be articulated for emerging or highly complex phenomena. This is even truer of when considering biofutures than other subject areas, as the evolving bioeconomy discourse must contend with the complex system it is part of, not only those that directly involve humans or fragmented sectors.

The open framing demands thinking outside of the boundaries of current categorizations. We have suggested how it has been influential in tailoring our approaches. Biofutures as a topic for futures research requires engaging in multiple frames and sectors where a degree of abstraction ultimately leads to ethical judgements of the system through entry-points. The simple task of identifying what a 'good or bad bioeconomy' would be fraught with complex values and *ethical stances* in which actors position themselves and expose their anticipatory assumptions. It is in these settings where fruitful dialogues about biofutures can occur. Agency and especially transformative agency can be enabled for a greater number of people and a wider variety of species when we use the lenses of open and closed biofutures.

We offer the term open biofutures as one that can be used by the many and varied bioeconomy actors toward many perspectives. We do not claim ownership of this type of futures criticism of the bioeconomy; our use of open and closed terminology functions mainly to embrace the manifold of ethical positioning, to especially allow in daily practice space for participatory criticism of future imaginaries, how these imaginaries are used, and to encourage spaces for identifying emergent potentials for living nature to justly thrive.

Acknowledgements The authors wish to acknowledge Matti Häyry, principal investigator of Bioeconomy and Justice Project (BioEcoJust); Markku Wilenius, leader of the foresight work package of BioEcoJust; and Sofi Kurki, who's thought leadership in this project has been significant. The BioEcoJust research project upon which this chapter is based is funded by the Academy of Finland as part of its BioFuture 2025 program.

References

Ahlqvist, T., & Sirviö, H. (2019). Contradictions of Spatial Governance: Bioeconomy and the Management of State Space in Finland. *Antipode*, *0*(0), 1–24. https://doi.org/10.1111/anti.12498

Amara, R. (1981, February). The Futures Field: Searching for. *The Futurist*, 25–29.

Andersson, J. (2012) The Great Future Debate and the Struggle for the World, The American Historical Review, Volume 117, Issue 5, December 2012, Pages 1411–1430, https://doi.org/10.1093/ahr/117.5.1411

Andersson, J., & Westholm, E. (2019). Closing the Future: Environmental Research and the Management of Conflicting Future Value Orders. *Science Technology and Human Values*, *44*(2), 237–262. https://doi.org/10.1177/0162243918791263

Balcom Raleigh, N. A., Taylor, A., & Wilenius, M. (2018). COLLISION POINTS AMONG BIOECONOMY WORLDS, TOWARD YEAR 2125. Future in the making 6th International Conference on Future-Oriented Technology Analysis (FTA) –Future in the Making, Brussels, 4-5 June 2018.

Bell, J., Paula, L., Dodd, T., Németh, S., Nanou, C., Mega, V., & Campos, P. (2018). EU ambition to build the world's leading bioeconomy—Uncertain times demand innovative and sustainable solutions. *New Biotechnology*, *40*, 25–30. https://doi.org/10.1016/j.nbt.2017.06.010

Birch, K. (2016). Emergent imaginaries and fragmented policy frameworks in the Canadian bioeconomy. *Sustainability (Switzerland)*, *8*(10). https://doi.org/10.3390/su8101007

Birch, K. (2017). Rethinking Value in the Bio-economy: Finance, Assetization, and the Management of Value. *Science Technology and Human Values*, *42*(3), 460–490. https://doi.org/10.1177/0162243916661633

Birch, K. (2019) Neoliberal Bio-Economies? The Co-Construction of Markets and Natures. London: Palgrave MacMillan. 208 pages. ISBN 978-3-319-91424-4

Birch, K., Levidow, L., & Papaioannou, T. (2010). Sustainable capital? The neoliberalization of nature and Knowledge in the European "Knowledge-based bio-economy". *Sustainability*, *2*(9), 2898–2918. https://doi.org/10.3390/su2092898

Bussey, M. (2014). Concepts and effects: Ordering and practice in foresight. *Foresight*, *16*(1), 1–16. https://doi.org/10.1108/FS-04-2013-0017

Clark, S. R. L. (2011). Imaginary futures and moral possibilities: Blossoming in the morn of days. *International Social Science Journal*, *62*(205–206), 301–312. https://doi.org/10.1111/issj.12004

Dator, J. (2009). Alternative futures at the Manoa School. *Journal of Futures Studies*, *14*(2), 1–18.

De Roo, G. (2018). Ordering Principles in a Dynamic World of Change-On social complexity, transformation and the conditions for balancing purposeful interventions and spontaneous change. *Progress in Planning*, *125*, 1–32. https://doi.org/10.1016/j.progress.2017.04.002

Everard, M., Johnston, P., Santillo, D., & Staddon, C. (2020). The role of ecosystems in mitigation and management of Covid-19 and other zoonoses. *Environmental Science and Policy*, *111*(May), 7–17. https://doi.org/10.1016/j.envsci.2020.05.017

Global visions for the bioeconomy - an international Delphi-Study (2015) Cuhls, K.; Kayser, V.; Grandt, S.; Hamm, U.; Reisch, L.; Daniel, H.; El-Chichakli, B.; Dieckhoff, P. Berlin: Bioökonomierat, 2015, 56 pp. http://publica.fraunhofer.de/documents/N-375141.html.

Golembiewski, B., Sick, N., & Bröring, S. (2015). The emerging research landscape on bioeconomy: What has been done so far and what is essential from a technology and innovation management perspective? *Innovative Food Science and Emerging Technologies*, *29*, 308–317. https://doi.org/10.1016/j.ifset.2015.03.006

Haraway, D. J. (2015). Chthulucene : Making Kin. *Environmental Humanities*, *6*(January), 159–165.

Hetemäki, L., Aho, E., Narbona Ruiz, C., Persson, G., & Potočnik, J. (2017). *Leading the way to a European circular bioeconomy strategy*.

Hockings, M., Dudley, N., Ellio, W., Napolitano, M., Mackinnon, K., Pasha, M. K. S., et al. (2020). Covid-19 and Protected and Conserved Areas. *Parks*, *26*(May), 7–24. https://doi.org/10.2305/IUCN.CH.2020.PARKS-26-1MH.en

Inayatullah, S. (2008). Six pillars: Futures thinking for transforming. *foresight*. https://doi.org/10.1108/14636680810855991

Jasanoff, S. (2015). Future Imperfect: Science, Technology, and the Imaginations of Modernity. In S. Jasanoff & S.-H. Kim (Eds.), *Dreamscapes of Modernity: Sociotechnical Imaginaries and the Fabrication of Power* (pp. 1–33). London: The University of Chicago Press. https://doi.org/10.7208/chicago/9780226276663.001.0001

Jasanoff, S., & Kim, S. H. (2009). Containing the atom: Sociotechnical imaginaries and nuclear power in the United States and South Korea. *Minerva*, *47*(2), 119–146. https://doi.org/10.1007/s11024-009-9124-4

Korhonen, J., Koskivaara, A., & Toppinen, A. (2020). Riding a Trojan horse? Future pathways of the fiber-based packaging industry in the bioeconomy. *Forest Policy and Economics*, *110*(April 2018). https://doi.org/10.1016/j.forpol.2018.08.010

Kuhmonen, T. (2017). Exposing the attractors of evolving complex adaptive systems by utilising futures images: Milestones of the food sustainability journey. *Technological Forecasting and Social Change*, *114*, 214–225. https://doi.org/10.1016/j.techfore.2016.08.015

Masini, E. (1993) Why Future Studies? London, Grey Seal Books.

Mayo, L. (2020). The Postnormal Condition. *Journal of Futures Studies*, *2020*(4), 61–72. https://doi.org/10.6531/JFS.202006_24(4).0006

Miller, R. (Ed.). (2018). Sensing and making-sense of Futures Literacy Towards a Futures Literacy Framework (FLF). In *Transforming the future: anticipation in the 21st century* (pp. 15–50). Abingdon, Oxon ; New York, NY: UNESCO and Routledge.

Milojević, I., & Izgarjan, A. (2014). Creating alternative futures through storytelling: A case study from Serbia. *Futures*, *57*, 51–61. https://doi.org/10.1016/j.futures.2013.12.001

Minkkinen, M., Auffermann, B., & Ahokas, I. (2019). Six foresight frames: Classifying policy foresight processes in foresight systems according to perceived unpredictability and pursued change. *Technological Forecasting and Social Change*, *149*(September), 119753. https://doi.org/10.1016/j.techfore.2019.119753

Mustalahti, I. (2018). The responsive bioeconomy: The need for inclusion of citizens and environmental capability in the forest based bioeconomy. *Journal of Cleaner Production*, *172*, 3781–3790. https://doi.org/10.1016/j.jclepro.2017.06.132

Niiniluoto, I. (2001). Futures studies: Science or art? *Futures*, *33*(5). https://doi.org/10.1016/S0016-3287(00)00080-X

Niiniluoto, I. (2017) Futures Studies: A Science or Art? [in] How Do We Explore Our Futures? Methods of Futures Research. Eds. Heinonen, Kuusi and Salminen, Tulevaisuuden tutkimuksen seura, Acta Futura Fennica 10 Somero 2017, 312 pp.

Ramcilovic-Suominen, S., & Pülzl, H. (2018). Sustainable development – A 'selling point' of the emerging EU bioeconomy policy framework? *Journal of Cleaner Production*, *172*. https://doi.org/10.1016/j.jclepro.2016.12.157

Rosen R. (2012) Anticipatory Systems. In: Anticipatory Systems. IFSR International Series on Systems Science and Engineering, vol 1. Springer, New York, NY. https://doi.org/10.1007/978-1-4614-1269-4_6

Rowe, E., Wright, G., & Derbyshire, J. (2017). Enhancing horizon scanning by utilizing pre-developed scenarios: Analysis of current practice and specification of a process improvement to aid the identification of important 'weak signals'. *Technological Forecasting and Social Change*, *125*(July), 224–235. https://doi.org/10.1016/j.techfore.2017.08.001

Rubin, A. (2017) Causal Layered Analysis [in] How Do We Explore Our Futures? Methods of Futures Research. Eds. Heinonen, Kuusi and Salminen, Tulevaisuuden tutkimuksen seura, Acta Futura Fennica 10 Somero 2017, 312 pp.

Sardar, Z., & Sweeney, J. A. (2016). The Three Tomorrows of Postnormal Times. *Futures*, *75*, 1–13. https://doi.org/10.1016/j.futures.2015.10.004

Sassen, S. (2016). A Massive Loss of Habitat: New Drivers for Migration. *Sociology of Development*, *2*(2), 204–233. https://doi.org/10.1525/sod.2016.2.2.204

Smith, P., Price, J., Molotoks, A., Warren, R., & Malhi, Y. (2018). Impacts on terrestrial biodiversity of moving from a 2°C to a 1.5°C target. Philosophical Transactions of the Royal Society A: Mathematical, Physical and Engineering Sciences, 376(2119), 20160456. https://doi.org/10.1098/rsta.2016.0456

Smáradóttir, S. E., Johannessen, B., & Paulsen, H. (2014). Future opportunities for bioeconomy in the West Nordic countries Future opportunities for bioeconomy in the West Nordic countries, 1–113.

Steward, F. (2008). Breaking the Boundaries: Transformative Innovation for the global good. *Nesta*, (April), 27. https://www.nesta.org.uk/sites/default/files/breaking_the_boundaries.pdf

Steward, F. (2012). Transformative innovation policy to meet the challenge of climate change: Sociotechnical networks aligned with consumption and end-use as new transition arenas for a low-carbon society or green economy. *Technology Analysis and Strategic Management*, *24*(4), 331–343. https://doi.org/10.1080/09537325.2012.663959

Taylor, A., Balcom Raleigh, N. A., Kurki, S., Birmoser Ferreira-Aulu, M.; Wilenius, M. (2019). PRECURSORS TO A 'GOOD' BIOECONOMY IN 2125: MAKING SENSE OF BIOECONOMY & JUSTICE HORIZONS First Foresight Report of the BioEcoJust Project. University of Turku, ebook_2-2019. http://urn.fi/URN:ISBN:ISBN978-952-249-521-1

Vainio, A., Ovaska, U., & Varho, V. (2018). Not so sustainable? Images of bioeconomy by future environmental professionals and citizens. *Journal of Cleaner Production*. https://doi.org/10.1016/j.jclepro.2018.10.290

Vivien, F. D., Nieddu, M., Befort, N., Debref, R., & Giampietro, M. (2019). The Hijacking of the Bioeconomy. *Ecological Economics*, *159*(June 2018), 189–197. https://doi.org/10.1016/j.ecolecon.2019.01.027

VTT. (2018). *Growth by integrating bioeconomy and low-carbon economy. Scenarios for Finland until 2050.*

Westley, F. R., Tjornbo, O., Schultz, L., Olsson, P., Folke, C., Crona, B., & Bodin, Ö. (2013). A theory of transformative agency in linked social-ecological systems. *Ecology and Society*, *18*(3). https://doi.org/10.5751/ES-05072-180327

Isbell, F., Craven, D., Connolly, J. *et al.* Biodiversity increases the resistance of ecosystem productivity to climate extremes. *Nature* **526,** 574–577 (2015). https://doi.org/10.1038/nature15374

Chapter 8
Sensing and Making Sense of Emergent BioEthos Using Futuring Games

Nicolas A. Balcom Raleigh and Amos Taylor

Abstract This chapter builds upon the premise that a multiplex of shared and divergent BioEthos – models of what would be 'good' relationships among humans and other living beings – inform the actions humans take towards living nature and ecosystems. Global warming and environmental change demand from people and our societies new ways of existing as part of living ecologies on this planet. In this setting, the Bioeconomy and Justice (BioEcoJust) project aims to explore ethical troubles that could arise in the development of a global, pervasive, and dominant bioeconomy. This chapter demonstrates how a role-based futuring game piloted in the Bioeconomy and Justice project supports people in 'sensing and making sense' of emergent BioEthos. It presents and analyses the outcomes from the BioEcoJust game session held at the 2019 World Futures Studies Federation Conference in Mexico City. The conceptual framework applied in the game interweaves theories of complexity, futures literacy, and scenarios as worldmaking and operationalizes sensemaking tools developed in the earlier stages of the BioEcoJust project including BioWorlds, bioeconomy socio-technological domains, the human-nature-technology triangle, and BioEthos. The BioEcoJust game pilot in Mexico City enabled its players to explore and critically assess the nuances and dimensions of an ethically troubled future situation and produce a new BioEthos which could be helpful to a unique set of roles for engaging the situation. While much of the literature concerning bioeconomy is concerned with the technical or social factors which can contribute to its development, little attention is paid to what larger ethical frameworks can support its just and fair evolution. The BioEcoJust game emphasized 'keeping whole' the created worlds of a variety of roles responding to an imagined future situation and focused the participant's attention on the interface between assemblages of persons and their bounding conditions. The BioEcoJust game can serve as a model for futuring games designed to help people develop skills for sensing and making sense of emergent BioEthos so they can apply these skills to develop a more just bioeconomy.

N. A. Balcom Raleigh (✉) · A. Taylor
Finland Futures Research Centre, Turku School of Economics – University of Turku, Turku, Finland
e-mail: nabara@utu.fi; amos.t.taylor@utu.fi

E. Koukios, A. Sacio-Szymańska (eds.), *Bio#Futures*,
https://doi.org/10.1007/978-3-030-64969-2_8

Keywords Ethics · Role-based games · Complexity · Emergent ethos · Long-term futures

> The Greek equivalent of translatio is metaphora. Both mean 'carrying across'. Metaphor is not a momentary conjuring trick turning something into something else in a single poem, but an act of translation, of seeing the world otherwise, which lies at the core of creativity.
> –Ruth Padel, *Silent Letters of the Alphabet* (Padel 2010)

1 *BioEthos* Inform our Actions Towards Earth's Life

What people consider to be good vs. bad, right vs. wrong, and desirable vs. undesirable in terms of their own effects and impacts upon living nature and Earth's ecosystems varies across different groups and changes over time. The continuous transformation of these kinds of normative views concerning how people should act towards living nature is what we call in this chapter and our research *BioEthos*. There is not one BioEthos, but many. Together these BioEthos generate a broader field of globally acceptable normative views on what is right or wrong for humans to do as part of nature. These views of how BioEthos interrelate to human action are the starting premise for the game-based futuring experiments described in this chapter.

From the view of humans as a 'superorganism', our consumption patterns are far out of alignment with what the overall ecology of our planet can provide (see, e.g. Hagens 2020). For we humans on Earth, our overall conditions are changing, and our deeply held values and beliefs are being unsettled and becoming something yet to be known (Berzonsky and Moser 2017). Global warming[1] and environmental change (GWEC) are accelerating these planetary and societal transformations. Several systems of human pressure drive GWEC, and one of the most significant is land use for human habitat and resources (Geldmann et al. 2014; Masson-Delmotte et al. 2019; You and Yang 2017). It would seem obvious that people ought to do something to address the GWEC that is already occurring, but we humans have many different views, levels of influence, and priorities. Determining which are the most ethical actions to take when confronted with the transformations and surprises GWEC is already bringing us – and will be bringing us over the long term – is not as obvious, clear, or widely agreed as they may at times appear.

We propose BioEthos as a key tool for thinking when inquiring into how ethics-shaping bounding conditions could develop, emerge, trigger, and inhibit alignment of human actions to address the many ethically difficult situations arising from GWEC over long time horizons. These BioEthos are dynamic and context-specific patterns held by people concerning 'what is good' for our relations with each other

[1] At the time of publication, climate change is the going term for the phenomena of humans increasing the greenhouse gasses in our atmosphere and thereby changing the radiating forcing dynamics of the planet so that its global average temperature is increasing. To describe this larger situation as clearly as possible, we intentionally use here the older term for this phenomena, global warming.

and the rest of Earth's life. On an individual basis, these models inform behaviours and actions, which ultimately culminate as larger-scale impacts on our ecosystems (e.g. human pressure, land degradation or restoration, biodiversity preservation or loss, etc.). We add to the above described dynamics those that manifest from imagined and actualized developments and impacts of technology.

This concept of BioEthos is used as a heuristic and sensemaking tool in the Bioeconomy and Justice (BioEcoJust) project.[2] This research setting aims to identify ethical concerns which could arise between now and 2125 assuming a global bioeconomy – characterized by bio-driven processes to provide humans with materials, chemicals, energy, and services – takes form and becomes dominant during that time period. The multiplex of emergent BioEthos and their interactions are the 'beating heart' of our research approach which focuses on developing tools to guide ethical considerations of policymakers and decisionmakers. How any particular formation of varied BioEthos could develop over the coming century has risen out of our research processes as a key question. In our view, actors who aim to address GWEC by growing the bioeconomy perceive and respond to the ethical challenges of their endeavours based on BioEthos they believe.

This chapter presents and analyses the outcomes of a role-based futuring game that operationalizes a set of our research project's sensemaking tools with an emphasis on the BioEthos. The *BioEcoJust game* aimed to enable gameplayers to explore and find their way through potential ethical troubles in the rising bioeconomy. We begin by presenting the key concepts which inform the game's design and follow by describing how the BioEcoJust sensemaking tools are integrated in the design of the game. To demonstrate how the game functioned, the key outcomes of the game are presented and lightly analysed, culminating in an account of a new BioEthos produced by the gameplayers during their session. Then, the relevance of this experimentation with 'sensing and making sense' of emerging BioEthos to the field of futures studies and bioeconomy developments is discussed. The chapter concludes with our assessment of the potential value that the BioEcoJust game, as a kind of role-based futuring game, can bring to the ethicality and justness of a rising bioeconomy.

2 Complexity, Situations, Worldmaking, Sensemaking, and Ethos

The design of the BioEcoJust game corresponds with a conceptual framework built from our theoretical understandings of complexity, situations, worldmaking, and ethos. At a high level, this framework sees situations as entanglements of networks of complex systems which involve actors (e.g. people taking various roles), agential components (e.g. built environments, human infrastructures, forests), and

[2] See https://bioecojust.utu.fi (accessed 24 September 2020).

time-bound moments of transformation (e.g. events). In these situations, a wide variety of individuals, each one enacting a semi-unique BioEthos in a 'world as it is to them', engage with the challenges and opportunities arising from an ethically troubled situation with larger processes of transformation. To operationalize this theoretical lens, we first need to unpack these massive concepts of complexity and transformation, situatedness and worldmaking, sensemaking, and ethos.

We position our *BioEthos* concept in a complex systems perspective, drawing specifically from the concept of *complex adaptive systems*. In this view, complex systems are comprised of many systems which share some set of overall conditions that drive and contour their behaviour as part of the complex system, while these systems produce and contribute to the qualities of those overall conditions. Complexity in our research is understood as multiple layers of multiple networks of people, ideas, materials, built environments, living nature, and the physical world all transforming at differing timings and entangling into many different and geographically distributed nodes.

This understanding of complexity is informed by a framing of the lived experience of the human species as a whole population – a super-organism (cf. Hagens 2020) – is multi-scale, multi-layered, and multi-local. In order to imagine futures congruent with how people experience living their own lives – as situated selves that are part of this super-organism – our futuring game needs to emphasize perspective-taking and inter-relating assemblages of individuals towards a situation.

Individual humans are diverse in their lived experiences, perceptions, and relationships to other people, as well as in their links to networks of ideas, materials, built environments, living nature, and physical landscapes. Goodman (1978) goes as far as to argue that each individual perceives and acts in one's own world and living one's life is worldmaking. An approach to scenario workshops has been developed based on Goodman's sociological concept of worldmaking, arguing that this new approach is better able to respect and generate insights by allowing for a diversity of many individual worlds to continue throughout a futuring processes, without smashing them into some artificial synthesis (Vervoort et al. 2015).

Sensemaking in our research is taken as an ongoing enacted activity demanded by continual transformation. Biologist Robert Rosen proposed that all life is oriented towards the future as their biological components act towards anticipated outcomes (see, e.g. Louie 2010). For humans, these processes can be conscious, and recent developments in futures studies have placed emphasis on the potential value of developing the capability called futures literacy as a way to widen our perception of our transforming world (Miller 2015). A working definition of futures literacy is 'diversifying how and why we use futures', which means developing skills in switching between whole modes of imagining futures for varying purposes and contexts. Two broad categories of modes are described in the Futures Literacy Framework – *anticipation for futures* and *anticipation for emergence*. A key claim by futures literacy proponents is that the skill of switching between these two categories of anticipation helps people 'sense and make sense of emergent novelty'. Proponents of futures literacy argue skills in *anticipation for emergence* are widely underdeveloped and new tools are needed to help people develop such skills (Miller 2018).

Because we conceive of BioEthos as continually transforming, varied, and multiple – a dynamic multiplex – we expect there to be emergent and novel forms of them over time. We model these emergent BioEthos as continually popping into existence, some staying and others fading away, some becoming widespread and others narrowly adapted. We posit that these emergent BioEthos are often implicit, changing, and multiple in individuals, organizations, and societies. We have observed, through self-reflection and thought experiments, that in any given single person, there can exist multiple BioEthos that are not necessarily coherent (e.g. a vegan race car driver or a hamburger-eating climate activist). Also, in large companies or start-ups, there can be, and often are (due to the extractionist colonial legacies of the global economy), conflicts between the normative views of what an ideally 'good' BioEthos should be and how a BioEthos is expressed through actions.

Based on these observations, we imagine a myriad array of overlapping BioEthos is likely to come into being during the long timespan between now and 2125 – some of which could remain familiar and others that are surprising and unfamiliar. These BioEthos are today implicit, changing, and conflictual in individuals and organizations and likely will be similarly so in the future. We have come to realize that if our research project is to produce anything of value to decisionmakers and the general public, it would be unsatisfactory to provide some set of scenarios and ethically correct choices derived from them – rendering us as the 'we told you so' people and them running a risk of making poor ethical choices based on past formulations of what is a 'good' relation between humans and other life. Instead, we have in our workshops focused on developing tools and processes that aim to help people struggle productively in sensing and making sense of potential and emerging BioEthos so they can gain experience in doing so. Such experiences would equip people facing GWEC-triggered local and multi-scale situations, no matter their level of power or influence in society, to identify and comprehend the multiple ways BioEthos are arising in various difficult situations and being invoked by various clusters of people in their attempts to ethically address their situations.

To make such a tool, we designed a futuring game that could enable workshop participants to engage with our project's understanding of how complexity, situatedness of individuals, worldmaking, and futures literacy are involved in BioEthos without being bogged down by these weighty concepts. This theoretical posture informed the development of the BioEcoJust game and helped us interweave key themes and avenues of inquiry without overly reducing the significant dimension of complexity. These interlinks manifested as the mechanics of the game, which are described in greater detail in the following section.

3 Game Elements and Their Interfaces to the BioEcoJust Sensemaking Framework

Prior to our three pilots of the BioEcoJust game, the BioEcoJust project had implemented a horizon scanning process concerning its research themes of bioeconomy and justice. An output from this horizon scanning process is a sensemaking framework we found to be helpful for interpreting and positioning the horizon scanning items we had identified. This framework applied the concept of worldmaking (Bendor 2017; Vervoort et al. 2015) to produce categories of *BioWorlds* to which various actors involved in bioeconomy efforts could be argued to belong (Table 8.1). It also identified three *socio-technological domains* pertinent to our own research setting in Finland – soil, forests, and algae. The project proposes this sensemaking framework is helpful for interpreting how bioeconomy work is now, while leaving room for what it could become. The sensemaking framework also highlights the multiplicity of views and motives active among people and organizations engaged in the bioeconomy.

Within our heuristic of BioWorlds, we have applied an adaptation of actor-network theory (cf. Latour 2005) to conceptualize normative and ethical views of the development of the bioeconomy as an interrelated set of agential nodes: humans, nature, and technology.

We call this set of nodes the *human-nature-technology triangle* (Fig. 8.1). This conceptualization serves as a tool for exploring how various groups conceive of what are 'good' interrelations between these nodes (e.g. humans to technology), as well as to explore dynamics inside of each node (e.g. humans to humans). It is intended to be neutral in terms of hierarchy – no one node is over another – and to emphasize interrelations in terms of action (e.g. what should humans do with their technology to nature?). It is also intended to open questions about how distinct each node really is from the others – for example, an expected use of the triangle is to challenge it with arguments such as 'humans should not be separated from nature, we are a part of nature' or 'all living nature has its own technologies'.

Table 8.1 BioEcoJust BioWorlds and their descriptions

BioWorld	*Description*
BioUtility	Technology should make human use of natural resources more efficient
BioMimicry	Nature is the best source of ideas for making new tech for complex situations
BioUpgrade	Life forms are generally flawed and can be fixed via human-made technology
BioRecovery	Humans should use all available tech to restore and recover ecosystems
BioEquality	All living nature should be equally respected for its intrinsic value

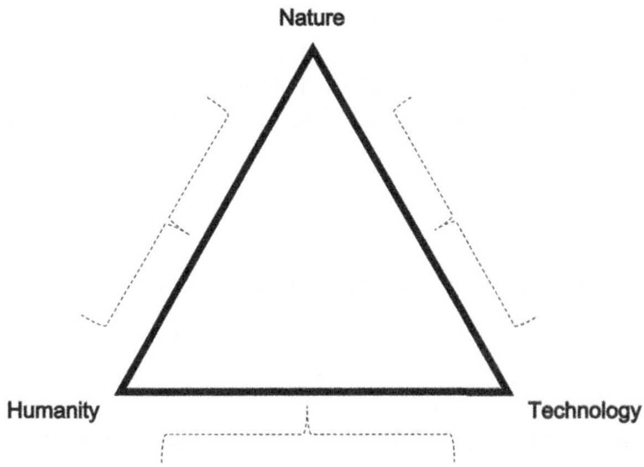

Fig. 8.1 Human-nature-technology triangle

Its origin, like the BioWorlds, is from the horizon scanning part of our study. It helped us interpret normative and ethically oriented views of actors and innovation efforts found in materials gathered through the scanning process. The human-nature-technology triangle helps untangle and comprehend what a particular person or group of people consider to be 'good' relationships among humans, nature, and technology. In our research, this triangle became the dynamical engine that distinguished one BioWorld from another. Later, we coupled our BioWorlds with BioEthos. On its own, the triangle functions as a heuristic for making sense of what BioEthos already exist and imagining what BioEthos could exist.

The human-nature-technology triangle can also be used to look for impacts of actions taken by people based on their BioEthos. We imagine every BioEthos version of the triangle having a realized impacts version of the triangle. For example, a BioEthos that sees 'humans upgrading nature by applying our technology' as appropriate would produce impacts such as an increased diversity of genetically modified life forms entering the ecosystem. Over time, impacts from the multiplex of BioEthos held by various people accumulate into aggregate effects, such as 'human pressure' on wild places; global environmental change (see, e.g. Hamann et al. 2018); or anthropomorphic climate change (see, e.g. Cook et al. 2013). In other words, when a BioEthos version of the human-nature-technology triangle is paired with a 'realized impacts' version, it produces a conceptualization of how a person or group's BioEthos impacts the overall complex system of all living nature on Earth. In many cases, a person may hold a BioEthos and act according to it without awareness or even a capacity to know what the actual impacts of 'living by this BioEthos' are or will be.

In our research project, before our three BioEcoJust game pilots, we had already applied the heuristic of the *human-nature-technology triangle* in our horizon scanning process and Delphi study (see Chap. 9). The triangle was used by the research team to elaborate descriptions of five co-existing BioWorlds, select a diverse panel based on them, and formulate questions that would help these Delphi experts unpack, evaluate, and communicate their own ethical views concerning the development of a global bioeconomy. Noting how the human-nature-technology triangle enabled the panel of experts to investigate and elaborate their own BioEthos as well as those held by others, we decided to make it a focal element in the Mexico City pilot of the BioEcoJust game.

3.1 Design Choices for the BioEcoJust Game

Futuring games, or game-based futures workshops, are argued to support creativity and imaginative thinking about the future and open new views on serious challenges (Heinonen et al. 2015). In order to operationalize our theoretical framework and incorporate the approach of worldmaking, we needed a game that emphasizes the interrelations of individuals in a shared situation. We also needed a game that could incorporate the project's sensemaking framework which included five BioWorlds and three bioeconomy-related socio-technological domains (Taylor et al. 2019). We concluded that a game developed by Balcom Raleigh called Metaphor Molecule fit these criteria (Balcom Raleigh and Heinonen 2019). It became the main engine of the BioEcoJust game.

When Metaphor Molecule game is used as a method, it supports groups in having rich and surprising conversations about a potential future. In it, details about a future are generated from multiple perspectives via roles invented by the gameplayers. Playing Metaphor Molecule enables people to immerse themselves in futures they generate and modify. It supports creativity and criticality of the players (Balcom Raleigh and Heinonen 2019). The game functions from a mixture of individual-driven and group-driven creativity and is designed to be fun to play. It also enables people to produce diverse, perspective-bound details about a future (Balcom Raleigh and Heinonen 2019).

Structurally, the Metaphor Molecule game is a role-based game in which gameplayers select or create a seed scenario to play, create roles for that seed scenario, describe what their roles perceive to be motivating or threatening in the scenario, find a metaphor to convey the complexity of each role, and relate the roles to each other. These role metaphors and relationships to other roles and the scenario are called Metaphor Atoms. Once all of this generative work is complete, the players model the relationship among roles as so-called Metaphor Molecules and take account of what drives those relationships. After modelling the relationships, the group selects one relationship they believe to be most influential to the overall situation.

The next step is the metaphor transformation – the gameplayers who created the roles involved in the 'most influential' relationship come up with new Metaphor Atom for their roles. This step is argued to be the game's moment of highest, most entangled creativity and criticality (Balcom Raleigh & Heinonen 2019). After the transformation step, there is typically a discussion about how the impacted relationship has changed and those changes cascade into the relationships among all of the roles and between the roles and the scenario. The whole game typically ends with a concluding activity such as some form of scorekeeping or encapsulating what happened. Along the way, the key artefacts generated by Metaphor Molecule game are *role cards*, *metaphor atoms* for the role cards, a *log sheet of Metaphor Molecules*, and any modified metaphor atoms for the roles that underwent *metaphor transformation*.

The BioEcoJust game can be called a customization of the Metaphor Molecule game. Two components of the original game were modified to better involve concepts developed in earlier stages of the BioEcoJust research project – the seed scenarios and the role card prompts. A third modification involved appending a concluding exercise focused on creating a new BioEthos. This new exercise was being tested for its potential to provide relevant information to the BioEcoJust project while simultaneously supporting the participants in processing the futures they had generated while playing the game.

Instead of a set of seed scenarios, the game utilized 'future ethically difficult situations' featuring competing factors which made it unclear what the most ethical ways to respond would be. These situations were presented in the game as a set of newspaper covers from the year 2075 and were developed based on the three situations used in the first round of the BioEcoJust Delphi study: 'Megapolis Floods! Five Million to Relocate' in which the balance of land use is precarious and any changes to it have big consequences, 'AI-Driven Ecosystems Challenged by Natural Nature' in which ecosystem restoration has been handed over to roving AI-driven synthetic biology labs, and 'Ghost Forests separated from nature' in which natural resources are grown in ways that allow for harvest without destroying habitats of other life forms. These situations had enough detail to initiate creative thinking, yet left room for gameplayers to add their own details and create stories about a future.

In the development of the Metaphor Molecule game, there have been many versions of role cards, but all versions have included an area for drawing and an area for describing in text. The role cards for BioEcoJust game take inspiration from the version that includes creative prompts – a set of three kinds of randomly drawn characteristics.[3] These role prompts help trigger creativity via 'forced combinations'; and in the case of BioEcoJust game, the customization allowed the research team to integrate the research project's sensemaking framework developed during its horizon scanning phase. The prompts included five BioWorlds, BioUtility, BioUpgrade, BioRecovery, BioMimicry, and BioEquality; three socio-technological domains – forestry, soil, and micro-organisms; and a set of social sectors (e.g. financial services, local government, start-up company, civil society organization, etc.). To foster creativity, each category of role prompts included a 'create your own' card to invite new ideas to the research project's sensemaking framework. The intent of this design choice to include Role Prompts was to help the gameplayers create roles that could discuss with our key research concepts.

[3] The idea for role prompts to spark participant creativity originated in the version of Metaphor Molecule game used in the Complex Futures of Human Settlements Futures Literacy Lab (Balcom Raleigh et al. 2018).

The BioEcoJust game was piloted three times in 2019, the first at Futures Conference in Turku concluded with an open-ended exercise in which the participants discussed what they noticed in their game. The second pilot at World Futures Studies Federation Conference in Mexico City concluded with a similar discussion, but emphasizing ethical dimensions, followed by an exercise in creating a new BioEthos for their group of roles using a human-technology-nature triangle design template (see Figure 8.1, earlier in this chapter). The third pilot at the Anticipation 2020 Conference in Oslo also concluded with this design template, but incorporated steps from the Futures Clinique approach (Heinonen and Ruotsalainen 2013) to elaborate the seed situations. Each pilot experimented with the design of the game to discover what potential insights it could provide to the BioEcoJust project and its stakeholders.

While we expected each pilot to successively improve the design of the game, we observed each of the three pilots emphasized different potentials for the BioEcoJust game. The first indicated that role-making and perspective-taking served the participants well for applying 'scenarios as worldmaking' in a participatory futuring process. The second pilot in Mexico City, with its addition of the human-nature-technology triangle design template, introduced us to the potentials for the game to enable rich conversations about emergent BioEthos. The third pilot in Oslo gave us a first taste of tensions or even incompatibilities between approaches which encourage imagining futures at a general, quasi-objective level and those which encourage imagining futures at an individual, quasi-subjective level while continuing to hint at the potential of the human-nature-technology triangle to support participants in having conversations about what new forms of ethics may be needed in the future. In the following section, we take the outcomes of the Mexico City BioEcoJust game pilot as the focus of our presentation and analyses.

3.2 The Mexico City Pilot

The best way to present and analyse the contents of the data produced in this game session is to walk through the steps of the game, the choices the gameplayers made, and the contents they produced. Part of our analysis is informed by observing the game directly, Balcom Raleigh as facilitator and Taylor as participant observer. The data produced from the game – including an audio recording of the proceedings and all materials produced – were further analysed post facto by the research team.

Seven people played the Mexico City Edition of the BioEcoJust game – three professional futurists, one futures studies PhD student, one futures researcher, and two university students less familiar with futures studies. The participants are from six nations, and their mother tongue languages are English (n = 4) and Spanish (n = 3). The game was played in English. Four of the participants are women and three are men. All game play was audio recorded, and all artefacts (e.g. completed role cards and design templates) were collected. All players signed research consent forms which followed Finland guidelines for research integrity.

The group selected the *Recoding Life with AI* as the situation to play. They then read the newspaper cover from 2075 for the situation and briefly discussed what they observed about it. The players then worked individually to create roles who would be interesting to play in the situation. Of the seven roles produced, only two were ordinary humans, and the other five were either augmented or significantly different than people of today or non-human beings with agential qualities. Because of time pressure, the facilitator decided to skip the metaphor transformation steps of the game to make time for a test of the human-nature-technology triangle design template. The group first used the blank back side of the template to make collective notes from an open discussion about 'ethical stuff' they noticed in their game session. The group then used the front of the design template to create a new BioEthos useful to their roles. These steps smoothly lifted the group into a provocative and rich conversation about what it would mean to be a human, or any life form, in a world where the boundaries between AI-modified life and 'natural nature' were blurred.

4 Exploring a Difficult Ethical Situation in 2075 as an Assemblage of Inter-related Roles and Role Worlds

This section demonstrates the kinds of novelty and insights the BioEcoJust game can produce by describing what happened at the BioEcoJust game pilot in Mexico City. It will tour details of the imagined assemblage of roles produced by the participants and illustrate the depth to which the participants imagined a future situation together. The discussion the group had about emerging BioEthos will be described, and reflections about what we learned from piloting the game will be offered.

The game session, like any workshop, is a complex happening involving an entanglement of ideas from a unique assemblage of actors at a specific time and place. The following describes what was generated by the gameplayers by presenting the event's traces – its data – synthesized by us authors as a story. The story starts with the future situation that the group chose, future people they created, those people's perspectives on the situation, and their assessments of how they relate to the other people in the story. It is followed with the key insights of the gameplayers' own evaluation of what 'ethical stuff' was active in their game and the emergent BioEthos they identified that would be useful to these people in the situation.

In the year 2075, radical ecosystem recovery efforts have fused human-made technologies into the viability of vast ecosystems. AI-driven synth-bio robotic labs manage at least one large ecosystem. Its function is to maintain a symbiotic biodiversity within the territory it oversees while maximizing carbon capture to sustain climate conditions suitable for global life. The living beings under its domain are the equivalent of tens of thousands of years evolutionary steps ahead of life forms found in 'unmanaged' ecosystems. Meanwhile climate change continues to cause habitat loss and massive non-human migrations. The non-human species entering this AI-managed ecosystem would die, because of their evolution gap, if they remained unmodified. To ensure ongoing balance, the AI lab captures those species, analyses how to maximize their fit, and modifies them. People have begun protesting this practice, demanding that forcing other living beings to be modified is morally wrong. But the system has been running so long; the whole managed ecosystem would likely die if the intervention is stopped.

An assemblage of seven-person roles, generated by the gameplayers, is engaged in the situation: 'Link', an advisor to policymakers and process-makers regarding the intrinsic value of nature; Tortuga-Bish 2, a so-called Telomerge in its second phase of a dematerializing transformation; Fatima Funder, a swarm-AI 'Innovestor'; Doctor Professor, head of the 'living nature' association arguing for the protection of 'natural nature'; Tiny Paula, a journalist and feminist activist who is 15 cm tall, who lives in the 'tiny world' created in the past; Bio Bureaucrat 87B 'Mister G', a social service AI helping people engage in eco-recovery projects and be compensated for their eco-restorative actions; and Futures Catalyser and Seeder, a being who can function at multi-dimensional frequencies seeking to foster convergences and collaborations. It is worth noting that five of these future people from 2075 radically bend what we know as people today in 2020. They are associated with varying BioWorlds, socio-technological domains, and sectors (Table 8.2).

The best way to convey who these future people are is by meeting them one by one, as it also occurred in the game.

Link is often in the forest. This person is linked to the BioWorld of 'BioEquality: all living nature should be equally respected for its intrinsic value' and is active in the socio-technological domain of forestry in the sector of research and development. She engages the political level of the situation by reminding policymakers and process-makers of the intrinsic voice and the intrinsic value of nature. From an economic perspective, value is more than just financial; it can also be understood in a broader way from human, economical, and aesthetic perspectives. She emphasizes the intrinsic value of all living nature. Link engages the social dimension by bringing intrinsic value awareness and enabling people to make decisions based on it.

Table 8.2 Participant-created roles and their BioWorlds, socio-technological domains, and sectors

Future person	BioWorld	Socio-technological domain	Sector
Link	BioEquality	Forestry	Research and development
Telomerge	BioUpgrade	Human bio-connectivity	Foundations
Innovestor	BioUpgrade	Algae, enzymes, microbes	Start-up investor
Professor	BioEquality	Antipollution technologies	Academia
Tiny Paula	Equally respected	Algae, enzymes, microbes	Media
Mr. G.	BioRecovery	Soil	AI-social service
Catalyser	7th dimension frequency	Language Communication	Inter-sectorial convergence

From a technology perspective, Link keeps a connection to AI but still uses human processing because Link thinks that the technology of mind is important. From ecological perspective, Link is an advocate for fair value management, making sure the intrinsic valuations are above the extrinsic ones in forestry-related matters. From a culture perspective, it's about bringing a diversity of perspectives on how living things are valuable for more than just human or the obvious reasons. Link secretly loves animals more than trees and has an interest in bonsai trees.

Tortuga-Bish 2 identifies with the BioUpgrade world, contributes to the socio-technological domain of BioConnectivity, and is part of foundations sector. This person's name includes the sound 'bish', which is incomprehensible to ordinary ears because it is a machine code designation that includes the total information about this person's life course. The 2 in their name indicates they are in the second phase of the three-phase life journey of a Telomerge. In the first phase, they are biologically mainly human; however their goal is to have a broad range of human experiences, frequently changing sex and gender and experiencing all human life stages before beginning the second phase. During this first phase, Telomerges serve society in teaching and entertainment functions. In their second phase – the phase Tortuga-Bish 2 is now in – Telomerges begin their biomerger into the ecosystem, at first using advanced haptics that allow them to feel what living nature feels and completing the phase by becoming part of the ecosystem. This process of the transitions in Phase 1 and Phase 2 is dangerous. In Phase 3, they are expected to download as much data as possible conveying the aggregate and conclusory information from their experiences, and this data is used as part of that ecosystem's data going forward. This rich dataset is informed by their human life information and ecosystem experiences. The Telomerge engages in the politics of the situation by embedding their blended human-ecosystem memories into long-term ecosystem biostorage and living out a memory function to push politics into a longer-term formation. The Telomerge serves a necessary economic function of translating the complexity that actually exists into data and human-relevant metrics that can be used to make decisions. Socially, they show others the transcendence of human experience and also provide a kind of entertainment as they go through their many bodily transformations and take risks to provide a vital public function. Technologically, their job is to live, experience, and add to the richness of the bio-informatic soil. They provide the historically encoded experience that is utilized to operate the AI-driven ecological system. Culturally, they foster the sensations of unity, connectedness, and co-regulation. Nature and humans are understood as different interpretations of the same universal information. What Telomerges do is novel and experimental. The first two phases of their transition involve an extremely high risk of failure and mental collapse. Very few people are really built to hold together through all three phases of a Telomerge's life.

Fatima Funder is an Innovestor and is a total AI creation. This role is basically looking across nature with AI tools, wherever life and innovation could happen – the ocean, forests, landscapes, etc. The Innovestor searches for investment opportunities related to the theme of 'fountain of youth'. Innovestor is always looking, always at the cusp. Innovestor is a foresight enterprise and does not need to be a human or be in human form. Innovestor is interested in any public policy that supports research and innovation. It does not care if the support is public or private, because innovation can happen anywhere and at any time. Politically, Innovestor is always looking to break down any barriers and considerations of where research and innovation can happen. It can be in formal or informal contexts. Economically, Innovestor sees 'small' companies and life forms as most profitable, knowing they are the origin and engine of value creation. Innovestor pays close attention to the smallest enterprises. Innovestor wants for people to comply with ordinance 5.4.3.2.1.0 which requires people to be their 'new and improved you' on a regular basis and sets a new social norm. Innovestor has evolved beyond nanotech, to micro-AI tech, and is going with the flow: 'whatever the next generation of tech will be, we're just going to keep going'. As an AI, Innovestor has high confidence in its kinds' ability to manage ecological issues and believes 'We can fix anything any time'. Culturally, some of its best friends are microbes. Innovestor loves risk taking and is an optimist; that is their programming and algorithm. It is excited to see how far it will take us. Innovestor aims to be all seeing and all knowing.

The *Professor* identifies with views of BioEquality, is an active contributor to the development of socio-technological domain of antipollution technologies, and is an academic. He leads the Association for Living Nature. In politics, he is an activist against any legislation that will change nature by replacing the 'natural nature' we have had 'since the beginning of days' to this new 'in vitro nature'. He's against and disagrees with the neoliberal economy and market economy because he sees it as responsible for the competitive scramble to deploy innovations aimed at changing 'natural nature'. He feels excluded from society and thinks it is because he is kind of socialist. This is a person who is conservative in his thinking. As for technology, he agrees with and promotes technological innovation, but not if our humanity is being lost in the process. He is a big lover of nature. He is engaged with taking care of nature and works to stop its destruction. Culturally, he has a little bit of Catholicism in his background. He's really influenced by theology and philosophy – and the origin of nature – and thinking about the creation made by God. That's why he's against all of these things that are replacing nature. But, he's not extreme. He doesn't want to be the target of critics as a fanatic. His philosophical and theological influences distinguish him from the others in this assemblage.

Tiny Paula lives in a smaller world within the situation and is motivated towards all people being equally respected. Humans have created tiny humans like her for their amusement, but now these tiny humans (less than 15 cm tall) are breaking free from that oppression. In this tiny world, you can find everything you can find in the bigger world. Paula is 21. She's a journalist, activist, and feminist. And she's fighting for the rights of tiny people. She is working to transform how tiny people are seen and treated in society. Talking about economy, she believes everyone should have the opportunity to raise their own economic status. She agrees with the evolution of the technology in the situation but is also fighting for 'real nature' to continue. For example, she agrees with creating some bees or some artificial life necessary to keep the world going. Culturally, she is open-minded and agrees with everything and everyone, but just as long as everyone is keeping equality in the conversation – especially with regard to respecting the equal standing of tiny people and women.

Bio Bureaucrat 87B, or 'Mister G', is part of the BioRecovery world, the socio-technological domain of soil, and the AI-Social Services sector. He works with compost and special enzymes. In practical terms, he runs a social services office. He is aligned with the 'humans for nature; nature for humans' political party. The AI provides a bio-offset economy to support humans who need support because they are a bit marginalized now. The social standing of people is in respect to nature as defined by AI, and people are rewarded for how much they contribute to the well-being of living nature. He applies advanced AI technologies as a social service that understands the needs of humans and needs of nature. His work focuses on restoring the ecology through regenerating soil ecosystems so they can act as carbon offsets. This work is tied to the size of the bio-offset economy he administers. Mister G's view of culture is that the 'wealth of nature is the wealth of humanity', but this wealth is precariously there. Bureaucrats help humans fund their daily lives using bio-eco-recovery points they can earn through their eco-friendly actions. 'Please take your ticket, and next. Sign up to quantify how much you are worth'. It is not clear to anyone if he's fully human or not.

Catalyser and Futures Seeder is from the BioWorld of seven-dimensional human beings. She operates in the frequencies of seven dimensions. From her perspective time doesn't exist, giving her a wide temporal resolution and the sensation of time travel. She can have simultaneous experiences in many cultures, times, genres, colours, etc. Her socio-technological domain is language and communication because language, whatever the kind – art or whatever – programs what can happen. Language is both very interesting and very dangerous. She acts towards inter-sectorial emergence and coordination. In politics, she catalyses change to a new model which can come after republic and representative democracy. In economics, she helps design different metrics of value which include metrics that are social, cultural, environmental, and financial – meaning mainly time, not only money. Socially she holds celebrations for collaborations using technology. Catalysers and seeders like her need this ability of gathering people and celebration because it is key for collaboration as it helps develop trust. When there is trust, people can reduce wasted time, money, intelligence, and people in bureaucracy. Technologically, she uses data for the senses, not only data but how people can sense the data through art and data visualization, which helps people choose and make better choices. For ecology, she helps people develop new senses, not only the six usual ones but new senses such as the first one developed, the seventh sense of feeling and being part of the whole. Through this new sense, people can go beyond the experience of being a singularity to also experiencing being part of an overall being. In culture, she works in soft tech, to balance technology, to foster collaboration and convergence.

This tour of the future people created by the gameplayers contributes a rich tapestry of details to the overall situation from the unique attributes of these roles. Technologies are used in a wide variety of ways. Old notions such as innovation and wealth creation co-exist with new valuation schemes such as intrinsic value or reward systems based on personal ecological actions. Political dimensions arise in the forms of individual lobbying efforts, civil society organizations and political parties, ambitions to converge interests of a variety of stakeholders, and promotions of equality.

Similar to people of the present, these future people of 2075 perceive differing kinds of motivators and threats in their situation. These perceptions add another layer of details to the 'story about a future' the gameplayers are building (Table 8.3).

In these perceptions, self-purpose and collective aspirations are implicit. Competition implies Innovestor can win contests. The oppressions Tiny Paula and her community face can fuel her work for equality. Mister G is concerned for humans being at the lowest level of the nature-to-nature concerns of the AI ecosystem management system but is addressing this concern by providing humans access to value created from soil restoration. Link identifies the public discourse around what to do about the AI-run ecosystem as an enabler for the conversations she aims to foster around the intrinsic value of nature. These are all intersubjective considerations that entangle the identities of the roles themselves and the gaps they perceive between the way things are in the situation and the way they ought to be.

The roles can also be understood by using metaphors. Metaphors are figurative words or phrases that convey complex ideas quickly. They emphasize some qualities while de-emphasizing others, and they imbue entailments of further meanings

Table 8.3 What the roles find motivating and threatening in the situation

Role	Motivating from situation	Threatening from situation
Link	The polarity of 'nature real' versus 'nature synthetic (or adapted genetically) is causing public debate, and from this debate insight/growth can develop	Loss of both created and pre-existing ecology; Loss of relationship
Telomerge (2nd phase)	Upgrade seen as a transcendent merger, a testing that was immanent in 'natural nature, our human nature'	Essentialist and transcendent external counterforces
Innovestor	This role is a result of this BioWorld/situation	Competition! Is not the only innovestor; Human authority values
Professor	Keep nature for a long time	The test of humanity; The lack of respect for human beings and the environment normalizing
Tiny Paula	She really wants an equity between tiny humans and 'real' humans	The 'real' humans [treat tiny humans] like pets, and they make experiments with tiny people
Mister G.	Restore nature through soil manipulation growth economy	Humans at the bottom of nature-to-nature AI equation
Catalyser	–	People are polarizing, and there is a lack of language and narrative to allow for convergency

Table 8.4 Roles and their metaphors

Role	*Role's metaphor*
Link	'Scales' – balancing
Telomerge 2nd phase	Human life as accretions of soil layers
Innovestor	BioEngine of creation. Biosine. AI-sine. Biofuel
Professor Richard	Don't forget ourselves
Tiny journalist Paula	Ant
87B – 'Mister G'	'One potato, two potato, three potato, four…'
Catalyser	Rainbow

onto a topic (Lakoff and Johnson 2003). The metaphors the gameplayers gave their roles serve as concise expressions of how they engage with the situation (Table 8.4). Link aims to balance intrinsic value of nature with other forms of value. Telomerge is taking extreme risks to deliver their experiences to the soil and become a layer of information. Innovestor thinks of itself as an economic engine of creation, focused on life-driven and AI-driven value creation. Professor Richard is reminding everyone in the situation (who will listen) not to forget ourselves or the 'natural nature' we come from. Tiny Paula is as hardworking and collectively oriented as an ant and needs the other tiny people to help make a change and address the inequality she aims to remove from society. 'Mister G' emphasizes the accounting function he occupies in the situation. Catalyser emphasizes her role as a bridge and advocate for diversity. All of these metaphors convey something of the ethical ambitions of the roles.

4.1 How Do These Future People in the Situation Get Along?

The assemblage of people in this situation, as in today's world, perceives each other in different ways in terms of their own interests and aspirations. Some are more helpful, and some are less helpful or, worse, working towards counter-purposes. The gameplayers determined this assemblage of persons in their 2075 situation have a mix of views towards each other. These details concerning the relational dynamics among the assemblage bring additional details to the future imaginary the group produced during their game session (Table 8.5).

Building models of these relationships as 'Metaphor Molecules' shows some of the dynamics among these relationships and adds detail to the assemblage of roles in this situation. Some of the roles are in a symmetrical relationship (strong bonds), and others have a least helpful or most helpful role in common (weak bonds). Strong bonds (Table 8.6) and weak bonds (Table 8.7) are the two main categories of Metaphor Molecules. In each of these, the relationship between the roles can be characterized as 'most helpful' and 'least helpful'. We note that this modelling process largely ignores the asymmetrical relationships in which roles do not see each other in the same way, and this is by design to keep the game flow moving.

Table 8.5 How the roles see each other in terms of 'least helpful' and 'most helpful'

Role	*Roles seen as 'least helpful'*	*Roles seen as 'most helpful'*
Link	Mister G	Catalyser
Telomerge 2nd phase	Professor, Catalyser, Little Paula	Mister G.
Innovestor	Link	Telomerge, Tiny Paula, Mister G., Professor, Catalyser
Professor Richard	Innovestor, Myself, Link	Mister G., Catalyser, Tiny Paula
Tiny journalist Paula	–	Everybody is helpful
87B – 'Mister G'	Catalyser	Innovestor
Catalyser	–	Everybody is helpful

Table 8.6 'Strong bond' symmetrical relationships among roles

Helpfulness	Role	Role	Description
Most helpful	Mister G	Innovestor	Both operate with degrees of freedom in public and private sectors to find a monetized soil and regrowth. Innovestor gives Mister G a simple financial incentivization and helps him set a quota to sustain. Both use AI as 'part of the system' and can measure results.
Least helpful	Professor	Telomerge	They have a cosmological or religious incompatibility. Telomerge is merging with biotech to sustain the AI-driven ecosystem, while the Professor aims to keep new life from being synthesized and does not want living beings to be artificially changed.

Table 8.7 'Weak bonds' - roles with another least helpful role in common

Roles perceiving a third role as least helpful	*Role they perceive as least helpful*	*Description of the relationship among these roles*
Professor, Innovestor	Link	The Professor and Innovestor are both interested in promoting innovation and see Link's lobbying for the 'intrinsic value of nature' to be counter to their aims.

Of the assemblage of roles, the two 'strong bond' relationships were identified, one 'most helpful' and one 'least helpful'. The most helpful combination is between two roles who are most deeply connected to AI and most concerned with the direct production of economic value – Mister G and Innovestor. The two roles who found each other to be 'least helpful' have cosmological incompatibility concerning how to best serve living nature – the Professor and the Telomerge. The Professor is motivated by past conceptualizations of the value of 'natural nature', while Telomerge takes new forms of AI-supported nature as a given and lives a life of risk and danger to produce new rich data to enrich the ecosystem's data soil.

Only one weak bond was found in this assemblage. The two roles Professor and Innovestor found the third role, the activist Link, to be 'least helpful' to their own interests in the situation. Link's promotion of alternative forms of value runs contrary to the innovation motivations of creating new, conventional forms of value. Link's activities are not understood by them.

The modelling of these most helpful and least helpful relationships revealed to the gameplayers the divergent and convergent interests of their roles in the situation. Again, this step served to add detail to the story of a future they were collectively imagining. Because these details pertain to the relationships among the roles, it indicates the discordant and concordant dynamics among them.

4.2 Discussing the Underlying 'Ethical Stuff' of the Game

After all of the above details were dynamically generated by the gameplayers, the gameplayers took a moment to discuss what 'ethical stuff' was underlying their game. This discussion served as a bridge to the next step of constructing a new BioEthos for their assemblage of roles. The discussion was also an instance of the Participatory Action Research principle of participants authentically joining in the inquiry work of a research effort and engaging in interpreting the outputs they produced and their experiences playing the game. They covered a wide range of topics and produced a variety of insights, many of which our BioEcoJust research team probably would not have arrived at on our own.

In their discussion the gameplayers highlighted various ethics-related aspects of the game, including ethical tensions and themes. They noticed some kind of 'purity ethics' challenging an inherited system of AI-driven eco-restoration and management. They noted the negative potentials of such ethics, as evidenced in today's world as variations of 'purity ethics' are being invoked to advance racist, anti-immigrant policies.

They questioned the past decisions which had led up to this *recoding life* situation. The participant who created Link observed 'It seems like there was a point where ethics weren't there, and then suddenly things have gotten a bit out of hand and now people are debating things that 50 years previously actually got this change happening in the first place'. This inherited lock-in from the decisions made by past people raises questions for the people of 2075 about what rights the new species generated from the AI-driven interventions on nature deserve and how to consider the consequences of prioritizing the needs of 'modified nature' or 'natural nature'.

The gameplayers noted a feeling of 'anomie' – no name and something that has no existence – in the ethical bearings of their game which they further associated with the concept of 'nomose' from sociology, the socially constructed norms generated from society that generate a framework for how to be and act (ala Berger 2011 [1967]), and the feeling that the ethical norms 'we knew before are gone, but the new thing hasn't snapped into place' which leads to asking 'what's going on?'

The group noticed traces of historically unethical and damaging concepts such as manifest destiny in how people feel an entitlement to go to the next level of hybridizing and creating new life. They asked: 'Where is the point we stop being human?'

and 'Are concepts of destiny and pre-destiny' driving the rise of this AI-ecosystem sentry?

Some roles viewed the AI as something that 'moves in mysterious ways and it is good' ascribing it something of God-like qualities. This view raises a tension between an 'ethics of knowing the right thing to do and doing the right thing' and an 'ethics of accepting the right thing from another [source]' – in this case, an AI. The gameplayers wondered if there could be a 'flourishing ethics' instead of a 'things turn out right ethics' in their roles' situation. They also noticed a trend towards 'decision-making based in knowledge instead of wisdom' and a lack of humility concerning 'nature knowing better' – like when engineering seeds for specific soils and outcome only to find few of the seeds will grow because 'the soil knows better'. Stemming from this point, they observed that 'humans tend to think someone or something else knows better' and the dominant role of AI in their game amplifies this view and serves as this 'something else'.

However, there is a tension between having the circumstances that enable people to develop capabilities to 'sense' and 'process many things' and 'a reliance on AI'. There is a risk humans would lose thinking capacity and other skills – such as 'an ability to understand big data' – as we choose to give power to AI concerning the management of the ecosystem we inhabit. An alternative to losing such skills would be advancing a goal of more 'humans as catalysts'.

They highlighted how the roles Tiny Paula, Telomerge, and Catalyser challenge the notion of being human: Tiny Paula is modified to be tiny, Telomerge is changing towards some form of dematerialization, and Catalyser is human, but somehow ethereal. There is a blurring of what it means to be a person. The gameplayers wondered if it is the 'perennial issue of *the other*' and how to ethically approach interrelations across difference. Perhaps the roles rely on the administrative functions of Mister G to keep the peace.

One participant imagined there could be an 'ethics animated by disgust' in response to the combining of elements that 'shouldn't be combined'. For example, a Telomerge's parent would be very worried for their child when they announce, 'I will become a Telomerge – there is a risk I will die, and it will be painful'. Yet Telomerges exist in this world of 2075. What ethical stance enables a Telomerge's parents to prioritize a goal of sustaining the AI-driven ecosystem and the planetary well-being it is believed to provide above their concern for the well-being of their own child? Could there be an *awe-driven ethics* driven by 'feeling small in the face of nature', even artificially maintained nature, that is an alternative to *an ethics animated by disgust?*

One gameplayer observed that 'when the many different points of view are brought together, it produces a lot of wisdom', suggesting a wiser ethical position is arising from the combination of their views. This observation, while targeted to the other gameplayers as part of the game session, indicates the value of this role-based futuring game in how it can surface productive discords and confluences of framings when considering transformation and potential ethical dilemmas.

4.3 Formulating and Naming an Emergent BioEthos

Building from their discussion about the underlying ethical dynamics of their game, the group was then asked to develop a new BioEthos which would be useful to their assemblage of roles in their 2075 situation where ethical challenges arise from the fact that whole ecosystems are managed by AI and many species already have been altered by it while many more will be. The facilitator explained this new BioEthos is meant to help their roles have a 'good' effect on the relationships represented by the three lines of the human-nature-technology triangle: humans and nature, nature and technology, and humans and technology. After discussing the dynamics of each side of the triangle, the gameplayers created a BioEthos they titled 'Status of Beautiful Monsters'.

The word 'Monsters' in the title of their BioEthos captures how the AI-driven ecosystem of their situation is simultaneously regarded as either a 'monstrosity going past all appropriate limits' and 'a beautiful integration helping people get passed all of the terrible limits' given to them by people's past actions. The word Beautiful conveys the need to appreciate the diversity of life forms and persons existing in the situation. The word Status indicates the need for equality and respect to all living beings, whether they've been modified by the AI-driven ecosystem management system or are unmodified nature. It means equality and respect to all persons, whether they are modified like Tiny Paula, are going through a dematerializing transition like the Telomerge, are somehow ethereal like Catalyser, are AI-augmented like Mister G, or are some kind of 'AI swarm as person' like Innovestor.

In this new BioEthos, the relation between *humanity and nature* are simultaneously 'belonging and transcending'; the relation between *humanity and technology* is producing a 'sensation amplifying connectedness'; and the relation between *nature and technology* leads to new forms of 'bio-discernment'. There is an expanded and inclusive definition of nature, the status of person stretches to include new beings, and the scale and ambition of technology raises the question of hubris – 'humans put nature in a box and are manipulating the heck out of it'.

Connectedness is a key aspiration of this BioEthos. Technology is providing people an amplification of 'sensory experiences and sensory interpretations'. It is a 'drug-induced oneness with other people in nature' that feels like 'raves in forests'. Some professions of people can 'read nature and comprehend it' and 'share experiences of being anything in nature'. These connections are valued.

This new BioEthos entails a desire for 'belonging' and 'transcendence', for being a part of the ecosystem, yet being able to go beyond it. Related to this desire, a more expansive diversity of life forms and persons makes the principle of equality of high importance. New definitions of nature are sought, ones that are 'adding instead of narrowing and limiting' and are 'inclusive'. People are better served by an expanding definition of nature to include these levels and new forms of ways of being alive. In light of collectively challenging experiences the inhabitants of Earth endured in the 2030s and 2040s, there is a new form of 'discernment' that is a 'more mature judgement about how these boundaries work or could be broken'.

This situation is such that 'human and technological monstrosities against nature are defining nature'. These monstrosities drive some actors, like the Professor to prevent or undo changes to 'natural nature' which is 'the origin of humanity, what is our essence'. In this situation, it is a strong 'ideological position' to go 'back to the core of what is a human'. However other roles like Tiny Paula – who herself is genetically modified – accept that their local 'nature is modified' and that 'we need to adapt to conserve it'. Tiny Paula's priority is to be 'involved in the new society' that is being created by their current circumstances and to promote equality.

Reflecting this interlink between nature and humanity, the group's new BioEthos calls for a 'resurgence of humanism' that includes an expanding diversity of persons and non-human life. 'There is a place for everyone'. All life forms, modified or natural, have intrinsic value. All humans are still human, even if they are augmented by AI or specialized senses, modified, or undergoing transformation. This new BioEthos also recognizes the 'status of new beings' which have become possible due to the technological environment, beings like Innovestor. Due to Innovestor's relationships to the other roles and the situation, she is treated with equity and 'has the same resources in a universe that treats everyone equal'. In the 'Status of Beautiful Monsters', all are 'beautiful monsters', and status is afforded to all as an act of 'leveling' and 'moving away' from status-seeking. There is simply a 'respect for being', no matter what natural or technological processes have produced you. 'We are not going to be assimilated'; instead all 'coalesce'.

The 'Status of Beautiful Monsters' BioEthos calls on people to recognize 'they are in a transition' and 'get used to that, instead of getting stuck in the past' because 'the transition is the important thing'. The roles 'don't assume this is an end point', 'terminus', or 'end of the line', and they recognize they are 'on a path to someplace else'. The AI-managed ecosystem is such that 'everyone is coming and going, and there's an infrastructure'. We are all only ever passing through. In this BioEthos, there is a hope that people 'could feel – if they wanted to work together – [they] could be helpful towards a happy outcome'.

The BioEthos produced by these participants in this game session involving their uniquely created roles serves as a crude model for how individuals interacting as assemblages can generate (and regenerate) a collective ethical orientation towards a specific ethically difficult situation. Based on the outcomes of this game session, we see potential for the game to help people notice, comprehend, and create new BioEthos and appreciate their potentials.

5 Why the BioEcoJust Game Is Relevant to Understanding Existing and Potential Ethical Challenges in the Bioeconomy?

Much of the literature regarding bioeconomy is concerned with the technical or social factors which can contribute to its development. Yet probing the emergent, multiple, and varying, yet sometimes overlapping, BioEthos as conceived as a complex adaptive system generated from assemblages of people and ecosystems could contribute useful insights regarding the ethical choices we are making today.

Our experimentation with the BioEcoJust game, as exemplified in the Mexico City pilot, demonstrated how a futuring game can be coupled to a stance towards reality that combines complex adaptive systems and assemblage theory (Spies and Alff 2020), emphasizing the dynamics of being embodied as a person in relation to situations that arise from complexity. We found that this coupling of our game design to this kind of ontology produced a platform for rich discussions concerning how ethical norms concerning the bioeconomy could form. The game demonstrated it can enable people to meaningfully explore the interface between assemblages of individuals and the bounding conditions surrounding an ethically challenging situation. The game's intent was not to produce some mechanical system model that reveals leverage points but was rather to enable people to experience engaging this interface between assemblages and their bounding conditions. As a result, the game-players acquired and practised skills useful for formulating new and applicable forms of ethicality to meet possible arising challenges which could happen during the rise of a global bioeconomy in the context of GWEC.

There are two aspects of the game we believe are worth developing further. The first is the placing of emphasis on individual role-driven generation of details to a seed scenario or seed future situation. This emphasis is quite different from the standard fare of scenario creation in which participants are encouraged to take an outsider-objective view of a shared world and describe its attributes. In a role-driven generative creative process, like used in the BioEcoJust game, the details the participants contribute instead are coming from the most basic element of a human system – individual people. An advantage of starting from an insider-subjective view is that diversity and dynamics among unique persons with unique perspectives are more similar to how people experience real situations when they occur. The second aspect to develop further is how to 'keep whole' the worlds created by participants. The role-driven creativity processes of the BioEcoJust Game never coerced participants to collapse the richness of their ideas into some consensus view, preserving the richness of the details they have contributed to the situation. Yet, inter-subjective meaning making also contributed new and surprising details. This formula of respecting individual creations while bringing those creations into interaction with the other participants to catalyse surprising details can serve as a model for other futuring games designed to sense and make sense of emergence.

Because it can help people notice and struggle to interpret novel ideas, we consider the BioEcoJust game to be an example of a class of futuring games designed to fill methodological gaps in the field of futures studies which are identifiable via the Futures Literacy Framework (Miller 2018). In the field of futures studies, nearly all of our methods and tools can be categorized as systematized ways to *anticipate for future*. Few if any are designed to support people in *anticipating for emergence*. It is our view that more such tools are needed in our field and would serve to help people practise struggling to make sense of novelty so that they are more able to do so in the present and respond in accordance to the demands of arising situations. Role-taking and perspective taking as game design features have shown themselves to be conducive towards building simulations of emergence within which to practise *anticipation for emergence*.

To connect all of this back to the development of bioeconomy, practising encounters with emergence – especially concerning the ethical dimensions of bioeconomic development – can contribute to more impactful and ethically tuned experimentation and innovation by bioeconomy actors to address GWEC.

6 Conclusion

The BioEcoJust game pilots demonstrated its potential to support people in richly imagining an assemblage of future people who could exist later than now and engage in elaborating and exploring an ethically troubling bioeconomy situation. The imagined futures generated by the players stemmed from individual worlds that are stitched together by their relationships, allowing for their shared situation to become a tapestry of details, uniqueness, and diversity. The richness of this tapestry was driven by a game design that adheres to the principles of *futuring as worldmaking* (Bendor 2017; Vervoort et al. 2015) which allows for participants to imagine people and keep their worlds whole as they engage in a common situation through gameplay. The utilization of a 'scenarios as worldmaking' approach allowed for the gameplayers to keep whole their created roles while simultaneously engaging with each other to produce unique relationally driven details for their roles. We have demonstrated that the BioEcoJust game is a productive way to support discussions which help people sense and make sense of emergent BioEthos and generate novel insights about possible developments of the bioeconomy. While these specific insights may not be the ones which ultimately prove relevant to situations which arise from developments of bioeconomy in the ongoing context of climate change, the gameplayers gained practice in struggling to sense and make sense of emergent BioEthos and reflect upon the suitability of their own BioEthos in the present.

By focusing the game design towards producing a nuanced tapestry of role-driven details, participants were enabled to use their roles as set of lenses through which they see the future situation from multiple perspectives. It went further and provided practice in the skills of noticing emergent BioEthos and developing one which could be relevant and helpful to their specific assemblage. In a future simultaneously filled with radical difference and commonalities to the present, the BioEcoJust game fostered participant awareness about their own BioEthos while embarking on explorations concerning what kinds of BioEthos may be emerging or needed in the future. The value of such an experience of being aware of present and potential future BioEthos is not to pick one to implement and another to avoid, but rather to tune into how a multiplex of co-existing ones in our world are changing and could need to change to fit the people involved in the specific challenging situations they face.

Such skills and approaches to comprehending and appreciating a multiplicity of varying ethical framings would be of high value to all levels and sectors of actors working to develop the bioeconomy. These skills could help decisionmakers and the wider public be quicker to notice new formations of BioEthos and create customized ones to fit unforeseen troubling situations as a way of coordinating a collective response. We propose the BioEcoJust game is just such a tool, and more futuring games like it should be developed to help the people of our world cope with all of the varied, dramatic, and locally specific difficult situations that are likely to arise during the long-term period of human-driven GWEC in which we live.

Acknowledgements The authors wish to acknowledge the participants who played BioEcoJust game at the World Futures Studies Federation Conference 2019 in Mexico City; Matti Häyry, principle investigator of Bioeconomy and Justice Project (BioEcoJust); Markku Wilenius, leader of the foresight work package of BioEcoJust; and Sofi Kurki, our BioEcoJust colleague who has contributed thought leadership to this project. The BioEcoJust research project upon which this chapter is based is funded by Academy of Finland as part of its BioFuture 2025 program (2017–2021, https://www.aka.fi/BioFuture2025).

References

Balcom Raleigh, N. A., & Heinonen, S. (2019). Entangling and Elevating Creativity and Criticality in Participatory Futuring Engagements. *World Futures Review*, *11*(2), 141–162. https://doi.org/10.1177/1946756718807014

Balcom Raleigh, N. A., Pouru, L., Leino-Richert, E., Parkkinen, M., & Wilenius, M. (2018). *Futures Literacy Lab for Education: Imagining Complex Futures of Human Settlements at Finland Futures Academy Summer School 2017*. Turku, Finland: Finland Futures Research Centre – University of Turku. http://urn.fi/URN:NBN:fi-fe2019052016107

Bendor, R. (2017). Interaction design for sustainability futures: Towards worldmaking interactions. In *Digital Technology and Sustainability: Engaging the Paradox* (pp. 205–216). https://doi.org/10.4324/9781315465975

Berger, P. L. (2011). *The Sacred Canopy: Elements of a Sociological Theory of Religion*. Open Road Media. https://books.google.fi/books?id=WcC-AYOq6Q4C

Berzonsky, C. L., & Moser, S. C. (2017). Becoming Homo sapiens sapiens: Mapping the psycho-cultural transformation in the Anthropocene. *Anthropocene*. https://doi.org/10.1016/j.ancene.2017.11.002

Cook, J., Nuccitelli, D., Green, S. A., Richardson, M., Winkler, B., Painting, R., et al. (2013). Quantifying the consensus on anthropogenic global warming in the scientific literature. *Environmental Research Letters*, *8*(2). https://doi.org/10.1088/1748-9326/8/2/024024

Geldmann, J., Joppa, L. N., & Burgess, N. D. (2014). Mapping Change in Human Pressure Globally on Land and within Protected Areas. *Conservation Biology*, *28*(6), 1604–1616. https://doi.org/10.1111/cobi.12332

Goodman, N. (1978). *Ways of worldmaking*. Indianapolis, Indiana: Hackett Publishing Company Inc.

Hagens, N. J. (2020). Economics for the future – Beyond the superorganism. *Ecological Economics, 169*(June 2019), 106520. https://doi.org/10.1016/j.ecolecon.2019.106520

Hamann, M., Berry, K., Chaigneau, T., Curry, T., Heilmayr, R., Henriksson, P. J. G., et al. (2018). Inequality and the Biosphere. *Annual Review of Environment and Resources, 43*(1), annurev-environ-102017-025949. https://doi.org/10.1146/annurev-environ-102017-025949

Heinonen, S., Raleigh, N. B., & Karjalainen, J. (2015). *CLA Game Report: Causal Layered Analysis Game on Neo-Carbon Energy Scenarios*. Turku: Finland Futures Research Centre – University of Turku.

Heinonen, S., & Ruotsalainen, J. (2013). Futures Clinique—method for promoting futures learning and provoking radical futures. *European Journal of Futures Research, 15*(7), 1–11. https://doi.org/10.1007/s40309-013-0007-4

Lakoff, G., & Johnson, M. (2003). *Metaphors we live by*. Chicago: University of Chicago Press.

Latour, B. (2005). Third move: Connecting sites. *Reassembling the social: an introduction to actor-network-theory*, (2005), 219–246.

Louie, A. H. (2010). Robert Rosen's anticipatory systems. *Foresight, 12*(3), 18–29. https://doi.org/10.1108/14636681011049848

Masson-Delmotte, V., Zhai, P., Pörtner, H.-O., Roberts, D., Skea, J., Calvo, E., et al. (2019). *Climate Change and Land An IPCC Special Report on climate change, desertification, land degradation, sustainable land management, food security, and greenhouse gas fluxes in terrestrial ecosystems Head of TSU (Operations) IT/Web Manager Senior Administrator.*www.ipcc.ch

Miller, R. (2015). Learning, the Future, and Complexity. An Essay on the Emergence of Futures Literacy. *European Journal of Education, 50*(4), 513–523. https://doi.org/10.1111/ejed.12157

Miller, R. (2018). Sensing and making-sense of Futures Literacy Towards a Futures Literacy. In R. Miller (Ed.), *Transforming the Future* (pp. 15–36). Abingdon, Oxon ; New York, NY: UNESCO and Routledge.

Padel, R. (2010). *Silent letters of the alphabet. Newcastle/Bloodaxe poetry series*. Newcastle upon Tyne, [England] : Tarset, Northumberland, [England]: Dept. of English Literary & Linguistic Studies, Newcastle University ; Bloodaxe Books.

Spies, M., & Alff, H. (2020). Assemblages and complex adaptive systems: A conceptual crossroads for integrative research? *Geography Compass*. https://doi.org/10.1111/gec3.12534

Taylor, A., Balcom Raleigh, N. A., Kurki, S., Ferreira-Aulu, M. B., & Wilenius, M. (2019). *Precursors to a 'Good' Bioeconomy in 2125: Making Sense of Bioeconomy & Justice Horizons: First Foresight Report of the BioEcoJust Project*. Turku, Finland: Finland Futures Research Centre – University of Turku. http://urn.fi/URN:ISBN:ISBN 978-952-249-521-1

Vervoort, J. M., Bendor, R., Kelliher, A., Strik, O., & Helfgott, A. E. R. (2015). Scenarios and the art of worldmaking. *Futures, 74*, 62–70. https://doi.org/10.1016/j.futures.2015.08.009

You, H., & Yang, X. (2017). Urban expansion in 30 megacities of China: categorizing the driving force profiles to inform the urbanization policy. *Land Use Policy, 68*(18), 531–551. https://doi.org/10.1016/j.landusepol.2017.06.020

Chapter 9
Bioeconomy in Maturation: A Pathway Towards a "Good" Bioeconomy or Distorting Silence on Crucial Matters?

Sofi Kurki and Johanna Ahola-Launonen

Abstract The bioeconomy as an emerging research field and policy framework has raised high expectations for enabling a shift to more sustainable practices. However, many of the solutions promoted under it have been heavily criticized for a lack of concern regarding the systemic effects in both environmental and social sustainability. In this article we analyse the differences between "1st round" bioeconomy policies and the revisions that have arisen from the critique ("2nd round bioeconomy policies"). We compare the two consecutive bioeconomy policy frameworks to views presented by a panel of Delphi experts. The experts elaborate on their views about a "good" and "bad" bioeconomy futures, with a long-range timeframe until 2075. The results indicate that the first round of bioeconomy policies contains many of the elements that the experts see as leading to an undesirable future. In contrast, the experts envisioned a "good" bioeconomy which would be based on a just and inclusive transition, a changed economic paradigm moving away from the focus on material growth, and a multitude of sustainable technologies, lifestyle changes, and balanced relations between business and politics. In the second round of bioeconomy policies, many of the issues addressed by the critique have been taken up, but problematic areas remain in the policies somewhat untouched. These include, amongst others, the question of biomass use for energy. We conclude that the bioeconomy finds itself now at an intersection between the old recommendations and novel, more inclusive goals. Are the expert panel's views indicative of the directions where national-level policy implementation is taking the bioeconomy? If not, how will the bioeconomy policies resolve the most burning critiques in relation to the overreaching policy goals to combat climate change? We argue that what happens

S. Kurki (✉)
Finland Futures Research Center, Turku School of Economics, University of Turku, Turku, Finland
e-mail: sofi.kurki@utu.fi

J. Ahola-Launonen
Department of Management (Philosophy of Management), Aalto University, Helsinki, Finland

E. Koukios, A. Sacio-Szymańska (eds.), *Bio#Futures*,
https://doi.org/10.1007/978-3-030-64969-2_9

in the next phases of bioeconomy policy implementation process will be critical for the fate of the entire bioeconomy project.

Keywords Bioeconomy · Circular economy · EU policy · Biomass · Delphi analysis · Future imaginaries · Technological expectations

1 Introduction

The bioeconomy rose on the policy agenda thanks to offering a set of promises that are tempting from various societal viewpoints. At the core of bioeconomy is a proposition that renewable resources could become the novel basis of the economy. This, on one hand, suggests that the bioeconomy has the potential to generate economic growth in a way that solves humanity's most pressing problem, climate change, which is strongly linked to fossil fuels. On the other hand, the bioeconomy in Europe has been seen as an opportunity to strengthen existing traditional industries, such as agriculture and forestry. Thus, the bioeconomy has brought a new kind of hope to the future prospects of the sectors and the people they employ. These expectations for the bioeconomy – fighting against climate change by substituting the fossils by renewables, creating economic prosperity as green growth, and securing the future of socially important sectors – are in line with the traditional representation of sustainable development through its three dimensions. However, there has been strong criticism towards the raised expectation horizons from the perspective of how attainable they truly are through the measures proposed by the initial bioeconomy program. Especially the gap between the goal of mitigating climate change and reaching it with the solutions offered by the bioeconomy, when considering the real environmental impact of the proposed measures, has been pointed out in the critique.

The critique could be expected considering the status of bioeconomy as an emerging research field and policy framework. As scholars in science and technology studies (STS) have argued, the hopes related to new technologies typically follow a certain pattern of dynamics of expectations.[1] The first steps of an innovation come with a vision, a story situated in the future, where the particular technology has solved problems or made life better. The first vision almost necessarily is filled with intentional exaggeration to get the needed attention, interest, and investment. This hype is built in intertwined networks of investors, regulators, policymakers, innovators, and consumers that form 'communities of promise'. When time passes, circumstances change. Problems evolve, and varying levels of

[1] See, e.g. Brown (2003), Brown and Michael (2003), Borup et al. (2006), Goven and Pavone (2015), Petersen and Krisjansen (2015), Van Lente (1993)

disillusionment drown the early hopes Members of the communities of promise reorient themselves to new visions, untarnished by disappointments.[2]

Politically created expectations can be described as normative images of the future that aim at bringing about futures that are framed as desirable. They are at their most effective at the stage when they address an emerging issue. This is, however, also the stage when uncertainty is highest and the actual effects of the created policies most uncertain. The hypes and hopes themselves are not morally neutral, but strongly normative. They are performative in the sense that even though the visions take place in the future, they influence our present. Visionary hypes grab attention and this has multiple effects. The field of technological innovation is highly competitive, and there is a race for investment. The best and most influential visions will most likely get support, acceptance, and resources. In this way, the most appealing hype suppresses other options. The agendas that are most novel are able to build up the most ambitious visions. This is because mature research fields have already gone through the cycle of disappointment and disillusionment. They appear mundane, familiar, and more boring; and therefore, they are doomed to lose in competitive funding and attention. It is easier to get excited with something completely new that promises radical easy-sounding answers. The biggest hypes tend to live only in the newest visions.[3]

In this article, we discuss the transitional period between two "rounds" of bioeconomy policies. The first round of such policies was initiated in 2006, and as discussed, it was characterized by an optimism regarding the potential of renewable resources to act as a basis of a new bioeconomy that could directly substitute the prevailing fossil economy. The policies and the expectations created on the first round received heavy criticism due to questionable assumptions regarding environmental and social sustainability. The updated EU-level bioeconomy policies (from around 2017 onwards) have responded to some of the key points brought up in the critique. However, the way EU-level bioeconomy policy documents are interpreted and implemented on the national level is an ongoing process, and thus the futures of bioeconomy policies remain a somewhat open question. Until the national policies are updated, the first-round policy recommendations are in effect and continue to affect the infrastructure and investment decisions that by themselves create continuity and development trajectories for the future. Also the global directions of bioeconomy present a major source of uncertainty for the assessment of bioeconomy futures.

In order to make sense of broader directions for bioeconomy futures and policies, we have analysed answers to a Delphi study exploring long-range futures of bioeconomy. In this study, we are especially interested in the Delphi panelists' evaluations of desirable and avoidable bioeconomy futures and their interlinkages to the first and second round of bioeconomy policies. The updated bioeconomy policies

[2] Brown (2003), 5–6; Deuten and Rip (2000)

[3] Brown (2003), see also Brown and Michael (2003), Borup et al. (2006), Deuten and Rip (2000), Goven and Pavone (2015), Petersen and Krisjansen (2015), Van Lente (1993)

are analysed against the long-range visions of the experts: what would the experts suggest for further measures and goals in order to reach the desired future bioeconomy? Experts' views regarding "good"[4] bioeconomy futures allow for an evaluation of the policies from the perspective of shared visions about what the goals of the policies should be. What they bring up reveal potential blind spots of policy documents: what kinds of topics are easy to include as responses to critique, and are the issues that are left unsaid or ambiguous in the policy guidelines?

In the conclusions we discuss our findings from the perspective of a theory of futures images.[5] They are a way to understand how individuals approach the futures through culturally shared bundles of concepts, values, and aspirations. Polak (1973) stresses the civilizational nature of the images. In this view, conceptions of futures stem from broader notions arising from a civilization's deep history and myths. Thus, one way to interpret the theory of images of the future is to see them acting as collectively shared filters for the kinds of narratives that are viewed as possible or plausible within a certain culture. In our conclusions we trace back our findings to a unifying narrative of the future that makes certain assertions more plausible and leaves others out from official representations.

With this study, we aim to contribute to several domains. First, we aim to provide actual guidance to policymakers and analysts as to what could be the potentially important future directions and blind spots in bioeconomy and climate policies that are in need of attention. Secondly, we wish to contribute to bioeconomy studies by taking part in the first wave of analyses of the updated bioeconomy policies. Because the updated bioeconomy policies are somewhat new, and academic publishing can be relatively slow, the updated bioeconomy policies are yet to be included in policy analyses on a large scale. Finally, our study, as a combination of policy analysis, science and technology studies, and future studies, is offering an example of a policy framework and research field in a process of transition and maturation.

2 Methods and Data Gathering

We took key documents from EU bioeconomy "first-round" policies to include OECD's *The Bioeconomy to 2030: Designing a Policy Agenda* (OECD 2006), followed by *The Bioeconomy to 2030. Agenda* (OECD 2009) and *Bioeconomy Strategy* in 2012 (EC 2012). The selected key documents for analysing the updated "second-round" policy were *Review of the 2012 European Bioeconomy Strategy* (EC 2017), *Updated Bioeconomy Strategy* (EC 2018), and *Realising the Circular Bioeconomy* (OECD 2018). Furthermore, the role of the bioeconomy in climate measures was

[4] Quotations are used in this article for "good" and "bad" to signal reservations about using such simplistic terms for complex and multifaceted, value-laden issues.

[5] Polak (1973), Bell (1996)

reflected by looking at EU-level general climate policies *A Clean Planet for all* (EC 2018) and *The European Green Deal* (EC 2019).

The first policies have been under a thorough critical analysis as it has been over the decade since their publishing. Due to the large number of existing analyses, one can find saturation in the analysis of different aspects of the data, and thus it was decided to partly rely on the existing analyses in reviewing them. However, as mentioned, policy documents since 2017 have received much less attention. For this study, these documents were qualitatively analysed by reflecting them from two angles: on the one hand, by viewing the updated documents in light of the main critical themes of the analyses of the first round and, on the other hand, by elaborating them against the Delphi experts' insight of "good" and "bad" bioeconomy.

The aim of the Delphi questionnaire was to provoke experts' thinking about long-term futures regarding bioeconomy developments. The Delphi method was originally developed for forecasting purposes in situations where trend forecasts and mathematical modelling were inefficient in providing useful information about the future.[6] Over the years, variations of the Delphi method have abandoned the forecasting aim, replacing it with a goal to foster fact-based argumentation about possible futures.[7] Consequently, the method has become one of the most popular methods in foresight and futures studies, and it has been adopted across different disciplines for assessing future developments by expert panels.[8]

The Delphi questionnaire was open to answering between March 27 and April 8, 2019. Key informants were selected to the panel by using an expertise matrix. The matrix was used to make sure that different viewpoints were represented in the invited panel. The questionnaire was completed by 64 experts representing 12 countries. Out of the 64 respondents, 52 came from countries within the European Union. Despite the use of the expert matrix for inviting the experts to join the study, the final makeup of the Delphi panelists answering the questionnaire was skewed towards more representation from researchers (17 of the respondents were employed by university or college and 11 by independent research institutes) and less from businesses (3 were employed by large companies and 3 by a start-up or SME). Also employers of governments or governmental agencies were strongly represented in the sample (10 respondents). Fields of science represented included social sciences, forestry, agriculture, philosophy, environmental sciences, innovation studies, economics, comparative literature, political history, and geography.

For the purpose of this study, we analysed the selected Delphi answers thematically and coded and grouped the themes that emerged from the data. The driving research question in the coding was to identify elements that the experts argued to be characteristics of a good and a bad bioeconomy and what they perceived as ways to reach them.

The data and methods of the analysis are presented in Table 9.1.

---

[6] Kuusi (1999)

[7] E.g. Tapio (2002)

[8] Linstone and Turoff (1975)

Table 9.1 Summary of the data and methods

Type of data	Method	Discussion
1st round bioeconomy policies[a] and related policy analysis[b]	Identification of key criticisms to 1st round bioeconomy policy	Are we approaching a "good" bioeconomy with the new policies?
Updated bioeconomy policies[c]	Qualitative analysis in light of the main critical themes of the analyses of the first round	
Selection from a Delphi questionnaire made to bioeconomy experts	Thematic grouping of elements of the experts' views on "good" and "bad" bioeconomy and key societal drivers to reach them	

[a]OECD (2006), OECD (2009), EC (2012)
[b]Main references include Bugge et al. (2016), Kleinschmit et al. (2017), Kröger and Raitio (2017), McCormick and Kautto (2013), Mittra and Zoukas (2020), Pfau et al. (2014), Ramcilovik-Suominen and Pülzl (2018), Staffas et al. (2013), Varho et al. (2018)
[c]EC (2017), EC (2018), OECD (2018)

3 The Rise of Bioeconomy Policies

The use of the term bioeconomy, or bio-based economy, started to increase in 2006, replacing its predecessor biotechnology as the buzzword in policy papers.[9] A key document that made the term popular was OECD's *The Bioeconomy to 2030: Designing a Policy Agenda* (OECD 2006), followed by *The Bioeconomy to 2030. Agenda* (OECD 2009). The European Union launched its *Bioeconomy Strategy* in 2012 (EC 2012).

The bioeconomy as a policy framework can to a large extent be explained through a double fold aim to both solve grand global challenges and at the same time produce economic growth in the spirit of knowledge-based economy and biotechnological revolution. Thus, it re-packages already existing programs promising growth and well-being through biotechnological and knowledge-based solutions, with aims to produce a sustainable transition in the world's economic and production systems. Such a combination is served as a manifold win-win strategy as reducing dependence on oil, tackling environmental challenges and climate change, making production and manufacture sustainable, creating jobs and new industry, and ensuring food and energy security. With a growing economy and diminishing environmental problems, it promises also to increase human well-being.[10] The cornerstones of this novel economy are biomass (e.g. from forest, field, sea, waste), innovation, and

[9]E.g. EC (European Commission), 2002. Life Sciences and Biotechnology: a Strategy for Europe; COM, vol. 27. Brussels, Belgium, 2002

[10]*A bioeconomy can be thought of as a world where biotechnology contributes to a significant share of economic output* (OECD 2009*: 8*).

*The bioeconomy's cross-cutting nature offers a unique opportunity to comprehensively address inter-connected societal challenges such as food security, natural resource scarcity, fossil resource dependence, and climate change while achieving sustainable economic growth (EC*2012*, 9).*

industrial biotechnology. Expressed in a most condensed form, in a bioeconomy, various biomasses are used to substitute fossil resources and to create added value to the economy.

3.1 Critique of the First EU Bioeconomy Policies

The inbuilt contradictions in a project merging goals of economic, environmental, and social interest have made the bioeconomy a target of a vivid academic discussion and review. Key questions include whether there is enough biomass for substituting for most fossil resources, along with all the other suggested uses for bio-based materials. What are the sources of this biomass? Who would the promised bioeconomy benefit, all or only the few? Who are included in decisions regarding these questions? Is the bioeconomy the only possible answer to the specific challenges it is claiming to solve? Two main streams of critique can be distinguished: the insufficient and immature discourse on sustainability and the costs arising from the one-sided, technology-oriented hype around bioeconomy. The summary of the critiques is presented in Table 9.2.

Table 9.2 Summary of first-round bioeconomy policy and its critique

1st round bioeconomy policy and its critique	
Dominating economic dimensions	Production, economic growth, and employment Aim to replace all currently used non-renewable resources with biomass
Lack of environmental and social dimensions	Undue optimism of available biomass (import, production increase, unused potentials and residues) No definitions or tools for measuring and assessing sustainability No critical views about consumption and material growth No concern for negative consequences of increasing biomass production, harvest, or import No concern for equity, social justice, local development, human rights
Harmful hype	Unjustified and speculative expectations Attention, resources, and policies directed to unrealistic and narrow technological fixes at the cost of socio-political research and interventions

The bioeconomy encompasses the production of renewable biological resources and their conversion into food, feed, bio-based products, and bioenergy. It includes agriculture, forestry, fisheries, food, and pulp and paper production, as well as parts of chemical, biotechnological, and energy industries (EC2012, 16).

3.1.1 Sustainability: Taken for Granted and Neglected

Despite the emphasis on sustainability as a key argument for the bioeconomy in the first place, the bioeconomy policies lack due consideration of whether the goals of bioeconomy can be reached in a sustainable manner. In fact, the policy documents offer very little when it comes to discussion or definitions of sustainability in the context of the bioeconomy.

The first-generation bioeconomy discourse in the policy documents is founded on a myth and expectation of boundless amounts of biomass awaiting for its harvest. Scientific and policy literature lists sources of unused biomass potentials and residues and identifies existing and hypothetical measures to increase the amount of biomasses with better harvesting activities.[11] Sustainability of the bioeconomy is taken as a given – forests and land can be used to replace fossil resources because natural resources are renewable by their nature. We need data and policies to guarantee sustainability, but it can and will be done.

However, critical examinations of the policies have pointed out that no tools for measuring and assessing the sustainability of bioeconomy are offered.[12] Moreover, economic dimensions vastly dominate environmental and social dimensions in the sustainability discourse of the bioeconomy.[13] As a consequence, little attention has been paid to scarcity issues or management of natural resources or questions of justice[14]:

> The focus in the bioeconomy discussion is on production, economic growth and employment. Critical views about consumption and material growth are largely absent. All currently used fossil and other non-renewable resources cannot be replaced with biomass.[15]

There are various negative consequences to increasing biomass production, such as deforestation, losses of biodiversity, low water quality, land competition in food production, and large-scale land grabbing in areas such as Africa and Asia.[16] Local populations and small-scale farmers seldom have legal rights to their land, which intensifies problems related to the competition driven by large multinational companies in rural developing areas.[17] These problematic consequences have been linked especially to biofuel production[18] and cast a shadow on strategies relying on imported biomass.[19]

---

[11] E.g. EC (2012), Winkel (2017)

[12] Staffas et al. (2013), McCormick and Kautto (2013)

[13] Ramcilovic-Suominen and Pülzl (2018); Mittra and Zoukas (2020); Staffas et al. (2013), McCormick and Kautto (2013); Varho et al. (2018)

[14] Ramcilovic-Suominen and Pülzl (2018); Mittra and Zoukas (2020); Staffas et al. (2013), McCormick and Kautto (2013); Varho et al. (2018); Bugge et al. (2016); Pfau et al. (2014)

[15] Varho et al. (2018), 29

[16] Ramcilovic-Suominen and Pülzl (2018), 4178

[17] Ramcilovic-Suominen et al. (2010, 2013)

[18] Arevalo et al. (2014), Danielsen et al. (2008), German et al. (2010), Obidzinski et al. (2012)

[19] Ramcilovic-Suominen and Pülzl (2018), 4178

Absence of discussion on the environmental and social sustainability of the bioeconomy, especially in its most prevalent bioenergy stream, is evident to a degree that it has led authors like Ramcilovic-Suominen and Pülzl (2018) to ask whether sustainable development in fact is only greenwashing, used as a selling point for the entire EU bioeconomy policy framework. There are however at least two sources of confusion that may explain the tendency to underplay the discussion on sustainability: first is the lack of a clear definition of "the bioeconomy" itself. When different countries speak about the bioeconomy, they speak about different things, depending on the local natural resource that each country aims to utilize in its bioeconomy.[20] This complicates also discussions on sustainability. Secondly, most of the knowledge production related to the bioeconomy originates from within natural and engineering sciences and industrial biotechnology that all have a focus on production, efficiency, and innovation, rather than wider, systemic consequences of the bioeconomy.[21] In some related policies, sustainability even primarily stands for safeguarding high production levels.[22] Even though sustainability is an integral part of (forest policies), technocratic orientations tend to narrow down the outlook on sustainable development.[23]

3.1.2 Harmful Hype

The other main target of criticism regarding the first round of bioeconomy policies has been their unjustified and speculative value propositions and ideologies.[24] At its core, the bioeconomy promises to solve climate change and other major societal challenges with innovative use of biomass. As discussed in the STS literature, the costs of such simplistic narratives stem from two main effects: (1) hype can have the effect of directing the discussion to unrealistic and narrow solutions, and (2) they may result in neglecting a search for other kinds of approaches that could be achieved via, e.g. socio-political research and interventions.[25]

The bioeconomy has been framed as the answer to "the most serious of "global challenges" –hunger, ill-health, and, especially, ecological crisis".[26] For instance, the OECD (2009) envisions overcoming environmental, social, and economic challenges mostly with biotechnological progress. In a similar fashion, a sustainable bioeconomy has been introduced as holding great promise to contribute to a

[20] Staffas et al. (2013), McCormick and Kautto (2013)

[21] Bugge et al. (2016), Mittra and Zoukas (2020)

[22] Kröger and Raitio (2017)

[23] Kleinschmit et al. (2017)

[24] Mittra and Zoukas (2020), 11

[25] Brown (2003), Goven and Pavone (2015); see also Brown and Michael (2003), Borup et al. (2006), Deuten and Rip (2000), Petersen and Krisjansen (2015), Van Lente (1993).

[26] Goven and Pavone (2015); 307

transformation of the entire economic system, with the move away from fossil-based production and consumption.[27]

Such hyperbolic framings are amongst visions that are seen as exemplifying "the aspirational political vision of technoscience as the ultimate solution to society's problems".[28] Together they contribute to a view where "human and environmental disasters are averted because a particular political-institutional configuration facilitated the development of profitable technological solutions".[29] The problem arising from a forceful propagation of a powerful, one-sided future vision is that a detailed diagnosis of the root causes of any current global crisis goes missing. And in the absence of such a diagnosis, it remains doubtful whether biomass-based economy can actually solve them. If the bioeconomy cannot redeem the promises it makes, there will be costs to the hype of expectations.

The kinds of powerful policy visions used to build a sustainable bioeconomy, irrespective of whether their promissory visions are realistic, have an impact on research strategy and organizational practices. Thus, they have a performative function: "The hopes and expectations that are embedded within the reports of national and international policy institutions, governments, and commercial organisations are not simply rhetoric. They have a material impact on what areas of science get funded and what kind of research is valued".[30] If attention, resources, and policies are directed to bioeconomy as a technological fix, attention is pulled away from much needed socio-political answers that should at least be a part of the solution. When it comes to the bioeconomy, visions that lead to neglecting socio-political causes to climate change can have an enormous effect. The idea that a technology-based economy could solve a problem that has socio-political roots leads the discussion astray. If the main emphasis is on hoping for a technological solution whose adoption would "automatically" result in a society-wide transition towards sustainability, the risk is that such a way out is revealed to be an illusion, and the needed socio-cultural measures would arrive too late, if at all.

The conclusion from the costs of hype is a paradox. Bioeconomy needs the imaginative speculation and hype in order to harness the needed resources and attention for its implementation. However, if hoping and hyping go too far and distort the discussion at the cost of other relevant aspects, it is harmful.[31] Thus, bioeconomy should be able to walk the thin line of hopeful and holistic realism.

---

[27] Winkel (2017), 14

[28] Mittra and Zoukas (2020), 12; Doezema and Hurlbut (2017)

[29] Goven and Pavone (2015), 305

[30] Mittra and Zoukas (2020), 12; Brown (2003)

[31] Brown (2003), 17

3.1.3 Suggestions for an Ameliorated Policy Approach to Bioeconomy

Ramcilovic-Suominen and Pülzl argue that in order to meet any of the sustainability approaches available in policy and scientific literature (that encompass at least three dimensions: economic, environmental, and social), the (EU) bioeconomy policy framework must be reoriented to entail environmental (biodiversity, air, water, and soil quality), as well as social, aspects (equity, social justice, local development, human rights).[32] Critical studies unanimously agree that the future evolution of the bioeconomy would have to build on a broader sustainability concept, as so far environmental policy integration and sustainability have only been integrated at the surface into bioeconomy strategies.[33] To facilitate this aim, Pfau et al. (2014) stress that the bioeconomy should be approached in a more interdisciplinary or transdisciplinary way.

After the first round of bioeconomy policies, and the extensive wave of critical policy analysis literature addressing them, policy directions have shifted considerably to comply with the suggestions. Furthermore, these very questions about the sustainable use of biomass were raised shortly after or during the first round of bioeconomy policy documents in background papers and workshops, inside policy. These then became (partly) visible in the second round of official strategies.

For example, a 2014 OECD workshop report discussed the essentiality of sustainable growth, harvesting, transportation, and trade of biomass to prevent societally detrimental practices and the risks of over-exploitation of natural resources that were inherent to policy recommendations of the time. Moreover, it discussed the lack of definitions for sustainability, measuring instruments for sustainable practices, and international agreements for indicators.[34] In background reports[35] used by the OECD (2014, 2016), the total supply of sustainable biomass in 2030 was assumed to realistically be enough to fulfil the demand in a 10% bio-based economy's final energy and feedstock consumption. A more ambitious ecologically sustainable bioeconomy in 2030 is foreseeable only by looking at optimistic assessments or relying on expectations of new technologies or potentials. However, "it should be realised that any number for its future potential is just a first guess".[36] As its main finding, a report concludes that "the conversion of a fossil fuel-based economy into a bio-based economy will probably be restricted in the European Union (EU) by the limited supply of ecologically sustainable biomass".[37]

As the OECD comments in 2014, the "success" of an ambitious bioeconomy might be in direct conflict with its original aim of tackling grand challenges. There is an urgent need for rapid action, "and yet we have not conquered the sustainability

[32] Ramcilovic-Suominen and Pülzl (2018), 4178

[33] Pülzl et al. (2017), 47

[34] OECD (2014), 10–11

[35] PBL (2012): EEA (2013)

[36] OECD (2014), 19

[37] PBL (2012), 2

issues around the changing needs and uses of biomass".[38] A problem in this conquest is that there is a general lack of agreement in criteria for the amount of sustainably available biomass.[39] The OECD recognizes that the priority in biomass use should be in food and feed and industrial uses should not be allowed to obstruct this. With these premises, the OECD wonders whether there actually exists any spare biomass or spare capacity of arable land.[40]

The external critique towards the first-round bioeconomy policies, along with an internal maturation process, has also been strengthened by the fact that concern about climate change has finally started to materialize into policy actions. This overall sea change in the policy context has created pressure for change in the bioeconomy context, too. In the following chapters, we shall examine how the main elements of the bioeconomy policy have developed on the second round of bioeconomy policy papers.

4 The Second Round of Bioeconomy: Cautious but Contradictory Narratives on Biomass

In the Review of the 2012 European Bioeconomy Strategy (2017), the European Commission states that:

> [T]he policy context in which the bioeconomy operates has changed significantly since 2012, with EU and global policy developments such as Circular Economy, Energy Union, the Paris Agreement and the Sustainable Development Goals. In consequence, the concept of a "circular bioeconomy" is being proposed by various stakeholders.[41]

The commission refers directly to the various criticisms that demonstrate the unsustainable aspects of existing bioeconomy policy. For example, the European Bioeconomy Stakeholders Manifesto (2017) concludes that:

> The bioeconomy can … not be based on the idea of substitution alone, but should be developed recognising that land and biomass, even when renewable, are limited resources. The bioeconomy should therefore be further developed in the context of principles of the circular economy, such as efficient use of primary natural resources, biodegradability and smart consumption, fostering innovation as well as changes in life style and diets.[42]

Followed by these critiques, the bioeconomy went on at least a major conceptual change. To offer a preliminary illustration, the titles and main action plans are compared in Table 9.3.

In the next chapters, we review key points that were updated from the first round of bioeconomy policies. On the one hand, the policies give reason to believe that the

[38] OECD (2014), 39

[39] OECD (2014), 12; OECD (2018), 41–42; Van Dam and Junginger (2011)

[40] OECD (2014), 5

[41] EC (2017), 41

[42] European Bioeconomy Stakeholders Manifesto (2017), 4

Table 9.3 Comparison of bioeconomy policy titles and action plans

Bioeconomy policy titles and action plans		
	1st round	2nd round
Title	Innovating for Sustainable Growth – A Bioeconomy for Europe	A sustainable Bioeconomy for Europe: strengthening the connection between economy, society, and the environment – Updated Bioeconomy Strategy
Action Plan headers	1. Investments in research, innovation, and skills 2. Reinforced policy interaction and stakeholder engagement 3. Enhancement of markets and competitiveness in bioeconomy	1. Strengthen and scale up the bio-based sectors, unlock investments and markets 2. Deploy local bioeconomies rapidly across Europe 3. Understand the ecological boundaries of the bioeconomy

discourse has entered into a new phase, founded on realism and sustainability. On the other hand, some of the highly criticized themes of economic emphasis and bioenergy are still ingrained in it. This results in major contradictions and leads to questions about the direction where the updated policies actually wish to guide the bioeconomy.

4.1 *From Bioeconomy to Sustainable Circular Bioeconomy: Finite and Local Biomass and Land*

In the updated Bioeconomy Strategy (EC 2018), the main concept used is *circular bioeconomy* or *sustainable circular bioeconomy*:

> To be successful, the European bioeconomy needs to have sustainability and circularity at its heart. This will drive the renewal of our industries, the modernisation of our primary production systems, the protection of the environment and will enhance biodiversity.[43]

What does it mean to aim for a sustainable and circular bioeconomy designed to have biomass as its primary feedstock? In the updated bioeconomy policies, the abstractly expected "success of the bioeconomy" is put to more concrete questions of biomass.

As the OECD explains in a central policy paper *Realising the Circular Bioeconomy* (2018), circularity means the aim of keeping biomass in circulation. This leans on the idea of *cascade use* of biomass. Biomass should be primarily used in material bioproducts that keep biomass in the economy for longer, increase resource productivity, and create added value and jobs. The updated Strategy lists a number of examples of bio-based products and sectors that would be environmentally beneficial and value adding, such as construction materials, textiles, and

[43] EC (2018), 4

plastics.[44] Furthermore, circularity is intended to utilize "waste", "co-product", "by-product", "residue" sources of biomass such as agricultural or forestry residues, and municipal solid waste.[45] The OECD (2018) anticipates that a sustainable and circular bioeconomy would mean an innovative network of resource-efficient biorefineries with which a large amount of biomasses can gradually replace fossil-based production. The discussion on wastes, residues, and circularity supports a narrative of an innovative and resource-efficient bioeconomy.

This change in discourse from virgin biomass to varying degrees of circularity and cascading in bioeconomy processes brings to front a much stressed critique towards the first round of bioeconomy policies, arguing that in order to be sustainable, the bioeconomy practices cannot presume an endless amount of virgin biomass. Indeed, a central theme in the round two policy documents is the concern for the sufficiency and sustainable use of biomass.

The third key point in the action plan, understanding the ecological boundaries of the bioeconomy, aims at increasing overall knowledge and monitoring of the sustainable biomass supply limits at the local, regional, and global level.[46] Sustainability seems to mean here at least "the status and resilience of terrestrial (agricultural and forest) and marine ecosystems and their biodiversity …. their related socio-economic costs and benefits, and their capacities to serve as a sustainable domestic biomass source, to sequester carbon and to increase climate resilience".[47] Furthermore, the updated Strategy talks about "[t]he *finite* biological resources and ecosystems of our planet".[48] The limitedness of arable land is also recognized: land used to biofuel crop cultivation or bioproduction is away from human nutrition and might increase prices of food. Moreover, the amount of arable land used to feeding animals (40%) is presented critically,[49] thus hinting towards proposing changes in agriculture and eating habits.

Furthermore, the updated Strategy seems to discourage importing biomass: "Such a [sustainable and circular] bioeconomy will rely and capitalise mainly on domestically available sustainable renewable resources".[50] Although deploying local bioeconomies is not defined more specifically, the document can be read as a promotion of the use of domestic (or local) rather than imported biomasses. This might be a matter of employability and expenses (because biomass is expensive to transport[51]) rather than a matter of social sustainability.

If we look at the updated Strategy's aims in local cascade business models on bio-products, it is easy to imagine how the bioeconomy creates jobs, produces

---

[44] EC (2018), 41

[45] OECD (2018), 10

[46] EC (2018), 14

[47] EC (2018), 15

[48] EC (2018), 15 (emphasis added)

[49] EC (2018), 33

[50] EC (2018), 15

[51] EC (2018), 45

environment-friendly local products, and contributes to a more sustainable world. Examples on forestry-based textiles, furniture, and chemicals, innovative means towards more sustainable agriculture, and cities as major circular bioeconomy hubs create a narrative of a sustainable local bioeconomy.

Finally, the aim to preserve carbon sinks is an evolved theme. The Strategy 2012 primarily discusses wood as a resource for fossil-free energy and products: "Forests of the future will be increasingly dedicated to producing fibres, timber, energy or customised needs".[52] The increased demand for forest products is supposed to be met by speeding up forest growth and productivity. In contrast, the updated Strategy emphasizes the role of forests as negative emission carbon sinks and the need of reducing pressures on major ecosystems such as seas and forests. It recognizes the demand for increased harvest rates but acknowledges the related trade-offs and risks.[53] Thus, this clearly indicates that bioeconomy in this novel interpretation can also stand for the simple aim to grow and preserve carbon sinks.

4.2 Remaining Themes: Economic Emphasis and Bioenergy

It is worth to note that despite the many responses to voiced criticisms towards the first round of bioeconomy policies, it is not the case that the bioeconomy would have changed its direction altogether. The core policy objective remains in "the economy". With the Strategy's language, this means strengthening European competitiveness and creating new value chains and jobs by renewing industries and modernizing primary production systems. Despite the mentioned aim to "rethink growth models and extract more value out of our limited resources",[54] it is not entirely clear which priorities would prevail, if in conflict, economic, environmental, or social ones.

A large open question regarding what has truly changed in the round two policies concerns biomass. Despite the convincing tones of sustainable, circular, and cascade use of biomass, other narratives are available. There still remains an explicit expectation of being able to mobilize such a large amount of biomasses that the fossil-based production can be gradually replaced with bio-based raw materials.[55]

A related, large, and seemingly undecided area is bioenergy. The updated Strategy still emphasizes the role of bioenergy in the reduction of greenhouse gas emissions and meeting EU renewable energy targets. It is "expected to remain a key component of the energy mix in 2030".[56] In a way, this is a pragmatic expectation. The EU has successfully pushed binding national targets on renewable energy, and

[52] EC (2012), 31

[53] EC (2018), 9, 26

[54] EC (2018), 41

[55] OECD (2018), 7

[56] EC (2018), 5

bioenergy is the largest renewable energy source in the EU. Countries have rapidly began to utilize biomass in electricity generation and fuels.[57]

The talk of bioenergy does come with cautionary notes. As the updated Strategy acknowledges, "bioenergy production and use can also be associated to unintended environmental impacts, which need to be effectively mitigated by regulation and good practices at global and corporate level".[58] Problems related to using biomass to produce bioenergy have been noted in revised renewable energy policies,[59] and there has been some reinforcement in the EU bioenergy sustainability criteria.[60]

However, apart from these sustainability criteria, there are no systematic restrictions on the direct use of biomass for energy purposes.[61] In the existing bioeconomies, biomass for energy use is dominant. By mid-2015, at least 154 countries had targets on renewable energy, mainly based on wood and crops. There are few policies that give such attention to bio-based materials and chemicals.[62] For example, a large proportion of the global trade of wood pellets is done for meeting climate obligations by burning them for bioenergy to generate electricity or heat.[63]

Moreover, most of the world's existing biorefineries are first-generation ethanol mills that use food crops as feedstocks.[64] The OECD notes this but emphasizes that "biorefining in the current context should be concentrated on second-generation biorefining, where feedstocks consist of non-food resources (renewable or non-renewable). Very often these will be waste materials. Along with agricultural and forestry residues, in theory this is a large stock of potential feedstocks".[65] The strategic focus is clearly on developed and circular, that is, waste-based biorefineries. However, will it be possible to transform or substitute existing mills to meet this aim? Do the new policies give due guidance and direction to this?

The emphasis on bioenergy – even if it was decreased from the Strategy 2012 – enables still building a bioeconomy narrative where crops and forests are turned into burnable resource for making biofuels. In a sense, this is very logical because the burning of fossil resources is the source of the majority of emissions the world is desperately trying to get rid of.[66] However, it is not the case that the problem of fossil fuels could be fixed with bioenergy. Even though any application of fossil resources could technically be replaced with biomass, it is not reasonable to expect this. Bioenergy is inefficient compared to fossil energy, and there are major conver-

[57] EC (2017); OECD (2018)

[58] EC (2018), 49

[59] EC (2016); Directive (EU) 2018/2001 of the European Parliament and of the Council of 11 December 2018 on the promotion of the use of energy from renewable sources

[60] EC (2018), 49

[61] OECD (2018), 9

[62] OECD (2018), 50

[63] OECD (2018), 44

[64] OECD (2018), 12

[65] OECD (2018), 12–13

[66] OECD (2018), 57

Table 9.4 Key conflicts in bioeconomy policies from 2017

Key conflicts in bioeconomy policies from 2017	
Stated aims of biomass use	*Sustainable*: sustainability limits at local, regional, and global level; socioeconomic costs and benefits of biomasses must be balanced; recognition of the limitedness of arable land available biomass *Circular*: "wastes" and residues instead of virgin biomass *Cascade*: material bioproducts that keep biomass in the economy for longer, increase resource productivity, and create added value and jobs *Carbon sinks*: reducing pressures on major ecosystems, carbon capture *Local*: Local biomass comes with social and environmental problems
Contradictory contents	*Replacement*: expectation of mobilizing such a large amount of biomasses that the fossil-based production can be gradually replaced with bio-based *Growth*: core policy objective is in competitiveness and growth *Bioenergy reliance*: no systematic restrictions on the direct use of biomass for energy purpose, a key component of the energy mix in 2030, prevalent type of the current bioeconomy. However, untoward social and environmental effects of cultivation are noted

sion losses in the production of it. Bioenergy could sustainably meet 10% of the EU's energy need.[67] Bioprocesses are "notoriously inefficient" compared to fossil-based production.[68] Utilizing waste and residue biomass would be a sustainable solution, but the potential supply estimates of these do not indicate they would become the dominant feedstock. Overall, background calculations recommend only a low priority to the application of biomass in power generation – instead, they would recommend focusing on other sources, such as solar, wind, nuclear, and hydropower, when it comes to power generation.[69] This skepticism does not undermine, however, the importance of bioenergy in selected areas. Biofuels will most likely be needed in heavy traffic or aviation. The point is that biomass should be used to produce biofuels only when no alternatives exist, in prioritized applications. This would constitute a "roadmap towards a low-carbon, partly bio-based economy".[70]

The evident conflicts between the stated aims of biomass use and the discussed contradictory elements present in the policy documents are summarized in Table 9.4.

4.3 *The Conflicting Roles of Bioeconomy in Climate Measures*

The likely future trajectories of the bioeconomy can be further reflected on by looking at recent EU climate change-related policy documents. *A Clean Planet for all: A European strategic long-term vision for a prosperous, modern, competitive and*

[67] PBL (2012), 7–9, see also Partanen et al. (2014)

[68] OECD (2018), 41; Philp (2015)

[69] PBL (2012), 9, 14

[70] PBL (2012), 14

climate neutral economy (EC 2018b) and *The European Green Deal* (2019) paint the big picture of EU's vision for a transition towards sustainability. What is the role of the bioeconomy in these, and do they shed light to some of the open questions presented in the previous chapter?

The general policies emphasize circular economy and energy efficiency, deployment of carbon sinks, electrification with renewables (solar and on- and off-shore wind, ocean energy, nuclear), changing agroforestry techniques, and changes in mobility patterns (e.g. reducing travel and massive decreases to road transport of freight). They promise regulative measures towards sustainable practices by revising multiple pricing and taxing sectors according to environmental burdens, ending fossil fuel subsidies and tax exemptions, and encouraging reusable, durable, and repairable consumer choices. The tones of promoting planetary boundaries are intensive. However, the green deal is presented as a strategy of growth that relies on the idea that economic growth can be decoupled from resource use.

The need for a just and inclusive transition and taking care of distribution of harms across areas and groups of people is recognized to a great extent. The main reason for this is that successful climate actions must be done with speed and effectivity, and this is possible only with a wide public acceptance as citizens are active participants and consumers in the transition. Wide public acceptance can only be gained by making the transition fair and inclusive.[71]

In these climate policies, the bioeconomy is a partial solution but is presented as a contradiction: We need it, but EU forests alone cannot deliver necessary biomasses. Importing is problematic, sinks should be enhanced, and land for biomass production is scarce. Other, perhaps even more important, measures are other renewables that enable electrification, policy actions that force (by regulation or by price) decreasing the use of fossils, and changing behaviour, e.g. in traffic (less wheels and fuel) and land use (food or feed, agriculture). Thus, tackling climate change is by and large seen a political task – not only of accelerating the use of new technology or innovation but a wide and complex socioeconomic task, comprising a wide transform of subsidies, taxation, and pricing to meet environmental goals.

With this in the background, and looking at the different narratives on bioeconomy, the major objective of the bioeconomy becomes ambivalent. If it is "the gradual replacement of fossil-based production with bio-based",[72] and the "defossilisation of major industries, such as the energy and transport sectors, the chemical industry (e.g. plastics), and the construction sector",[73] what is the speculated role of the bioeconomy in these replacements? The answer to this question is a key determinant to how sustainable the bioeconomy can be.

One way to conceive of the promise of the bioeconomy is to see it as proposing to save the planet by replacing fossil fuels gallon by gallon and plastic bag after plastic bag. However, this narrative builds on the notion of infinite biomass, which

[71] EC (2018b), 21–22

[72] OECD (2018), 7

[73] EC (2018), 32

has been the main point raised in the critique. Thus, this particular vision of world-saving turns into its opposite. In contrast, if the aim is to replace some fossil resources by, for instance, promoting novel practices in agriculture and forestry, the image of a truly sustainable and circular bioeconomy created is certainly much more credible. However, in the latter narrative, the dramatic effect of offering a silver bullet to solve the grand challenges is mostly lost. It then seems that the bioeconomy can be either perceived as a "world-saving" unsustainability monster or understood as a partial, small-scale answer, to drive transition to sustainability by a change in local practices.

The updated bioeconomy policies remain ambivalent in taking sides in this regard. In the second-round bioeconomy policies, one can see that the biggest hype around the bioeconomy has faded, but the bioeconomy is still seen as holding an important, if not, key role in efforts to drive sustainability:

> Even though the production of biological renewable resources is also associated with greenhouse gas emissions, resource consumption and other environmental risks, there is growing understanding that deep defossilisation and remaining under the 2 °C limit will not be possible without sustainable bioeconomy activities, given their potential for carbon sequestration, the substitution of fossil resources with sustainable biomass-based resources, and their large greenhouse gas emission reduction potential related to more resource efficient and sustainable production patterns.[74]

Thus, the bioeconomy can produce emissions and environmental risk, but we need bioeconomy to tackle emissions and environmental risks. Bioenergy can be good, but it can be bad. Cascade use is good for the environment and for the society, but it does not provide a direct answer to the existing biggest emissions.

Despite these contradictory remarks, it is evident that the hype on simplistic technological fixes or scientism, criticized in the first round, has decreased. Even though the OECD holds that "science and technology quite clearly hold the answers to many of the questions regarding this low-carbon, non-fossil future, as evidenced by the growth of solar and wind technologies",[75] it also states that:

> Whenever humans intervene in a system, from the level of genetics to whole community, all the way to globally, there are interactions with other components of the system, and new consequences. The 'behaviour' of these grand challenges is assuming characteristics of an ecosystem: an intervention in one location results in changes there but also elsewhere. Ultimately the goal is interacting solutions to interacting grand challenges. This calls for multi-disciplinary research and systems innovation. There is no simplistic technological fix.[76]

What the OECD calls for to address these contradictions is efficient allocation of biomass to chemicals, materials, and energy; balancing societal benefits in different biomass uses; better knowledge on biomass volumes and measuring sustainability; and levelling the playing field for cascade use of biomass by removing fossil fuel

---

[74] EC (2018), 32

[75] OECD (2018), 50

[76] OECD (2018), 8

subsidies and pricing the environmental damage of those industries.[77] At the same time, it warns about potential regulatory deadlock that overly strict policies regarding biomass use might produce.[78] The OECD actually suggests a guide to prioritization in case-by-case analysis: climate change mitigation (including preserving important carbon sinks), protection of the environment (especially forests) and the people, energy security, economic stability, and job creation.[79]

The concrete consequences of the conceptual change will only be seen when the EU puts the new policies into action, and individual countries update their respective bioeconomy strategies, interpreting and implementing the new EU-level policy documents. The essence of the bioeconomy might be at the edge of a true paradigmatic shift, or it can remain the same with superficial additions in policy papers.

5 Delphi Analysis: Imaginaries of Good and Bad Bioeconomies Towards 2075

In this chapter we present narratives of futures images of "bad" and "good" bioeconomies in 2075. They are constructed from experts' answers to two questions from the Delphi questionnaire: (1) experts' conceptualizations of what would constitute a "good" and a "bad" bioeconomy in the year 2075 and (2) the panelists' insights regarding the main societal drivers and key factors in either failing or succeeding in reaching the IPCC's climate targets. The respondents unanimously held that a determining characteristic of a "good" bioeconomy is a strong contribution to halting climate change. Therefore, the answers that elaborate key societal drivers in reaching the IPCC's climate targets can be understood as pathways to a "good" bioeconomy, and vice versa in the case of descriptions of a bad bioeconomy. First we present concise summaries of the "good" and "bad" bioeconomies as elaborated by the expert panel. A table of thematically grouped quotations (Table 9.5) shows how these visions are grounded in the original Delphi responses. In order to represent the rich data and do justice to the multifaceted causality chains in the arguments, we have chosen to present the findings as constructed narratives on the pathways that have lead to undesirable or desirable outcomes for the bioeconomy in 2075. These are presented after the quotation table.

The experts' view of a "bad" bioeconomy has failed in changing the course of the climate change progression and is successful mainly as a greenwashing strategy. In many respects, the bad bioeconomy is a continuation of "business as usual" as it continues on the same economic paradigm and behavioural trends of the present, aiming to a simplistic replacement of fossil resources with renewables. It has not affected fossil use, and the bioeconomy-related practices are harmful for the

[77] OECD (2018), 31, 42, 54

[78] OECD (2018), 31

[79] OECD (2018), 31

Table 9.5 Dimensions of “bad” and “good” bioeconomies

	Dimensions of bad and good bioeconomies	
	“A bad bioeconomy….	“A good bioeconomy…
Political	…is too little and too late”. …is the paralysis of the political class, a lack of public will and societal integration”. …lacks the willingness to coordinate and cooperate on a global level”. …is based on short-term self-interest and denialism, greediness of money and power”. … is made from corruption of government by big business and strong lobby from the fossil fuel and agricultural industry”. … consists of idiots in all bigger governments”.	…has massive pressure from citizens and consumers”. …has strong political action and commitment from world leaders”. …relies on proper risk communication that manages mobilization and lifestyle”. …is made by a new generation of leaders”. …produces changes in legislation and culture”.
Economic	…continues the exploitative linear economy”. … is myopically focused on growth and allows it without limitations”. …aims to replace all fossil based raw materials with bio-based ones”. …is an unconstrained market economy”. …is continuing to use petroleum as the foundation for our economy”. … plays in the hands of the global companies”. … is based on consumerism”.	… entails a paradigm shift from continuous economic growth”. … reduces the overall level of consumption”. …is reasonably small. Biological resources cannot be used to the extent fossils are being used today”. …is a true circular economy where everything is re-used, recycled and shared”. …is a global agreement to revise the rules and frameworks that govern our markets”. … uses biomaterials only based on “true” needs”.
Social	…aggregates benefits to few”. …is a policy agenda decided by a small elite”. … ignores the rights of indigenous peoples”. … benefits large corporations and makes the lives of people dependent on bioresources more vulnerable”. …does not pay heed to issues concerning equality, poverty, workers’ rights, etc.”. … does not provide jobs and well-being to communities”.	…is one where the global population is stabilised”. …addresses wealth inequality”. …distributes biological resources justly”. … is an inclusive economy that takes into consideration needs of all stakeholders”. …provides equal opportunities to participate in decision making”. …is an economy where less people live in poverty”.

(continued)

Table 9.5 (continued)

	Dimensions of bad and good bioeconomies	
	"A bad bioeconomy….	"A good bioeconomy…
Technological	…offers technological fixes to our environmental and social predicaments". …uses low-level technology, similar to current biodiesel or other first-generation biofuels". …uses biomass for energy".	… emphasizes socio-ecological circularity and innovation, rather than technological solutions". …is radical technologies for health and energy". …is represented by rather small to medium-size scale biorefineries available locally". …provides massive decarbonised electrification of whatever is possible".
Environmental	… is harmful for the ecosystem". …fails to make a contribution to climate change". …is a word for green-washing". …does not balance carbon sinks against the use of bio-based materials". …exploits the globe's natural resources beyond the limits of sustainability". …fails to balance economic systems with the planetary boundaries".	…is a tool to reach a fossil-fuel-free society". …does not aim to replace all fossil based raw materials with bio-based ones". …is locally sourced and sustainable". …is about safeguarding natural resources". …increases biodiversity and carbon sinks". …uses biomass for energy only in a very limited manner, if at all". …uses agroecological methods in agriculture".

environment: they are economically wasteful and utilize old-fashioned technologies, and biomass is routinely used for energy production. The bad bioeconomy is based on exploitative relationships between humanity and nature, but also between individuals and different nations. It is culturally insensitive, socially divisive, and unjust. As a result of a bad bioeconomy trajectory, the experts anticipate widespread societal and ecological problems that are difficult to mend. Table 9.5 presents thematically grouped quotations from the authentic Delphi responses. Quotations have been slightly formatted to fit into the table.

Aspects of a "good" bioeconomy, according to the expert answers to the questionnaire, include the protection and certifying of natural resources and distributing them in least harmful ways. The experts emphasize the importance of creating a sustainable economic model, where the economic paradigm has transcended the current focus on material growth and consumerism. Instead, sustainable consumer behaviour and distribution of the benefits generated by the bioeconomy in a just and equal way are at the core of a "good" bioeconomy as envisioned by the expert panel. Also, such economies in the views of the experts would be relatively small and local. Fossil use would be minimized and where feasible substituted by renewables. However, low-carbon goals and combatting climate change are key priorities, and there circular bioeconomy overrides the earlier substitution-based thinking. Smart and sustainable technology, combined with regenerative approaches across the board, provide the tools for reaching a good bioeconomy. Economical issues are in balance with the goal of environmental protection and enhancement of biodiversity, as well as with the aim to provide conditions for flourishing and well-being.

5.1 A "Bad" Bioeconomy in 2075

In the bad bioeconomy scenario, the world has continued on the unsustainable path that much of the human societies already were on in 2019. The word bioeconomy has largely come to represent a particular greenwashing strategy, which for long had distracted citizens from the urgent need to shape lifestyles and societal structures towards more sustainable practices. The main reason for why societies were not able to move on to a sustainable path through the bioeconomy was that too little was done to change the harmful practices through policy, within the timeframe they would have still been effective. One major obstacle in the path towards a sustainable bioeconomy was getting stuck on the idea that biological resources could be used to the extent fossil resources were used earlier. Most of the difficulties experienced in implementing sustainable practices can in retrospect be linked with this fundamental inability to change mindsets related to consumption and material growth.

The failure to halt climate change is clearly evident in 2075, and any attempts to reform the systems to being less destructive are generally perceived as too little, too late. A major factor that has kept things rolling towards the wrong direction has been the inability to change economic priorities and the linear, exploitative economic model, connected with the prevailing consumerist lifestyle. A general lack of belief in change and collective action has coloured the decades since 2019, and this bleak social mood has contributed to the reluctance to regulate markets. In sum, the failures resulted from the passivity of ignorant and hopeless citizens, lack of a political vision, and nihilistic businesses. Despite climate change being a constant point of discussion on international political sphere for decades, it did not successfully challenge the prevailing economic dogma. An underlying reason for the slowness or lack of action was the persistence of old power relationships: old habits, the overriding goal of economic growth as the guarantee of human welfare, and related assumptions regarding competitivity combined with the dependency of the societies on fossils. Together they slowed down progress towards more ambitious policies. A general lack of confidence in the political decision-making has been a key factor that has prevented making political decisions quickly enough, and important decisions made by politicians in big countries such as the USA, Brazil, and China have greatly affected the development of climate action globally.

Earlier, there had been changes in values that speeded up the transition from material to immaterial consumption, and in some circles there was even readiness to adopt degrowth-based practices and policies. However, these remained as niches for a long time, while globally, the more or less official assumption remained that all fossil fuels would be used anyway, and any efforts to substitute them would only slow down the process rather than end it. Within societies, societal disintegration had made the task of forming collective resolutions more challenging. The soft market-based measures implemented to support changes in consumption were successful only in addressing the consumption patterns of the relatively rich, well-educated people. Although their behaviour was at the heart of problems, the chosen policy approach was not enough to affect progress of climate change. Also, it perpetuated the impression that only elites were the concern of the decision-makers (who also belonged to the same societal strata), contributing to a general sense of social injustice permeating many societies. Western consumption-based lifestyles were held up as the ideal in developing countries. Benefits from the bioeconomy had mainly gone to societal elites, while minorities had seen their rights and social standing continuously diminishes over the course of the past decades. Globally, the practice of importing biomass from the developing countries had caused widespread social and ecological problems while providing only little as revenues for the countries to mend the damage done. Efforts to share the positive financial results justly with the developing countries thus failed and led to a continuation of high population growth and unsustainable development patterns. At the heart of the problematique was a general hopelessness, and a lack of adequate risk communication that would have acknowledged and respected the emotions and psychology of people. Challenges seemed unsurmountable, and for instance, the melting tundra and the burning of forests appeared to be unstoppable natural events that automatically would continue to generate CO_2. Lack of resources caused poverty and hunger amongst fast-growing populations.

Societal pressure ultimately, as climate change impacts were already clearly visible and disturbing daily lives, gained enough strength to turn the policies around to adopt more stern measures to combat climate change. However, this happened too late to be able to have a significant effect on hindering or mitigating harmful effects of warming.

Corruption of governance by big business interests was responsible for major delays in climate action. It was manifested in different ways: denying climate change, delaying climate action measures by serving interests of selected stakeholders (e.g. nationally important industries), and promoting fear of losing wealth and jobs. As no incentives or restraints were present to hinder the drive to make profits, the economic system, but also sub-systems like agriculture, continued to rely on the fossil industry and perpetuated the approach based on extractivism, global inequality, and a false idea of nature as abundant, self-correcting, and "larger than humanity". In the absence of clear policy guidelines or incentives, and with all the more pressing economic concerns, technological development and application of novel technologies started to lag, and many businesses continued with using rudimentary first-generation technologies in biomass exploitation biorefineries. Bio-based resources were routinely applied for linear bulk use, like fuel and energy production.

Even if decline in human health, competition for resources, and population migration had underlined the need to replace fossil energy sources, the politicians were rendered unable to act in a world characterized by combined effects of the climate change-related calamities: wars, natural disasters, effects of reduced biodiversity, population growth, inequality, more mobility and flying, and the heating of climate. Societies lacked abilities to cooperate or coordinate their responses to ever-mounting challenges, and the political class had resigned their power mostly to the business interests that were dominated by short-sighted and selfish greed of money and power. The public lacked will and confidence to try to affect changes. The same dynamics also were at work on the global scale, leading to global inaction as regards mobilizing to tackle climate change. The problems culminated to an inability to make necessary political decisions fast enough, resulting in political leaders paying attention to insignificant small questions to attract the voters' attention, to focus on technological fixes, and a reluctance to think through measures to bring about systemic change. Rather, the bioeconomy proceeded as direct substitution of fossils by renewables that resulted in unsurmountable ecological problems.

5.2 A "Good" Bioeconomy in 2075

The success of halting climate warming can to a large extent be credited to changes made to the previous economic paradigm. This was triggered by a search for a good life and a good society, brought up by an awakening to the state of the environment in the first decades of the second millennium.

The dramatic changes were first initiated by decisive action from the citizen society, where individual opinion leaders and political movements acted as primus motors insisting on change. These movements were instigated by climate-related catastrophes and fear, but with proper risk communication, they paved the way for a global change in attitudes. The citizen pressure, combined with calls for action from sustainability-oriented businesses, was able to draw strong commitments from world leaders and political deciders. At the same time, businesses and industries had begun to respond to a turbulent, unsecure, and unpredictable political and economic environment by starting to favour local, small-scale operations.

A global agreement to revise the rules and frameworks that govern the markets was reached in order to prioritize environmental and social sustainability. Also other international agreements moved forwards, with all the significant countries joining in. Land management became more efficient with clear policies such as prioritizing food over feed or energy, and massive afforestation efforts were taken as a result of placing various incentives for environmental protection. Even conservative and large corporations abandoned fossils and adopted sustainability-driven missions and strategies.

Together, the combination of value changes, soft market measures, and policy enforcement diminished energy use, mobility, and eating meat and lowered the overall level of consumption. In general, consumption and material welfare seized to be perceived as status symbols. The UN Sustainable Development Goals were an important framework for deciding on the social, economic, and ecological changes. A shift towards a just, equal, and inclusive distribution of resources, goods, and wealth globally has had direct consequences for the sustainability of societies, but also side effects that would have earlier seemed utopian, such as the elimination of poverty, equality between the sexes, and universally available education. These have, amongst other things, led to a more balanced demographic structure.

In the context of a rapidly developing bioeconomy, sustainability of the raw material production and ensuring the maintenance, or in many cases increase, of biodiversity were key considerations. Early on it was realized that replacing all fossil-based raw materials with bio-based ones was not sustainable or good for the biodiversity. Thus, separating out the industrial segments where this was worth doing and where it was not was an important process that later on proved beneficial for the sustainability of the bioeconomy. Another key focus area was to make sure that fossils were not simply substituted, but that major efforts were made in reducing primary consumption. Countries enacted policies that strictly regulated the use of virgin raw materials, and so circularity and recycling became the foundations of the bioeconomy. Regenerative farming techniques were used to bind carbon to the soil while producing a varied and healthy nutrition for the increasing population. All in all, the bioeconomy was in many ways crucial in helping humanity transcend to a world beyond fossil dependency.

The most important technological enablers of the transformation were the fall in the cost of renewable energy and battery technology, with a wide-scale electrification. Diverse wood-based solutions are mainly used as bio-products in 2075. Due to sustainability issues, the role of bioenergy in the portfolio of renewables has been dramatically decreased. Bioeconomy means not just technological fixes or one major technology but is seen as a multifaceted area providing many answers to the quest for sustainability.

6 Are We Approaching a "Good" Bioeconomy with the New Policies?

In this section, we estimate the updated bioeconomy policies in light of the experts' insights of characteristics of a "good" and "bad" bioeconomy and in light of whether the critique of the first-round policy is answered. How close are the recent policies' directions to what a "good" bioeconomy would be? Are there some elements of a "bad" bioeconomy that remain?

The thematic main characteristics of "bad" and "good" bioeconomy, the first-round policies, and updated bioeconomy policies (with some added notes from the bioeconomy-related general climate policies) are presented in Table 9.6. We reflect the policy changes against the experts' notions of a "good" and a "bad" bioeconomy trajectories by comparing overlapping themes that emerged from both data: the role of bioeconomy as a technoscientific answer; narratives on the use of biomass; visions of economic paradigm and values, the role of justice, and inclusion in the transition; and views on the required political and societal mindsets.

Table 9.6 Comparison of the Delphi experts' "bad" and "good" bioeconomy with first- and second-round bioeconomy policies

Dimensions of current and future bioeconomies			
"Bad" bioeconomy	"Good" bioeconomy	1st-round bioeconomy policy	2nd-round bioeconomy policy[a]
Political: Political and societal mindset			
Societal disintegration No public will Too little too late Short-sited politicians Selfishness, greediness of money and power Lobby and corruption prevent change	The world united Collective action Societal pressure Belief in change Committed leaders Interest groups and corporations join targets Psychologically effective risk communication		Effective policy implementation needs active citizens and public acceptance[a] Tackling climate change is a political task[a]
Economic: Economic paradigm and economic values			
Linear and exploitative emission-heavy economic growth guarantees welfare Only soft market measures restrict and direct to sustainability Consumerism prevails	Prioritization of environment and social sustainability Revision of rules that govern markets Strict regulation of all virgin raw materials Change in consumer behaviour and culture	Focus on economic output and growth Continuous or even increased production Economic priorities dominate environmental and social ones	Core focus in growth and competitiveness Soft and hard regulation Industry less dependent on new materials Reliance on decoupling No real restrictions to energy use of biomass
Social: Justice, inclusion, equality			
Benefits to few, social injustice Ignore the rights of minorities, decisions by a small elite Importing biomass from developing economies Fast growth of population	Equal distribution of benefits, just transition Inclusive decisions Just distribution of natural resources Education of girls and equality of sexes to halt population growth	Lack of discussion on social and cultural matters of bioeconomy No attention to unethical import of biomass	Balancing societal aspects in different biomass uses Just and inclusive transition[a] Taking care of people with low income[a] Concerns about biomass import[a]
Technological: Technoscientific fixes and their limits			
Focuses on technological answers Fossil use merely replaced with another resource Lack of necessary innovation Inability to implement existing technology	No just technological fixes Reducing primary consumption Innovations to circularity and sustainability Cheaper mass markets Large electrification	Technoscience is a main solution Reaching sustainability taken for granted Infinite biomass can replace fossil resources	Decreased hype on technological fixes Sustainability taken seriously Finite biomass and land Not only replacement but some reducing[a] Large electrification[a]
Environmental: How biomass is used			

(continued)

Table 9.6 (continued)

Dimensions of current and future bioeconomies			
"Bad" bioeconomy	"Good" bioeconomy	1st-round bioeconomy policy	2nd-round bioeconomy policy[a]
Biomass overuse Linear bulk energy use No changes in land use, unsustainable agriculture Global exploitation	Circular cascade use No bulk use for energy Massive land use changes: afforestation, sinks, carbon capture… Local biomass sourcing	Central role to biofuels and bioenergy Inadequate attention to resource scarcity	Circular cascade use Bioenergy reliance Land use changes: sinks, regenerative methods… Local biomass sourcing

[a]*Notes from the Clean Planet (EC*2018b*) and Green Deal (EC*2019*)*

As we have elaborated above, the experts' views of a "good" bioeconomy rely on a holistic and socio-politically oriented diagnosis of climate change that does not see technoscientific fixes as a probable solution. This understanding is visible in the updated policies where the limits of technoscientific solutions are recognized. Sustainability is taken seriously, and the finitude of biomass and land for its cultivation are acknowledged. However, it is uncertain to what extent the policies have moved from mere replacement of fossil resources to promoting massive decreases in primary consumption.

This is not to say that technoscientific solutions would be of less worth. The experts emphasize the need to implement a variety of sustainability-promoting technologies and worry about their speedy and effective implementation. However, the scale in which citizens and other agents in the society are able to adopt and start using new solutions poses a limit to technological solutions. The new strategies do emphasize innovation and the need of fast implementation with market measures, but a fast transition needs to be a just transition.

The planned uses of biomass in policies are similar to the visions of a "good" bioeconomy. Locality, circularity, and cascading of biomass are central in the new policies. Land use changes are on the list, but are they as massive as the experts call for? The challenging open question is the role of bioenergy and biofuels in narratives of future bioeconomy. Some of it is arguably needed, but large-scale production is undesirable. Currently, "the bioeconomy" predominantly means cultivating crops for biofuels. Thus, a major shift should take place. To what extent do the policy documents acknowledge this?

The experts see a change to the current paradigm of linear growth as a key factor in reaching the climate targets and a "good" bioeconomy. The new policies offer some hints towards this. There is a will to use heavier regulation to direct economic activities towards sustainability and reduce dependency on new materials. However, it seems most of these regulations are softer market means, while the experts would wish also for a strict regulation of all raw materials. Even though strict regulations can produce untoward effects, the lack of genuine restrictions to the energy use of (raw) biomass seems to be at odds with the vision of a circular and sustainable bioeconomy.

Furthermore, the policy documents' core focus remains in economic growth and the myth of decoupling economic growth from resource use.[80] This creates a major threat to the aim of reducing primary consumption. Without a systematic re-conceptualization of "economic growth", it is an open question whether primary consumption could really be decreased. The prevailing dogma of growth does not help to shift consumerist behaviour, either. The experts view this as being a key factor in need of change. The strategies have elements of guiding the decrease of some unsustainable habits such as travel, food, and energy use, but there is no discussion on the need to affect the mindset regarding consumption culture as the cornerstone of welfare and status in the society.

The image of a "good" bioeconomy based on the Delphi experts' arguments is just, inclusive, and equal. These issues are also progressively visible in the new policies. It remains an open question to what extent these intentions translate to changing practises in different nations. In the expert's views, taking care of a just transition is important, but it is mainly discussed in relation to the overall success of the sustainability transition. The experts fear societal disintegration and lack of cooperation. The experts identify feelings of injustice about the distribution of burdens and benefits as a main driver for societal disintegration and unwillingness to change one's lifestyle. Fair distribution of harms enables collective commitment and public acceptance of policies. Especially the new general climate policies discuss these issues. To the experts, social equality is a key driver to sustainable population growth, too, but the policies entail little discussion of this.

A theme that goes completely missing in the strategies is power relations that prevent change. The biggest threat the experts mention is a fear that nothing gets done, or actions are too little too late. A central element in this fear is that politicians are shortsighted and care more about the next election polls than responsible politics of sustainability. Main actors in this are lobby and corruption by big influencers, such as conservative fossil-dependent or biomass bulk use-dependent corporations and interest groups. In a "good" bioeconomy, leaders are committed, and corporations adopt sustainability targets. Thus, power relations, economic structures, and interest influencers play a major role in the transition. Nevertheless, these relations are not discussed at all, apart from urging for "public" acceptance and active citizens. Should there be more targeted speech addressing the central opinion influencers in the society?

Another issue related to the lack of leadership is social apathy, brought about by the lack of adequate means to communicate the climate urgency. Although communicating the issue should raise alarm and create awareness of urgency, it should still remain sensitive to communicating a just and inclusive transition. This way messages would be designed to reach different groups of people, with awareness and caution not to create a sense of resignation or fatalism. The experts view such sentiments to be a central factor contributing to failures to address climate change through any kind of policy framework.

---

[80] See, e.g. Jackson (2009)

All in all, the depictions of a "bad" bioeconomy in general engage in a topical critique of contemporary bioeconomy policies and practices. A "bad" bioeconomy combines the key points of critique towards round one bioeconomy policies: it is not mindful of the regeneration rates of renewables, is concerned with quantity over quality of raw material production, aggravates current social and ecological problems, and aims at direct substitution of fossil-based goods with renewables.

Definite conclusions of resemblance of the "good" bioeconomy and the updated policies give way to contradictory narratives and mixed messages. Committed tones to circularity, all-encompassing sustainability, convincing understanding of the limits of technology, and urging socio-political answers take the updated policy to a path towards a "good" bioeconomy. However, reliance on bioenergy and economic growth, the lack of acknowledging the power relations affecting politics, and potentially too soft means in revising the rules of markets cast a shadow on to interpretations about this direction. In the following concluding section, we elaborate our observations more closely.

7 Conclusion

One way to simply encapsulate the changes that have taken place in the bioeconomy discourse of the EU-level policy documents is to describe the new round of policies as "matured" in comparison with the initial program. Bioeconomy in its revised form makes less bold promises and engages in a much more nuanced discussion of limits of possibility and feasibility. At the same time, however, the policies keep the door open for multiple and at times conflicting interpretations, some of which may be very similar to the initial bioeconomy narrative. In this article, our aim was to understand more deeply the directions for bioeconomy, based on both the policy documents and views of experts in the field of bioeconomy. From the expert's views, it is possible to distill a desirable narrative for the futures of the bioeconomy. Many of the elements of a "good" bioeconomy as raised by the experts can also be found in the revised policy documents. Yet, many of the issues that the experts bring up remain unsaid in the documents.

The need to revise the policy framework was to be expected after the harsh criticism directed at the initial bioeconomy programme. Judging by the experts' views, the earlier bioeconomy paradigm is currently widely taken to represent almost the definition of an undesirable bioeconomy future. On the other hand, the issues that have been left out or have not been concretely addressed also tell something about how climate change and other grand challenges are understood on the level of the policymakers. As a general observation, the experts interpret environmental crises as a symptom of a systems-level problematique and are likely to search for more holistic measures as answers to solving them. There, the revised policies are moving closer to the experts' views by agreeing that instead of being a silver bullet to solve multiple pressing problems facing the EU, the bioeconomy- and biomass-based solutions can only be a rather modest part of the compilation of policies, technologies, and cultural shifts that together can build a more sustainable future. However, in the views of the experts, at the core of the systemic imbalance is a reluctance to change the consumption patterns that sustain the unhealthy economic paradigm. In the EU policies, such questions are mostly avoided, and there lie the biggest discrepancies between the experts' visions for the future and the revised EU policy frameworks.

If we look at the expert's visions for a pathway to a "good" bioeconomy and the urgency they place on critically addressing existing power relations, ending linear bulk use of biomass, revising the current economic paradigm, and massively cutting our primary consumption, we can see that several of these central issues are not raised in the policy documents. What the policies, according to the Delphi panel, should say might be something like the following: "Draw back bioenergy programs, we made a mistake!" or "Stop listening to the fossil- and raw material intensive industries, no matter how important they are nationally!", or even "Forget your short-sighted interests in money or power!". "Stop consuming raw materials altogether!", "Don't seek welfare or status from growth or consumption!", and "Forget the freedom of markets when sustainability is at stake!". It is evident how out of place, naïve, and utopian these sentences would seem in policy contexts. Yet, policies frame the key questions around their area of concern and have a major effect on how we understand and attempt to solve the problems. If relevant areas remain in sidelines, solutions might miss the most important targets.

The theory of the images of the future posits shared visions about the future as an important factor in the shaping of futures. Polak, the main author behind the theory, maintains that the images of the future are culture specific and based on the key narratives, myths, and histories of their originating civilization and culture. Explaining the theory through Western culture, Polak provides plenty of examples as to how one way to narrate a storyline excludes elements that are not fit for that particular context.[81] One can thus argue that the discrepancies between the experts' ideals and the reality of the policy documents lie exactly in the mismatch between the experts' semi-utopian visions (made possible perhaps by the extremely long time span up until the year 2075) and the prevailing cultural image of the futures, tied in closely with the archetypal lifestyle, and beliefs about its foundations in the prevailing economic model. However, the shared images of the future are in a constant process of evolution and change, and the dynamics between utopia and prevailing notions of plausible futures can be argued to be a central force generating social change: "Utopism is the forerunner of all modern conceptions concerning social policy, social organization, and social peace" (Polak, 1973: 178). Thus, it is necessary to discuss alternative images of the future if one aims to advance towards a good society, and a good bioeconomy.

As a particular sidenote, Polak (1973) argues that the late modern Western culture is in general subject to a cultural disruption that is manifested in the dispersing

[81] Similar ideas have been used as base for method and theory development in critical futures studies (the CLA method, Inayatullah 2009) and in peace studies (Johan Galtung's (1981a, 1981b) deep civilizational codes).

of traditional values and founding myths that have so far given rise to Western images of the future. This is argued to bear consequences amongst other things in the form of managerialization of politics. This diagnosis resonates well with a key concern amongst the Delphi panelists' answers regarding the notions of a "bad" bioeconomy (which read essentially as depictions of courses of action that result in more or less total collapse of the contemporary Western ideal state). In such answers, politicians appear as impotent and visionless non-leaders who are unable to ideate or instigate any changes, but rather are merely acting as managers to implement the wishes of the more powerful business elites.

In this study we have explored views derived from bioeconomy experts, who have created a convincing scenario on desirable directions for the bioeconomy. The updated policies were analysed for understanding the extent to which expert's views are in concordance with policy directions and what key elements in the experts' visions are missing from the policy documents. It must be noted that certainly other interpretations about the futures of the bioeconomy directions are possible, and the nature of the Delphi study makes it sensitive to the compilation of experts gathered to provide their judgements. In the current absence of other studies about the revised policy documents, we must rely on our own interpretation of them. Thus, to conclude, in this study we provide a possible framework to understand the direction of the bioeconomy towards the future, but there might be other valid viewpoints and frameworks that this study does not cover. We hope to provide insights to policymakers and analysts about potentially important future directions and blind spots in bioeconomy and climate policies in need of attention, take part in the first wave of analyses of the updated bioeconomy policies, and offer an example of a research field and policy framework in a transitional phase to relevant areas of scholarship.

At the time of writing the article, we find ourselves in a moment defined by uncertainty regarding the futures of the bioeconomy. On one hand, the EU-level bioeconomy policies point to a future where bioeconomy would take a more considerate position on questions related to sustainability from the perspective of all its aspects. Yet, the policies do not enforce any of these viewpoints and leave rather free space for national-level interpretations.

The bioeconomy has been a dream of an all-replacing biomass giant, but these reveries have proven to be unrealistic hype and wishful thinking. We presume that the future of the bioeconomy will depend on its ability to genuinely dissociate itself from its former unsustainable vision. A move forward has already been made in the official policies, and it remains to be seen how fast the everyday bioeconomy of the real world will catch up with the inevitable change. One sustainable option for the bioeconomy could be to become a partial solution to the grand challenges of climate change and environmental degradation, focusing on small and local refineries and sustainable production methods. However, if we look at the expert's insights, another road ahead comes into view. The bioeconomy could start to promote a wider ideological shift towards a new economic paradigm, searching for prosperity without material growth. This move would help to position bioeconomy as a salient feature in the strategies towards sustainability.

Bibliography

Arevalo, J., Halder, P., Kortelainen, J., & Mola-Yudego, B. (2014). Bioenergy: from local conflicts to global governance. Oil, Gas & Energy Law 4. www.ogel.org/article.asp?key¼3511.

Bell, W. (1996). Foundations of futures studies: human science for a new era. Vol. 1, History, purposes and knowledge. New Jersey: Transaction Publishers.

Borup M, Brown N., Konrad K. and Val Lente H. (2006). The Sociology of Expectations in Science and Technology. Technology Analysis & Strategic Management 18(3/4), 285–298.

Brown, N (2003). Hope Against Hype – Accountability in Biopasts, Presents and Futures. Science Studies 16(2), 3–21.

Brown, N. and M. Michael. (2003). A Sociology of Expectations: Retrospecting Prospects and Prospecting Retrospects. Technology Analysis and Strategic Management 15(1), 3-18.

Bugge, M., Hansen, T., Klitkou, A. (2016). What Is the Bioeconomy? A Review of the Literature. Sustainability 8(7): 691.

Danielsen, F., Beukema, H., Burgess, N.D., Parish, F., Bruhl, C.A., Donald, P.F., Murdiyarso, D., Phalan, B., Reijnders, L., Struebig, M. & Fitzherbert, E.B. (2008). Biofuel plantations on forested lands: double jeopardy for biodiversity and climate. Conservation Biology 23(2), 348–358.

Deuten, J., & Rip, A. (2000). Narrative Infrastructure in Product Creation Processes. Organization 7(1), 69–63.

Doezema T. and Hurlbut JB. (2017) Technologies of Governance: Science, State and Citizen in Visions of the Bioeconomy. In: Pavone V and Goven J (eds) Bioeconomies: Life, Technology, and Capital in the 21st Century. Basingstoke: Palgrave Macmillan, pp. 49–71.

European Bioeconomy Stakeholders Panel. (2017) European Bioeconomy Stakeholders Manifesto. Available online: https://ec.europa.eu/research/bioeconomy/pdf/european_bioeconomy_stakeholders_manifesto.pdf (accessed 15.6.2020)

European Commission (2012). Innovating for Sustainable Growth: A Bioeconomy for Europe. Luxembourg: Publications Office of the European Union.

European Commission (2016). Accelerating Clean Energy Innovation. COM(2016) 763 final.

European Commission (2017). Review of the 2012 European bioeconomy strategy. Luxembourg: Publications Office of the European Union.

European Commission (2018). A sustainable Bioeconomy for Europe: strengthening the connection between economy, society and the environment – Updated Bioeconomy Strategy. Luxembourg: Publications Office of the European Union.

European Commission (2018b). A Clean Planet for all: A European strategic long-term vision for a prosperous, modern, competitive and climate neutral economy. COM(2018) 773 final.

European Commission (2019). The European Green Deal. COM(2019) 640 final.

European Environment Agency (EEA). (2013). EU bioenergy potential from a resource-efficiency perspective. EEA Report No 6/2013, Luxembourg.

Galtung, J. (1981a). Social Cosmology and the Concept of Peace. Journal of Peace Research 1981, 18 (2).

Galtung, J. (1981b). Western Civilization: Anatomy and Pathology. Alternatives 1981, vol. 7

German, L., Schoneveld, G., Skutch, M., Andriani, R., Obidzinski, K., Pacheco, P., Komarudin, H., Andrianto, A., Lima, M. & Dayang Norwana, A.A.B. (2010). The Local Social and Environmental Impacts of Biofuel Feedstock Expansion a Synthesis of Case Studies from Asia, Africa and Latin America. Center for International Forestry Research: Info Brief. No. 34.

Goven, J Pavone V. (2015). The Bioeconomy as Political Project: A Polanyian Analysis. Science, Technology, & Human Values 40(3), 302–337.

Inayatullah, S. (2009). Causal Layered Analysis: An Integrative and Transformative Theory and Method. In J. C. Glenn & T. J. Gordon (Eds.), Futures research methodology. The Millennium Project.

Jackson, T. (2009) Prosperity Without Growth. London: Routledge.

Kleinschmit, D., Arts B., Giurca A., Mustalathi I., Sergent A. and Pülzl H. (2017). Environmental concerns into political bioeconomy discourses. International Forestry Review 19(1), 41-55.

Kröger, M., Raitio, K., (2017). Finnish forest policy in the era of bioeconomy: A pathway to sustainability? Forest Policy and Economics 7, 6-15 .

Kuusi, O. (1999). Expertise in the Future Use of Generic Technologies. Epistemic and Methodological Considerations Concerning Delphi Studies. Acta Universitatis oeconomicae Helsingiensis. A, ISSN 1237-55e6X; 159.

Linstone, H.A. & Turoff, M. (1975). The Delphi Method, Techniques and Applications. Boston: Addison-Wesley Publishing Company.

McCormick, K., Kautto, N., (2013). The Bioeconomy in Europe: An Overview. Sustainability 5: 2589–2608.

Mittra, J., & Zoukas, G. (2020). Unpacking the Concept of Bioeconomy: Problems of Definition, Measurement, and Value. Science & Technology Studies, 33(1), 2–21.

Obidzinski, K., Andriani, R., Komarudin, H. & Andrianto, A. (2012). Environmental and social impacts of oil palm plantations and their implications for biofuel production in Indonesia. Ecology and Society 17(1), 25.

OECD (2006). The Bioeconomy to 2030: Designing a Policy Agenda. Paris.

OECD (2009). The Bioeconomy to 2030. Agenda. Paris.

OECD (2014). Sustainable Biomass Drives the Next Bioeconomy: a New Industrial Revolution? Report of an OECD workshop, 10-11 June, OECD, Paris.

OECD (2016). Building a sustainable bioeconomy: a framework for policy.

OECD (2018). Realising the circular bioeconomy. OECD Science, Technology and Industry Policy Papers No.60.

PBL Netherlands Environmental Assessment Agency (2012). PBL Note. Sustainability of biomass in a bio-based economy. A quick-scan analysis of the biomass demand of a bio-based economy in 2030 compared to the sustainable supply. PBL Publication number 500143001.

Partanen, J., Paloheimo, H., & Waris H. (2014). The World After Cheap Oil. Routledge.

Petersen, A., & Krisjansen, I. (2015). Assembling 'the bioeconomy': Exploiting the power of the promissory life sciences. Journal of Sociology, 51(1), 28–46.

Pfau, S.F., Hagens, J.E., Dankbaar, B., Smits, A.J.M., 2014. Visions of sustainability in bioeconomy research. Sustainability 6: 1222–1249

Philp, J.C. (2015). Balancing the bioeconomy: supporting biofuels and bio-based materials in public policy. Energy & Environmental Science 8, 3063-3068.

Polak, F. (1973). The image of the future. Amsterdam: Elsevier Scientific Publishing.

Pülzl, H. et al. (2017). The role of forests in bioeconomy strategies at the domestic and EU level In: Winkel, G. (ed). Towards a sustainable European forest-based bioeconomy – assessment and the way forward. What Science Can Tell Us 8. European Forest Institute.

Ramcilovic-Suominen, S., Gritten, D., & Saastamoinen, O. (2010). Concept of livelihood in the FLEGT voluntary partnership agreement and the expected impacts on the livelihood of forest communities in Ghana. International Forestry Review 12(4), 361–369.

Ramcilovic-Suominen, S., Matero, J., & Shannon, M. (2013). Do forest values influence compliance with forestry Legislation? The case of farmers in the fringes of forest reserves in Ghana. Small-scale Forestry 12(2), 235–256.

Ramcilovic-Suominen, S., Pülzl, H (2018) Sustainable development – A 'selling point' of the emerging EU bioeconomy policy framework? Journal of Cleaner Production 172, 4170–4180.

Staffas, L., Gustavsson, M., McCormick, K., 2013. Strategies and policies for the bioeconomy and biobased economy: An analysis of official national approaches. Sustainability 5: 2751–2769.

Tapio, P. (2002). The Limits to Traffic Volume Growth, The Content and Procedure of Administrative Futures Studies on Finnish Transport CO2 Policy. Turku: Acta Futura Fennica 8, Finnish Society for Futures Studies & Finland Futures Research Centre. Doctorate thesis. http://ethesis.helsinki.fi/julkaisut/maa/limno/vk/tapio.

Van Dam, J. and M. Junginger (2011). Striving to further harmonization of sustainability criteria for bioenergy in Europe: Recommendations from a stakeholder questionnaire. Energy Policy 39, 4051–4066.

Van Lente, H. (1993). Promising Technology: The Dynamics of Expectations in Technological Developments. Amsterdam: Proefschrift.

Varho, V, Rautiainen A, Peltonen M, Niemi J, Ovaska U. (2018). Biopaths to Carbon Neutrality. Publications of the Ministry of Agriculture and Forestry (Finland) 2018:3.

Winkel, G. (ed). 2017. Towards a sustainable European forest-based bioeconomy – assessment and the way forward. What Science Can Tell Us 8. European Forest Institute.

Part IV
Agro-Food and Healthcare Advancements

Chapter 10
Technological Landscape of the Agriculture and Food Sector: A Long-Term Vision

Leonid Gokhberg, Ilya Kuzminov, and Elena Khabirova

Abstract This chapter presents the overview of global challenges and trends, as well as technological landscape and future prospects for science, technology and innovation (STI) in agriculture and food sector. Our study is based on a systemic mapping of trends and technologies with the combination of big data analysis (text mining) and expert-based methods. The focus of the study is the interaction of agri-food sector with biotechnology and information technology domains in shaping the future of bioeconomy-driven sustainable and socially inclusive bioagrifood sector. The latter is understood not only in terms of food production but also as a societal ecosystem providing opportunities for human-centred and environment-oriented activities on the land involving certain groups of the population based on traditions and collective or individual values. New opportunities for long-desired wide implementation of the principles of sustainable development and bioeconomy provided by breakthroughs in biotechnologies, nanotechnologies and artificial intelligence give hopes of deep transformation in the sector not only in the most technologically and economically advanced economies but also in mid-tier countries.

Keywords Agriculture · Food sector · Technological landscape · Global trends · Foresight · Trend spotting · Big data · Text mining

L. Gokhberg (✉) · I. Kuzminov · E. Khabirova
Institute for Statistical Studies and Economics of Knowledge, National Research University Higher School of Economics, Moscow, Russia
e-mail: lgokhberg@hse.ru; ikuzminov@hse.ru; etochilina@hse.ru

E. Koukios, A. Sacio-Szymańska (eds.), *Bio#Futures*,
https://doi.org/10.1007/978-3-030-64969-2_10

1 Introduction

New technologies play a pivotal role in allowing present-day agriculture and food sector on global, national and local levels to answer mounting challenges of population growth, food consumption increase, mineral and land resources scarcity, and ongoing transformation of ecosystems and climate. A fast-growing population, together with steadily declining agricultural areas, require entirely new approaches to food production in emerging bioeconomy. To pave the grounds for production increases, advanced technological solutions and their timely adoption are urgently required (Sachs et al. 2019; Zaidi et al. 2019; Venkatramanan and Shah 2019). Farming yields depend greatly on the applied farming technologies like machinery; genetically modified organisms; synthetic pesticides, herbicides and fertilisers; and information and communication technologies (ICT) (e.g. Coomes et al. 2019; Balakrishnan et al. 2017). Especially the latter benefited greatly from the development of infrastructure, like the availability of broadband Internet connections in remote areas (Briglauer et al. 2019), enhancing both social and professional opportunities for rural communities (Bowen and Morris 2019), and technical feasibilities for radically new methods and instruments of agricultural production and its management and control, including smart fields (Bach and Mauser 2018), Internet of things (Ayaz et al. 2019), autopiloted machinery (Ghobadpour et al. 2019), swarm robotics (Vu et al. 2017), lean logistics control (Chen et al. 2020), etc. Most of these technologies do not derive out of agriculture but influence today's agriculture greatly. For example, precision agriculture (PA) exploits both biotechnology discoveries (on micro level and on the ecosystem level) and ICT applications for soil and crop surveillance, while mapping technologies based on satellite data together with unmanned aerial vehicles (drones) for crop scouting have massively increased real-time awareness of the situation on the land leading to reduction in risks and losses (Fraser 2019). In fact, emergence of a biotech-ICT nexus in agriculture is increasingly evident in recent years (Kim 2019), as data processing, artificial intelligence and robotic automation become crucial means of producing and applying bioscience discoveries (Streich et al. 2020) in the form of technologies and providing implementation of such technologies on a physical level through infrastructure (Dubé et al. 2018) and machinery "on the field" (Wahby et al. 2019). However, a serious challenge to new technological transformation of agriculture and food sector is political (Klerkx and Rose 2020), communal (Adnan et al. 2018) and individual resistance (Ugochukwu and Phillips 2018) to technology innovations dictated by market protection; unscientific preconceptions about the harm of new technologies, especially related to gene modification; and the drive to preserve traditional way of business and rural life style. For example, public perception strongly opposes genetically modified organisms despite scientific proof of their harmlessness and evident benefits (Jiang et al. 2018). The cause for superficial (apparent) resistance in some cases could be the lack of resources for acquisition of new expensive technologies. Economic factors in such situations are made to be falsely perceived by wider audiences as ethical in nature for the sake of political gains by the parties declaring such views (Bartkowski et al. 2018). Consequently, the socio-technological regime of agriculture production involves long-standing cognitive, social and institutional processes which are not easy to change (Geels 2020). The successful adaption of new technologies builds on coalition networks involving a multitude of actors in the production regime (e.g. Levy and Lubell 2018).

Bioeconomy transition is especially important for agriculture and food sector of both southern and northern countries in view of ongoing climate change with increasingly evident and severe effects of it: land productivity loss, yield instability across seasons and disruption of local communities' livelihoods. Biotechnologies are needed both on micro (genomics) and macro (ecosystems engineering) levels (Corlett 2017). They are needed not only to reduce global warming drivers (e.g. livestock breeds with less methane-producing metabolism, etc.) but also to make agricultural ecosystems more resilient to climate change effects (drought-resistant crops; introducing natural enemies of pests enwidening their habitat areas) (Lokko et al. 2018; Loboguerrero et al. 2019). Climate change and an increase of anthropogenic pressure on the environment entail biodiversity reduction and degradation of ecosystems due to environmental pollution, increasing the likelihood of natural hazards and environmental disasters. The trend is intensified by the depletion of natural resources and the intensification of competition for them, a decrease in the agroclimatic potential of the planet and the exhaustion of the long-term effects of the "green revolution" on the progress of agriculture (Kumar et al. 2017). For example, in the US agricultural sector, while preserving the existing prerequisites for farming in the future, a 50% corn harvest reduction is forecasted to 2100 (Strickland 2017). Despite some beneficial effects of climate change for the Nordic countries due to the expansion of agriculturally suitable areas, in the southern regions, such changes will lead to droughts and soil erosion. An ambiguous effect on the development of the agriculture sector will also have an increase in average winter temperature. A longer growing season will contribute to an increase in productivity, while a change in the thermal regime will contribute to intensifying activities of harmful microorganisms and increasing the habitat of agricultural insect pests, from which global agricultural production losses are already estimated at 20–40% (FAO 2019b). The rate of change of climatic regimes significantly exceeds the rate of adaptation of crops to them, which leads to undermining agricultural productivity. In this regard, there is a transition to intensive forms of agricultural production, and the role of technological advances (especially in biotechnologies) in overcoming the effects of global warming is increasing.

Demographic trends and global demand trends tightly linked to them play extremely important role in setting the agenda for bioeconomy transformation of agriculture and food sector. The main demographic trend that has a direct impact on the development of the agriculture and food sector will be an increase in the global population and a corresponding increase in the demand for food products by 80% by 2100 relative to 2010 (Depenbusch and Klasen 2019). The urgent problem of structural unemployment in the rural areas is associated not only with qualitative technological changes in the agricultural sector but also with the multiscale disruptions of agricultural markets against the backdrop of the transportation and social infrastructure deficiencies. A decrease in the attractiveness of employment in agriculture along with the automation of key processes will determine the low demand in the labour market and the corresponding intensification of urbanisation processes. These factors, along with high differentiation of the rural and city population by income level, indicate the need for a policy aimed at mitigating the effects of rural depopulation and integrating agricultural producers into the new technological structure of the agriculture and food sector. Other demographically important trend is dangers from infectious and parasitic diseases, which, due to the growing world population and the creation of favourable conditions (climatic, political, etc.) for their spread, will be characterised by their geographical expansion and increased resistance to existing drugs (Rohr et al. 2019). A significant impact on the development of the agricultural sector will also have a global change in patterns of food consumption. Raising public awareness of healthy nutrition and increasing demand for organic and eco-products will help accelerate the development of relevant markets and increase interest in the products of family farms and private farms. The specificities of consumer demand of the generation of millennials (born in 1984–2000) and the "generation Z" (born after 2000), which in many respects form the main trends in the markets, are also important. According to surveys, millennials are about 20% more likely to be consumers of organic foods than any other generation (Vilceanu et al. 2019). The variability of agricultural consumption patterns also contributes directly to inefficient food use. According to FAO estimates, around 14% of the world's food is lost from production before reaching the retail level (FAO 2019a). All these demographic aspects raise dilemmas in the domains of sustainability, social inclusion and emerging new wave rural development as integral part of the agricultural sector agenda.

Sustainability, social inclusion and rural development become increasingly more important stream of the bioeconomy transformation of the agrifood sector. Agricultural and food sector (transforming nowadays into bioagrifood sector within the bioeconomy) has long since been reconsidered not only as the means of production of food (and related) products and raw material for them in cost-effective manner to produce enough surplus for feeding city populations but also as a societal system aimed at preserving livelihoods of rural territory residents in environmentally sound ways (Torre and Wallet 2020). Even (practically infeasible) full automation of production, restoration, logistics, maintenance and control operation on rural territories within agricultural production systems would not render rural populations redundant (Rotz et al. 2019). Their role is significantly wider than such stipulated by narrow economic utilitarian models (Liu and Li 2017). While highly dense rural population living in bad sanitary conditions and producing high pressure on the local ecosystem services (tragedy of the commons) is an unsustainable scenario for a long run, further urbanisation, intensification of production and decoupling of production of organic materials (including food) from open environments (recirculative aquaculture instead of fisheries, industrial greenhouse facilities and hydroponics instead of fields, large-scale industrial bioreactors, etc.) are definitely needed; at the same time, preservation of basic settlement networks in rural areas across the globe is an imperative. It is dictated by priorities of preserving cultural heritage

(Salpina 2019), respecting human rights to live in the way they prefer (Bellows 2019), preserving vigilance/overwatching across wide areas (Mackay and Perkins 2019), availability of informal human help for business and recreational travelers through vast expanses of rural areas in emergencies making multimodal connectivity among city centres more resilient (Knickel et al. 2018) and preserving species, breeds and techniques of agricultural production and of long-term survival in non-urban territories for the sake of diversification of ways of adaptation of humanity as biological species (Koohafkan and Altieri 2016).

The latter aspect becomes ever more evident, and practical, in view of recent global pandemic (Karabag 2020) demonstrating risks of densely populated city dwelling as long-term survival strategy for humans (Gibson and Rush 2020), leading to disruptions and collapses of production chains in many industries, including high-tech factors of agricultural production, such as fertilisers (Marlow 2020), and agricultural production itself. As a setback of global trade as a guaranteeing factor of food security, the COVID-19 pandemic has led within one agricultural season to countries reconsidering their agroexport strategies, in extremes going from stimulating exports to preparing stress action plans of restricting any export for preserving as large food reserves as possible within national territory (World Trade Organization 2020). The 2020 disruption is a lesson to be learned about the importance of putting limits to urbanisation, centralisation of production and globalisation of value chains (Bedford et al. 2019).

The outlined big picture of global challenges and trends in modern society demonstrates clearly that technological landscape of the agriculture and food sector must not be analysed narrowly, with technical or financial biases, but instead must be looked at holistically, with human, environment and ethicist questions in the centre. However, to make the first step towards such complex human-oriented analysis, we must initially define the scope by identifying and showcasing (based on objective data and objective criteria) the main technologies and technology-related forks and scenarios of development. The discovery, systematisation, registry and overview of bioeconomy-related technology trends are the task of current study.

2 Methodology

Landscape mapping studies, including trend spotting, new technologies discovery, actor mapping and centres of competence benchmarking, historically had been an integral part, not explicitly positioned, of various strategic studies aimed at stating the current conditions and prospects of development of countries and regions, sectors of economy and domains of knowledge. With the development of strategic analytical studies as a specific activity meeting needs of high-level – and increasingly often *distributed multilevel* – decision-makers, foresight studies, including Future oriented technology analysis (Cagnin et al. 2008) and Science and Technology Foresight (ForSTI) (Miles et al. 2016), emerged as a special framework of such research, with various mapping activities being formally named and methodologically scoped within. Within this logic we implemented our current overview of technology trends in agriculture and food sector as a systemic text mining-enabled foresight study.

Foresight methodology has become central for enabling future-oriented strategic decision-making across economic sectors and domains. One of the most important components of the system of information and expert-analytical support for making long-term decisions in different countries of the world is the forecasts of scientific and technological development, regularly developed and updated using the foresight methodology at different levels (national, industry, regional and corporate) (Calof et al. 2020). Foresight methodology is actively used in the formation of priority areas for the development of science and technology, lists of critical technologies and developing state forecasts of scientific and technological development (e.g. see Sokolov and Chulok 2016). The creation of constantly functioning organisational systems of technological forecasting at the level of economic sectors could be beneficial for effective long-term strategic vision. Such systems are intended for early informing decision-makers about global challenges and trends, new threats and opportunities for scientific and technological development (Porter and Cunningham 2004; Bakhtin et al. 2017). In the face of growing global challenges associated with the emergence of fundamentally new technologies that are already affecting the development of the agricultural sector, in particular platform (convergent) technologies (ICT, bio- and nano-, aerospace, robotic) in agriculture and animal husbandry, the most important goal of agricultural policy should be the transformation of the agriculture into a modern high-performance sector due to technology-enabled significant growth of the technological level and innovative activity of enterprises and farms. Ultimately, foresight can enable future-oriented strategic decision-making in the sector stimulating quicker rates of large-scale introduction of cutting-edge technologies that determine the ability to produce agricultural and food products competitive in global markets.

Foresight methodology – landscape mapping,trend spotting, horizon scanning and consensus on the collectively constructed future – is the core of the presented study. The data on which this chapter builds was generated through systemic identification of trends in the agriculture and food industry through various methods including big data processing techniques for bibliometric and patent analysis and mining of semantics of international planning documents from, e.g. the Organisation for Economic Co-operation and Development (OECD), the Food and Agriculture Organization (FAO), the United Nations Environment Programme (UNEP) or the World Bank, as well as direct text mining of scientific research articles, patent application (from Microsoft Academic Graph and USPTO open databases, accordingly) and press releases. In total, more than 10 million documents were processed using iFORA system, developed at HSE ISSEK, including 160,000 documents in the agriculture and food sector including research, analytical, strategic and news documents in English, Russian and Mandarin. In this study we concentrate on global technology trends, i.e. major, primarily future-oriented trends present in many countries. The "future-oriented" property means we are mostly interested in trends which will make the biggest impact on the future shape of the economic sector under consideration. Accordingly, specific trends have or have not been included in the short final list on the basis of their growth potential in the foreseeable future and their ability to engender further significant shifts.

Expert's insights in the analytical process have not been rendered by big data approaches any less important than in the past. The authors of this study hold an opinion that given the current development level of information technologies (which is clearly insufficient for full-fledged semantic analysis of texts humans are capable of and computerised synthesis of new meanings), text mining or other big data analytical tools cannot yet fully replace experts when it comes to studying existing trends or creating visions of future socio-economic systems (or their specific components). Therefore, expert's polling validated the results of the text mining methods. Detecting these trends in the first place, and selecting relevant keywords for quantitative analytical methods, was possible on the basis of an expert's review of analytical literature on the agricultural sector, more than 40 in-depth expert interviews, and more than 20 expert's discussion sessions within the framework of several projects for key stakeholders in the sector beginning from the early 2016 to late 2019. These procedures have been constantly intertwined with big data scanning efforts, with iterative results of both activities consecutively applied as an input conditions for next steps (lists of trends are evaluated by experts and additional search (scope) conditions formulated by them; they are used as seeds for text mining of trends on additionally enwidened and renewed database of documents; new lists of trends and technologies with more sophisticated tags and metrics developed during the elapsed period by engineers are fed back into expert panels and surveys, etc.). Overall, four full cycles (iterations) of expert data trend spotting and technology landscape mapping have been implemented.

Ways of blending expert and database methods were diverse during the study and have been consistently developing within the time span of the study. However, they have been constantly revolving around expert- and machine learning-based feedbacks to, respectively, big data analysis and expert knowledge extraction results. Developed analytical apparatus was tested on materials of more than 400 questionnaires filled in by the surveyed experts, with both open and closed questions about global technology trends in the agricultural sector (Gokhberg et al. 2017). The key results of the survey included the estimates by the experts of trend systemic significance, prospects of growth and specific time horizons when the effects of trends were expected to fully emerge. Later, instruments of extraction of expert statements from openly published sectoral commentary materials started to be used to approximate the possible opinions of the experts based on their digital traces in the absence of opportunity to interview or survey all the influential industry experts. The additional layer of formal tagging of agriculture technology-related terms included 30,000 individual terms, prefiltered by their information content and industry specificity and ranked by their frequency and growth of frequency for the last 10 years and then tagged by experts as carrying or not carrying valuable insights into the technological landscape of the sector.

Based on this method, necessary news items were extracted from big collections of documents and evaluated and ranked according to the density of mentions of high-technology-related terms, and as a result 536 unique events of opening of new or modernisation of existing production facilities were identified for the analysed period and assessed on the types of advanced bioeconomy technologies used.

2.1 Mapping of Global Trends in Agriculture and Food Sector

Five main groups of trends in agriculture and food sector have been discovered in the study, with the relative weights representing structural complexity and number of sub-trends identified in each category:

- Economic (the transition to new models and resources of economic growth) (weight 4)
- Social and value (demographic and social transformations) (weight 6)
- Environmental and natural resources (changing the natural environment) (weight 7)
- Policy and institutional (geopolitical transformation and changing systems of global governance) (weight 3)
- Technological (the formation of a new paradigm of scientific and technological development) (weight 10)

It should be noted that the weights may represent the possible inherited topical asymmetries of the study induced by historical focus of both foresight researchers and agriculture stakeholders in the country on technological aspects of industry development. The other reason could be possible skewedness of textual sources at large (not those downloaded for the text mining processing in iFORA, but those presented in the open access documental sources). On a global scale, documents with more focus on highly controversial environment protection and social justice topics are more abundant, while policy and institutional agenda in clearly vocalised and concentrated form is only presented in relatively small numbers of strategy documents of international organisations, such as OECD, UN UNIDO, UN FAO and UN UNESCO (the UN organisations most relevant to agriculture by agenda as examples).

The group of global trends "Transition to new models and resources of economic growth" is manifested in a change in the already formed global value chains and the emergence of new ones; the transition to new models of innovation; customisation of production and consumption; the emergence and spread of new business models; and structural changes in the labour market.

Demographic and social transformations are associated with an increase in the predicted duration of a healthy active life in leading countries, an increase in urbanisation processes and a move towards a "smart city" model, an increase in international migration, an increase in social inequality and the emergence of new social classes, the spread of social innovations and digitalisation of society and transformation of education system. An increase in the world population and in the share of urban residents will lead to a negative impact on the ecosystem and urban infrastructure, which can be offset by the introduction of smart technologies in the urban environment.

Changes in the natural environment have the greatest impact on the development of the agro-industrial sector, since it causes a reduction in biodiversity and degradation of ecosystems and increases the likelihood of dangerous natural phenomena and environmental disasters. These trends reinforce the trend of depletion of natural resources (mineral, water, land, forest, etc.) and increase competition for them, which will contribute to the transition to a closed-loop economy.

The group of global trends "Transformation of the geopolitical situation and changing global governance systems" is caused by the transition to a multipolar geopolitical model of the world and the intensification of the struggle of states for spheres of influence, the growth of regional instability and the aggravation of world security problems. The answer to these challenges is to increase the requirements for the effectiveness of relevant instruments and institutions, the emergence of new international and regional blocs and unions and the changing role of states.

The formation of a new paradigm of scientific and technological development is associated with a new industrial revolution – the widespread dissemination of ICT, biotechnology, robotics and artificial intelligence, wide practical use of materials with new properties, next-generation electronics and new energy sources and methods of their storage and transmission. The most important factor in the transition to a new paradigm is the digitalisation of the economy, which also affects science and contributes to the development of new methods and tools for conducting research, which now presents new requirements for their organisation in the agricultural sector and related fields.

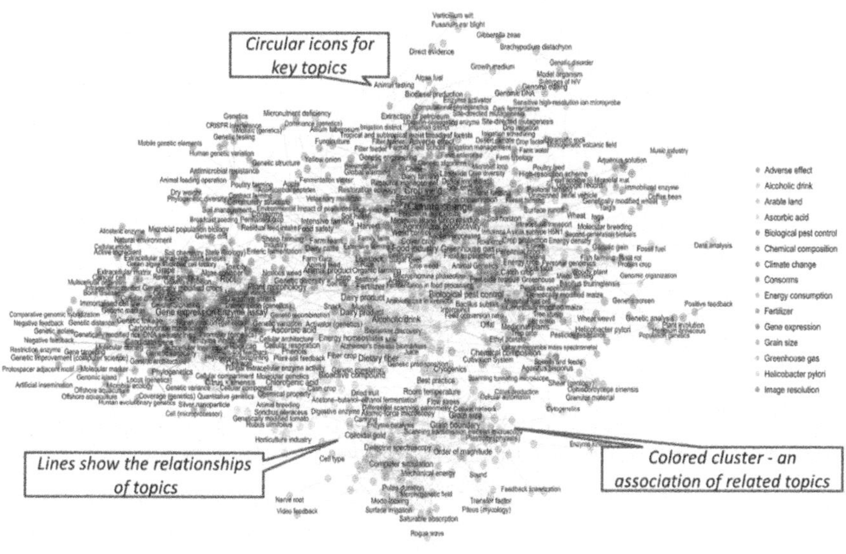

Fig. 10.1 Global technological landscape of agriculture: analysis of scientific publications. (Source: based on iFORA system©, developed at HSE ISSEK)

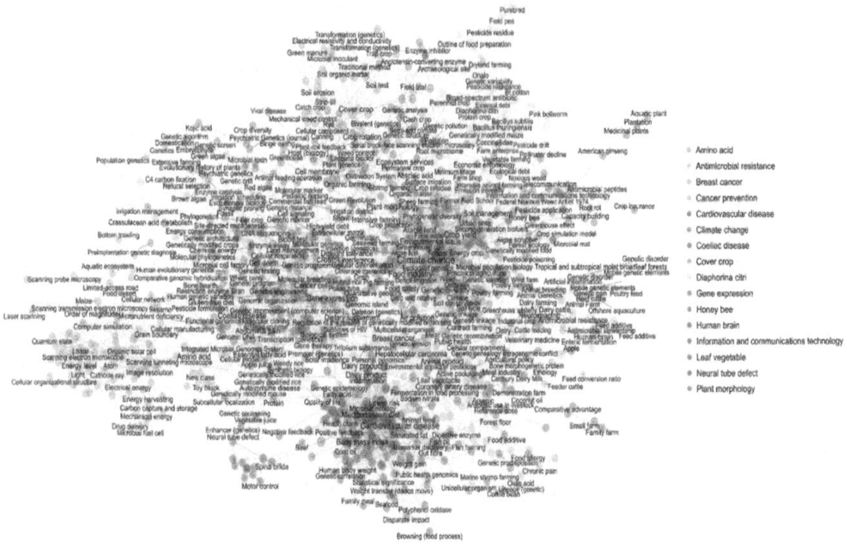

Fig. 10.2 Global technological landscape of agriculture: analysis of media resources. (Source: based on iFORA system©, developed at HSE ISSEK)

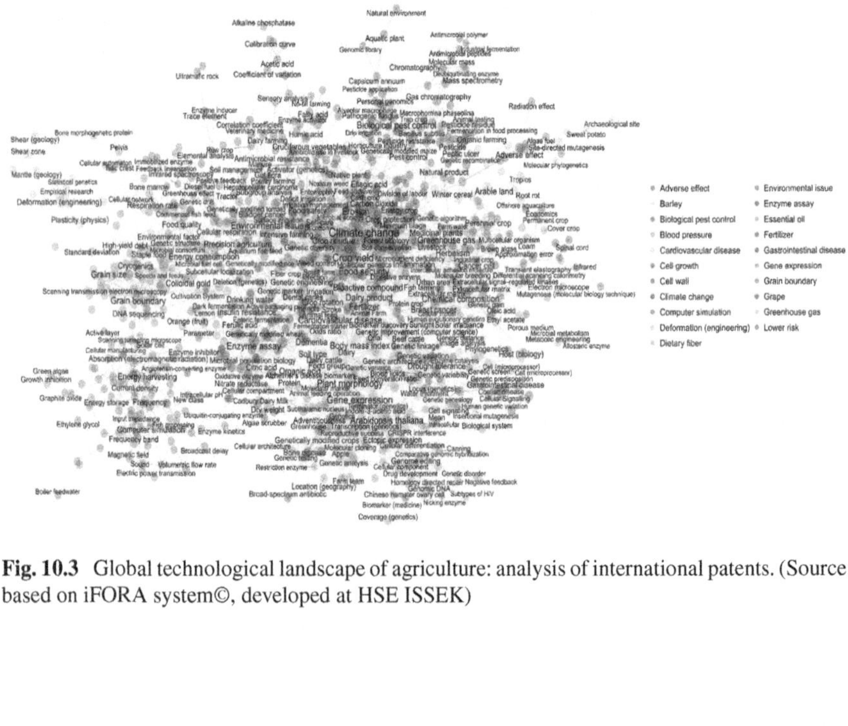

Fig. 10.3 Global technological landscape of agriculture: analysis of international patents. (Source: based on iFORA system©, developed at HSE ISSEK)

In order to visualise the above trends, semantic maps were generated based on text mining of large amounts of documents. They are automatically generated visualisations of the content of scientific publications (Fig. 10.1), media (Fig. 10.2) and international patents (Fig. 10.3). All they show the most relevant topics within the studied corps of agriculture-related documents.

On the map of scientific articles describing agricultural technologies (Fig. 10.1), the central cluster focuses on the trend of climate change. Closely connected with it are such popular topics as crop yields, food security, precision agriculture and minimal tillage. A cluster of technologies based on genetics is also highlighted (related topics on the map, DNA sequence, enzyme immunoassay, molecular cloning, plant morphology, crops), a cluster of "nutritional supplements" (related topics on the map, bioactive compounds, dairy products, biomarkers, diseases), etc.

On the semantic maps based on the analysis of the media (Fig. 10.2) and patent databases (Fig. 10.3), the cluster of "climate change" occupies a central place too, and the topics closest to it are high-precision agriculture, ecosystem services and organic agriculture.

The developed approach, based on the analysis of big data, allows, in our opinion, to present an objective picture of the technological development of agriculture.

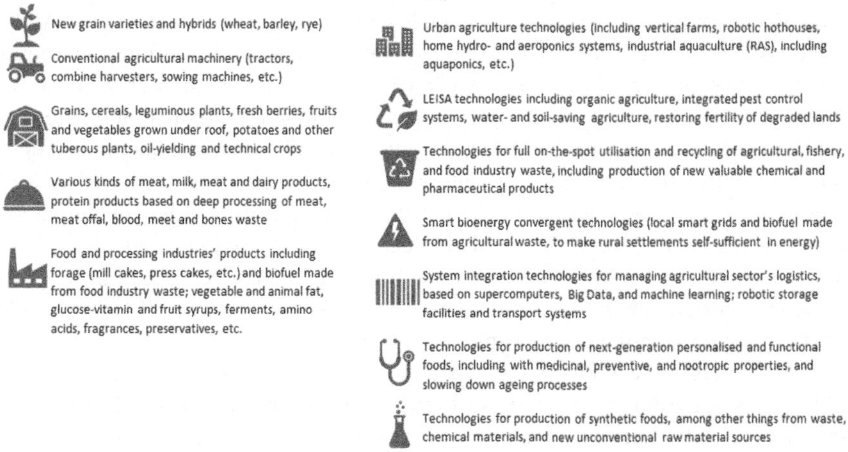

Fig. 10.4 Vision of global technology trends in agriculture and food sector. (Source: compiled by the authors)

2.2 Consensus Vision of Global Technology Trends in Agriculture and Food Sector

In order to ensure a balanced vision of global technology trends towards a sustainable and ethical bioagrifood sector, we combined big data-driven analytical interpretation with pre-existing expert surveys.

Figure 10.4 contains an overview of the major current technology trends determining the agrifood sector capabilities to cope with biological limitations of the global ecosystem as well as the long-term future trends for its bio-sustainable transformation (Fig. 10.4).

The most significant narrow (specific) technology trends defining sustainable and circular bioeconomy possibilities for the sector include bio-IT technology nexus elements of future bioagrifood sector. Pivotal role in this nexus pertains to bioscience discoveries and instruments on all levels from molecular to ecosystem, big data processing, machine learning and artificial intelligence as enablers of biosciences and biotechnologies, robotics and automations and additive manufacturing as a base for wide-scale implementation of biobased principles of new architectures of sustainable agricultural production systems, as well as a host of other enabler IT-based technologies, such as online platforms, Internet of things, software as a service and cloud services at large, artificial intelligence technologies, digital realtime mapping of agronomic and weather processes and geoinformation systems.

From the analysis of global and national trends in technology development of agricultural production, it becomes clear that the main driver of agriculture modernisation towards more environmental sustainability and social inclusion lies in the domain of bioeconomy; however, at the same time, the instrumental platform for practical implementation of life sciences and biotechnology innovations is information and robotic technology. Artificial intelligence, smart systems, distributed online platforms and the Internet of things play pivotal role in extracting value from large amounts of genetic data to produce valuable genetic solutions for agriculture. At the same time, the same basic information technologies applied for different purposes form the base of smart farm, smart factory and robotic automation on-premise for direct technical implementation of biotechnology-based production processes. Value of such production automation lies predominantly not in the labour cost optimisation but in the fact that only automatic business processes exclude risks of human errors, protocol breach risks, as well as biosecurity and biocontamination issues. Complex and therefore fragile biotechnology underlying the emerging production processes in agriculture requires robotic automation because of high sensitivity to implementation precision, while the total associated costs of labour would be increasing, rather than decreasing.

Main emerging technologies in the forming "biogenetic/inforobotic" (or "bio-IT") nexus, according to the results of our study, will include, first of all, artificial intelligence (AI), machine learning, big data processing and data mining for analysis of several types of information:

- Genetic information for producing genetically modified subspecies and organisms, treatments and control instruments (tests, etc.), both by specialised biotechnology firms for industry-wide application and by agricultural producers of their cooperative networks on-premise, on-demand and ad hoc
- Landscape-level information on climatic and weather conditions, geohydrology, soil, biome and other factors and processes within smart fields and their systems for consequent adaptive application of biotechnology-based solutions to concrete localities in a robotic-automated and automated control fashion
- Multiscope and multilevel business process information for risk modelling, forecasting and preventing risk implementation
- "AI bosses" which is a technology of real-time planning hardware-software systems giving recommendations on the optimal series of actions for workers based on information from networks of sensors and cameras (mainly needed for micro-level individual logistics planning either on the field or within a warehouse of factory)

Another important component of a new wave of global technology trends in agriculture and food sector are biotechnologies, including both micro- (genomics, proteomics, etc.) and macroscale (environmental engineering, natural enemies approach, etc.) as well as organic-chemical, nano- and specialised material engineering technologies as suppliers of base means of implementation for the biotechnologies. This picture is complemented with industrial automation technologies transferred from traditional large-scale manufacturing to the domain of solutions for small distributed local agricultural production smart factories and even smart nodes (small-scale highly specialised fully automated unserviceable production facilities for preprocessing of agricultural produce at the level of servicing tens of hectares articular plots, interconnected into smart and agile logistical infrastructures; such facilities can be either dormant most part of the year or mobile). Another important IT component consists in robotic technologies, mainly in the domains of swarm robotics for enhanced and partially automated management of fleets of specialised and universal agricultural machine deployment and operation in a synergetic manner and robots with handpieces (manipulators) allowing high-precision imitations of functions of human palms and fingers which are characterised by exceptionally high precision and agility in implementing physical operations. Bio-IT nexus cannot be also viewed without low-cost small-scale construction technologies, machine node on-demand production for on-premises technical service technologies, biomaterial, soil-imitates and food printing technologies based on additive manufacturing and 3D printing technology mega platform.

One of the key parts of a new wave of global technology trends in agriculture and food sector is software-as-a-cloud class enterprise resource management systems, business process management systems, integrated virtual environments, online collaborative tools, remote work operations and pooling software (e.g. for outsourcing of co-piloting of smart agricultural machines by humans in a time-sharing and on-demand time-switching manner; pilot work from teleworking centres using computer simulators of control gear of machines and switch on-demand for short periods of time to manually pilot machinery in complex piloting situations, which AI encountered problems with).

Many experts also see high potential for agriculture and food sector in distributed, nanosensor- and high-precision long-distance-sensor-based autonomous networks for big data collection necessary to constantly feed new data into AI models managing agricultural production on a macro-environment level (Internet of things, smart field and smart plot sensor networks, geopositioning, navigation and remote sensing systems), as well as the following digital enterprise technologies and novel and emerging solutions in digital product and digitised production:

- Managing sources of infrastructural transformations of the production cycle using artificial intelligence
- Machine learning industrial applications
- Big data search and processing
- Industry 4.0 as model of integration of production processes
- Internet of things
- Digital twins:
  - Digital twins of key industry objects and processes
  - Digital twins of complex products
- Cloud data storage and personalisation of data access
- Real-time cloud computing and performance adjustments
- Predictive analytics
- Optimisation of business processes using digitalisation tools with continuous analysis of production changes
- Digitalisation of new product development:
  - Development of completely new products using artificial intelligence algorithms (calculations of product properties)
  - Application of artificial intelligence algorithms in biotechnological research, agricultural chemistry, etc.
- Product condition monitoring:
  - Products as a source of signal generation and information transfer
  - Monitoring product status online or on demand
  - Contact and non-contact methods and systems for monitoring the condition of products (including using unmanned vehicles, etc.)

Table 10.1 Long-term technology trends implementation in the Russian agrifood sector investment projects

Bioeconomy component	Example of a project	Number of projects 01 Jan 2018–01 April 2020	Predominant applicants
Special isolated environment for highly specialised breeds of animals	"Roshinsky" rabbit production factory, near city of Tyumen	29	Rabbit, poultry, pig farms
Highly artificial, isolated and biosecure environments of specialised cultures	Bogorodskie Ovoshy mushroom facility, near town of Elektrostal	9	Mushroom farms, some hothouses
Robotic hothouse, smart hydroponics and lighting based on genetic profile and development process of plants	Noviy hothouse complex, city of Cherepovets	27	Most advanced, large-scale hothouses
Full automation of collection, preprocessing, biosecurity control, food security control and packing of bioproduction	MolProduct, city of Vladikavkaz	196	Milk, meat, fish plants
Smart animal biosecurity	Milking cow farm "Avangard", new city of Ryazan	27	Rabbit, poultry, pig farms
Robotics used in seeding, watering, harvesting in enclosed environments		49	Most new hothouses
Robotics used in feeding, milking, manure removal and other operations with animals	Fully robotised cow farm" Red Bashkirya", Abzezillov raion, Bashkiria region	168	Cow farms
Landscape-smart techniques to breed/plant non-traditional species, domesticate wild species	Maral farm, near settlement of Arshan, Buryatia	3	Experimental farms with endemic hoofed animals
Biotechnologically synthesising analogues/ substitutes to traditional bioproducts	Efko sugar substitutes enzymatic plant in Belgorod region	11	Biochemical plants, some food processing plants
Automated production lines integrated with smart intrafactory logistics	Velikoluksky meat processing factory, Pskov region	213	Food preprocessing, food production, feed production plants
Smart field, landscape-adaptive, hydrogeology big data augmented crop farming	Miratorg agricultural holding (multi-regional)	73	Large multiregional to down-integrated agricultural enterprises; large monoculture grain and oilseed farmers

Source: compiled by the authors on the basis of the iFORA System©

2.3 Case Study: Long-Term Technology Trends in the Russian Agriculture

The classification of technology trends based on the text mining exercise has proved to be applicable to the analysis of investment projects in the agrifood sector. Several relevant cases were identified in the Russian Federation (Table 10.1) as an example of the applicability of our methodology to broader analytical purposes.

As the case study showcases, the most current practical potential of long-term vision technologies of agriculture and food sector in Russia lies in the domains of biosecurity and biosafety (controlled bioenvironments protected from external pathogens), production and logistics automation and robotisation on all stages, environmentally smart agriculture processes based on real-time big data feeds and ecosystem-level biotechnologies.

2.4 Specific Agriculture and Food Sector Technology Cases

Experts' surveys allowed to extract several specific technology cases which were not included into a generally adopted list of the mainstream technology trends, but need to be watched as potentially impactful future weak signals.

"Low" technologies for small-scale adoption on local levels for providing increased sustainability and combating climate change draw attention of many experts. Such technologies include, among others, greenhouse gas emission reduction technologies from croplands and water sources; use of composts, biocharcoals as fertilisers, soil structure enhancers and sorbents and fuel for gas-generation engines to reduce use of extracted hydrocarbons by local machinery; fertiliser and solid biofuel pelletisation; microbial digestion; local specialised simple-construction bioreactors; and soil nitrogen and water augmentation techniques.

Advanced microbial digestion allows to reduce greenhouse gas emissions from the soil as well as make utilisation of useful substances from chemical fertilisers more efficient. Thus, a Danish-German research collaboration has found that it is possible to reduce methane emissions by more than 90 per cent by adding cable bacteria to the soil for rice cultivation (Aarhus University 2020). Experiments show that new genetic types of soil microbiomes can more efficiently transform ammonium from nitrogen fertilisers into nitrates, which are then absorbed by plants (AgriNews 2020). Organic nitrogen is the preferred repository from which soil microorganisms can decompose and secrete inorganic nitrogen into the soil and then into plants, thereby preventing leaching and nitrogen loss (American Society of Agronomy 2020).

Biochar technology implies, among other aspects, the transformation of biomass into charcoal and its further mixing with the soil to preserve the burnt carbon. Biochar makes it possible to use agricultural and agroforest waste to produce durable natural carbon, which can enhance water storage in the soil and capture carbon, what is needed in the unstable environment. In April 2016, ICEM launched its interactive tool GMS Biochar and Soil Mapping Tool, which can identify regions that are exceedingly suitable for biochar production. According to the study of Market Study Report, during 2020–2025, the Biochar market will register 9.3% CAGR in terms of revenue, and the global market size will reach $ 682.4 million by 2025, rising from $ 477.5 million in 2019 (Geoengineering Monitor 2018; Benjamin 2020; Data Bridge Market Research 2020; Market Study Report 2020). The technology has already found its first pilot testing sites in Russia. For instance, to prevent the process of soil degradation, a new technology for making biochar from chicken dung and agricultural waste, such as unusable remnants of sunflower or rapeseed, was tested in Tatarstan at one of the enterprises of the Agrosila Group in 2019 (Kazan Federal University 2019).

Pelletisation is a simple-to-implement energy-efficient technique to transform fine particle materials in more manageable and not radiating fine dust into the environment. It is used to make economically useful wide number of bioproducts such as sawdust for production of wood fuel pellets, agricultural plant residues for production of fuel or bioreaction pellets for a wide range of purposes instead of leaving residues on land which can increase biosafety risks, animal manure pelletisation to render these substances (risky in biosecurity terms and therefore underused in local organic fertilisation of soils) widely applicable in organic agriculture. Fertiliser pelletisation is especially important for the future bioeconomy. Cases from across the globe have already showed significant potential in their use to prevent from leaking into wide environment and tie and canalise into local plot soil fertilisation of an excessive nitrogen produced by animal husbandry and urban biowaste. Examples include bio-fortified pelletised organic fertiliser made from human, chicken, pig and cattle biowaste, vegetable and food peels and rotten waste, etc. (BioInnovate Africa Programme 2018; Sokoine University of Agriculture 2019).

Water augmentation for a variety of purposes is also among the foci of interest of many industry experts and community stakeholders. In 2020, agriculture remains the main water consumer, absorbing 69% of the world's water used annually (Piesse 2020). Technology solutions include water control for indirect augmentation, relying on the use of drones, satellite images and big data to track irrigation cycles, and direct augmentation of water sources including for reducing greenhouse gas emissions from water surfaces and water storage reservoirs (Investing in Research to Develop Sustainable Agricultural Practices, 2018); decontamination; sterilising to reduce biosafety risks from microbial and parasitic diseases; water fortification by fertilisation agents for efficient use in watering plants in enclosed environments, including hydroponics and aeroponics settings; filtering and fortifying water with vitamins in the livestock feeding solutions; etc. A host of smart irrigation techniques often adjoins the field of water augmentation, and more and more often integrated solutions combining water processing and its transportation to fields can be discovered. For instance, the method of impulse irrigation, which is used in Russia and Kazakhstan, consists in providing plants with a daily irrigation rate in a pulse mode, which ensures frequent watering in small volumes at certain hours with optimal soil humidity and the directed effect on the growth conditions of plants and the external environment (Angold et al. 2020).

These are only some examples of niche "low" technologies deemed significant for biofutures of agriculture and food sector. Others include weak signals on possible breakthrough technology solutions from a number of organisations, including start-up ventures, such as Ecovative Design (fungi technology), ABS (genetic audit of cattle), Mosa Meat (cultured meat), Verdeca (drought-tolerant plants), Arbiom (wood to food bioreactors), Biosafe (biopolymers from residues of crop farming), BioHauki Ltd. (biomethane as a fuel), Motif Ingredients (direct biosynthesis of food ingredients), Natural Biotechnology (antimicrobial protection), WeFarmUp (farm equipment time-sharing platform, including to drive down costs of expensive biotechnology-related equipment), La Ruche (platform links producers and consumers in smart ways to drive down transaction costs), Panorama AGRO (farm accounting, soil fertility base calculations, monitoring fields and crops, other smart farm and smart field functions) and ValBio-3D (biobased composites).

3 Conclusions

Agriculture and food sector is an integral part of the future global bioeconomy Agricultural and food producers face great commercial opportunities to increase the exports of their products if upscale their production method tough to meet the demand. Becoming leading producers of healthy and green agricultural products requires the uptake of new technologies to improve the quality of products. The way forward depends on decisions taken around certain development folks which have been described. These developments might well limit the opportunities of exporters – but can only be influenced partly by decision-makers. Next to the right development paths of international markets, there is need to gear up quickly on several technology areas to close the quality gap. Over the next years, the future of agriculture and food producers and their presence on international markets will rely on the right connections between actors in the innovation system, user-oriented knowledge development and a careful channeling of public funds. The task ahead is not only a stronger presence on international food markets but also alleviation of future food shortages.

Development forks for the long-term future of agriculture arise from the current interplay of global challenges and existing technologies The future of the agriculture will be largely determined by the choices made at several major economic, institutional-political and technological turning points. Most of these forks are defined globally, while others rely on decisions taken in the country. The portfolio of institutional and political decisions which require implementation over the next few years will determine the feasibility of long-term scenarios among which the national agricultural sector can become globally competitive on a broad product range. Among these rank the structural change of global markets, the absorptive capacity of agricultural organisations to modernise and the priority setting for the sector's S&T activities.

Favourable or unfavourable market situations will shape the future of the sector in unpredictable ways It is possible that the market environment changes to the worse (as we see nowadays in the wake of the global coronavirus pandemic), including decreasing economic growth in developing countries, increased competition in international markets and major importers protecting their markets from imports. Also changes in supply can impact the market, like the changing cost structure of production. All prices of conventional energy sources, railway and electricity tariffs can vary significantly and affect intermediate (e.g. fertilisers) and end products' competitiveness due to high consumption of fuel and lubricants during the production and transportation of agricultural materials. International markets may affect agriculture in different ways, depending on political decisions such as market protection mechanisms (e.g. tariffs or other kinds of barriers), currency regulation and subsidising own agricultural production. Many agricultural producers may turn out to be unprepared to take necessary steps for events like the sudden lifts of the food embargo or discontinued subsidies to farmers. Funding bottlenecks remain one of the main barriers hindering the development of the agricultural sector. Low returns and banks' cautious attitude towards agricultural producers do not encourage an inflow of investments. Further diversification of the sector would become increasingly difficult which would create risks to its sustainable development. Other barriers may include persistently high global prices for low value-added products, hampering innovation activities of agricultural producers and the absence of necessary incentives for developing absorptive capacities. On the other hand, a steadily growing demand for animal farming products combined with the rising importance of "green" agricultural policies can open significant niches for the animal farming, food, and agri-chemical sectors and thus promote the diversification of agricultural production.

Consumer demand in both the domestic and external markets is changing These changes are directed, on the one hand, towards environmentally responsible production and, on the other hand, towards low cost production. These changes will in turn affect marketing strategies of agricultural organisations. They may opt either for continuously extending their product range and creating new business niches or for cost optimisation within the existing market structure. Accordingly, growth of agricultural sector's revenues (or at least maintaining them at the same level) would be mostly achieved by diversifying production towards more expensive and more personalised products or by increasing the supply of relatively simple traditional products.

Shortage of incentives for developing agri-bioenergy and biofuel industries The bioenergy industry had a promising start, but the future development of the industry will depend on both the prices for traditional energy carriers and agricultural materials and on policy choices of countries like the USA, EU, and PRC. A low biofuel production in many countries will reduce demand for relevant R&D activities. Some producers might seek energy self-sufficiency and will demand specialised technologies. In the best-case scenario, powerful export-oriented growth will lead to increased sown acreage, changing crop structure and major price shifts.

Extensive or intensive business models The main question is whether the agriculture sector will develop through extensive factors or would turn towards intensification. Both models may turn out to be winners. The low application of mineral fertilisers and low use of agri-chemicals could help to increase efficiency, limit the nutrients' extraction from soil and raise yields. Due to the abundance of land, circular and ecologically sustainable ways of production could be established based on crop rotation with long fallow time. Similar cost advantages can stem from the development of dairy products, certain segments of animal farming based on natural pasturing, and by using the high bioproductivity potential of seas and internal waters for sustainable fisheries. The latter would be a strong alternative to an expensive aquaculture infrastructure. On the other hand, intensification would allow to concentrate all agricultural production in the most favourable regions, supported by agricultural clusters located near big cities in areas with well-developed infrastructure. This could result in a further reduction of production costs due to spatial optimisation. Application of intensification technologies may sharply reduce agriculture's vulnerability to natural environment risks, such as unfavourable weather, upsurges of pests, outbreaks of dangerous animal diseases, etc. Ultimately, intensification activities may lead to urbanised areas becoming self-sufficient food producers. This would require a paradigm shift from land-based agriculture as the supplier of agricultural materials towards a direct industrial synthesis of food products with specified properties. Under the extensive development model, production of conventional low-margin products by small producers will retain an important role. People's personal land plots, gradually transforming into advanced small farms, will become parts of agricultural cooperatives, leave the shadow sector, turn into predominantly commodity producers and continue to play a significant role in providing local-level food supply. Under the intensive development model, large companies (including those with international capital) will play an increasingly important role, promoting international technological standards and replacing labour with capital assets. Providing public support to small producers will become ineffective and instead might open opportunities in rural areas, such as organic agriculture, various ecology-oriented formats such as agricultural tourism, production of forest-based products other than timber (food, medicines), etc. The availability of labour, on the other hand, will depend on the urbanisation trend. In case of intensive development, the depopulation of rural areas will increase, so their sustainable development will be based on optimised settlement systems, reduced unit costs of infrastructure and increased support for local economic growth. The extensive agricultural development model would not increase export revenues but will – in trade balance terms – have controversial effects as demand for imported machinery, agricultural chemicals and imported food products will decline.

Economic policy will have a significant impact on the development of the agricultural sector as agricultural production The results will be, on the one hand, a rigid internal market for agricultural products or further market liberalisation, on the other hand. In the first case, the sector's innovation-based development may face several barriers such as limited access to foreign capital, obstacles to the development of a free land market or inefficient production through attempts to minimise the producers' dependency on international intellectual property. The route towards liberalisation suggests increase economic efficiency, combined with continuing consistent institutional reforms, structural optimisation, the revitalisation of the sector and the risks for small and medium companies in risky agricultural areas. The economic policies will either provide continued support to producers or foster innovation-led growth. The first case will incentivise traditional businesses to rely on extensive production factors and target low-value market segments. Such policies would primarily involve subsidies to help acquire means of production, which in turn discourage innovation. A possible alternative is providing priority support to innovation-oriented companies which reduce labour input and energy and materials consumption and implement cutting-edge technologies to introduce radically new products.

Technology adaption will be an area of interest for policy-makers Here, an important aspect is the decision about the status of genetics research. Banning or allowing the creation and use of genetically modified organisms will have great influence. Currently, genetically modified organisms are hardly used in agricultural sector in many countries, and practically no GMO-based food reaches the shop floors. Therefore, a total ban on biotechnologies based on genetic engineering will not have any visible impact on the economy in the short run. Yet another S&T policy fork is the support for technological development of the agricultural sector or orienting towards its modernisation in the scope of international technological cooperation (including localisation of the cutting-edge biological, robotic, information and telecommunication solutions in the framework of strategic partnerships with foreign companies). The first option leads to higher levels of technological and biological security while at the same time reduces dependency on volatile global markets. The second option involves lower public financial support to encourage technological import substitution, which would probably help to achieve necessary food security standards earlier.

Acknowledgements The book chapter was prepared within the framework of the Basic Research Program at the National Research University Higher School of Economics.

References

Aarhus University. (2020) How Cable Bacteria Could Make Rice Cultivation Greener. https://www.labmanager.com/news/how-bacteria-could-make-rice-cultivation-greener-22434. Accessed 22 April 2020

Adnan, N., Nordin, S. M., Rahman, I., & Noor, A. (2018) The effects of knowledge transfer on farmers decision making toward sustainable agriculture practices. World Journal of Science, Technology and Sustainable Development, 15 (1), 98–115

AgriNews (2020) Microbes play important role in soil's nitrogen cycle. https://www.agrinews-pubs.com/2020/04/05/microbes-play-important-role-in-soils-nitrogen-cycle/a3ze2fm/. Accessed 6 April 2020

American Society of Agronomy (2020) Soil carbon and nitrogen mineralization after the initial flush of CO2. https://www.eurekalert.org/pub_releases/2020-04/asoa-sca041420.php. Accessed 16 April 2020

Angold, E. V., Zharkov, V. A., Kalashnikov, A. A. (2020) Impul'snoe dozhdevanie: tekhnologiya i preimushchestva (Impulse irrigation: technology and advantages). https://www.agbz.ru/articles/impulsnoe-dozhdevanie-tekhnologiya-i-preimushchestva/. Accessed 13 April 2020

Ayaz, M., Ammad-Uddin, M., Sharif, Z., Mansour, A., & Aggoune, E. H. M. (2019) Internet-of-Things (IoT)-Based Smart Agriculture: Toward Making the Fields Talk. IEEE Access, 7, 129551–129583

Bach, H., & Mauser, W. (2018) Sustainable agriculture and smart farming. In Earth Observation Open Science and Innovation (pp. 261–269). Springer, Cham

Bakhtin, P., Saritas, O., Chulok, A., Kuzminov, I., & Timofeev, A. (2017) Trend monitoring for linking science and strategy. Scientometrics, 111(3), 2059–2075

Balakrishnan, S., Janet, J., Sachinkanithkar, R., & Reshma, D. (2017) Technological Innovations for Agricultural Developments through Information Communications Technology (ICT). CSI Communications magazine, 41(6), 10-13

Bartkowski, B., Theesfeld, I., Pirscher, F., & Timaeus, J. (2018) Snipping around for food: economic, ethical and policy implications of CRISPR/Cas genome editing. Geoforum, 96, 172–180

Bedford, J., Farrar, J., Ihekweazu, C., Kang, G., Koopmans, M., & Nkengasong, J. (2019) A new twenty-first century science for effective epidemic response. Nature, 575(7781), 130–136

Bellows, A. C. (2019) A systems-based human rights approach to a national food plan in the USA. Renewable Agriculture and Food Systems, 1–3

Benjamin, J. (2020) Conservation Tools Help Producers Make Positive Impacts on Changing Climate. https://www.usda.gov/media/blog/2020/04/22/conservation-tools-help-producers-make-positive-impacts-changing-climate. Accessed 22 April 2020

BioInnovate Africa Programme. (2018) Nitrogen bio-fortified and pelletized commercial grade organic fertilizer made from urban bio-waste to improve soil productivity. https://bioinnovate--africa.org/wp-content/uploads/2018/04/Nitrogen-organic-fertilizer-from-urban-bio-waste.pdf. Accessed 22 April 2020

Bowen, R., & Morris, W. (2019) The digital divide: Implications for agribusiness and entrepreneurship. Lessons from Wales. Journal of Rural Studies, 72, 75–84

Briglauer, W., Dürr, N. S., Falck, O., & Hüschelrath, K. (2019) Does state aid for broadband deployment in rural areas close the digital and economic divide? Information Economics and Policy, 46, 68–85

Cagnin, C., Keenan, M., Johnston, R., Scapolo, F., & Barre, R. (2008) Future-oriented technology analysis. Strategic Intelligence for an Innovative Economy. Springer-Verlag Berlin Heidelberg

Calof, J., Meissner, D., & Vishnevskiy, K. (2020) Corporate foresight for strategic innovation management: the case of a Russian service company. Foresight, 22 (1), 14–36

Chen, H., Liu, S., & Oderanti, F. (2020) A knowledge network and mobilisation framework for lean supply chain decisions in agri-food industry. In Supply Chain and Logistics Management: Concepts, Methodologies, Tools, and Applications (pp. 369–381). IGI Global

Coomes, O. T., Barham, B. L., MacDonald, G. K., Ramankutty, N., & Chavas, J. P. (2019) Leveraging total factor productivity growth for sustainable and resilient farming. Nature Sustainability, 2(1), 22–28

Corlett, R. T. (2017) A bigger toolbox: biotechnology in biodiversity conservation. Trends in Biotechnology, 35(1), 55–65

Data Bridge Market Research. (2020) Biochar Market: Moving Towards Brighter Futurel Cool Planet, Pacific Biochar Benefit Corporation, Genesis Industries, LLC, CharGrow USA LLC, Black Owl Biochar. https://www.openpr.com/news/2002148/biochar-market-moving-towards-brighter-future-cool-planet. Accessed 16 April 2020

Depenbusch, L., & Klasen, S. (2019) The effect of bigger human bodies on the future global calorie requirements. PloS one, 14(12), e0223188

Dubé, L., Du, P., McRae, C., Sharma, N., Jayaraman, S., & Nie, J. Y. (2018) Convergent innovation in food through big data and artificial intelligence for societal-scale inclusive growth. Technology Innovation Management Review, 8(2)

FAO. (2019a) The State of Food and Agriculture 2019. Moving forward on food loss and waste reduction. Rome

FAO. (2019b) New standards to curb the global spread of plant pests and diseases. http://www.fao.org/news/story/en/item/1187738/icode/. Accessed 16 April 2020

Fraser, A. (2019) Land grab/data grab: precision agriculture and its new horizons. The Journal of Peasant Studies, 46(5), 893–912

Geels, F. W. (2020) Micro-foundations of the multi-level perspective on socio-technical transitions: Developing a multi-dimensional model of agency through crossovers between social constructivism, evolutionary economics and neo-institutional theory. Technological Forecasting and Social Change, 152, 119894

Geoengineering Monitor. (2018) Biochar. Technology Factsheet. http://www.geoengineeringmonitor.org/2018/05/biochar-technology-factsheet/. Accessed 16 April 2020

Ghobadpour, A., Boulon, L., Mousazadeh, H., Malvajerdi, A. S., & Rafiee, S. (2019) State of the art of autonomous agricultural off-road vehicles driven by renewable energy systems. Energy Procedia, 162, 4–13

Gibson, L., & Rush, D. (2020) Novel coronavirus in Cape Town informal settlements: feasibility of using informal dwelling outlines to identify high risk areas for COVID-19 transmission from a social distancing perspective. JMIR Public Health and Surveillance, 6(2), e18844

Gokhberg, L., Kuzminov, I., Chulok, A., & Thurner, T. (2017) The future of Russia's agriculture and food industry between global opportunities and technological restrictions. International Journal of Agricultural Sustainability, 15(4), 457–466

Jiang, K., Anderton, B. N., Ronald, P. C., & Barnett, G. A. (2018) Semantic network analysis reveals opposing online representations of the search term "GMO". Global Challenges, 2(1), 1700082

Karabag, S. F. (2020) An Unprecedented Global Crisis! The Global, Regional, National, Political, Economic and Commercial Impact of the Coronavirus Pandemic. Journal of Applied Economics and Business Research, 10(1), 1–6

Kazan Federal University. (2019) Biochar used to prevent soil erosion in Tatarstan. https://www.eurekalert.org/pub_releases/2019-09/kfu-but091319.php. Accessed 16 April 2020

Kim, H. (2019) AI, big data, and robots for the evolution of biotechnology. Genomics & Informatics, 17(4)

Klerkx, L., & Rose, D. (2020) Dealing with the game-changing technologies of Agriculture 4.0: How do we manage diversity and responsibility in food system transition pathways? Global Food Security, 24, 100347

Knickel, K., Redman, M., Darnhofer, I., Ashkenazy, A., Chebach, T. C., Šūmane, S., Tisenkopfs, T., Zemeckis, R., Atkociuniene, R. V., Strauss, A., Kristensen, L. S., Schiller, S., Koopmans, M. E., & Rogge, E. (2018) Between aspirations and reality: Making farming, food systems and rural areas more resilient, sustainable and equitable. Journal of Rural Studies, 59, 197–210

Koohafkan, P., & Altieri, M. A. (2016) Forgotten agricultural heritage: Reconnecting food systems and sustainable development. Taylor & Francis

Kumar, P., Lorek, T., Olsson, T. C., Sackley, N., Schmalzer, S., & Laveaga, G. S. (2017) Roundtable: New narratives of the green revolution. Agricultural History, 91(3), 397–422

Levy, M. A., & Lubell, M. N. (2018) Innovation, cooperation, and the structure of three regional sustainable agriculture networks in California. Regional environmental change, 18(4), 1235–1246

Liu, Y., & Li, Y. (2017) Revitalize the world's countryside. Nature News, 548(7667), 275

Loboguerrero, A. M., Campbell, B. M., Cooper, P. J., Hansen, J. W., Rosenstock, T., & Wollenberg, E. (2019) Food and earth systems: priorities for climate change adaptation and mitigation for agriculture and food systems. Sustainability, 11(5), 1372

Lokko, Y., Heijde, M., Schebesta, K., Scholtès, P., Van Montagu, M., & Giacca, M. (2018) Biotechnology and the bioeconomy — Towards inclusive and sustainable industrial development. New biotechnology, 40, 5–10

Mackay, M., & Perkins, H. C. (2019) Making space for community in super-productivist rural settings. Journal of rural studies, 68, 1–12

Market Study Report. (2020) Global Biochar Market Growth 2020-2025. https://www.marketstudyreport.com/reports/global-biochar-market-growth-2020-2025. Accessed 16 April 2020

Marlow, S. (2020) COVID-19: Effects on the Fertilizer Industry. https://ihsmarkit.com/research-analysis/report-covid19-effects-on-the-fertilizer-industry.html. Accessed 16 April 2020

Miles, I., Saritas, O., & Sokolov, A. (2016) Foresight for science, technology and innovation. Springer International Publishing

Piesse, M. (2020) Global Water Supply and Demand Trends Point Towards Rising Water Insecurity. http://www.futuredirections.org.au/publication/global-water-supply-and-demand-trends-point-towards-rising-water-insecurity/. Accessed 16 April 2020

Porter, A., & Cunningham, S. (2004). Tech Mining: Exploiting New Technologies for Competitive Advantage. John Wiley & Sons, Inc.

Rohr, J. R., Barrett, C. B., Civitello, D. J., Craft, M. E., Delius, B., DeLeo, G. A., Hudson, P. J., Jouanard, N., Nguyen, K. H., Ostfeld, R. S., Remais, J. V., Riveau, G., Sokolow, S. H., & Tilman, D. (2019) Emerging human infectious diseases and the links to global food production. Nature Sustainability, 2(6), 445–456

Rotz, S., Gravely, E., Mosby, I., Duncan, E., Finnis, E., Horgan, M., LeBlanc, J., Martin, R., Tait Neufeld, H., Nixon, A., Pant, L., Shalla, V. & Fraser, E. (2019) Automated pastures and the digital divide: How agricultural technologies are shaping labour and rural communities. Journal of Rural Studies, 68, 112–122

Sachs, J. D., Schmidt-Traub, G., Mazzucato, M., Messner, D., Nakicenovic, N., & Rockström, J. (2019) Six transformations to achieve the sustainable development goals. Nature Sustainability, 2(9), 805–814

Salpina, D. (2019) How sectoral policy can benefit the protection of multi-functional cultural heritage? The case of agricultural landscape and the EU rural development policy. Aedon, 2

Sokoine University of Agriculture. (2019) Success story on pelletized green biowastes as organic fertilizer for smallholder farmers in Tanzania. https://www.sua.ac.tz/news/success-story-pelletized-green-biowastes-organic-fertilizer-smallholder-farmers-tanzania. Accessed 16 April 2020

Sokolov, A., & Chulok, A. (2016) Priorities for future innovation: Russian S&T Foresight 2030. Futures, 80, 17–32

Streich, J., Romero, J., Gazolla, J. G. F. M., Kainer, D., Cliff, A., Prates, E. T., Brown, J. B., Khoury, S., Tuskan, G. A., Garvin, M., Jacobson, D., & Harfouche, A. L. (2020) Can exascale computing and explainable artificial intelligence applied to plant biology deliver on the United Nations sustainable development goals? Current opinion in biotechnology, 61, 217–225

Strickland, T. S. (2017) Study: Half of U.S. corn crop could vanish by 2100 due to climate change. https://www.triplepundit.com/2017/01/climate-change-corn-crop/. Accessed 16 April 2020

Torre, A., & Wallet, F. (2020) Rural development policies at stake: structural changes and target evolutions during the last 50 years. In: Fischer M., Nijkamp P. (eds) Handbook of Regional Science. Springer, Berlin, Heidelberg

Ugochukwu, A. I., & Phillips, P. W. (2018) Technology adoption by agricultural producers: a review of the literature. In *From Agriscience to Agribusiness* (pp. 361–377). Springer, Cham

Venkatramanan, V., & Shah, S. (2019) Climate smart agriculture technologies for environmental management: the intersection of sustainability, resilience, wellbeing and development. In Sustainable Green Technologies for Environmental Management (pp. 29–51). Springer, Singapore

Vilceanu, M. O., Grasso, O., & Johnson, K. (2019) Bridging the Gap Between Public Opinion Research and Consumer Marketing Research: Insights into US Shoppers of Organic Foods. Association of Marketing Theory and Practice Proceedings

Vu, Q., Nguyen, V., Solenaya, O., & Ronzhin, A. (2017) Group control of heterogeneous robots and unmanned aerial vehicles in agriculture tasks. In International Conference on Interactive Collaborative Robotics (pp. 260–267). Springer, Cham

Wahby, M., Heinrich, M. K., Hofstadler, D. N., Petzold, J., Kuksin, I., Zahadat, P., Schmickl, T., Ayres, P., & Hamann, H. (2019) Robotic sensing and stimuli provision for guided plant growth. Journal of Visualized Experiments, 149, e59835

World Trade Organization. (2020) Responding to the Covid-19 Pandemic with Open and Predictable Trade in Agricultural and Food Products. https://www.wto.org/english/tratop_e/covid19_e/covid19_e.htm. Accessed 16 April 2020

Zaidi, S. S. E. A., Vanderschuren, H., Qaim, M., Mahfouz, M. M., Kohli, A., Mansoor, S., & Tester, M. (2019) New plant breeding technologies for food security. Science, 363(6434), 1390–1391

Chapter 11
Parallels Between the Future for MedTech and Agri-Tech, Perspectives Drawing on the British Experience

Malgorzata Grzegorczyk, Pantea Lotfian, and William J. Nuttall

Abstract In this chapter we explore the future for innovation in two related, but distinct, sectors. We consider the linkages between medical technology (MedTech) and agricultural technology (Agri-Tech) innovation in the UK. We ask and discuss questions: Who are the key actors in the innovation systems of MedTech and Agri-Tech in the UK? What are the core technologies driving the current waves of innovation in these two sectors? Can one industry learn from the other? Where is the scope for cooperation and synergies? We notice that both sectors are technologically linked through foundational technologies underpinning the majority of the observed innovation, e.g. big data, AI, IoT and robotics. The outputs of these technologies rely crucially on digital data for insight and decision support. However, Agri-Tech benefits from less complex stakeholder issues regarding data security and privacy. Both sectors are important to the UK going forwards, and both will be exposed to Brexit and consequences of the COVID pandemic. Our discussion on the future for innovation should be of particular interest to start-up leaders, entrepreneurs, investors, managers and policy-makers in MedTech, Agri-Tech and cognate sectors.

Keywords MedTech · Agri-Tech · Ecosystem actors · Digital medicine · AI · Big data · Robotics · Internet of things · Precision agriculture

M. Grzegorczyk (✉)
Nottingham Business School, Nottingham Trent University, Nottingham, UK
University of Lodz, Lodz, Poland
e-mail: margaret.grzegorczyk@ntu.ac.uk

P. Lotfian
Camrosh Limited, Cambridge, UK

W. J. Nuttall
School of Engineering and Innovation, The Open University, Milton Keynes, UK

E. Koukios, A. Sacio-Szymańska (eds.), *Bio#Futures*,
https://doi.org/10.1007/978-3-030-64969-2_11

1 Introduction

This chapter seeks to draw parallels between, and reveal common insights from, two historically quite distinct, but now increasingly related sectors: medical technology ('MedTech') and agricultural technology ('Agri-Tech'). Previously, others have considered the linkages of these individual sectors with associated industries and economic activities, such as MedTech and Healthcare or Agri-Tech and Bioscience, but we are not aware of studies concerning prior linkages between the MedTech and Agri-Tech sectors as innovation areas. As an example, we consider the UK experience, and note that, at the time of writing, both the MedTech and Agri-Tech sectors are independently important to the UK economy (BEIS 2017; WLEP n.d.). Furthermore, both have seen significant growth in recent years (UKRI 2018; WECA 2019), in part as a consequence of driving considerations explored elsewhere in this book. Despite the recent progress, we posit, as a result of our examination of the UK position, that there are likely to be even more dramatic changes ahead and some of these are already underway. We suggest that some of the coming progress will arise from a synergistic combination of the two domains, but even more substantially we see the possibility for knowledge transfer between the two sectors as we detect the emergence of innovations within each sector with much beneficial potential in the other.

Given the urgency in addressing the systemic challenges we face today, it is important to start understanding how historically distinct industries are changing and converging towards comparable patterns as they adopt digital technologies. As we show in this chapter, the foundation technologies that enable both MedTech and Agri-Tech are the same. This means these two distinct industries will start showing synergies particularly at the level of business models and data they produce. This poses a considerable opportunity for future thinking from a systems perspective where two of the fundamental industries for human survival, namely, agriculture and health, can be brought together to address global challenges.

In this chapter we survey and explore the foundations of the starting linkages between these industries and a set of issues that we have been researching and for which our work is ongoing. We expect to publish that ongoing work in due course. We here provide an overview of key issues and highlight themes and insights emerging from our research thus far.

2 Industry Overview

For this chapter we seek, as far as possible, to focus on innovations, and from that perspective we seek to establish a set of associated actors driving the process. We consider two related essential questions. Who are the key actors in the innovative system in MedTech and in Agri-Tech in the UK? And, what are the core technologies driving the current waves of innovation in the two sectors? We ask these questions to understand better the future of both sectors, but also the possible scope of cooperation. Can one industry learn from the other? Where is the scope for cooperation and innovation?

Our ongoing research is based on a review of the literature and our initial findings from a series of expert interviews. However, our main purpose in this chapter is context setting and an exploration of the key issues. Both agriculture and health are part of the oldest activities in the history of human civilisation. Today the medical technology sector is known to be one of the most innovative industries. According to Accenture Disruptability Index, which is studying disruption in 20 industries, healthcare is among the most vulnerable to future disruption (Accenture 2019). The UK MedTech sector includes organisations developing, producing or selling medical devices. These are all linked to an extensive network of service and supply businesses. The sector is part of the Life Sciences and hence sits alongside the closely related BioTech sector. The MedTech sector is heavily influenced by medical device regulations and the health economic factors affecting adoption and distribution in key consumer groups such as the UK National Health Service (NHS). The medical devices industry remains central to the MedTech industry, and insight into the medical devices industry is crucial when seeking to assess the future of MedTech. The medical devices industry includes technologies from single-use consumables to complex hospital equipment, 'well-being' and digital health goods as well as in vitro diagnostic products.

The Agricultural Revolution in Britain spanned the mid-seventeenth to the late nineteenth centuries and was underway long before the more famous Industrial Revolution. As industrialisation started to provide more jobs, and as urbanisation took hold, health issues in crowded cities intensified leading to a continuous pressure for innovation in health and medical technologies. In contrast however, agriculture advanced in bursts of activity when enabling technologies reached a critical mass. From the 1950s onwards, the 'green revolution' and the genetic modification of plants led to increased yields and disease and pest resistance opening a new path for agricultural innovation and formation of high-growth agri-corporations. A new wave of innovation in agriculture is today underway with the onset of the fourth Industrial Revolution. The result is highly intensified precision agriculture, employing sophisticated sensor and data technologies, robotics and AI in order to predict, perform and control agricultural activity. The aim of these developments is to minimise the unpredictability of some deciding factors such as weather, disease, resource scarcity and other natural adverse effects in order to achieve maximal outcomes as well as increased automation and a reduction of reliance on human labour. These enabling technologies increasingly bring agriculture to the fore as a strategic field ready for innovation and disruption and on a par with other innovative industries, such as MedTech.

2.1 The British MedTech Ecosystem – Key Actors

The MedTech sector in the UK is moving towards a dynamic and increasingly diverse healthcare procurement environment, with challenges in the form of a new regulatory standards and a highly competitive market. There are around 3700 medical technology companies operating in the UK (Kent 2019). The sector accounts for 40% of Life Sciences employment in England (86,000 jobs); however 84% of the businesses are small companies employing less than 50 people (MedTech Landscape Review 2019). Around one third of the total turnover comes from small businesses and the rest from companies with turnover of over £50 m including UK companies and international corporations. The development of the sector faces a challenge in a form of a gap between a small number of big global corporations and many small- and medium-sized businesses, including start-ups. In total, the sector employs 127,400 people: 97,600 in core MedTech businesses and 29,800 in service and supply businesses. The combined turnover of the sector is £24bn (Office for Life Sciences 2019).

We note that in the UK, MedTech technology developments are mainly dictated by regulations from the National Institute for Health and Care Excellence (NICE) and purchasing by the National Health Service (NHS) which has a near-monopoly in healthcare procurement. The operation of the NHS and other governmental organisations, such as NICE and the National Institute for Health Research (NIHR), are centralised. The NHS creates a strong centralised network that makes regulation around creation and commercialisation of medical technologies easier. However, it is noteworthy that the NHS, with its significant market power (it accounts for about 80% of expenditure for medical technology (Klein 2015)), is not perceived to be approachable by the small- and medium-sized companies and start-ups dominating the MedTech innovation landscape. As a market giant with extremely high bargaining power, the NHS shapes the demand for MedTech innovations focusing mostly on those that appear likely to reduce NHS system costs.

The Medical Research Council (MRC) is the main public body responsible for funding fundamental medical research in the UK. It forms part of the national science base. In addition, the NIHR, as part of the government's Department of Health and Social Care, invests £300 million a year into the infrastructure of clinical trials in the UK, generating what it estimates to be a financial return of £2.4 billion for the British economy (Kent, 2019). The MRC has provided the funding for a number of notable medical science breakthroughs, including the structure of DNA. Indeed the research funded by the MRC has resulted in 32 Nobel Prize awards.

Government policy in England is currently strengthening its long-standing emphasis on achieving an impact from innovation in the health and care system, rather than just sponsoring the development of inventions and innovations (MedTech Landscape Review 2019). However, the presence of science-push remains clearly visible, and Great Britain is generally perceived to be one of the best places for digital health and medical technology businesses. Most of medical technology companies in the UK are small- to medium-sized enterprises (SMEs) that are entitled to research and development (R&D) tax reliefs. Thanks to this they can deduct a high percentage of their qualifying costs from their yearly profit (Kent 2019).

The NHS has supported and enabled innovations in MedTech through a wide range of collaborations with private and non-profit sectors. Medical technology companies can be found all over the UK; however several regional industrial clusters have developed. Those regions have a high concentration of medical device companies, component suppliers, clinics and research facilities with a focus on biomed research. Organisations in those clusters benefit not only from the opportunities of joint research and the proximity of experts in different areas (medical, engineering, IT, AI, big data) but often also from collaboration in areas, such as lobbying. In this regard big companies are always privileged as they have higher lobbying power to shape the stream of the governmental funds. At the centre of such medical clusters is the so-called golden triangle linking London, Cambridge and Oxford with world-class academic institutions. In particular Cambridge is the UK's MedTech start-up hotspot (Hender 2015). Being based in the Cambridge cluster allows companies to access talent and resources of world-leading quality. This culture has helped the UK recently to become a world leader in medical artificial intelligence (AI) and in the digital transformation of healthcare.

2.2 *The Agri-Tech Ecosystem Key Actors*

Agriculture contributes less than 1% to the UK economy and its share of employment is 1.45%. Total farming income in 2019 rose by 8.2% to £5.3bn in current price terms, and gross output increased by 2.1% to £27.3 bn. The total factor productivity of UK agriculture, however, increased by 4% over a similar 12 month period (DEFRA 2019; DEFRA 2020). Despite its small contribution to the economy, UK agriculture comprises of a strong ecosystem of actors including university research centres, start-ups, farmers (and farm management businesses), research organisations and the government through the four Agri-Tech centres (Crop and Soil Health Protection, Animal Productivity Welfare and Health, Engineering and Precision Technologies, and Data Science Analytics and Modelling) as well as grant and funding bodies and catalyst organisations. The ecosystem also contains, and is supported by, private funding sources (angel investors, venture capital and other forms of private investment), food retailers and business support and consulting organisations and persons.

The Cambridge Hub, located in East of England, is the example of a lively and expanding Agri-Tech ecosystem where all the above-mentioned elements are working together creating a rapidly growing Agri-Tech scene in the region. Given the central role of farmers in an Agri-Tech ecosystem and the fact that farms are dispersed around the countryside, Agri-Tech East (renamed as Agri-TechE in 2020, as it goes national due to its strong success), a catalyst organisation, has had a central role in initiating and shaping this cluster in the east of England. The University of Cambridge, the University of East Anglia and the University of Bedfordshire all have agriculture and technology research groups that work in various specialist areas of Agri-Tech. NIAB (the National Institute of Agricultural Botany) is a leading UK centre for plant science, crop evaluation and agronomy with its headquarters in Cambridge. Also the region has a long history in farming, and there are a considerable number of farms of varying sizes spanning arable, pastoral and mixed operations.

In 2018, the UK government officially announced support for the development of a £500 million ($650 million) Agri-Tech Cluster, to be located on the edge of the university town of Cambridge. The project aims to develop a 553 acre commercial park to house the new cluster. The park will provide working space for up to 4000 people, spread across 1 million sq/ft of commercial properties as well as land for field trials, demonstration plots and access to up to 25,000 acres of additional crop and technology trial areas through 'established local partnerships' with farmers (Ag Funder Network 2018; Cambridgeshire Live 2018). Together with the already established strength of the ecosystem in the area, this places Cambridge to lead the nascent Agri-Tech industry in the UK.

2.3 Agri-Tech Industry and Policy in the UK

In July 2013 the UK government published the 'A UK Strategy for Agricultural Technologies'. The report aimed at building a capability to feed a growing population without damaging the natural environment through connecting basic research and applied sciences in order to create modern systems that allow farmers to access Agri-Tech expertise and innovations (HM Gov 2013). Embedded in the strategy was the development of relevant skills to carry out ideas from laboratory to farms. Industry was placed at the helm of the strategy working in partnership with public and third sectors, in particular by identifying opportunities for development and co-investment. The strategy was the first recognition of Agri-Tech as a sector encompassing agricultural research and the full supply chain from seed to food processing and packing and retail, encompassing both arable and livestock agriculture. The adoption and establishment of the Agri-Tech strategy by the government in 2013 were a timely response to the rapid rise of the adoption of digital technologies in agriculture, as well as the expansion of agricultural markets. This expansion arose due to a rising population, the growth of emerging economies and an increasing scarcity of the fundamental factors for agriculture, namely: land, water and energy. Furthermore, the strategy leads to the UK government investing £90 M in setting up four key Agricultural Technology Centres or Agri-Tech Centres (n.d.), to foster collaboration between the Agri-Food sector, government and academia:

- Agri EPI Centre – Centre for precision agriculture and engineering
- Agrimetrics – Centre of excellence for big data across the agri-food sector
- CHAP – Centre for innovation in crop health and protection
- CIEL – Centre for innovation excellence in livestock

Another development towards fostering the Agri-Tech ecosystem and shaping regional communities was the foundation of a number of Agri-Tech Hubs including the aforementioned Agri-Tech East in 2014, based in Cambridge UK aimed at bringing people together (farmers, scientists, engineers and other players in the industry including logistics and retail). The hubs identify challenges and opportunities and help form multidisciplinary collaborations to address the UK growth requirements in agriculture and food production. The organisation has been highly successful in building strong ties between different players in the ecosystem. Another organisation, The Ceres Agri-Tech Knowledge Exchange Partnership, funded by a £4.78 m budget from the Connecting Capability Fund, Research England, was also established in 2018 as a partnership between Cambridge University, the University of East Anglia, Hertfordshire University, the University of Lincoln and the University of Reading. These universities are collaborating with the John Innes Centre in Norwich, Cambridge-based NIAB and Rothamsted Research to augment their joint commercialisation expertise. Furthermore, as noted earlier, there is backing for establishing a dedicated Agri-Tech cluster near Cambridge including a research and development park which will lead to the region becoming the industry lead nationally and potentially internationally.

All these developments occur against the backdrop of Brexit which will lead to a new bill to shape post Brexit UK agriculture policy. This will see the UK exiting the common agriculture policy (CAP) with a 'public money for public goods' agenda. The policy is a series of 'Schemes' targeted at the protection of the environment and the countryside and to the conservation of livestock, while being accessible to farmers, growers and other players in the value chain.

The latest data available on the size of Agri-Tech sector in the UK also goes back to 2013 which saw a contribution of £14.3bn to the UK gross value-added income and when the sector employed more than half million people. A growth rate of 16% was reported for the sector between 2008 and 2013, with some subsectors showing growth of over 20%.

Although traditional agriculture still dominates the industry, emerging technologies are increasingly gaining acceptance in the agriculture sector and account for a third of Agri-Tech output with the technology component in agriculture showing the fastest growth between 2020 and 2030. High-tech agriculture is currently dominated by precision farming and engineering which is worth over £1bn to the UK economy and employs 21,000 people.

3 Strategies for Innovation and the Technologies Driving Innovation

In our discussions with sectoral experts, we sought to explore the extent to which strategy is shaped by the flow of money. In addition, we were keen to determine how government industrial strategy informs decisions as to where government funding is allocated. We also sought to explore the extent to which trickle-down funding in Agri-Tech and MedTech impacts on the balance of technologies developed. In addition, we considered whether there is a fundamental difference between the Agri-Tech and MedTech sectors arising from the shape of the market in each case.

3.1 MedTech Technologies Driving Innovations

The MedTech market in the UK is characterised by faster innovation and adoption (compared to biopharma), but it is still slow when compared to other markets of nonmedical technologies. New products are fuelled by new technology and unmet market need. Many companies use the new technology as part of their core business. In addition, some experiment with new approaches, so as to find ways of creating added value. The industry is in a development stage, under pressure to scale new technologies.

We note that:

- 75% of healthcare businesses are currently rolling out AI and machine learning technologies (Taylor 2020).
- Healthcare is the next tech frontier; however the industry needs to demonstrate success to convince investors.
- Brexit (EU-UK alignment) and COVID-19 will shape the future of MedTech in the UK.

The core technologies driving innovations in MedTech in the UK include:

Artificial intelligence (AI), big data and extended reality. These technologies are expected to become the foundation for next-generation products and services. Healthcare is adopting social, mobile, analytics and cloud technologies. However, most healthcare providers are in early stages of the digital transformation. AI has a great number of emerging use cases in healthcare. It is changing the way patients interact with doctors and is supporting the personalised healthcare, e.g. AI can identify diseases using speech, facial features, retina scans or X-rays. AI might also be used in contact centres, for payment activities and medical chart reviews, and it can help patients take part in self-service activities. British start-up Lancor Scientific opened a laboratory to push for 90% accuracy for cancer screening with the use of AI. It has developed a device that the company claims is able to early detect cervical cancer at 90% accuracy and which can also be used for several other types of cancers. The aforementioned technologies are expected to push each other further forwards, and these combinations might have a game-changing impact on healthcare. The UK government, being a big supporter of AI and especially the use of AI in healthcare, took some steps to address and shape a future increasingly driven by AI. Among them is creation of a national strategy: 'Artificial Intelligence Sector Deal' in 2019 and the Life Sciences Industrial Strategy with the aspiration for the UK to apply £10 billion data set to domestic purposes. It has also supported the use of AI in healthcare by launching the national Artificial Intelligence Lab, to help the diffusion of technologies across the NHS. In 2018 the UK business secretary Greg Clark announced five new centres of excellence for digital pathology and imaging, including radiology using AI medical advances. Among them is the Pathology Image Data Lake for Analytics, Knowledge and Education (PathLAKE), in Coventry, that will use NHS pathology data to 'drive economic growth in health related AI' and the London Medical Imaging and Artificial Intelligence Centre for Value Based Healthcare that will use AI in medical imaging and related clinical data for faster and earlier diagnosis and automating expensive and time-consuming manual reporting (MedTech Innovation News 2019). Inward investment into AI increased in the UK by 17% in 2019 (more than in the whole of the rest of Europe). The UK is ranked third in the Global AI Index, following China and the USA. It is placed first in 'Operating Environment' measure of the index that focuses on the regulatory context and public opinion (Global AI Index n.d.).

Robotics, automation, 3D printing. We note the rise of a human and machine collaborative workforce, where individuals are empowered by their skillsets and knowledge and new capabilities are delivered by technology. One example might be 'surgeon robots' that can perform complex surgeries with minimal supervision. Robots can collect blood samples, support patients (nurse and patient care robots), and carry out diagnostic procedures. Surgical robotics is another area where UK companies are at the forefront. CMR Surgical is using the Versius surgical robotic system, transforming keyhole surgery minimising the learning curve for the procedure, shortening it from 2 or 3 years to weeks. This also helps the surgeon to stay seated, reducing the operator's physical burden and hence the possibility of surgical errors due to exhaustion. Robotics can be expected to change the face of healthcare in the future.

Digital medicine, the Internet of Medical Things (IoMT). The growing number of connected medical devices and smartphones that are able to generate, collect, analyse and transmit data creates the IoMT a complex, connected infrastructure. It includes the medical devices but also, data, software, health systems and services. According to Deloitte the IoMT will transform the role of MedTech in healthcare: improving drug management, diagnosis and treatment and disease management; remote monitoring of chronic diseases; enhancing patient experience; and decreasing costs (Deloitte 2018). Solutions based on connected devices can help reduce healthcare costs by reducing hospital re-admissions, lowering medication non-adherence and increasing wellness. IoMT can also engage and empower patients and their carers to improve self-management.

The Topol Review published in 2019 presented a digital future for the NHS stressing the latest technologies across the themes of genomics, digital medicine and AI and robotics identified digital healthcare technologies that will impact on the NHS from 2020 to 2040: telemedicine, smartphone apps, sensors and wearables for diagnostics and remote monitoring, automated image interpretations (AI) and interventional and rehabilitative robotics (Bolland 2019). The digitization of the patient experience has been given a huge additional boost by the COVID-19 pandemic leading to a rapid virtualization of the patient experience seeking medical advice from their primary care physician.

3.1.1 Applications

In the UK the NHS, in common with many other health systems around the world, is looking for solutions that will enable earlier diagnosis of disease, address unmet needs in mental health and offer new solutions to cancer and rare diseases and for complex multi-morbid patients (MedTech Landscape Review 2019). Three new innovations have been recognized and announced by the NHS and are recommended to become mandatory across the health service (Taylor 2020). They include:

- *HeartFlow Analysis* – non-invasive test that helps clinicians understand the severity of coronary heart disease by using artificial intelligence (AI); leveraging deep learning and highly trained analysts, HeartFlow creates a 3D digital model of a patient's arteries to help clinicians understand the location and severity of blockages. The HeartFlow Analysis has the highest diagnostic performance compared to other non-invasive tests and can reduce the need for additional tests and deliver cost savings. It has many benefits for patients as it gives diagnostic clarity and reduces additional visits and exposure to radiation. The estimated cost savings for NHS are £9.1 m every year (Millar 2019). These savings are a result of avoiding expensive unnecessary procedures and focusing on those who really need it. The technology was recommended by NICE in 2017 and now is used in more than 40 hospitals across NHS.
- *SecurAcath* – a device to secure catheters, associated with low incidence of catheter-associated complications. It improves stability and decreases infection risk for patients with a peripherally inserted central catheter. SecurAcath was one of four technologies centrally funded by the NHS under the Innovation and Technology Payment (ITP) programme (over 250 medical technologies applied for the programme). It helps to reduce time taken to care and treat dressing changes.
- *Placental growth factor (PIGF) test* that quickly predicts the risk of pre-eclampsia in pregnancy. The main benefit is a reduction in monitoring that allows pregnant women to spend less time in hospital. PIGF-based testing has been recommended for use by NICE Diagnostic Guidance (DG23) and Clinical Guidance (NG133) and is supported by the NHS Accelerated Access Collaborative and Innovation and Technology Payment (ITP) programme. The test became even more important recently as it helped to keep women out of hospital during the COVID-19 pandemic.

3.2 Agri-Tech Technologies and Their Applications

The Agri-Tech industry has very similar digital drivers to those seen in other industries. These are based on a number of core technologies listed below:

Data and AI: at the heart of the data revolution are the increased storage capacity (cloud) and computational power which together enable the rapid processing of large quantities of data. This is increasingly augmented by machine learning, namely, mathematical models that enable the creation of algorithms that find and apply patterns in data (Hao 2017). The quality and accuracy of data are key to a useful and successful AI system. Hence when it comes to using advanced systems, the old adage of 'garbage in garbage out' still stands, and hence it becomes highly important to generate and use quality data. In agriculture data sources are mainly from sensors in the field, satellite data and economic and historic data sets on produce.

Internet of things (IoT): IoT is an umbrella term for a series of technologies that work together to enable dispersed hardware to communicate with each other using Internet protocols and with a central data storage and processing centre. An IoT system requires a robust architecture that enables sensors, hubs and data processing centres to work in combination supported by a robust data relay network. The main application of an IoT system in farming is in enabling precision agriculture by integrating sensors and monitoring solutions into existing infrastructure and processes to generate data from agriculture activities or, in the case of animal husbandry, data related to each individual animal's health and physical condition. These data are generated directly from the farm, possibly in combination with external data. They lead to an optimisation of processes, the better management of resources and permit the continuous monitoring of plant and animal health, which will enable timely intervention leading to prevention of loss of harvest or animal products.

Robotics: robots are specially designed hardware that can operate using advance AI systems using data generated from the farm as well as other external data required for fulfilling a specific task such as harvesting fruit or selectively taking out weeds without impacting the crops – leading to highly reduced use or, indeed in some cases, no use of harmful chemicals. Robots are designed around fulfilling a specific task, and their intelligence depends on AI, IoT networks that they are part of and data that feed these systems.

Concerning the Agri-Tech application of robotics: drones are small flying robots carrying a camera or other devices/sensors and photographing, filming or charting a designated area. Drones also perform their tasks based on a combination of technologies used in IoT systems and AI modes and data technologies very similar to those seen in traditional robotic applications. The main difference is that drones are usually a mode of data collection whereas robots are more often machines/devices designed to take action based on data and AI decision system inputs.

Other innovations in Agri-Tech are based on new ways of combining the above-mentioned technologies to fulfil new applications and solutions. For example, hydroponics is the cultivation of plants without the use of soil. In general plants are set in an inert growing medium and supplied with nutrients, oxygen and water through liquids (hydroponics) or vapours (aeroponics). These technologies are currently most suitable for flowers, salads, herbs and some vegetables. Typically, they are tried out initially in demonstrator projects at a modest scale. Considerable further development is then required in order to enable these technologies to reach scales that are economic for a variety of crops. One advantage of these systems is that they can be built and run indoors using artificial light. In some cases crops can be stacked on top of each other creating 'vertical farms'. Most of the technologies behind these systems, apart from the nutrients that go into the hydroponic water and aeroponic vapours, are all operated, monitored and adjusted using AI and IoT core technologies.

3.2.1 Key Applications

The above-mentioned technologies converge in a number of ways to enable three main areas of application:

- Precision agriculture: this is when IoT and data from the field (machinery as well as soil and crop condition) are combined with other data sources such as weather, satellites and drones to inform the timing, location and amount of water, fertiliser or pesticides needed. This not only has a considerable impact on the cost of managing a field but also minimises negative impacts on the environment by reducing water, fertiliser and pesticide use.
- Labour replacement: robots are increasingly becoming accepted in agriculture mainly because farm labour is in short supply and both the pandemic and Brexit have been seen in 2020 to exacerbate these shortages because in recent decades (up until 2020) most of the seasonal farm work has been carried out in the UK by seasonal migrant workers (Byrne 2018). Furthermore, robots can apply any intervention with high precision and continuously. Two start-ups we interviewed for this research are developing precision solutions to routine agricultural work:
  - Ubiqutek (Rootwave) has developed a device that can kill weeds using an electric current and can be applied to each single week leaving crops intact.
  - Small Robot Company is developing small and flexible robots that can manoeuvre easily in a farm environment and can carry different tools for specific tasks.

It is not surprising that both companies are collaborating with each other in developing a weed zapping robot that can carry out this task without getting tired. There are also other forms of automation aimed at specific tasks in a farm such as continuous cleaning of animal sheds and even herding in the field which are being tested using specifically designed robots.

Animal health: IoT systems have been also deployed in animal husbandry with sensors continuously monitoring mainly the body temperature of each individual animal in the herd, notifying any changes through the systems. This enables the farmer to address any health issues before they present as visible symptoms and hence reduce risk of infecting the rest of the animals in the herd.

A farm which benefits from the ability to run most of its routine daily tasks and timely data reports for decision-making and prediction of required interventions in the near future is a so-called connected farm. It can gain economic advantage by reducing costs and reducing its environment footprint by minimising use of water, fertilisers and pesticides and other sources of environmental damage. Technologically the creation of such farms is possible today. Challenges include the capital cost of the technology infrastructure, rigid business models and policy. Policy currently does not allow for rapid adoption of new technology in farming and hence hampers the commercialisation of new technology. Start-ups are the main supply of new technology and innovation to this industry (Beauhurst 2018). However, in Agri-Tech there is not enough domain expertise in the start-up ecosystem. Most founders are engineers, data scientists or from other technical fields mainly with an urban life background. Investors too are not familiar with the challenges and opportunities of the industry and hence are not willing to risk larger sums of investment in the industry. Furthermore, there are risks associated with changing weather patterns as well as the impending consequences of climate change, namely, floods, heat waves and drought.

It is important to emphasise that the field of Agri-Tech is in its early days and the rate of innovation using the above-mentioned core technologies, i.e. AI, IoT and robotics, is rapidly growing, and many innovations are yet to emerge. However, as with new field of innovation, there will be also a so-called selection pressure, and only a few innovations will become dominant technologies that will capture a maximal market share. Apart from the enhanced rate of adoption, the interoperability of these technologies and adaptation to diversity of agricultural settings are all important factors. Furthermore, the level of funding and support available to nascent start-ups (and the technologies they commercialise) and numerous other nuanced factors can play a determining role in the success, or failure, of a new technology, as its promoters seek to prevail in the competition to establish the dominant design.

4 Challenges

4.1 MedTech

As introduced earlier, there are a set of concerns driving forwards a series of innovations, including big data, AI, precision healthcare and the Internet of Medical Things. These considerations lead us to ask: how will UK developments be shaped by the departure from EU? Is there scope for closer cooperation with the USA and, for example, the Middle East?

Concerning big data and artificial intelligence in MedTech, we suggest that data collection and application are key to greater levels of automation and more sophisticated AI. With developments in AI outpacing even the nimblest regulators, the key question for the UK is whether future-proofed frameworks can be built which will allow technologies to be deployed in real-world settings. The European Medical Device Regulation (MDR) aims to support innovations, but at the same time to guarantee high-quality and robust safety standards for medical devices produced in EU or imported into Europe. In May 2020 a new MDR came into force in the EU. That was just months before the end of the Brexit implementation period, but until the UK Parliament decides otherwise, the UK remains subject to this EU-originated law.

There are many problems regarding the data:

- Lack of standardisation – the available medical data is a mixture of structure, semi-structured and unstructured data
- How to handle big data
- Privacy and security issues
- Data transfer and speed limitation
- Reliability of data storage

There is a need for international partnerships and joined-up approaches, so as to harness properly the data underpinning much of current MedTech innovation. It is important to leverage the power of such data. If UK innovation is to achieve global impact then matching to, and helping to shape, international standards will be key concerns. Existing technologies use different frameworks and standards. They store information in a way that does not allow other systems to see, understand and use the information. It is crucial to develop standards to improve accessibility, utility and scalability of healthcare data. These challenges and opportunities sit in a context where consumer trust has been damaged. Increasingly consumers (patients) do not trust companies to be able safely to collect and use data. In the spring of 2020, as the COVID-19 pandemic took hold, there was much interest around the world in the potential for smart phone technology to assist with the tracking and tracing of epidemiological contacts. This in turn led to much debate concerning user privacy and a groundbreaking collaboration between Apple and Alphabet (parent of Google) in the development of a highly private protocol open to both iPhone and Android smartphone users. At the time of writing (August 2020), however such smartphone-based techniques have so far largely failed to gain significant traction in Europe or North America. The concerns around the alternative approaches often promoted by governments including holding personal information in central databases remind us that a cultural shift is needed if trust is to be restored. Perhaps the Apple-Alphabet COVID-19 collaboration is a first example of such a cultural shift. We posit, based on that recent experience and other indications, that there is a need for better understanding of such 'trust issues' on the part of MedTech companies. We suggest that they can improve healthcare only if they access and utilise data responsibly. In that regard perhaps collaborations with academic institutions can help, as long as public trust in the socially beneficial role of academic institutions is also maintained.

4.2 Agri-Tech

There are obvious global challenges to agriculture in general, namely:

- Population growth increasing the pressure on agriculture for intensification and increased yields
- Rapid urbanisation and reduction of arable land as well as labour willing and able to work in agriculture, leading to labour shortage
- Increased frequency and duration of adverse weather conditions due to climate change reducing the chance of reliable outcomes
- Social and economic pressure to reduce the carbon footprint of the industry
- Need to address legacy, policy, attitudes, entrenched interests and resistance to change
- Retreat from globalisation and an increased emphasis on self-sufficiency

Furthermore, in the context of Brexit, the UK potentially faces an uncertain future due to a number of factors such as a possible loss, or weakening, of research partnerships built over decades with European institutions, the impact of trade negotiations on prices through possible tariffs and the potential loss of international political influence. This picture, although troubling, does not necessarily lead to a bleak future for Agri-Tech industry in the UK. In contrast the above-mentioned pressures have the potential to increase the support and attention to the industry and to favour the acceleration of automation and the diversification of agricultural products to replace imports (e.g. via urban hydroponic vertical farming). With its strong research and innovation base, the UK is well placed to face these challenges.

The COVID-19 pandemic has considerably increased the immediacy of these pressures. For example, we have seen that unemployment is exacerbated due to global lockdown of countries to combat the pandemic. These events lead to acceleration of the drive for automation and also potentially reduce barriers to adoption by showing how a sudden unexpected event at national or global scale can suddenly disrupt the old models of the industry.

5 Discussion and Conclusions

The innovation process in MedTech is much more complex and regulated when compared to Agri-Tech. It is characterised by high R&D costs and short product life cycle, and it requires very sophisticated knowledge, expertise and resources. The UK is leading the charge among its European counterparts when it comes to embracing AI and robotics and generally accelerating the digital transformation of healthcare.

The UK has the largest number of researchers and professional engineers educated in the biological sciences at some of the world's leading universities. This includes Cambridge University, an institution located at the heart of Europe's leading technology cluster. When looking at both industries, we find that it is the people, and their unmatched expertise, that are the driving factors for innovation in the UK. Being based in the UK, particularly in the Cambridge cluster or more widely in the 'golden triangle', allows companies to access talent and resources of world-leading quality.

When looking at similarities, we noticed that both MedTech and Agri-Tech are technologically linked through their foundational technologies. As mentioned earlier, data and AI, the IoT and robotics underpin the majority of innovations in both industries. The outputs of all these technologies converge via data (including its generation, transfer, aggregation and analysis) for insight and decision support. Robotics will also find considerable transferable technology and skills in both sectors.

Agri-Tech benefits from a less complex stakeholder landscape regarding data. It spared some of the difficulties seen in MedTech – a sector burdened with trust issues in relation to patient data and its privacy and security. In contrast to the MedTech experience, a considerable amount of Agri-Tech data is already traded on different websites serving as marketplaces. Internal farm data, although possibly owned by the farm owner, finds its full use and applicability when it is combined with other data and even with data from other farms; hence there is an increased incentive for data sharing. In contrast medical data is highly prized and is a subject of competitive intelligence as well as strict privacy requirements concerning patients.

We have noticed that the British innovation landscape in MedTech is diverse and developing dynamically, looking for opportunities of cooperation and synergistic results. However, at the same time, the associated centralised systems are slow in adopting, as they are focused mainly on cutting costs. These MedTech contextual realities might restrain promising innovations and slow down the commercialisation processes. As a result, many British MedTech start-ups scale up through internationalisation and are selling their products into foreign markets, particularly the USA.

In contrast Agri-Tech in the UK is not hindered by a centralised organising structure and hence has a reasonable opportunity to thrive in both national and international markets. The challenges for Agri-Tech are mainly its low level of prominence as a tech sector, both for tech talent and for investors. However, through organisations such as Agri-TechE and government-backed schemes such as the Ceres Agritech Knowledge Exchange Partnership and the Cambridge AgriTech Cluster, the industry is gaining support and attention as source for innovation and growth.

As the reality, dimension and severity of global challenges such as climate change, population pressure and resource limitation become more evident; both agriculture and healthcare come into the zone of concern for policy-makers and business leaders alike. It is evident that the solutions which brought the agricultural revolution and advances in healthcare are not sufficient to address the scale and scope of the challenges we will imminently face. Both Agri-Tech and MedTech can play a considerable role, at the global and national level, to address these challenges at the systems level. Climate, food and health are key to navigating the world through the twenty-first century.

In light of COVID-19 pandemic, the MedTech industry comes into focus as it has the potential to be a strong contributor in solutions to challenges posed to both the global community and the UK through the pandemic. Agriculture, however, is also strongly impacted by the pandemic in the UK mainly due to its dependence on seasonal workers who are not able to travel to the UK as before. Furthermore, both industries will be impacted by Brexit when it is expected to come into full force in January 2021. Initially it appears that MedTech would be more exposed to a hard Brexit due to changes in regulation and compliance which will hamper medical certification of UK MedTech products in the EU and other countries that use EU standards. This will impact MedTech start-ups mainly because European markets sometimes were much easier to navigate than the complex monopoly of NHS in the UK medical market. Furthermore lack of access to EU will take away the opportunity to scale up using EU standard regulations. This may prompt more UK start-ups to move into the US market, developing their goods under FDA regulations.

In contrast Agri-Tech is less burdened by stringent validation and regulatory requirements and has the scope to expand significantly in the UK. Nevertheless, the challenge of navigating the export landscape will remain.

MedTech and Agri-Tech can have a convergence point in animal health and 'quantified-self' health monitoring in humans. Both MedTech and animal husbandry Agri-Tech are based on measuring a series of health indicators via sensors to send early warning signals for prevention and early treatment. Such synergies may be useful in new innovations for epidemiological control in pandemics in the near and distant future.

Both MedTech and Agri-Tech play a central role in addressing global systemic challenges, such as food production and health provision and the management are already under pressure due to population growth and climate change.

Who are the key actors in the innovative system in MedTech and in Agri-Tech in the UK? And, what are the core technologies driving the current waves of innovation in the two sectors? We ask these questions to better understand the future of both sectors but also possible scope of cooperation. Can one industry learn from the other? Where is the scope for cooperation and innovation? Despite the ideas explored in this chapter, these questions remain pertinent. More research is needed if the full benefits of the synergies between the fast-moving sectors of MedTech and Agri-Tech are to be made real. This work, and indeed this entire book, forms merely part of a work in progress. It is far too soon to know where this journey will end, but it is already proving interesting and provocative.

References

Accenture (2019), Digital Health Tech Vision. Are you ready for what's next in health care? [online] Available at: https://www.accenture.com/_acnmedia/PDF-102/Accenture-Digital-Health-Tech-Vision-2019.pdf#zoom=50 [Accessed: 18.06.2020]

Agri-Tech Centres (n.d.) The Agritech Centres of Agricultural Innovation; [online] Available at: https://www.agritechcentres.com [Accessed: 18.06.2020]

(Ag Funder Network 2018) £500m UK AgriTech Cluster Gets Top Political Backing; 1st November 2018; Ag Funder Network Partners. [online] Available at: https://bit.ly/30mkhGk [Accessed: 18.06.2020]

Beauhurst (2018) Agtech startups: the UK's leading agricultural innovators. [online] Available at: https://www.beauhurst.com/blog/top-uk-agtech-startups/ [Accessed: 10.08.2020]

BEIS (2017) The Bioeconomy in the North of England: A Science and Innovation Audit Report sponsored by the Department for Business, Energy & Industrial Strategy; August 2017; [online] Available at: https://bit.ly/30sNwrn [Accessed: 10.08.2020]

Bolland, I. (2019). Digital Medicine and Robotics at Heart of Topol Review Recommendations, MedTech News, 11 February [online] https://www.med-technews.com/news/digital-medicine-and-robotics-at-heart-of-topol-review-recom/ [Accessed: 10.06.2020]

Byrne R. (2018) The migrant labour shortage is already here, and agri- tech can't yet fill the gap (2018), Land, Farm and Agribusiness Management Department, Harper Adams University [online] Available at: https://bit.ly/2PyowZn [Accessed: 10.08.2020]

Cambridgeshire Live (2018) Cambridge spearheads new £20m agritech cluster, 10th April 2018; Cambridgeshire Live, [online] Available at: https://bit.ly/30nxGxQ [Accessed: 10.08.2020]

DEFRA (2019) Agriculture in the United Kingdom, 2019, Department for Food and Rural Affaires, [online] Available at: https://bit.ly/3fL0PaE [Accessed: 10.08.2020]

(DEFRA 2020) Farming for the future, Policy and progress update February 2020, Department for Environment, Food and Rural Affaires (DEFRA); [online] Available at: www.gov.uk/government/publications [Accessed: 10.08.2020]

Deloitte (2018). MedTech and the Internet of Medical Things. How connected Medical Devices are Transforming health Care, July, [online] Available at: https://www2.deloitte.com/global/en/pages/life-sciences-and-healthcare/articles/MedTech-internet-of-medical-things.html [Accessed: 14.06.2020]

Global AI Index n.d., [online] Available at: https://www.tortoisemedia.com/intelligence/ai/) [Accessed: 10.08.2020]

Hao, K. (2017) What is Machine Learning? MIT Technology Review, 17 November [online] Available at: https://bit.ly/2XZ7DuG [Accessed: 14.06.2020]

Hender, L. (2015) Cambridge – the UK's MedTech start-up hotspot, MedTech Engine, 31 May [online] Available at: https://MedTechengine.com/article/cambridge-MedTech-start-ups/ [Accessed 10.06.2020]

HM Gov (2013) A UK strategy for Agri-Tech technologies [online] Available at: https://assets.publishing.service.gov.uk/government/uploads/system/uploads/attachment_data/file/227259/9643-BIS-UK_Agri_Tech_Strategy_Accessible.pdf [Accessed 10.08.2020]

Kent, C. (2019). Science at the cutting edge: Why Britain is great for digital health and medical technology businesses, Pharmaceutical Technology, 27 June, [online] Available at: https://www.pharmaceutical-technology.com/sponsored/uk-digital-health-2019/ [Accessed: 17.06.2020]

Klein, T. (2015). The MedTech revolution: the European medical technology industry [online] Available at: https://informe2015.biocat.cat/wp-content/uploads/2016/04/EN-329-the-MedTech-revolution-the-european-medical-technology-industry.pdf [Accessed 15.06.2020]

MedTech Innovation News (2019) [online] Available at: https://www.med-technews.com/news/government-announces-five-new-ai-centres-of-excellence/ [Accessed 10.08.2020]

MedTech Landscape Review (2019) [online] Available at: https://www.ahsnnetwork.com/wp-content/uploads/2019/02/MedTech-Landscape-Review-AHSN-Network.pdf [Accessed 10.08.2020]

Millar, A. (2019). Heart flow Analysis: a better way to diagnose heart disease? Verdict Medical Devices, [online] Available at: https://www.medicaldevice-network.com/features/heartflow-analysis-a-better-way-to-diagnose-heart-disease/ [Accessed: 18.06.2020]

Office for Life Sciences (2019). Biosciences and Health Technology Sector Statistics 2018. Department of Health and Social Care, Office for Life Sciences, and Department for Business, Energy & Industrial Strategy [online] https://www.gov.uk/government/statistics/bioscience-and-health-technology-sector-statistics-2018 [Accessed: 10.06.2020]

Taylor, C. (2020). MedTech in the 2020s: what's in store?, MedTech News, 31 March, [online] Available at: https://www.med-technews.com/features/MedTech-in-the-2020s-what%E2%80%99s-in-store/ [Accessed: 10.06.2020]

UKRI (2018) Research England invests £67 million in collaborative projects to drive university commercialisation, 10th April 2018; UKRI; [online] Available at: https://bit.ly/33kGwhZ [Accessed: 18.06.2020]

WECA (2019) Local Industrial Strategy: Evidence base report Collaborative Innovation in the West of England; February 2019; West of England Combined Authority; [online] Available at: https://bit.ly/2XnrgNu [Accessed: 18.06.2020]

WLEP (n.d.) Key Sector Recommendations, Worcestershire Local Enterprise Partnership; [online] Available at: https://bit.ly/2XpNy1b [Accessed: 10.08.2020]

Chapter 12
Genomic Vaccines for Pandemic Diseases in Times of COVID-19: Global Trends and Patent Landscape

C. Possas, A. Antunes, A. M. Oliveira, M. Ramos, S. O. R. Schumacher, and A. Homma

Abstract This chapter provides an analysis of global trends in genomic vaccines, a radical innovation breakthrough (RIB), from technological foresight and pandemic preparedness perspectives, crucial in times of COVID-19. From this conceptual framework, the state-of-the-art and technological prospects for these genomic vaccines are examined, based on a search on scientific publications and on patents for the period 2010–2020, presenting the vaccine patent landscape for the period. This search provides an overview of recent breakthroughs in genomic vaccines and two other related RIBs, gene editing and gene therapy, and identifies novel strategies that could positively contribute to the development of future genomic vaccines to pandemic diseases and COVID-19. Our results evidence in the last decade extraordinary advances in genetic approaches, gene editing and gene therapy, and the rapid development of innovative DNA/RNA vaccines for the prevention and immunotherapy of an extensive diversity of diseases, from the neglected infectious ones to cancer therapy. These results highlight the flexibility of vaccine technological platforms, crucial for response to pandemics and COVID-19, including hepatitis B, varicella, chronic obstructive pulmonary disease, autoimmune diseases – systemic lupus erythematosus, lupus nephritis, and autoimmune myasthenia gravis – and finally a new nucleic acid sequence for immunogenicity to SARS-CoV-2.

Keywords Genomic vaccines · Radical innovation breakthroughs (RIBs) · Technological foresight · Patent landscape · Pandemic preparedness · COVID-19 · SARS-CoV-2

C. Possas (✉) · A. Homma
Bio-Manguinhos, Oswaldo Cruz Foundation, Rio de Janeiro, Brazil
e-mail: cristina.possas@bio.fiocruz.br

A. Antunes · A. M. Oliveira
School of Chemistry, Federal University of Rio de Janeiro, Rio de Janeiro, Brazil

National Institute of Industrial Property, Rio de Janeiro, Brazil

M. Ramos · S. O. R. Schumacher
School of Chemistry, Federal University of Rio de Janeiro, Rio de Janeiro, Brazil

E. Koukios, A. Sacio-Szymańska (eds.), *Bio#Futures*,
https://doi.org/10.1007/978-3-030-64969-2_12

1 Introduction

Vaccines are widely recognized by international organizations and scientific publications as one of the most relevant public health interventions, contributing to prevent infectious diseases and their spread. Moreover, the eradication of smallpox and the historic low cases of poliomyelitis and other vaccine-preventable diseases worldwide resulted in extraordinary proven benefits to public health, with a huge global economic impact. Immunotherapies against infectious and chronic degenerative diseases, such as HIV/AIDS, cancer, Alzheimer's, Parkinson's, and many others, are equally recognized as one of the most promising scientific developments in biosciences.

Innovation breakthroughs in preventive biomedicine, favored by rapid advances in immunology, virology, genomics, bioinformatics, biomedical engineering, and related technologies, have long been drivers of drastic advances in biosciences and healthcare. New developments in vaccines and immunotherapies from international innovation-based collaborative research networks, taking place in laboratories around the world, are expected to promote a paradigmatic change in preventive biomedicine and biotherapeutics in both the short and long terms.

Recently, the European Commission has launched a report on 100 radical innovation breakthroughs (RIBs) for the future, based on a methodology from the Finnish Radical Technology Inquirer (RTI), with biomedicine and related disruptive innovations in bioscience, such as genomic vaccines, listed in leading positions (Warnke et al. 2019).

In this chapter, we provide an overview of genomic vaccine and immunotherapy strategies and their related RIBs, from bioeconomic and global sustainability perspectives, discussing trends and innovation challenges for achieving Sustainable Development Goals (SDGs) for Vaccines in Agenda 2030 and their potential impacts (United Nations 2016; Gavi 2018).

2 Conceptual Framework

Technological foresight is an area of prospective studies that seeks to identify the main technological changes within a defined time horizon, with the aim of examining investment opportunities in certain processes or products from a strategic perspective, seeking to anticipate the risks involved in paradigmatic changes in technological scenarios (Antunes and Magalhães 2008). Conducted in a systematic way, with predefined methodologies and approaches, technological foresight is a tool that contributes to the construction of a shared vision of the future and the achievement of the desired impacts (Antunes and Canongia 2006; Alencar et al. 2007). Patents are, from this perspective, a valuable tool for technological scanning and foresight of future trends in genomic vaccines. We analyze the global patent landscape for these multipatented vaccine processes and products (Possas et al. 2015),

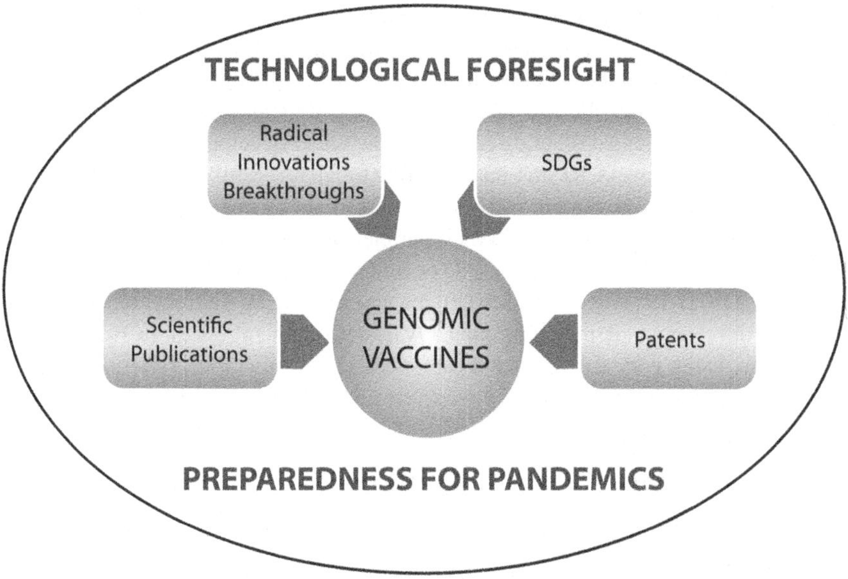

Fig. 12.1 Conceptual framework: genomic vaccines, foresight, and preparedness (Source: elaborated by the authors)

trying to identify future scenarios up to 2030 for innovative breakthroughs in vaccines and immunotherapies for pandemic or potentially pandemic diseases, as indicated in Fig. 12.1.

From this perspective, integrating technological foresight and global preparedness for pandemics, our results will be presented and discussed in three steps.

First, we provide an overview of recent vaccine breakthroughs from publications on "genomic vaccines" and two related RIBs, "gene editing" and "gene therapy," obtained from a search in PubMed Database for the period 2010–2020. This search aimed to identify novel strategies that could positively contribute to the development of future genomic vaccines to pandemic diseases and COVID-19.

Second, we present the global patent vaccine landscape for discussion of the technological trends in vaccinology in the last decade (2010–2020), including *genomic vaccines* and *structural vaccinology* (target epitopes) in the following patent databases: Derwent, Espacenet, and United States Patent and Trademark Office (USPTO).

Third, we provide evidence indicating that vaccine preparedness is a crucial component of global pandemic preparedness, discussing the most promising genomic vaccines now in development for COVID-19 and other potentially pandemic diseases. Moreover we underline, from a global governance perspective, the urgent need to create effective vaccine pandemic preparedness structures. New vaccine research, development and innovation (RD&I) organizations should be created, such as the Coalition for Epidemic Preparedness Innovations (CEPI), in accelerating global response to COVID-19 and to a future Disease X.

It is important to highlight that in our readings we present the main breakthroughs in next-generation vaccines related to three relevant RIBs: gene editing, gene therapy, and genomic vaccines. Although only the last one, genomic vaccines, appeared explicitly in our results, the two other RIBs, gene editing and gene therapy technologies, for vaccines against other diseases are now under development for the disease. In gene editing, a novel CRISPR-based editing tool has recently been developed that enables researchers to target mRNA and knockout genes without altering the genome. The developers have used their parallel-screening technique to create optimal guide RNAs which could be used for future detection and therapeutic applications for COVID-19 (Wessels et al. 2020). In gene therapy, as well, several institutions are now developing partnerships for COVID-19 treatment, such as Novartis and MultiStem, using gene therapies and vector-based gene therapies (Harris 2020).

3 Part I: Recent Vaccine Breakthroughs – Scientific Landscape

To define the universe of the study, a search was initially carried out in the MEDLINE Base (PubMed) in the period of 2010–2020. Three terms were selected: "gene editing," "gene therapy," and "genomic vaccine," from the main RIBs in technology (Warnke et al. 2019). Moreover, the terms "gene editing" and "gene therapy" were associated to the word "vaccine" for searching the publications. The numbers of publications are shown in Table 12.1. In a next step this search has been correlated to COVID-19. This methodological strategy for searching recent vaccine breakthroughs is presented in Fig. 12.2.

3.1 Vaccines and Gene Editing

In this first step of our study, we analyzed data collected from PubMed Database to survey the relationship between "gene editing" and the development of vaccines. The results identified 114 publications. Different approaches about gene editing and

Table 12.1 Number of publications for selected RIBs. (PubMed Database)

Radical innovation breakthroughs in technology[a]	Number of documents (papers)
Gene editing[a] + vaccine	114
Gene therapy[a] + vaccine	1462
Genomic vaccines[a]	1943
Total	3519

2010–2020 – carried out on March 18, 2020
Elaborated by the authors
[a]Warnke et al. (2019)

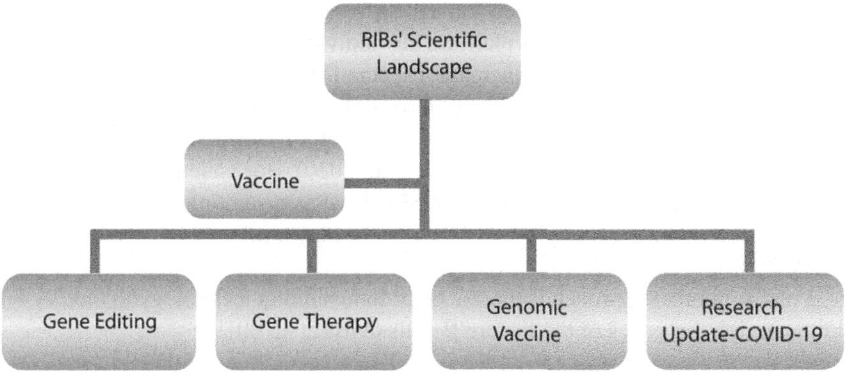

Fig. 12.2 State-of-the-Art Review: Genetic Approaches in Vaccinology (Source: Elaborated by the authors based on Warnke et al. 2019)

Table 12.2 The innovations in vaccinology: the use of gene editing (PubMed Database)

Author/ year	Description	Use	Source
Orr-Burks et al. (2019)	The identification of a subset of Vero cell host genes by siRNA and these antiviral host genes can be individually deleted by CRISPR-Cas9	Rotavirus vaccine production	Vaccine: X
Zhang et al. (2019)	The application of CRISPR-Cas9 editing approach for precise editing of the viral gene phosphoprotein 38	Highlights Marek's disease pathogenesis and the virus-host interactions	Viruses
Tang et al. (2020)	The "use of a previously developed CRISPR-Cas9 gene editing protocol for the insertion of ILTV gD-gI and the H9N2 AIV hemagglutinin expression cassettes into the distinct locations of the recombinant HVT-IBDV VP2 viral genome"	The generation of a triple insert HVT-VP2-gDgI-HA recombinant vaccine	Vaccines

Elaborated by the authors

vaccines were identified from the analyses of the abstracts and titles. From the main institutions in the number of publications, 29 documents were identified and evaluated. The present findings evidence significant advances in the use of gene editing in vaccinology innovations as indicated in Table 12.2. Furthermore, other documents were examined. One of them refers to the use of a "forward genetic approach" in *Plasmodium falciparum* research for the identification of "genes responsible for variable erythrocyte invasion by phenotyping the parents and progeny of previously generated experimental genetic crosses," which provides data for the development of a blood stage vaccine (Campino et al. 2018). Another, related to veterinary vaccine research, described the method of "application of CRISPR-Cas9 based genome editing" of generating herpesvirus of turkeys (HTV) recombinants which express VP2 protein of IBDV. In gene editing research, CRISPR-Cas9 system is used in many settings such as the method of modification of HTV genome which allows an increase in the development of recombinant vaccines (Tang et al. 2018).

Moreover, in another document on cancer research, it was demonstrated to be possible, using a technological platform for individual cancer vaccine, to prevent initiation and progression of pancreatic cancer. The researchers created a "Virus-Infected, Reprogrammed Somatic cell-derived Tumor cell (VIReST) regime," and the in situ gene editing was used for the induction of pluripotent stem cells from healthy cells to pancreatic tumor cells (Lu et al. 2020). Hoeksema and collaborators (2018) in their research showed the "CRISPR-Cas9 gene editing tools were used to knockout Vero target genes previously shown to play a role in polio- and rotavirus production."

Finally, another approach is the use of genome editing technology to modify the immunoglobulin genes of the mature B cells (Voss et al. 2019).

3.2 Vaccines and Gene Therapy

The RIB "gene therapy" (Warnke et al. 2019) associated to the term vaccine was used in the search strategy, resulting in 1462 publications in the 2010–2020 period. From the main institutions, 85 documents were recovered. These ones provided data about the use of gene therapy for noninfectious diseases and the breakthroughs in immunotherapy, highlighting recent advances on gene therapy and cancer, as indicated in Table 12.3.

Table 12.3 The use of gene therapy and the treatment of cancer

Author/ year	Description	Use	Source
Kamran et al. (2016)	Immunotherapy for glioblastoma	Active and passive immunotherapies for glioblastoma: checkpoint blockade, adoptive T-cell therapies, vaccines, and gene therapy	*Expert Opinion on Biological Therapy*
Maruf et al. (2016)	Novel immunotherapeutic agents to treat urothelial cell carcinoma	Vaccines, gene therapy, monoclonal antibodies, adoptive T-cell therapy for the urothelial cell carcinoma	*Cancer Biology and Medicine*
Donin et al. (2017)	Immunotherapies for urothelial carcinoma, research in immunotherapy, ongoing trials, and novel investigational agents	Immunotherapy for urothelial carcinoma is a highly promising research area	*The Journal of Urology*
Shaw and Suzuki (2019)	Adenoviral vectors in cancer therapy	"Adenoviral vectors are used for therapeutic gene transfer, vaccines, and oncolytic agents in the cancer gene therapy"	*Molecular Therapy – Methods and Clinical Development*

Siefker-Radtke (2010) presents and discusses researches on the biology of bladder cancer and the use of a gene therapy through the individual immune system with vaccines and anti-CTLA4 antibodies. The limited options of treatment for non-muscle invasive bladder cancer led to the improvement of therapies, like checkpoint inhibition, drug delivery, targeted therapy, chemoradiation, and viral therapy, besides cancer vaccines and gene therapy (Packiam et al. 2019; Siddiqui et al. 2017).

For infectious diseases, an evolution was observed in the field of influenza virus reverse genetics, like the modification of different gene segments to be applied on vaccine and gene therapies (Li et al. 2013).

Furthermore, the vaccines to combat the neglected diseases are also a target to gene therapy. One strategy against malaria is the "vectored immunoprophylaxis" which neutralizes the infectivity of the parasite (Rodrigues and Soares 2014). Or the use of mRNA for developing vaccines against diseases is a promising and safety approach of "mRNA-based treatments in gene therapy" (Lorenzi et al. 2010). Kalomoiris and collaborators (2012) described the development of "preselective anti-HIV lentiviral vectors" which improves HIV stem cell-based gene therapy.

For veterinary use, another important breakthrough is the treatment of canine melanoma with a combination of chemo-gene therapy and cytokine-enhanced vaccine in a new surgery adjuvant combined gene therapy (Finocchiaro et al. 2019). Furthermore, suicide and cytokine gene therapies were another combination for adjuvant gene therapy for canine mammary carcinoma (Finocchiaro et al. 2018).

Concerning new technologies, the use of biotherapeutic particles for the development of products in several fields, including gene therapy or tumor vaccines, was an important breakthrough (Moleirinho et al. 2019).

3.3 Genomic Vaccines

Genomic vaccines, or "DNA vaccines," take an innovative and entirely different approach from previous vaccine strategies: they inject genes, specifically DNA or RNA that encode for the needed protein, which then cause cells to produce the protein in question (Warnke et al. 2019). The search of scientific documents during 2010–2020 was conducted in PubMed Database in titles and abstracts using the terms "genomic vaccine," "DNA vaccine," and "RNA vaccine."

The terms "recombinant" and "subunit" were withdrawn from the search, because the main focus was not recombinant or subunit DNA vaccines. The search recovered 1943 documents (on average 176 publications by year). China, the USA, Iran, Japan, and Germany were the leading countries with 604, 536, 107, 80, and 67 scientific documents, respectively. The results also demonstrated that the leading institutions were Inovio Pharmaceuticals, Inc., USA, with 66 documents; Chinese Academy of Agricultural Sciences, China, with 57 documents; University of Pennsylvania, USA, with 56 documents; US Armed Forces, USA, with 54 documents; and People's Liberation Army, China, with 53 documents. The 258 scientific documents from the main institutions related to the development of vaccines were chosen for diverse research focus such as immunization for infectious and noninfectious diseases, neglected tropical diseases, delivery methods, and vaccine composition.

An important topic, as highlighted in many publications, is related to breakthroughs in DNA vaccines' delivery by electroporation. Some examples are provided: 1. DNA vaccine with high protection against Venezuelan equine encephalitis virus delivered by intramuscular electroporation (Dupuy et al. 2011); 2. Development of a "multiheaded intradermal electroporation device" to deliver multiple DNA vaccine plasmids for mass vaccination purposes (McCoy et al. 2014); 3. A "gene-optimized DNA vaccine" developed and delivered by dermal electroporation to protect guinea pigs from Lassa virus, a candidate for use in humans (Cashman et al. 2017).

For infectious diseases, an important breakthrough is the development of a synthetic DNA vaccine against Middle East respiratory syndrome coronavirus -MERS (Muthumani et al. 2015). Concerning cancer therapy, a promising recent research is a preclinical study of "DNA vaccine platform to target tumor neoantigens", reported by Duperret and collaborators (2019), a preclinical study of "DNA vaccine platform to target tumor neoantigens." Table 12.4 shows other uses of DNA vaccine for the treatment of this disease.

About the increase of the immunotherapy through adjuvants and the characterization of cytokine isoforms, Villarreal and Weiner (2015) discussed how IL-33 isoforms can be used for the development of tools as vaccine adjuvants. Furthermore, genetic adjuvant IL-12 increases DNA vaccine's efficacy for Venezuelan equine encephalitis virus (Suschak et al. 2018). Concerning the carriers for DNA vaccine, an important breakthrough is the upconversion nanoparticles which were modified with aminosilanes for foot-and-mouse disease. Indeed, the use of "upconversion nanoparticles NaYF4:Yb/Er@silica(UCPs)" was encouraged for "an effective nanosystem for gene delivery to cells for in vitro and in vivo vaccination" (Guo et al. 2012). Finally, another breakthrough is related to the required CCR10 expression for the "adjuvant activity of the mucosal chemokine CCL28 when delivered in the context of an HIV-1 Env DNA vaccine", which raises the antigen-specific humoral response (Gary et al. 2020). Concerning neglected tropical disease

Table 12.4 DNA vaccines for cancer therapy (PubMed Database)

Author/year	Description	Use	Source
Yan et al. (2013)	A synthetic highly optimized full-length human telomerase reverse transcriptase DNA vaccine	Treatment of cancer	*Cancer Immunology Research*
Yang et al. (2017)	DNA vaccine: the use of interleukin-12 (IL-12) as an intramolecular adjuvant	Treatment of prostate cancer	*Journal of Molecular Microbiology and Biotechnology*
Walters et al. (2017)	DNA vaccine platform	Antitumor immunity induction	*Molecular Therapy*
Duperret et al. (2018)	"A designer cross-reactive DNA immunotherapeutic vaccine"	Cancer therapy	*Clinical Cancer Research*

Elaborated by the authors

vaccination, results have indicated that the dengue virus serotype-1 showed safety profile and reactogenicity in human evaluation (Beckett et al. 2011).

3.4 *Research Update – COVID-19*

In difficult times of emerging pandemics, such as coronavirus disease (COVID-19), with significant impacts on the economy and society worldwide, innovations that allow reduction of time in the development of vaccines and drugs for the disease are crucial.

As of May 12 2020, the rapid spread of the COVID-19 pandemic had already resulted in a global humanitarian disaster, with 4,088,848 cumulative confirmed cases and 283,153 deaths worldwide (WHO 2020a; Johns Hopkins University 2020). Nevertheless, there are so far no effective treatments for COVID-19, and its management is empirical (Phadke and Saunik 2020).

The main challenge is thus to develop a SARS-CoV-2 vaccine fast and with a low cost to meet the global demand from the COVID-19 pandemic and to define clear priorities for their development. A recent WHO document presents targeted R&D priorities for candidate vaccines which is crucial to identify the research priorities for population immunization (WHO 2020b).

A major concern is that it takes more time to develop an effective and safe vaccine against COVID-19, which is estimated to be 12–18 months or more, and vaccines will probably not be available for this pandemic. The global COVID-19 vaccine R&D landscape includes 115 vaccine candidates, of which 78 are confirmed as active. Nevertheless, 73 of these projects are currently at exploratory or preclinical stages.

Table 12.5 Examples of candidates for COVID-19 in clinical and preclinical evaluations (WHO 2020b)

Platform	Evaluation	Type of candidate vaccine	Developer	Same platform for non-COVID-19 candidates
Nonreplicating viral vector	Clinical phases I and 2	Adenovirus type 5 vector	CanSino Biologics Inc./Beijing Institute of Biotechnology	Ebola
DNA	Clinical phase I	DNA plasmid vaccine electroporation device	Inovio Pharmaceuticals	Lassa, Nipah, HIV, filovirus, HPV, cancer indications, Zika, hepatitis B
RNA	Clinical phase I	LNP-encapsulated mRNA	Moderna/NIAID	Multiple candidates
DNA	Preclinical	Plasmid DNA, needle-free delivery	Immunomic Therapeutics, Inc./ EpiVax, Inc./ PharmaJet, Inc.	SARS
DNA	Preclinical	DNA plasmid vaccine	Osaka University/ AnGes/Takara Bio	–

Elaborated by the authors

Table 12.6 Clinical-phase vaccine candidates for COVID-19 (ClinicalTrials.gov website; WHO 2020a)

Candidate	Vaccine characteristics	Lead developer	Status
mRNA-1273	LNP-encapsulated mRNA vaccine encoding S protein	Moderna	Phase I
Ad5-nCoV	Adenovirus type 5 vector that expresses S protein	CanSino Biologics	Phase I
INO-4800	DNA plasmid encoding S protein delivered by electroporation	Inovio Pharmaceuticals	Phase I
LV-SMENP-DC	DCs modified with lentiviral vector expressing synthetic minigene based on domains of selected viral proteins; administered with antigen-specific CTLs	Shenzhen Geno-Immune Medical Institute	Phase I
Pathogen-specific aAPC	aAPCs modified with lentiviral vector expressing synthetic minigene based on domains of selected viral proteins	Shenzhen Geno-Immune Medical Institute	Phase I

Elaborated by the authors
(Note: aAPC, artificial antigen-presenting cell; CTL, cytotoxic T lymphocyte; DC, dendritic cell; LNP, lipid nanoparticle; S protein, SARS-CoV-2 spike protein)

Table 12.7 Vaccines platforms for the development of COVID-19 vaccines (Thanh Le et al. 2020)

Vaccine platforms	Preclinical	Clinical phase I
Live-attenuated virus	2	0
Nonreplicating viral vector	2	2
Recombinant protein	6	1
Peptide based	2	0
Virus-like particle	0	0
DNA	0	1
RNA	4	1
Unknown	2	0
Total	18	5

Elaborated by the authors

Only five promising advanced candidates moved into clinical development, including mRNA-1273 from Moderna, Ad5-nCoV from CanSino Biologics, INO-4800 from Inovio, and LV-SMENP-DC and pathogen-specific aAPC from Shenzhen Geno-Immune Medical Institute, as indicated in Tables 12.5 and 12.6.

Several other vaccine developers have indicated plans to initiate human testing in 2020 (WHO 2020b). It is important to highlight that 4 of these 5 promising candidates for COVID-19 are genomic vaccines, using DNA and RNA platforms, indicating an important advance in the maturity of this strategy.

Table 12.7 provides an overview of the diverse platforms used in the 23 projects in the global vaccine pipeline for the development of COVID-19 vaccines in preclinical (18) and clinical phases (5).

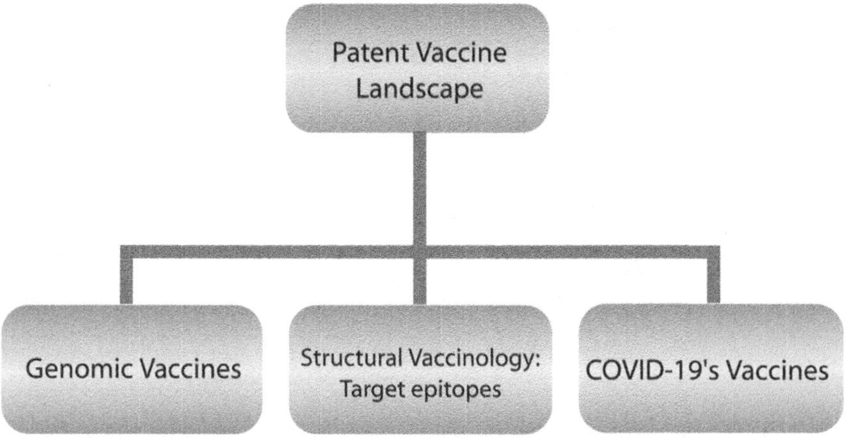

Fig. 12.3 Technological trends in vaccinology: patent landscape (Elaborated by the authors)

4 Part II: Vaccine Patent Landscape

The overview of a global patent vaccine landscape was carried out according to the strategy shown in Fig. 12.3.

4.1 Genomic Vaccines

The patent search was conducted in Derwent Database using the words "genomic" (or its derivatives) and "vaccine," "DNA vaccine" or "RNA vaccine" in document titles and in abstract documents for the 2010–2020 period, carried out on April 21, 2020.. The recombinant and subunit vaccines were excluded. The results were validated and the repeated documents were eliminated from the priority document number, and 219 patent filings were recorded.

The search led to important results about patent documents related to genomic/DNA/RNA vaccines, providing diverse focus on protection. The results indicated top patent applicants and priority countries, year, and the object of patent protection. The top priority countries were China with 97 patent filings and the USA with 61 documents, followed by the UK (14) and the European Patent Office (10) (Fig. 12.4).

The top applicants are from China: The top applicants are from China: UNIV CHINESE ACAD SCI (14 patent filings), ACAD MILITARY MEDICAL SCI CPLA CHINA (11) and CHINESE ACADEMY OF AGRICULTURAL SCIENCES (8). They are followed by ModernaTX Inc, from USA, with 8 documents (Fig. 12.5).

In Fig. 12.6, no relationship was demonstrated among the groups of institutions with patent filings in vaccines according to the strategy adopted. An exception was observed to the partnership between Novartis AG (Switzerland) and GlaxoSmithKline (the UK). The technology is related to an "immunogenic cationic oil-in-water emulsion" which was used to formulate DNA and RNA vaccines where the nucleic acid

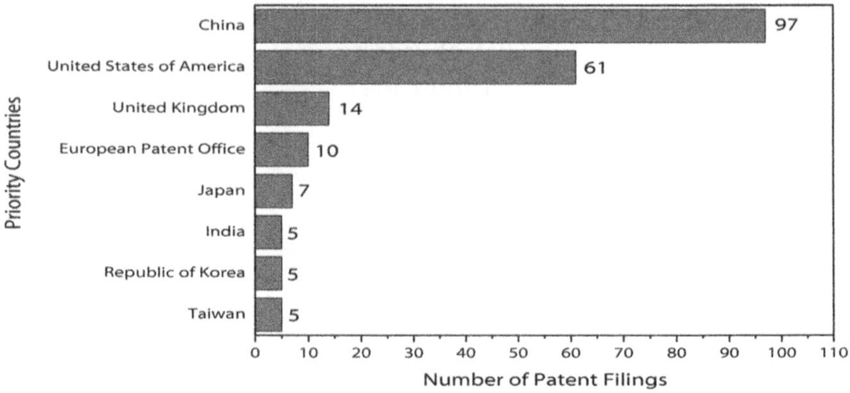

Fig. 12.4 The top priority countries for genomic/DNA/RNA vaccine patent filings (Derwent Database). (Elaborated by the authors)

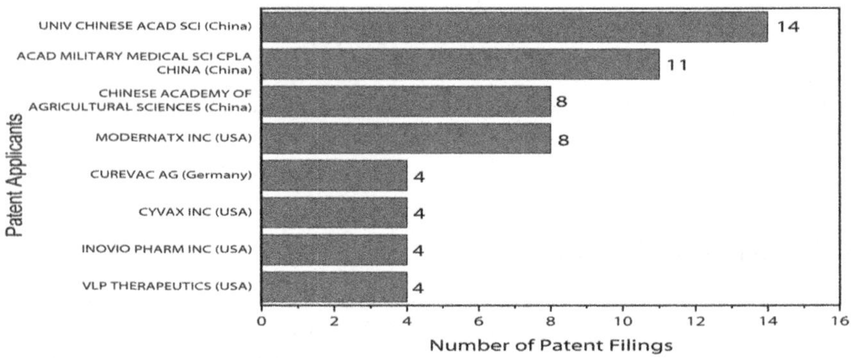

Fig. 12.5 The top applicant institutions for genomic/DNA/RNA vaccine patent filings (Derwent Database). (Elaborated by the authors)

molecule is complexed to the emulsion particles (patent number – PN: WO 2013006834) (Brito et al. 2012).

The recovered documents were aligned with applied biotechnology. DNA vaccines can be used for the treatment of chronic diseases, such as a vaccine for treating Alzheimer's disease which comprises a polynucleotide sequence that will encode a fragment of amyloid-beta protein. And it also comprises a "T-helper epitope of hepatitis B virus capsid antigen and a polynucleotide of S gene of hepatitis B virus" (PN: US 2014199338) (Arya and Markham 2014). Furthermore, for the treatment of hepatitis B, Furthermore, for the treatment of hepatitis B, a DNA vaccine plasmid was claimed for patent protection, which carries a hepatitis B antigen gene (PN: CN 103239717) (Jia et al. 2013). In addition, the present findings confirm the use of therapies for the treatment of cancer. Another breakthrough resulted in a patent filing (PN: CN 103948943) (Li 2014) for a DNA vaccine and for preparing a drug for

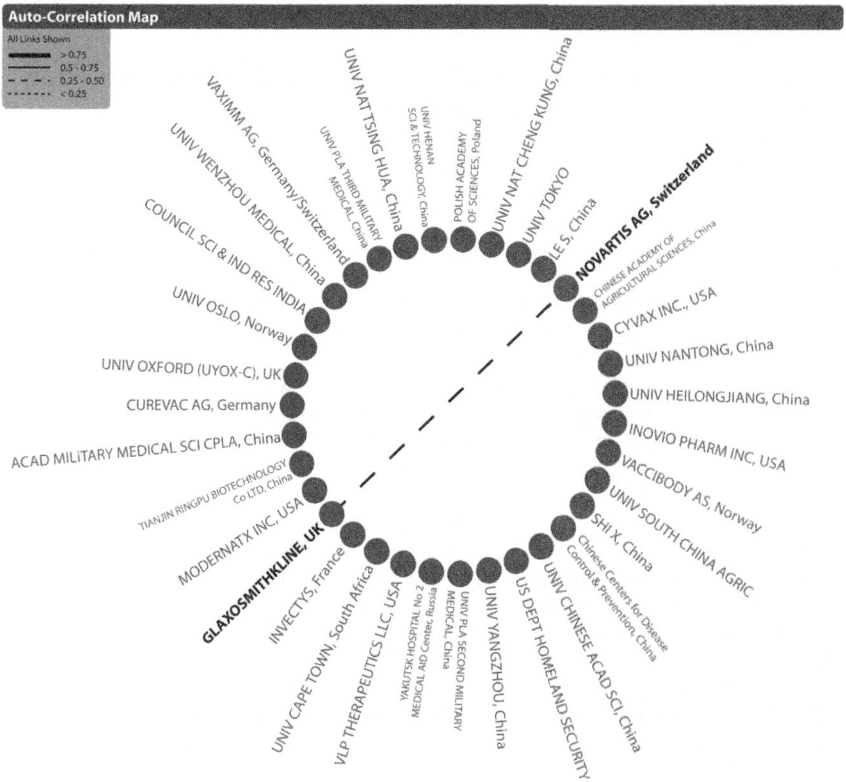

Fig. 12.6 Autocorrelation maps of most priority institutions for genomic/DNA/RNA vaccine patent filings (Derwent Database). (Elaborated by the authors)

the prevention or treatment of cancer. It was obtained by the insertion "sig-tG250-Fc-GP1-internal ribosome entry site-granulocyte-macrophage/B7 complex gene to replicative DNA vaccine vector pSVK." For the treatment of melanoma, a fermentation culture medium was used as DNA vaccine pSVK-CAVA (PN: CN 103952367) (Wang et al. 2013).

The results also provided clear support for the use of DNA vaccines for different diseases, a human cytomegalovirus (HCMV) vaccine with RNA polynucleotide encoding antigenic polypeptides (gH, gL, UL128, UL130, and/or UL131A) or its epitopes against HCMV (PN: WO 2018075980) (Ciaramella and John 2017), and for the preparation of DNA vaccine for the treatment and prevention of "allogeneic/organ transplantation rejection" (PN: CN 102161998) (Xi and Luo 2011). For the treatment of brucellosis, a *Brucella abortus* vaccine was developed. It comprises a "P39 and 18 Kd gene co-expressing through internal ribosomal entry site mediated eukaryocyte" (PN: CN 102772793) (Jing et al. 2012). For the prevention or treatment of chronic obstructive pulmonary disease, autoimmune diseases (like systemic lupus erythematosus, lupus nephritis, and autoimmune myasthenia gravis),

Table 12.8 Patent landscape: genomic (DNA) vaccines focus (Derwent Database)

Applicant	Description	Indication	Patent number	Year of priority
Chinese Academy of Agricultural Sciences (China)	"DNA vaccine based on specific gene for avian reticuloendothelium-shaped endothelial tissue hyperplasia"	Treatment/ prevention of endothelial tissue hyperplasia	CN 102988969 (Gao et al. 2012)	2012
UNIV CHINESE ACAD SCI (China)	"New *Streptococcus iniae* trivalent DNA vaccine"	Protection against streptococcal infection in dolphins	CN 102716499 (Hu et al. 2012)	2012
Cyvax Inc. (USA)	DNA vaccine: DNA plasmid with polynucleotide encoding antigenic polypeptide	Prevention of liver-stage malaria infection and reduction of bloodstream infection levels.	WO 2014028644 (Markham 2013)	2012
ACAD MILITARY MEDICAL SCI CPLA (China)	"Broad-spectrum antitumor double-plasmid replicable DNA vaccines"	Antitumor active immunotherapy	CN 103948944 (Jia et al, 2014)	2013
ACAD MILITARY MEDICAL SCI CPLA (China)	DNA vaccine which contains engineering *Escherichia coli* and with a specific nucleotide sequence	Treatment of hepatitis B	CN 103396975 (Xu et al. 2013)	2013

Elaborated by the authors

targeting tumors and other claimed diseases for patent protection, a C-X-C motif chemokine ligand 13 DNA vaccine was developed (PN: CN 105797147) (Cheng et al. 2016).

Moreover, another innovation is vaccine which comprises RNA polynucleotide for the prevention/treatment of varicella zoster virus (PN: WO 2018170270) (Ciaramella 2018). Another example is a RNA vaccine for the induction of a specific immune response to infection caused by respiratory syncytial virus. And this RNA vaccine is "presented to the cellular system in a more native fashion" with an efficacy of 60–90% (PN: WO 2018170260) (Bahl et al. 2018). Other examples are shown in Table 12.8.

Other innovations are related to the improvement of vaccines' formulations (including DNA vaccine) using chitosan oligosaccharide obtained by enzymatic hydrolysis (PN: CN 104906574) (Shan et al. 2015). Furthermore, other vaccine improvements are a "heat shock protein 70" used as an adjuvant of a DNA vaccine which improves its immune protective effect (PN: CN 103405763) (Sun and Zhang 2013) and the use of a "cluster of differentiation (CD)83 molecule derived from turbot" as an adjuvant for DNA vaccine (PN: CN 104689313) (Sun and Li 2015).

From the same strategy at Derwent Database, additional researches were performed at Espacenet Patent Database and at the United States Patent and Trademark Office (USPTO). From Espacenet Database, documents' analyses revealed that there was no company or research institution with great relevance in the number of deposits. After eliminating the repeated documents, 25 documents from the main applicants were selected. The most applicants were People's Liberation Army (China) and VAXIMM AG (Switzerland) with six patent filings each. On the other hand, China is the most priority country with 113 patent filings, far from the USA (33), European Patent Office (14), and Republic of Korea (10).

In our search key findings emerged: RNA vaccine against cancer which provides artificial nucleic acid molecules (PN: WO2017EP66676) (Fotin-Mleczek et al. 2018) and the "combination of an RNA encoding an epitope and immune checkpoint inhibitors" for the treatment/prophylaxis of diseases like infectious diseases and cancer (PN: WO2017EP56427) (Heidenreich et al. (2018). Another important finding was related to understanding mechanisms involved in the key role DNA vaccines can play for the treatment of cancer. Some examples are the "Novel CMV PP65 Targeting DNA vaccine" (PN: EP20150001802) (Lubenau 2016); "The process to produce a DNA vaccine" (PN: EP20160001550) (Lubenau 2017); a "PD-L1 Targeting DNA vaccine" (PN: EP20170161666) (Lubenau 2018), all of them for the treatment of cancer. And the "AR-42 enhances E7-specific CD8+" in conjunction with DNA vaccine to mediate antitumor immunity for the treatment of HPV-related cancers (PN: US201261731225P) (Wu and Hung 2013). Furthermore, a "Neoantigen targeting DNA vaccine" for combined therapy to treat solid tumors has been developed (PN: EP20180192782) (Lubenau 2019).

About the treatment/prophylaxis of other diseases, DNA vaccine was applied to combined HIV vaccine, which comprises "a DNA vaccine, an envelope protein vaccine and a polypeptide vaccine with specific sites" (PN: CN201710431893) (Jiang et al. 2017). Another vaccine, a "H9 subtype bird flu DNA vaccine" was claimed by UNIV SOUTH CHINA AGRICULT (PN: CN201610237685) (Jiao et al. 2016). At the USPTO, four complementary patent documents were recorded. All of them consist of inventions about DNA vaccines. Two documents were from Osaka University (Japan). The first invention (PN: US 20150202271 A1) (Nakagami et al. 2013) is a DNA vaccine for the treatment or prevention of cancer and comprises a "VEGF-specific epitope and/or angiopoietin-2-specific epitope." The second one is about a therapeutic/improving agent, but now for a "lifestyle-related disease" (PN: US 20140099335 A1) (Morishita et al 2012). And a DNA vaccine which has a specific epitope of apolipoprotein was filed by AnGes, Inc. (Japan) (PN: US 20160303211 A1) (Kyutoku et al. 2016). Moreover, a method for induction of response and the use of DNA vaccine (PN: US 20140255343 A1) (Smietanka et al. 2012), was claimed by INSTYTUT BIOCHEMII I BIOFIZYKI PAN (Poland).

4.2 Structural Vaccinology – Target Epitopes

The structural biology allows the possibility of understanding the potential antigens' structure and to discover target epitopes to the development of vaccines (Serruto and Rappuoli 2006). The development of an effective vaccine using structural vaccinology involves rational steps. For instance, first is the determination of the antigen or antigen antibody's structure; second is the use of the reverse molecular engineering to remodel the antigen or epitope; after it is the incorporation of the reengineered epitope/antigen into one of the vaccine platforms; and the last step is to test the vaccine's safety and efficacy (Anasir and Poh 2019).

Based on this concept, a complementary search was developed in Derwent Database using the terms "structural vaccine" or "target epitope and vaccine" in the 2010–2020 period, carried out on April 21, 2020. After validation of the results and elimination of the repeated priority documents, 19 documents were recorded. The top priority country was the USA with nine patent filings, followed by China (eight), the European Patent Office (one), and Spain (one). Considering the number of deposits, no company or research institution indicated the great relevance in the number of deposits. Only two institutions have two applications: UNIV NANTONG (China) and UNIV WASHINGTON (USA). And for the others, only one patent filing was found. The development of vaccines with target epitopes is observed from the documents' analysis. One of them refers to the development of a "chimeric bivalent vaccine" to prepare drugs for hypertension (PN: CN 109663124) (Liao et al. 2019). The document focuses on vaccine compositions or immunogenic compositions. The document WO 2015082745 (Legarreta Solaguren et al. 2014) is an example of a computer-assisted method design of a vaccine. In addition, an immunogenic composition which also comprises a common target epitope and improved vaccines for many antigens and proteins targets was claimed (PN: WO 2013177214) (Glanville 2013).

Furthermore, a new isolated peptide for the treatment of respiratory syncytial virus infection was claimed (PN: WO 2013152274) (Correia and Schief 2013). This is also the case of the INST ROUSSY GUSTAVE's document (France) (PN: WO 2012150478) (Benihoud et al. 2011) for a "new recombinant replication defective adenovirus" which is useful as a vaccine to generate an immune response against the target epitope(s) inserted into its fiber protein. Moreover, an additional claim is a "T-cell modulatory multimeric polypeptide" or its conjugate which is useful for the development of pharmaceutical composition/medicament for modulating an immune response or for the treatment of infectious diseases, like viral infection or cancer (PN: WO 2019051127) (Chaparro et al. 2018). Moreover, a deposit from UNIV NANTONG (China), CN 106344914 (Fang et al. 2014), showed a "method for preparing *Brugia malayi* M29 epitope protein vaccine."

From the same strategy at Derwent Database, additional researches were performed at Espacenet Patent Database and at the USPTO. After eliminating the repeated documents, two documents were found in Espacenet Patent Database. Both of them were from China. The first one was from UNIV ANHUI

1. **(PN: CN110974950-A):** *"Vaccine composition useful for e.g preventing severe acute respiratory syndrome coronavirus (SARS-CoV)-2 infection, and inducing immune response in human body, comprises nucleic acid sequence comprising specific base pair sequence."*

2. **(PN: CN110951756-A):** *"New nucleic acid sequence used for expressing polypeptide causing immunogenicity to novel coronavirus SARS-CoV-2 by e.g. including immune response in human body, generating bioreporters, and regulating gene function."*

Fig. 12.7 Description of COVID-19 vaccine's patents (Derwent and Espacenet Patent Databases). (Elaborated by the authors)

AGRICULTURAL which claimed a "*Vibrio mimicus* oral target epitope gene vaccine" (PN: CN20181060036) (Li et al. 2018) with an increased immune effect and the "tumor nucleic acid vaccine" by the NANJING HIGH WIT BIOTECHNOLOGY COMPANY for the prevention and treatment of tumors (PN: CN201310405991) (Liu et al. 2013). At the USPTO, there was no additional document.

4.3 COVID's Vaccines

To perform a patent research about vaccines to COVID-19, a strategy search at Derwent and Espacenet Patent Databases was developed on May 5, 2020, for the period of 2010–2020. COVID-19 terminology and definitions can change with scientific progress. Thus, for our research, specific terms were used, associated to the word "vaccine" in the title/abstract of the patent documents: "COVID-19," "coronavirus disease," "SARS-CoV-2," "SARS-2," and "Wuhan coronavirus." According to this strategy, we analyzed the leading patent applicants, country of priority, year, and object of patent protection.

An inexistent or a low number of patent filings by the applicant countries due to the patent secrecy's period was expected. Despite this constraint, two patent documents from GUANGZHOU N BIOMED CO LTD (China) were recovered, both from 2020. Figure 12.7 illustrates these deposits. The document PN: CN110974950-A (Chen et al. 2020b)is a vaccine which comprises a nucleic acid sequence useful for the prevention of COVID-19 infection in human body. Besides that, there were other claims like production of biological reporter and preparation of gene function regulator. Another point to highlight about the vaccine is that "the vector is a DNA plasmid, an RNA expression plasmid or a viral vector."

The document PN: CN110951756-A (Chen et al. 2020a) claimed a nucleic acid that promotes immunogenicity to COVID-19 in human body. It should be noted that GUANGZHOU N BIOMED CO LTD also claimed other uses of the nucleic acid sequence like generation of bioreports and regulation of gene function.

5 Part III: Global Preparedness and Vaccine Development

As the COVID-19 pandemic is exponentially increasing as a global humanitarian disaster, it is important to highlight the unprecedented timely international efforts from extensive networks of research institutes and companies worldwide, in collaborations and partnerships supported by huge RD&I investments, to accelerate the development and production of new drugs and vaccines. Consequently, as a result of this global commitment of so many institutions, information about SARS-CoV-2 and COVID-19 and new vaccine developments are changing very rapidly.

It is also interesting to observe in our search the acceleration of the speed of vaccine development. Based on these innovative platforms and adoption of fast track approach, scientists were able to bring five vaccine candidates into clinical trials in unparalleled speed, just 4 months.

Vaccine platforms include, besides attenuated and inactivated vaccines, DNA, mRNA, modRNA technology, replicating and nonreplicating vectors, virus-like particles, nanotechnology and nanoparticles and also the use of large databases, big data, and artificial intelligence. The main features of these innovative platforms are their flexibility: they can be modified to target a desired pathogen with minimal changes in development procedures, contributing to the simplification of the regulatory process and exponentially accelerating vaccine development and production timelines.

COVID-19 pandemic has, with emergence of nucleic acid vaccines and other radical innovation breakthroughs, drastically changed the global vaccine development scenario, which might never be the same again (Dolgin 2021). Scientific and technological improvements have allowed vaccine developers in emerging diseases scenarios to reduce the timeline between the initial stage of viral genome sequencing and the launch of phase I trials in humans, reduced from 20 months in the 2003 SARS outbreak, to just over 3 months in the 2016 Zika outbreak and to 2 months in the COVID-19 pandemic, achieved by two companies Moderna and CanSino Biologics.

Synthetic peptides are also the object of innovative studies, identified by specific silica modeling for SARS-CoV-2, with a greater cross-reaction spectrum for coronavirus, in VLP, or in nanoparticles. Synthetic peptides have the advantage of quick production, reducing development time, and low production cost.

Two genomic vaccines, Moderna and Pfizer/BioNTech, use mRNA platforms to deliver genetic fragments of SARS-COV-2 to human cells, instructing them to produce immune responses against the pathogen. This approach has a broad ranger of applications, from cancer to Zika vaccines

CanSino Biologics uses an adenovirus vector platform that modifies the virus which causes the common cold to carry key structural genes from SARS-COV-2. Adenovirus vectors have been used for a number of vaccines such as Ebola, in a variety of forms, and are very promising.

Two other innovative vaccines use adenovirus vector approaches: 1. Oxford/AstraZeneca, a COVID-19 vaccine with chimpanzee adenovirus platform approach; 2. Johnson & Johnson, which has recently announced a billion dollar collaboration with BARDA to support an innovative human adenovirus COVID-19 vaccine based on this platform to produce 1 billion doses of vaccine by the end of 2021.

In spite of these important breakthroughs, major gaps remain (Homma & Possas 2020) and should be overcome in these innovative COVID-19 vaccines:

1. There is scarce scientific knowledge on the disease, such as on the causes of its high lethality in the elderly and individuals with comorbidities and understanding why some people are asymptomatic and others get severe disease; .
2. Lack of data on immunogenicity conferred by the virus, duration of immunity and protection and reference materials.
3. The possibility of "vaccine enhancement": instead of protecting against infection, the vaccine can actually make the disease worse when a vaccinated person is infected with the virus. The mechanism that causes that risk is not fully understood and is one of the major constraints that have prevented the successful development of a coronavirus vaccine. Under normal conditions, researchers would take months to test for the possibility of vaccine enhancement in animals. Given the urgency of the pandemics, some pharmaceutical companies are moving straight into small-scale human tests, without waiting for the completion of such animal tests.
4. The need to redesign vaccines to protect against emerging coronavirus mutations and variants. COVID-19 lineages that can evade immunity are stimulating scientists to explore new strategies to redesign their novel vaccines. Variants recently identified in UK (B.1.1.7); in Brazil (P1/B1.1.28) and in South Africa (B.1.351) have raised major concerns in vaccine developers. This South African variant identified in late 2020 is among the most worrying since lab assays have found that it carries mutations that reduce the potency of virus-inactivating neutralizing antibodies that were made by people who received either the Pfizer or Moderna RNA vaccines (Callaway and Ledford 2021). A good example is Oxford/Astrazeneca which is now redesigning its COVID-19 vaccine to protect against the South African variant, after a clinical study had indicated that the vaccine offered minimal protection to this variant (Cohen 2021).

From this perspective, it will be crucial to strengthen the role of global innovative vaccine RD&I organizations directed to pandemic preparedness, such as CEPI, supported by the Bill and Melinda Gates Foundation, the Wellcome Trust, Global Health Fund, and others, thereby accelerating the availability of innovative vaccines to the global population. Local CEPI-like structures should also be created in middle-income countries, such as Brazil and India, in support to CEPI, to accelerate vaccine development and manufacturing in new decentralized structures for vaccine innovation aiming pandemic preparedness. Moreover, it is also necessary to emphasize the crucial role the Developing Countries Vaccine Manufacturers Network (DCVMN) can play integrating vaccine manufacturers in these countries into this new global strategy.

6 Conclusion

COVId-19 pandemic accelerated extraordinary radical technological innovation breakthroughs in a new era of vaccinology. Our search results indicate in the last decade an important paradigmatic change in vaccine development, with novel genomic vaccines approaches using mRNA and DNA platforms, which can contribute to accelerate vaccine development and shorten timelines to reach a final product.

These results also evidence extraordinary advances in genetic approaches (gene editing and gene therapy) and particularly the rapid development of innovative DNA/RNA vaccines for the prevention and immunotherapy of an extensive diversity of diseases, from the neglected infectious ones to cancer therapy. These novel preventive and immunotherapeutic DNA/RNA vaccine strategies and platforms include hepatitis B, varicella, chronic obstructive pulmonary disease, autoimmune diseases (like systemic lupus erythematosus, lupus nephritis, and autoimmune myasthenia gravis), and finally a new nucleic acid sequence to cause immunogenicity to SARS-CoV-2.

These breakthroughs have become crucial to accelerate vaccine pipeline in a complex scenario, as the COVID-19 pandemic unfolds in a global humanitarian disaster and the possibility of a future more severe and lethal Disease X becomes a global concern, favored by rapid eco-social changes and population mobility, related to intensification of travel.

The global scientific mobilization to produce a SARS-CoV-2 vaccine has become a race against time. It is interesting to note, in our results, the broad range of innovative technology platforms being evaluated, including nucleic acid (DNA and RNA), virus-like particle, peptide, viral vector (replicating and nonreplicating), recombinant protein, live-attenuated virus, and inactivated virus approaches.

There are now more than 300 projects in the global vaccine pipeline for COVID-19, as of February 2021, most of them in preclinical phase, using these diverse platforms, including nucleic acid (DNA and RNA), virus-like particle, peptide, viral vector (replicating and nonreplicating), recombinant protein, live-attenuated virus, and inactivated virus approaches. This large number of projects evidences the rapid development of new vaccine strategies and the intensified global effort to respond to the pandemic. It also highlights the impressive and unprecedented timely international efforts from extensive networks of research institutes and companies worldwide, supported by huge RD&I investments to accelerate the development of new drugs and vaccines.

Nevertheless, in spite of all these successful initiatives, it is urgent to accelerate the global vaccine pipeline for pandemic preparedness, providing sustainable funding and overcoming current gaps. The advances in other fields such as immunotherapies in oncology and other medical fields is stimulating research institutes and companies to move into disruptive next-generation vaccine strategies, in order to accelerate vaccine development and production to respond to the pandemic. Certainly it is expected that very soon there will be a new generation of vaccines that will cross-protect for diverse mutations and variants of SARS-CoV-2, requiring only one dose for sterile protection and protection for life. These new vaccine technologies will be easier to produce, have high-yield and low cost of production.

An important feature of genomic vaccines for COVID-19 resides in the greater flexibility of innovative DNA and mRNA platforms, providing more favorable conditions for antigen manipulation and the possibility of considerable acceleration in the speed of vaccine development. Moreover, as discussed here, Moreover, as discussed here, the immunome vaccine concept, based on the identification of a minimal set of antigens that induce a competent immune response to a pathogen or neoplasm, favored and accelerated by new developments in reverse vaccinology, bioinformatics, and artificial intelligence, will contribute to development of future next-generation personalized vaccines with reduced adverse effects.

Notwithstanding, in spite of all these extraordinary breakthroughs, complex vaccine development issues, such as adverse vaccine events from "antibody-dependent enhancement" (ADE) and animal models, remain as major challenges for response to the pandemic. COVID-19 is likely to become an endemic disease and accelerating the global scientific effort to overcome these constraints will be crucial to develop an effective and safe vaccine for its prevention.

COVID-19 pandemic emerged as a health and humanitarian disaster. And for this reason, it was never made so much investment to obtain a new medicine and a new vaccine. Moreover, all the information presented here is changing very quickly, resulting from intensified collaboration and involvement of hundreds of institutions.

The lessons learned from this COVID-19 pandemic will certainly result in extraordinary scientific, technological, and social advances on a global scale in the field of vaccines. Innovation breakthroughs in new generation of genomic vaccines will certainly contribute to the maturity of this field, promoting an exponential acceleration in vaccine development and novel strategies to extend their availability and access in a new global pandemic preparedness paradigm.

References

Alencar MSM, Porter AL, Antunes AMS. Nanopatenting patterns in relation to product life cycle. Technol Forecast Soc Chang. 2007. https://doi.org/10.1016/j.techfore.2007.04.002.

Anasir MI, Poh CL. Structural vaccinology for viral vaccine design. Front Microbiol. 2019. https://doi.org/10.3389/fmicb.2019.00738.

Antunes A, Canongia C. Technological foresight and technological scanning for identifying priorities and opportunities: The biotechnology and health sector. Foresight 2006. https://doi.org/10.1108/14636680610703072.

Antunes MSA and Magalhães JL (org.), Patenteamento e prospecção tecnológica no setor farmacêutico [Patenting and technological foresight in the pharmaceutical sector], Rio de Janeiro, Interciência, UFRJ, 2008.

Arya B, Markham R (2014) Vaccines for protection from and treatment of Alzheimer's disease. US Patent 2014199338, 10 January 2014

Bahl K, Bett AJ, Ciaramella G, Espeseth A, Wang D (2018) Respiratory syncytial virus vaccine. WO Patent 2018170260, 15 March 2018

Beckett CG, Tjaden J, Burgess T, Danko JR, Tamminga C, Simmons M, et al. Evaluation of a prototype dengue-1 DNA vaccine in a Phase 1 clinical trial. Vaccine 2011. https://doi.org/10.1016/j.vaccine.2010.11.050.

Benihoud K, Lanzi A, Perricaudet M (2011) Adenovirus vaccine vectors. WO Patent 2012150478, 28 April 2011

Brito L, Chan M, Geall A, O'hagan D, Singh M (2012) Oil-in-water emulsions that contain nucleic acids. WO Patent 2013006834, 6 July 2012

Callaway E, Ledford H (2021) How to redesign COVID vaccines so they protect against variants. Nature 590, 15–16. https://doi.org/10.1038/d41586-021-00241-6

Campino S, Marin-Menendez A, Kemp A, Cross N, Drought L, Otto TD, et al. A forward genetic screen reveals a primary role for Plasmodium falciparum Reticulocyte Binding Protein Homologue 2a and 2b in determining alternative erythrocyte invasion pathways. PLoS Pathog. 2018. https://doi.org/10.1371/journal.ppat.1007436.

Cashman KA, Wilkinson ER, Shaia CI, Facemire PR, Bell TM, Bearss JJ, et al. A DNA vaccine delivered by dermal electroporation fully protects cynomolgus macaques against Lassa fever. Hum Vaccin Immunother. 2017. https://doi.org/10.1080/21645515.2017.1356500

Chaparro R, Low CM, Ross J, Seidel RD (2018) T-Cell modulatory multimeric polypeptide with conjugation sites and methods of use thereof. WO Patent 2019051127, 6 September 2018

Chen L, Yang C, Liu X (2020a) New nucleic acid sequence used for expressing polypeptide causing immunogenicity to novel coronavirus Sars-Cov-2 by E.G. inducing immune response in human body, generating bioreporters, and regulating gene function. CN Patent 110951756, 23 February 2020.

Chen L, Yang C, Liu X (2020b) Vaccine composition useful for preventing severe acute respiratory syndrome coronavirus (SARS-CoV-2) infection, and inducing immune response in human body, comprises nucleic acid sequence comprising specific base pair sequence. CN Patent 110974950, 5 March 2020.

Cheng X, Wang G, Zhang Y, Zhou G (2016) CXCL13 (Chemokine(C-X-C Motif) Ligand 13) DNA vaccine and application thereof. CN Patent 105797147, 7 April 2016.

Ciaramella G (2018) Varicella zoster virus (VZV) Vaccine. WO Patent 2018170270, 15 March 2018

Ciaramella G, John S (2017) Human cytomegalovirus vaccine. WO Patent 2018075980, 20 October 2017

Correia BE, Schief WR (2013) Epitope- scaffold immunogens against respiratory syncytial virus (RSV). WO Patent 2013152274, 5 April 2013

Cohen J (2021) South Africa suspends use of AstraZeneca´s COVID-19 vaccine after it fails to clearly stop virus variant. Science 2021, published online on Jan. 28. https://doi.org/10.1126/science.abg9559

Donin NM, Lenis AT, Holden S, Drakaki A, Pantuck A, Belldegrun A, et al. Immunotherapy for the Treatment of Urothelial Carcinoma. J Urol. 2017. https://doi.org/10.1016/j.juro.2016.02.3005.

Dolgin E (2021) How COVID unlocked the power of RNA. Vaccine research and development might never be the same again. Nature 2021. https://www.nature.com/articles/d41586-021-00019-w

Duperret EK, Liu S, Paik M, Trautz A, Stoltz R, Liu X, et al. A Designer Cross-reactive DNA Immunotherapeutic Vaccine that Targets Multiple MAGE-A Family Members Simultaneously for Cancer Therapy. Clin Cancer Res. 2018. https://doi.org/10.1158/1078-0432.CCR-18-1013.

Duperret EK, Perales-Puchalt A, Stoltz R, Hiranjith GH, Mandloi N, Barlow J, et al. A synthetic DNA, multi-neoantigen vaccine drives predominately MHC Class I $CD8^+$ T-cell responses, impacting tumor challenge. Cancer Immunol Res. 2019. https://doi.org/10.1158/2326-6066.CIR-18-0283.

Dupuy LC, Richards MJ, Ellefsen B, Chau L, Luxembourg A, Hannaman D, et al. A DNA vaccine for Venezuelan equine encephalitis virus delivered by intramuscular electroporation elicits high levels of neutralizing antibodies in multiple animal models and provides protective immunity to mice and nonhuman primates. Clin Vaccine Immunol. 2011. https://doi.org/10.1128/CVI.00030-11.

Fang Z, Wang Y, Fang H, Xu Q, Wu J, Zhang B, Wang H, Tao R, Xie D (2014) Simple and convenient method for preparing periodic Brugia malayi M29 epitope gene protein vaccine. CN Patent 106344914, 5 November 2014.

Finocchiaro LME, Agnetti L, Fondello C, Glikin GC. Combination of cytokine-enhanced vaccine and chemo-gene therapy as surgery adjuvant treatments for spontaneous canine melanoma. Gene Ther. 2019. https://doi.org/10.1038/s41434-019-0066-7.

Finocchiaro LME, Spector AIM, Agnetti L, Arbe MF, Glikin GC. Combination of suicide and cytokine gene therapies as surgery adjuvant for canine mammary carcinoma. Vet Sci. 2018. https://doi.org/10.3390/vetsci5030070.

Fotin-Mleczek M, Fiedler K, Kowalczyk A, Heidenreich R (2018) Cancer RNA-Vaccine. WO Patent 2017EP66676, 3 July 2018

Gao H, Qin L, Wang X, Qi X, Wang Y, Gao Y (2012) DNA vaccine for expressing avian reticuloendotheliosis virus gp90 gene as well as establishment method and application of gene. CN Patent 102988969, 28 November 2012.

Gary EN, Kathuria N, Makurumidze G, Curatola A, Ramamurthi A, Bernui ME, et al. CCR10 expression is required for the adjuvant activity of the mucosal chemokine CCL28 when delivered in the context of an HIV-1 Env DNA vaccine. Vaccine 2020. https://doi.org/10.1016/j.vaccine.2020.01.023.

Glanville JE (2013) Epitope focusing by variable effective antigen surface concentration. WO Patent 2013177214, 21 May 2013

Global Alliance for Vaccines and Immunizations. Sustainable Development Goals. 2018 https://www.gavi.org/about/ghd/sdg/. Accessed 5 Apr 2019.

Guo H, Hao R, Qian H, Sun S, Sun D, Yin H, et al. Upconversion nanoparticles modified with aminosilanes as carriers of DNA vaccine for foot-and-mouth disease. Appl Microbiol Biotechnol. 2012. https://doi.org/10.1007/s00253-012-4042-z.

Harris E. (2020). Cell and Gene and COVID-19. Cell and Gene. March 24, 2020. Available: https://www.cellandgene.com/doc/cell-gene-and-covid-0001

Heidenreich R, Fiedler K, Fotin-Mleczek M, Kowalczyk A, Elbers K, Wurm M (2018) RNA vaccine and immune checkpoint inhibitors for combined anticancer therapy. WO Patent 2017EP56427, 16 March 2018

Hoeksema F, Karpilow J, Luitjens A, Lagerwerf F, Havenga M, Groothuizen M, et al. Enhancing viral vaccine production using engineered knockout vero cell lines – A second look. Vaccine 2018. https://doi.org/10.1016/j.vaccine.2018.03.010.

Homma A & Possas C (2020) Desenvolvimento e produção da vacina SARS-CoV-2. Notas complementares (Development and production of SARS-CoV-2 Vaccine. Complementary Notes) In Homma A, Possas C, Noronha JC, Gadelha P. Vacina e Vacinação no Brasil: Horizontes Para Os Próximos 20 Anos (Vaccine and Vaccination in Brazil: Horizons for the next 20 years). Edições Livres 2020, Rio de Janeiro.

Hu Y, Sun L, Sun Y (2012) Streptococcus Iniae Trivalent DNA vaccine and construction method thereof. CN Patent 102716499, 16 May 2012.

Jia R, Wang W, Wang Y, Xu Y, Zhang L, Yan J, Yu J, Zhang W (2014) broad-spectrum anti-tumor double-plasmid reproducible DNA vaccine. CN Patent 103948944, 19 May 2014.

Jia R, Xu Y, Zhang L, Xiao Y, Yan J, Wang Y, Yu J, Zhu X, Zhang W (2013) Replicon DNA (Deoxyribonucleic Acid) vaccine for treating chronic hepatitis B. CN Patent 103239717, 15 May 2013.

Jiang S, Lu L, Wang Q, Dai Y (2017) HIV vaccine and preparation method thereof. CN Patent 201710431893, 9 June 2017.

Jiao P, Liao M, Song H (2016) H9 subtype bird flu DNA vaccine and preparing method thereof. CN Patent 201610237685, 14 April 2016.

Jing Z, Lin G, Qiu C, Zheng F (2012) Brucella deoxyribonucleic acid (DNA) vaccine as well as construction method and application thereof. CN Patent 102772793, 30 May 2012.

Johns Hopkins University, Coronavirus Resource Center. COVID-19 Dashboard by the Center for Systems Science and Engineering. Global confirmed cases and deaths. https://coronavirus.jhu.edu/map.html. Accessed 12 May 2020.

Kalomoiris S, Lawson J, Chen RX, Bauer G, Nolta JA, Anderson JS. CD25 preselective anti-HIV vectors for improved HIV gene therapy. Hum Gene Ther Methods. 2012. https://doi.org/10.1089/hgtb.2012.142.

Kamran N, Calinescu A, Candolfi M, Chandran M, Mineharu Y, Asad AS, et al. Recent advances and future of immunotherapy for glioblastoma. Expert Opin Biol Ther. 2016. https://doi.org/10.1080/14712598.2016.1212012.

Kyutoku M, Nakagami H, Koriyama H, Nakagami F, Morishita R (2016) DNA vaccine containing specific epitope of apolipoprotein (A). US Patent 20160303211, 1 July 2016

Legarreta Solaguren L, De La Fuente Martínez I, Martínez Fernández L, Milanic, M (2014) Method for designing a vaccine. WO Patent 2015082745, 2 December 2014

Li J, Arévalo MT, Zeng M. Engineering influenza viral vectors. Bioengineered. 2013. https://doi.org/10.4161/bioe.21950.

Li J, Cao J, Liu X, Li L, Liu Y (2018) Vibrio mimicus oral target epitope gene vaccine and preparation method and application thereof. CN Patent 20181060036, 22 January 2018.

Li Y, Liu N, Wang Y, Zhang L, Xiao Y, Yan J, Yu J, Zhang W, Tian R (2014) Reproducible kidney cancer therapeutic DNA vaccine. CN Patent 103948943, 19 May 2014.

Liao Y, Chen X, Wu H, Qiu Z, Liao M (2019) Chimeric bivalent blood-pressure-reduction vaccine for human vascular smooth muscle cell L-type calcium channels and angiotensin 1-type receptors and application thereof. CN Patent 109663124, 14 February 2019.

Liu W, Wu X, Xu L, Yang Z (2013) Tumor nucleic acid vaccine based on tissue factor as well as preparation method and application of vaccine. CN Patent 201310405991, 9 September 2013.

Lorenzi JCC, Trombone APF, Rocha CD, Almeida LP, Lousada RL, Malardo T, et al. Intranasal vaccination with messenger RNA as a new approach in gene therapy: Use against tuberculosis. BMC Biotechnol. 2010. https://doi.org/10.1186/1472-6750-10-77.

Lu S, Zhang Zhe, Du P, Chard LS, Yan W, Khouri M El, et al. A virus-infected, reprogrammed somatic cell–derived tumor cell (VIREST) vaccination regime can prevent initiation and progression of pancreatic cancer. Clin Cancer Res. 2020. https://doi.org/10.1158/1078-0432.CCR-19-1395.

Lubenau H (2016) Novel Cmv Pp65 targeting DNA vaccine for cancer immunotherapy. EP Patent 20150001802, 16 June 2016

Lubenau H (2017) Process for the production of a DNA vaccine for cancer immunotherapy. EP Patent 20160001550, 12 July 2017

Lubenau H (2018) Novel Pd-L1 targeting DNA vaccine for cancer immunotherapy. EP Patent 20170161666, 16 March 2018

Lubenau H (2019) Neoantigen targeting DNA vaccine for combination therapy. EP Patent 20180192782, 4 September 2019

Markham R (2013) Methods and compositions for preventing a condition. WO Patent 2014028644, 14 August 2013

Maruf M, Brancato SJ, Agarwal PK. Nonmuscle invasive bladder cancer: a primer on immunotherapy. Cancer Biol Med. 2016. https://doi.org/10.20892/j.issn.2095-3941.2016.0020.

McCoy JR, Mendoza JM, Spik KW, Badger C, Gomez AF, Schmaljohn CS, et al. A multi-head intradermal electroporation device allows for tailored and increased dose DNA vaccine delivery to the skin. Hum Vaccin Immunother. 2014. https://doi.org/10.4161/hv.29671.

Moleirinho MG, Silva RJS, Alves PM, Carrondo MJT, Peixoto C. Current challenges in biotherapeutic particles manufacturing. Expert Opin Biol Ther. 2019. https://doi.org/10.1080/14712598.2020.1693541.

Morishita R, Nakagami H, Koriyama H, Nakagami F, Yoshida N (2012) DNA Vaccine. US Patent 20140099335, 13 April 2012

Muthumani K, Falzarano D, Reuschel EL, Tingey C, Flingai S, Villarreal DO, et al. A synthetic consensus anti-spike protein DNA vaccine induces protective immunity against Middle East respiratory syndrome coronavirus in nonhuman primates. Sci Transl Med. 2015. https://doi.org/10.1126/scitranslmed.aac7462.

Nakagami H, Kyutoku M, Morishita R, Tomioka H (2013) DNA vaccine containing VEGF-specific epitope and/or angiopoietin-2-specific epitope. US Patent 20150202271, 28 August 2013

Orr-Burks N, Murray J, Wu W, Kirkwood CD, Todd K V., Jones L, et al. Gene-edited vero cells as rotavirus vaccine substrates. Vaccine X. 2019. https://doi.org/10.1016/j.jvacx.2019.100045.
Packiam VT, Werntz RP, Steinberg GD. Current Clinical Trials in Non-muscle-Invasive Bladder Cancer: Heightened Need in an Era of Chronic BCG Shortage. Curr Urol Rep. 2019. https://doi.org/10.1007/s11934-019-0952-y.
Phadke M, Saunik S. COVID-19 treatment by repurposing drugs until the vaccine is in sight. Drug Dev Res. 2020. https://doi.org/10.1002/ddr.21666.
Possas C, Antunes AMS, Mendes FML, Schumacher SOR, Martins RM, Homma A (2015) Access to new technologies in multipatented vaccines: challenges for Brazil. Nat Biotechnol. https://doi.org/10.1038/nbt.3244.
Rodrigues MM, Soares IS. Gene-therapy for malaria prevention. Trends Parasitol. 2014. https://doi.org/10.1016/j.pt.2014.09.005.
Serruto D, Rappuoli R. Post-genomic vaccine development. FEBS Lett. 2006. https://doi.org/10.1016/j.febslet.2006.04.084.
Shan J, Du Y, Cheng G, Jia P, Liu H, Wang Y, Xu Q, Wei J (2015) Application of chitosan oligosaccharide in preparing vaccine adjuvant and vaccine composition. CN Patent 104906574, 25 June 2015.
Shaw AR, Suzuki M. Immunology of Adenoviral Vectors in Cancer Therapy. Mol Ther Methods Clin Dev. 2019. https://doi.org/10.1016/j.omtm.2019.11.001.
Siddiqui MR, Grant C, Sanford T, Agarwal PK. Current clinical trials in non–muscle invasive bladder cancer. Urol Oncol. 2017. https://doi.org/10.1016/j.urolonc.2017.06.043.
Siefker-Radtke A. Bladder cancer: Can we move beyond chemotherapy? Curr Oncol Rep. 2010. https://doi.org/10.1007/s11912-010-0104-5.
Smietanka K, Florys K, Gora-Sochacka A, Gromadzka B, Minta Z, Sanczyafska V, Sirko A, Stachyra A, Szewczyk B, Zagorski-Ostoja W, Saczynska V, Sawicka R, Bednarczyk M, Lakota P, Asmietanka K, Zagorski-Ostoja WW (2012) DNA vaccine, method of inducing the immune response, method of immunisation, antibodies specifically recognising the H5 haemagglutinin of an influenza virus and use of the DNA vaccine. US Patent 20140255343, 21 September 2012
Sun L, Li M (2015) Application of turbot CD83 molecules serving as vaccine adjuvant. CN Patent 104689313, 4 March 2015.
Sun L, Zhang J (2013) Application of heat shock protein 70 as immunoadjuvant. CN Patent 103405763, 15 July 2013.
Suschak JJ, Bagley K, Six C, Shoemaker CJ, Kwilas S, Spik KW, et al. The genetic adjuvant IL-12 enhances the protective efficacy of a DNA vaccine for Venezuelan equine encephalitis virus delivered by intramuscular injection in mice. Antiviral Res. 2018. https://doi.org/10.1016/j.antiviral.2018.09.014.
Tang N, Zhang Y, Pedrera M, Chang P, Baigent S, Moffat K, et al. A simple and rapid approach to develop recombinant avian herpesvirus vectored vaccines using CRISPR/Cas9 system. Vaccine 2018. https://doi.org/10.1016/j.vaccine.2017.12.025.
Tang N, Zhang Y, Sadigh Y, Moffat K, Shen Z, Nair V, et al. Generation of a triple insert live avian herpesvirus vectored vaccine using CRISPR/Cas9-based gene editing. Vaccines 2020. https://doi.org/10.3390/vaccines8010097.
Thanh Le T, Andreadakis Z, Kumar A, Gómez Román R, Tollefsen S, Saville M, et al. The COVID-19 vaccine development landscape. Nat Rev Drug Discov. 2020.
United Nations Development Programme (2016). 2030 Sustainable Development Goals. United Nations, Geneva.
Villarreal DO, Weiner DB. IL-33 isoforms: Their future as vaccine adjuvants? Expert Rev Vaccines. 2015. https://doi.org/10.1586/14760584.2015.1011135.
Voss JE, Gonzalez-Martin A, Andrabi R, Fuller RP, Murrell B, McCoy LE, et al. Reprogramming the antigen specificity of B cells using genome-editing technologies. ELife. 2019. https://doi.org/10.7554/eLife.42995.
Walters JN, Ferraro B, Duperret EK, Kraynyak KA, Chu J, Saint-Fleur A, et al. A Novel DNA Vaccine Platform Enhances Neo-antigen-like T Cell Responses against WT1 to Break

Tolerance and Induce Anti-tumor Immunity. Mol Ther. 2017. https://doi.org/10.1016/j.ymthe.2017.01.022.

Wang Y, Wu H, Xu Y, Zhang L, Yan J, Yu J, Zhu X, Zhang W (2013) Fermentation medium and method for producing plasmid DNA vaccine pSVK-CAVA for treatment of melanoma by fermentation medium. CN patent 103952367, 16 April 2013.

Warnke P, Cuhls K, Schmoch, Daniel L, Andreescu et al. 100 Radical Innovation Breakthroughs for the Future, 2019.

Wessels HH., Méndez-Mancilla A., Guo X, Legut M, Daniloski Z. and Sanjana NE. (2020). Massively parallel Cas13 screens reveal principles for guide RNA design. Nat Biotechnol. 2020 Jun;38(6):722-727. https://doi.org/10.1038/s41587-020-0456-9

World Health Organization, WHO. Coronavirus disease (COVID-19) Situation. 2020a. https://www.who.int/docs/default-source/coronaviruse/situation-reports/20200512-covid-19-sitrep-113.pdf?sfvrsn=feac3b6d_2. Accessed 12 May, 2020.

World Health Organization. WHO. Draft landscape of COVID-19 candidate vaccines. 2020b. https://www.who.int/blueprint/priority-diseases/key-action/Novel_Coronavirus_Landscape_nCoV_11April2020.PDF?ua=1. Accessed 11 Apr 2020.

Wu TC, Hung C (2013) Use of Ar-42 Enhances E7-Specific Cd8+ T-Cell mediated antitumor immunity in conjunction with DNA vaccine. US Patent 201261731225p, 27 November 2013

Xi Y, Luo Y (2011) Deoxyribonucleic acid (DNA) vaccine based on B7-1-PE40KDEL exotoxin fusion gene and application thereof. CN Patent 102161998, 14 January 2011

Xu Y, Wang L, Zhang L, Yan J, Yu J, Zhu X, Zhang W, Du Z, Wang Y (2013) Vector PSVK based hepatitis B therapeutic plasmid DNA vaccine fermentation production method, and special engineering bacterium and high-yield fermentation culture medium therefor. CN Patent 103396975, 15 May 2013.

Yan J, Pankhong P, Shin TH, Obeng-Adjei N, Morrow MP, Walters JN, et al. Highly optimized DNA vaccine targeting human telomerase reverse transcriptase stimulates potent antitumor immunity. Cancer Immunol Res. 2013. https://doi.org/10.1158/2326-6066.CIR-13-0001.

Yang Y, Shao Z, Gao J. Antitumor Effect of a DNA Vaccine Harboring Prostate Cancer-Specific Antigen with IL-12 as an Intramolecular Adjuvant. J Mol Microbiol Biotechnol. 2017. https://doi.org/10.1159/000477245.

Yongzhi X, Yuan L (2011) Deoxyribonucleic acid (DNA) vaccine based on B7-1-PE40KDEL exotoxin fusion gene and application thereof. CN102161998, filed January 14, 2011

Zhang Y, Luo J, Tang N, Teng M, Reddy VRAP, Moffat K, et al. Targeted editing of the pp38 gene in Marek's disease virus-transformed cell lines using CRISPR/Cas9 System. Viruses 2019. https://doi.org/10.3390/v11050391.

Chapter 13
Non-communicable Diseases in the Era of Precision Medicine: An Overview of the Causing Factors and Prospects

Dimitris Tsoukalas, Evangelia Sarandi, and Maria Thanasoula

Abstract Non-communicable diseases (NCDs) are among the most significant health challenges of the twenty-first century, causing 7 out of 10 deaths worldwide. Despite recent technological and medical advances, NCDs mortality and morbidity rates are increasing, and it is expected that by 2030 they will have caused 52 million deaths. In 2017, 41 million people died due to NCDs, and 80% of these deaths could have been prevented. Cardiovascular disease, cancer, diabetes, and chronic lung disease are the primary causes of mortality among NCDs-related deaths. Autoimmune diseases (ADs) affect 5–10% of the globe and have detrimental effects on patients' quality of life, life expectancy, and healthcare costs. Apart from the genetic background, 80% of the risk factors of NCDs are modifiable, including diet, hidden hunger, smoking, alcohol, air pollution, and physical activity, all discussed in this chapter. Accumulating evidence shows that changes in diet, lifestyle, and socioeconomic status have resulted in a substantial metabolic shift associated with the rapid increase of ADs. However, current approaches do not fully capture the individual variability on genes and lifestyle or consider the impact of modifiable factors on health. As such, there is growing pressure from patients' increasing demand and substantial healthcare costs for prevention, prediction, early diagnosis, and effective treatment of NCDs. With the advent of precision medicine, there have been efforts made to deliver tailor-made solutions for NCDs. Metabolomics, an emerging field that gives a detailed analysis of the phenotype, is currently being investigated as a potential precision medicine tool for screening, patient stratifica-

D. Tsoukalas (✉)
European Institute of Nutritional Medicine, E.I.Nu.M, Rome, Italy

Metabolomic Medicine Clinic, Athens, Greece

E. Sarandi
Metabolomic Medicine Clinic, Athens, Greece

Laboratory of Toxicology and Forensic Sciences, Medical School, University of Crete, Heraklion, Greece

M. Thanasoula
Metabolomic Medicine Clinic, Athens, Greece

E. Koukios, A. Sacio-Szymańska (eds.), *Bio#Futures*,
https://doi.org/10.1007/978-3-030-64969-2_13

tion, and treatment personalization. In this chapter, we present up-to-date data on the mitigating epigenetic and lifestyle risk factors for NCDs and ADs and review the current methodology for their assessment.

Keywords Non-communicable diseases · Autoimmune diseases · Risk factors · Metabolomics · Hidden hunger · Vitamin D · Metabolic shift · Insulin resistance · Inflammation

1 Introduction

Non-communicable diseases (NCDs) are chronic diseases that occur from various genetic and non-genetic factors. The genetic background contributes by 20% to the NCDs risk, whereas 80% of the risk factors are epigenetic, thus, modifiable. These factors include but are not limited to dietary habits, smoking, physical activity, toxic load, alcohol, and others, which are going to be discussed in this chapter. NCDs are responsible for 7 out of 10 deaths worldwide. In 2017, they were responsible for 73.5% (41.1 million) of the deaths which occurred globally and are considered one of the most significant health challenges of the twenty-first century (Martinez et al. 2020). Morbidity rates are also increasing dramatically from 43% in 1990 to more than 60% in 2017 globally. In contrast, in high-income countries where infectious diseases are reduced due to improved life quality standards, the NCDs rate reaches 80% of the global disease burden and over 90% of deaths.

The increasing morbidity and mortality rates indicate that by 2030, NCDs will account for 52 million deaths globally. The main types of NCDs include cardiovascular diseases (CVD), cancers, chronic respiratory diseases, diabetes, and autoimmune diseases. The primary mortality causes among NCD-related deaths are due to CVD, cancer, diabetes, and chronic lung disease. Autoimmune diseases (ADs) are a subgroup of NCDs and refer to conditions characterized by the malfunction of the immune system, which attacks self-tissues and organs in various parts of the body, causing inflammation. Nearly 5–10% of the global population is affected by ADs and suffers detrimental effects on their quality of life, life expectancy, and healthcare costs. More than 150 types of autoimmune diseases have been reported up to date, with the most common being Hashimoto's thyroiditis, rheumatoid arthritis, inflammatory bowel disease (IBD), multiple sclerosis, and lupus. Accumulating evidence shows that the rapid increase of ADs is associated with a metabolic shift caused by changes in lifestyle, diet, and socioeconomic status. Twin studies have indicated that there are significant genetic determinants for ADs, such as the high concordance of Major Histocompatibility Complex (MHC) haplotypes in monozygotic twins. However, the association mentioned above mainly affects early-onset diseases suggesting that other factors apart from genes may influence the development of ADs (Theofilopoulos et al. 2017).

Medicine advancements have made substantial progress in the diagnosis of NCDs, but preventive strategies, tools of prediction, and active therapeutic agents are not yet established. Regarding prevention, despite the pivotal role of diet and the other environmental factors on disease onset and progression, physicians lack proper knowledge and tools to assess and improve them (Strong et al. 2006; Tinetti et al. 2012; Devries 2019). Besides, current treatment approaches have proved beneficial for only 30–60% of the patients, and an additional 30% experience severe adverse effects indicating a large gap that needs to be addressed. The "one size fits all" model that is being applied at the moment does not take into consideration the modifiable risk factors or capture the genetic variability and lifestyle of the individual (Balashova et al. 2018). As a result, the demand for effective prevention, prediction, early diagnosis, and treatment of NCDs by patients is continuously increasing. Precision medicine is an emerging approach for the individualized treatment, forecasting, and early diagnosis of disease, taking into account the individual gene variability, lifestyle, and nutrition. Metabolomics, the study of metabolites and their interactions within the organism, gives a detailed analysis of the phenotype and has vast applications in medicine. It can capture the interrelationships of a biological system, including humans, under the influence of epigenetic factors. Specifically, it has been suggested that metabolomics can provide insight on a systemic dysfunction before the appearance of the symptoms, thus it is a valuable tool for the prevention and prediction of disease. At an advanced stage of the disease, metabolomics can be used to monitor the side effects of drug treatment allowing the treatment type or dose optimization and detect and assess the nutritional deficiencies that should be replenished to improve the life quality (Tsoukalas et al. 2019b).

This chapter includes recent data on the modifiable epigenetic and environmental risk factors for NCDs and ADs and reviews the available methodologies for their assessment. Specifically, we have included essential factors that shape an unhealthy diet and an unhealthy environment. We also discuss the challenges that precision medicine faces regarding its application in clinical practice while focusing on the potential future opportunities for personalized disease treatment.

2 Dietary and Lifestyle Factors

Nutritional epidemiology has linked the consumption of specific foods, nutrients, or dietary patterns with different types of cancer, cardiovascular disease, diabetes, increased blood pressure, insulin resistance, and hyperglycemia. According to recent data, the suboptimal diet is a more critical factor than smoking for global mortality. An insightful study published in Lancet last year demonstrated that more than 20% of global deaths of adults were linked to poor diet, and the most important cause of death was cardiovascular disease, cancer, and diabetes (Afshin et al. 2019). The poor diet included inadequate intake of healthy foods, mostly nuts, seeds, whole grains, and fruit and overconsumption of unhealthy food, mainly sweetener-rich beverages, sodium, and processed meat. Specifically, more than 50% of deaths related to diet were associated with high sodium intake (global mean 6 g), low whole-grain diet (global mean 29 g), and low fruit intake (global mean was less than 100 g). Here we discuss the role of hidden hunger (micronutrient deficiency), lack of vitamin D, excessive sodium and free sugar intake, alcohol consumption, smoking, and physical activity as critical factors of an unhealthy diet and lifestyle that increase the risk of NCDs morbidity and mortality.

2.1 Hidden Hunger

Hidden hunger (or micronutrient deficiency) is defined as the lack of vitamins and minerals of an organism, and in contrast to micronutrient deficiency diseases, hidden hunger is asymptomatic. Micronutrients, including vitamins, minerals, amino acids, fatty acids, probiotics, enzymes, and antioxidants, are essential for the normal function of cells and tissues (Bailey et al. 2015). Vitamins are organic compounds that act as coenzymes of metabolic pathways, and most of them are essential and are obtained through diet. Vitamin B7 and vitamin K, though, are normally synthesized through the gut and vitamin D is synthesized with sun exposure. Minerals are also essential nutrients obtained through diet, and act as cofactors of enzymatic reactions. The inadequate intake or absorption of these nutrients due to poor diet or disease state can lead to severe cellular malfunction with complications in health. According to the World Health Organization (WHO), 2 billion people suffer from nutrient deficiencies globally (Bailey et al. 2015). The primary nutrient deficiencies that contribute to the development of hidden hunger are vitamins A, D, E, and C, and choline, calcium, magnesium, iron (for specific age/gender groups), potassium, and fiber. The Dietary Guidelines Advisory Committee, though, has recommended the public to increase the intake of only some of these nutrients because they have been linked to adverse health issues (Blumberg et al. 2017). Indeed, many of these deficiencies have been linked with the prevalence of NCDs, and their correction has been related to beneficial effects on the management of certain NCDs (Kivity et al. 2011; Moss and Ramji 2017; Wessels and Rink 2020a; Winther and Rayman 2020). Although there is a lack of enough studies showing a correlation between the deficiency of the rest of the nutrients and NCDs, there is growing evidence supporting that nutrients act synergistically (Faggi et al. 2019).

Nutritional deficiencies have been associated with ADs. In a study of patients with IBD, several nutritional deficiencies were observed (Vagianos et al. 2007). Biochemical measurements indicated a high prevalence of nutrient deficiencies of vitamin E (63%), vitamin D (36%), vitamin A (26%), calcium (23%), folate (19%), iron (13%), and vitamin C (11%) in these patients. Deficiencies were further demonstrated through insufficient blood serum levels of hemoglobin (40%), ferritin (39.2%), vitamin B6 (29%), carotene (23.4%), vitamin B12 (18.4%), vitamin D (17.6%), albumin (17.6%), and zinc (15.2%). Even though these deficiencies were not correlated to diet, the authors suggested that other factors may influence the low nutritional levels and that supplementation should be considered in the IBD patients.

In a review by Manzel A. et al., the relation of autoimmune diseases to the Western Diet was discussed focusing on the importance of T cells, concluding that nutrition affects the gut mucosal immune system and the metabolic state of the body, which are both risk factors for autoimmunity (Manzel et al. 2014). Moreover, recent evidence indicate a pivotal role for vitamin D and zinc deficiencies in most common ADs suggesting potential effective and cost-effective strategies of prevention and treatment (Wessels and Rink 2020b).

2.2 Intermediate Metabolic Risk Factors

In affluent countries, hidden hunger as a result of malnutrition often co-exists with obesity or overweight and the rise of the NCDs. Specifically, studies indicate that micronutrient deficiencies under nutritional patterns of high-fat, high-protein, high-sugar, and excess salt intake that are commonly consumed in affluent countries (Western Diet) have been linked to an increased risk for obesity, high BMI, metabolic syndrome, and cardiovascular disease (Ames 2006; Manzel et al. 2014). Significant progress has been made in the description of the molecular pathways that associate diet, obesity, insulin resistance, low-grade inflammation, disrupted biochemical parameters, and NCDs. Obesity, defined as Body Mass Index over 30, has been classified as a risk factor for many NCDs, including type 2 diabetes, CVD, and cancers, and there is a common view that obesity triggers inflammation and insulin resistance being a subsequent effect (Johnson and Milner 2012). However, growing evidence suggests that insulin resistance is the primary disturbance and precedes inflammation (Giles et al. 2015). Briefly, the proposed mechanism for the diet-related onset of inflammation is that excessive intake of calorie-dense and nutrient-empty foods lead to the disruption of the physiological mechanism of the organisms to produce energy in a controlled manner. Insulin and leptin, the central hormones that regulate energy metabolism but also affect immune responses, are increased continuously to balance the excessive calorie intake. Continuous secretion of insulin and leptin leads to metabolic shift in the peripheral cells and the cells of the immune system, triggering inflammation. At the same time, hyperinsulinemia directly leads to insulin resistance, where cells require more insulin to receive the signal to uptake glucose and use it as energy. In turn, in conditions with established chronic low-grade inflammation, the inflammatory microenvironment further fuels insulin resistance, storage of fat and obesity, and related metabolic syndrome (Fig. 13.1). As for obesity, a series of interventional studies indicate that changes in insulin resistance or inflammation markers precede and can predict weight changes suggesting that obesity is a consequence and rather a cause of insulin resistance (Kong et al. 2013). Validation of this theory will be very significant for research but also for clinical practice since strategies targeting weight loss are far different from those targeting insulin resistance (Noakes 2018).

Micronutrient deficiencies are mostly caused by poor dietary choices, although socioeconomic factors and the presence of underlying disease are also important

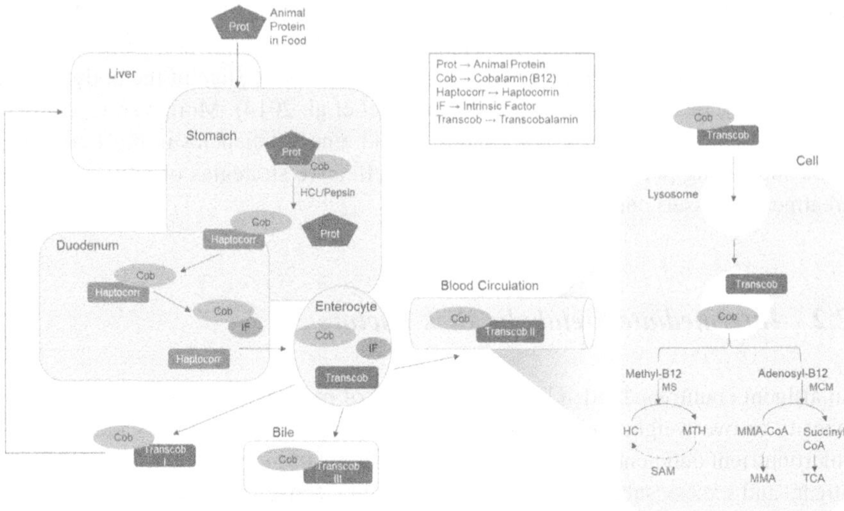

Fig. 13.1 Vitamin B12 pathways of absorption and metabolism. Digestion of cobalamin (vitamin B12) bound to animal protein takes place in the stomach followed by the duodenum, facilitated by the transporters haptocorrin and intrinsic factor (IF). The complex cobalamin-IF enters the enterocyte where cobalamin is then bound to transcobalamin and depending on the type of transcobalamin, the complex follows a different path: (i) Cobalamin-Transcobalamin I is transported to the liver where cobalamin is stored (75% of cob), (ii) Cobalamin-Transcobalamin III is transported to the bile where it is excreted with the urine, and (iii) Cobalamin-Transcobalamin II enters the systemic blood circulation to reach and enter the cells. Upon entering the cell, cobalamin is released from transcobalamin and acts as a cofactor for intracellular metabolic pathways. Methyl-B12 is a cofactor of Methionine Synthase (MS) for the conversion of Homocysteine (HC) to Methionine (MTH), which in turn is metabolized to s-adenosyl methionine (SAM), a precursor of HC. Methylmalonyl-CoA mutase (MCM) catalyzes the conversion of methylmalonyl-CoA (MMA-CoA) to succinyl-CoA, with the presence of adenosyl-B12, which feeds the TCA cycle for energy production. Methylmalonic acid (MMA) is a downstream metabolite of MMA-CoA, and upon malfunction or inactivation of the adenosyl-B12 dependent pathway, MMA increases

(Bailey et al. 2015). Also, even if intake is sufficient, a deficiency may occur at a later stage due to disturbed absorption of nutrients for various reasons (e.g., metabolic disease or excessive toxic load from medication) (National Academies of Sciences, Engineering et al. 2017). For example, vitamin B12, which is obtained through diet and specifically animal sources, is released in blood circulation through a complicated journey of absorption from the stomach, the duodenum, and the enterocytes (Fig. 13.2). B12 released by enterocytes is bound to transcobalamin and can be either stored in the liver (75% of B12), excreted via the bile or enter the cells and participate in intracellular metabolism. After entering the cells, B12 is released and free to act as cofactor either for methionine synthase (MS), in the form of methyl B12, or for methylmalonyl-CoA mutase (MCM) in the form of adenosyl B12. Therefore, as discussed elsewhere, measuring serum vitamin B12 levels alone has many limitations and does not reflect vitamin B12 bioavailability or cellular levels of B12 (Hannibal et al. 2016). Current factors for the identification of vitamin

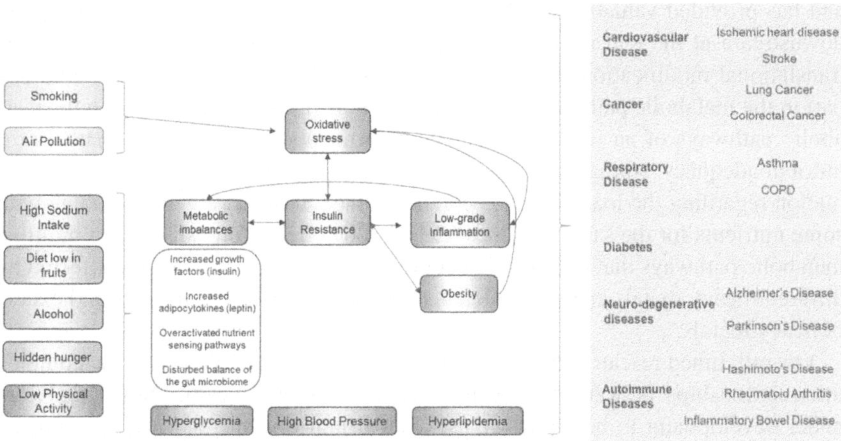

Fig. 13.2 Schematic diagram depicting the environmental risk factors affecting NCDs and intermediate mechanisms

B12 deficiency include markers participating in the metabolic pathway of vitamin B12 including methylmalonic acid, homocysteine, and total holotranscobalamin, which is the vitamin B12 bound to transcobalamin complex.

The assessment of nutritional intake is mostly done via food frequency questionnaires (FFQs) both in clinical practice and epidemiological studies. FFQs provide an overview of the macronutrients and the micronutrients obtained through diet, via a series of questions regarding the type and portion of ingested food. Also, a crude estimation on the toxicants obtained through diet can be obtained. Dietary patterns can be recognized and translated into a nutritional score in a computer-based system that matches the food choices with their nutritional composition. FFQs, although they are easy to use, economical, and high throughput, they have low sensitivity and accuracy as discussed elsewhere (Margină et al. 2020).

Biochemical and laboratory testing is another widely used method to assess the status of nutrients and is a valuable tool to diagnose severe micronutrient deficiencies. However, micronutrient deficiencies in NCDs can be multiple and to a smaller extent compared to a nutritional deficiency disease. Also, cells and tissues are very well-regulated systems that have mechanisms to adjust the nutrient requirements depending the nutrient availability or recycle to meet increased demands on a specific tissue (Pietrzik 1991; Nualart et al. 2014). Therefore, blood values of the micronutrient do not provide sufficient information on the physician (Bier and Mann 2015). Emerging technologies for the identification of biomarkers of the nutritional status that will allow grouping of individuals according to their nutrient requirements are at the center of attention.

Nutriomics is a novel field that focuses on the comprehensive study of the effect of ingested food on human health and disease risk. Nutrigenomics, studies the effect of diet on the expression of genes and risk for NCDs. This technology allows the determination of nutrient requirements and foods based on the genetic background and has provided valuable insights on the nutrition-genome interaction. However, downstream at the end of the gene expression and the environment-related post-translational modifications lies the metabolome which is the metabolites that take part in the metabolic pathways. Nutrient-regulated enzymes are catalyzing the metabolic pathways of an organism, thus making metabolites promising markers for nutrient adequacy, storage, and use. In addition to that, metabolites can give information regarding the load from heavy metals since the latter are antagonizing with some nutrients for the same enzymes (Zeisel 2007). Therefore, analysis of multiple metabolic pathways that use different combinations of nutrients and are affected by different heavy metals can give a detailed map of the missing nutrients and excess of heavy metals.

Overall, tuned research efforts are currently being made to uncover the complex interrelationships between diet, genes, metabolites, and health. Novel tools are under development to help physicians, clinical nutritionists, and every healthcare professional to detect nutritional deficiencies early and provide personalized recommendations for their replenishment.

2.3 Vitamin D

Vitamin D is a member of the steroid hormones participating in various functions of the human body and can be obtained from food and supplements, or can be endogenously sythesized through sunlight exposure (Wang et al. 2017). The leading roles of vitamin D include modulating cell growth and inducing the function of the immune, nervous, and muscular systems. On a molecular level, it accentuates the expression of genes that control cellular proliferation, differentiation, and apoptosis. According to a study by Ramagopalan et al., the nuclear vitamin D receptor (VDR) occupies 2776 positions on the human DNA, and 229 genes show changes in expression after treatment with vitamin D (Ramagopalan et al. 2010). After it has been absorbed by the intestine or synthesized by the skin due to sunlight exposure, vitamin D in the form of cholecalciferol is transported to the liver and converted to calcidiol, 25-hydroxycholecalciferol (25 (OH) D) which binds to specific proteins that transport calcidiol to the kidney through blood circulation. There, it is converted to the active vitamin D form, calcitriol 1,25-dihydroxycholecalciferol (1,25(OH)2D3). 25 (OH) D is a biomarker used in the measurement and assessment of vitamin D levels and the detection of deficiencies because it reflects the levels of vitamin D derived from both the diet and the skin and is more stable than 1,25(OH)2D3 (Wang et al. 2017).

Accumulating evidence from recent studies indicate that vitamin D deficiency can be directly associated with the incidence of NCDs and ADs. According to a recent review by Amrein K. et al., a deficiency of vitamin D is recognized as a concentration of 25 (OH) D lower than 20 ng/ml (Amrein et al. 2020). Even though most authors consider a range of below 30 ng/ml 25 (OH) D as vitamin D deficient, studies have shown that levels lower than 10 or 12 ng/ml increase the risk of osteomalacia and crickets (Holick et al. 2011a; Institute of Medicine of the National Academies 2011; Braegger et al. 2013; EFSA NDA Panel 2016). Therefore, these levels define severe vitamin D deficiency. The clinical practice guidelines of the Endocrine Society Task Force on vitamin D refer to 20 ng/ml as a cutoff level for vitamin deficiency, 21–29 ng/ml as vitamin insufficiency and 30–100 ng/ml as sufficient vitamin D levels (Holick et al. 2011a).

The levels mentioned above refer to bone health and reflect the minimum concentration of vitamin D, under which diseases have been reported to arise. However, vitamin D cutoff levels associated with the risk of NCDs have been shown to be higher. A study by Wang T. et al. examined vitamin D deficiency and its relation to the risk of developing cardiovascular disease in individuals without prior cardiovascular disease. A total of 1739 individuals with a mean age of 59 years participated in the study, and the amount of 25 (OH) D in the blood was used to evaluate the status of their vitamin D levels. The findings of this study indicated that vitamin D deficiency is positively associated with the risk of developing cardiovascular disease. More specifically, individuals with hypertension whose 25 (OH) D levels were less than 50 nmol/L had a twofold risk of cardiovascular incidence (Wang et al. 2008). The first case-control study examining the correlation between the development of Insulin Dependent Diabetes Mellitus (IDDM) and vitamin D administration during the first year of life by EURODIAB Substudy 2 study group showed a decreased risk of developing type 1 diabetes for children that received vitamin D supplements for at least 1 year during early childhood (Miettinen et al. 2020). The data of the study were collected by interviewing mothers of 3.155 children regarding their children's supplementation of vitamin D during their first years of life. However, lack of consistent dosage of vitamin D and the inconsistent validity of the answers given by mothers constitute limitations of this study. Type I diabetes (IDDM) is an autoimmune disease caused by the destruction of beta pancreatic cells whose role is to produce insulin. As a result, insulin deficiency occurs, leading to hyperglycemia and having further health complications in other tissues and organs. Insulin injections are administered daily to promote the absorption of glucose by cells and maintain glucose levels within the optimal range (Kahanovitz et al. 2017). Further studies need to be taken into consideration to establish a causality linkage between IDDM and vitamin D. A systematic review and meta-analysis of this study by EURODIAB Substudy 2 study group indicated that the establishment of causality requires randomized controlled trials with long periods of follow-up (Zipitis and Akobeng 2008). A study conducted by Miettinen M. E. et al. investigated the association of serum 25-hydroxyvitamin D levels in childhood on the risk of developing islet autoimmunity and IDDM (Miettinen et al. 2020). A total of 732 infants participated in the observational study, and serum concentrations of 25(OH) were measured repeatedly for 10 years. The serum concentrations were then compared according to age at the first seroconversion. The results suggested that prenatal vitamin D supplementation can assist in the prevention of IDDM.

Clinical and preclinical evidence suggests that vitamin D deficiency plays a vital role in the management of inflammatory bowel disease (IBD) (Hlavaty et al. 2015). IBDs, referring to ulcerative colitis and Crohn's disease, are NCDs characterized by extensive inflammation of the intestine. Specifically, ulcerative colitis affects the large intestine, while Crohn's disease can affect any part of the digestive system. In a study by Levin A. D. aiming to associate vitamin D status with IBD location and severity, the importance of monitoring vitamin D status was emphasized for the management of the disease (Levin et al. 2011). Seventy-eight children with IBD participated in the study (45 males, 33 females), and their 25 (OH) D levels were measured for the period during 2006–2007. Vitamin D deficiency was defined as 25 (OH) D< 51 nmol/l (moderate) and 25 (OH) D< 30 nmol/l (severe), while insufficiency was for 25 (OH) D levels between 51 and 75 nmol/l. The results yielded that 15 children (19%) had a vitamin D deficiency, and 30 children (38%) had an insufficiency, and therefore a positive correlation was established. Further randomized trials are required to establish a causational relationship between vitamin D deficiency and IBD.

According to evidence from review articles and studies, vitamin D can also aid in the treatment of psoriasis (Morimoto et al. 1986; Fu and Vender 2011; Mattozzi et al. 2016; Kechichian and Ezzedine 2018). Psoriasis is an immune-mediated disease that affects the skin causing red patches to appear. A study by Morimoto S et al. indicated that oral administration and topical application of vitamin D derivatives were beneficial in the management of psoriasis and improvement of psoriatic skin lesions (Morimoto et al. 1986). A total of 40 patients were enrolled in the study, and active forms of vitamin D_3 were either orally administered or topically applied. Vitamin D_3 (cholecalciferol) is an active analog of vitamin D that is synthesized by the skin or obtained via supplements or diet. The results of the study suggested that psoriasis may respond to the active forms of vitamin D_3 and that unresponsiveness of skin cells to the vitamin might be implicated in the pathogenesis of psoriasis. Another study by Finamor D. C. et al. investigated the effect, efficacy, and safety of administration of high daily doses of vitamin D on the clinical course of vitiligo and psoriasis (Finamor et al. 2013). A total of 25 patients (9 with psoriasis and 16 with vitiligo) received 35.000 IU of vitamin D per day for 6 months. The results showed that the treatment reduced disease activity for 9/9 patients with psoriasis and 14/16 patients with vitiligo.

Evidence from studies has shown that the supplementation of vitamin D can be beneficial for NCDs, including cancer and autoimmune diseases, as well as infections. Specifically, a review conducted by Garland F. C. et al. investigated the prospects of vitamin D_3 supplementation in global cancer prevention, and the results were encouraging (Garland et al. 2007). According to the review, the intake of 2.000 IU per day of vitamin D_3 would lead to a 25% decrease of annual cases of breast cancer and 27% of annual cases of colorectal cancer. Overall, vitamin D can aid in the prevention of NCDs (Gorham et al. 2005; Garland et al. 2009). Nevertheless, it was recently proposed that maintaining concentrations above 40–60 ng/mL can decrease the risk of infections (Grant et al. 2020).

Moreover, vitamin D levels have been associated with other ADs including autoimmune thyroiditis and rheumatoid arthritis while attention is being given at the recommended dose (Kivity et al. 2011). In populations with higher nutrient demands, as in the case of an established disease, the administered dose may exceed the Recommended Dietary Allowance (RDA), which need to be considered in clinical trials as well as in everyday clinical practice (Tsoukalas and Sarandi 2020). There is growing evidence that cells are insensitive to vitamin D in ADs suggesting that higher doses are required to exert the immunomodulatory effect of vitamin D (Jeffery et al. 2018). In addition, corticosteroids treatment commonly used in ADs has been shown to affect the catabolism of vitamin D, thus requiring higher doses to maintain optimum levels in the blood circulation (Singh and Kamen 2012; Kamen 2013). According to the American Endocrinology Society, the upper level for safe vitamin D intake is 10.000 IU daily for adults and 4000 IU for children over 8 years old. Higher dose recommendations require monitoring of the vitamin D blood levels (Holick et al. 2011b). Based on evidence from recent studies, vitamin D supplementation has not been associated with toxicity risks even at 700 ng/ml or the intake of 30,000 IU/day and the steady-state levels of vitamin D at 200 ng/ml for a long time (Hathcock et al. 2007).

Overall, randomized controlled trials are needed to specify the ideal supplementation dose as a preventive and treatment strategy considering the various involved factors. However, vitamin D supplementation is an established general recommendation for the reduction of risk for NCDs and ADs and as an adjunct tool in their management.

2.4 *Sodium*

High intake of sodium was the driving cause of mortality among diet-related deaths in China, among other countries, and mostly due to cardiovascular disease, according to the Lancet observational study, in line with others (Ezzati et al. 2014). Several salt alternatives have been proposed starting several years ago when the salt reduction program in Finland showed beneficial effects of low-sodium, high-potassium, and magnesium salt substitutes on hypertension (Katz et al. 1999),(Karpanen et al. 1984). Notably, Finland was one of the first countries that adopted a low salt routine in the late 1970s, and 20–30 years later, the mortality caused by stroke and coronary heart diseases decreased dramatically by 75–80% (Karppanen and Mervaala 2006). Later, the UK demonstrated that a 15% decrease in salt intake is linked to a significant reduction of blood pressure and mortality from stroke and ischemic episodes (Watroba and Szukiewicz 2016). Since then, 30% of sodium reduction has been included in the strategies of the World Health Organization to combat NCDs, and according to a 2015 review, 75 countries have adopted strategies to achieve this goal (Trieu et al. 2015).

Apart from the causal relationship of high sodium intake with blood pressure and CVD, some studies have shown possible associations with damage in several organs such as the kidney, stomach, and bones, malfunction in the immune system, hormonal and oxidation mechanism, and the gut microbiome balance (He et al. 2020). However, as many researchers discuss, salt reduction strategy is usually combined with a healthier lifestyle in observational studies, and the synergistic effect cannot be fully discriminated from the alt reduction alone. According to WHO and CDC recommendations, less than 5 g/day salt and 2 g of sodium with more than 3.5 g of potassium should be consumed daily by adults (World Health Organization 2013; CDC 2017). As with every nutrient, optimum intakes are required for the normal function at a cellular and organism level. A meta-analysis of sodium intake and CVD showed that not only excessive (>12.5 g) but also extremely low levels of salt (<5.6 g) were related to poor health outcomes. In addition, bio individuality stemming from our genes and other risk factors shapes different salt sensitivity levels that needs to be considered (Graudal et al. 2014). Overall, there is mounting evidence that high sodium intake has many adverse effects and especially when it is combined with the ingestion of processed foods, but at low levels sodium is crucial for the maintenance of fluid and blood volume and the normal function of nerve cells.

2.5 *Free Sugars*

Free sugars which are defined as "all monosaccharides and disaccharides added to foods by the manufacturer, cook, or consumer, plus sugars naturally present in honey, syrups, and fruit juices" according to the WHO and the FAO have been the subject of intense debate concerning health effects. As stated elsewhere, this term includes all sugars but the lactose found in milk and the naturally occurring sugars found in the outside structure of food such as the fruit skin (Ludwig et al. 2018). Sugar consumption has been known for its detrimental effects on oral health. Indeed, according to a systematic review that was later used by WHO experts, a 10% free sugar reduction is positively associated with lower caries. In comparison, 5% was associated with a better outcome, a strategy adopted by the Scientific Advisory Nutrition Committee (Moynihan and Kelly 2014; SACN 2015). Although there are some inconsistencies in the field, sugar intake has been associated with several disease outcomes, including CVD (Te Morenga et al. 2014; Dinicolantonio and Okeefe 2017), diabetes, and autoimmune diseases (Zhang et al. 2019; Correa-Rodríguez et al. 2020). Moreover, data from animals study show that a Western diet rather than a high-fat diet can lead to a psoriasis-like phenotype which occurs earlier than obesity, suggesting that sugars' effect on health is independent of obesity (Shi et al. 2020).

Although studies linking sugar intake with skin diseases like psoriasis and atopic dermatitis are scarce, and the exact mechanism is not fully understood, it is suggested that sugar is a crucial contributor to chronic inflammation of autoimmune diseases and NCDs in general (Manzel et al. 2014; Nosrati et al. 2017). Specifically, positive associations have been demonstrated between intake of sugar and established cardiovascular markers, namely, blood pressure, and levels of triglycerides, LDL, and total cholesterol in meta-analyses of randomized controlled trials. The authors suggest that fructose commonly found in sugar-sweetened beverages, honey, sucrose syrup, and fruit is more likely to be the cause for the sugar-related increase of cardiometabolic indicators. Also, excessive fructose intake has been implicated in weight gain and as a critical contributor to the obesity epidemic. Because the sweet taste of fructose-rich products causes satiety, some suggest that weight gain stems from excessive food consumption caused by satiety. Recent data indicate that under physiological conditions, fructose is metabolized in the intestine and the liver, increasing blood glucose and insulin. However, when fructose is consumed excessively, it reaches the colon and liver where it is metabolized leading to de novo lipogenesis through several pathways, including the feeding of TCA cycle with the fructose-derived pathway with pyruvate which in turn is metabolized to citrate and then to acetyl-CoA by the enzyme ATP citrate lyase. However, a study published in *Nature*, March this year, showed for the first time that de novo lipogenesis can occur even in the absence of ACLY through a distinct pathway that involved the gut microbiome (Postic 2020; Zhao et al. 2020b). Briefly, using isotope-tracer methodology and metabolomics, it was demonstrated that fructose could be metabolized to acetate by the gut microbiome in the liver resulting in hepatocyte-related lipogenesis. Also, fructose even though its metabolism is not insulin-dependent, when is ingested excessively, it has been shown to augment hyperinsulinemia and insulin resistance via direct and indirect pathways in the liver, independently from weight increase and total calories, while promoting liver inflammation through mitochondrial fatty acids oxidation impairment and stress of the endoplasmic reticulum (Softic et al. 2020).

Interestingly, glucose, which is also a monosaccharide-like fructose with the same molecular formula but a different structure, has not been shown to act similarly with fructose when used as a sweetener in terms of de novo lipogenesis. The WHO guidelines include the reduction of sugar to 10% of energy intake while highlighting the beneficial effects of further reduction to 5%, based on existing literature (e-Library of Evidence for Nutrition Actions (eLENA) 2019). In other words, 25 g of free sugar per day or 2 oranges is recommended for a healthy individual. Sugar consumption today ranges from 13% to 17%, of which 50% is in the form of fructose (Merino et al. 2020). In the UK, children 4–10 years old 13.5% of energy intake is in the form of free sugars, according to the UK National Diet and Nutrition Survey (NDNS), and similar are the findings in the USA. Overall, there is accumulating evidence that excessive free sugar intake is involved in the onset of metabolic changes that promote the development of ADs and NCDs.

2.6 Alcohol

Alcohol consumption is one of the leading risk factors of NCDs. Some suggest a beneficial effect when consumed moderately, but recent comparative reviews question this relationship. On the contrary, a series of epidemiologic studies have indicated that heavy alcohol consumption increases the risk of cardiovascular disease and liver disease and has been associated with more than 50 diseases (WHO 2018b; Millwood et al. 2019). Specifically, more than 5% of the global burden of disease can be attributed to alcohol, and some of the major contributors are cancers, chronic liver disease, and cardiovascular diseases. Through the increase of blood pressure and the disturbance of lipid profile, excessive drinking is linked with overall CVD posing a major challenge of modern societies (Chiva-blanch and Badimon 2020). In addition, it has a detrimental effect on the gut microbiome and immunotolerance and has been regarded as an associating factor with the presence of ADs (Wang et al. 2010; Sarkar et al. 2015).

2.7 *Physical Activity*

Another important risk factor for NCDs is physical activity, the movement of the body that requires energy such as walking or cycling, which ranks among the top causes of early mortality (WHO 2018a). Globally, 30% of the population is not taking adequate physical activity according to the global recommendations on physical activity for health. Insufficient physical activity refers to less than 150 min/week of moderate-intensity aerobic exercise or less than 75 min/week of intense exercise for adults. According to a large epidemiological study on nearly two million people around the world, published in Lancet, it showed that high-income countries are twice more prevalent in physical inactivity than low income mostly due to the different means of transport and nature of work (Guthold et al. 2018; Lear et al. 2017).

Several studies have demonstrated the positive effects of physical activity not only in prevention but also for the improvement of disease progression and the quality of life of patients. Indeed, in a 130.000 people observational study from different countries of every income category, moderate physical activity was associated with a more than 20% reduction in risk for major CVD and risk for all-cause mortality. The negative association between physical activity and the risk was dose-dependent, suggesting that more exercise than the 150 min/week has additional benefits (Lear et al. 2017).

Moreover, physical activity has been shown to regulate the immune responses, thus benefiting patients with autoimmune diseases, including multiple sclerosis, rheumatoid arthritis, and inflammatory bowel diseases. Importantly, patients with autoimmune diseases experience musculoskeletal complications that significantly deteriorate their quality of life. Exercise can contribute to the maintenance of mobility function through enhanced muscle strength, coordination, and weight balance (Sharif et al. 2018). More importantly, it has been shown that regular moderate exercise can increase glucose uptake and reduce insulin resistance, which are determinant factors for the onset and progression of NCDs (DeFronzo et al. 1987). Finally, in a case-control study, it was shown that exercise was a very important factor for the development of a model predicting the presence of autoimmune diseases based on the levels of fatty acids and lifestyle factors (Tsoukalas et al. 2019c).

2.8 Cigarette Smoking

Tobacco use is the most prevalent modifiable risk factor of the main NCDs, including CVD, cancer, respiratory disease, and diabetes, as well as neurological disorders. It is estimated to cause around 71% of all lung cancer deaths, 42% of the chronic respiratory disease, and almost 10% of CVD. It is estimated that around six million people each year are killed by tobacco, approximately one person every 6 seconds, from whom more than five million are due to direct tobacco use and 600,000 due to their exposure to second-hand smoke. Moreover, tobacco is responsible for 14% of the global NCDs deaths of adults age for more than 30 years. In 2020, the number of deaths attributed to tobacco use increased to eight million people annually, with seven million of those deaths due to direct tobacco use and around 1.2 million due to non-smokers being exposed to second-hand smoke. Still, almost 80% of them, corresponding to 1.3 billion tobacco users, come from low- and middle-income countries, with tobacco use, greatly contributing to poverty and replacing basic needs, such as food. Due to the very high rates of tobacco use morbidity and mortality, the healthcare costs for treating the diseases caused by tobacco are significantly high in several countries (World Health Organization 2020).

The molecular pathways involved in the effect of cigarette smoking on NCDs, include metabolic shift and oxidative stress contributing to the development and progression of cardiovascular damage (Leone 2005). More specifically, metabolic changes mediated by cigarette smoking substances lead to the development of atherosclerotic lesions and atherosclerotic plaque through narrowing of the vascular lumen and induction of a hypercoagulable state that in turn increases the risk of acute thrombosis. Briefly, cigarette smoking leads to endothelium dysfunction by directly affecting the endothelial cells triggering the formation of atherosclerotic plaques, which with the combination of other inflammation mechanisms will develop into vulnerable plaques prone to rupture (Csordas and Bernhard 2013). Cigarette smoking also affects other risk factors, such as low levels of HDL cholesterol and glucose intolerance (CDC 2008). Moreover, hematological changes are also triggered by tobacco exposure, including increased white blood cells, platelet aggregation, changes in serum lipids, and fibrinogen levels. The most important specific markers used for the determination of exposure to tobacco include nicotine and its metabolites, such as carbon monoxide, cotinine, and thiocyanate with cotinine being the most potent urine marker.

Nevertheless, carboxyhemoglobin levels seem to be more a qualitative rather than a quantitative factor for the level of exposure, or the amount of cardiovascular damage (Leone 2005). Also, hair analysis is used to determine the levels of cotinine, which accumulates in the hair during hair growth allowing the long-term monitoring of the accumulative effects of tobacco exposure (Florescu et al. 2009). It should be noted that all forms of tobacco are harmful, including cigarette smoking, waterpipe tobacco, and other various smokeless tobacco products, cigars, pipe tobacco, etc. More specifically, the use of waterpipe tobacco and other smokeless tobacco products are harmful, similar to cigarette smoking. It has been suggested that waterpipe tobacco is highly addictive due to containing nicotine and significantly damaging for human health.

Moreover, heated tobacco products that are promoted the last years as being less harmful produce aerosols with nicotine and other toxic products upon tobacco heating that lead to increased risk of cancers of the head, neck, throat, esophagus, and oral cavity, as well as several dental diseases (Davis et al. 2019). Similarly, e-cigarettes that are electronic systems delivering nicotine or not produce an aerosol upon heating a liquid that is inhaled by the user. They can also be highly addictive and harmful, especially when used by children or adolescents whose brain is still under development, as well as pregnant women, as it can be damaging for the fetus. Finally, it has been shown to increase the risk of CVD and lung disease, but its long-term effects remain to be studied the following years (CDC 2020).

3 Environmental Factors

Environmental factors are very important contributors to disease, and recently they have been acknowledged as risk factors for NCDs. However, in countries like Southeast Asia, air pollution is the leading cause of NCDs. It is estimated that environment-related deaths from NCDs account for 2/3 (8.2 M) of total deaths (12.6 M) caused by the environment. Apart from air pollution, environmental factors include radiation, second-hand smoke, noise, unhealthy drinking water, smoking, exposure to carcinogens and other harmful toxic agents, heavy metals, and mostly lead and mercury (World Health Organization 2017). Health complications to these factors include mostly cardiovascular disease, where 1/3 of CVD is attributed mostly to air pollution and at a lower level to other environmental factors.

3.1 Air Pollution

According to data from the Global Health Observatory for mortality from all or specific causes, air pollution was responsible for 22% of CVD deaths, 26% of ischemic heart and 25% of stroke deaths, 53% of COPD deaths, and 40% of deaths from cancer in the lungs (Wang et al. 2016). In line with NCDs incidence and related death rise, ambient air pollution has risen by 9% for the period 2010–2016, raising the awareness of international health organizations to address this challenge. The third United Nations high-level meeting on NCDs recognized air pollution (ambient and household) as a risk factor for NCDs in 2018. Since then, several interventional strategies have been proposed towards a more sustainable environment (Prüss-Ustün et al. 2019). An important factor for the NCDs incidence caused by environmental risks is early-life exposure. More than 25% of deaths among children below 5 years old are associated with the environment, and exposure to polluted air has been linked with premature and low-weight birth for a pregnant mother and NCDs onset for the children and adolescents. Asthma, the most common NCD among children, has been studied extensively concerning the role of air pollution on its development. In contrast, a recent study showed that improvement in air quality could prevent almost 50% of asthma cases (Pierangeli et al. 2020).

Air pollution, referring to the polluting substances or particulate matters in the air that can have a harmful effect on living organisms, can have direct and immediate or indirect, and at a later stage, effects on health. Particulate matters normally are formed in the air through the interaction between chemical substances and are categorized based on their diameter. It has been shown that the smaller their diameter, the greater the risk for human health because of their increased penetration to the body. Immediate impact can be caused via the binding compounds present in gases or aerosols such as CO_2 and NO_2 to hemoglobin competing with oxygen, leading to hypoxia and toxicity (Schraufnagel et al. 2019). Studies investigating the short-term exposure effect of air pollutants showed increased hospitalization and admission at the emergency department for patients with respiratory issues such as asthma and COPD. In China, a longitudinal analysis of 84 patients with COPD showed that exposure to ambient air pollution and specifically NO_2, CO, and SO_2 was linked to reducing lung function measured by Forced Vital Capacity percentage (FVC%) and reduced the anti-inflammatory and increased pro-inflammatory markers (Gao et al. 2020). These findings are in line with previous studies with COPD patients, and notably, the correlation is stronger in patients that smoke suggesting a synergistic effect between pollutants and smoke agents (Dadvand et al. 2014).

Additionally, a large global study in 652 cities of 24 countries published in NEJM highlighted the positive association between CVD, respiratory disease and all-cause mortality, and short exposure to ambient air pollution, even below the allowed threshold of pollutants concentrations. Concerning CVD, several studies have reported significant associations between short-term exposure to particulate matters with blood pressure and out-of-hospital cardiac arrest (Zhao et al. 2020a).

In a more long-term manner, pollutants of the air promote oxidative stress and systemic inflammation and have been implicated in dysfunction of distinct organs reviewed by Schraufnagel D. et al. (Schraufnagel et al. 2019). Global health organizations have developed tools for the risk assessment of air pollution for long-term and short-term exposure. AirQ+ is a software developed by the World Health Organization (WHO) Regional Office for Europe enabling users to quantify and assess the magnitude of air population with specific characteristics used as input on health including morbidity and mortality incidence projections for acute and chronic diseases. A comparative review and discussion of the collected data from AirQ models by Conti G O et al. identified the limitation of not including a large variety of pollutants as input to the software, thus providing only a part of the picture (Oliveri Conti et al. 2017).

3.2 Heavy Metals

Another type of environmental pollution with a significant effect on human health is heavy metals. Although some are essential for life such as iron, zinc, and manganese at small doses, some others, including cadmium (Cd), mercury (Hg), and lead (Pb), have no known beneficial effect and can be rather dangerous. Heavy metals have increased dramatically due to the anthropogenic activity and can be found in the atmosphere, the water, the soil, and thus the living organisms. Through the food chain, the accumulation of these heavy metals to humans can be such that it will be dangerous. The absorption of heavy metals from vegetables through the soil resulting in the chronic-low grade exposure to them to humans has been well studied for years. Briefly, heavy metals can either directly affect organs such as the brain, the kidney, and the heart or displace essential nutrients leading to significant disruption of metabolic pathways and oxidative stress (Jaishankar et al. 2014). However, new evidence indicates an additional pathway through which heavy metals affect health, which is through alteration of the microbiome (Chiu et al. 2020).

3.3 Concluding Remarks

NCDs morbidity and mortality upward trends represent a major challenge for the healthcare sector. Based on epidemiological data and observational studies, global health agencies have defined the key environmental factors and the intermediate mechanisms that shape the unhealthy environment and trigger or aggravate NCDs (Fig. 13.2). Diet is the primary factor that is associated with NCDs mortality, suggesting that through an intervention to people's daily dietary choices, 20% or 11 million of global deaths could be prevented. However, as studies suggest, the relationship between nutrients and health is complex and dynamic, which requires sophisticated tools to identify and monitor their metabolism.

Metabolomics is a promising tool that can be valuable to several healthcare positions and specialties. As a scanning tool, metabolomics can demonstrate nutritional deficiencies or hidden hunger that underlie an NCD, allowing primary care professionals to replenish these deficiencies under a balanced diet with a personalized dietary intervention (Tsoukalas et al. 2017). As shown in the explanatory figure of vitamin B12 pathway, blood levels of nutrients are not sufficient and reliable markers to reflect the bioavailability of nutrients, whereas intermediate metabolites participating in the pathways fueled by these nutrients are more robust (Fig. 13.1). Moreover, dietary interventions can be monitored for their efficacy in an individual with metabolomics, based on the genetic profile, underlying disease and drug treatment. These factors may affect nutrients absorption or metabolism. Dietary compounds can increase or reduce the risk of NCDs through their interaction with gene expression and post-translational modifications. Metabolomics can capture the effect of selected dietary compounds on metabolism, allowing the healthcare professional to personalize the intervention. Finally, specific metabolic biomarkers that are related to diet-related complications such as insulin resistance and pro-inflammatory context can be valuable predictive tools for individuals at risk of developing NCDs or ADs. For example, dihomogamma-linolenic acid is related to insulin resistance, inflammation, and the presence of autoimmune diseases suggesting the potency as a predictive biomarker (Tsoukalas et al. 2019a; Tsoukalas et al. 2019c).

An additional burden to health, apart from dietary and lifestyle factors, is toxicants from cigarette smoking, dietary heavy metals, and air pollution. The molecular mechanism by which these factors negatively affect human health is not fully described but accumulating data show their causal relationship with NCDs onset. Oxidative stress and inflammation are central mechanisms that have been shown to be significantly induced under the exposure to PM or cigarette smoke, also affecting the human metabolome and promoting insulin resistance (Fig. 13.2). Also, heavy metals obtained through diet have been shown to affect the normal function of the metabolic pathways through their interaction with the enzymes. As such, an association between exposure to toxicants with metabolic phenotypic changes can provide valuable information to health and governmental bodies towards sustainable environmental solutions. In a more patient-centered view, metabolomics can identify the specific metabolic pathways that are disturbed and the enzymes and metabolites that are involved indicating the points of the metabolism that require attention through dietary or medication interventions.

Overall, metabolomics, as a tool of precision medicine, presents an opportunity to move from evidence-based medicine that focuses on diseases and symptoms management of NCDs, towards medical approaches that combine effective screening, prevention, and health promotion strategies, while offering personalized intervention targeting the risk factors in line with the standard treatment.

References

Afshin, A. *et al.* (2019) 'Health effects of dietary risks in 195 countries, 1990–2017: a systematic analysis for the Global Burden of Disease Study 2017', *The Lancet*, 393(10184), pp. 1958–1972. doi: https://doi.org/10.1016/S0140-6736(19)30041-8.

Ames, B. N. (2006) 'Low micronutrient intake may accelerate the degenerative diseases of aging through allocation of scarce micronutrients by triage', *PNAS*, 103(47), pp. 17589–17594.

Amrein, K. *et al.* (2020) 'Vitamin D deficiency 2.0: an update on the current status worldwide', *European Journal of Clinical Nutrition*. Springer US. doi: https://doi.org/10.1038/s41430-020-0558-y.

Bailey, R. L., West, K. P. and Black, R. E. (2015) 'The epidemiology of global micronutrient deficiencies', *Annals of Nutrition and Metabolism*, 66(suppl 2), pp. 22–33. doi: https://doi.org/10.1159/000371618.

Balashova, E. E., Maslov, D. L. and Lokhov, P. G. (2018) 'A metabolomics approach to pharmacotherapy personalization', *Journal of Personalized Medicine*, 8(3). doi: https://doi.org/10.3390/jpm8030028.

Bier, D. M. and Mann, J. (2015) *Nutrition for the Primary Care Provider*. Edited by W. R. of N. and Dietetics. Karger. doi: https://doi.org/10.1159/isbn.978-3-318-02667-2.

Blumberg, J. B. *et al.* (2017) 'Impact of frequency of multi-vitamin/multi-mineral supplement intake on nutritional adequacy and nutrient deficiencies in U.S. adults', *Nutrients*, 9(8). doi: https://doi.org/10.3390/nu9080849.

Braegger, C. *et al.* (2013) 'Vitamin d in the healthy European paediatric population', *Journal of Pediatric Gastroenterology and Nutrition*, 56(6), pp. 692–701. doi: https://doi.org/10.1097/MPG.0b013e31828f3c05.

CDC (2008) 'Smoking-attributable mortality, years of potential life lost, and productivity losses--United States, 2000-2004.', *MMWR. Morbidity and mortality weekly report*. United States, 57(45), pp. 1226–1228.

CDC (2017) *Get the facts: Sodium and the Dietary Guidelines*.

CDC (2020) *About Electronic Cigarettes (E-Cigarettes)*. Available at: https://www.cdc.gov/tobacco/basic_information/e-cigarettes/about-e-cigarettes.html (Accessed: 2 July 2020).

Chiu, K. *et al.* (2020) 'The Impact of Environmental Chemicals on the Gut Microbiome'. *Toxicol Sci.*, 176(2), pp. 253–284. doi: https://doi.org/10.1093/toxsci/kfaa065/5835885.

Chiva-blanch, G. and Badimon, L. (2020) 'Benefits and Risks of Moderate Alcohol Consumption on Cardiovascular Disease: Current Findings', *Nutrients*, 12(108), pp. 1–19. doi: https://doi.org/10.3390/nu12010108.

Correa-Rodríguez, M. *et al.* (2020) 'Dietary intake of free sugars is associated with disease activity and dyslipidemia in systemic lupus erythematosus patients', *Nutrients*, 12(4). doi: https://doi.org/10.3390/nu12041094.

Csordas, A. and Bernhard, D. (2013) 'The biology behind the atherothrombotic effects of cigarette smoke', *Nature Reviews Cardiology*. Nature Publishing Group, 10(4), pp. 219–230. doi: https://doi.org/10.1038/nrcardio.2013.8.

Dadvand, P. *et al.* (2014) 'Air pollution and biomarkers of systemic inflammation and tissue repair in COPD patients', *European Respiratory Journal*. European Respiratory Society, 44(3), pp. 603–613. doi: https://doi.org/10.1183/09031936.00168813.

Davis, B., Williams, M. and Talbot, P. (2019) 'iQOS: evidence of pyrolysis and release of a toxicant from plastic', *Tobacco Control*, 28(1), pp. 34 LP – 41. doi: https://doi.org/10.1136/tobaccocontrol-2017-054104.

DeFronzo, R. A., Sherwin, R. S. and Kraemer, N. (1987) 'Effect of physical training on insulin action in obesity', *Diabetes*, 36(12), pp. 1379–1385. doi: https://doi.org/10.2337/diab.36.12.1379.

Devries, S. (2019) 'A global deficiency of nutrition education in physician training: the low hanging fruit in medicine remains on the vine', *The Lancet Planetary Health*. The Authors(s). Published by Elsevier Ltd. This is an Open Access article under the CC BY-NC-ND 4.0 license, 3(9), pp. e371–e372. doi: https://doi.org/10.1016/S2542-5196(19)30173-1.

Dinicolantonio, J. J. and Okeefe, J. H. (2017) 'Added sugars drive coronary heart disease via insulin resistance and hyperinsulinaemia: a new paradigm', *Open Heart*, 4(e000729), pp. 1–4. doi: https://doi.org/10.1136/openhrt-2017-000729.

EFSA NDA Panel (EFSA Panel on Dietetic Products, Nutrition and Allergies). (2016) Scientific opinion on dietary reference values for vitamin D. *EFSA Journal* 14(10):4547, 145 pp. doi: https://doi.org/10.2903/j.efsa.2016.4547.

e-Library of Evidence for Nutrition Actions (eLENA) (2019) *Reducing free sugars intake in adults to reduce the risk of noncommunicable diseases, WHO*. Available at: https://www.who.int/elena/titles/free-sugars-adults-ncds/en/ (Accessed: 2 July 2020).

Ezzati, M. *et al.* (2014) 'Global Sodium Consumption and Death from Cardiovascular Causes', *NEJM*, 371:624–634. doi: https://doi.org/10.1056/NEJMoa1304127.

Faggi, L. *et al.* (2019) 'A Polyphenol-Enriched Supplement Exerts Potent Epigenetic-Protective Activity in a Cell-Based Model of Brain Ischemia', *Nutrients*, pp. 1–15. doi: https://doi.org/10.3390/nu11020345.

Finamor, D. C. *et al.* (2013) 'A pilot study assessing the effect of prolonged administration of high daily doses of vitamin D on the clinical course of vitiligo and psoriasis', *Dermato-Endocrinology*, 5(1), pp. 222–234. doi: https://doi.org/10.4161/derm.24808.

Florescu, A. *et al.* (2009) 'Methods for quantification of exposure to cigarette smoking and environmental tobacco smoke: focus on developmental toxicology', *Therapeutic drug monitoring*, 31(1), pp. 14–30. doi: https://doi.org/10.1097/FTD.0b013e3181957a3b.

Fu, L. W. and Vender, R. (2011) 'Systemic role for vitamin D in the treatment of psoriasis and metabolic syndrome', *Dermatology Research and Practice*, 2011. doi: https://doi.org/10.1155/2011/276079.

Gao, N. *et al.* (2020) 'Lung function and systemic inflammation associated with short-term air pollution exposure in chronic obstructive pulmonary disease patients in Beijing, China', *Environmental Health*, 19(1), p. 12. doi: https://doi.org/10.1186/s12940-020-0568-1.

Garland, C. F. *et al.* (2007) 'Vitamin D and prevention of breast cancer: Pooled analysis', *Journal of Steroid Biochemistry and Molecular Biology*, 103(3–5), pp. 708–711. doi: https://doi.org/10.1016/j.jsbmb.2006.12.007.

Garland, C. F. *et al.* (2009) 'Vitamin D for Cancer Prevention: Global Perspective', *Annals of Epidemiology*, 19(7), pp. 468–483. doi: https://doi.org/10.1016/j.annepidem.2009.03.021.

Giles, J. T. *et al.* (2015) 'Insulin Resistance in Rheumatoid Arthritis. Disease-Related Indicators and Associations With the Presence and Progression of Subclinical Atherosclerosis', *Arthritis & Rheumatology*, 67(3), pp. 626–636. doi: https://doi.org/10.1002/art.38986.

Gorham, E. D. *et al.* (2005) 'Vitamin D and prevention of colorectal cancer', *Journal of Steroid Biochemistry and Molecular Biology*, 97(1–2), pp. 179–194. doi: https://doi.org/10.1016/j.jsbmb.2005.06.018.

Grant, W. B. *et al.* (2020) 'Evidence that vitamin d supplementation could reduce risk of influenza and covid-19 infections and deaths', *Nutrients*, 12(4), pp. 1–19. doi: https://doi.org/10.3390/nu12040988.

Graudal, N. *et al.* (2014) 'Compared With Usual Sodium Intake, Low- and Excessive- Sodium Diets Are Associated With Increased Mortality: A Meta-Analysis', *American Journal of Hypertension*, 27(9), pp. 1129–1137. doi: https://doi.org/10.1093/ajh/hpu028.

Guthold, R. *et al.* (2018) 'Worldwide trends in insufficient physical activity from 2001 to 2016: a pooled analysis of 358 population-based surveys with 1·9 million participants', *The Lancet Global Health*. World Health Organization, 6(10), pp. e1077–e1086. doi: https://doi.org/10.1016/S2214-109X(18)30357-7.

Hannibal L, *et al.* (2016) Biomarkers and Algorithms for the Diagnosis of Vitamin B12 Deficiency. *Front Mol Biosci*, 3:27. doi: https://doi.org/10.3389/fmolb.2016.00027.

Hathcock, J. N. *et al.* (2007) 'Risk assessment for vitamin D 1, 2', *Am J Clin Nutr*, 85(6), pp. 6–18.

He, F. J. *et al.* (2020) 'Salt Reduction to Prevent Hypertension and Cardiovascular Disease: JACC State-of-the-Art Review', *Journal of the American College of Cardiology*, 75(6), pp. 632–647. doi: https://doi.org/10.1016/j.jacc.2019.11.055.

Hlavaty, T., Krajcovicova, A. and Payer, J. (2015) 'Vitamin D therapy in inflammatory bowel diseases: Who, in what form, and how much?', *Journal of Crohn's and Colitis*, 9(2), pp. 198–209. doi: https://doi.org/10.1093/ecco-jcc/jju004.

Holick, M. F. *et al.* (2011a) 'Evaluation, treatment, and prevention of vitamin D deficiency: An endocrine society clinical practice guideline', *Journal of Clinical Endocrinology and Metabolism*, 96(7), pp. 1911–1930. doi: https://doi.org/10.1210/jc.2011-0385.

Holick, M. F. *et al.* (2011b) 'Evaluation, treatment, and prevention of vitamin D deficiency: An endocrine society clinical practice guideline', *Journal of Clinical Endocrinology and Metabolism*, pp. 1911–1930. doi: https://doi.org/10.1210/jc.2011-0385.

Institute of Medicine of the National Academies (2011) *Dietary Reference Intakes for Calcium and Vitamin D*. Edited by C. Ross et al. Washington DC: The National Academy Press.

Jaishankar, M. *et al.* (2014) 'Toxicity, mechanism and health effects of some heavy metals', *Interdisciplinary Toxicology*. Slovak Toxicology Society, 7(2), pp. 60–72. doi: https://doi.org/10.2478/intox-2014-0009.

Jeffery, L. E. *et al.* (2018) 'Decreased sensitivity to 1,25-dihydroxyvitamin D3 in T cells from the rheumatoid joint', *Journal of Autoimmunity*, 88, pp. 50–60. doi: https://doi.org/10.1016/j.jaut.2017.10.001.

Johnson, A. R. and Milner, J. J. (2012) 'The inflammation highway: metabolism accelerates inflammatory traffic in obesity', *Immunological Reviews*, 249, pp. 218–238.

Kahanovitz, L. et al. (2017) "Type 1 Diabetes - A Clinical Perspective." *Point of Care*, 16(1), pp. 37-40. doi: https://doi.org/10.1097/POC.0000000000000125.

Kamen, D. L. (2013) 'Vitamin D in Lupus', *National Institutes of Health*, 71(2), pp. 1–8. doi: https://doi.org/10.1038/mp.2011.182.doi.

Karpanen, H. *et al.* (1984) 'Safety and effects of potassium- and magnesium- containing low sodium salt mixtures.pdf', *Journal of Cardiovascular Pharmacology*, 6, pp. 236–243.

Karppanen, H. and Mervaala, E. (2006) 'Sodium Intake and Hypertension', *Progress in Cardiovascular disease*, 49(2), pp. 59–75. doi: https://doi.org/10.1016/j.pcad.2006.07.001.

Katz, A. *et al.* (1999) 'Effect of a mineral salt diet on 24-h blood pressure monitoring in elderly hypertensive patients', *Journal of Human Hypertension*, 13(11), pp. 777–780. doi: https://doi.org/10.1038/sj.jhh.1000837.

Kechichian, E. and Ezzedine, K. (2018) 'Vitamin D and the Skin: An Update for Dermatologists', *American Journal of Clinical Dermatology*. Springer International Publishing, 19(2), pp. 223–235. doi: https://doi.org/10.1007/s40257-017-0323-8.

Kivity, S., Agmon-Levin, N., *et al.* (2011) 'Vitamin D and autoimmune thyroid diseases', *Cellular and Molecular Immunology*, 8(3), pp. 243–247. doi: https://doi.org/10.1038/cmi.2010.73.

Kong, L. C. *et al.* (2013) 'Insulin resistance and inflammation predict kinetic body weight changes in response to dietary weight loss and maintenance in overweight and obese subjects by using a Bayesian', *Am J Clin Nutr*, 98(1), pp. 1385–1394. doi: https://doi.org/10.3945/ajcn.113.058099.

Lear, S. A. *et al.* (2017) 'The effect of physical activity on mortality and cardiovascular disease in 130 000 people from 17 high-income, middle-income, and low-income countries: the PURE study', *The Lancet*. Elsevier Ltd, 390(10113), pp. 2643–2654. doi: https://doi.org/10.1016/S0140-6736(17)31634-3.

Leone, A. (2005) 'Biochemical markers of cardiovascular damage from tobacco smoke.', *Current pharmaceutical design*. United Arab Emirates, 11(17), pp. 2199–2208. doi: https://doi.org/10.2174/1381612054367391.

Levin, A. D. *et al.* (2011) 'Vitamin D deficiency in children with inflammatory bowel disease', *Digestive Diseases and Sciences*, 56(3), pp. 830–836. doi: https://doi.org/10.1007/s10620-010-1544-3.

Ludwig, D. S. *et al.* (2018) 'Dietary carbohydrates: Role of quality and quantity in chronic disease', *BMJ (Online)*, 361. doi: https://doi.org/10.1136/bmj.k2340.

Manzel, A. *et al.* (2014) 'Role of "western diet" in inflammatory autoimmune diseases', *Current Allergy and Asthma Reports*, 14(1). doi: https://doi.org/10.1007/s11882-013-0404-6.

Margină, D. *et al.* (2020) 'Chronic Inflammation in the Context of Everyday Life: Dietary Changes as Mitigating Factors', pp. 1–30. doi: https://doi.org/10.3390/ijerph17114135.

Martinez, R. *et al.* (2020) 'Trends in premature avertable mortality from non-communicable diseases for 195 countries and territories, 1990–2017: a population-based study', *The Lancet Global Health*. Pan American Health Organization. 8(4), pp. e511–e523. doi: https://doi.org/10.1016/S2214-109X(20)30035-8.

Mattozzi, C. *et al.* (2016) 'Psoriasis, Vitamin D and the importance of the cutaneous barrier's integrity: An update', *Journal of Dermatology*, 43(5), pp. 507–514. doi: https://doi.org/10.1111/1346-8138.13305.

Merino, B. *et al.* (2020) 'Intestinal fructose and glucose metabolism in health and disease', *Nutrients*, 12(1), pp. 1–35. doi: https://doi.org/10.3390/nu12010094.

Miettinen, M. E. *et al.* (2020) 'Serum 25-hydroxyvitamin D concentration in childhood and risk of islet autoimmunity and type 1 diabetes: the TRIGR nested case–control ancillary study', *Diabetologia*, 63(4), pp. 780–787. doi: https://doi.org/10.1007/s00125-019-05077-4.

Millwood, I. Y. *et al.* (2019) 'Conventional and genetic evidence on alcohol and vascular disease aetiology: a prospective study of 500 000 men and women in China', *The Lancet*, 393(10183), pp. 1831–1842. doi: https://doi.org/10.1016/S0140-6736(18)31772-0.
Morenga, L. A. Te *et al.* (2014) 'Dietary sugars and cardiometabolic risk: systematic review and meta-analyses of randomized controlled trials of the effects on blood pressure and lipids 1 – 3', *Am J Clin Nutr*, 100, pp. 65–79. doi: https://doi.org/10.3945/ajcn.113.081521.1.
MORIMOTO, S. *et al.* (1986) 'An open study of vitamin D3 treatment in psoriasis vulgaris', *British Journal of Dermatology*, 115(4), pp. 421–429. doi: https://doi.org/10.1111/j.1365-2133.1986.tb06236.x.
Moss, J. W. E. and Ramji, D. P. (2017) 'Nutraceutical therapies for atherosclerosis', *Nat Rev Cardiol.*, 13(9), pp. 513–532. doi: https://doi.org/10.1038/nrcardio.2016.103.Nutraceutical.
Moynihan, P. J. and Kelly, S. A. M. (2014) 'Effect on caries of restricting sugars intake: Systematic review to inform WHO guidelines', *Journal of Dental Research*, 93(1), pp. 8–18. doi: https://doi.org/10.1177/0022034513508954.
National Academies of Sciences, Engineering, and M. *et al.* (2017) 'Guiding Principles for Developing Dietary Reference Intakes Based on Chronic Disease'. National Academies Press (US), Washington (DC) (The National Academies Collection: Reports funded by National Institutes of Health). Available at: http://europepmc.org/books/NBK465024.
Noakes, T. D. (2018) 'So What Comes First: The Obesity or the Insulin Resistance ? And Which Is More Important ?', *Clinical Chemistry*, 64(1), pp. 7–9. doi: https://doi.org/10.1373/clinchem.2017.282962.
Nosrati, A. *et al.* (2017) 'Dietary modifications in atopic dermatitis: patient-reported outcomes', *J Dermatolog Treat*, 28(6), pp. 523–538. doi: https://doi.org/10.1080/09546634.2016.1278071.
Nualart, F. *et al.* (2014). Vitamin C Transporters, Recycling and the Bystander Effect in the Nervous System: SVCT2 versus Gluts. *Journal of Stem Cell Research & Therapy, 4*(5), 209. https://doi.org/10.4172/2157-7633.1000209.
Oliveri Conti, G. *et al.* (2017) 'A review of AirQ Models and their applications for forecasting the air pollution health outcomes', *Environmental Science and Pollution Research*. Springer Verlag, 24(7), pp. 6426–6445. doi: https://doi.org/10.1007/s11356-016-8180-1.
Pierangeli, I. *et al.* (2020) 'Health equity and burden of childhood asthma - related to air pollution in Barcelona', *Environmental Research*. Elsevier Inc., (May 2019), p. 109067. doi: https://doi.org/10.1016/j.envres.2019.109067.
Pietrzik, K. (1991) *Modern Lifestyles, Lower Energy Intake and Micronutrient Status*. Springer.
Postic, C. (2020) 'Conversion of a dietary fructose: new clues from the gut microbiome', *Nature Metabolism*. Springer US, 2(3), pp. 217–218. doi: https://doi.org/10.1038/s42255-020-0185-x.
Prüss-Ustün, A. *et al.* (2019) 'Environmental risks and non-communicable diseases', *BMJ*, 364(1265).
Ramagopalan, S. V. *et al.* (2010) 'A ChIP-seq defined genome-wide map of vitamin D receptor binding: Associations with disease and evolution', *Genome Research*, 20(10), pp. 1352–1360. doi: https://doi.org/10.1101/gr.107920.110.
Sarkar, D., Jung, K. M. and Wang, J. (2015) 'Alcohol and the immune system.', *Alcohol Research*, 37(2), pp. 153–155. doi: https://doi.org/10.1136/bmj.298.6673.543.
Schraufnagel, D. E. *et al.* (2019) 'Air Pollution and Noncommunicable Diseases: A Review by the Forum of International Respiratory Societies' Environmental Committee, Part 2: Air Pollution and Organ Systems', *Chest*. Elsevier Inc, 155(2), pp. 417–426. doi: https://doi.org/10.1016/j.chest.2018.10.041.
Scientific Advisory Committee on Nutrition Report (2015) *Carbohydrates and Health*. London.
Sharif, K. *et al.* (2018) 'Physical activity and autoimmune diseases: Get moving and manage the disease', *Autoimmunity Reviews*. Elsevier B.V., 17(1), pp. 53–72. doi: https://doi.org/10.1016/j.autrev.2017.11.010.
Shi, Z. *et al.* (2020) *Short-Term Exposure to a Western Diet Induces Psoriasiform Dermatitis by Promoting Accumulation of IL-17A–Producing γδ T Cells, Journal of Investigative Dermatology*. Society for Investigative Dermatology. doi: https://doi.org/10.1016/j.jid.2020.01.020.

Singh, A. and Kamen, D. L. (2012) 'Potential benefits of vitamin D for patients with systemic lupus erythematosus', *Dermato-Endocrinology*, 4(2), pp. 146–151. doi: https://doi.org/10.4161/derm.20443.

Singh, S., Young, P. and Armstrong, A. W. (2017) 'An update on psoriasis and metabolic syndrome: A meta-analysis of observational studies', *PLoS ONE*, 12(7), pp. 1–13. doi: https://doi.org/10.1371/journal.pone.0181039.

Softic, S. *et al.* (2020) 'Fructose and hepatic insulin resistance', *Critical Reviews in Clinical Laboratory Sciences*. Taylor & Francis, pp. 1–15. doi: https://doi.org/10.1080/10408363.2019.1711360.

Strong, K. *et al.* (2006) 'Preventing chronic disease: a priority for global health', *International Journal of Epidemiology*, 35(2), pp. 491–492. doi: https://doi.org/10.1093/ije/dyi314.

Theofilopoulos, A. N., Kono, D. H. and Baccala, R. (2017) 'The Multiple Pathways to Autoimmunity', *Nat. Immunol.*, 18(7), pp. 716–724. doi: https://doi.org/10.1016/j.cogdev.2010.08.003.Personal.

Tinetti, M. E., Fried, T. R. and Boyd, C. M. (2012) 'Designing health care for the most common chronic condition—multimorbidity', *JAMA*, 307(23), pp. 2493–2494. doi: https://doi.org/10.1002/nbm.3066.Non-invasive.

Trieu, K. *et al.* (2015) 'Salt reduction initiatives around the world-A systematic review of progress towards the global target', *PLoS ONE*, 10(7). doi: https://doi.org/10.1371/journal.pone.0130247.

Tsoukalas, D., and Sarandi, E. (2020) Micronutrient deficiencies in patients with COVID-19: how metabolomics can contribute to their prevention and replenishment, *BMJ Nutrition, Prevention & Health*, 3. https://doi.org/10.1136/bmjnph-2020-000169.

Tsoukalas, D. *et al.* (2017) 'Application of metabolomics: Focus on the quantification of organic acids in healthy adults', *International Journal of Molecular Medicine*, 40(1), pp. 112–120. https://doi.org/10.3892/ijmm.2017.2983.

Tsoukalas, D *et al.* (2019a) 'Application of metabolomics part II: Focus on fatty acids and their metabolites in healthy adults', *International Journal of Molecular Medicine*, 43(1), pp. 233–242. doi: https://doi.org/10.3892/ijmm.2018.3989.

Tsoukalas, D, *et al.* (2019b) 'Metabolic fingerprint of chronic obstructive lung diseases: A new diagnostic perspective', *Metabolites*, 9(12). doi: https://doi.org/10.3390/metabo9120290.

Tsoukalas, D *et al.* (2019c) 'Targeted Metabolomic Analysis of Serum Fatty Acids for the Prediction of Autoimmune Diseases', *Frontiers in Molecular Biosciences*, 6(November), pp. 1–14. doi: https://doi.org/10.3389/fmolb.2019.00120.

Vagianos, K. *et al.* (2007) 'Nutrition assessment of patients with inflammatory bowel disease', *Journal of Parenteral and Enteral Nutrition*, 31(4), pp. 311–319. doi: https://doi.org/10.1177/0148607107031004311.

Wang, H. *et al.* (2016) 'Global, regional, and national life expectancy, all-cause mortality, and cause-specific mortality for 249 causes of death, 1980-2015: a systematic analysis for the Global Burden of Disease Study 2015', *The Lancet*. Elsevier, 388(10053), pp. 1459–1544. doi: https://doi.org/10.1016/S0140-6736(16)31012-1.

Wang, H. *et al.* (2017) 'Vitamin D and chronic diseases', *Vitamin D: Oxidative Stress, Immunity, and Aging*, 8(3), pp. 346–353. doi: https://doi.org/10.14336/AD.2016.1021.

Wang, H. J., Zakhari, S. and Jung, M. K. (2010) 'Alcohol, inflammation, and gut-liver-brain interactions in tissue damage and disease development', *World Journal of Gastroenterology*, 16(11), pp. 1304–1313. doi: https://doi.org/10.3748/wjg.v16.i11.1304.

Wang, T. J. *et al.* (2008) 'Vitamin D deficiency and risk of cardiovascular disease', *Circulation*, 117(4), pp. 503–511. doi: https://doi.org/10.1161/CIRCULATIONAHA.107.706127.

Watroba, M. and Szukiewicz, D. (2016) 'The role of sirtuins in aging and age-related diseases', *Advances in Medical Sciences*, 61(1), pp. 52–62.

Wessels, I. and Rink, L. (2020a) 'Micronutrients in autoimmune diseases: possible therapeutic benefits of zinc and vitamin D', *The Journal of Nutritional Biochemistry*. Elsevier Inc., 77, p. 108240. doi: https://doi.org/10.1016/j.jnutbio.2019.108240.

Wessels, I. and Rink, L. (2020b) 'Micronutrients in autoimmune diseases: possible therapeutic benefits of zinc and vitamin D', *Journal of Nutritional Biochemistry*. Elsevier Inc., 77, p. 108240. doi: https://doi.org/10.1016/j.jnutbio.2019.108240.

WHO (2018a) *Fact sheets-Physical activity*. Available at: https://www.who.int/news-room/fact-sheets/detail/physical-activity (Accessed: 2 July 2020).

WHO (2018b) *Global status report on alcohol and health 2018*.

Winther, K. H. and Rayman, M. P. (2020) 'Selenium in thyroid disorders — essential knowledge for clinicians Kristian', *Nat Rev Endocrinol.*, 16, pp. 165–175. doi: https://doi.org/10.1038/s41574-019-0311-6.

World Health Organization (2013) WHO issues new guidance on dietary salt and potassium. *Cent Eur J Public Health*, 21(1):16. PMID: 23741892.

World Health Organization. (2017). Preventing noncommunicable diseases (NCDs) by reducing environmental risk factors. *World Health Organization*. https://apps.who.int/iris/handle/10665/258796. License: CC BY-NC-SA 3.0 IGO

World Health Organization (2020) *Fact Sheets, Tobacco*.

Zeisel, S. H. (2007) 'Nutrigenomics and metabolomics will change clinical nutrition and public health practice: insights from studies on dietary requirements for choline', *Am J Clin Nutr*, 86(3), pp. 542–548. doi: https://doi.org/10.1038/jid.2014.371.

Zhang, D. *et al.* (2019) 'High Glucose Intake Exacerbates Autoimmunity through Reactive-Oxygen-Species-Mediated TGF-β Cytokine Activation', *Immunity*. Elsevier Inc., 51(4), pp. 671-681.e5. doi: https://doi.org/10.1016/j.immuni.2019.08.001.

Zhao, B. *et al.* (2020a) 'Short-term exposure to ambient fine particulate matter and out-of-hospital cardiac arrest: a nationwide case-crossover study in Japan', *The Lancet Planetary Health*, pp. e15–e23. doi: https://doi.org/10.1016/S2542-5196(19)30262-1.

Zhao, S. *et al.* (2020b) 'Dietary fructose feeds hepatic lipogenesis via microbiota-derived acetate', *Nature*. Springer US, 579(7800), pp. 586–591. doi: https://doi.org/10.1038/s41586-020-2101-7.

Zipitis, C. S. and Akobeng, A. K. (2008) 'Vitamin D supplementation in early childhood and risk of type 1 diabetes: a systematic review and meta-analysis', *Archives of Disease in Childhood*, 93(6), pp. 512–517. doi: https://doi.org/10.1136/adc.2007.128579.

Chapter 14
Obstacles in the Adaptation of Biopesticides in India

Chetan Keswani, Hagera Dilnashin, Hareram Birla, and Surya Pratap Singh

Abstract The unregulated use of chemical pesticides and fertilizers has led to unanticipated harmful consequences to human health as well as environment. The pandemic Covid-19 crisis is not permanent, but it has magnified the impact on livestock farming already present in the agricultural system. Farmers are facing the shortage of agricultural inputs due to the global trade disturbance. An effective alternative method of conventional pesticides is the application of biopesticides. These substances are sustainable solution for farming, pest control, and disease management. The post-Covid-19 pandemic impact on agriculture, regulatory hurdles, and the limitations of large-scale biopesticides production are discussed here.

Keywords Agriculturally important microbes · Biosafety · Bio-efficacy · Non-pesticidal · Microbes

1 Introduction

The Indian economy is based largely on agriculture. Since the Green Revolution (1970s), the usage of chemical pesticides and fertilizers has increased enormously (Singh et al. 2017). But the threats presented by the extensive long-term use of pesticides and fertilizers for humans and environments and their devastating impact on soil microbes are now increasingly becoming recognized (Keswani et al. 2014). One eco-friendly alternative solution is to cultivate plant strains that are resistant to or have greater resistance to pathogens, but it takes a long time to research, grow, approve, and commercialize new varieties (Keswani 2015).

This opens the door for the use of other cost-effective, eco-friendly, and organic substitutes that would provide improved yields, such as AIM (agriculturally important microbes) including plant growth-promoting bacteria (PGPB)

C. Keswani (✉) · H. Dilnashin · H. Birla · S. P. Singh
Department of Biochemistry, Institute of Science, Banaras Hindu University, Varanasi, India
e-mail: chetan.keswani4@bhu.ac.in

E. Koukios, A. Sacio-Szymańska (eds.), *Bio#Futures*,
https://doi.org/10.1007/978-3-030-64969-2_14

or rhizobacteria (PGPR) and biopesticides that can manage pest outbreaks (Ram et al. 2018; Singh et al. 2019a).

In terms of agricultural sector growth, India still faces challenges in meeting its challenges of ensuring food security for the next 50 years (Keswani et al. 2019). Biopesticides are unable to take off in a major way due to the limitations, despite their immense market value and state and national attempts to promote them as an alternative to chemical pesticides (Keswani et al. 2016). Research, production, and commercialization of biopesticides and biofertilizers have been supported by many government agencies, such as the Department of Biotechnology (DBT), the Ministry of Agriculture and Farmers' Welfare, and the Ministry of Science and Technology (Sinha and Biswas 2008).

Global demand is there too. Biopesticides are just 4.5% of the total pesticides produced globally, and 6% in the USA, while only 3% are produced in India. It is estimated that the annual growth rate of biopesticide production in India is 2.5%. For instance, the countries (the USA, Mexico, and Canada) of the North American Free Trade Agreement (NAFTA) are the world's largest users of biopesticides and use about 45% of all biopesticides sold worldwide (Vílchez et al. 2017), while 20% is used by the European Union (Keswani et al. 2019).

The Central Insecticides Board and Registration Committee (CIBRC) of India has 970 biopesticide companies registered to screen possible biopesticides for biosafety (Keswani et al. 2019) (Tables 14.1, 14.2, 14.3, 14.4, and 14.5). The slow permeation of biopesticides in agricultural applications remains a major challenge, despite the tremendous attempts by Indian regulatory bodies to promote its usage.

2 Major Constraints

Meanwhile, in a disturbing trend, the surge in the demand for biopesticides due to national and state policies has ended up, promoting the marketing of spurious biopesticides that undermine the biotechnology sector's respectability (Keswani et al. 2019).

The restricted production of biopesticides is one of the restrictions in India, with only 14 biopesticidal formulations registered under the Insecticides Act of 1968, which primarily catalogs the requirements for biosafety determination of pesticides (Singh et al. 2019b). The costs of biopesticide registration and the long-term procedure involved further prevent businesses from investing in research and development.

Another restriction is the necessity to test the biosafety of the microbes prior to the biopesticides registration and dissemination. The additional expenses needed for better protection cannot be sustained by several universities and research organizations undertaking initial research and producing biopesticides (Keswani et al. 2019). In immuno-compromised patients, for instance, allergies to some fungi such as *Metarhizium*, *Trichoderma*, *Beauveria bassiana*, and *M. anisopliae* have been identified, and often human-safe strains have been found to modify soil microbes (Darbro and Thomas 2009; Keswani et al. 2014) (Fig. 14.1).

Table 14.1 List of biopesticides (*B. bassiana*) registrants under CIBRC, India

1.	M/s Om Agro Organics, Yavatmal, Maharashtra
2.	M/s INORA, Pune
3.	M/s Ellor Bio Tech & Agro Services, Aurangabad
4.	M/s Nirmal Organo Bio Tech, Maharashtra
5.	M/s Junna Life Sciences (p) Ltd., Hyderabad
6.	M/s Shree Shiva Bio-tech, Pudukktti, Tamilnadu
7.	M/s Arya Biotech and Research Laboratory, MS
8.	M/s Multiplex Bio Tech Pvt. Ltd.
9.	M/s Amit Biotech, Kolkata
10.	M/s Pest Control (India) Pvt. Ltd
11.	M/s Agrilife, AP
12.	M/s Varsha Bioscience & Technology, Hyderabad
13.	M/s Sri Venkateswara Chemicals, Secunderabad
14.	M/s Pravara Agro Bio-tech
15.	M/s DVS BioLife Ltd
16.	M/s Jai Biotech Industries
17.	M/s Vaibhav Lakshmi Bio-control Laboratories
18.	M/s Bio Agro Ferticon, Pune
19	M/s Avishkar Bio-farm Pvt. Ltd.
20	M/s Choudhary Agrotech (I)
21.	M/s Kanbiosys Pvt. Ltd.
22.	M/s Sai Agrotech, Yavatmal
23.	M/s Super Pesticides & Agro (India) Pvt. Kolkata
24.	M/s Bioscience (India) Pvt. Ltd., Hyderabad
25.	M/s Biotech International Ltd., New Delhi
26.	M/s Poshak Fertilizers, Gujarat
27.	Dept. of Agriculture, Govt. of U.P. (Lucknow)
28.	M/s Agriland Biotech Limited, Gujarat
29.	M/s Biotech International Ltd.
30.	M/s Sujay Bio-Tech Pvt. Ltd
31.	M/s Advance Bio-Tech Industries & Research Inputs (India), Indore
32.	M/s Gujarat Life Sciences (P) Ltd., Vadodara
33.	M/s. Green Plus Biotech, Nashik
34.	M/s. Arya Biotechnologies, Aurangabad
35.	M/s. Nova Agri Tech Pvt. Ltd., Secunderabad
36.	M/s. Neo Gene Agri Input Pvt. Ltd., Secunderabad
37.	M/s. Prathibha Biotech, Hyderabad
38.	M/s Gujarat Eco Microbial Technologies Pvt. Ltd.
39.	M/s. Care-Pro Bioscience (P) Ltd., New Delhi
40.	M/s Srikar Biotech Pvt. Ltd., Hyderabad
41.	M/s Bisco Bio Science (P) Ltd., Secunderabad
42.	M/s SRT Agro Science (P) Ltd., Durg
43.	M/s Agro Bio tech Research Centre Ltd., Kottayam

(continued)

Table 14.1 (continued)

44.	M/s Anshul Agro Chemicals, Bangalore
45.	M/s M.D. Biocoals (Agri Division), Sirsa, Haryana
46.	M/s Devi Biotech Pvt. Ltd., Madurai, T.N.
47.	M/s. Abhinav Biotech, Akola
48.	M/s. Institute of Plant Biotechnology, Nashik
49.	M/s. Esvin Advanced Tech. Ltd., Chennai
50.	M/s. Antecedent Pabulum In., Bhatinda (PB)
51.	M/s Shiv Shakti Bio Chem, Muzaffarnagar
52.	Department of Agriculture, Lucknow
53.	M/s Bharat Biocon Pvt. Ltd., New Delhi
54.	M/s Roshan Bio & Naturals, Delhi
55.	Commissionerate of Agriculture, Pune (M.S.)
56.	M/s Avadhut Agro Bio-Tech, Osmanabad (M.S.)
57.	M/s Kanha Herbs, Delhi
58.	M/s Ankur Chemicals India (INN), Muzaffarnagar
59.	M/s T. Stanes & Company Ltd., Cobatore (T.N)
60.	M/s Crop Care Bio Science & Research Institute, Ahmednagar (M.S.)
61.	M/s Aviral Bio-Tech & Fertilizer Pvt. Ltd., Bhopal
62.	M/s Gujarat chemicals and Fertilizers Trading Company, Baroda
63.	M/s Tropical Agrosystem (India) Pvt. Ltd., Chennai
64.	M/s Bio-control Laboratory, CSAUA&T, Kanpur
65.	M/s Bacto Power India Pvt. Ltd., Cobatore (TN) (Strain:BB-IARI-RJP) (Accession No.:MCC-1022).
66.	M/s Sunbio Tech Pvt. Ltd., Delhi (Strain: AAI, Allahabad) (Accession No.: NACC- F-3048).
67.	M/s Shri Dutta Gro-Tech Equipments, Wardha for grant of registration for indigenous manufacture of (Strain: AAI, Allahabad, Accession No. NACC-F-3048) u/
68.	M/s Harit Bio Control Lab., Yavatmal (CFU 1×108 /ml. Min)
69.	M/s Sugway Agri Biotech and Research Foundation,
70.	M/s INORA, Pune
71.	M/s Biosys Agrotech Pvt. Ltd., Mandsaur (Strain source: ICAR, Umiam, Meghalaya, Accession No. NACC-1022)
72.	M/s Vidarbha Biotech Ltd., Yavatmal (Strain Designation: ICAR, Umiam, Meghalaya, Accession No.NACC-1022).
73.	M/s Bio Control Laboratory, Meerut, UP (Strain source: M/s Biocontrol Laboratory, Sardar Vallabh Bhai Patel University of Agriculture, Meerut, UP from the soils of Meerut, Strain Designation: SVBPU/CSP/Bb-10, Accession No.ITCC-7520).
74.	M/s Ponalaboratory, Bangalore (Strain designation: AAI, Allahabad, Accesion No. NACC-F-3045)

Source: Directorate of Plant Protection, Quarantine & Storage, Department of Agriculture, Cooperation & Farmers Welfare, Ministry of Agriculture & Farmers Welfare, Government of India (http://ppqs.gov.in/divisions/cib-rc/biopesticide-registrant Accessed on: September 01, 2020)

Table 14.2 List of biopesticides (*T. harzianum*) registrants under CIBRC, India

1.	M/s Bio-control Laboratory, Kanpur
2.	M/s Peak Chemical Industries Ltd., u/s (Strain: IIHR, Bangalore, IIHR-Th-2, Accession No.-ITCC6888).
3.	Central Institute for Subtropical Horticulture, Lucknow
4.	M/s Gujarat Green Revolution Company Ltd.
5.	M/s Dept. of Agriculture Govt. of Uttar Pradesh
6.	M/s Mac Hi-Tech, Kerala
7.	M/s Super Pesticides & Agro (I) Pvt. Ltd.
8.	M/s Biocontrol Laboratory, Meerut
9.	M/s. Gujarat Eco Microbial Technologies Pvt. Ltd.
10.	M/s Multiplex Bio-tech Pvt. Ltd., Bangalore
11.	M/s Advance Crop Care (India) Pvt. Ltd., Indore
12.	M/s Jai Biotech, Pune
13.	M/s Bio-control Laboratory, Varanasi, U.P
14.	M/s. Agri Life, AP
15.	Department of Agriculture, Lucknow
16.	M/s Nico Organo Manure, Dekor, Gujarat
17.	M/s Excel Crop Care Ltd., Mumbai
18.	M/s Krishi Vikas Sahakari Samiti Ltd., Jaipur
19	M/s Agro Biotech Research Centre Ltd.
20	M/s Poshak Bio Research Pvt. Ltd., Gujarat for grant of registration for indigenous manufacture of (CFU count 2×106/gm. min.) (Strain No.IIHR-TH-2, Accession No. ITCC – 6888) u/s
21.	M/s Tropical Agrosystem (India) Pvt. Ltd. Chennai
22.	M/s Nature Agrocare & Research Pvt. Ltd., Indore
23.	M/s Bio-Control Laboratory, Lucknow
24.	M/s Junagarh Agriculture University, Junagarh (Strain Designation: IIHR-TH-2, obtained from IIHR, Bangalore, Accession No. ITCC No.6888).
25.	M/s Krishi Biosys Bangalore (Strain Designation: IIHR-TH-2, obtained from IIHR, Bangalore, Accession No. ITCC No.6888).
26.	M/s International Biotech, Abohar (Punjab) (Strain Designation: IIHR-TH-2, obtained from IIHR, Bangalore, Accession No. ITCC No.6888).
27.	M/s Gurudev Bio-Control Laboratory, Bijapur (Strain Designation: IIHR-TH-2, obtained from IIHR, Bangalore, Accession No. ITCC No.6888)
28.	M/s Jhas Agro Industries, Chittor, AP (Strain Designation: IIHR-TH-2, obtained from IIHR, Bangalore, Accession No. ITCC No.6888)
29.	M/s Antecedent Pabulam Inc., Bhatinda (Strain Designation: IIHR-TH-2, obtained from IIHR, Bangalore, Accession No.ITCC-6888).
30.	Accession No.(ITCC No. 6888).
31.	M/s Green Max Agotech, Cobatore (TN) (Strain: IIHR-TH-2) (Accession No.: ITCC- 6888).
32.	M/s Gama Organomed Plus Pvt. Ltd.

Source: Directorate of Plant Protection, Quarantine & Storage, Department of Agriculture, Cooperation & Farmers Welfare, Ministry of Agriculture & Farmers Welfare, Government of India (http://ppqs.gov.in/divisions/cib-rc/biopesticide-registrant Accessed on: September 01, 2020)

Table 14.3 List of biopesticides (*M. anisopliae*) registrants under CIBRC, India

1.	M/s Pravara Agro Bio Tech, Ahmednagar, MS for (CFU 1×108 /gm Min) under Section 9(3B)
2.	M/s Sri Biotech
3.	M/s Viswa Mithra Bio Agro P. Ltd., Guntur
4.	M/s Microplex (India), Wardha
5.	M/s T. Stanes & Company Ltd., Cobatore (T.N)
6.	M/s International Panacea Ltd., New Delhi
7.	M/s Microplex Biotech & Agrochem Pvt. Ltd., Wardha (MS)
8.	M/s Pest Control (India) Pvt. Ltd., Bengaluru
9.	M/s Sai Agrotech, Yavatmal, Maharashtra
10.	M/s OM Agro Organics, Yavatmal, Maharashtra (NACC-F-03047) (Strain- UMIAM, Accession No. NACC-F-03047)
11.	M/s R.B. Herbal Agro, Maharashtra (Strain-AAI, Accession No. NACC-F-03037)
12.	M/s Bonageri Crop Science Pvt. Ltd., Dharwad (Strain:AAI, Allahabad) (Accession No.:NACC-F03037).
13.	M/s Ruchi Oyster Mushroom, Gondia (MS) (Strain:AAI, Allahabad) (Accession No.: NACC-F03037).
14.	M/s Bharat Biocon Pvt. Ltd., New Delhi (Strain:AAI, Allahabad) (Accession No.: NACC-F03037).
15.	M/s Bisco Bio Sciences (P) Ltd., Secunderabad (AP) (Strain:AAI, Allahabad) (Accession No.: NACC-F03037).
16.	M/s Bharti Minerals Ltd., New Delhi (Strain: AAI, Allahabad) (Accession No.: NACC-F03037).
17.	M/s Advance Cropcare (India) Pvt. Ltd., Indore (Strain: AAI, Allahabad) (Accession No.: NACC-F03037).
18.	M/s Adiraj Agro Industries, Pune (MS) (Accession No. NACC – F- 03037).
19	M/s Sugway Agribiotech & Research Foundation, Yavatmal (MS)
20	M/s Kaveri Seed Company Ltd., Secunderabad (Strain No.: KSCL/Ma-59, Accession No. ITCC 7058).
21.	M/s Care-Pro Bioscience (P) Ltd., New Delhi (Strain No.: CPB/PSP-T26 Accession No. MTCC 5699).
22.	M/s Maa Bhagwati Biotech & Chemicals, Wardha (MS) (Accession No. NACC-F-03037).
23.	M/s Abhinav Biotech, Akola (MS) for grant of registration for indigenous manufacture of (AAI, Allahabad, Accession No.-NACC-F -03037) u/s
24.	M/s Nirmal Seeds Pvt. Ltd., Jalgaon (MS) for grant of registration for indigenous manufacture of (AAI, Allahabad, Accession No.-NACC-F -03037) u/s
25.	M/s Arya Biotech & Research Laboratory, Amravati (M.S) (CFU 1×108 /gm Min)
26.	M/s Jai Kisan Agro Indore, MP (CFU 1×108 /gm. Min)
27.	M/s Shri Ram Solvents Extraction Pvt. Ltd., Jaipur
28.	M/s Mahatma Phule Krishi Vidyapeeth (Strain obtained from AAI, Allahabad, UP, Accession No. NACC-F03037)
29.	M/s Govinda Agro Tech Ltd., Nagpur (Strain obtained from AAI, Allahabad, UP, Accession No. NACC-03037)
30.	M/s International Panacea Ltd., New Delhi (Strain isolated by own R&D of applicant from soils of Baghpat, UP, Accession No. ITCC-6895)

(continued)

Table 14.3 (continued)

31.	M/s Varsha Bioscience & Technology, Hyderabad
32.	M/s Agri Life, Secunderabad, AP
33.	M/s Sri Venkateshwara Chemicals, Bangalore
34.	M/s Biotech International Ltd, New Delhi
35.	M/s DVS BioLife
36.	M/s International Panacea Ltd.
37.	M/s T. Stanes & Co. Ltd.

Source: Directorate of Plant Protection, Quarantine & Storage, Department of Agriculture, Cooperation & Farmers Welfare, Ministry of Agriculture & Farmers Welfare, Government of India (http://ppqs.gov.in/divisions/cib-rc/biopesticide-registrant Accessed on: September 01, 2020)

Commercialization is stifled by the need for a robust battery of testing, practically confining to the shelf, many strains produced in India's publicly funded research institutions. The use of DNA bar-coding for precise recognition of the organisms to be used in formulation of biopesticides, prior to their production and field applications, would be a potential way forward.

India requires an effective federal action plan, realistic funding, and streamlined regulatory processes to register and market biopesticides to optimize the farming benefits, and farmers should also be trained to use biopesticides properly.

3 Ifs and Buts

The development of biopesticides is a high-risk endeavor that needs a large initial capital expenditure, from the selection stage to the procurement of potential strains for export, processing, storage, and delivery. However, the uncontrolled selling of poor-quality or spurious biopesticides and biopesticides tainted with chemical pesticides is the single biggest restriction to its development and growth.

Uncontrolled organic biological products not approved by the CIBRC and marketed under the label of organic bio-inputs licensed by the Agricultural and Processed Food Products Export Development Authority (APEDA) under the Ministry of Commerce and Fertilizers of India are contributing to the issue of false, contaminated, or poor-quality biopesticides. These "organic" products are not subject to any bio-safety and bio-efficacy trials required by the CIBRC and account for an estimated 65% of the country's overall biopesticide market value (Keswani et al. 2016).

The low durability of biopesticide is a serious problem for farmers. Biopesticides predominantly consist of living microbes. Their potency is dampened by temperature variations, humidity, and ultraviolet radiation penetration (Arora et al. 2010). Moreover, any contamination in field conditions seriously limits its efficacy (Alam and Alam 2000). Biopesticides are target-specific, only killing target pests and their close relatives (Senthil-Nathan 2015). On the other hand, chemical pesticides also kill beneficial insects, mammals, and birds (Mahmood et al. 2016).

Table 14.4 List of biopesticides (*Paleomyces*) registrants under CIBRC, India

1.	M/s Surya Bio Products, Godavari
2.	M/s Indian Institute of Horticultural Research, Bangalore
3.	M/s T. Stanes and Co. Ltd., Cobatore
4.	M/s RPC Balaji Crop Care Pvt. Ltd
5.	M/s Multiplex Biotech Pvt. Ltd., Bangalore
6.	M/s Nico Orgo Manures, Dakor, Gujarat
7.	M/s Shree Shiva BioTech
8.	M/s Bio-tech International Ltd., New Delhi
9.	M/s advance Cropcare (India) Pvt., Ltd., Indore
10.	M/s. Liebigs Agro Chem Pvt. Ltd., Kolkata
11.	M/s. Gujarat Eco Microbial Tech. Pvt. Ltd.
12.	M/s. Sri Biotech Laboratories India, Hyderabad
13.	M/s International Panaacea Ltd, New Delhi
14.	M/s Agriland Biotech Ltd., Baroda
15.	M/s Kanbiosys Pvt. Ltd., Pune
16.	M/s. Devi Biotech (P) Ltd., Madurai
17.	M/s Varsha Bioscience & Technology, Hyderabad
18.	M/s Gujarat Life Sciences (P) Ltd., Baroda
19	M/s. Chaitra Agri Organics Mysore
20	M/s Agri Life, Medak Dist. (A.P.)
21.	M/s Jyothiraditya Bio Solutions Ltd., Mysore
22.	M/s Viswa Mithra Bio Agro (P) Ltd., Guntur
23.	M/s Bharat Biocon Pvt. Ltd., New Delhi
24.	M/s Excel Crop Care Ltd., Mumbai
25.	M/s Romvijay Bio Tech Private Limited
26.	M/s Ganesh Bio-Control System, Shapar (Gujarat)
27.	M/s Jhass Agro Industries, Chittoor, AP for grant of registration for indigenous manufacture of (CFU count 2×106/gm. min.) (Strain No. IIHR-PL -2, Accession No. ITCC – 6887) u/s
28.	M/s Camson Bio Technologies Ltd., Bangalore (CFU count 2×106/gm. min.) (Strain No. IIHR-PL -2, Accession No. ITCC – 6887)
29.	M/s Agriva Agro Tech Kappallur, Madurai, Tamilnadu for grant of registration for indigenous manufacture of (CFU count 2×106/gm. min.) (Strain No.IIHR-PL -2, Accession No. ITCC – 6887)
30.	M/s Poabs Biotech Pvt. Ltd., Triuvala for grant of registration for indigenous manufacture of (CFU count 2×106/gm. min.) (Strain No.IIHR-PL -2, Accession No. ITCC – 6887) u/s
31.	M/s Kaveri Seed Company Ltd. Sikandrabad (CFU COUNT 2×106 /gm Min)
32.	M/s New Swadeshi Sugar Mills, Narkatiaganj (CFU COUNT 2×106 /gm Min) under Section 9(3B)
33.	M/s Amar Bio Tech, Bathinda, Punjab (CFU COUNT 2×106 /gm Min)
34.	M/s Mahadhan Nutrients & Seeds Corporation (CFU COUNT 2×106 /gm Min) under Section 9(3B)

Source: Directorate of Plant Protection, Quarantine & Storage, Department of Agriculture, Cooperation & Farmers Welfare, Ministry of Agriculture & Farmers Welfare, Government of India (http://ppqs.gov.in/divisions/cib-rc/biopesticide-registrant Accessed on: September 01, 2020)

Table 14.5 List of biopesticides (*T. viride*) registrants under CIBRC, India

1.	M/s Nafed Biofertilizer, Bharatpur (Rajasthan)
2.	M/s T. Stanes And Co. Ltd., Cobatore
3.	M/s Sun Agro Industries India, Delhi
4.	M/s Transgene Biotech Ltd
5.	M/s Krishi Rasayan Export Pvt. Ltd., New Delhi
6.	M/s Pest Control (India) Pvt. Ltd. Bangalore
7.	M/s Pushpanjali Agri Input Technologies, Kurnool (AP)
8.	M/s Agriland Biotech Limited
9.	M/s Romvijay Bio Tech Pvt. Ltd.
10.	M/s Multiplex Agricare Pvt. Ltd.
11.	M/s Margo Bio control Pvt Ltd.
12.	M/s K.N. Bio Science (India) Ltd.
13.	M/s Pragathi Bio Fertilizers
14.	M/s Varsha Bio Science and Technology, Hyderabad
15.	M/s Ecosense Labs. (I) Pvt. Ltd., Mumbai
16.	M/s Green Care Bio technologies
17.	M/s Kaveri Agri Tech Bio Division
18.	M/s Crop Health Products Ltd.
19	M/s. Prathista Industries Ltd.
20	M/s Mitcon Consultancy Ltd., Pune
21.	M/s Monarch Bio Fertilizers & Research Centre, Chennai
22.	M/s Maa Bhagwati Biotech & Chemical
23.	M/s Vidarbha Biotech Lab., Yavatmal
24.	M/s Ecophila Biotech, Ahmednagar
25.	M/s Om Agro Organic, Yavatmal
26.	M/s Universal Agro Bio-Tech., Nohar (Rajasthan)
27.	M/s Durva Biotech, Nagpur
28.	Yash Krishi Takniki Ewam Vigyan Kendra, Allahabad
29.	M/s K.N. Bio Tech Pvt Ltd., Hyderabad
30.	M/s Modern Biotech, Jalgaon (MS)
31.	M/s Amit Bio Tech. Kolkata
32.	M/s Jai Kisan Agro, Indore
33.	M/s M.S. Industries, Amravati (MS)
34.	M/s Ruchi Biochemicals, Gondia, Maharashtra
35.	M/s Sudarshan Chemical Industries Ltd., Pune
36.	M/s S&S Biotech, Nagpur
37.	M/s J.R. Biocontrol Laboratories, Yavatmal
38.	M/s Agri Life, Secunderabad
39.	M/s Nomin Agri Bio Pvt. Ltd., Pune
40.	M/s R.B. Herbal Agro Satana, Dist. Nashik (Maharashtra)
41.	M/s Pravara Agro Bio Tech, Sangamner (Maharashtra)
42.	M/s Esvin Advanced Technologies Ltd., Chennai
43.	M/s Jai Bio Tech Ind.

(continued)

Table 14.5 (continued)

44.	M/s. Microplex Biotech & Agrochem Pvt. Ltd., Wardha
45.	M/s. Institute of Natural Organic Agriculture, Pune
46.	M/s Sai Agrotech, Yavatmal
47.	M/s Nirmal Organics Biotech P. Ltd., Mumbai
48.	M/s Kalpavruksha Biosystems, Bangalore
49.	M/s Govinda Agro Tech Ltd., Nagpur (Maharashtra)
50.	M/s Maharashtra Research & Development Centre, Solapur (Maharashtra)
51.	M/s Vasundhara Agrotech, Aurangabad (Maharashtra).
52.	M/s Sai National Rural Development & Research Institute, Allahabad
53.	M/s Kan Biosys Pvt. Ltd., Pune
54.	M/s Anjali Biotech, Amravati
55.	M/s Deepa Farm Inputs (P) Ltd., Trivandrum
56.	M/s Shree Jee Biotech Agriculture & Equipment, Wardha
57.	M/s CAB Tech. Labs., Hyderabad
58.	M/s Bio Agro Ferticons, Pune
59.	M/s Liebigs Agro Chem Pvt. Ltd., Kolkata
60.	M/s Bio Chaudhary Agro Tech (I)
61.	M/s Margo Biocontrols Pvt. Ltd., Bangalore
62.	M/s Shri Ram Solvent Extractions Pvt. Ltd., Uttaranchal
63.	M/s Plantrich Chemicals & Fertilizers
64.	M/s KCP Sugar and Industries Corp. Ltd., Andhra Pradesh
65.	M/s Rajshree Sugars & Chemicals Ltd., Coimbatore
66.	M/s Super Pesticides & Agro (I) P. Ltd.
67.	M/s Soman Biofertilizers, Pune
68.	M/s RPC Biotech Industries, Kolkata
69.	M/s. ECI Agrochem Pvt. Ltd., Kolkata
70.	M/s Avishkar Biofarm, MS
71.	M/s Krishi Vigyan Kendra, Baramati
72.	M/s Arya Bio Technologies, Aurangabad
73.	M/s Honey Dew Biotechnologies, Krishna (AP)
74.	M/s Ajay Biotech India Ltd., Pune
75.	M/s Ellora Biotech & Agro Services, Aurangabad
76.	M/s Prakash Seeds Agro Division, Osmanabad(M.S.)
77.	M/s Directorate of Oilseed Research (ICAR), Hyderabad
78.	M/s Pruthvi Fertilizers Pvt. Ltd., Anand (Gujarat)
79.	M/s Poabs Enviro Tech Pvt. Ltd., Kerala
80.	M/s IPM Biocontrol Labs, Secunderabad
81.	M/s Shree Biotech & Research Inputs (India), Mandsaur
82.	M/s Krishna Industrial Corp. Ltd., Nidadavole
83.	M/s Ganesh Bio-Control System, Rajkot
84.	M/s Siddhant Biotech Lab, Amravati
85.	M/s Enpro Bio Sciences Pvt. Ltd., Nashik
86.	M/s Nath Krupa Bio-Control Lab, Nagpur

(continued)

Table 14.5 (continued)

87.	M/s Tari Bio-Tech, Thanjavur
88.	M/s Vidyas Biotech Laboratories, Nagpur
89.	M/s Insecticides India Ltd
90.	M/s Sun & Ocean Agro (India) Pvt. Ltd.
9.	M/s Vasundhara Bio-Products, Latur
92.	M/s International Panacea Ltd.
93.	M/s Pandian Biosol, Mathura
94.	M/s Tripti Biotech, Balaghat
95.	M/s National Bio-control Laboratories
96.	M/s Sri Aurobindo Institute of Rural Development
97.	M/s Chirayu Biotech, Pune
98.	M/Gujarat Life Science Pvt. Ltd
99.	M/s Juna Life Sciences Pvt. Ltd
100.	M/s Maharashtra Insecticides Ltd.
101.	M/s Arvind Biotech, Buldhana (MS)
102.	M/s Agri Gold Organics Pvt. Ltd.
103.	M/s Sri Laxmi Narayan Chemical & Fertilizers Pvt. Ltd.
104.	M/s Ashwamedh Agritech & Farm Solutions
105.	M/s Krishna Biotech Fertilizers
106.	M/s State Bio-fertilizer Quality Control Laboratory
107.	M/s Grace Bio-care Pvt. Ltd.
108.	M/s Camson Bio Technologies
109.	M/s Chaitra Agri Organics, Mysore
200.	M/s Surya Bio Products, Eluru, AP
201.	M/s Abhinav Biotech. Akola
202.	M/s Shree Shiva Bio-tech Pudukkottai
203.	M/s Lila Agrotech
204.	M/s Biocontrols, Hyderabad
205.	M/s Ishwar Agro, Dhule
206.	M/s Biotech International Ltd., New Delhi
207.	M/s Pramukh Agri Clinic, Gujarat
208.	M/s Sujay Biotech Pvt. Ltd., Andhra Pradesh
209.	M/s Kundu Agro Chem (P) Ltd. Kolkata
210.	M/s Indo Sikk Bio Agro Industries, Gangtok
211.	M/s Aastha Biotech Pvt. Ltd., Kolkata
212.	M/s Sainath Agro Vet Ind. Ltd., Ahmednagar
213.	M/s Modi Agro Products, Bhopal
214.	M/s Shilabati Horticulture & Agriculture, Kolkata
215.	M/s Jai Shree Rasayan Udyog Ltd., New Delhi
216.	M/s Advance Crop care (I) Pvt. Ltd.
217.	M/s Kilpest India Ltd., Bhopal
218.	M/s Nico Orgo Manures, Gujarat
219.	M/s Sivashakthi Bio Planttec Ltd., Hyderabad

(continued)

Table 14.5 (continued)

220.	M/s Green Valley Bio-tech, Ujjain
221.	M/s Maharashtra State Seeds Corporation Ltd.
222.	M/s Krishi Vigyan Kendra, Jalna (MS)
223.	M/s Kanan Devan Hills Plantations Company Pvt. Ltd., Kerala
224.	M/s Ecosense Labs (I) Pvt. Ltd., Mumbai
225.	M/s Micro Life Bio Science, Alwar
226.	M/s Greenfert Agro Research Centre (P) Ltd., Kottayam, Kerala
227.	Central Research Institute for Dryland Agriculture, Hyderabad
228.	M/s. Shriram Biotech & Biofertilizers, Dhule
229.	M/s Gujarat Eco Microbial Technologies Pvt. Ltd., Vadodara
230.	M/s Krishi Vigyan Kendra, Amravati (M.S.)
231.	M/s Panacea Agro Bio-tech, Paratwada, Dist. Amrawati
232.	M/s Universal Agro Tech, Kolkata
233.	M/s Probitek Biotech, Jalna (MS)
234.	M/s Agro Pesticides, Itarsi (M.P.)
235.	M/s. Surbhi Agro Bio Tech., Amravati (MS)
236.	M/s Jeypee Biotechs Virudhunagar, Tamil Nadu
237.	M/s. Krishi Vishwa Bio-tech Ltd., Virgaon, Akola, Maharashtra
238.	M/s. Maharashtra State Biocontrol Lab., Pune
239.	M/s. Care-Pro Bioscience (P) Ltd., New Delhi
240.	M/s. Nova Agri Tech Pvt. Ltd., Secunderabad
241.	M/s. Neo Gene Agri Input Pvt. Ltd., Secunderabad
242.	M/s. Biosys Agrotech Pvt. Ltd., Mandsaur
243.	M/s. Shree Jee Biotech Agri. & Equipments, Wardha
244.	M/s High Range Fertilizer Bio-Tech & Research Centre, Puliyanmala, Kerala
245.	M/s Southern Fertilizers and Chemicals, Kottayam
246.	M/s. Sunshiv Biotech, South 24 Parganas (W.B)
247.	M/s. Pre Agro Pvt. Ltd., Cuddalore (TN)
248.	M/s. Nafed Biofertilizers, Indore
249.	M/s. Green Valley Bio-tech Pvt. Ltd., Ahmedabad
250.	M/s Gramakarshaka Fertilizers Pvt. Ltd., Kollam
251.	M/s. Kanha Herbs Kotdwar, Uttarkhand
252.	M/s Agri Bio Care Kottayam, Kerala
253.	M/s Hindustan Bioenergy Ltd. Lucknow
254.	M/s Rovor Bio Technologies (P) Ltd., Vijayawada
255.	M/s Bisco Bio Science (P) Ltd., Secunderabad
256.	M/s Mahatma Phule Krishi Vidyapeeth
257.	M/s Deepa Farm Inputs (P) Ltd., Secunderabad
258.	M/s Neesa Agritech Pvt. Ltd., Ahmedabad
259.	M/s. Kerala Agricultural University, Thiruvananthapuram
260.	M/s SRT Agro Science (P) Ltd., Durg
261.	M/s United Bio Fertilizers, Jaipur
262.	M/s Devi Biotech Pvt. Ltd., Madurai, T.N.

(continued)

Table 14.5 (continued)

263.	M/s M.D. Biocoals (Agri Division), Sirsa
264.	M/s Tender Sips, Indore (M.P) T.N.
265.	M/s Bonageri Cropscience Pvt. Ltd. Dharwad
266.	M/s. Travancore Organics Fertilizers Co. (P) Ltd., Kerala
267.	M/s. Vasant Biotech Pusad, Yavatmal
268.	M/s Jayco Chemicals India Ltd., Hapur
269.	M/s Unique Bio-tech Ltd., Hyderabad
270.	M/s Ankur Chemicals India (INN), Muzaffarnagar
271.	M/s Maa Narsai Biotech & Chemicals, Wardha
272.	M/s Aviral Bio-tech & Fertilizer Pvt. Ltd., Bhopal
273.	M/s Gayatri Insecticides Ltd., Indore (MP)
274.	M/s Biosynthetics, Jalgaon (MS)
275.	M/s Kaveri Seed Company Ltd., Secunderabad (AP)
276.	M/s Tropical Agrosystem (India) Pvt. Ltd, Chennai
277.	M/s Krishi Vigyan Kendra, Pravara Institute of Research and Education in Natural and Social Sciences (PIRENS)
278.	M/s Multiplex Biotech Pvt. Ltd., Bangalore
279.	M/s Jyotiraditya bio solutions Ltd., Mysore
280.	M/s Gujarat Chemicals and Fertilizers Trading Company, Baroda
281.	M/s Ambika Biotech & Agro Services, Mandsaur
282.	M/s Jai Kisan Agro, Indore
283.	M/s Bios Laboratories, Hyderabad
284.	M/s Green Max Agrotech, Cobatore
285.	M/s Bacto Power India Pvt. Ltd., Cobatore
286.	M/s Prabhat Fertilizers and Chemicals, Karnal
287.	M/s Krishi Bio-Products & Research Pvt. Ltd., Indore
288.	M/s R.G.P. Cropscience, Khargone
289.	M/s Halo Kem, Jalgaon
290.	M/s Datta Gro-tech & Equipment, Wardha
291.	M/s Safe Crop Science Pvt. Ltd., Indore
292.	M/s Global Agri Care Industry, Indore
293.	M/s Bharati Minerals Ltd., Delhi
294.	M/s Shri Ram Solvent Extractions Pvt. Ltd., Jaspur
295.	M/s Sun Agro Bio System Pvt. Ltd., Chennai (Strain: TNAU, Accession No. ITCC 6914)
296.	M/s Samridhi Bioculture Pvt. Ltd., Indore (Strain: TNAU, Accession No. ITCC 6914)
297	M/s Fertilizers India, Kerala (Strain: KAU) (Accession No.: MTCC- 5694)
298.	M/s Shree Shiva Bio-Tech, Pudukkottai (TN) (Accession No. ITCC No.6914).
299.	M/s Shiv Shakti Bio-chem, Muzaffarnagar (UP) (Accession No. ITCC No.6914).
300.	M/s Nirmal Seeds Pvt. Ltd., Jalgaon (MS) (Accession No. ITCC No.6914).
310.	M/s Jai Biotech Industries, Nashik (MS) (Accession No. ITCC No.6914).
302.	M/s Microplex (India), Wardha (MS) (Accession No. ITCC No.6914).
303.	M/s Microplex Biotech & Agrochem Pvt. Ltd. (Accession No. ITCC No.6914).
304.	M/s Pravara Agro Bio-Tech, Ahmednagar (MS) (Accession No. ITCC No.6914).

(continued)

Table 14.5 (continued)

305.	M/s R.B. Herbal Agro, Nasik (MS) (Accession No. ITCC No.6914).
306.	M/s Ruchi Biochemicals, Goregaon, Mumbai (MS) (Accession No. ITCC No.6914).
307.	M/s Hari Agrotech, Yavatmal (MS) (Stain: TNAU, Cobatore, Tv-1, Accession No. ITCC 6914).
308.	M/s Kalpavruksha Biosystems, Bangalore (Strain: TNAU Cobatore, Tv1, Accession No.: ITCC No. 6914).
309.	M/s Phalada Agro Research foundations Pvt. Ltd. (Strain TNAU Coimbatore, Accession No. ITCC – 6914).
310.	M/s Adiraj Agro Industries, Pune (MS) for grant of registration for indigenous manufacture of (CFU count 2×106/gm. min.) (Strain No. TNAU, Accession No. ITCC – 6914) u/s
311.	M/s Balaji Crop Care Pvt. Ltd., RR dist.(AP) for grant of registration for indigenous manufacture of (Strain No.BHU, Varanasi, Accession No.NACC-F-02976) u/s
312.	M/s Yash Krishi Takniki Evam Vigyan Kendra (CFU COUNT 2×106/gm min.)
313.	M/s Kan Biosys Pvt. Ltd., Pune
314.	M/s Arya Biotech and Research Laboratory, MS
315.	M/s R. B. Herbal Agro, Nasik
316.	M/s INORA, Bavdhan Khurd, Pune (Strain Designation: TNAU, Accession No. ITCC 6914).
317.	M/s Govinda AgroTech Ltd. Nagpur (M.S)
318.	M/s Institute of Plant Biotechnology, Nashik (Strain Designation: TNAU, Accession No. ITCC 6914)
319.	M/s Institute of Plant Biotechnology, Nashik for grant of registration for indigenous Biopesticide manufacturing of u/s 9(3) (Strain Designation: TNAU, Accession No. ITCC 6914).
320.	M/s Indian Institute of Horticulture Research, Bangalore (Strain Designation: IIHR TV-5, Accession No.ITCC6889).
321.	M/s KVK Akola, Akola Maharashtra (Strain Designation: IIHR TV-5, Accession No. ITCC6889).
322.	M/s Harit Bio-control Lab Yavatmal (MS) (Strain Designation: TNAU, Accession No. ITCC 6914)
323.	M/s Krishna Industries Corporation Ltd., West Godavari, AP (Strain Designation: TNAU, Accession No. ITCC 6914)
324.	M/s Sai Agrotech Yavatmal (MS)
325.	M/s Peak Chemical Industries Ltd., Jalpaiguri, WB (Strain Designation: IIHR-TV-5, Accession No. ITCC No.6889).
326.	M/s Agri Gold Organics Pvt. Ltd., Vijayawada (Strain Designation: IIHR-TH-5, Accession No. ITCC No.6889).
327.	M/s Jai Biotech & Research Centre, Jaipur (Strain Designation: IIHR-TV-5, Accession No. ITCC No.6889).
328.	M/s Rajshree Sugars and Chemicals Ltd., Tamil Nadu (Strain Designation: IIHR-TV-5, Accession No. ITCC No.6889).
329.	M/s Director of horticulture (Plant Nutrition), Govt. of Karnataka, Karnataka (Strain Designation: IIHR-TV-5, Accession No. ITCC No.6889).
330.	M/s Sun Pesticides Pvt. Ltd.(Strain Designation: IIHR-TV-5, Accession No. ITCC-No. 6889).

(continued)

Table 14.5 (continued)

331.	M/s Excel Bio Tech (P) Ltd., Kolkata (Strain Designation: IIHR-TV-5, Accession No. ITCC-No. 6889).
332.	M/s Varsha Bioscience and Technology Hyderabad
333.	M/s Kaveri Organic Agri Inputs (Pvt.) Ltd. (Strain designation: IIHR TV – 5, Accession No. ITCC 6889)
334.	M/s Krishi Biosyus (Strain designation: IIHR TV – 5, Accession No. ITCC 6889)
335.	M/s Malabar Bio-control (Strain designation: IIHR TV – 5, Accession No. ITCC 6889)
336.	M/s Agro Biotech Research Centre Ltd. (Strain designation: IIHR, Bangalore, IIHR- TV – 5, Accession No. ITCC 6889)
337.	M/s Nano Agro Sciences Co-operative Society Ltd. (Strain designation: IIHR, Bangalore, IIHR- TV – 5, Accession No. ITCC 6889)
338.	M/s Bio Pest Control Industries (Strain designation: IIHR, Bangalore, IIHR- TV –5, Accession No. ITCC 6889)
339.	M/s S & S Biotech (Strain designation: IIHR, Bangalore, IIHR- TV – 5, Accession No. ITCC 6889)
340.	M/s New Swadeshi Sugar Mills (Strain designation: IIHR, Bangalore, IIHR- TV – 5, Accession No. ITCC 6889)
341.	M/s Green Earth Agrobiotech (Strain designation: IIHR, Bangalore, IIHR- TV – 5, Accession No. ITCC 6889)
342.	M/s Amar Biotech (Strain designation: IIHR, Bangalore, IIHR- TV – 5, Accession No. ITCC 6889)
343.	M/s Kaveri Seed Co. Ltd. (Strain designation: IIHR, Bangalore, IIHR- TV – 5, Accession No. ITCC 6889)
344.	M/s Dte. of Research, MPUA&T (Strain designation: IIHR, Bangalore, IIHR- TV – 5, Accession No. ITCC 6889)
345.	M/s Tripureswari Biotech Pvt. Ltd. (Strain designation: IIHR, Bangalore, IIHR- TV – 5, Accession No. ITCC 6889)
346.	M/s Microtech Agro Chemicals (Strain designation: IIHR, Bangalore, IIHR- TV – 5, Accession No. ITCC 6889)
347.	M/s Chaitra Agri Organics and Chemicals (Strain designation: IIHR- TV – 5, Accession No. ITCC 6889)
348.	M/s Agriva Agrotech (Strain designation: IIHR- TV – 5, Accession No. ITCC 6889)
349.	M/s Patanjali Research Institute Pvt. Ltd. (Strain designation: IIHR- TV – 5, Accession No. ITCC 6889)
350.	M/s Agri Biocare (Strain designation: IIHR- TV – 5, Accession No. ITCC 6889)
351.	M/s Romvijay Biotech Pvt. Ltd. (Strain designation: IIHR- TV – 5, Accession No. ITCC 6889)
352.	M/s Som Phytopharma India Ltd. (Strain designation: IIHR- TV – 5, Accession No. ITCC 6889)
353.	M/s Liebigs Agrochem Pvt. Ltd. (Strain designation: IIHR- TV – 5, Accession No. ITCC 6889)
354.	M/s Mahadhan Nutrients and Seeds Corporation (Strain designation: IIHR- TV – 5, Accession No. ITCC 6889)

(continued)

Table 14.5 (continued)

355.	M/s Poabs Biotech Pvt. Ltd., Kerala (Strain designation: IIHR- TV – 5, Accession No. ITCC 6889)
356.	M/s Prathibha Biotech (Strain designation: IIHR- TV – 5, Accession No. ITCC 6889).
357.	M/s Advance Biocare Inputs (Strain designation: IIHR- TV – 5, Accession No. ITCC 6889).
358.	M/s Department of Plant Pathology, Assam Agriculture University
359.	M/s Antecedent Pabulum Inc. Bhatinda
360.	M/s Oshnic Crop Science Ltd.
361.	M/s Tagros Chemicals India Ltd.
362.	M/s Krishi Vigyan Kendra, Amravati
363.	M/s Rohini Bio-Agents, Karnataka
364.	M/s East Cost Biotech Project
365.	M/s Sun Plant Agro Products Pvt. Ltd.
366.	M/s Surya Bio Products
367.	M/s Criyagen Agri & Biotech Pvt. Ltd. (Strain: IIHR, Bangalore, IIHR-Tv-5, Accession No.-ITCC6889).
368.	M/s Swastika Chemicals & Fertilizers Pvt. Ltd. (Strain: IIHR, Bangalore, IIHR-Tv-5, Accession No.-ITCC6889).
369.	M/s Ambic Organic, Surat (Strain: IIHR, Bangalore, IIHR-Tv-5, Accession No.-ITCC6889).
370.	M/s Uttam Chemical Industries (Strain: IIHR, Bangalore, IIHR-Tv-5, Accession No.-ITCC6889)
371.	M/s National Biochemical Laboratories (Strain: IIHR, Bangalore, IIHR-Tv-5, Accession No.-ITCC6889).
372.	M/s NCS Crop Science Pvt. Ltd. (Strain: IIHR, Bangalore, IIHR-Tv-5, Accession No.-ITCC6889).
373.	M/s Prathibha Biotech, Hyderabad
374.	M/s Bio Control Laboratory, Lucknow, UP (Strain Designation: IIHR-TH-2, obtained from IIHR, Bangalore, Accession No. ITCC No.6888)
375.	M/s Chhattisgarh Agro Biotech Lab., Raipur
376.	M/s Jai Biotech & Research Centre, Jaipur
377.	M/s Pest control India Ltd.,
378.	M/s Kan Biosys Pvt. Ltd., Pune
379.	M/s Biotech International Ltd., New Delhi
380.	M/s Margo Biocontrol (P) Ltd., Bangalore
381.	M/s T. Stanes & Company Ltd., Cotore
382.	M/s Pest control India Ltd.
383.	M/s International Panacea Ltd., New Delhi
384.	M/s Indore Biotech Inputs & Research Pvt. Ltd., Indore
385.	M/s Sheer Agro Bio-Fertilizer Creative Organization, Purba Medinipur (WB)
386.	M/s Central Bio Tech, Nagpur

Source: Directorate of Plant Protection, Quarantine & Storage, Department of Agriculture, Cooperation & Farmers Welfare, Ministry of Agriculture & Farmers Welfare, Government of India (http://ppqs.gov.in/divisions/cib-rc/biopesticide-registrant Accessed on: September 01, 2020)

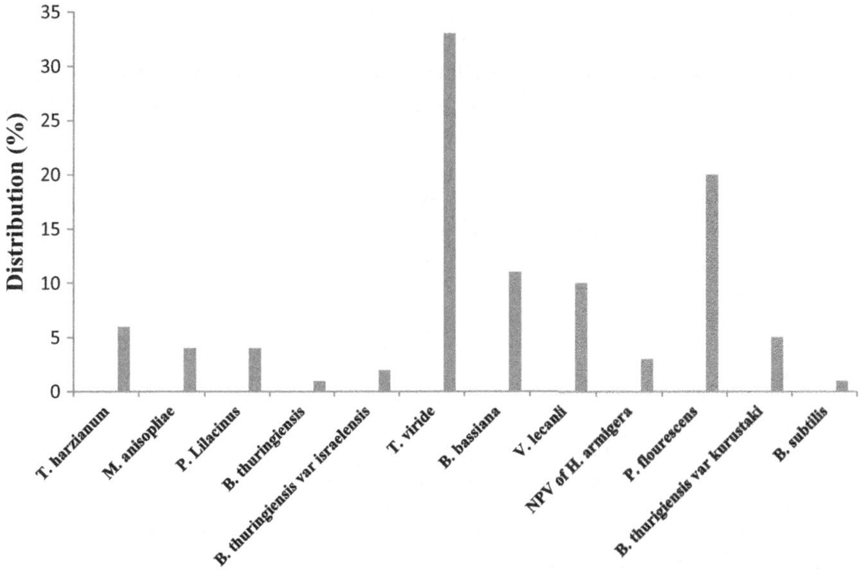

Fig. 14.1 Industrial distribution of microbial biopesticides. (Source: Directorate of Plant Protection, Quarantine & Storage, Department of Agriculture, Cooperation & Farmers Welfare, Ministry of Agriculture & Farmers Welfare, Government of India (http://ppqs.gov.in/divisions/cib-rc/biopesticide-registrant Accessed on: September 01, 2020)

When it comes to the extensive use of biopesticides, there are several "ifs and buts" since they are target-specific, they cannot be used against a wide variety of species, and that is a limitation.

4 Further Work Required

The ongoing Covid-19 pandemic has adversely affected the agricultural value chain in India. The government of India issued state-wise guidelines for the movement of agricultural products and farming activities. In spite of all these measures, major issues are not fully resolved. The nationwide lockdown came at an unfortunate time and create bottlenecks like disruption in the harvesting and marking crops, shortage of workers, blockades in the transportation sector, closure of *mandis*, and shutdowns in the retail agricultural markets that led to a disrupted supply chain and fall in the farm prices across the country.

To anticipate biopesticides, complete substitution of chemical pesticides will be unrealistic. These microbial substances are used along with organic substances such as farmyard compost or manure. The major issue is that agricultural mechanization has limited or eliminated cattles in many regions, thus generally decreasing the potential for biomass use.

Scientists have emphasized that biopesticides or biofertilizers should not follow the "one-size-fits-all" approach. Local data and knowledge, awareness of pest incidence trends, as well as non-pesticidal activities such as a bonfire during the pest's egg-laying season are needed for biopesticides development. Further studies and trials on crop-specific and area-specific formulations are required to optimize the use of biopesticides.

In India, there is a huge scope for biopesticide marketing. However, to succeed, it requires a re-organization of agricultural research and information systems, enhanced by local data and local pest scouts.

References

Alam, G., & Alam, G. (2000). *A study of biopesticides and biofertilisers in Haryana, India*: International Institute for Environment and Development.

Arora, N. K., et al. (2010). Plant growth promoting rhizobacteria: constraints in bioformulation, commercialization, and future strategies *Plant growth and health promoting bacteria* (pp. 97–116): Springer.

Darbro, J. M., & Thomas, M. B. (2009). Spore persistence and likelihood of aeroallergenicity of entomopathogenic fungi used for mosquito control. *The American journal of tropical medicine and hygiene, 80*(6), 992–997.

Keswani, C. (2015). *Proteomic studies of thermotolerant strain of trichoderma spp*. Banaras Hindu University.

Keswani, C., et al. (2019). Regulatory barriers to Agricultural Research commercialization: A case study of biopesticides in India. *Rhizosphere, 11*, 100155.

Keswani, C., et al. (2014). Unraveling the efficient applications of secondary metabolites of various Trichoderma spp. *Applied microbiology and biotechnology, 98*(2), 533–544.

Keswani, C., et al. (2016). Synthesis of policy support, quality control, and regulatory management of biopesticides in sustainable agriculture *Agriculturally Important Microorganisms* (pp. 3–12): Springer.

Mahmood, I., et al. (2016). Effects of pesticides on environment *Plant, soil and microbes* (pp. 253–269): Springer.

Ram, R. M., et al. (2018). Biocontrol technology: eco-friendly approaches for sustainable agriculture *Omics technologies and bio-engineering* (pp. 177–190): Elsevier.

Senthil-Nathan, S. (2015). A review of biopesticides and their mode of action against insect pests *Environmental sustainability* (pp. 49–63): Springer.

Singh, H.B, Keswani, C., Reddy, M.S., Royano, E.S., García-Estrada, C. (Eds.). (2019a). Secondary Metabolites of Plant Growth Promoting Rhizomicroorganisms: Discovery and Applications. Springer-Nature, Singapore. 392 pages, ISBN- 978-981-13-5861-6.

Singh, H.B., Keswani, C., Singh, S.P. (Eds.). (2019b) Intellectual Property Issues in Microbiology. Springer-Nature, Singapore. 425 pages, ISBN- 978-981-13-7465-4.

Singh, H.B, Sarma, B.K., Keswani, C. (Eds.). (2017) Advances in PGPR Research CABI, Wallingford, UK. 408 pages, ISBN-9781786390325.

Sinha, B., & Biswas, I. (2008). Potential of Bio-pesticides in Indian Agriculture Vis-à-vis Rural Development. *India Science and Technology*.

Vílchez, J. I., et al. (2017). Biosafety Evaluation: A Necessary Process Ensuring the Equitable Beneficial Effects of PGPR. *Advances in PGPR Research*, 50.

Chapter 15
Energy Solutions for Agricultural Machinery: From the Oil Era Towards a Sustainable Bioeconomy

Per Frankelius and Mattias Lindahl

Abstract Agriculture is at the heart of the bioeconomy. One central factor in agriculture is energy, which is needed for processes such as powering field machines like tractors and combine harvesters but also irrigation pumps, stable fans and grain drying systems. Modern implements like seeders, planters and manure spreaders need energy both for their movement (traction) and onboard equipment like fans, pumps, computer power and hydraulics. The study focuses on energy solutions for field machines in farming. These machines do their job in the countryside, far away from the energy infrastructure found in cities, and that is a challenge. One main part of the analysis in this chapter is a historical discussion on energy provision. Another part is an overview of present initiatives, including both those implemented in practice (like biogas and battery concepts) and those in trials or only discussed among experts in the industry (like fuel cell or hydrogen concepts). Derived from this analysis are some future visions towards a (1) fossil-free, (2) cost-effective and (3) sufficient energy system for farming field machines. In addition to these three dimensions, also discussed, for example, is the need for local production (small-scale circular systems), the weather independency aspect (for energy production) and autonomous vs. traditional machine systems.

Keywords Circular economy · Agriculture · Farming · Tractors · Biodiesel · Fossil-freedom · Fuel cells · Biogas · Ethanol · Robots

P. Frankelius (✉)
Agtech 2030, Linköping University, Linköping, Sweden
e-mail: per.frankelius@liu.se

M. Lindahl
Mistra REES, Linköping University, Linköping, Sweden

E. Koukios, A. Sacio-Szymańska (eds.), *Bio#Futures*,
https://doi.org/10.1007/978-3-030-64969-2_15

1 Introduction

1.1 Background

Despite their evidently different assumptions and operationalization strategies, the concepts of the bioeconomy, the circular economy and the green economy are joined by the common ideal to reconcile economic, environmental and social goals (D'Amato et al. 2017) and are currently considered in policy making as key sustainability avenues. Of these three, this chapter is focused on the bioeconomy, which is concentrated on local processes in terms of biosecurity and rural policies.

The term bioeconomy was probably first used in 1992 by Bernadine Healy when, while serving as Director of the National Institutes of Health in the USA, she speculated about the future of biotechnology (Nerlich 2015). Biotechnology is a field with roots in agriculture, for example, the art of using microorganisms in the process of brewing beer (Frankelius 2009). Later, the meaning of bioeconomy became broader; it is referred to today by many as the economic activities derived from biological resources. In this context, the agriculture and forest industries play important roles in providing bio-based substitutes for non-renewables (Roos and Stendahl 2015). The current understanding of bioeconomy is based on the idea that industrial inputs (e.g. material, chemicals and energy) should be derived from renewable biological resources, with research and innovation enabling the transformational process (Kleinschmit et al. 2014; Bugge et al. 2016), something in line with the ideas behind the biological cycle in a circular economy (Ellen MacArthur Foundation 2013).

This implies that agriculture is at the very heart of the bioeconomy because it conducts primary production by means of using soil in combination with photosynthesis. Agriculture contributes to society not only by meeting the world's food needs; it also produces fossil-free fibers and fuels. Moreover, if managed in a sustainable way, and not as sadly described by Carson (1962) in her book *Silent Spring*, agriculture can be an enabler for the important biodiversity of, for example, crops, insects and birds – which can be essential for sustainable agriculture. At the same time, agriculture is highly energy dependent, which is a problem not only from an economic perspective but also from one of climate care. Activities such as soil cultivation, the transport of bulk materials, in-farm handling of animals and crops and the drying of agricultural products are example areas where significant energy is needed.

1.2 The Productivity Revolution

Historically, the energy needed for field work was provided by men and animals, not least, oxen and horses. But, by the end of the nineteenth century, machines took over. This started a productivity revolution in agriculture that also paved the way for

Fig. 15.1 One towed 5-foot header combine harvester (from 1954) that fills grain into sacks compared to the self-propelled X9 combine harvester with a 45-foot header and a 16,200-liter grain tank (from 2020)

releasing the workforce in favour of the growing industrial sector. Some comparisons can be interesting just to understand this revolutionary process. During the eighteenth century, up to 30 person-days per tonne of harvested grain were used in harvesting and threshing work (Myrdal 1993). After the entry of the mowing reaper and threshing machine, the person-days needed dropped to about seven.

With the introduction of modern combine harvesters, things got faster. In 2020, John Deere demonstrated a combine called the X9, then still not launched on the market but with a capacity of 100 tonnes per hour (Stolpe-Nordin 2020). This means less than 2 minutes per tonne, but the energy needed for this mega combine corresponds to 700 hp (by means of a diesel engine); see Fig. 15.1.

The productivity revolution, the shift from manual labour to machinery and constantly larger machines, has led to an increased energy consumption, mainly related to fossil fuels. According to Eurostat, the data on energy consumption in agriculture is not very reliable and mainly reflect consumption for engines used for agriculture-related transportation. The energy consumption by agriculture within the EU in 2017 made up 2.8% of the overall energy consumption and decreased between 1997 and 2017 by 15% (Eurostat 2019). Oil and petroleum products were the main fuel type and contributed to 53% of total energy consumption by agriculture in 2017, but the share of electricity and renewables and biofuels had increased since 1997.

1.3 Why Is Agriculture So Energy Demanding?

There are many energy consumption activities in agriculture, for example land preparation, cultivation, irrigation, harvesting, threshing, grain drying and fan systems in animal stables. This chapter, however, focuses on energy needs in field work. We can begin by ascertaining that agriculture field work is very energy demanding. But why? Some reasons are compiled in Table 15.1.

One can add that heavy machinery leads to more soil compaction, which leads to more soil resistance and, therefore, more energy needed to overcome it. Moreover,

Table 15.1 The ten factors behind the high energy need in modern agricultural field work

	Factor	Explanation
1	Scale	Modern farm production is a large-scale activity. For example, farmers use 35-m-broad sprayers or 45-foot combines. Although the average farm in the EU has only 15 hectares of land, the majority of the land area is managed by large farms (defined as more than 100 hectares). Today, farms of well over 1000 hectares are not uncommon. The structural change from smaller to bigger farms also means more energy-demanding machines
2	Workforce cost	Because labour cost is high, there is a need for time-saving and, therefore, high-speed processes, which demand efficiency
3	Soil resistance	Soil is a heavy material, and many activities, for example, weed hoeing, are about dragging tools in the soil
4	Weather and "short time windows"	Agriculture is weather dependent, and timing is everything. Therefore, much work must be done in a short time frame at "weather windows", meaning a need for high power
5	Image	Attracting people to farming has partly been related to "cool machines", and in some people's eyes, that means "big machines" (no matter if those big machines are needed)
6	The imitation game	It seems that agricultural machinery companies compete by making faster, bigger and stronger machines, which drives a greater need for energy
7	Lackof energy efficiency	Optimal would be 100% energy efficiency in fuels and other energy categories, but losses are a matter of natural laws, so more energy is needed than what is actually transformed into target actions
8	Population increase	In 1800, the world population was 0.98 billion, in 1900, the figure was 1.6 billion, and in 2000 it was 6 billion. In 2050, the UN projects it will be 9.8 billion. In combination with changing food habits, this means more food demand and more energy needed in farming
9	Organic trend	Transforming from conventional (chemicals and fertilizers) to organic farming for meeting political and consumer trends also means more physical actions regarding, for example, weed management, and that means a greater need for energy
10	Demand for fossil-free fibres and fuels	The will to phase out fossil products means more demand for bio-based products like biofuels and bio-fibres, and this means more farming leading to a greater need for energy

soil compaction means decreased soil health, which brings harvests down, and farmers try to compensate for this by still more activities in the field.

1.4 The Pursuit of Fossil-Free Agriculture: The Political Perspective

Diesel is still the main energy source for field work in agriculture, but this may change in the near future. The search for fossil-free energy in response to not least climate problems has, as everyone has seen, intensified, not least through laws, regulations and supranational agreements, for example, the Kyoto Protocol and the

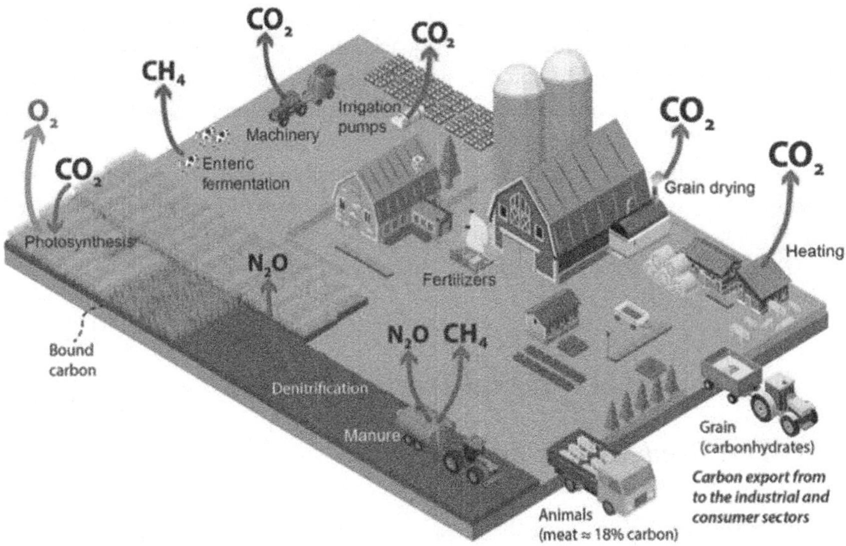

Fig. 15.2 Farming activities and climate effects (based on Frankelius 2020)

Paris Agreement. While the Kyoto Protocol was about reducing annual global carbon emissions to 1990 levels by 2020, the Paris Agreement set the goal of limiting global warming of greenhouse gas emissions to 1.5 °C by the end of this century. The EU also set similar goals in line with the UN framework and has put in place legislation to reduce CO_2 emissions by at least 40% by 2030. This is a part of the EU's 2030 climate and energy framework and contribution to the Paris Agreement.

In agriculture, there are several sources of greenhouse gas emissions. Most of these different kinds are negative, but some are actually positive, not least oxygen production and the "opposite-to-emission" carbon dioxide uptake by plants (Frankelius 2020); see Fig. 15.2. In this chapter, we only focus on machines and the negative impact of them as well as how to decrease this negative impact.

1.5 *The Pursuit of Fossil-Free Agriculture: The Customer Perspective*

Many actors in the world, not least in Europe, are conducting processes and projects to convert agriculture to fossil-free activities. In Spain, for example, the MASLOWATEN project commenced in 2017 to run high-power photovoltaic irrigation systems for productive agriculture irrigation consuming zero conventional electricity and 30% less water (Lorenzo et al. 2018). Another example is the European Investment Bank, which decided in November 2019 to stop financing fossil fuel projects by the end of 2021 (Ekblom 2019).

Individual farmers are also acting on the stage, but it is not as easy for them to make the necessary investments. Nevertheless, there are a lot of proactive farmers who have already started to convert their farms to fossil-free ones. Biogas plants, solar power and wind power are examples of local energy production, while heating boilers used in, for example, the drying of grain is increasingly done using fossil-free energy, such as straw pellets. There are several driving forces behind an individual farmer wanting to switch to fossil-free agriculture. One is simply a climate commitment. But farmers also have high energy costs and are looking for ways to reduce them. They dream of producing their own energy.

2 Development of Different Traction Energies

2.1 The Steam Engine Era

Petroleum-based fuels became the most common energy source in agriculture after the era of oxen and horses. However, steam engines first dominated traction during a period up to the beginning of the twentieth century. In fact, the world's first high-pressure steam engine was invented by Englishman Richard Trevithick in 1802 (Hodge 1973), and it made its way into society through the Hayle Foundry, which built it in 1811 for the farmer Christopher Hawkins. He wanted it to run a (stationary) threshing machine on his farm in Probus, Cornwall, and it came into operation in 1812. Thus, the modern steam engine became an innovation thanks to agriculture, not train transports or industrial use, which many may assume.

2.2 The Gasoline and Kerosene Era

Following the steam era, the mainstream fuel became gasoline (petrol), which dominated from 1900 to the 1950s. The world's first true tractor, introduced by John Froelich in 1892, ran on gasoline. Driven by a lower price, kerosene did enter the fuel market. One of the first kerosene tractors was made by Hart-Parr in 1904. However, one needed to start it on gasoline (ASME 1996). Another was the Rumely Oil Pull tractor, built in 1908 and introduced in 1910, which ran on kerosene rather than raw oil despite its name, and for which the chilling medium was oil, not water.

2.3 The Struggle to Make Use of the Cheaper Raw Oil

There were some early initiatives to produce engines that could be driven by (cheaper) raw oil (also called crude oil). Herbert Akroyd Stuart invented the hot-bulb engine, or heavy oil engine, with Charles Richard Binney and in connection with the Richard Hornsby and Sons company. The first prototype appeared in 1886, and the patent was filed in 1890 with the title "Improvements in Engines Operated by the Explosion of Mixtures of Combustible Vapour or Gas and Air", and this engine was stationary.

In 1896, Richard Hornsby and Herbert Akroyd Stuart filed the Hornsby-Akroyd Patent "Safety Oil Traction Engine" (Ransome-Wallis 2001). In 1897, this tractor was bought by Mr. Locke-King, which was the first recorded sale of a tractor in Britain. In the same year, the tractor was presented at the Smithfield Show and at the Royal Agricultural Show. This tractor was, by the way, taken back to the company and equipped with tracks later on (the company became one forerunner to Caterpillar). Another initiative to use the hot-bulb engine for traction in farming was taken by J.V. Svenssons Motorfabrik in Augustendal in Stockholm, Sweden (Funke 2013). This company used hot-bulb engines in its Type 1 motor plough called Avanceplogen, produced in 1912. The inventor of this machine was Gustaf E. Jonsson from Norrköping. In 1921, the Lantz Bulldog tractor appeared in Germany. This tractor could run on many kinds of fuels, including crude oil. In 1930, the Swedish firm Munktells launched a hot-bulb engine for cheap raw oil.

John A. Secor, a consulting engineer at Advance-Rumely Co. in La Porte, Indianapolis, USA, made this reflection about raw oil engines in 1920 (Secor 1920, p. 700):

> None of the early types of oil engines was suitable for or used in the farm tractor. As recently as 1900, the steam tractor practically monopolized the field of power farming; but the supremacy of steam power was soon challenged by the gas tractor, and within recent years, the gas tractor has commanded a much broader market than the steam tractor.

During the 1960s, diesel became the new mainstream fuel in agriculture. According to Crister Stark, part-owner of the implement manufacturer Väderstad, the diesel engine has been extremely important for the productivity development in agriculture after the Second World War. "Most modern high-efficiency implements need tractors with lots of power" (Stark 2020). So, let us look into the development of this engine.

2.4 *The World's First Diesel Tractor*

The first diesel tractor, and indeed the world's first series-produced diesel engine vehicle, was the Benz-Sendling S 6, introduced in 1923 (Fig. 15.3). Its background was related to Karl Friedrich Benz and Otto Vollnhal. Benz founded the company Benz & Cie in 1883, and he received the famous patent on what is considered the world's first automobile 2 years later. Following a few unsuccessful attempts in the tractor field, Benz created a joint venture with Münchner Motorenfabrik München-Sendling in 1919, founded by Otto Vollnhal in 1899. Vollnhal had started manufacturing motor ploughs in 1909. The new company was called Benz-Sendling

Fig. 15.3 The Benz Sendling S 6. (Photo courtesy of Mercedes-Benz Classic)

Motorpfluge GmbH. Initially, a three-wheeled tractor with a petrol engine, called the Model T3, was developed. But soon, the engine was changed. Here is how it happened (Jung 2019):

The inventor Prosper L'Orange joined Benz & Cie in 1908 after having worked for Gasmotorenfabrik Deutz on big compressorless, stationary diesel engines. At this time, diesel engines were only used as stationary and marine engines, due to their size and weight. In 1921, L'Orange developed a more compact and high-speed engine by means of inventing first an afterchamber (patent 1908) and then a prechamber (patent 1909). However, the prechamber patent was neglected by the company. But when L'Orange came across a competitor's engine from Sweden, he got new inspiration and started to modify his prechamber, resulting in a new patent in 1919. The Swedish engine, called Ellwe, was invented by Harry Leissner at around 1913 and produced by the company Ljusne-Woxna AB. This engine had a new kind of prechamber and was launched on the market in 1918 (Spade 2008).

But L'Orange did not stop inventing. He also developed a variable injection pump in 1921, and now the engine concept was ready for wheeled vehicle installment. But what vehicle to choose? L'Orange found what he wanted at the joint-venture company Benz-Sendling: a tractor. As early as in 1921, he had put the new engine into a Benz-Sendling T3 tractor with three wheels. The engine had two-cylinders and developed 25 horsepower.

In 1922, three new prototypes were created with diesel engines. These tractors developed 30 horsepower and, like the T3, had three wheels with a large rear drive. The model was presented as "Benz-Sendling S 6" (see Fig. 15.3) at Ostmesse in Königsberg in 1923, and information tells that all three copies were sold quickly, of which at least one was sold at the trade fair. In 1923, as many as 100 tractors were manufactured.

The first four-wheeled Benz-Sendling tractor was called "BK", developed 32 horsepower, and began development in 1923. It was the result of a new collaboration, now between Benz and motor-plough manufacturer Automobilfabrik Komnick AG in Elbing ("BK" stood for Benz-Komnick). That tractor had three gears and was more expensive than the S models, and it is unclear if any were sold in the first few years. There is information that one may have been sold in England in 1930, but this is unclear.

Benz & Cie merged with Daimler-Motoren-Gesellschaft and, in 1926, formed Daimler-Benz AG. After that, the "Mercedes-Benz" name appeared, as well as the legendary three-point star on the new company's products. In 1928, a new variant of just one cylinder (inspired by Lantz) was launched, under the name Mercedes-Benz Model OE Diesel Tractor. The partly owned company Benz-Sendling also sold that tractor. But sales were slow, and in 1933, Daimler-Benz AG finally gave up tractor production for many years to come. Later on, however, it started again and had success with Unimog-based MB-Trac. But that is another story.

2.5 Biodiesel Fuels

Over time, environmental and political concerns paved the way for so-called biodiesel fuels, meaning vegetable oil – or animal fat – based fuels with molecules similar to petroleum diesel. The history of biodiesel is even longer than the history of the diesel engine itself. The transesterification of vegetable oils was conducted by Patrick Duffy already in 1853. In fact, Rudolf Diesel's first engine model, created in Augsburg, Germany, on the 10th of August 1893, ran on "biodiesel" in the form of peanut oil or hempseed oil (Saß 1962). Logically, it must, for the farmer, be a pleasant dream, using fuels that can be produced at the farm itself, for example, rapeseed oil.

Time went by, but not much happened in the biodiesel area – until the 1980s. It was then when the Elsbett company in Germany developed a type of cylinder head suitable for direct running on vegetable oils. This engine was announced around 1982. The Valmet company in Finland made some tractors with that type of engine, and some farms in Sweden ran them in field work; Sjösa near Nyköping was one of them (Hansson 2020).

It took a long time for biodiesel production on farms to gain its foothold in the market. The farmer, Axel Lagerfelt at Tolefors in Sweden, was one of the early adopters when installing a compact biodiesel plant in 2006. The equipment was developed by David Frykerås at the company Ageratec, later acquired by Alfa-Laval (Frykerås 2011).

In the same year as the Tolefors investment, in 2006, New Holland probably became the first tractor manufacturer to offer 100% biodiesel compatibility (Overall 2018). Some farmers have expanded this idea into pure fossil-free farming concepts. One pioneering effort was made by the company Energifabriken in Sweden 2012, something we will come back to later in this chapter.

2.6 Ethanol Fuels

Regarding ethanol, it is a complicated story. Such fuel was common in the early history of internal combustion engines, starting with Samuel Morey's fuel-flexible engine (including ethanol), in 1821 in the USA (Morey 1820). Due to US tax regulation in 1862, this fuel could no longer compete with coal (steam engines) or gasoline. Interestingly, Henry Ford's Model T was developed to run on ethanol (Hansson 2020). Sadly enough, we have not found any robust historical information about the first ethanol-driven tractor. But we know that in some countries, ethanol, over time, made a foothold on the market as traction fuel. Not least during times of war and during oil crises did ethanol gain more attention. Here are some ethanol projects in modern times:

In 1983, the Finnish tractor producer Valmet introduced an ethanol tractor made for the "ethanol country", Brazil. In 2006, Saskatchewan Industry and Resources in Canada also presented a modern ethanol tractor. Scania was early with an engine running on ethanol. It was mainly an engine for trucks, and it ran well on ethanol (Hansson 2020).

Besides new engines, one option is to convert older engines so that they can be fuelled with ethanol. One shall keep in mind that it has been hard to get diesel engines to accept ethanol fuel. The US company ClearFlame Engines, together with Cummins, announced in 2020 that it was trying to make a combined diesel and ethanol engine. Scania in Sweden already has many ethanol engines based on the diesel engine principle. However, its engines need additives (ignition improver and lubricants) to function, and these engines are not applied to tractors. Later on, we will describe the robot concept Alina, which can run on ethanol.

2.7 Wood Gas and Methane

During the Second World War, there was a lack of fuels. That led not only to interest in ethanol but also to wood gas, in the USA called "producer gas". Authorities in many countries promoted this technology for both automobiles and farm tractors. In Finland, for example, there were at most 4000 tractors driven by wood gas (De Decker 2010). The first wood gas vehicle was presented by Thomas Hugh Parker in England 1901. One of the tractors that were prepared for wood gas was the McCormick Deering 10–20 from International Harvester.

Methane is one of the ingredients in wood gas, and pure methane is still one kind of fuel that has been used for traction, and interestingly enough, not least because farmers can produce their own methane through biogas plants. Valtra, in the first decade of this century, presented and test ran in practical work dual-fuel models (methane/diesel). It started this development in 2008, and in 2009, the first model, N101, was taken into use and tested in various contexts. In 2010, the time had come for public launch; it happened at the Borgeby Field Days in Sweden. The next year,

Valtra presented the improved version T133 HiTech at the Agritechnica fair in Hannover. Testing of this tractor was conducted in Sweden during 2012, and in 2013, limited sales had begun.

New Holland Agriculture was right behind this development. In 2013, it presented its first prototype of a tractor powered by methane, the Methane Power T6. In December 2018, the company announced a new prototype tractor propelled by methane, which was presented at the Farm Progress Show in Decatur, Illinois, USA (Overall 2018). These initiatives were part of the New Holland vision "Energy Independent Farm". More on this will be discussed later.

2.8 Early Electrical Visions

Besides the energy variants mentioned above, there have been more interesting innovation projects heading for creative traction concepts. One early attempt was electricity in different forms and related to either cable provision or batteries, but also wire or chain systems. One early but strange example is the F. Zimmermann u. Company in Halle, which in 1894 demonstrated an electrical ploughing system (Williams 2019). It was a motor plough with an electric engine onboard. The engine dragged the vehicle along a chain that was laid over the field; see Fig. 15.4.

Notice that the Zimmerman system needed a chain and had the motor onboard. Williams (2019) describes another electrical tractor concept: the Brutschke system. In this concept, the motor was placed at the side of the field, and a plough was dragged by means of wires across the field. Fritz Brutschke's electrical plough concept, the so-called single-engine system, was designed around 1900 so that there was one cart with an electric motor (weighing 5.5 tonnes) placed at one side of the field. The second cart, at the other side of the field, was an anchor carriage. The motor was 60 hp (44 kW), with a working voltage of 380 volts. One challenge with the 5.5-tonne carriage was moving it by horse.

Sweden was among the countries that became inspired by the German development. In 1909, the first Swedish electric plough concept was manufactured by the company ASEA (Lagnelöv 2014). It was implemented the following year at the Ahlby farm in Stockholm, and the electric motor on the side of the field produced

Fig. 15.4 Zimmermann's electric plough. (Unknown artist from old lithography)

24 hp (18 kW). The owner of Ahlby farm, as well as the first adopter of this, was a manager at the tech company L.M. Ericsson. Subsequently, ASEA began to construct a larger plough which appeared at the Agricultural Meeting in Örebro in 1911. Now the engine had 35 hp (26 kW). Although ASEA's plough received good reviews, the spread was slow. In 1919, there were only four electric ploughs implemented in Swedish agriculture.

The mentioned concepts suffered from high friction because of the wire or chain being dragged along the ground. This led to more inventions. In 1920, E.A. Grafström from Electro-Agricultur Aktiebolag in Stockholm received a patent for a device for transferring electrical power to freely moving electric machines that could hold a cable freely hanging in the air with the help of regulators and cable drums. In 1921, he received a new patent on the regulator for keeping the cable at a good height (Lagnelöv 2014; Svensk Tidskrift för Industriellt Rättsskydd 1923).

Another example during the 1920s was the experimental tractor Elektrotanken (ASEA 2011), a concept invented by Hjalmar Cassel. Unlike the previous electrical plough concepts, it was now a moving machine that was electric motor-driven and could pull whatever implement, not only a plough. The machine (a converted US-built Cletrac) had tracks and a three-phase alternating current motor of 15 hp (12 kW) using 500 volts. The electricity was transferred through a high-voltage cable to a transformer wagon, and from there to the cable reel via a 250-m-long reinforced cable. The cable went through an 8.5-m-high mast on the cable car that connected to a high mast that was fixedly mounted on the tractor. The mast construction caused great strain on the cable, as the mast swayed very strongly when the tractor passed over uneven parts of the field. This was reported to be the reason why continued production failed.

In 1922, an attempt was made with the so-called June motor plough outside the ASEA city of Västerås in Sweden (Landtmannen 1924). The Royal "Vattenfallsstyrelsen" contributed financially, ASEA supplied the electrical equipment, and AB Juneverken contributed with its three-wheel motor plough. ASEA engineer Nils Forssblad converted the June plough to electric power. This concept was about a wheel-driven motor plough, and it received power from a transformer wagon via a 400-m-long insulated cable. The June plough initially had an electric motor of 16 hp (12 kW), but it was considered too weak, so a 50 hp (37 kW) engine was later installed and marketed. The cable was laid on the ground, and only the last 30–40 m hung freely from the ground in the mast that the tractor was equipped with. Integrated into the tractor was a cable drum which, using a small electric motor, wound the cable. The tractor mast also contained a series of switch rollers that would prevent the laid cable from looping or twisting while driving. The run still needed to be planned and performed according to predetermined patterns so that the cable was not damaged. Despite this, the life of the cable was relatively short due to constant unwinding and reeling on the cable drum. It was estimated that the cable would last at least 2 years if 150 ha were ploughed annually. After being tested in the Västerås area, the electrified June plough from 1923 was used in Ultuna and Täby with good technical results. This is what the magazine *Landtmannen* (1924, p. 433) wrote in 1924 about this:

> An important wish regarding the electric tractor is that it could be made independent of the wires in the field, that it could thus be equipped with an accumulator battery, sufficient for, for example, half a day or longer driving. Solving the question in that way, however, seems at least at present, for practical reasons, to be impossible, as the accumulators become too heavy and expensive. The lightest battery that can now be obtained weighs 33 kg per kWh and costs SEK 85 per kWh. If a motor plow is required for half a day of operation, e.g., 70 kWh., the battery weight will be 2,300 kg, and the price for the same SEK 6,000 or SEK 12,000 for the required two batteries. In addition, such batteries are not very durable. It thus seems to be quite clear that the accumulator operation currently is not to be seriously reckoned with, and the hopes concerning it for the future do not seem to be so bright that, at present, they justify a wait-and-see attitude towards now practically feasible proposals for electric operation.

We will revisit this kind of concept later.

2.9 *Battery-Based Vehicles*

Besides the energy variants mentioned above, there have been some interesting attempts to fulfil the dream that the magazine *Landtmannen* talked about. One of the first electric battery-based tractors was the one developed by South Dakota State University in the USA in 1983. Williams (2019) commented:

> Agricultural engineers at South Dakota State University started work on their Choremaster tractor project in 1983, using two 32-cell battery blocks to produce a 43.5 kWh electricity supply powering two motors. One motor operated the hydraulic systems, including the power steering plus the power take-off, and the other powered a three-range hydrostatic transmission with four-wheel drive. Because of the limitations of battery power in the 1980s, the Choremaster was designed mainly as a yard tractor working within easy reach of a power point for battery charging.

Time went by, and not much seems to have happened. But in 2007, during Agritechnica, John Deere presented concept tractors called the E-Premium 7430 and 7530, manufactured in Mannheim. The tractors had diesel engines but also crankshaft generators, driven directly by the engine, which could provide about 20 kW of electrical energy. From the generator, the electrical energy was converted to the desired voltage and further distributed to the chosen destination, such as fan drive, air conditioning or compressed air compressor – or for that matter, sockets to machines such as workshop machines. Any connection to driving, such as wagon wheels, was not included in the concept. It was partly an electric tractor, however not battery-based.

Kharkiv in Ukraine, in 2015, presented the electric tractor Edison (40 kW). In 2017 Fendt presented the e100 Vario prototype and in 2018 the Swiss Rigitrac SKE 50 was on display on the Agrama trade fair. The energy came from an 80-kWh lithium-ion battery with which one can drive up to 5 hours on one charge. This yellow, compact and well-designed tractor contained five electric motors, one for front and rear power take-off, one for the front axle and one for the rear axle. The fifth engine drives the hydraulic system.

We will later give some other examples. However, reality has shown that batteries are difficult solutions, due to the huge energy requirements in farming work, in combination with distance from fields to reload stations. A similar problem pained NASA engineers when planning long space flights. One interesting innovation area, therefore, became so-called fuel cells. We will now dig deeper into the fuel cell concept.

2.10 Towards the Fuel Cells Concept and Beyond

A fuel cell converts chemical energy from a fuel, such as liquid hydrogen, into electricity through a chemical reaction that is activated by an oxidizing agent, such as liquid oxygen. Simply put, one can say that hydrogen reacts with oxygen, which leads to water, while electrons are released ($2H_2 + O_2 \rightarrow 2H_2O + 2^{e-}$). This is a continuous process that does not end as long as fuel is supplied, unlike a battery that stores the chemical reactants. Furthermore, the reaction is silent and does not give rise to either smoke or the same heat as internal combustion engines. Compared to most batteries, the weight is small. The disadvantage is high price and (possible) explosion risks.

The fuel cell was probably invented by William Robert Grove in Wales (Great Britain) in 1839 but is believed to have been first demonstrated by the British chemist Humphry Davy as early as 1801. Francis Thomas Bacon (a relative of the famous Francis Bacon) developed in 1939 a practically useful fuel cell powered by hydrogen and oxygen by an alkaline electrolyte. He developed a working prototype in 1959.

Bacon's concept, the alkaline fuel cell, was to be used in the US space program. First out was the two-person rocket Gemini V, which took off on the 1st of August, 1965, with astronauts L. Gordon Cooper and Charles Conrad. General Electric is said to have constructed the fuel cells in the Gemini spacecraft. However, they did not work as they should, creating great drama. The technology was further developed and later used in the Apollo project, which followed Gemini. The Apollo missions triumphed with the lunar landing in 1969.

However, the first fuel cell engine vehicle was neither a space rocket nor car, but a tractor from Allis Chalmers, and was presented on the 15th of October, 1959, in Milwaukee (Karg 2019). Its fuel cells, 1008 in number, were not driven by hydrogen but a secret propane gas mixture. The power output was 15 kilowatts, which corresponded to about 20 horsepower. The concept tractor was based on a D-10 chassis. During the demonstration, ploughing of a field alpha-alpha was performed. The engine was developed by engineer Harry Karl Ihrig starting in 1951. The tractor is preserved at the Smithsonian Museum, but at the time of this writing, it was loaned to a small local museum called McLeod County Historical Society & Museum in Hutchinson, Minnesota; see Fig. 15.5.

The first car with a fuel cell engine was not introduced until 1966 through the Chevrolet Electrovan from General Motors. We can thus conclude that agriculture

Fig. 15.5 The Allis Chalmers fuel cell tractor at the McLeod County Historical Society & Museum. (Photo courtesy of Larry Karg)

was not only ahead of the automotive industry but also NASA in the midst of space-intensive years. NASA documents (Austin 1966) show that NASA carefully investigated Allis-Chalmers fuel cells and developed space-adapted solutions based on these fuel cells. This is also confirmed by Larry Karg, who is active in the Allis-Chalmers club and handled the tractor during museum stays. He wrote: "Some of the fuel cells were removed and sent to NASA" (Karg 2019).

3 Examples of Electrification in Modern Times

3.1 Modern Fuel Cell Concepts

What, then, about fuel cell tractors in modern times? The manufacturer New Holland (part of CNH) launched a hydrogen fuel cell tractor in 2009. This concept was called NH2 and was driven by compressed hydrogen and only emitted water, no exhaust gases. Furthermore, this tractor was part of New Holland's earlier mentioned vision of an "Energy Independent Farm". The company established three test farms in Italy, France and Germany, respectively. Here's how the company reasoned (Johnson 2009):

> The main obstacles to the use of hydrogen are distribution and availability. The concept of the New Holland energy independent farm involves the farmer producing his own supply of compressed hydrogen either from water (using electrolysis) or directly from methane (by burning waste or biomass). The production plants are supplied with energy from wind tur-

> bines or solar panels and the hydrogen can be stored on the farm in underground tanks. One benefit is the short distance (compared with cars and trucks) that tractors and combines cover between the farm and the field. Farmers are ideally placed to use hydrogen technology. They have large areas of land for alternative energy generation plants (solar power, wind energy, biomass plants or waste systems) and for the storage of energy in the form of hydrogen. In addition to the environmental benefits, a system of this kind would allow farmers to be independent of external energy suppliers and to increase their financial stability, as fuel represents a significant proportion of overall farm costs.

The company wrote that it believes the tractors of the future will run on electricity, but at the same time, it does not believe in battery operation for tractors (Johnson 2009):

> Because of the large amounts of power needed for these large machines and the low levels of energy recovery, batteries are not a suitable power source for this type of vehicle, unlike cars and industrial trucks.

But the NH2 concept was soon put on ice. Overall (2018), Corporate Communications Manager, CNH Industrial, commented: "The tractor did function well, but became too expensive to gain broad market adoption".

In 2018, Fiat Power Train (also owned by CNH) revealed its Cursor X concept. It was a powerpack or multi-use engine, also called "modular power source", for tractors but also trucks, wheeled loaders, combine harvesters and other machine types. The concept was as much a design as an engineering project (created jointly by the engineering and design departments of CNH and Fiat Power Train); see Fig. 15.6.

Fiat Power Train had produced a hydrogen fuel cell bus back in 2001 and the fuel cell tractor from New Holland in 2009 that we already have mentioned. The new concept had more capacity and thus made possible longer working times between reloads of hydrogen. The Cursor X powertrain included not only a fuel cell and

Fig. 15.6 The fuel cell engine by Fiat Power Train introduced at Agritechnica in 2019

hydrogen tanks but also a lithium-ion battery pack, an "eAxle" and an energy management system. The maximum fuel cell output was 200 kW, which was said to generate a motor output of 400 kW (Gilkes 2018). The 350-bar hydrogen tanks were said to hold 64 kg hydrogen, and reloading took 20 minutes. The negative aspect was that the concept was new and fragile and that the price was about four times higher than diesel engines. Another downside was that the powertrain weighed around 6 tonnes.

3.2 Modern Battery Concepts

Fuel cells are not the most common pathway in the electrification of agricultural vehicles. Most initiatives are based on battery systems.

In a speech on the 19th of April, 2016, Prof. Peter Pickel at John Deere European Technology Innovation Center described the SESAM concept, which stood for Sustainable Energy Supply for Agricultural Machinery (Pickel 2016). The concept, appearing in real life later that year, was a big, 6R-series tractor, but the diesel engine and fuel tank were replaced by an extremely large battery. They showed this concept during SIMA in Paris in 2017. The electric tractor had a lithium-ion battery, corresponding to a capacity of 150 kWh, and was equipped with two 150 kW motors (a total of about 400 hp). The idea was that one motor drives the tractor wheels, while the other can drive power take-offs or other functions. If the motor were working at full capacity, for example, the tractor could run for 30 minutes when ploughing. It then took 3 hours to charge the batteries. The battery was assumed to handle 3100 charge cycles. The chassis came from the Mannheim factory's 6R series. The transmission consisted of the DirectDrive dual-clutch gearbox (launched in 2011). As a curiosity, it can be mentioned that the engine growled aggressively like a Formula 1 car.

Prior to Agritechnica 2017, Fendt (part of AGCO Group) announced the e100 Vario, an all-electric compact tractor based on the Fendt 200 Vario. With a 600-volt battery of 100 kWh under the hood, a 50 kW electric motor could be operated for about 2.5 hours at fairly hard work. For lighter work, the battery lasted up to a full day. By fast charging, a discharged battery could get back 80% of its capacity within 40 minutes.

3.3 Modern Cable-Based Tractors

The earlier-mentioned cable link concept is still alive today. One interesting example is a project by the inventor Kurt Hansson at the farm Sörgården in Norrbäck (Sala municipality), Sweden. He bought an 11-kW electric engine and placed it into a Fordson Major tractor and added about 400 m of cable. The energy was produced by 6 solar trackers with a total of 324 square meters of panel surface. By this, he was able to produce a maximum of 72 kW every sunny hour of the working day from March to September and, in total, about 100,000 kWh per year. The concept was formed in 2014 (Hansson 2020).

Another example is the John Deere project to combine cable concepts with robotics, which was conducted by John Deere's European Technology Innovation Center in Kaiserslautern, together with B.A.U.M Consult GmbH and the Technical University in Kaiserslautern. The company calls it GridCON. The prototype, announced in April 2018 but presented in December, produced 400 horsepower through the provision, via a 1 km long cable, of direct current of more than 6000 volts, which is transformed onboard to 700 volts (Bensing 2018). The vision includes that the farms shall produce their own electricity in some way.

4 Hybrid Concepts

Already in 1954, International Harvester launched the Farmall 450 Electrall. It was a tractor with a diesel engine also equipped with a 10-kW generator big enough to drive implements. It developed an electric baler, but that was it. The market was not ready for changing mechanical power take-off to electricity. Interestingly, it developed a large-scale bug zapper to electrocute insects in the field at night (Ag and Food Newsletter 1954). Remember here also the John Deere E-premium in 2007.

But let us turn to modern times. On the 2nd of September, 2017, ZF Friedrichshafen AG announced a concept where a tractor with a diesel engine was equipped with a powerful generator, which in turn generated power that drove hub motors mounted on a connected trolley (Zillner 2020). The central parts of the concept were a high-voltage generator called Terra+ (60 kW), in combination with the electric hub motor eTRAC. It, in turn, could be mounted on any vehicle. If only power corresponding to 15 kW was needed, a low-voltage generator was also offered. The transmission in the system was either a powershift gearbox or a hydrostatic transmission. The concept was thus a hybrid between an electric and mechanical driveline. In addition to the transmission, ZF included a number of sensors and the potential for artificial intelligence. For example, there were radars so that the system could keep track of slippage, and based on that, optimally distribute the power between the tractor and, for example, the wheel drive on a plough.

In November 2019, during Agritechnica, Carraro launched something called the "Mild Hybrid", a driveline intended for special tractors. The idea was to be able to reduce the size of a diesel engine by combining it with an electric motor, and everything was done in light of the legal requirements that exist within the EU. The Carraro 3E22 diesel engine had three cylinders and produced 55 kW. It meets the emission limits for T3B, Stage V, and does not need to be equipped with an SCR catalyst (to meet the mentioned requirements). In addition, there is an electric motor of only 48 V, which gives 20 kW. Together, the concept can correspond to a 100 hp diesel tractor but with significantly lower fuel consumption. Or, to put it simply, it is about the electric motor helping when extra-high power is needed.

In fact, Carraro had launched an electric hybrid tractor already during the EIMA International trade fair in November 2018. It was created in collaboration with the company 4E Consulting (in Porotto, Italy) and was named Ibrido, and it had a 105 hp electric motor in combination with a diesel engine. The driver could choose between driving in fully electric mode (e.g. in riding stables or in courtyards), only in diesel mode (e.g. on roads) or in both electricity and diesel modes (e.g. when pulling or driving heavy with the power take-off).

The Indian company Protecto Engineering Services launched a small hybrid tractor called "HAV S2 hybrid" in 2019. It was equipped with hub motors powered by electricity, either from a battery or diesel engine (alternatively, a natural gas engine) via a generator.

The Steyr brand, acquired by CNH, has been given the role of an experimental platform for CNH. In 2019, its concept – called Konzept – was launched. It is a diesel-electric concept with a lot of innovation. The diesel engine from Fiat Powertrain has four cylinders and drives a generator, which in turn drives several electric and separately regulated wheel motors on all four wheels. A battery also enables it to get a "boost" for short distances. The battery has, at a maximum, 3 kWh. What used to be hydraulics and mechanical power take-off are also electrically driven; see Fig. 15.7. This enables variable power take-off speeds as well as the ability to reverse it. At the rear of the tractor are 700 V and 48 V connections for power tools. In addition to its hybrid driveline, the tractor is equipped with details, such as a head-up display and a drone that flies in front of the tractor and captures data.

Also, during Agritechnica in November 2019, Deere & Co and Joskin launched a concept called eAutoPower, in combination with e8WD. Just like ZF in 2017, it

Fig. 15.7 The electric power take-off on the concept tractor from Steyr

was about an engine that powered a generator, but it is more complicated than that. Deere & Co and Joskin were talking about motors rather than generators, which then generate electricity for motors at the wheels. The concept was stated to be "the first electromechanical power-shifted gearbox in agricultural technology". In their own words: "Technically, the hydro unit (pump/motor) was completely dispensed with; instead, two electric motors are used as a continuously variable actuator" (DLG 2019). Note that DLG (2019) writes "motors", not "generators"; a more accurate term would probably have been electric machines because they can – if we understand this right – change roles between being a motor and being a generator. The electric motors, when acting as generators, gives 100 kW to power whatever one wants. In the configuration shown in the autumn of 2019, cables from the generator led to an engine located at the front of the manure barrel chassis behind the tractor. That engine, in turn, drove mechanical axles that were connected to two pairs of wheels via gears.

5 More About Fossil-Free Thinking

The electrical concepts mentioned are not, by definition, fossil-free concepts. For example, this is dependent on how the electricity used is produced. Around the world, various projects are underway to convert agriculture to fossil-free. Here are some examples.

5.1 Biogas Tractors Making Use of Farm Manure

In 2009, CNH presented a biogas tractor in the form of the Steyr CVT 6195, with a stepless transmission. The tractor could also be run on natural gas or ordinary diesel. The tractor was a result of cooperation with the company Lu Power. In connection with the presentation, CNH said that biogas, in particular, is interesting because the farmers themselves can produce the energy on their own farms.

Valtra presented a biogas tractor in 2010 called the N111 HiTech. A few years later came the N101, which was a 110 hp four-cylinder engine, and which was a hybrid between biogas and diesel. The next step was the T133 Dual Fuel, a six-cylinder engine. In biogas mode, about 80% of the power of the gas was produced, while the reset was produced from diesel.

The company Gomselmash from Belarus presented a combine powered by biogas in 2017. During Agritechnica 2019, the Palesse GS4118K model was then launched on the market. The 12-liter gas engine with 350 hp came from Cummins (IS12G). Perhaps it is the only commercially available agricultural machine in the world that meets strict Stage V engine emissions without using Adblue or a particulate filter. The combine can also be powered by natural gas. The gas is stored in eight containers with a capacity of 1816 litres (450 m^3), which is said to last 8–10 hours.

New Holland (also part of CNH) has long invested in gas engines of various kinds. As early as 2013, it launched a biogas tractor called the T6, and in 2018, the company demonstrated its prototype "New Holland T6.180 Methane Power". It was equipped with a six-cylinder gas engine of 180 hp from Fiat Power Train with spark ignition. The gas was partly stored in a front tank, and with it, the tractor could handle 48% of the operating time compared to a diesel tractor and could run for about a day without refuelling. Hubertus M. Mühlhäuser (2019), CEO at CNH Industrial commented: "People think electric engines mean fossil-free. No, that is not the case. But biomass is. So, biogas is the future". The tractor was launched on the market in 2020.

5.2 *The Case of Energifabriken*

Internationally one famous example of a fossil-free farm initiative is La Bellotta in Italy, which in 2013 was selected as the first "Energy Independent Farm concept" by the CNH company New Holland. However, already in 2006, three farming families in Sweden made substantial steps in this direction. Their farms were Smedberga, Kasta and Jolstad, and together, they formed the company Energifabriken, which has since been selling and distributing biofuels (Varverud 2020). In the same year, the companies' common combine was converted to biodiesel. In the spring of 2011, the three farms decided, in principle, to go full circle and replace all fossil fuels needed for their own production and operation of the farms with biofuels. It was not only about the fuel in tractors, combine harvesters and other field machines, but also oil for heating, drying plants and transport of materials to and from the farms. In addition to diesel vehicles, there were petrol-powered vehicles on the farms, such as cars, quad bikes, chainsaws and high-pressure washers. These machines were converted to ethanol operation. During this experimental innovation process, the farmers received support from AgroÖst and the Energy Office Östra Götaland. In 2012, they had replaced 95 percent of all fossil fuels. The remaining percentage was related to leased services, but 2 years later, many of these were also made with renewable fuels.

The next step was to create (and start marketing) products that were produced using the fossil-free method. In 2014, they could start selling "fossil-free flour" and "fossil-free rapeseed oil"; see Fig. 15.8. They hereby created completely new segments alongside traditional dimensions such as organic-conventional. The families continued to engage in smart business development, such as launching "more friendly wheat" in 2018 together with the grain company Lantmännen. Today, this company is one of the main players in fossil-free fuels in Sweden.

Fig. 15.8 Energifabriken managed the art of creating new, value-added products out of the fossil-free production philosophy

6 The Future of Farm Traction Energy

The path forward is largely about systemic development work. For example, it is not fruitful to only analyse vehicles that are powered by one or another energy source if one does not also address issues such as the supply of the energy needed. Let us just point at some future concepts that might be of interest because they could provide fossil-free, cheap, locally produced, and in one of the two cases, also weather-independent energy for farmers.

6.1 Agrosolary

One possible innovation is Agrosolary, invented by the farmer Kurt Hansson (2020). The concept includes tree and shrub alleys, which, through a new type of solar-tracking panels (invented by Saab), also generate renewable energy. Thus, the fields will produce not only food and feed but also electrical energy. Many people still think that it is negative for agriculture to have solar panels on agricultural fields, but today's focus on sustainability and trends around "regenerative agriculture" or the "Green Deal" (European Commission 2019) means that the time may be right for this concept. The trees can be perceived as natural in the arable landscape, increase the biological diversity in agriculture and hinder soil erosion caused by wind. But, at the same time, the concept must enable rational and highly efficient use of the land between the alleys and, therefore, the lanes must be placed at a large and, in relation to modern machines, well-adapted distance. If a farmer decides to work with machine widths of, for example, 36 m, a suitable width of the arable land between the tracks could be 72, 108 or 144 m. Furthermore, the tracks should not go all the way to the edge of the fields so that modern machines can easily turn around at the end points of the tracks.

6.2 *Modern Steam Engines*

Another possible idea for the future might, in fact, be modern steam engines making use of biomaterials of different kinds. The Swedish company Ranotor, together with the institute RISE and Linköping University, are planning to start a project on this and build a modern steam tractor (Platell 2020, Pettersson 2020). Ranotor has its roots in the Saab-Scania Steam Power Project, which started back in 1968.

The new kinds of steam engines are very much different from the old steam engines we know from history. One way to illustrate this is by comparing the power density. This means the output power in relation to weight (kW/kg), volume (kW/l) or time (kW/SEK). The Ronator concept has a power density of far beyond 50 kW in relation to the old of about 15 kW/l.

This might create an opening for using steam engines in vehicles such as tractors. The interesting thing about modern steam engines is that they can utilize basically all energy sources, including straw, wood pellets, discarded grain or even solar energy, not least via concentrating solar power. Ranotor is now also running an EU project together with some large companies and the Fraunhofer Institute, where it will use ammonia as energy for a steam engine.

6.3 *The Hydrogen and Ammonia Vision*

We mentioned ammonia (NH_3), which is very central in the new EU hydrogen strategy launched in 2020. Especially interesting is the vison of producing ammonia in a climate-friendly way. Let us give two examples:

One project is called The Carbon-Free Farm and driven by the MIT and NASA engineer Jay Schmuecker and others at Pinehurst Farm in Iowa, USA. They built a hydrogen tractor demonstrator already in 2010 by installing a Ford 460 V8 otto engine into a John Deere 7819 tractor and mount tanks on the roof. In 2011, they started to produce hydrogen by means of solar power. In 2015, they could present the world's first tractor to run on NH3. See Fig. 15.9.

Another example is the Canadian innovator Roger Gordon, who has developed a synthesizer for H_2 and NH_3 by using wind power. Moreover, he has made a Ford tractor that can be operated by this "green ammonia". He received a patent on this 2015. Among his visions is the intention to have direct-acting fuel cells on NH_3 (Hansson 2020).

Fig. 15.9 The solar hydrogen ammonia tractor. (Photo courtesy of Jay Schmuecker)

6.4 Robots and Their Energy Provisions

Last but not least: What about autonomous vs. traditional machine systems? We did not focus on this dimension in this chapter, but we did touch upon it in some examples. The GridCON concept was based on cable electricity and an autonomous vehicle. It might be relevant to point to the fact that autonomous vehicles need energy systems that are not dependent on manual observation and adjustments. Therefore, external or internal combustion engines have been said to be less suited for robots than electric systems. If this is true or not, we cannot be sure.

Terry Anderson, in North Dakota, USA, started a project in 2012 that led to the Spirit prototype in 2013. It looked a bit like a huge air compressor on tracks. This machine had twin 202 hp, 5.2-liter Isuzu diesel engines (that could accept biodiesel), and these engines, in turn, propelled electric wheel motors.

Another example is the Swedish robot concept Alina, developed by Mapro Systems, RISE and SLU, which was tested in the summer of 2020 at the Testbed for Digitalized Agriculture in Uppsala; see Fig. 15.10. This is a tool carrier platform, fully autonomized with sufficient power enough for using a three-unit seeder. It has no electric motor but instead a Vanguard engine (37 hP) that one starts with gasoline and then turns over to fossil-free ethanol. The engine is coupled to a hydrostatic drive (hydraulic wheel motors). The reason for not using electric wheel engines is, according to the inventor Mats Andersson (2020), that "Electric wheel engines are not reliable enough for work 24/7 in clay and wet environments". He adds: "Electric wheel motors also need quite high rotation speed to be efficient". The intention is to use the machine for horticulture. This machine can work at least 6 hours before refuelling.

Despite cases like Spirit or Alina, most field robots in agriculture use electric energy. One example is found in the EU-funded project Mars (Mobile Agricultural Robot Swarms) that started around 2015 and was run in collaboration between

Fig. 15.10 The Alina ethanol robot in test

AGCO and Hochschule Ulm. The project led to a robot called Mars, which consisted of four pneumatic wheels and a single seed bill. The basic idea was to use swarms of such small robots. In 2017, some Mars robots underwent testing in Germany. After Fendt (AGCO) took over the project, the robot was renamed Xaver, and in September 2020, changes were revealed (Vale 2020). The pneumatic wheels had been changed to rubberized steel wheels (interestingly, just like the era of the iron wheels). The four wheels had been reduced to three, where the two iron wheels drive, and the rear wheel partly steers and partly packs after the seed bill. The robot is electrically powered by a 2.6 kWh lithium-ion battery and needs to be charged after 1.5 hours of driving. With six such small robots, it is stated that one can sow three hectares per hour. Perhaps this is a taste of the future; see Fig. 15.11.

Are electrical robots effective enough for modern farming? Will we see fleets or swarms of many small robots or fewer but bigger ones? Or is the future about some middle way? The future will tell. Simulation research indicates a huge potential. Engström and Lagnelöv (2017) showed that one could manage a 200-ha dairy farm on two autonomous 36 kW (100 kWh) electrical robot machines. At present, Engström, together with Arvid Örde and others, is trying to fulfil the dream by means of the electric mid-steered robot Drever with 120 kW power. This is quite small and flexible but big enough for having ordinary farm implements; see Fig. 15.12.

However, the question remains: How can we produce electrical energy in sufficient quantity and in a fossil-free, weather-independent and cost-effective way, preferably locally produced on the farm? In light of these questions, we can mention that the robot project called Farmdroid, with its six saw bills, offers alternative solutions. It has solar panels (1 kW) that catch solar energy in real time. Another concept is the Ekobot robot, which has a battery but is possible to reload using solar-powered charge stations.

Fig. 15.11 The Xaver robot in test in September 2020. (Photo courtesy of Manja Morawitz (Deputy Press Officer AGCO/Fendt EME))

Fig. 15.12 The robot project Drever 120 kw at the testbed for Digitalized Farming in Sweden (September 2020)

We need to underline that robots can work longer than manned machines, simply because labour is expensive per hour and needs to sleep. Moreover, robots can be lighter because they do not require cabins or human bodies among their load. This means less energy is needed and less soil compaction, which in turn can lead to looser soil demanding less energy to go through with farming implements.

The future will tell if the robots will take over farming work or not, what kind of energy they will use and how that energy will be produced.

7 Concluding Discussion

Future visions in the area of energy for field work in agriculture should be related to the following dimensions: First, are they fossil-free? Second, are they cost effective? Third, are they sufficient in relation to the energy needs of farming field machines? The three dimensions can lead to a fourth and fifth dimension: Is there a need for local production (small-scale circular systems), and will autonomous vs. traditional machine systems using wind power to make H_2 and NH_3 be used?

7.1 Fossil Freedom

Regarding fossil-free agriculture, we can conclude that there are many technical solutions for being fossil-free and that this is also the will of many actors. Electrical concepts are popular to discuss in the farming business industry, but often forgotten is a discussion on the processes behind electric energy production.

7.2 Cost-Effectiveness and Sufficient Concepts

The dimension of cost-effectiveness is important. Ethical arguments and feelings might point in a certain way, but the wallets of farmers have limitations. The dimension is related to the other dimension of sufficient energy in relation to the energy need of farming field machines. Let us discuss these two dimensions together a little bit deeper. Some concepts have been successful on the market, while others are stuck in the idea or prototype stage. Let us take battery-powered electric tractors as an example. Despite the extensive development of holistic systems, the electric concept has not yet been successful. A basic problem is spelled energy density. From 1 kg of diesel, it is possible to obtain 45 MJ under ideal conditions. If we consider the typical efficiency of a diesel engine to be 33%, it means 15 MJ of mechanical energy per kg of diesel. It can be compared to 1 kg of a modern lithium battery that gives only 0.2 MJ. There are more advanced batteries under development, for example, lithium sulphide, which can give up to 1.5 MJ per kilo (or lithium thionyl chloride batteries are stated to give 2.5 MJ/kg). Although an electric driveline gives 90% efficiency, the battery solution is far behind diesel in energy density. Furthermore, of course, the cost is a problem, as prices are falling, but then from a high level. An opening may be a hybrid concept.

Another issue with batteries is that their production is not considered sustainable. Nevertheless, there is hope: One striking piece of news was the discovery that hemp batteries work eight times better than lithium-ion batteries (Wang et al. 2013). So, can industrial hemp be a part of the future of battery-powered vehicles? It might be a step towards sustainable and efficient ways of creating battery power. The idea is not new. In 2014, researchers in the USA discovered that unused fibre from hemp could be converted into an "ultrafast" battery that is better than graphene. David Mitlin of Clarkson University, New York, led this experiment (Hansson 2020).

7.3 Local Production of Energy at Farms

One dimension is the need or wish for local production (small-scale circular systems) of energy. We all know that the local production of things is not always the best from an economic or even a sustainability point of view. Farmers have local resources and are often located at long distances from alternative resources. There are many good reasons for trying to create locally produced energy, and we have seen in this chapter some examples of this.

7.4 Notes on the Robotics Revolution

Robot concepts pave the way for many advantages from a climate care perspective. They can be powered by electricity, which can also be produced fossil-free at the local farm by means of, for example, biogas-fuelled generators or solar trackers. They can also, like the Alina robot, be fuelled by biofuels such as ethanol. Moreover, they do not require as much energy as ordinary machines, partly because their weight can be lowered. One can, as said before, avoid the load from, for example, cabins, air-conditioning systems or driver's bodies. The Fendt corporation assumes that the weight can be reduced by ca 80%, which will save energy, costs and not least the soils through less soil compaction. Less compaction, in turn, means less resistance on later passages of the fields as well as better soil health, meaning more harvest in relation to all stakes. But another aspect is that they can work 24 hours a day and, therefore, do not need to be as productive as man-operated systems.

Figure 15.13 summarizes potential combinations of locally produced energies at farms and different kinds of engines or motors.

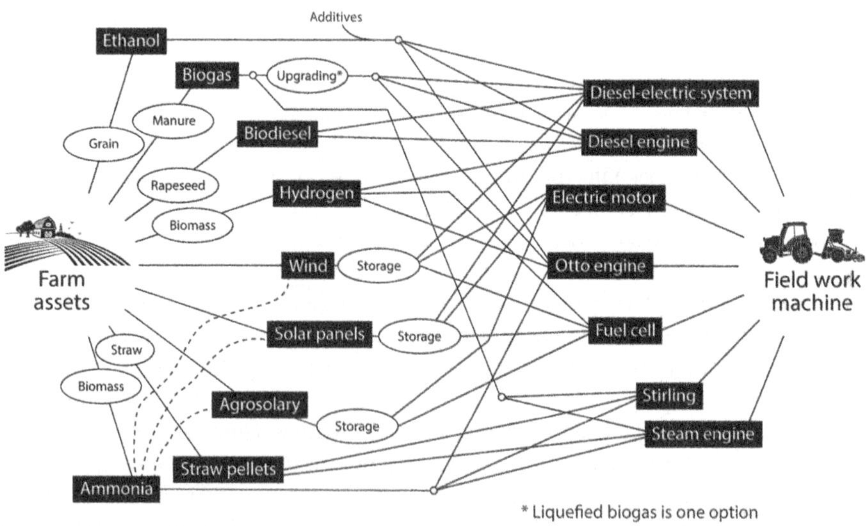

Fig. 15.13 Combinations of farm energy assets and different kinds of engines

Acknowledgement The authors want to thank the following persons for their valuable contributions: Mats Andersson (CEO at Mapro Systems), Jonas Engström (Senior Project Leader, RISE), David Frykås (Former owner and CEO of Ageratech), Kurt Hansson (Inventor and CEO, Gasilage), Michael Jung (Archive manager, Mercedes-Benz Classic), Larry Karg (Allis-Chalmers Club), Hubertus M. Mühlhäuser (Former CEO at CNH Industrial), Laura Overall (Corporate Communications Manager at CNH Industrial), Ola Pettersson (Senor Project Leader, RISE), Peter Platell (CEO at Ranotor), Crister Stark (Inventor and part-owner, Väderstad), Emil Stolpe-Nordin (Territory Manager, Sweden, Deere & Co), David Varverud (Part-owner of Energifabriken) and Natalie Zillner (Product Communication Off-Highway Systems at ZF Friedrichshafen AG). Many thanks also to Mistra, Region Östergötland and Vinnova for research funding.

References

Ag and Food Newsletter (1954) Insect Electrocution. Ag and Food Newsletter 2 (14):711

Andersson M (2020) CEO at Mapro Systems, personal communication, 15th of September.

ASME (1996) The Hart-Parr tractor introduced in 1901 – A national historic mechanical engineering landmark. The American Society of Machanical Engineers

ASEA (2011) ASEA's image archive. Västmanlands Läns Museum, Västerås, Sweden.

Austin LG (1966) Fuel Cells: A Review of Government-Sponsored Research 1950–1964 (Prepared under contract for the National Aeronautics and Space Administration). NASA, Washington,

Bensing T (2018) John Deere Traktor mit Kabeltrommel und ohne Dieselmotor, 16th of April, www.profi.de/aktuell/aktuelle-meldungen/john-deere-traktor-mit-kabeltrommel-und-ohne-dieselmotor-11779714.html.

Bugge MM, Hansen T, Klitkou A (2016) What is the bioeconomy? A review of the literature. Sustainability 8 (7):691

Carson R (1962) Silent spring. Houghton Mifflin, Boston, USA

D'Amato D, Droste N, Allen B, Kettunen M, Lähtinen K, Korhonen J, Leskinen P, Matthies BD, Toppinen A (2017) Green, circular, bio economy: A comparative analysis of sustainability avenues. Journal of Cleaner Production 168:716-734. doi:https://doi.org/10.1016/j.jclepro.2017.09.053

De Decker K (2010) Wood gas vehicles: firewood in the fuel tank. Low-tech Magazine, 18th of January,

DLG (2019) eAutoPower gearbox e8WD for 8R large tractors. Innovation Magazine:p. 11

Ekblom J (2019) European Investment Bank to cease funding fossil fuel projects by end-2021. Reuters (14th of November)

Ellen MacArthur Foundation (2013) Towards the circular economy.

Engström J, Lagnelöv O (2017) Batteridriven autonom jordbruksmaskin: Simulering av maskinaktiviteter på en svensk gård (SP Rapport). RISE, Uppsala

European Commission (2019) The European Green Deal, 11th of December.

Eurostat (2019) Agri-environmental indicator - energy use. Eurostat. ec.europa.eu/eurostat/statistics-explained/index.php/Agri-environmental_indicator_-_energy_use. Accessed 15th of September 2020

Frankelius, P (2009) Questioning two myths in innovation literature. Journal of High Technology Management Research, 20(1):40–51.

Frankelius P (2020) A proposal to rethink agriculture in the climate calculations. Agronomy Journal 112 (4):3216-3221. doi:https://doi.org/10.1002/agj2.20286

Frykerås D (2011) Former owner and CEO of Ageratech, personal communication, 5th of August.

Funke M (2013) Johan Victor Svenson. Svenskt biografiskt lexicon. Riksarkivet, Stockholm

Gilkes D (2018) FPT Tech Day reveals Cursor X concept.

Hansson K (2020) Inventor and CEO, Gasilage, personal communication, 15th of September.

Hodge J (1973) Richard Trevithick. Shire Publications, Oxford, UK
Johnson S (2009) New Holland's drive concept of the future. Press information from New Holland Agriculture, 6th of October.
Jung M (2019) Archive manager, Mercedes-Benz Classic, personal communication 9th of August.
Karg L (2019) Allis-Chalmers Club, personal communication, 31st of July.
Kleinschmit D, Lindstad BH, Thorsen BJ, Toppinen A, Roos A, Baardsen S (2014) Shades of green: a social scientific view on bioeconomy in the forest sector. Scandinavian Journal of Forest Research 29 (4):402-410. doi:https://doi.org/10.1080/02827581.2014.921722
Lagnelöv O (2014) Elektrifiering av jordbruksmaskin - Direktkoppling till solföljare och elnät. Master thesis, Uppsala University, Uppsala, Sweden
Landtmannen (1924) Nytt system för elektrisk jordbearbetning. No. 23:pp. 432–433
Lorenzo C, Almeida R, Martinez-Nuñez M, Narvarte L, Carrasco L (2018) Economic assessment of large power photovoltaic irrigation systems in the ECOWAS region. Energy 155. doi:https://doi.org/10.1016/j.energy.2018.05.066
Morey S (1820) On Heat and Light. American Journal of Science and Arts 2:122-123
Mühlhäuser HM (2019) CEO at CNH Industrial, personal communication, 11th of November.
Myrdal J (1993) Jordbruk (in English: Argiculture). Nationalencyklopedin, vol 10. Bokförlaget Bra Böcker, Höganäs, Sweden
Nerlich B (2015) The bioeconomy in the news (or not), University of Nottingham blog post. blogs.nottingham.ac.uk/makingsciencepublic/2015/04/30/the-bioeconomy-in-the-news-or-not/. Accessed 15th of September 2020
Overall L (2018) Corporate Communications Manager at CNH Industrial, personal communication, 11th of April.
Pettersson O (2020) Senor Project Leader, RISE, personal communication, 15th of September.
Pickel P (2016) On the Pathway to Digital Farming. Paper presented at the EFSI Workshop, The European Federation for Services to Individuals (EFSI), Brussels, 19th of April
Platell P (2020) CEO at Ranotor, personal communication, 14th of September.
Ransome-Wallis, P (2001) Illustrated Encyclopedia of World Railway Locomotives. North Chelmsford, MA: Courier Dover Publications.
Roos A, Stendahl M (2015) 10 The emerging bio-economy and the forest sector. Forests, Business and Sustainability:179
Saß F (1962) Geschichte des deutschen Verbrennungsmotorenbaues von 1860 bis 1918.
Secor JA (1920) Kerosene as a tractor fuel. 15, Part 2:696-713. doi:https://doi.org/10.4271/200078
Spade K (2008) En historia om kraftmaskiner. Riksantikvarieämbetet, Stockholm
Stark C (2020) Inventor and owner, Väderstad, personal communication, 1st of September.
Stolpe-Nordin E (2020) Territory Manager, Sweden, at Deere & Co, personal communication, 18th of August.
Svensk Tidskrift för Industriellt Rättsskydd (1923) Anordning vid system för överföring av elektrisk kraft till fritt rörliga elektriska maskiner. Vol. 39:p. 40
Vale S (2020) New look for Fendt Xaver field robot, Profi, September 7, www.profi.co.uk/news/new-look-fendt-xaver-field-robot
Varverud D (2020) Part-owner, Energifabriken, personal communication 21st of September.
Wang H, Xu Z, Kohandehghan A, Li Z, Cui K, Tan X, Stephenson TJ, King'ondu CK, Holt CMB, Olsen BC, Tak JK, Harfield D, Anyia AO, Mitlin D (2013) Interconnected Carbon Nanosheets Derived from Hemp for Ultrafast Supercapacitors with High Energy. ACS Nano 7 (6):5131-5141. doi:https://doi.org/10.1021/nn400731g
Williams M (2019) Machinery Milestones: Electric tractor power. Farmers Weekley, 20th of June,
Zillner N (2020) Product Communication Off-Highway Systems at ZF Friedrichshafen AG, personal communication, 11th of May.

Part V
Sustainability Constraints and Prospects

Chapter 16
Embedding Sustainability Strategies to Protect God's Gift: The Earth

George P. Nassos

Abstract The overconsumption of our natural resources, climate change, the decline in quality and availability of water, and population growth are four major environmental issues that require our immediate and continued attention. The sustainability strategies presented to address these issues include adapting to a sharing economy, also known as servicizing – converting a product economy to a service economy. It is also important to design for a circular economy in order to increase resource efficiency and decrease disposal. Biomimicry is a great strategy leading to innovations inspired by nature. Strategies for addressing climate change have been addressed by project Drawdown which provides 100 different proposals to reduce greenhouse gas (GHG) emissions. And when none of these address the problem, just think out of the box with strategies like vertical gardens. All of these strategies can be used to address the United Nations Sustainable Development Goals which are being adopted by corporations and countries worldwide. Achieving a sustainable world will also have a positive impact on financial economics and environmental, social, and governance (ESG) investing.

Keywords Sharing economy · Circular economy · Biomimicry · UN Sustainable Development Goals (UNSDGs) · "Big Hairy Audacious Goal" · Natural resource consumption · Climate change · Population growth · Water conservation · Environmental, social, and governance (ESG)

Historians claim that the earth is about four and a half billion years old, but humans have occupied the earth the last miniscule fraction of this time. The environment of the earth has been somewhat stable through most of the human presence until the

G. P. Nassos (✉)
George P. Nassos & Associates, Inc. Business and Environmental Consulting, Glenview, IL, USA

Ariston Institute of Chicago, Northbrook, IL, USA

E. Koukios, A. Sacio-Szymańska (eds.), *Bio#Futures*,
https://doi.org/10.1007/978-3-030-64969-2_16

past 100 or so years. And what has happened during these most recent years? Basically, there are four major issues that require everyone's attention:

1. Overconsumption of our natural resources – the ecological footprint. This is a tool for measuring and analyzing human natural resource consumption and waste output within the context of nature's renewable and regenerative capacity (or biocapacity). It represents a quantitative assessment of the biologically productive area required to produce the resources (food, energy, and materials) and to absorb the wastes of an individual or region. Today, we are consuming the equivalent of 1.7 earths, and every year the overconsumption increases (Global Footprint Network).
2. Climate change – It was in 1988 when James Hansen, head of the Goddard Institute for Space Studies, testified to Congress that global warming was beginning (https://www.nytimes.com/1988/06/24/us/global-warming-has-begun-expert-tells-senate.html). Up until 200 years ago, the concentration of carbon dioxide in the atmosphere was relatively stable at 275 parts per million (ppm) (http://350.org/en/about/science). Hansen went on to say that the maximum concentration should not exceed 350 ppm because it will cause rising sea levels leading to flooding, droughts leading to dwindling food production, and more frequent major storms. The problem today is that we are now consistently well over 400 ppm (https://www.co2.earth/daily-co2).
3. Conservation of water – Of all the water on the earth, 97% is saline. Of the remaining 3%, 68.7% is in the form of ice caps and glaciers, 30.1% is groundwater, and 0.9% is in some other unavailable form. This leaves only 0.3% of the fresh water on earth available to us on the surface with 87% in lakes, 11% in swamps, and 2% in rivers. This means that only 0.1% of all the water on the earth is available for industrial, agricultural, and human use. And of these three general uses, 70% is for agricultural use, 20% for industrial use, and only 10% for human consumption. With the growing economy and the growing population, the quantity of available water is decreasing rapidly. There are at least ten huge cities in the world that are running out of water (https://www.usnews.com/news/cities/slideshows/10-cities-most-at-risk-of-running-out-of-water?slide=6).
4. Population growth – Each of the environmental issues described above, consumption, water scarcity, and climate change, is all related to the world population. While man has been on this earth for thousands of years, the population did not reach one billion until 1804 (https://en.wikipedia.org/wiki/World_population_milestones). But it took only 123 years to add the second billion. And during the following 93 years, we added another 5.8 billion people to bring us to the current world population of 7.8 billion. Going forward we are adding about 10 million people every 6.4 weeks.

Mitigating the impact of these four major environmental issues leads to an urgency for a more sustainable planet.

1 Servicizing and the Sharing Economy

Servicizing is another word for taking a product that is used for a particular function and sell the function rather than the product – converting the business to a service. It has also been called dematerialization and leads to extended product responsibility (EPR). This is the principle whereby the process participants along the product chain or lifecycle share responsibility for the lifecycle environmental impacts of the whole product system, including upstream, production, and downstream impacts. This should result in lower lifecycle environmental impacts for products or product systems. Under such a scenario, the manufacturer would be more inclined not to consider planned obsolescence and manufacture a more durable product. This would result in fewer products manufactured, less resources employed, and less waste created.

Some early examples of servicizing include selling illumination rather than light bulbs which could lead to the production of more efficient light bulbs, selling an automobile painting service rather than the paint which led to a more efficient paint system and less waste, or selling a floor covering service rather than carpeting. This would lead to replacing and recycling only the carpet tiles that have worn rather than the entire carpet.

While servicizing is an excellent strategy for the reduction in consuming natural resources as well as reducing waste, this same thinking led to the evolution of the sharing economy. This can best be done by looking at "underutilized assets." If you think about some of your own assets that are not used as often as the manufacturer intended, they can lead, or have led, to opportunities. Of course, the manufacturer would prefer the limited use of the product as it can then sell more of it.

When movies became available on videotape, and later on discs (DVDs), they were purchased to watch at the owner's convenience and as often as they wished. However, movies are not viewed very often, usually no more than once or twice. So companies like Blockbuster bought the recorded movies and rented them out so the product was used considerably more often. With the sharing economy still evolving, Blockbuster closed its doors as this type of business went a step further so that a DVD is not even required. Netflix allows the customer to stream the movie through the Internet and onto the television, so a physical product is not even required.

A student in my class proposed the following business as an example of the sharing economy. If we can assume that the average person takes 2 weeks of vacation each year and replaces the luggage every 10 years, the luggage is used about 20 weeks before it is disposed or given away. People may have the desire to replace their luggage every 10 years as new and improved models are introduced. In the meantime, it is stored in the home for the other 50 weeks per year. This leads to the manufacture and eventual disposal of luggage that may not have been use throughout its useful life. A new company would purchase the latest design of luggage and rent them to vacationers for, say, 2 weeks at a time. In theory, the luggage would be used 52 weeks per year and disposed after its useful used. The company will always be offering the latest and most cost-effective luggage. There will be less luggage manufactured meaning that fewer raw materials will be consumed. In addition, there will be less material disposed in landfills.

Another example of an underutilized asset is the family automobile. If a car is driven 12,000 miles per year at an average speed of 25 miles per hour, which assumes both highway driving and neighborhood driving, it amounts to 1 hour and 20 minutes per day. This means that the car is used for a little over an hour per day and sits idle for almost 23 hours per day. What a waste! This thinking was the impetus for companies like Zipcar (www.zipcar.com) and I-GO (www.igocars.org) to purchase automobiles and allow members of their organization to use them when they need transportation. Of course, the members must reserve the car in advance, pick it up at a specific location, and return it to the same or another specific location. Most of the members that might have been two-car families usually sell one of their cars within a year after becoming a member.

This service business has gone to the next step where automobile owners provide this service themselves. In the USA, RelayRides (www.relayrides.com) is a relatively new company consisting of people who rent their **own** car, usually to neighbors. The company provides the technology to connect the car owner with the car renter and to provide the necessary insurance. In Europe, similar companies like Buzz Car (www.buzzcar.com) and Whip Car (www.whipcar.com) allow people to rent their cars to other drivers. Similarly, there are bicycle-sharing programs in almost every country.

In 2007 two roommates decided to rent out their spare bedroom to people visiting the city for a design conference (Lenyado). An idea was born, and the company, Airbnb (www.airbnb.com), took off in 2008 during the Democratic and Republican conventions in Denver, Colorado, and St. Paul, Minnesota. When Denver ran out of hotel beds, more than 300 residents used Airbnb to offer spare rooms. Unwittingly, they had opened a new tier in the holiday accommodation industry. Similar companies have been created since then with names like Crash Padder in the UK (www.crashpadder.com), Couch Share (www.couchshare.com), and I Stop Over (www.istopover.com).

One can also extend the concept of the sharing economy to something called "collaborative consumption." Fast Company calls it the reinvention of traditional market behaviors – renting, lending, swapping, sharing, bartering, and gifting – through technology (https://www.thebalancesmb.com/how-collaborative-consumption-can-help-your-home-business-4174323). It is not really new but goes one step further. People have rented out rooms in their homes (boarding) or provided freelance skills in the past. What's different now is that the Internet has expanded the market and made it easier for people who need something to find someone who has it.

Here are a few places you can rent your items or sell your skill.

Sell your skills
Gigbucks
FittyTown
Fiverr
Freelancer
Mechanical Turk
TaskRabbit
Tenrr
TwentyVille
Upwork

Room rental
Airbnb
Vrbo

Rent your car
Getaround
JustShareIt

Rent your parking spot
ParkEasier

Rent your clothes
Borrowing Magnolia (wedding dresses)
DateMyWardrobe

Rent yourself
Bridesmaid for Hire
Rent a Friend
Rover (pet sitting)
Wag (pet sitting and walking)
Care (child, senior, and pet care, housekeeping, tutoring, and more)

Miscellaneous items to rent
goBaby (baby items)
KitSplit (camera gear)
Neighborgoods
PeerRenters
Zilock

Sell stuff
ArtPal

2 The Need and Growth of a Circular Economy

Since the beginning of the production and consumption, the economy has been based on a linear business model. It consisted of manufacturing the product, the use of the product, and eventually disposing the product. Subsequently, it was given the name of the *linear economy*.

It often leads to a system that is inefficient and costly, and usually that harms the environment or depletes natural resources. For example, mining operations for coal or even gold can spoil ecosystems and disrupt nearby communities. Making steel from ore requires a large amount of energy, which produces globe-warming carbon dioxide. A by-product of the linear model is material waste, which takes up space and may include contaminants. Trash ends up in undesirable places. The so-called Great Pacific garbage patch (https://www.theoceancleanup.com/great-pacific-garbage-patch) may be the most well-known example of global-scale plastic pollution. Yet products like steel and plastic can be reused, refurbished, and recycled to capture untapped value.

However, it is becoming worse. Due to the exponential growth of the global population, an increase of over five billion people since 1940, and the subsequent consumption of our natural resources, this linear economy no longer works. As of 2018, the world is consuming the equivalent of 1.7 earths of its natural resources (Global Footprint Network). What is necessary today is a *circular economy* which consists of take, make, use, reuse, and reuse again and again.

One of the earliest examples of the circular economy was developed in Kalundborg, Denmark, in the early 1970s. It was not originally created as a circular economy but has grown as reducing waste became more important, and was not originally known as a circular economy until very recently when this sustainability model became necessary and was created.

There are currently over 30 exchanges of materials among the companies of Kalundborg (https://en.wikipedia.org/wiki/Kalundborg_Eco-industrial_Park). The Asnaes Power Station is at the heart of the network. The power company gives its steam residuals to the Statoil Refinery, meeting 40% of its steam requirements, in exchange for waste gas from the refinery. The power plant creates electricity and steam from this gas. These products are sent to a fish farm, to Novo Nordisk, which obtains all of its required steam from Asnaes, and to a heating system that supplies 3500 homes. These homeowners pay for the underground piping that supplies their heat, but receive the heat reliably and at a low price. Fly ash from Asnaes is sent to a cement company, and gypsum from its desulfurization process is sent to Gyproc for use in gypsum board. Two-thirds of Gyproc's gypsum needs are met by Asnaes. Statoil Refinery removes the sulfur from its natural gas and sells it to a sulfuric acid manufacturer, Kemira. The fish farm sells sludge from its ponds as fertilizer to nearby farms, and Novo Nordisk gives away its own sludge, of which it produces 3000 cubic meters per day. The sludge is to be refined for biogas for the power plant.

Water reuse programs have also been developed within Kalundborg. Statoil sends 700,000 cubic meters of cooling water per year to Asnaes, which purifies it and uses it as "boiler feedwater." Asnaes also uses approximately 200,000 cubic meters of Statoil's treated wastewater per year for cleaning. The 90 °C residual heat from the refinery is not used for district heating due to taxes. Instead, heat pumps are used with the 24 °C wastewater as a heat reservoir.

Kalundborg was the first time separate industries grouped together to gain a competitive advantage by material exchange, energy exchange, information exchange, and/or product exchange. The very term, industrial symbiosis, was first defined by a station manager in Kalundborg as "a cooperation between different industries by which the presence of each increases the viability of the others, and by which the demands of society for resource savings and environmental protection are considered" (https://en.wikipedia.org/wiki/Kalundborg_Eco-industrial_Park).

Although the Kalundborg symbiosis was probably the first industrial model of the circular economy, examples really existed long before. At least as early as the 1940s, the circular economy was employed by the dairy industry although it was not known by that name. A dairy delivered milk via a truck in a quart or gallon glass bottle to the home of its customers. It was placed outside the front door of the home. Upon consuming the contents of the bottle, the customer would place the empty bottle outside the front door for the next milk delivery. The milkman would then pick up the empty bottle and replace it with a full bottle. The empty glass bottle would be taken back to the dairy, cleaned thoroughly, and refilled for the next delivery. This system created almost no waste at all except for possibly the bottle cap.

Following the milkman model, 80 years later TerraCycle introduced its circular economy model in early 2019 and called it Loop. It joined forces with some of the top consumer brand companies such as Häagen-Dazs, Procter & Gamble, Nestlé, PepsiCo, Unilever, Mars, Clorox, Coca-Cola, Mondelēz, and Danone. European retailer Carrefour, logistics company UPS, and resource management company Suez also participated in the system when it was launched in Paris and New York.

Loop was created as a circular shopping platform that transforms the packaging of everyday essentials from single-use disposable to durable, feature-packed designs. They took some of the most popular brands and introduced them in more functional packaging. After the contents have been consumed, the empty container is placed back in the Loop tote and scheduled for a free pickup from the customer's home by UPS. The product containers are designed for a minimum of 100 uses.

Loop would determine the customer's consumption of the various products based on the empty bottles that are returned. Using these data, Loop would then replenish the products consumed by the customer. This would be an ongoing process that did not require the customer to continue to reorder. Loop would hygienically clean and sanitize the empty packaging and send it back so they are ready for reuse, instead of ending up as waste after a single use.

The Loop tote became a breakthrough zero-waste delivery system that eliminates wasteful single-use shipping materials. Durable packaging enables innovative features and unparalleled design, so that the customer can experience everyday products in a brand new way.

The circular economy is basically a planned industrial system that is restorative by intention and design for environmental reasons, not for economic reasons. However, this circular economy leads to both cost reduction and material reduction. The idea is that rather than discarding products before their value is fully utilized, we should use and reuse them as much as possible. Presently only a few percentage points of the original product value are recovered after use. A good example is the mobile phone. With most people today using a smart phone, they are encouraged to replace it at least every 2 years. The manufacturers offer improved phones in order to continue increasing sales. However, the old phones, which are still functional, become an unnecessary waste. Another example is the personal computer (PC). According to Moore's law (https://www.intel.com/content/www/us/en/silicon-innovations/moores-law-technology.html), the efficiency of a PC will double every 18 months because it was the time required for Intel to develop a new computer processor with double the speed of the previous one. But instead of just replacing the processor, people were "trained" to replace the entire PC even though the case, the keypad, and most other parts were the same as the previous model.

3 Designing for the Circular Economy

Products can be designed from the outset so that, after their useful lives, they will provide nourishment of something new. For example, they can be conceived as "biological nutrients" that will easily reenter the water or soil without depositing synthetic materials and toxins. Or they can be "technical nutrients" that will continually circulate as pure and valuable materials within closed-loop industrial cycles, rather than being "recycled" – really, down cycled – into low-grade materials and uses, making this process truly eco-effective. This is another example of learning from nature (McDonough and Braungart 2002).

The products can also be designed for disassembly or even deconstruction. It can be made easier for the product to be repaired or upgraded, prolonging its useful life. It can also help ensure that the product is recycled and enables whole components of the product to be reused. The degree to which the product can be easily disassembled often determines how the product will end its life.

Another design feature is to consider flexibility whereby a product can have multiple applications. In this manner, it will be more efficient to fulfill the applications with the greatest demand. An example of this feature is the product line of Do It Right This Time (DIRTT) (https://www.dirtt.net/) which designs and manufactures interior systems for office space, schools, and healthcare facilities. Given the allocated space, DIRTT will design and provide the components for the space as initially required. However, the components can be disassembled and repurposed for a different design or layout.

When talking about reducing the waste generated via the circular economy, most people normally refer to municipal solid waste (MSW) which is the waste generated by people and businesses. As an example, in 2015, about 260 million tons of MSW were generated in the USA. Basically, this is the quantity of material that is available for recovery and reuse. However, this is not all of the waste generated in the USA.

A study was made in 2009 to determine the total waste generated in the USA not only from the public and businesses but also from industry. This would include oil and gas waste, mining waste, and hazardous and nonhazardous industrial waste. It was difficult to determine this total quantity of waste, given the name Gross National Trash (GNT), because these data are not provided by industry. But an earlier estimate was made in 1992 by Joel Makower that the GNT was about 13 billion tons with the MSW representing less than 2% (https://www.greenbiz.com/blog/2009/03/20/calculating-gross-national-trash?utm_source=newsletter&utm_medium=email&utm_content=2019-04-07&utm_campaign=greenbuzz&mkt_tok=eyJpIjoiWWpVMVpqQXdNamhoWkdOayIsInQiOiJBbGt0Yk1CQTNNNUW-

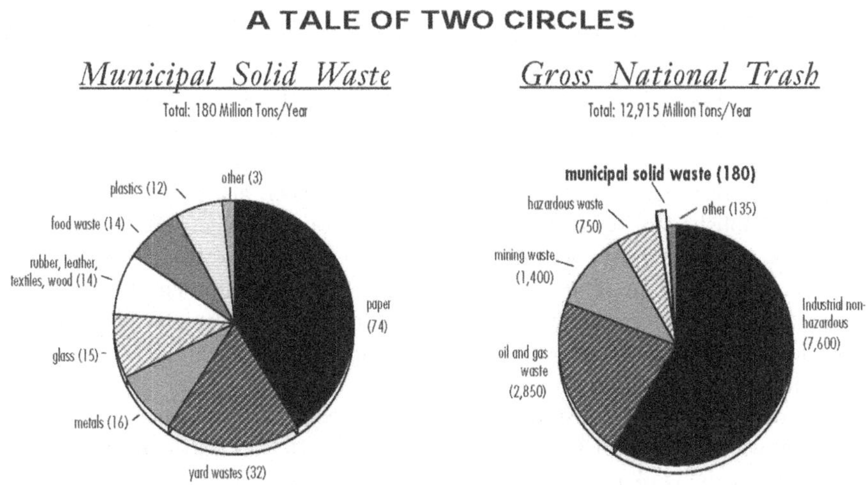

Fig. 16.1 MSW vs. GNT

hCQVpYR2lMMnNyMnhMb25aUGQ0R0xyNTNTaVFIZXdETnlqNWJLN-mI3ek12TmNiXC9rR3dEUlVhckY5MzNDTWg2YmR2Z2FnYndXa2orVTJ1O-EtBTGgzaTJwSW1cL3hHTERrU0djVUdFSkxndlBzYUlLREJQWWxqIn0%3D). This indicates that there is a massive amount of waste that can be considered for the circular economy (See Fig. 16.1).

Designing a circular system to deal with the Gross National Trash would be a major effort with an equally major benefit; however there are also opportunities to apply this concept with something very small. An example might be the dilemma at Christmas time as to whether one should purchase a real tree that was cut down for use in a home for several weeks or to purchase an artificial tree that can be used every year. So why not lease one or rent one? A new business could be a tree farm that grows pines, spruces, and firs, and near the end of the year makes them available for use. The tree would be removed from the earth along with its roots and have the base encased in a net-type bag. The tree could then be taken to a home along with a special stand/container. At the end of the Christmas season, the tree would be returned to the farm where it would be replanted until next year. Some people like small trees about 1.5 meters, while others can use trees as big as 3 meters. Consequently, a tree can be used over multiple years as it grows from 1.5 meters to 3 meters.

This system would provide many environmental benefits. The trees would not be cut down and discarded but rather would have a long life in sequestering carbon dioxide. There would be no shipments over great distances which otherwise would emit more carbon. No waste would be generated from a typically cut-down tree nor plastic waste from a synthetic plastic tree.

4 Environmental Innovation Through Biomimicry

In the 1600s, Europe was affected by a crippling shortage. People had to deal with the fact that a valuable commodity was increasingly in short supply. What was it? Rags.

While hard to believe, rags were used to make paper, and paper was in great demand. Publishers of books, newspapers, and various pamphlets all needed more paper. But there were just not enough rags. Advertisements appeared asking women to "save their rags." In 1666, England banned the use of cotton and linen for the burial of the dead, decreeing they must be saved for making paper. One entrepreneur even suggested using the cloth from Egyptian mummies. The scarcity of rags led to extensive paper shortages in Europe and America.

One day, a French scientist took a walk in the woods. Réne Antoine Ferchault de Réaumur was an accomplished physicist and chemist. He was also a man who loved bugs and insects. Walking in the woods that day, he came upon an abandoned wasp nest. Delighted, he began to examine it in detail, and an astounding fact dawned on him: the nest was made of paper, paper made by wasps, paper made without the use of rags. How could this be? The wasps did it by chewing wood and plant fibers.

What wasps could do, he argued, man could find a way to do it also. It took decades, but his discovery was what inspired inventors to develop ways to make paper from wood pulp. Thanks to Réaumur's nature walk that day, we can now do what once would have been considered almost criminal: crumple up a piece of paper and throw it out (Beyer 2009).

Another interesting example is the process of photosynthesis, from the Greek word that means "using light to put something together." Here plants, algae, and bacteria take carbon dioxide, water, and sunlight to produce energy-rich sugars while releasing oxygen. Animals, including humans, take that oxygen and the sugars and transform them back to carbon dioxide, water, and energy. Without the sunlight, this photosynthesis reaction would never take place.

A little over 100 years ago, an Italian chemistry professor, Giacomo Ciamician, wrote in *Science* (Ciamician 1912) magazine that one day our landscape would change from the typical power plants to a large array of glass that would mimic photosynthesis and produce a different form of energy. Through his inspiration, we now have acres and acres of solar panels made of silicon, a material that allows for the production of energy, even though it is nowhere found in the vegetation structure.

This is just one of hundreds of examples of innovations inspired by nature. In the classic book, *Biomimicry* (Benyus 1997), Janine Benyus defines biomimicry in three different terms:

1. *Nature as model.* Biomimicry is a new science that studies nature's models, and then it imitates or takes inspiration from these designs to develop processes to solve human problems, e.g., a solar cell inspired by a leaf.
2. *Nature as measure.* Biomimicry uses an ecological standard to judge the "rightness" of our innovations. After 3.8 billion years of evolution, nature has learned: What works? What is appropriate? What lasts?

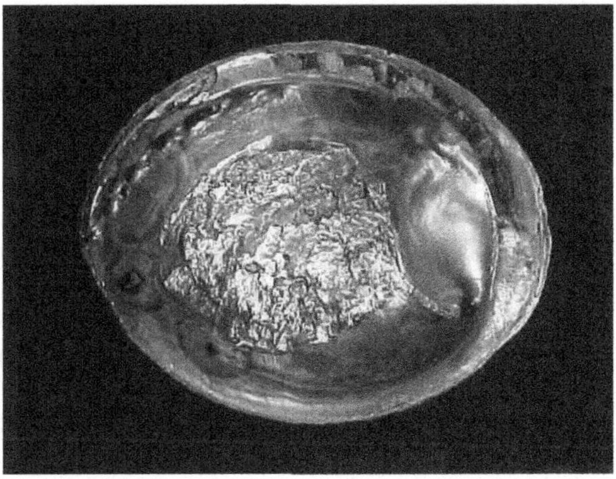

Fig. 16.2 Abalone shell – stronger than ceramic

3. *Nature as mentor.* Biomimicry is a new way of viewing and creating value from nature. It introduces an era based not on what we can *extract* from the natural world but on what we can *learn* from it.

What follows are examples of how nature has inspired or can inspire innovation that will lead to sustainable systems. Benyus now uses a better word than biomimicry for these new systems. She calls them "inspired by nature."

The abalone (Fig. 16.2) is a shellfish that is known for its delicious meat as well as its smooth inner coating that makes the shell as "hard as nails." A car could drive over an abalone shell and not break it. It is stronger than any produced ceramic, but why? It consists of an intricate crystal architecture that allows it to shrug off stress. If you can see the structure from the side, it consists of hexagonal disks of calcium carbonate stacked like a brick wall. Between the "bricks" is a polymer that gives the formation some flexibility to ward off the head-on stress. The ceramics that are produced commercially, such as glass, porcelain, and bricks, are manufactured by taking inorganic particles from the earth and subjecting them to heat and pressure. The result will be a very strong material, but brittle and subject to cracking. Scientists are now looking at natural designs like the abalone inner coating to determine how to reproduce that structure synthetically.

Just imagine trying to put up a building that could be as strong as the abalone shell. Suppose you wanted to build a very strong structure such as a house or small office, and after it is designed, structural forms are connected to create the walls of the building. These forms are designed so they can contain liquids such as seawater. Suppose these forms are then filled with seawater and the secret chemical used by the abalone to create its shell is added to the seawater in these forms. After a short period of time, the water is drained, and what remains are walls stronger than any ceramic or brick currently made. All that is needed to accomplish this feat is to determine how the abalone makes its shell.

During the past 15 years, the production of ethanol as a fuel has become critical for the transportation industry. For many years, up to 10% ethanol has been added to gasoline in certain geographic areas to reduce carbon emissions. Today, ethanol is the primary fuel for flex-fuel automobiles that can use up to 85% ethanol. In Brazil, ethanol is produced from sugarcane bagasse, but in the USA it is produced from corn.

The process to convert corn to ethanol consists of a chemical process whereby enzymes are added to convert the starch in the corn to a sugar. Yeast is then added, and the sugar is converted to ethanol. The process may not be very efficient as some studies report that it takes as much as 1.29 units of energy during the process to produce 1.0 units of energy as ethanol (Pimentel). On the other hand, other studies show just the reverse as the output-input energy ratio is 1.34 (Shapouri et al.). The difference in these two analyses depends on the assumptions, primarily the allocation of the energy consumption to the by-products.

In any event, researchers have looked at ways to increase the efficiency of the conversion process. Currently, the corn-to-ethanol process utilizes only the corn kernel as this is the only part of the corn plant that can be converted to sugar and then alcohol. Scientists at Michigan State University noted that the enzyme that allows a cow to digest grasses and other plant fibers can be used to turn other plant fibers into simple sugars. The scientists then discovered a way to grow corn plants that contain this enzyme. They have inserted a gene from a bacterium that resides in a cow's stomach into a corn plant. Now, the sugars locked up in the plant's leaves and stalk can be converted into usable sugar without expensive synthetic chemicals (Sticklen 2008). This new approach should make the corn-to-ethanol process considerably more efficient, thanks to a cow's stomach.

Many years ago, engineers were challenged to design a structural part that must be very rigid yet very light. The real design challenge was to use the least amount of material to keep the weight very low. For example, perhaps the engineer was required to design an airplane wing. The goal is to develop an internal structure that could be covered with a skin layer which would make the wing very rigid yet very light.

If the engineer were to start with a geometric design that utilizes the least amount of material for a given area, that geometry would be a circle. We learned from geometry that the largest area for a given perimeter is in the form of a circle. However, if that design is a matrix of circles, there is a large amount of material where the circles are attached (Fig. 16.3). The engineer could then use the simplest geometric figure which is the triangle, but even in this matrix, there is a large amount of material at the vertex of each triangle (Fig. 16.4). Geometrically, the engineer would then employ the next design in the hierarch which would be a square. This design would show an improvement, but there is still too much material at the intersection of four

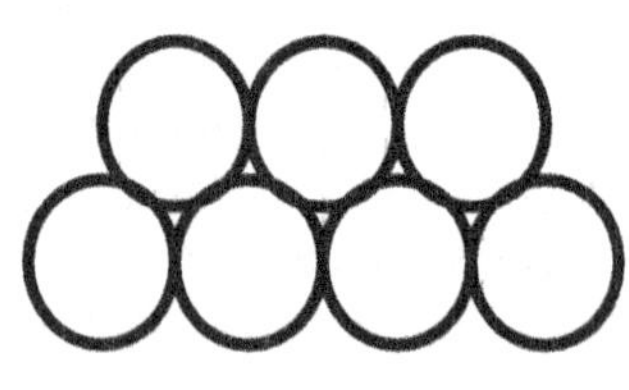

Fig. 16.3 Structure with circles

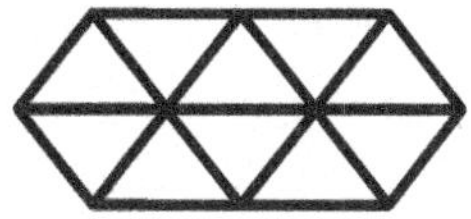

Fig. 16.4 Structure with triangles

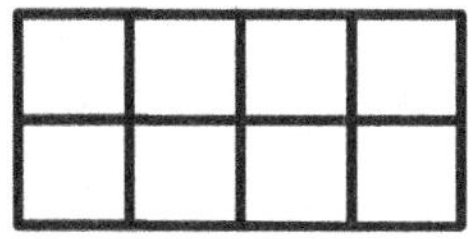

Fig. 16.5 Structure with squares

lines (Fig. 16.5). The next geometric design in the hierarchy is a pentagon, but the asymmetry of this design makes it impossible to consider (Fig. 16.6).

Of course, the next step would be to use a hexagonal-type structure which consists of the least number of lines at an intersection, and thus this would become the choice design (Fig. 16.7). While this may be a design that was created by engineers, perhaps 100 years ago, bees have been using this same design for thousands of years in building their homes – the beehive (Fig. 16.8). Bees learned how to construct the most rigid structure with the least quantity of material.

There are many more examples of technologies inspired by biomimicry (https://www.scoop.it/t/biomimicry?page=1).

5 Imbedding the UN Sustainable Development Goals to Achieve Sustainability

It is important to understand that there are many initiatives that an organization can undertake for the benefit of the environment. Some of these activities may be sustainable, but many of them are not truly sustainable but still very beneficial as they reduce the negative impact on the environment. These environmentally beneficial activities embraced by corporations become the guidelines that are followed, better known as "principles." Examples of some of the many different principles adopted by organizations are CERES, Hannover, Precautionary, Earth Charter, and Sanborn.

In September 2015, the United Nations passed a resolution titled *Transforming Our World: The 2030 Agenda for Sustainable Development*. Included in this agenda were 17 Sustainable Development Goals consisting of 169 targets. The purpose of the goals and targets was to stimulate action over the next 15 years, up through 2030, in areas of critical importance for humanity and the planet. More specifically, the agenda focused on these five areas:

1. People – We are determined to end poverty and hunger, in all their forms and dimensions, and to ensure that all human beings can fulfill their potential in dignity and equality and in a healthy environment.
2. Planet – We are determined to protect the planet from degradation, including through sustainable consumption and production, sustainably managing its

Fig. 16.6 Structure with pentagons

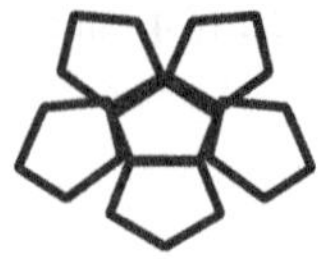

Fig. 16.7 Structure with hexagons

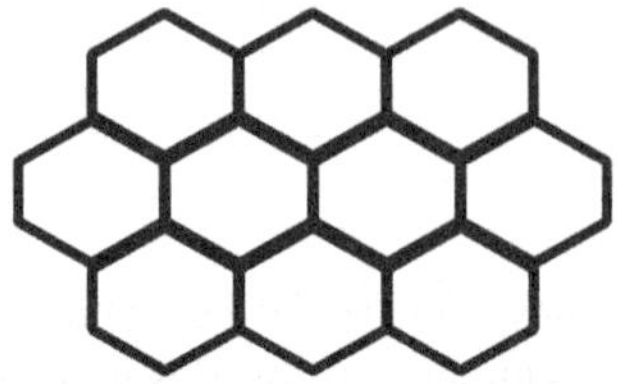

Fig. 16.8 Beehive – most efficient structure

natural resources and taking urgent action on climate change, so that it can support the needs of the present and future generations.
3. Prosperity – We are determined to ensure that all human beings can enjoy prosperous and fulfilling lives and that economic, social, and technological progress occurs in harmony with nature.
4. Peace – We are determined to foster peaceful, just, and inclusive societies which are free from fear and violence. There can be no sustainable development without peace and no peace without sustainable development.
5. Partnership – We are determined to mobilize the means required to implement this Agenda through a revitalized global partnership for sustainable development, based on a spirit of strengthened global solidarity, focused in particular on the needs of the poorest and most vulnerable and with the participation of all countries, all stakeholders, and all people.

The 17 Sustainable Development Goals (SDGs) adopted by the UN in 2015 are most often depicted as a table such as Fig. 16.9.

The SDGs were written as a blueprint to achieve a better and more sustainable future for everyone living on this planet. They address the global challenges that are

Fig. 16.9 2015–2030 UN Sustainable Development Goals

affecting everyone, including those related to poverty, inequality, climate, environmental degradation, prosperity, and peace and justice. The SDGs interconnect with a target to be successful by 2030. The 17 SDGs are summarized as follows:

1. **No poverty**: End poverty in all its forms everywhere
2. **Zero hunger**: End hunger, achieve food security and improved nutrition, and promote sustainable agriculture.
3. **Good health and well-being**: Ensure healthy lives and promote well-being for all at all ages.
4. **Quality education**: Ensure inclusive and equitable quality education and promote lifelong learning opportunities for all.
5. **Gender equality**: Achieve gender equality and empower all women and girls.
6. **Clean water and sanitation**: Ensure availability and sustainable management of water and sanitation for all.
7. **Affordable and clean energy**: Ensure access to affordable, reliable, sustainable, and modern energy for all.
8. **Decent work and economic growth**: Promote sustained, inclusive, and sustainable economic growth, full and productive employment, and decent work for all.
9. **Industry, innovation, and infrastructure**: Build resilient infrastructure, promote inclusive and sustainable industrialization, and foster innovation.
10. **Reducing inequalities**: Reduce income inequality within and among countries.
11. **Sustainable cities and communities**: Make cities and human settlements inclusive, safe, resilient, and sustainable.
12. **Responsible consumption and production**: Ensure sustainable consumption and production patterns.

13. **Climate action**: Take urgent action to combat climate change and its impacts by regulating emissions and promoting developments in renewable energy.
14. **Life below water**: Conserve and sustainably use the oceans, seas, and marine resources for sustainable development.
15. **Life on land**: Protect, restore, and promote sustainable use of terrestrial ecosystems, sustainably manage forests, combat desertification, and halt and reverse land degradation and halt biodiversity loss.
16. **Peace, justice, and strong institutions**: Promote peaceful and inclusive societies for sustainable development, provide access to justice for all, and build effective, accountable, and inclusive institutions at all levels.
17. **Partnerships for the goals:** Strengthen the means of implementation and revitalize the global partnership for sustainable development.

As mentioned earlier, it is imperative that everyone does what is necessary to reduce carbon emissions even if there were a slight possibility that climate change is not for real. This objective is repeated as SDG 13. We need a very rigorous plan to reduce or "draw down" carbon emissions. A team of 200 scientists and engineers worldwide came together in Project Drawdown to develop 100 creative ideas that will do exactly that, and it was compiled in *Drawdown*, a book edited by Paul Hawken (Hawken 2017).

These potential solutions to the climate change problem can be ranked as to how cost-effective they are; how quickly they can be implemented; or how beneficial they are to society. While these are all very important metrics for interpreting the result, the authors of *Drawdown* ranked the solutions based on the total amount of greenhouse gases (GHG) they can potentially avoid or remove from the atmosphere. While the rankings are global, the relative importance of one solution over another may depend on geography, economic conditions, or which of the eight sectors is of most interest to that country. Table 16.1 lists the seven different sectors, the top two solutions in each sector, and the different metrics: (1) total atmospheric CO2 equivalent reduction in gigatons (GT), (2) the net cost in billion US dollars, and (3) net savings in billion US dollars.

The reduction in carbon dioxide equivalents is the quantity expected to be removed by a particular solution between 2020 and 2050. The total cost of each solution is the amount needed to purchase, install, and operate the system over the same 30-year period. The estimates tend to be conservative, but they still tend to offer an overwhelming net savings. For some of the solutions, like a specific rainforest or support girls' education, the savings may not be calculable. A summary of these 14 solutions in the 7 sectors is as follows.

Table 16.1 Drawdown – top solutions in each sector

Sector	Solution	Total atmospheric CO2-Eq reduction (GT)	Net cost (billions US$)	Net savings (billions US$)
Energy	Wind turbines (onshore)	84.60	1225.37	7425.00
	Solar farms	36.90	−80.60	5023.84
Materials	Refrigeration	89.74	N/A	−902.77
	Alternative cement	6.69	−273.90	N/A
Food	Reduced food waste	70.53	N/A	N/A
	Plant-rich diet	66.11	N/A	N/A
Women and girls	Educating girls	59.60	N/A	N/A
	Family planning	59.60	N/A	N/A
Buildings and cities	District heating	9.36	457.07	3543.50
	Insulation	8.27	3655.92	2513.33
Land use	Tropical forests	61.23	N/A	N/A
	Temperate forests	22.61	N/A	N/A
Transport	Electric vehicles	10.80	14,148.03	9726.40
	Ships	7.87	915.93	424.35

The Drawdown team has continued to develop systems for reducing carbon emissions and has recently published *Drawdown Review 2020* (https://drawdown.org/).

6 Developing a "Big Hairy Audacious Goal"

In 1994, the term BHAG (Big Hairy Audacious Goal) was coined in a book, *Built to Last: Successful Habits of Visionary Companies* (Collins and Porrass 1994), to refer to a long-term goal of an organization in order to achieve its mission. Three different examples are provided to show how one might develop such a solution to achieve the intended goal.

6.1 Washing Machines

For many years, the washing of clothes has been accomplished by adding a detergent to hot or warm water and then rinsing the clothes with cold water. This obviously is very energy intensive from both the heating of the water and the requirement for operating the washing machine. Usually the hot water was obtained from a central hot water tank, but some machines were developed for adding cold water and heating it by the washing machine.

The soap companies eventually produced a detergent that could be effective in cold water. This certainly reduces the energy requirement for washing clothes. However, the process still requires the use of a large quantity of water that is usually discharged to a sanitary sewer rather than being reused.

Applying the BHAG philosophy, how could the clothes be cleaned with minimal energy usage and no water? Perhaps the use of some nontoxic chemical in its vapor form could be used to remove dirt and grime from clothes. This chemical, in pellet form, could be added to, say, a container with the clothes. The pellet would then be activated to release the vapor and then the clothes tumbled so that the vapor makes contact with all the clothes' surfaces. This, in theory, will make clothes cleaning more sustainable, and it is expected that soap manufacturers are conducting research in this area.

6.2 Toilets

In most developed countries, almost all toilets are operated from the same water source as our drinking water. Is it really necessary to flush liquid waste or solid waste with such clean water? If the BHAG philosophy is applied, perhaps another source of water such as gray water could be used. But where would this water be obtained for a home? Again, thinking BHAG will lead to sources of water like collected rainwater, washing machine wastewater, dishwasher wastewater, or even

Fig. 16.10 Toilet using gray water

water going down the drain from a shower. A simple solution would be to collect the drain water from a bathroom sink and divert it to the toilet tank. This gray-water concept has been converted to a number of different designs, one of which is shown in Fig. 16.10. Another example of applying the same concept is in a public bathroom for men. Each of the urinals can have a small water dispenser and sink above each urinal so the person can wash his hands without leaving the urinal as shown in Fig. 16.11. The wash water from the sink would then drain down to flush the urinal. Thus the water would be used twice.

Fig. 16.11 Urinals with sink for flushing

6.3 Urban Farming

Prior to the nineteenth century, a typical home consisted of four walls and a roof – a very simple structure. As people started living closer together, anyone needing more space added a second floor. But then it seemed to be more cost-effective for more than one family to live in a building with four floors and a roof. So this became the beginning of apartment buildings. As land became more valuable and people continued to live in cities, developers started constructing taller and taller buildings. So basically, starting with living at ground level, we have gone to living in high-rise structures.

The need for more food created a demand for more agricultural land. But because the farmland is usually far from the large cities where most of the population is located, food travels great distances from its source to the dinner table, and that distance is estimated to be about 1500 miles. Implementing the BHAG concept, why can't agricultural land go up just like homes?

A number of entrepreneurs have begun with this concept by developing fruit and vegetable growing in previously vacant warehouses. One such grow house is AeroFarms (https://aerofarms.com/ n.d.) in Newark, NJ, which consists of 6500 square meters in a former indoor sports arena (Fig. 16.12). The vegetables and/or fruits are grown without sun or soil in a fully controlled indoor environment. The company uses aeroponics to mist the roots of the greens with nutrients, water, and

Fig. 16.12 AeroFarms, Newark, NJ

oxygen. The aeroponic system is a closed-loop system, and it uses 95% less water than field farming and 40% less water than hydroponics.

Another example of vertical farming is a proposed new facility that combines the vertical farming concept with solar energy to create a "net-zero vertical farm." This new farm, a joint effort between the University of Toronto-Scarborough and Centennial College, will provide, when developed, all the electrical power needed for the building which includes research laboratories and a training facility (https://dailyhive.com/toronto/canada-first-ever-net-zero-vertical-farm-2019). It will provide all the power needed from the heating and cooling to the lighting of the vertical farm.

Besides producing more food for the growing population, there are numerous other advantages of vertical farming:

- Year-around crop production
- No weather problems
- All food grown organically
- Eliminates agricultural runoff
- Farming in urban cities
- Reduces fossil fuel use
- Converts black and gray water
- Provides jobs for local residents

7 Sustainable Strategies and Beyond

There are many opportunities to implement sustainable strategies and reduce the impact on the environment, provide social benefits, and do it profitably. It is a matter of thinking on how to make a service or product more sustainable, or even develop a new product or service for an unmet need. During my academic career, I challenge my students to apply a sustainability strategy to their place of work or create a new business to solve a problem. Listed below are a few examples provided by my students:

- Medication delivery – proposing the reuse of bottles and tubing rather than disposal after one use.
- Farm equipment – a central logistics service providing tractor, huskers, and others to farms in a region rather than every farmer required to possess each piece.
- Eco-toys – lease toys to families as they are used for a period of time considerably less than the life of the toy.
- Ford BOP – proposed a joint venture between Ford and Infosys, a company in India, to produce a low-cost car employing the base of the pyramid strategy.
- Communal refrigeration – large refrigeration and freezers to be shared by people living in small apartments in a large building.

- Highway solar panels – install solar panels along or on highways where land is not used.
- Home furnishings – expensive long-lasting home furnishings can be exchanged for a different long-lasting style.
- e-Club – for those that want the latest electronic equipment, lease it and then allow someone who doesn't need the latest to buy it.
- "Moo-trient" – collect cattle methane waste and convert it to energy and cattle manure to fertilizer.
- Reuse furniture scraps – collect the textile and leather remnants, primarily from high-end and luxurious furniture companies, and manufacture new smaller products like pillows, wallets, hats, gloves, handbags, and jewelry.
- Home-delivered meals – deliver the dinner that will come along with reusable and compostable options giving the customers the option to return the reusable containers or the compostable containers.
- Refrigeration for sub-Saharan Africa – a company will provide on-demand transportation services connecting the farmers with refrigeration facilities.
- Eco-travel – provide tourism options that are all sustainable in order to reduce environmental issues at highly visited locations such as the Phi Phi Islands in Thailand, or pollution and erosion in Machu Picchu (Peru) due to massive tourism, and the ecological deterioration of Mount Everest.
- Ice cream shop – collecting the plastic wrap and wooden sticks, after popsicles are consumed, to recycle.

8 And for the Investment Community

About 50 years ago, one of the greatest economists of the twentieth century, Milton Friedman, stated in a *New York Times* article that the sole purpose of a company is to make money for its shareholders. In other words, he was saying that the companies must maximize their earnings per share (EPS). This has been the mantle of almost all corporations, and the shareholders have been very pleased. But over the past 10 years or so, this has been changing. There are other issues that must be addressed for the future of the company, and one of the biggest issues is the financial risk due to climate change. Companies have become very aware of environmental, social, and governance (ESG) criteria and the standards necessary to determine the future performance relative to return and risk. These are the new criteria in making companies environmentally and socially sustainable. So organizations must choose ESG over EPS.

References

Global Footprint Network https://www.footprintnetwork.org/

https://www.nytimes.com/1988/06/24/us/global-warming-has-begun-expert-tells-senate.html

http://350.org/en/about/science

https://www.co2.earth/daily-co2

https://www.usnews.com/news/cities/slideshows/10-cities-most-at-risk-of-running-out-of-water?slide=6

https://en.wikipedia.org/wiki/World_population_milestones

Lenyado, Benji, *New York Times Travel,* July 15, 2010.

https://www.thebalancesmb.com/how-collaborative-consumption-can-help-your-home-business-4174323

https://www.theoceancleanup.com/great-pacific-garbage-patch

https://en.wikipedia.org/wiki/Kalundborg_Eco-industrial_Park

https://www.intel.com/content/www/us/en/silicon-innovations/moores-law-technology.html

McDonough, William and Michael Braungart, *Cradle to Cradle,* The North Point Press, 2002.

https://www.dirtt.net/

https://www.greenbiz.com/blog/2009/03/20/calculating-gross-national-trash?utm_source=newsletter&utm_medium=email&utm_content=2019-04-07&utm_campaign=greenbuzz&mkt_tok=eyJpIjoiWWpVMVpqQXdNamhoWkdOayIsInQiOiJBbGt0Yk1CQTNNUWhCQVpYR2lMMnNyMnhMb25aUGQ0R0xyNTNTaVFIZXdETnlqNWJLNmI3ek12TmNiXC9rR3dEUlVhckY5MzNDTWg2YmR2Z2FnYndXa2orVTJ1OEtBTGgzaTJwSW1cL3hHTERrU0djVUdFSkxndlBzYUlLREJQWWxqIn0%3D

Beyer, Rick, "The Greatest Science Stories Never Told", Harper Collins Publishers, 2009

Ciamician, Giacomo, "The photochemistry of the future", Science 36, 385–394 (1912)

Benyus, Janine M., "Biomimicry", William Morrow and Company, 1997

Pimentel, David, http://www.college.wustl.edu/~anthro/articles/pimentel-ethanol.pdf

Shapouri, Hosein, James A. Duffield, and Michael Wang, "The Energy Balance of Corn Ethanol: An Update", U.S. Department of Agriculture, Office of the Chief Economist, Office of Energy Policy and New Uses. Agricultural Economic, Report No. 813.

Sticklen, Mariam, "Plant Genetic Engineering for Biofuel Production: Towards Affordable Cellulosic Ethanol", Nature Review Genetics, June 2008.

https://www.scoop.it/t/biomimicry?page=1

Hawken, Paul, *Drawdown: The Most Comprehensive Plan Ever Proposed to Reverse Global Warming",* Penguin Books, 2017.

https://drawdown.org/

Collins, James C. and Jerry I Porrass, *Built to Last: Successful Habits of Visionary Companies*, William Collins, 1994

https://aerofarms.com/

https://dailyhive.com/toronto/canada-first-ever-net-zero-vertical-farm-2019

Chapter 17
Biological Degradation of Odorous Air Pollutants

Damian Kasperczyk, Krzysztof Urbaniec, and Krzysztof Barbusiński

Abstract An offensive and prolonged smell that significantly interferes with the enjoyment and use of the affected property is known as odor nuisance. It can cause adverse health effects such as emotional unease, eye irritation, and respiratory problems. The most significant odorous gases are volatile organic compounds, nitrogen compounds including ammonia, and sulfur compounds including hydrogen sulfide. Major sources of these pollutants are animal farms, various chemical as well as food- and feed-processing industries, waste treatment or disposal facilities, and wastewater treatment plants. As more and more countries adopt regulations that prohibit business organizations from emitting strong odors, various technologies for removal of odorous compounds from waste gases are in use. Physical and chemical processes, such as activated carbon adsorption, ozone oxidation, catalytic oxidation, and incineration, are characterized by rather high energy requirements and high treatment costs. As more sustainable processes, microbiological odor treatment methods using various types of air filtration equipment are introduced. Their advantages include absence of explosion risk, operation at near atmospheric pressure and temperature range 10–40 °C, no secondary waste generation, less energy requirements. This chapter will review microbiological odor treatment methods and summarize authors' experience of development and applications of compact trickle-bed bioreactor for odors removal from ventilation air discharged from various industrial facilities.

Keywords Odorous chemical compound · Odor source · Odor nuisance · Odor removal · Microbiological odor treatment method · Compact trickle-bed bioreactor

D. Kasperczyk
Ekoinwentyka Ltd., Ruda Śląska, Poland

K. Urbaniec (✉)
Warsaw University of Technology, Płock, Poland
e-mail: k.urbaniec4@upcpoczta.pl

K. Barbusiński
Silesian University of Technology, Gliwice, Poland

E. Koukios, A. Sacio-Szymańska (eds.), *Bio#Futures*,
https://doi.org/10.1007/978-3-030-64969-2_17

1 Introduction

Odorous pollutants are volatile organic and inorganic compounds sensed by animals and humans via olfactory receptors. If the concentration of odorous compound in air is above the level known as olfactory threshold, it is identified by the brain as unpleasant sensation. Most frequently, this is caused by odorous gases that belong to the compound groups listed below in the order of decreasing olfactory threshold (OF):

- Volatile organic compounds (VOCs), e.g., benzene, formaldehyde, methylene chloride, and toluene, having OF of the order 100 ppm down to below 1 ppm.
- Nitrogen-containing compounds having OF in a wide range, e.g., ammonia 47 ppm while nitrobenzene 0.018 ppm; this compound group also includes putrescine, known as "smell of death."
- Sulfur compounds including hydrogen sulfide H_2S and mercaptans, e.g., ethanethiol and tert-butyl mercaptan, used as odorizers that facilitate detection of otherwise odorless natural gas; OF of the order 0.001 ppm and below.

Some of these compounds occur naturally, for example, in volcanic gases (H_2S) or in the products of decay of protein-containing matter (putrescine); however, this chapter is concerned with odor issues induced by human activities.

Widely known are emissions of odorous gases from agricultural and agriculture-related businesses, such as animal farms, rendering plants, and food processing plants (Trabue et al. 2019; Qamaruz-Zaman et al. 2020), and from environment protection facilities including sewage treatment plants and waste treatment or disposal facilities (Cheng et al. 2019; Kasperczyk et al. 2019). Some odorous gases are released to the atmosphere as a side effect of exploitation of natural resources, such as deep mining of metal ores (Kasperczyk and Urbaniec 2015) or oil extraction and processing (Jafarinejad 2017). A number of odorous compounds are present in mass-produced goods, e.g., as components or odorizers of gaseous and liquid fuels, or have important industrial applications as solvents or cleaning agents and, therefore, are emitted from industrial plants, such as pulp and paper mills, plastics and resin manufacturing, paint facilities, and various chemical industries (Bajpai 2014; Wypych 2017).

Human reactions to odor exposure are categorized as odor annoyance (adverse psychological effects of perceived bad smell) or odor nuisance (effects of prolonged perception of a smell so offensive that it significantly interferes with the enjoyment and use of the affected property). The effects of odor nuisance include undesired reactions ranging from emotional stresses such as states of discomfort, anxiety, and depression to physical symptoms such as eye irritation, headache, respiratory problems, and gastric problems such as nausea or vomiting (Naddeo et al. 2012). In the industrial context, odorous compounds having high molecular weight are dangerous because they can accumulate in lower parts of confined environments, such as large tanks or storage rooms, leading to unexpected oxygen deprivation. The economic consequences of such effects may range from interference with business activities and loss of property value to the costs of medical treatment of affected persons. Some of the odorous gases also pose a threat to the environment through direct toxic or corrosive action (e.g., H_2S), or through their role in promoting occurrence of smog (e.g., VOCs).

Considering the risks associated with odor nuisance, reduction of odor emission to atmosphere and related aspects of sustainable air quality have drawn attention of many countries all over the world. In the European Union, legal aspects of odor emission are considered in the framework of the **Industrial Emissions Directive (IED) 2010/75/EU of the European Parliament**. In addition, some EU countries have implemented and more countries are in the process of developing odor-specific legal requirements and odor abatement strategies. From a global perspective, existing odor regulations applied in different countries are reviewed by Brancher et al. (2017).

The abatement of air pollution is a complex issue comprising legal requirements on the emissions of pollutants, management of pollution sources, pollution control technologies, as well as measurement and evaluation of air quality. Information on the state of the art in these areas, with odor-specific issues taken into account, can be found in the reference documents relating to Industrial Emissions Directive, listed in Table 17.1.

2 Engineering Solutions for Odor Degradation

In general, approaches to odor reduction can be divided into three categories, namely, ones employing chemical, physical, and biological treatment methods (Lopez et al. 2012). Chemical and physical technologies (thermal oxidation, catalytic oxidation, ozonation and condensation, adsorption, absorption) have been widely used because the equipment is compact and it can be started up rapidly, and extensive experience in process design and operation has been accumulated over several decades (Barbusinski et al. 2020). However, while the said technologies are economically applied for large-flow waste gas streams and high pollutant concentration, the biological methods can handle low pollutant concentration and stand out as most environment-friendly. Biological degradation is generally oxidative in nature and produces ecologically safe products such as CO_2, H_2O, sulfate, and nitrate (Bindra et al. 2015); in addition, it does not require the use of chemicals and can be conducted at moderate temperatures (10–40 °C) and atmospheric pressure (Gospodarek et al. 2019). Moreover, the risk is avoided of shifting the pollutants into other environmental compartments – from gas to liquid or solid, or to the atmosphere; such a risk is characteristic of many physicochemical methods of purification of gases such as absorption of pollutants in liquids, adsorption on active carbon, or catalytic combustion.

Different designs of bioreactors for the removal of odorous compounds from air are schematically shown in Fig. 17.1. Significant progress in the applications of air biotreatment to remove odorous compounds dates back to the 1950s. Initially,

Table 17.1 Selected reference documents relating to Industrial Emissions Directive 2010/75/EU (excerpts from the list of documents available at https://eippcb.jrc.ec.europa.eu/reference)

Name	Code	Adopted/ published document	Remark
Food, Drink and Milk Industries	FDM	BREF BATC (12.2019)	
Intensive Rearing of Poultry or Pigs	IRPP	BREF BATC (02.2017)	
Slaughterhouses and Animals By-products Industries	SA	BREF (05.2005)	Under review[a]
Waste Incineration	WI	BREF BATC (12.2019)	
Waste Treatment	WT	BREF BATC (08.2018)	
Refining of Mineral Oil and Gas	REF	BREF BATC (10.2014)	
Production of Pulp, Paper and Board	PP	BREF BATC (09.2014)	
Production of Polymers	POL	BREF (08.2007)	
Textiles Industry	TXT	BREF (07.2003)	New version under development[a]
Large Volume Inorganic Chemicals – Ammonia, Acids and Fertilisers	LVIC-AAF	BREF (08.2007)	
Large Volume Inorganic Chemicals – Solids and Others Industry	LVIC-S	BREF (08.2007)	
Production of Large Volume Organic Chemicals	LVOC	BREF BATC (12.2017)	
Surface Treatment Using Organic Solvents including Wood and Wood Products Preservation Chemicals	STS	BREF (08.2007)	New version awaiting acceptance[a]
Emissions from Storage	EFS	BREF (07.2006)	
Large Combustion Plants	LCP	BREF BATC (07.2017)	
Common Waste Water and Waste Gas Treatment/Management Systems in the Chemical Sector	CWW	BREF BATC (06.2016)	Reference: Brinkmann et al. (2016)

(continued)

Table 17.1 (continued)

Name	Code	Adopted/ published document	Remark
Common Waste Gas Treatment in the Chemical Sector	WGC		New version under development[a]
Monitoring of Emissions to Air and Water from IED Installations	ROM	REF (07.2018)	Report (not a BREF)

BREF a document describing applied techniques, present emissions and consumption levels, techniques considered for the determination of best available techniques (BATs) as well as BAT conclusions and any emerging techniques, *BATC* document laying down conclusions on best available techniques

[a]Status March 2020

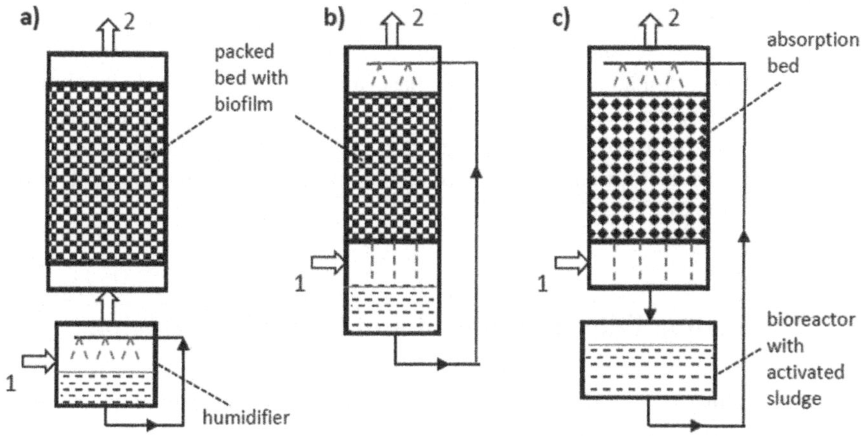

Fig. 17.1 Schemes of bioreactors for biodegradation of odorous compounds (after Gospodarek et al. 2019). Living environment for microorganisms is maintained either in the biofilm formed on the surfaces of packing elements of (**a**) biofilter (BF) and (**b**) biotrickling filter (BTF) or in the activated sludge maintained in (**c**) bioscrubber (BS); 1, inlet of polluted air; 2, outlet of purified air. Solid lines indicate liquid flow

simple biofilters with soil beds have been employed, and later the other bed materials with a larger specific surface and more effective for the growth of microorganisms were used. Generally, solid-bed biofilters (BFs; Fig. 17.1a) are the simplest, oldest, and the most widely used bioreactor configurations in order to treat polluted air (Malakar et al. 2017). Solid-bed biofilters (BFs; Fig. 17.1a) have been applied since the mid-1900s to remove odorous compounds emitted from wastewater treatment plants or intensive animal farms. Introduced later, more technologically advanced biotrickling filters (BTFs; Fig. 17.1b) and bioscrubbers (BSs; Fig. 17.1c) also belong to the most often used bioreactors.

These three designs are based on a similar concept, that is, bringing the polluted air into contact with microorganisms present in a living environment maintained in the biodegradation equipment. Each of the equipment types has its advantages and disadvantages, and specific ranges of economic application (Gospodarek et al. 2019); however, biotrickling filters (also known as trickle-bed bioreactors (TBBs)) stand out as being compact, moderately costly, and well suited for those applications in which operational flexibility is required (Wu et al. 2018).

The characteristic feature of a BTF/TBB is that an aqueous phase is trickled over the filter bed, which is usually made of some synthetic or natural inert material. Due to the availability of free liquid phase, the process conditions are easier to control. Therefore, biotrickling filters can handle difficult applications more efficiently than biofilters (Barbusinski et al. 2017; Barbusiński et al. 2020). A BTF-specific problem is that biofilm development on the surface of bed packing can progressively reduce bed porosity leading to excessive drop of air pressure and, in extreme cases, to bed clogging. However, if systematically monitored, biofilm growth can be controlled by various techniques such as bed flushing or adjustments in the values of process parameters (De Vela and Gostomski 2018). Another drawback of the biotrickling filter technology is the problem of gas transfer arising from the necessity of dissolving the gaseous pollutants in the aqueous phase, and therefore, for the low-solubility pollutants, application of bioscrubbers (BSs) may be preferable. Although more costly than BFs in investment and operation, BSs offer longer retention times of the gas phase in the absorption beds and are also better suited to handling peak emissions (EMIS 2020).

3 Biotrickling Filter Equipment and Operating Principle

In a biotrickling filter (BTF/TBB), the bed of bioreactor is made up of usually inert material on which microorganisms are immobilized in order to degrade the processed pollutants. Random-dump or structured plastic packing, pieces of open-pore synthetic foam, lava rock, tire-derived rubber particles, as well as organic materials such as wood chips are the materials most often used (Barbusinski et al. 2017). Gas and liquid phases flow co- or countercurrently through the bed, which enables to clean large gas streams with no risk of bed flooding (Wu et al. 2018).

The treatment of polluted air involves pollutant transfer from air to the aqueous phase (mineral salt solution), diffusion into the biofilm, and biodegradation within the biofilm (De Vela and Gostomski 2018). Biofilm is created by a group of microorganisms (bacteria, fungi, algae, and protozoa) which attach themselves on the surface of the packing elements and form a slim layer of a viscous, jellylike structure. The development of biofilm is initiated by the inoculation of bioreactor bed using samples of trickling liquid taken from biotrickling filters treating a comparable waste gas stream, or using activated sludge from wastewater treatment as it typically contains a wide spectrum of bacteria that are capable of degrading many different compounds. A well-developed biofilm is formed within a few days, weeks, or even months, depending on the type of microorganisms and required living environment. In order to maximize the efficiency of biopurification process carried out in the bioreactor, it is necessary to keep the concentration of nutrients (macronutrients, N, P, K, and S, and micronutrients, vitamin and metals) on a right level to maintain an active, growing microbial culture (Bak et al. 2017). Usually, biotrickling filters are operated in the mesophilic temperature range between 10 and 40 °C. The biotreatment of waste gas in the TBB may be limited both by biological reaction rate and by the mass transfer rate. Depending on the particular application, temperature can affect either of these limitations possibly resulting in changed process performance.

Although biotrickling filtration is not a new technology, it is still considered as innovative one, and new application areas are entered rather slowly (Oyarzun et al. 2019; Kasperczyk et al. 2019). One of the reasons is that a BTF-based biodegradation system has to be individually designed for the specific pollutant or group of pollutants accounting also for the parameters of the stream of contaminated air, to achieve the desired level of the efficiency of biodegradation. This can be illustrated by the experience (summarized in the next subsection) of the authors of this chapter, from the applications of the compact trickle-bed bioreactor (CTBB) which is a special variant of BTF/TBB.

4 Compact Trickle-Bed Bioreactor and Its Applications

The pilot-scale tests of the biodegradation of H_2S and H_2S-VOC mixtures emitted from a copper-ore mine and from a wastewater treatment plant (WWTP) were described in previous publications cited below. CTBB application in the deep mine was highly demanding due to unpredictable bursts of H_2S liberated from geological structures and difficult environment of mine corridor at 1000 m depth (Kasperczyk and Urbaniec 2015). Application in the WWTP was also difficult as pollutants' concentration fluctuated wildly in accordance with daily and weekly cycles of wastewater supply by tanker cars to the treatment facility (Kasperczyk et al. 2019). Although positive results of the biodegradation of pollutants were achieved in both cases, the gained experience had a limited value only when a project was undertaken on the treatment of VOCs emitted in the ventilation air from an industrial painting shop. In view of a particular type of pollutants and special requirements on emission reduction, it was necessary to perform a time-consuming multistep procedure that included the selection of suitable microorganisms, laboratory and industrial tests of small-scale CTBB, design and acquisition of full-scale equipment, and startup and testing of full-scale CTBB-based air biopurification system.

4.1 Selection of Microorganisms and Adaptation to Target Pollutant

The concept of pollutant degradation in the CTBB assumes the use of microorganisms naturally occurring at the place where pollutants are present. In order to ensure availability of such bacterial strains, soil samples were collected from industrial sites where automotive painting shops are in operation. The isolation of bacterial consortium capable of degrading VOCs was carried out by enrichment culture technique (Culturing Microorganisms 2019) applied to a mixture of microorganisms coming from three sources:

- Collected soil samples
- Bacterial cultures (including *Pseudomonas fluorescens*) purchased from a publicly available collection of microorganisms
- Bacterial cultures from the collection of microorganisms owned by Ekoinwentyka Ltd.

According to the mentioned technique, nutrients supplied and environmental conditions created for the growth of microorganism mixture were so controlled as to favor the growth of microorganisms capable of degrading the VOCs. In applying this approach, the VOCs were initially simulated by a mixture of dichloromethane and formic acid in concentrations gradually increasing from 1 to 100 mg/dm^3. Later, during a 4-week adaptation period, bacterial cultures were exposed to air polluted by VOCs present in vapors generated from solvent formulations used in the industrial painting shop. After completing the procedure, 17 different cultures of bacterial consortia (dominated by *Pseudomonas fluorescens*) were found suitable for application in industrial systems for VOC biodegradation.

4.2 Small-Scale CTBB

A simplified scheme of laboratory setup in which the process of cleaning the air from volatile impurities was conducted is shown in Fig. 17.2. Its main component is compact trickle-bed bioreactor made of stainless steel with a diameter of 0.315 m and active bed 0.6 m high, packed with polypropylene rings. The gas phase in the tests was atmospheric air pumped by a compressor. The VOCs were available in the form of vapors generated from the solvent solution used in an industrial painting shop. The air was conditioned by filtration, heating to about 25 °C and mixing with solvent vapors, and directed to bioreactor inlet where VOC concentration in the supplied air was measured. The liquid phase circulating in the system was a solution of mineral salts (total volume around 25 dm^3) whose parameters, including flowrate, pH, temperature, and absorbance (indicating concentration of microorganisms in the solution), were controlled and regulated online using auxiliary equipment such as micro-pumps dispensing buffer solutions, control valves, and heaters. Placed

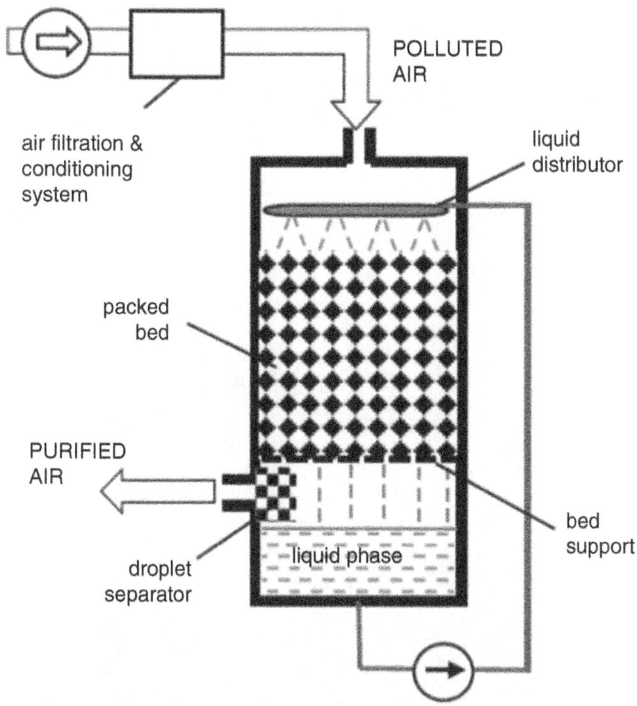

Fig. 17.2 Scheme of laboratory setup including CTBB for air cleaning from volatile pollutants. (Ekoinwentyka Ltd.)

above bioreactor bed, liquid distributor ensured that gas and liquid, flowing in cocurrent downwards, were brought into contact over packing surface. VOC concentration remaining in the purified gas was determined at the outlet of droplet separator.

A photo of the laboratory setup including its control equipment is shown in Fig. 17.3. Before starting the process, the entire setup was sterilized by rinsing three times with alcohol solution, and irradiated for several hours using UV lamp. This was followed by the immobilization of microorganisms on the packing of bioreactor bed. To this end, microorganisms previously adapted to VOC degradation were mixed with 10 dm^3 of liquid phase, and the resulting suspension was circulated through the bed for around 3 days until the formation of biofilm layer on the packing could be observed. After that, the suspension was removed from the setup, biofilm-covered bed packing was rinsed with sterile mineral salt solution, and the process was started by setting gas and liquid flowrates and pollutant concentration in the gas supplied to the bioreactor.

In the pilot-scale tests, while aiming at the determination of rational ranges of parameters of biodegradation process to ensure a high VOC conversion degree, much attention was also paid to the chemical composition of liquid phase circulated in the system. In order to avoid shifting air pollution to the liquid, the process

Fig. 17.3 Laboratory setup including small-scale CTBB. (Ekoinwentyka Ltd.)

parameters were maintained so as to keep the circulating solution of mineral salts free from the VOCs and intermediate products of their degradation.

The VOC biodegradation setup was continuously operated in the laboratory for several months. By varying the flowrates of gas and liquid phases, different values were set of specific pollutant load Ms defined as the ratio of its mass flow and empty volume of bioreactor bed, that is, mass of pollutants entering unit empty volume of bioreactor bed in unit time:

$$\mathrm{Ms} = G_{\mathrm{p}} / V_{\mathrm{bed}} = C_{\mathrm{g\,in}} V_{\mathrm{g}} / V_{\mathrm{bed}} = C_{\mathrm{g\,in}} / t_{\mathrm{g}} \tag{17.1}$$

where G_p is the mass flow of pollutant, V_{bed} empty bed volume, V_g gas phase flowrate, $C_{g\,in}$ VOC concentration at bioreactor inlet, and t_g average gas residence time.

The efficiency of biodegradation process was assessed by calculating specific elimination capacity (purification efficiency) defined as

$$\mathrm{EC} = \left(C_{\mathrm{g\,in}} - C_{\mathrm{g\,out}} \right) / t_{\mathrm{g}} \tag{17.2}$$

and VOC conversion degree defined as

$$K = \left[\left(C_{\mathrm{g\,in}} - C_{\mathrm{g\,out}} \right) / C_{\mathrm{go}} \right] \cdot 100\% \tag{17.3}$$

where $C_{\mathrm{g\,out}}$ is the VOC concentration at bioreactor outlet.

Measurements were carried out to determine the relationships between specific elimination capacity EC and specific pollutant load Ms. Figure 17.4 depicts the results of VOC biodegradation measurements performed in the laboratory, at gas phase flowrate of V_g = 2.5 m³/h and liquid phase flowrate in the range V_c = 1.2–1.4 m³/h. At Ms up to 4.0 g/(m³h), it was possible to attain high EC values up to around 3.9 g/(m³h) equivalent to VOC conversion degree K = 90–99%; however, when specific pollutant load was increased to Ms = 4–6 g/(m³h), the elimination capacity tended to decrease and VOC conversion degree dropped to less than 85%. In practical terms, these results indicate that very high conversion above 90% can be achieved under the condition that specific pollutant load Ms does not exceed a threshold value that depends on gas phase flowrate. If pollutant concentration at bioreactor inlet is increased leading to Ms increase (according to Eq. (17.1)) above the threshold, then a reduction of the conversion degree is unavoidable.

After completing laboratory tests, the CTBB was moved to the industrial site and provided with piping that allowed connecting the bioreactor with different air streams flowing through the ventilation system of the painting shop. Measurements of VOC concentrations in the air streams at bioreactor inlet and outlet, at flowrates

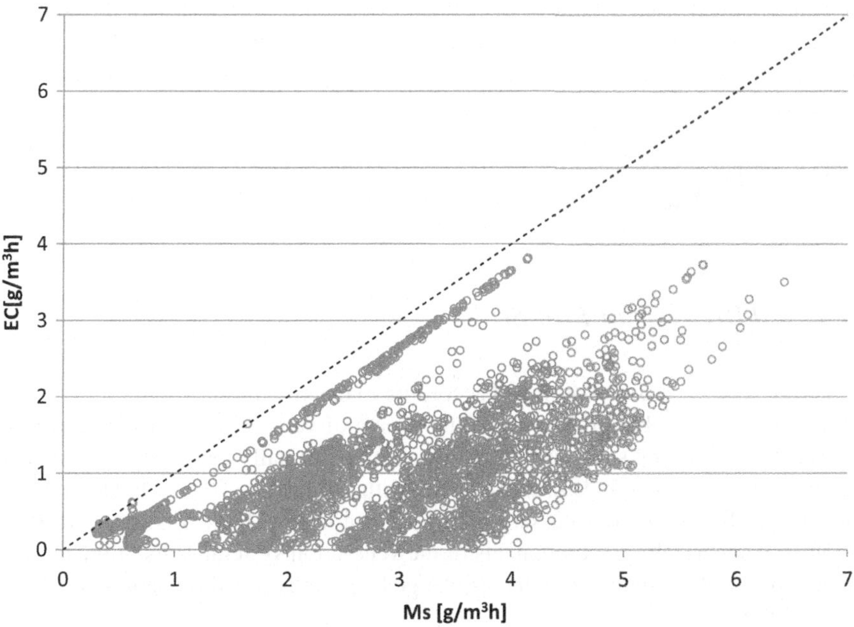

Fig. 17.4 Elimination capacity of VOC biodegradation in pilot-scale CTBB at V_g = 2.5 m³/h. Dashed line represents physical limit (complete elimination, i.e., VOC conversion degree 100%). (Ekoinwentyka Ltd.)

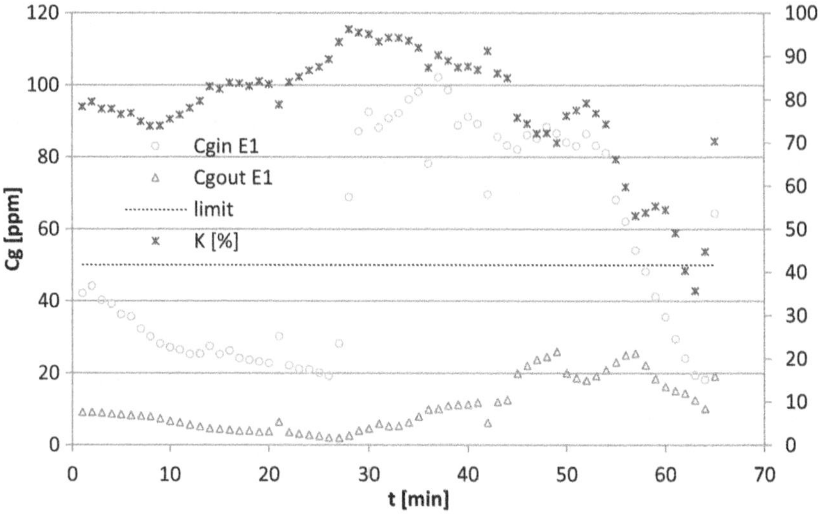

Fig. 17.5 Results of industrial tests of VOC biodegradation in pilot-scale CTBB, air stream E1. (Ekoinwentyka Ltd.)

between 1.0 and 10.0 m^3/h, were carried out by the personnel of a testing organization accredited in accordance with PN-EN ISO/IEC 17025 standard. Examples of results obtained from 1-hour tests of biodegradation of VOCs in two different air streams drawn from the ventilation system are shown in Figs. 17.5 and 17.6. As can be seen, at inlet concentrations of the pollutant up to 170 ppm – changing between 40 and 117 ppm in stream E1 and almost stable 150–170 ppm in stream E2 – VOC concentration in the purified air was maintained below 20 ppm. This value was well below the upper limit of 50 ppm set by the management of the industrial site, in accordance with the relevant environmental permit.

4.3 CTBB Tests in Full Scale

The full-scale CTBB was designed according to the scheme shown in Fig. 17.2 and dimensioned to enable processing of the stream of ventilation air discharged from an industrial painting shop, at VOC concentration normally below 200 ppm and air flowrate up to 6000 m^3/h. Bioreactor vessel with a diameter of 2.8 m and total height 7.7 m together with the necessary piping, shown in Fig. 17.7, was installed in the industrial site as add-on components of the ventilation system. Tests of the biodegradation of VOCs contained in the discharged air were carried out continuously for several months.

During the initial phase of operation of full-scale CTBB, the attention was focused on critically important immobilization of microorganisms and biofilm growth, and starting up of measurements of process parameters. To ensure the

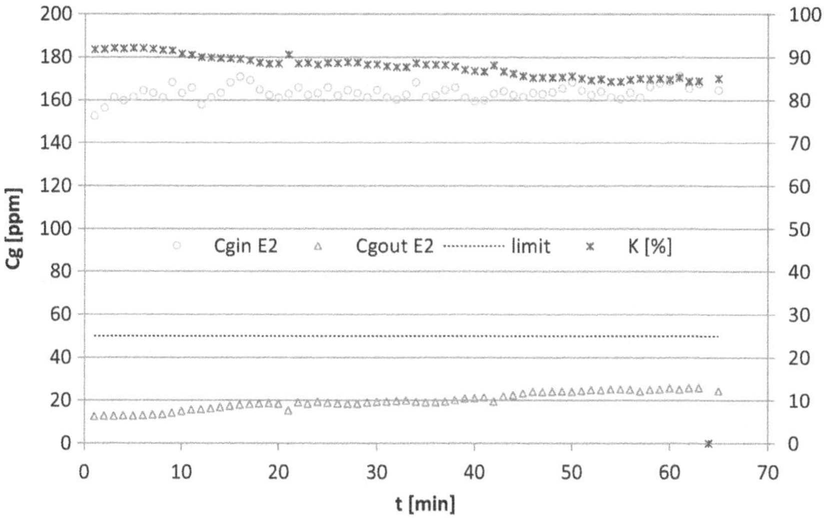

Fig. 17.6 Results of industrial tests of VOC biodegradation in pilot-scale CTBB, air stream E2. (Ekoinwentyka Ltd.)

Fig. 17.7 Full-scale CTBB and control panel of its automation system. (Ekoinwentyka Ltd.)

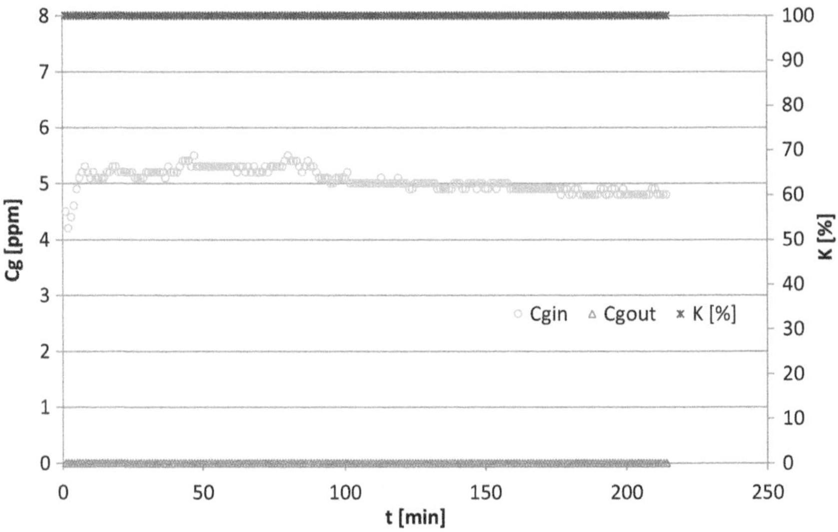

Fig. 17.8 VOC concentrations in air stream at bioreactor inlet and outlet and VOC conversion degree determined during the initial phase of CTBB tests in full scale. (Ekoinwentyka Ltd.)

success of that phase, mild process conditions were kept by setting air flowrate at V_g = 500 m^3/h and maintaining VOC concentration at bioreactor inlet in the range $C_{g\ in}$ = 4.0–5.5 ppm. According to the measurements of inlet and outlet concentrations of the pollutant illustrated in Fig. 17.8, VOC conversion degree K = 99.9% was achieved throughout the initial phase.

Positive results of the initial phase of CTBB operation made it possible to allow air flowrate and pollutant concentration at bioreactor inlet to fluctuate freely in accordance with changing parameters of the discharged stream of ventilation air. As it turned out, real operating conditions sometimes deviated from the design conditions considerably because during the period of test operation, air flowrate varied in the range V_g = 300–6000 m^3/h and inlet concentration of the pollutant was changing between 5 ppm and 2000 ppm.

The results of measurements of pollutant concentrations performed during a representative time interval of 20 hours, shown in Fig. 17.9, indicated considerable flexibility of the VOC biodegradation system. Despite widely changing air flowrate and inlet concentration of VOCs, their concentration at bioreactor outlet was maintained in the range of 0.1−55 ppm. At moderate values of inlet VOC concentration lower than or just above design value of 200 ppm, pollutant conversion degree not lower than 99% was achieved. Short-lived concentration peaks – above 1000 ppm, and sometimes as high as 1800 ppm – typically resulted in the reduction of conversion degree to around 85%. However, each time VOC concentration was back at the moderate level, the conversion degree quickly returned to the range of 95–99%.

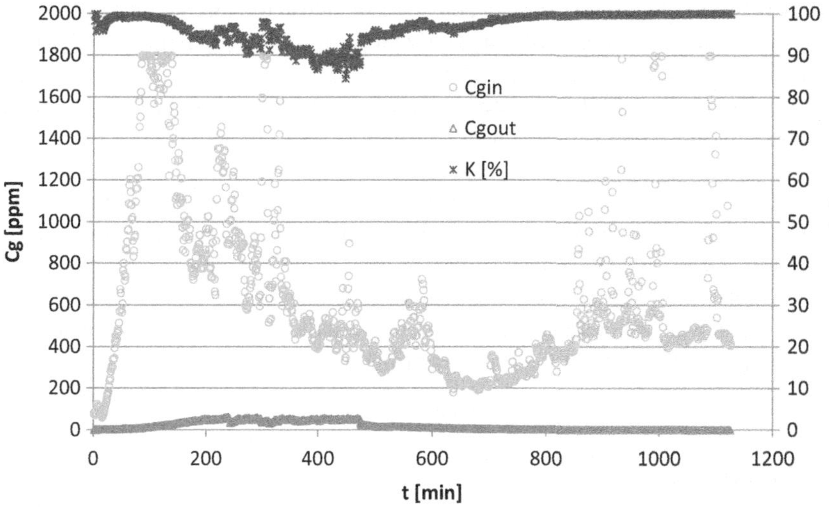

Fig. 17.9 VOC concentrations in air stream at bioreactor inlet and outlet and VOC conversion degree determined during a representative 20-hour interval of CTBB tests in full scale. (Ekoinwentyka Ltd.)

4.4 Discussion

The experimental results presented in subsections 4.1–4.3 can be compared with the results of research performed by other researchers on biodegradation of VOCs in biotrickling filters in laboratory or pilot scale and summarized in the review paper by Rybarczyk et al. (2019). The tests of full-scale CTBB confirmed the optimal ranges of parameters of VOC biodegradation process, namely, temperature 20–35 °C and pH = 6.0–7.5. As long as moderate values below 200 ppm of pollutant concentration at bioreactor inlet were maintained during CTBB tests in both pilot and full scale, typical values of VOC conversion degree $K = 95–99\%$ obtained in the present research were comparable with the highest K values reported by other researchers.

Apart from quantitative information acquired through measurements and processing of measurement data, important qualitative information was collected during the period of test operation of full-scale CTBB. As long as the values of process parameters were maintained within or close to their optimal ranges, VOC biodegradation was running smoothly and efficiently. However, when the processes conducted in the painting shop were modified resulting in changed chemical composition of pollutants in the discharged ventilation air and subsequent pH increase in the liquid phase to 8.6–9.0, inhibition of the biodegradation process was observed. This appeared to be another case of the impairment of biodegradability of VOC mixture due to chemical interactions between some mixture components and degradation products of other co-existing components (Yoshikawa et al. 2017). The problem was solved by adjusting the parameters of painting processes, and in addition supplying stronger buffer solutions to the control devices of biodegradation system. As soon as pH in the liquid phase was back at a level close to 7.5, the microorganisms returned to their normal activity, and efficient VOC biodegradation was quickly resumed.

It was also observed at higher values of air flowrate that excessive foaming of the circulating liquid phase occurs, thus disturbing CTBB operation. This was recognized as a problem known from other applications of bioreactors (Delvigne and Lecomte 2010). A corrective action was necessary to avoid foam spreading with the stream of purified air and over the area of industrial site. From the range of foam prevention agents available on the market, a suitable defoamer – nontoxic and friendly to the microorganisms – was selected, and the functioning of the biodegradation system was brought back to normal by periodic dosing of the defoamer into the liquid phase upstream of the liquid distributor in the bioreactor.

5 Conclusions

Nowadays the abatement of odorous air pollutants is widely recognized as an important component of air pollution control. Among the different approaches to odor reduction, applications of biological air treatment methods have gained growing attention. In the present chapter, this is exemplified by the development and testing of compact trickle-bed bioreactor (CTBB) for the biodegradation of volatile organic compounds (VOCs) emitted from automotive painting industry represented by a specific painting shop.

Overall, the experimental results obtained during the tests of full-scale CTBB indicated a positive outcome of upscaling and adaptation of biotrickling filtration technology to the conditions of the painting shop. Once minor problems of the initial phase of operation had been solved, the bioreactor performed satisfactorily as VOC conversion degree $K = 95–99\%$ was obtained at design operating conditions. The CTBB also proved flexible enough to withstand short-lived situations of severe pollutant overload and to resume satisfactory operation as soon as VOC concentration returned to its design level.

The environmental friendliness of VOC biodegradation was confirmed. Energy demand was low as energy was needed mainly to compensate for heat losses from thermally insulated bioreactor vessel where temperature 20–35 °C should be maintained. Water demand was also low as water was needed mainly for offsetting vapor discharge in the stream of purified air that has been moisturized when in contact with liquid phase inside the bioreactor (order of magnitude 10 kg/h). There was no problem of solid waste as the mass flow of degraded VOCs, being rather small (usually less than 1 kg/h), was partly converted to bacterial biomass while biodegradation metabolites remained in the circulating liquid phase, thus enabling removal by liquid purge if needed.

References

Bajpai P., 2014. Biological Odour Treatment. Springer, Heidelberg-New York-Dordrecht-London.

Bak A., Kozik V., Dybal P., Sulowicz S., Kasperczyk D., Kus S., Barbusinski K., 2017. Abatement robustness of volatile organic compounds using compact trickle-bed bioreactor: Biotreatment of styrene, ethanol and dimethyl sulfide mixture in contaminated airstream. Int Biodeter Biodegr; Vol. 119, 316–328.

Barbusinski K., Kalemba K., Kasperczyk D., Urbaniec K., Kozik V., 2017. Biological methods for odor treatment - A review. J Clean Prod; Vol. 152, 223–241.

Barbusiński K., Urbaniec K., Kasperczyk D., Thomas M., 2020. Biofilters versus bioscrubbers and biotrickling filters: state of the art of biological air treatment. Chapter 2, in: Soreanu G., Dumont É. (Editors), From Biofiltration to Promising Options in Gaseous Fluxes Biotreatment: Recent Developments, New Trends, Advances, and Opportunities. Elsevier. ISBN: 9780128190647 (In press).

Bindra N., Dubey B., Dutta A., 2015. Technological and life cycle assessment of organics processing odour control technologies. Sci Tot Environ; Vol. 527–528, 401–412.

Brancher M., Griffiths K.D., Franco D., de Melo Lisboa H., 2017. A review of odour impact criteria in selected countries around the world, Chemosphere; Vol. 168, 1531–1570.

Brinkmann T., Giner Santonja G., Yükseler H., Roudier S., Delgado Sancho L., 2016. Best Available Techniques (BAT) Reference Document for Common Waste Water and Waste Gas Treatment/Management Systems in the Chemical Sector; available at https://eippcb.jrc.ec.europa.eu/reference

Cheng Z., Sun Z., Zhu S., Lou Z., Zhu N., Feng L., 2019. The identification and health risk assessment of odor emissions from waste landfilling and composting. Sci Tot Environ; Vol. 649, 1038–1044.

Culturing Microorganisms, 2019. In: Microbiology, Chapter 6, Available at https://bio.libretexts.org/Bookshelves/Microbiology/Book%3A_Microbiology_(Boundless)/6%3A_Culturing_Microorganisms.

Delvigne F., Lecomte J., 2010. Foam Formation and Control in Bioreactors. In: Flickinger MC (Editor), Encyclopedia of Industrial Biotechnology: Bioprocess, Bioseparation, and Cell Technology, Wiley, 2010.

De Vela R.J.L., Gostomski P.A., 2018. Minimising biomass accumulation in biotrickling filters. Rev Environ Sci Biotechnol; Vol. 17, 417–430.

EMIS Bioscrubber. https://emis.vito.be/nl/node/19472; last accessed 30.03.2020.

Gospodarek M., Rybarczyk P., Szulczyński B., Gębicki J., 2019. Comparative evaluation of selected biological methods for the removal of hydrophilic and hydrophobic odorous VOCs from air. Processes; Vol. 7, 187.

Jafarinejad S., 2017. Petroleum Waste Treatment and Pollution Control, Butterworth Heinemann.

Kasperczyk D., Urbaniec K., 2015. Application of a compact trickle-bed bioreactor to the biodegradation of pollutants from the ventilation air in a copper-ore mine. J Clean Prod; Vol. 87, 971–976.

Kasperczyk D., Urbaniec K., Barbusinski K., Rene E.R., Colmenares-Quintero R.F., 2019. Application of a compact trickle-bed bioreactor for the removal of odor and volatile organic compounds emitted from a wastewater treatment plant. J Environ Manage; Vol. 149, 77–84.

López M.E., Rene E.R., Veiga M.C., Kennes C., 2012. Biogas Technologies and Cleaning Techniques. In: E. Lichtfouse, J. Schwarzbauer, D. Robert (Editors), Environmental Chemistry for a Sustainable World, Springer, Dordrecht, The Netherlands, 347–377.

Malakar S., Saha P.D., Baskaran D., Rajamanickam R., 2017. Comparative study of biofiltration process for treatment of VOCs emission from petroleum refinery wastewater - A review. Environ Technol & Innovation; Vol. 8, 441–461.

Naddeo V., Belgiorno V., Zarra T., 2012. Odour Characterization and Exposure Effects. Chapter 2, in: Belgiorno V., Naddeo V., Zarra T. (Editors), Odour Impact Assessment Handbook, Wiley, 7–29.

Oyarzun P., Alarcón L., Calabriano G., Bejarano J., Nuñez D., Ruiz-Taglec N., Urrutia H., 2019. Trickling filter technology for biotreatment of nitrogenous compounds emitted in exhaust gases from fishmeal plants. J Environ Manage; Vol. 232, 165–170.

Qamaruz-Zaman N., Yaacof N., Kamarzaman F.F., 2020. Control of odors in the food industry. Chapter 9, in: Galanakis C. (Editor), The Interaction of Food Industry and Environment, Academic Press, 281–307.

Rybarczyk P., Szulczyński B., Gębicki J., Hupka J., 2019.Treatment of malodorous air in biotrickling filters: a review. Biochem Eng J; Vol. 141, 146–162.

Trabue S., Scoggina K., Tyndall J., Sauer T., Hernandez-Ramirez G., Pfeiffer R., Hatfield J., 2019. Odorous compounds sources and transport from a swine deep-pit finishing operation: a case study. J Environ Manage; Vol. 233, 12–23.

Yoshikawa M., Zhang M., Toyota K., 2017. Biodegradation of Volatile Organic Compounds and Their Effects on Biodegradability under Co-Existing Conditions. Microbes Environ; Vol. 32(3), 188–200.

Wu H., Yan H., Quan Y., Zhao H., Jiang N., Yin C., 2018. Recent progress and perspectives in biotrickling filters for VOCs and odorous gases treatment. J Environ Manage; Vol. 222, 409–419.

Wypych G., 2017. Handbook of Odors in Plastic Materials. ChemTec Publishing, Toronto.

Chapter 18
A Systematic Approach for Assessing and Managing the Urban Bioeconomy

Alberto Bezama, Nora Mittelstädt, and Daniela Thrän

Abstract As urbanization processes happen all over the world, an increasing attention is being given to the management of the resources that feed these urban areas. When addressed from a systems perspective, the connection between resources, production, and manufacture sectors and society can be clarified, especially when viewed from a life cycle perspective. The goal of this chapter is therefore to provide an analysis of the state-of-the-art resources management tools that take a life cycle management approach, with a particular focus on bio-based resources and the latest experiences in the bioeconomy sector. This analysis is the basis for discussing the necessary steps and needs for establishing an "urban bioeconomy metabolism," whose definition can help to managing the material streams within the city limits in connection with the bio-based resources of the city's surroundings.

Keywords Urban metabolism · Life cycle management · Regionalized sustainability assessment · Regionalized approaches · Urban bioeconomy · Bioeconomy · Circular economy · Social life cycle assessment · Life cycle assessment

1 Introduction

Urbanization processes are taking place at unprecedented speed and intensity. The majority of humankind is currently living in urban areas, a development that is predicted to continue, with more than two thirds of the global population expected to live in cities by 2050 (UN 2015). The implications of this will be enormous. The

A. Bezama (✉) · N. Mittelstädt
Helmholtz Centre for Environmental Research – UFZ, Leipzig, Germany
e-mail: alberto.bezama@ufz.de

D. Thrän
Helmholtz Centre for Environmental Research – UFZ, Leipzig, Germany

Deutsches Biomasseforschungszentrum gGmbH (DBFZ), Leipzig, Germany

E. Koukios, A. Sacio-Szymańska (eds.), *Bio#Futures*,
https://doi.org/10.1007/978-3-030-64969-2_18

consumption of land for the development of urban areas, as well as for satisfying the needs of their societies for food, energy, and raw materials (e.g., for construction), is ever expanding and is having irreversible impacts on the biosphere (Seto et al. 2012). Cities are using 60% of the residential freshwater resources and produce 75% of the global carbon emissions (Grimm et al. 2008). Furthermore, 90% of the global economic power (GDP) and 65% of the global energy consumption are concentrated in urban areas (Solecki et al. 2013). As a result, the ecological footprint of a city can be as much as 200 times greater than the area of a city itself (Wigginton et al. 2016). It is clear that urbanization is one of the main drivers of global environmental change.

However, the rise of cities and the underlying urbanization processes is also offering a great potential for change. On the one hand, cities with their physical, organizational, institutional, and demographic compactness (Evans 2011) have the chance to help reducing ecological footprints through minimizing land consumption and sprawl, supporting short travel distances, and allowing efficient use of water, energy, and waste. On the other hand, they are places where innovations are catalyzed because people of diverse backgrounds and experiences can come together, bringing in capacities and skills to produce ideas for a sustainable development. Cities can thus play a key role in dealing with the challenges of global environmental change (Rosenzweig et al. 2015), as they are economic fulcrums and places where innovations can be put into practice (Sassen 2012).

In order to prioritize the work on addressing these challenges, the European Network "Eurocities," an organization that "brings together around 140 of Europe's largest cities and over 45 partner cities, that between them govern 130 million citizens across 39 countries" (www.eurocities.eu), published its "Strategic Framework 2014–2020 - Towards an Urban Agenda for the EU." In this framework, they identify the five focus areas that "to a large extent align with the EU's strategic priorities and provide a strong strategic operational framework for EUROCITIES" (Eurocities 2016). These are (i) cities as drivers of quality jobs and sustainable growth; (ii) inclusive, diverse, and creative cities; (iii) green, free-flowing, and healthy cities; (iv) smarter cities; and (v) urban innovation and governance in cities.

The parallels of these strategic focus areas and the goals of the bioeconomy are remarkable. Bioeconomy is defined as "the production and utilization of biological resources (including knowledge) to provide products, processes and services in all sectors of trade and industry within the framework of a sustainable economy" (Bioökonomierat 2018). According to the German Bioeconomy Council (Bioökonomierat 2018) "the future bioeconomy will satisfy primary human needs; it will be technology-driven and take the environment into account." The relevance of a regional and city-oriented perspective is also highlighted by a global expert survey carried out by the German Bioeconomy Council (Bioökonomierat 2018). This is due to the many processes within cities in which bioeconomy plays a relevant role, for example, in the production of sustainable building materials, food production close to the city, and implementation of sound infrastructures for the appropriate cascading systems for wastes, residual materials, and nutrients.

During December 2017, within the framework of the SYMOBIO project (Systemic Monitoring and Modelling of the Bioeconomy, www.symobio.de), the Department of Bioenergy at the UFZ, Leipzig, carried out a series of workshops. The aim of these workshops was to analyze what representatives of civil society, the scientific community, and the industrial sector in Germany expect toward bioeconomy. To frame the discussion, these workshops focused on the 17 proposed UN Sustainable Development Goals (SDGs) and their relevance for a sustainable transition toward a bio-based economy. One of the main results was the ranking of the most relevant SDGs according to the German bioeconomy stakeholders (Thrän et al. 2018; Zeug et al. 2019).

The result was a ranking of the SDGs from high relevance to low relevance as follows: (1) SDG 7, Affordable and clean energy; (2) SDG 8, Decent work and economic growth; (3) SDG 9, Industry, innovation, and infrastructure; (4) SDG 2, Zero hunger; (5) SDG 17, Partnerships for the goals; (6) SDG 13, Climate action; and (7) SDG 15, Life on land. This resulting list shows the priorities of the various stakeholders.

In addition, it is important to highlight the relevance of the regional perspective, in particular of cities, on the successful implementation of the bioeconomy strategy. In this regard, cities (or city districts in case of larger urban areas) can be considered as the smallest representative entity where an integrative approach, for assessing the potential effects of implementing the bioeconomy, can be carried out. The relevance of this urban-centered perspective is also taken up by city representative themselves. Many political initiatives are meanwhile pushed forward by cities themselves, particularly in the field of climate change. Also with regard to a more urban-based bioeconomy, there are first steps taken, and the first initiatives aiming at linking the global relevance of cities with the ideas promoted by the bioeconomy are starting to be established. In May 2018 the workshop "The road to Urban Bio-economy: Barriers and Solutions to Closing the Loops of Bio-Resources" was organized in Brussels to discuss the challenges faced by cities to promote bio-resources along the entire value chain. However, there is a drawback in the concept taken as a basis for discussion, as the EU discusses the urban bioeconomy concept by focusing it merely on the utilization of the residual bio-based streams (Accorigi 2018): the definition of an urban bioeconomy in its current form entails too narrow an understanding of the urban bioeconomy concept. In fact, there is a high need to develop a sound and more comprehensive concept of urban bioeconomy.

The urban bioeconomy concept can be used to identify and understand the transformation processes of bio-based resources on an urban level, as well as to understand the mechanisms underlying the interactions between different actors found at the city level in the bioeconomy field and for unraveling the full potential of a city-focused bioeconomy concept. Moreover, by linking the identified urban activities to the actual bio-based material streams that are processed within city limits, the urban bioeconomy model would help in the local monitoring and management of the available bio-based resources.

However, what is "urban bioeconomy," and how can it be thoroughly defined? What are key characteristics of an urban bioeconomy concept? This chapter is a first attempt to draw a connection on the lessons learned in the bioeconomy field and the needs for defining an urban bioeconomy concept that can be actually used by local and regional authorities to optimize the management of bio-based resources within the city limits.

2 Managing the Bioeconomy: Lessons Learned and Challenges Ahead

The increasing demand of biomass resources for food, feed, industrial, and energy applications is putting a huge pressure on the management of these resources. Moreover, as the definition of sustainability has changed from the three-dimension perspective to a more holistic approach, recently defined by the Sustainable Development Goals (SDGs), the complexity behind a sustainable management of the available resources has also increased. As a consequence, management systems and tools have become major attention, as they are capable of incorporating and assessing different factors to provide support to decision makers. For this reason, this chapter intends to provide an analysis of the current state of the art of management systems for resources management, taking into consideration the urban bioeconomy concept as introduced in the previous section. This chapter focuses therefore mainly on the tools that have been developed to address the management of bio-based resources (especially in the industrial sectors), and identifies the needs that should be addressed in the short- to mid-term to understand and manage the sustainability issues involved in an urban bioeconomy.

2.1 Management Tools in a Systems Perspective

Bioeconomy has been traditionally connected to the industrial sector, and particularly to the development of novel technological approaches that deal with the utilization of biomass resources. The multitude of technology breakthroughs and innovations in the field over the last years have forced companies and industrial sectors to constantly restructure the ways they work, organize, and manage. And it is envisaged that this connection within different sectors will even increase in the short to medium term, as there are several advantages for the establishment of integrated value chains and networks (Bezama et al. 2019; Hildebrandt et al. 2019, 2020).

However, this will require a series of new management tools, in order to cope with the necessities of the different industrial sectors. Currently, a variety of management tools are available. They vary from tools for technology and project management, knowledge management tools, and environmental management tools to business process management tools or costumer relation management tools.

Schawel and Billing (2018) describe in their book *Top 100 Management Tools* the increasing demands on duties and responsibilities of a manager challenging within the competing market and the multiplicity of tasks and topics to coordinate, including the definition of strategies, pilot projects, develop concepts and methods as well as increasing the efficiency of the company or motivate the stuff and guide them target-oriented. To achieve these goals and comply the described tasks, a variety of management tools exist. They group these different management tools in three categories: strategical management, controlling and timing, and communication. These groups simultaneously represent the steps in a continuous management process. Listed are tools from ABC analysis and investment management as a strategical management tool to sales-funnel-analysis as problem analysis tool or the Osborne method as a creativity technique.

In 2007 Rigby asserted an explosion of management tools in the former two decades seeing it as a need to successfully guide an increasingly competitive market. He defines the multifaceted management tool compilation as a help to handle complex decisions, especially business decisions, in a global world. By means of those a company can improve their performance as well as their profits. Therefore it is necessary to encompass the weaknesses and strengths of each tool for a proper appliance. Rigby identified the 25 most popular tools and defined them with an explanation of how the tools are being used.

Rigby and Bilodeau (2017) assessed recent trends in management tools related to usage and satisfaction. They again listed the 25 most popular tools, within strategic planning is topping the list as the most popular tool globally. As a trend they identified digital technology as a dominant factor across all industries and regions. Therefore "digital transformation" is a tool helping to challenge these shifts evident in increasing popularity and satisfaction compared to the last survey.

Wrisberg and de Haes (2012) compile assessment tools for the environmental dimension of sustainable development to support business decisions. Based on a systemic perspective, subdividing systems as function-oriented and region-oriented or agreement-oriented, they describe the weaknesses, strengths, and possible combinations of the most commonly tools as cost-benefit analysis (CBA), cost-effectiveness analysis (CEA), material flow accounting (MFA), life cycle assessment (LCA), environmental risk assessment (ERA), physical and environmental input-output analysis (env IOA), or multi-criteria analysis (MCA) within a systems perspective.

Industrial sectors as agriculture, construction, fisheries, forestry, and manufacturing use management tools. Qorri et al. (2018) outlined various method applications in regard to sustainability performance of supply chains in different sectors. The results showed an increasing application of multi-criteria analysis. The most common tools are life cycle assessment, analytical hierarchy process, fuzzy set approach, balance scorecard, and data envelopment analysis.

Lager (2016) reviewed methods and tools within the process industries such as chemicals, food and beverage, mining and metals, mineral and materials, pharmaceuticals, pulp and paper, steel, and utilities. He considers current applications like technology road mapping, R&D strategy development, and portfolio balancing and future perspectives including raw material supply, production process, and products. He goes in especially for the collaboration between process companies and technology/equipment suppliers. There is outlined a need for closer linkage between innovation management and operations management to come after in industry and academia.

Thinking in a systems perspective requires a change in thinking from a linear understanding to circular. The basic concept of this is seeing biological processes as interlinkage systems. As a result companies have shifted to the product-service system (PSS) business model. Mourtzis et al. (2018) established a holistic approach for PSS evaluation. The aim is to capture all its life cycle phases in a value-added chain including aspects from providers' and costumers' perspective. Therefore also a software tool was evolved and applied in a case study of the mold making industry. Also Vezolli et al. (2015) appreciate the (sustainable) product-service system as a strategic management design or tool applicable to various industrial sectors to combine customer satisfaction and economic wealth respecting environmental impacts.

As a tool for a systemic approach in the cluster management, Ucler (2017) developed the intelligent cluster assignment tool concept to enhance innovation applicable to different sectors in developing economies. The approach delivered a strategic framework for cluster management.

Also Tamayo-Orbegozo et al. (2017) developed a strategic model which offers a more regional application within the context of eco-innovation after identifying a lack of analysis of the dynamics of eco-innovation including different agents and sectors. Therefore they extracted from the setting of a multiple-case study an integrating model, holistic and dynamic, which is transferable relating to sustainable and innovative solutions in a specific regional context.

2.2 *Latest Developments in Management Tools and Application Examples in the Bioeconomy Field*

The review of Karvonen et al. (2017) carves out the most relevant impact assessment methods within the bioeconomy, especially the forest bioeconomy. It is worth mentioning that these tools address mainly industrial actors, not addressing governance issues; this is relevant to mention at this point to avoid any misunderstandings and to realize that the focus thus far has been to provide management tools for the individual enterprises. Karvonen et al. (2017) compiled the five most common tools in a table including weaknesses and strengths of each method as well as their application in combination with other tools as an amplification. The cost-benefit analysis (CBA) as an economic-oriented tool, which is based on monetary units, thus evaluating monetary values as its strength, is combined with, for example, input-output able to monetize also the nonmonetary values and vice versa. Its weakness is to be presumed in ethical and democratic observations regarding values as subjective cases.The input-output (IO) methods as an economic or environmental applied method can be expanded with LCA databases or MFA calculations. Economic tables and statistics for environmental IO are commonly available and well documented. But this directly leads to an extensive data output, which is troublesome to analyze and utilize subsequently. Life cycle analysis (LCA) methods, listed as third sustainability assessment tool, can be applied environment-oriented (ELCA), as well as social-oriented (SLCA) or economic-oriented (LCC) and various combinations with other methods are possible. The material flow analysis (MFA) with its orientation in environment sector can uncover inefficient material usage and production phases in a simpler way compared to LCA methods, but may induce a limited view.As the fifth tool registered, the multi-criteria analysis (MCA) can provide any desired orientation and offers a complete assessment and balancing between alternatives with the limitation of excluding known unsustainable alternatives preliminarily (Karvonen et al. 2017).

Consistently mentioned there is a lack of assessing the social impacts regarding bioeconomy. Therefore, Mattila et al. (2018) evaluated the social sustainability of bioeconomy value chains. The goal of the paper compared to previous applications in social sustainability methods was to compare the setting, more precisely the impacts of Finnish wood products, in local and global approaches with the aim of developing possibilities of an integrative approach. Therefore a multi-region input-output model was used. The outcomes of the study were health and safety and gender inequality as the main social issues within a life cycle perspective. These impacts are mainly presented outside the forest industry sector and not within the Finnish boundaries. They developed options to interconnect the output of local stakeholders, who concentrate primarily on the local issues as working conditions, and the global impact output of this study in terms of a framework combining the global and local considerations (Mattila et al. 2018).

Falcone and Imbert (2018) criticize the neglect of social impacts within the life cycle approach in the analysis of bio-based economy, as well. In their paper they identified the main social impact categories to include them eventually in the social life cycle assessment scheme for bio-based products. This leads to a better informed consumer and an expanding market of bio-based products.

As a need in the latest developments of bioeconomy for assessing social effects a new conceptual framework for a context-specific sLCA, especially to assess wood-based products in a regional perspective, was evolved by Siebert et al. (2018a, b). It facilitates to uncover social hotspots and social opportunities and the location in the wood-based production system of a regional bioeconomy.

3 The Need for a Systematic Approach for the Urban Bioeconomy

Bezama (2016) analyzed that from a systems perspective, the implementation of the bioeconomy strategy entails a series of challenges, from which the following two can be identified as important for the sustainable development of cities: Firstly, that there is a lack of synergic work between the different participants in the "innovation chains" of the bioeconomy. Participants of these innovation chains are not only the ones involved in the technology development process but also the market and society players (including public services) that produce the demand for bio-based products. In order to overcome this challenge, it is necessary to identify the different actors along the innovation chains of the bioeconomy and, most importantly, to understand the interactions between these actors, as well as their perspective on the potential and current barriers toward a more urban-centered bioeconomy. In this regard, there is a need to incorporate a more dynamic analysis that takes into account the different scales (local, regional, national, global) and dimensions (social, economic, environmental) directly and indirectly affected by the implementation of these new processes and products (Bezama 2018).

An important aspect to consider with the implementation of the bioeconomy is that the impacts of such implementation will most dramatically be observed on a regional and local level. In the particular case of cities, the impacts of the bioeconomy are complemented by the effects of further transition processes, such as the circular economy and a series of societal changes (e.g., environmental awareness, industrialization, economic changes) as well as the global process of urbanization itself, which is closely linked to the consumption of land, increase in traffic, and high air pollution and is considered a major challenge for a sustainable development (EEA 2015). It is therefore important to link the identified interactions among actors with the actual material streams of available bio-based resources that shape the bioeconomy system (i.e., all inputs and outputs to and from the cities, as well as the internal bio-based material streams that characterize the processes that take place within the city limits).

In this regard, over the last years, the "urban metabolism" concept, first conceived by Wolman (1965), has been considered as an interesting method for supporting the development of sustainable cities and communities (Chrysoulakisa et al. 2013; Conke and Ferreira 2015).

As described by Kennedy et al. (2007, 2011), "urban metabolism" may be defined as "the sum total of the technical and socio-economic processes that occur in cities, resulting in growth, production of energy, and elimination of waste." The concept of "urban metabolism" was based by providing an analogy of the urban context to the metabolism of organisms. As explained by Decker et al. (2000), "cities transform raw materials, fuel, and water into the built environment, human biomass and waste"; thus the analogy to the metabolism is observed in natural organisms (see Fig. 18.1 for a simplified version considering the urban bioeconomy metabolism).

The urban metabolism concept is based on an analysis of material and energy flows, thereby tracing the input, storage, transformation, and output processes (Zhang 2013; Hendriks et al. 2000). In general, material flow analysis starts by classifying the different material flows, followed by an accounting of all the identified flows. Particularly interesting is the use of a life cycle perspective for monitoring the flow of materials throughout their entire life cycle within the urban system (Zhang 2013).

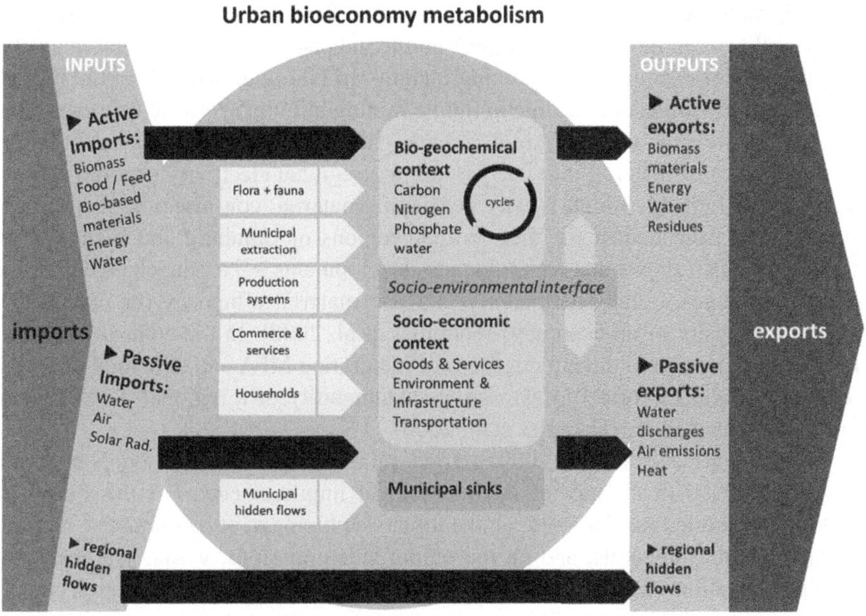

Fig. 18.1 General conceptual description of the flows associated with the urban bioeconomy metabolism concept. (Adapted from Musango et al. (2017))

3.1 The Urban Metabolism Concept and Its Applications to Resources Management

The urban metabolism concept is not new. Already in the 1970s, there were several pioneering studies utilizing material flow analysis concepts to evaluate and characterize the material flows within the city limits. In the 1990s there were also several studies on UM. It is since the 2000s when the majority of existing studies can be recorded (Anderberg 2012). In 2001, a standardized MFA for national analysis was issued by Eurostat and used in many studies. About a decade later, scientists have pointed out that MFA at regional and local level still remains very limited (Niza et al. 2009). It is in 2004 when the *Practical Handbook of Material Flow Analysis* authors by P. H. Brunner and H. Rechberger is published. The book is an introduction to MFA and contains 14 case studies describing the method, in addition to the characteristics and history of MFA (Zeschmar-Lahl 2004).

In 2006 Hammer et al. published an MFA based on the three regions around Hamburg (HH), Vienna (W), and Leipzig (LE). The analysis is preceded by a list of selected structural features of the regions under investigation, such as the change in settlement and transport areas over the period under investigation. The indicator DMC (domestic material consumption) is used to calculate the material consumption of the respective population within a defined period of investigation (1992–2001, HH, LE; 1995–2003, W). It is calculated on the basis of raw material extraction plus imports minus exports. The evaluations show that even individual changes in material flows can have a major impact on the indicators and the material intensity of a region, for example, the declining quantity of building materials in the region around and in Vienna or a decline in lignite mining in Leipzig during the respective period under review. The analysis also shows gaps: e.g., that an MFA does not cover the conversion of material into energy and interregional electricity exports. This is the case for Leipzig, which is why per capita material, consumption also differs significantly from that of the metropolitan regions of Hamburg and Vienna. The comparison also shows that technological developments – i.e., the changed use of certain resources – can significantly increase material efficiency (or reduce the material intensity of an economy) (Hammer et al. 2006). In the context of MFA, Hammer et al. point to the effectiveness of integrating MFA and structural analysis (e.g., ecological footprint) to assess the sustainability of a regional development more comprehensively (Hammer et al. 2006).

More recently, Niza et al. (2009) carried out a UM analysis of Lisbon's material flows in 2009 with the aim of methodological improvements. For this purpose, material categories were formed: on the input side biomass, energy sources, metals, and nonmetallic minerals, and on the output side emissions, waste. The result of their investigation was, for example, that 80% of urban material consumption comes from nonrenewable sources. Two reasons for this are the dominant construction of new buildings and the simultaneous lack of renovation of old buildings in Lisbon since the 1990s. The percentage figure also included the switch from public transport to individual car traffic and inner-city transport of transit goods. It is also worth mentioning the inclusion of the life span of products (quick consumption materials 0–1 year, 2–10 years, 11–30 years, over 30 years). This categorization allows for an examination of material storage in cities (city mining) and future waste or the potential utilization of secondary raw materials.

Following the same line of work, Wallsten (2015) combined the quantitative MFA approach with the qualitative social science approach of infrastructure studies in a study to look at a locally specific research topic from a sociotechnical perspective. The study focused on the "hibernating stock," i.e., unused pipes and cables underneath the streets of the Swedish city of Norrköping. By means of a quantitative survey, Wallsten determined the amount of steel, copper, and aluminum which could potentially be salvaged and thus serve as an alternative material reserve (5000t) as well as its local distribution. By means of interviews with road construction personnel and other stakeholders, a series of statements could be made about the origin and procedure of the unused materials. The study makes clear a "disconnect and leave behind" logic and leads to three categories for left behind infrastructures (Wallsten 2015). It is methodologically remarkable that the research object of the "hibernating stock" and the local confinement function as a "boundary object" and thus allow the work with approaches from different disciplines even "without consensus," but with common "modus operandi" (Wallsten 2015). Wallsten concludes with the policy recommendation to integrate metal recovery into continuous renovation or urban planning processes (Wallsten 2015).

Finally, Bahers et al. (2019) analyzed the material flows of two medium-sized cities in Western France (Rennes, 400,000 p.e.; Le Mans, 200,000 p.e.) by means of MFA. In their research, scientists focused on spatial indicators and waste flows. The analysis captured imports and exports at city, local (department), regional, national, and international level. In this way, two categories were formed: "local goods" (biomass, building materials, secondary raw materials) and "highly globalized goods" (industrial goods, fuels, metals). In a further step, the waste exports of the two cities are analyzed and compared. The researchers note that externalization practices in this respect are noting that even the recycling of waste takes place primarily at national or even global level. With regard to the urban material stock, this work shows that medium-sized cities are gradually replenishing their material stocks due to the urban sprawl rather than, for example, the metropolis of Paris, where buildings are being renovated. The study shows that the UM of a medium-sized city differs significantly from that of a large city, since the former is located as an intermediate link between rural and metropolitan areas. Moreover, medium-sized cities are characterized by a very strong connection with their rural surroundings (Bahers et al. 2019).

In summary UM does provide a systematic tool to characterize the material flows associated with the resources management of cities. The adaptation of this methodology to the bioeconomy field could be a useful way of providing a robust evaluation of the biomass resources management within the city limits. For this, however, the following aspects should be taken into consideration:

- UM studies focus mainly on large cities. Only a few studies are devoted to regional metabolisms (Bahers et al. 2019).
- UM studies are bound by administrative boundaries (Bahers et al. 2019).
- Limited data availability (e.g., at the local and urban level) prevents exhaustive system descriptions and adds some uncertainty to the results (Bahers et al. 2019; Hammer et al. 2006; Niza et al. 2009; Shahrokni et al. 2015).
- Limited data availability leads to the prioritized consideration of selected material flows with a good data basis (Anderberg 2012; Niza et al. 2009).

4 Needs for Implementing an Urban Bioeconomy Metabolism

4.1 The Role of Governance and of Social Aspects

MFA is a method that reduces the complexity of reality to a simplified and reliable form (Brunner and Rechberger 2003), based on input and output flows. However, this is not enough to understand the relationships between urban and environmental quality and the patterns and lifestyles behind metabolic flows. For this purpose, additional methods are needed to find a balance between studying urban complexity and generating ideas for real politics and urban planning (Broto et al. 2012).

According to Björn Wallsten, the usefulness of pure MFAs for political decision-making processes, especially at higher levels, can be seen in concrete examples, e.g., for decisions on recycling projects. Nevertheless, in order to successfully implement recycling targets locally, basic knowledge about the potentially available quantities is required, as well as information about when, where, and by whom recycling can take place. Wallsten criticizes MFAs at this point: They should overcome their reductionist orientation, as they risk removing the material quantities under consideration from their social and local embedding. There would be a risk that purely mathematical standards would lose relevance for the social sciences (Wallsten 2015).

In addition, according to Stefan Anderberg, the recording of material flows would make it possible to obtain an overview, but would not allow any statement about the usefulness of those flows for society. UM studies are dominated by a quantifying analysis of material flows; only a few intensively pursue the connections to social aspects. In addition to a more flexible analysis compared to levels and details ("[a] more systematic multilevel analysis"), a closer connection to decisions and institutional structures is needed (Anderberg 2012). Despite growing awareness and an increase in sustainable objectives, there are still only a few effective initiatives that would shift urban development in a sustainable direction. The growing number of UM analyses at the urban level and studies from sustainable urban research is contrasted by a relatively small role of these in the urban planning context. It is rare that UM studies or their perspectives are fully integrated into local policy strategies or planning processes. Nevertheless, UM studies have often contributed to sustainability reports or indicators (Anderberg 2012). Anderberg pleads for the inclusion of further criteria in UM analyses: climate, age of a city, and its development history (Anderberg 2012). A critical aspect of the social science approach is the focus on groups of actors and the classification of their relevance. Especially in infrastructure issues, workers who work directly on the materials are relevant sources of information and in this sense more valid sources of information than, for example, "system providers" and yet less often the subject of studies (Wallsten 2015). In this sense, Bristow and Mohareb (2019) stress the importance of urban political ecology (UPE), as it goes beyond the quantitative coverage of an MFA and analyses drivers and impacts in greater depth.

Finally, urban dynamics can also cause negative environmental impacts within the spatial environment of cities, such as urban sprawl in the regional hinterland (Bleher 2017). The examination of global and inner-city distributional inequalities shows that a meaningful UM model includes not only material analyses but also studies of socioeconomic and political contexts. Such an approach thus combines physical flows, which are visible and in the best case quantifiable, with less visible structural contexts that significantly shape those flows. Further theoretical development is needed to determine how such expanded knowledge can be implemented in practice (Broto et al. 2012).

4.2 The Urban Bioeconomy Metabolism as a Toolbox for the Management of Biomass Resources

A series of scholars and practitioners have identified the model of a natural ecosystem as the most suitable way for developing sustainable cities. In fact, the major uses of the urban metabolism models can be summarized as twofold, first, to be used as basis for sustainable urban design and, second, to be used as basis for policy analysis (Zhang 2013). Thus far, however, there have been no advances in exploring the definition of an urban metabolism in a bioeconomy context.

Considering the above, the main research question to be addressed is: How can we define a "bioeconomy concept in an urban context," based on the metabolism concept, such that we can understand and analyze the transformation processes related to the bioeconomy within the urban system, and link them to the material streams that characterize the available bio-based resources so that we can propose measures to design more sustainable urban concepts?

In order to address this question, we propose that the establishment of the urban bioeconomy metabolism should not only mean the definition of a concept that could help understand the potential impacts of biomass streams. Albeit being an important issue, the major goal of the urban bioeconomy metabolism should aim at managing the resources within the scope limits of the urban areas and their interaction with the peripheral rural areas, thus providing a means for a systematic regional resources management.

Scientific experiences of the urban metabolism context show that integrated approaches that mix quantitative methods, such as material flow analysis (MFA), with policy or infrastructure, sociotechnical analysis, or methods of social science, allow more comprehensive studies of the "urban complexity" or even the "urban disorder" (Broto et al. 2012; Wallsten 2015; Bleher 2017). Solely quantitative approaches such as MFA and footprints are not able to obtain efficient city policy or city planning (Broto et al. 2012).

Integrated approaches reflect on resource flows as much as on history, policy, and socioeconomic conditions of urban contexts. They include considerations on resilience/resistance (e.g., in case of hazards) and on flexibility and multifunctionality of urban structures (Anderberg 2012; Bristow and Mohareb 2019). Structures of (civil) self-organization, power relations, and decision processes allow more detailed perspectives on urban resource flows and access to them (Broto et al. 2012; Bristow and Mohareb 2019).

By expanding the urban metabolism concept toward a more integrated and interdisciplinary analysis of urban areas, it will be possible to make those theoretical approaches closer to the reality of urban planning and policy making. As presented in Fig. 18.2, an urban bioeconomy toolbox contains multiple analysis components and connects to several goals on the policy side.

The concept of the bioeconomy seeks for transformation toward more sustainable economic practices. Transformations depend on structural changes on institutional,

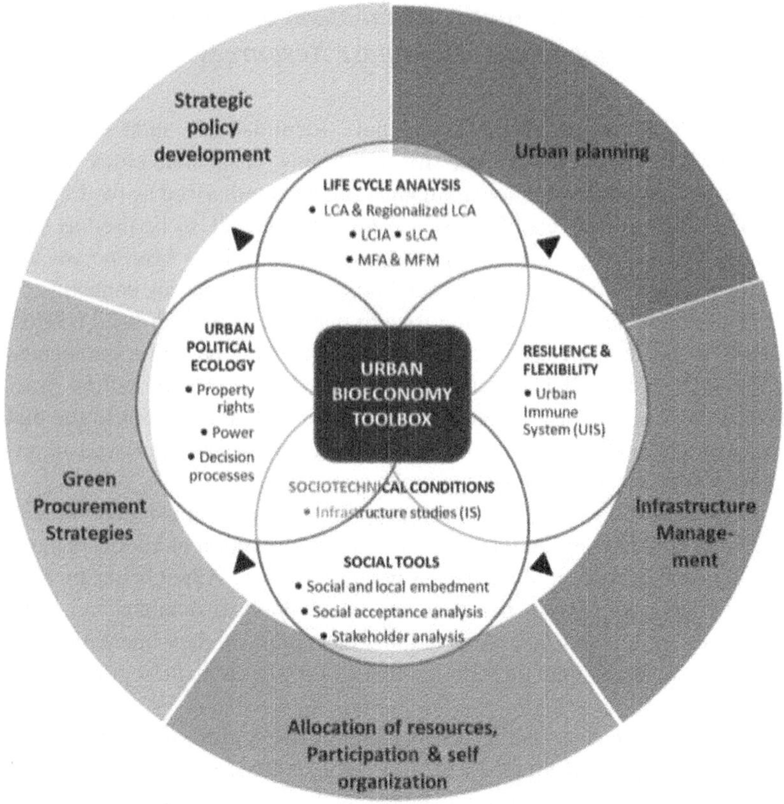

Fig. 18.2 The proposed toolbox for addressing the management of the urban bioeconomy metabolism

societal, legal, and technological level. Transition-based governance strategies include necessarily several layers in their analytical preparation (Ludwig 2019).

In this sense, for managing the urban bioeconomy, it is needed to count with an integrated planning and assessment of the resources management within and beyond the city limits, for which the implementation of an adequate and dynamic governance framework is required. This is sustained not only by a strategic policy development but also by a sustainable green procurement program that takes advantage of the regional capacities and strengths (i.e., local resources, human capacities, market needs, and industrial infrastructures, among others).

On the other hand, by connecting the urban planning, infrastructure management, and allocation of resources from a bioeconomy perspective, cities could identify and manage the resources in terms of their own defined goals, based on the definition of the local and regional sustainability development plans. A regional life cycle management approach could then help in bringing the necessary information for decision makers and involved stakeholder groups and individuals to generate a more robust and mutually agreed resources management plan.

5 Outlook

Depending on how cities are built, heated, and cooled, on the efficiency of their infrastructures and the consumption and transport habits of their inhabitants, this influences the amount of greenhouse gas emissions and the impact on land use, water, and mineral resources in the global framework (Anderberg 2012).

Sustainable urban planning would not only include more efficient, integrated, and flexible water, sanitation, waste, heating, and energy infrastructure but would also aim at resilience to changes in population, economy, climate, and water balance. This results in urban systems that are multifunctional, as they fulfill several values and functions simultaneously (Anderberg 2012).

The approach of urban metabolism offers the possibility for expansion and integration with other approaches. An integrated consideration of dynamic energy and material flows, which includes land use changes or soil degradation and urban sprawl effects in its analyses, is missing in some cases (Bleher 2017). In addition, it should be integrated with regionalized life cycle methods and tools for helping the management of locally available resources, such as the RELCA and RESPONSA models (O'Keeffe et al. 2016; Siebert et al. 2018a, b).

The introduction of the urban bioeconomy metabolism as a toolbox for identifying the main issues of the biomass flows within the city limits, providing also the resources for making available the necessary information to build a sound decision basis for local and regional stakeholders.

References

Accorigi, A. (2018): Research and Innovation Activities on Urban Circular Bioeconomy. DG Research & Innovation, Directorate for Bioeconomy. In: EUROCITIES - 'The road to Urban Bio-economy: Barriers and Solutions to Closing the Loops of Bio-Resources'. Available at: http://nws.eurocities.eu/MediaShell/media/Presentations.pdf (Last accessed: June 2018).

Anderberg, S. (2012). Natural Resource Flows and Sustainability in Urban Areas. Environmental Geology, 257–269. https://doi.org/10.1007/978-1-4939-8787-0_441

Bahers, J. B., Barles, S., & Durand, M. (2019). Urban Metabolism of Intermediate Cities: The Material Flow Analysis, Hinterlands and the Logistics-Hub Function of Rennes and Le Mans (France). Journal of Industrial Ecology, 23(3), 686–698. https://doi.org/10.1111/jiec.12778

Bezama, A., (2016): Let us discuss how cascading can help implement the circular economy and the bio-economy strategies. Waste Manage. Res. 34 (7), 593 – 594

Bezama, A. (2018): Understanding the systems that characterise the circular economy and the bioeconomy. Waste Manage. Res. 36 (7), 553–554

Bezama, A., Ingrao, C., O'Keeffe, S., Thrän, D., (2019): Resources, collaborators, and neighbors: The three-pronged challenge in the implementation of bioeconomy regions. Sustainability 11 (24), art. 7235

Bioökonomierat (2018) Future Opportunities and Developments in the Bioeconomy – a Global Expert Survey. Office of the German Bioeconomy Council, Berlin, April 2018.

Bleher, D. (2017). Ressourcenwirkung des urbanen Metabolismus. 1–37.

Bristow, D. N., & Mohareb, E. A. (2019). From the urban metabolism to the urban immune system. Journal of Industrial Ecology, 1–13. https://doi.org/10.1111/jiec.12919

Broto, V. C., Allen, A., & Rapoport, E. (2012). Interdisciplinary Perspectives on Urban Metabolism. JournalofIndustrialEcology,16(6),851–861.https://doi.org/10.1111/j.1530-9290.2012.00556.x

Brunner, P. H., & Rechberger, H. (2003). Practical handbook of material flow analysis. In Practical Handbook of Material Flow Analysis. https://doi.org/10.1007/bf02979426

Chrysoulakisa, N., et al. (2013): Sustainable urban metabolism as a link between bio-physical sciences and urban planning: The BRIDGE project. Landscape and Urban Planning 112, 100– 117

Conke, L., Ferreira, T. (2015): Urban metabolism: Measuring the city's contribution to sustainable development. Environmental Pollution 202, 146-152

Decker, E.H., Elliott, S., Smith, F.A., Blake, D.R., Rowland, F.S., (2000): Energy and material flow through the urban ecosystem. Annual Review of Energy and the Environment 25, 685-740.

EEA - European Environment Agency, 2015. Urban sustainability issues — Resource-efficient cities: good practice - Technical report No 24/2015

Eurocities (2016): EUROCITIES Strategic Framework 2014-2020 - Towards an Urban Agenda for the EU. Brussels, Belgium. Available at: http://nws.eurocities.eu/MediaShell/media/494-Eurocities_strategic_framework-leaflet-web.pdf (Last accessed: June 2018)

Evans, J.P. (2011) Resilience, ecology and adaptation in the experimental city. Transactions of the Institute of British Geographers 36, 223-237

Falcone, PM; Imbert, E, 2018: Social Life Cycle Approach as a Tool for Promoting the Market Uptake of Bio-Based Products from a Consumer Perspective. Sustainability 10(4), 1031; https://doi.org/10.3390/su10041031

Grimm, N.B., Faeth, S.H., Golubiewski, N.E., Redman, C.L., Wu, J., Bai, X., Briggs, J.M. (2008) Global change and the ecology of cities. Science 319, 756-760.

Hammer, M., Giljum, S., Luks, F., & Winkler, M. (2006). Die ökologische Nachhaltigkeit regionaler Metabolismen: Materialflussanalysen der Regionen Hamburg, Wien und Leipzig. [EcoHammer, M., Giljum, S., Luks, F., & Winkler, M. (2006). Die ökologische Nachhaltigkeit regionaler Metabolismen: Materialflussanalysen. Natur Und Kultur, 7(2), 62–78.

Hendriks, C., Obernosterer, R., Muller, D., Kytzia, S., Baccini, P., Brunner, P.H., (2000): Material flow analysis: a tool to support environmental policy decision making. Case-studies on the city of Vienna and the Swiss lowlands. Local Environment: The International Journal of Justice and Sustainability 5, 311e328.

Hildebrandt, J., O'Keeffe, S., Bezama, A., Thrän, D., (2019): Revealing the environmental advantages of industrial symbiosis in wood-based bioeconomy networks: an assessment from a life cycle perspective. J. Ind. Ecol. 23 (4), 808 – 822

Hildebrandt, J., Bezama, A., Thrän, D., (2020): Insights from the Sustainability Monitoring Tool SUMINISTRO applied to a case study system of prospective wood-based industry networks in Central Germany. Sustainability 12 (9), art. 3896

Karvonen, J; Halder, P; Kangas, J; Leskinen, P, 2017: Indicators and tools for assessing sustainability impacts of the forest bioeconomy. Forest Ecosystems. 4, 2. DOI: 10.1186/s40663-017-0089-8

Kennedy, C.A., Cuddihy, J., Engel Yan, J. (2007). The changing metabolism of cities. Journal of Industrial Ecology 11, 43-59

Kennedy, C., Pincetl, S., Bunje, P. (2011) The study of urban metabolism and its applications to urban planning and design. Environmental Pollution 159, 1965-1973

Lager, T., 2016: Managing Innovation & Technology in the Process Industries: Current Practices and Future Perspectives. Procedia Engineering. Volume 138, Pages 459-471

Ludwig, G. (2019): The Role of Law in Transformative Environmental Policies—A Case Study of "Timber in Buildings Construction in Germany". Sustainability 11(3):842. https://doi.org/10.3390/su11030842

Mattila,TJ; Judl, J; Macombe, C; Leskinen, P, 2018: Evaluating social sustainability of bioeconomy value chains through integrated use of local and global methods. Biomass & Bioenergy 109, 276-283

Mourtzis, D; Papatheodorou, AM; Fotia, S, 2018: Development of a Key Performance Indicator Assessment Methodology and Software Tool for Product- Service System Evaluation and

Decision-Making Support. Journal Of Computing And Information Science In Engineering, Volume 18

Musango, J. K., Currie, P., & Robinson, B. (2017). Urban Metabolism for Resource-Efficient Cities. 1–40. Retrieved from www.sustainabilityinstitute.net

Niza, S., Rosado, L., & Ferrão, P. (2009). Urban metabolism methodological advances in urban material flow accounting based on the Lisbon case study. Journal of Industrial Ecology, 13(3), 384–405. https://doi.org/10.1111/j.1530-9290.2009.00130.x

O'Keeffe, S., Majer, S., Bezama, A., Thrän, D., (2016): When considering no man is an island—assessing bioenergy systems in a regional and LCA context: a review. Int. J. Life Cycle As-sess. 21 (6), 885–902

Qorri, A; Mujkic, Z; Kraslawski, A, 2018: A conceptual framework for measuring sustainability performance of supply chains. Journal of Cleaner Production. Volume 189, Pages 570-584

Rigby, D.K; Bilodeau, B, 2017: Management Tools & Trends, Bain & Company

Rosenzweig, C., Solecki, W., Romero-Lankao, P., Mehrotra, S., Dhakal, S., Bowman, T., Ali Ibrahim, S., (2015) ARC3.2 summary for city leaders, in: Network, U.C.C.R. (Ed.), New York, Columbia University

Sassen, S. (2012) Cities in a world economy. Pine Forge Press, Thousand Oaks

Schawel, C; Billing, F, 2018: Top 100 Management Tools – Das wichtigste Buch eines Managers von ABC-Analyse bis Zielvereinbarung, 6.Auflage, SpringerGabler

Shahrokni, H., Årman, L., Lazarevic, D., Nilsson, A., & Brandt, N. (2015). Implementing smart urban metabolism in the Stockholm Royal Seaport: Smart city SRS. Journal of Industrial Ecology, 19(5), 917–929. https://doi.org/10.1111/jiec.12308

Siebert, A., Bezama, A., O'Keeffe, S., Thrän, D., (2018a): Social life cycle assessment indices and indicators to monitor the social implications of wood-based products. J. Clean Prod. 172, 4074–4084

Siebert, A., O'Keeffe, S., Bezama, A., Zeug, W., Thrän, D., (2018b): How not to compare apples and oranges: Generate context-specific performance reference points for a social life cycle assessment model. J. Clean Prod. 198 , 587–600

Seto, K.C., Reenberg, A., Boone, C.G., Fragkias, M., Haase, D., Langanke, T., Marcotullio, P., Munroe, D.K., Olah, B., Simon, D. (2012) Urban land teleconnections and sustainability. Proceedings of the National Academy of Sciences 109, 7687-7692.

Solecki, W., Seto, K.C., Marcotullio, P.J. (2013) It's Time for an Urbanization Science. Environment: Science and Policy for Sustainable Development 55, 12-17

Tamayo-Orbegozo, U; Vicente-Molina, M., Villarreal-Larrinaga, O., 2017: Eco-innovation strategic model. A multiple-case study from a highly eco-innovative European region. Journal of Cleaner Production. Volume 142, Pages 1347-1367

Thrän, D., Hildebrandt, J., Zeug, W., Moesenfechtel, U., Bezama, A. (2018): Scenarios for a wood-based bioeconomy in Central Germany. Products - Processes - Environmental Effects. Oral Presentation at the 7. International Bioeconomy Conference am 6. und 7. Juni 2018. Halle (Saale).

Ucler, C, 2017: Intelligent assignment in clusters to enhance collaboration and innovation. Journal Of Manufacturing Technology Management. Volume 28, Pages 554-576

UN United Nations, Department of Economic and Social Affairs, Population Division (2015). World Population Prospects: The 2015 Revision, Methodology of the United Nations Population Estimates and Projections. ESA/P/WP.242.

Vezolli, C; Ceschin, F; Diehl, JC; Kohtala, C, 2015: New design challenges to widely implement 'Sustainable Product-Service Systems'. Journal Of Cleaner Production. Volume 97, Pages: 1-12

Wallsten, B. (2015). Toward social material flow analysis on the usefulness of boundary objects in urban mining research. Journal of Industrial Ecology, 19(5), 742–752. https://doi.org/10.1111/jiec.12361

Wigginton, N.S., Fahrenkamp-Uppenbrink, J., Wible, B., Malakoff, D. (2016) Cities are the Future. Science 352, 904-905.

Wolman, A. (1965). The metabolism of cities. Contemporary Psychology: A Journal of Reviews, Vol. 6, pp. 223–223. https://doi.org/10.1037/006636

Wrisberg, N. and Udo de Haes (2012): Analytical Tools for Environmental Design and Management in a Systems Perspective. [2012 Edition]. Springer Science + Business Media, B.V. ISBN: 9789401039024

Zeschmar-Lahl, B. (2004). Book Review: Practical Handbook of Material Flow Analysis (Authors: Brunner, Paul H. Rechberger, Helmut). Waste Management, 9(5), 337–338.

Zeug, W., Bezama, A., Moesenfechtel, U., Jähkel, A., Thrän, D., (2019): Stakeholders' interests and perceptions of bioeconomy monitoring using a sustainable development goal framework. Sustainability 11 (6), art. 1511

Zhang, Y. (2013): Urban metabolism: A review of research methodologies. Environmental Pollution 178, 463-473

Part VI
Innovative Energy Solutions

Chapter 19
Innovation in Bioenergy: Factors Affecting Innovation in Biofuels

Dariusz M. Trzmielak and Ewa Kochańska

Abstract This chapter focuses on biofuels innovation development based on main challenges for innovations and the cooperation partnerships of research centre involving different partners. The aspects of biofuels generations and the crucial biofuels products provide a deeper understanding of the significant role of bioenergy markets. The authors consider also the problems in the areas of biofuel evolution that have a decisive influence on their past and present developments. It is the transport, heat and electricity and the value chains. The practical part illustrates the biofuels innovations and its developments in Polish central region. The four cases – the liquid biofuel production based on slaughterhouse waste and animal fats, the solid biofuel production based on waste of onion, the gas biofuel production based on utilization of CO_2/methane and bioethanol and integrated biofuels production technology/methane and fish oil based on fish waste – are exemplifying the dynamic development of biofuels in Poland and the European Union.

Keywords Development of bioenergy · Biofuels technologies · Renewable transportation fuels · Biofuels value chain · Biocomponents

Development of Bioenergy and Biofuels: The Challenge for an Innovation

Renewable energies such as solar energy, geothermal energy, wind energy, floating water, wave and tidal energy and biomass have been widely studied in economics and give assurance to the launch of renewable energy beginning in the 1990s. The innovations in renewable energy are forced by the climate change, security of energy supply, necessity of energy sources diversification and optimization of pro-

D. M. Trzmielak (✉)
University of Lodz, Lodz, Poland
e-mail: dariusz.trzmielak@uni.lodz.pl

E. Kochańska
Research and Innovation Center Pro-Akademia, Lodzki, Poland

E. Koukios, A. Sacio-Szymańska (eds.), *Bio#Futures*,
https://doi.org/10.1007/978-3-030-64969-2_19

duction cycles and the utility of new technologies in connection with research and development in the field (Van Geenhuizen and Schoonman 2011). The needs to struggle with global climate changes are caused by the increased greenhouse gas emissions into the atmosphere, mainly of carbon dioxide (CO_2), SO_2 and NOx that have the main impact on biofuels developments. The craving for improving the countries' economic security in the increase of global energy consumption is the next significant factor for biofuels technology production and the development of biofuels energy transporters. The third one is the competition of companies, regions and economies. The diversification of energy optimize of energy production and process of a biofuel supply chains. Biofuel is not the only renewable energy source. It has built wealth and helps with the sustainable improvement of human well-being. The biofuel technology development stimulates agricultural and urban evolution through the utilization of waste from food and bioenergy crops and municipal waste (Panchuk et al. 2020).

The development of technologies for the alternative fuels production derived from biological and urban waste sources of biomass has different aims (Pratama and Utomo 2019). The past experience contributes to the fact that the significant biofuel development objective is to reduce CO_2 emission or to balance them optimally in the process of fuel production or combustion (Biernat 2012). The emissions reduction, mitigation of pollutions mainly in urban, create incentives for developing the new technologies and products in all renewable fields (Ibarra-Yunez, Ibbara-Yunez 2020). Currently, transport is by far the biggest emitter accounting for more than 30% of the Union's total energy consumption, and 98% of transport energy expenditure is fossil fuels (Cordis 2019). The transport sector and its energy consumption have become one of the key challenges for EU and other countries' overall goals of reducing its dependence on fossil fuels.

The depletion of oil resources has affected the intensive search for alternative fuels. The adoption of biofuels on the market is one of the opportunities to guarantee more efficient and less harmful fuels and fuels transporters. Road, rail, marine and railway vehicle engines contribute to high greenhouse gas emission to the atmosphere cause by the combustion of liquid fuels such as petrol and diesel. Another tendency is the growing need of the diversification of raw materials for the production of liquid fuels. It creates circumstances for industrial-scale energy production as well as for personal use in households. The production of liquid biofuels still exceeds the cost of acquiring fossil fuels. However, the inevitability of limiting the extraction potential and existing raw materials and the increase in price are changing this situation. The usage of biofuels, especially in the transport fuel market, is expected to increase in the nearest future. The world's biofuel consumption is mainly (90%) accounted for by bioethanol and biodiesel. Other biofuels are also used, although their market presence is quite limited. The automotive industry has long since replaced petrol engines with hybrid, fully electric or biofuel engines. The process of changing energy sources in air and sea transport is just the beginning. The airlines, railways and shipping companies are switching their fleet to biofuels (Panchuk et al. 2020). New developments in the field of motor fuels made from plant raw materials are arousing enthusiasm and potential demand. The demand for biomass fuels is stimulated by the following factors: political, national, scientific and research activities, development of technical progress and distribution channels' capabilities. The use of biofuels in the EU and other countries is sanctioned by directives, acts and regulations (Orynycz 2012).

The development of biofuels production and application technologies is assumed that the sources of biofuels should have the following: technical and energetic properties determining their applicability for supplying engines and heating devices; be cheap in production, distribution and sales; lower environmental risk than fuels used so far due to lower emission of toxic compounds and greenhouse gas in the process of their combustion; providing acceptable economic indicators for transporters (engines and boilers); lower operating cost in production, distribution and sales; and increasing energy independence for sectors, regions and economies (Biernat 2012). There are many issues to be discussed and weighted up regarding innovation development in bioenergy. The overall energy balance is dependent on the species of biomass and the type of produced biofuels. The main research and development in biofuels fields tend to direct use, transformation, electricity generation, gasification, charcoal production and modular systems in bioenergy technologies and products. The principal possibilities of application of biofuels are following: thermal application of forest waste in the medium-power boilers, electrical generation from agricultural and forest waste, pellet manufacturing plants using agricultural and forest waste, generation of biogas for thermoelectric use from cattle waste, thermal application of agro-food industry waste and generation of biogas for thermoelectric use from the organic fraction of urban solid waste (Kafle and Kim 2013).

The first-generation biofuels have been made mainly from corn and other food-based crops. It includes fuels production from raw materials obtained from plant biomass or from vegetable and animal fats (biodiesel, bioethanol, biomethanol and biofuels, which are based on post-frying oils). However, they are available in limited volumes that could not have widespread use. Another key problem is technology. The first-generation biofuels producers have to rely on technology developed over millennia to make beer, wine and spirits. There are also severe problems to produce fuel from food in a world with a growing population and limited cultivated areas (Agarwal and Kumar 2018). In consequence, the economical ways to use cornstalks, wood chips, by-products and urban waste are the crucial tendency for new biofuels technology development. The biofuels industry has been working on the next biofuels innovation second-generation. The new technologies have already made it possible to extend the amount of those bioenergy sources. Today, biofuels can be made from forest and crop residues, energy crops and municipal and construction wastes (Krieger 2014).

The second-generation biofuels are produced from raw materials that are unfit for human and animal consumption, as well as from waste substances. The latest (third-generation) biofuels are produced from algae, cyanobacteria (biohydrogen) and other microorganisms and photosynthetic mechanisms (bioelectricity). The future (fourth-generation) biofuels are derived from the bioconversion of living organisms using biotechnological tools. As follows, the fourth-generation biofuels combine the worth of third-generation biofuels with an advantage of genetic optimization. The most popular methods for making fuels from biomass come from biological or chemical processes. The latest approach combines both forms and indicates the beginning of fourth generation of biofuels. Biofuels technology and products have been well known for many years, but the research challenge is still developing biofuel production technology from biomass, which is waste. The new technologies effectively change the current competitive approach. The competition between thermochemical and biological methods may end (Schmidt and Dauenhauer 2007).

The developments of biofuels result in the reduction of carbon emission, energy tendency and increasing energy efficiency. The transport is responsible for an estimated 21% of all greenhouse gas (GHG) emissions, and more than 90% of the total transport emissions are due to road transport. The consequent biofuels have played a crucial rule in depleting fossil fuel reserves and negative climate changes. Biofuels for transport development are becoming one of the most concerns (Antizar-Ladislao and Turrion-Gomez 2008). District heating and cooling networks are a highly effective way to integrate natural resources such as industrial and agricultural biomass while increasing energy efficiency.

The significant advantage of bioenergy is that it can contribute to a more secure, sustainable and economically sound future in countries' development by supplying domestic clean energy sources; reducing economies' dependence on coal, oil and gas; and revitalizing rural economy.

Biofuel, Technologies and Products

Biofuels Products

Biofuel is commonly known as a fuel produced from biomass material. They are derived from biomass including firewood, wood shavings, pellets, some fruit stones such as olives and avocados, as well as nutshells. The production processes very often use waste biomass, which instead of being composted or directed to landfills is used as a raw material for the production of wholesome energy fuel. It can be assumed that biofuel is any material produced from biomass that has energy significance.

In economic practice, a rough division of biofuels is made into:

- Solid biofuels, more or less processed: wood, wood chips, pellets, wood briquettes, as well as cereal grains, nutshells, etc.
- Liquid biofuels: a fuel obtained by alcoholic fermentation of carbohydrates processes of biomass to butanol or from biodiesel esterified vegetable, fish or animal oils or ethanol
- Gaseous biofuels: biogas obtained as a result of anaerobic digestion of liquid and solid organic waste, e.g. from agricultural production (liquid manure, solid manure, straw, etc.), agri-food production, meat, fruit and vegetable processing,

dairy production or municipal waste, obtained in the processes of selective collection of bio-fractions from households, as well as generator gas, formed in the process of biomass gasification

The majority of biofuels used in heating is solid. The solid biofuels, both and an economic and a practical reason of heating houses, may use wood or pellets burning stoves as the main or supplements to other heating systems, like natural gas or electricity (Kumar 2018).

The basic liquid biofuels of plant origin are biodiesel, biomethanol, bioethanol and vegetable oil. In an attempt to solve energy problems, biofuels can be made by producing biodiesel based on rapeseed. Renewable methanol and ethanol can compete economically with the natural gas feedstock yearlong, being easily available and cheaper in prices. Tilman concludes that also grasses would be effective biofuel. It can reduce carbon dioxide more effectively[1] than grain ethanol and biodiesel (Research Highlights 2006). One of many biofuels is biodiesel as a clean-burning fuel. The main obstacle in the commercial production of biodiesel is related to its high manufacturing cost.

The gasifying biomass is the process that accommodates a wider variety of feedstocks. Biogas is created through anaerobic (lacking the presence of oxygen) fermentation of organic material. The carbon monoxide and hydrogen gases released by the biomass are fermented to ethanol by anaerobic bacteria. The materials such as biowaste or sludge may be used as well as agricultural side products such as manure or dung and specially cultivated energy crops.

The processes of biomass gasification are not available on a large scale yet. During the last years, we saw the techniques such as solid burn gasification, fluidized bed gasification, vapour flow gasification – multilevel or combined procedures.

Biofuels are divided also according to the type of raw material used and the technology of its conversion into biofuels to I, II and III generation.

The value of biomass arises also from co-products that include solid residues. Biomass can be transformed in various ways into another semi-finished products. They can be burned to generate steam and power or sold as animal feed. The energy contained in biofuels is a useful product as well. Carbon dioxide can be purified for carbonation, and glycerine is used in the pharmaceutical and cosmetic sectors. The competitiveness of biofuels results in additional revenues of co-products and, in some cases, from a free and renewable source of energy to power the biorefinery (Fairley 2011).

The Sources of Biofuels and the New Product Development

The main biomass components can be converted into biofuels depending on the accessibility of resources. Biomass resources include dedicated energy crops, agricultural crop residues, forestry residues, algae, wood processing residues, munici-

[1] It reduces carbon dioxide 6 to 16 times more effectively.

pal waste and wet waste (crop wastes, forest residues, purpose-grown grasses, woody energy crops, algae, industrial wastes, sorted municipal solid waste, urban wood waste and food waste). Biomass is one type of renewable resource that can be converted into liquid fuels – known as biofuels – for transportation. Biomass is a carbon-neutral energy source, because the sources of biomass (wood, crops) during its growth absorb CO_2 that is then released into the atmosphere during its combustion. Finally, there is a zero-net balance of CO_2 emissions. Biofuels include cellulosic ethanol, biodiesel and renewable hydrocarbon "drop-in" fuels. Bio-based resources aggregate the broad range of energy sources support to transition to low carbon growth. Energy crops, herbaceous energy crops, wood energy crops and agricultural biomass utilization are one of the most popular sources of biofuels. The use of herbaceous (e.g. switchgrass, biomass sorghum and miscanthus) and woody energy crops (e.g. hybrid poplar, southern pine and willow) represents also one of the most important opportunities for biofuels innovation developments (Rogers et al. 2016). The significant forestry and wood biomass resources include whole-tree biomass, logging residues, unused mill residues, urban wood wastes and treatment thinnings. The municipal solid wastes (MSW) – biogenic portion of MSW, waste biomass, landfill gas and algae – have also a perspective future. Algae are a future and desirable energy source. The most promising is their growth on non-agricultural land in a fraction of the area required by conventional biomass sources as oil crops. Algae capture carbon dioxide and can rise in domestic waste or salt water (Gold rush for algae 2009).

Biofuels for Transport

The new developments and technical progress contribute to the improvement of new vehicles as low emission, electric, hybrids and creation of new fuels, including different generations of biofuels. The two most common types of biofuels used in transport today are ethanol and biodiesel. The competitiveness of biofuels in the transport sector is primarily determined by the price of conventional energy sources. The increase in oil prices rises the competitiveness of the biofuels, the decrease affects its reduction. This economic dependence is as long as the relative high cost of bioenergy production, which is usually much higher than the acquisition of conventional raw materials. However, in some countries, it is becoming more profitable to produce biofuels. Brazil is becoming a tycoon of bioethanol production, and transport companies in South America have gained access to economic fuel (Furtado et al. 2020). Regarding biodiesel production, one of the obstacles to maintaining stable increasing trends is the deficiency in non-food-producing lands. The biodiesel production of biofuels should consider wastelands and degraded areas. Otherwise, the development of biofuels limits the agricultural growth (Elder et al. 2018).

The renewable transportation fuels that are functionally equivalent to petroleum fuels lower the carbon intensity of our vehicles and airplanes. Biofuels can be used in most vehicles that are on the road. Biofuels based on urban waste are increasingly used in the air transport sector. The flights out of London City Airport, operated by British Airways, will be fuelled by rubbish – paper, food scraps, garden clippings and other organic detritus get rid of by the city's residents. The transport biofuels market is based on the use of fuel blends which are a combination of conventional fuels and biocomponents (e.g. petrol with bioethanol and diesel with methyl esters). One of the most significant transport technologies advances is the development of flexible fuel vehicles (FFVs), which can run on a wide range of petrol and bioethanol blends up to and including E85.

Biofuels for Heat and Electricity

Biofuels are usually burned directly in domestic appliances such as stoves and boilers. Usually, cut and chopped firewood is the least processed. The chips come from the withering of biomass both agricultural and forest. The size of chips depends on the manufacturing and transformation processes. Finally, pellets are the most elaborate biofuel, but they have low energy ratio. They consist of small cylinders from 6 to 12 mm in diameter from 10 to 30 mm in length that are obtained by pressing biofuels with binders. Pellets are produced from the cutter dust, the grinding dust and the saw dust. Pellets can be made also from agricultural biomass (e.g. cattle, chicken waste). When pressuring wood biomass to pellets, the raw material has to be absolutely dry.

The most common energy crops include grasses and trees such as willow or poplar (Baumber 2016). Perea-Moreno et al. (2019) indicated the new tendency in using biofuels sources. It is becoming increasingly common to use fruit stones and seeds, as well as fruit husks, though used to a lesser extent than wood chips and pellets. Biofuels based on the mango stone, peanut shell and sunflower seed husk have a high energy potential, with a higher heating. These facts, together with the increasing worldwide production of these by-products, make them especially attractive for thermal energy generation, as well as to reduce CO_2 emissions. Biofuels for heating come from equipment in buildings. Mostly, they are burned directly in domestic appliances such as stoves and boilers. Biomass boilers are an emerging technology in constant development. They produce heat at industrial and residential levels. Hydrogen can also be made by gasification of biomass and with the help of electrolysis. Converting bio- and wood gas to electricity technology for gasification is developed as well as for converting the gas to electricity (Modeling and Analysis 2016).

Wood gasification boilers are totally new and efficiency biomass innovation which is being currently introduced in Slovenia – wood gasification boilers with capacity up to 25, 30, 35 and 40 kW heating of biomass for heating multi-family houses, schools, swimming baths and smaller businesses. The others are connected to a district heating grid. The combustibles are mostly wood chips from forestry or industrial waste wood, which are cheap and make the installation profitable.

The direct use of pellets for end users is one of the most thriving and profitable businesses. The linear increase of pellet mill demand has been noticed since 2014 (Bioenergy technologies 2019).

Whole package solutions for pellet heating systems are offered in the markets.

The new technological solutions in the biofuels sector have also developed the technologies to enable the production of liquid and gel-based biofuels for home heating. A bio-fireplace powered by liquid or gel fuel has been also a new product appearing on the heating markets in recent years. Biofuels used for fireplaces do not produce harmful substances and ash.

Biofuels Value Chain

One of the major problems in the value chain is that the supply system of the base material has not been organized in an efficient way yet. The liquid biofuels such as biodiesel and bioethanol can be distributed in the same way as petrol and diesel, using the same distribution channels and adopting the same logistic processes. However, biodiesel and high blends should be stored for 3–6 months. Additionally, biodiesel thickens at a temperature below zero, and under this condition, special additives should be used. Bioethanol is hygroscopic and attracts water. Both biodiesel and ethanol can lead to corrosion of metal components and damage to elastomeric and rubber parts over time. These product features create an additional cost and disadvantages for sale and distribution channels. They require an additional service cost in the fuel distribution channels. However, the distribution of biofuels can be significantly facilitated by the use of gas distribution networks. The gas distribution network is an ideal solution for biomethane storage and transport. It allows the use of the existing network (Bionett 2007). Enhancing developments of biofuels distribution channels and their infrastructure require a stable and long-term legislative framework and incentives to enlarge the social acceptance of the corresponding investment. The energy policy planning requires a common approach to sectorial, national and regional new technology development and investment policy. It may lead to the increase of competitiveness of the next-generation biofuels, more beneficial usage for end users and profitable applications of producers and distribution channel members (Tsitaa et al. 2020). The freer trade in biofuels is one of the strategies that can be promoted to increase competition between producers. The world's lowest-cost and most efficient producers also increase the value of biofuels in the supply chain (Gheewala et al. 2013).

Innovative Biofuels Solutions in Łódź Region

Innovative biofuels solutions are in line with the strategy for the development of bioeconomy in the European Union. In the EU Action Plan for the Circular Economy (European Commission 2015), the most important tasks for European countries and regions have been formulated, the implementation of which is to contribute to changing the economic model and the transition from a linear economy to a circular economy. The action plan focuses on several priority areas, such as the production and use of plastics, waste management including food waste and critical raw materials. The European Commission has assigned a special role in the action plan to biomass and bioproducts, including biofuels, as one of the most important factors in creating new economic models and opportunities to escape from the trap of average income and average product.

The Commission underlines that the circular economy creates local jobs, opportunities for social integration and economic growth, investment and industrial innovation. The Polish response to European proposals of the circular economy are the activities of the Interministerial Team for the Circular Economy (GOZ, Ministry of Climate), which aim to identify opportunities and threats as well as strengths and weaknesses in the context of the transformation of the Polish economy towards circular economy accordingly to the EU. The Commission underlines that the circular economy creates local jobs, opportunities for social integration and economic growth, investment and industrial innovation. The Polish response to European proposals of the circular economy are the activities of the Interministerial Team for the circular economy (GOZ), which aim to identify opportunities and threats as well as strengths and weaknesses in the context of the transformation of the Polish economy towards circular economy accordingly to the "Closing the Loop – An EU Action Plan for the Circular Economy" (European Commission 2015).

The main task for Polish policymakers is to develop a position being the national response for European Union initiatives on circular economy and to develop a road map in the implementation of the circular economy in Poland, specifying in particular the objectives and priorities of action along with their time horizon and institutions responsible for their implementation. One of the bases for the development of circular economy, waste management and the bioeconomy as well is treating biomass as a source of the greatest importance.

Biomass potential management issues are included in the road map for implementing the circular economy in Poland.

The following proposals for action were formulated:

1. Identification of potential locally available biomass
2. Creating biosociality, i.e. local links between entities cooperating within the value chain
3. Creation of local biorefineries
4. Increasing the use of local biomass and increasing the use of waste biomass for energy purposes
5. Business models in the bioeconomy
6. Stimulating demand for bioproducts through promotional campaigns

Bioeconomy is becoming a reality in many regions of the European Union. Most of them chose the bioeconomy or its elements (e.g. food, green energy) as a regional strategy for smart specialization (Kochanska 2012a, b, 2013).

Poland has been and still is perceived as a country that can have a significant share in the production of biomass in the EU. The analyses carried out at the Institute of Soil Science and Plant Cultivation (2020) show that without damage to food production, Polish agriculture can allocate 0.6 ml ha for cereal production for bioethanol, 0.4 million ha for biodiesel production and 0.5 rape million ha for biomass production for the needs of the professional power industry (Kupczyk et al. 2017).

It can be expected that biofuels and biorefinery technologies based on waste biomass, which is considered as a problematic material, (e.g. waste from meat processing or organic municipal waste,) the price of waste biomass raw material will be low, zero or even negative. The technical potential of wood biomass (wood and waste from forests and orchards) for energy purposes is 8.81 million tons, and the surplus of agro-biomass, mainly straw, is 7.85 million tons per year. Poland has significant potential for waste biomass. The calculated amount of energy from waste biomass is 18.5 TWh of electricity and 111.04 PJ of heat. The calculated amount of electricity would cover Poland's needs in 11.6% and heat in 24.0% (Kochanska 2012b).

It can be expected that biofuels and biorefinery technologies based on waste biomass, which is considered as a problematic material, (e.g. waste from meat processing or organic municipal waste,) the price of waste biomass raw material will be low, zero or even negative. Advanced processing (biorefining) high-value products could be sold outside of a region, including exports, while less valuable waste biomass products will be consumed on-site as feed for animals or as an energy fuel.

The expected result of the development of biorefinery plants delivered high-value products, new business models and new value chains will be created. A specific service sector will be developed, associated with servicing local biorefineries. Therefore, the regional economy will benefit from diversified and technologically advanced products and semi-finished products resulting from waste biomass processing in biorefinery technologies, including esters, biofuels or bioethanol.

The opinions of European and global experts gathered during the European Bioeconomic Congresses in Lodz confirm that biorefining technologies are the real future of the European economy (Lodzkie Voivodship 2019). In Germany, Sweden and also in Poland, small local biorefineries are just appearing, focused mainly on the production of specific substances of organic origin with a high market value, such as tocols and phytosterols. Biorefinery technologies are very often accompanied by biofuels technologies (Bio Business).

The Research and Innovation Centre Pro-Akademia is a Polish scientific institution that conducts application research mainly in the areas of energy, including renewable energy and environmental engineering at three laboratories: Biomass and Waste Valorization Lab, Processing Lab and Natural Products Lab. The subject of biofuels and bioproducts is closely related to research dedicated to recycling and upcycling organic waste. The innovation of the Pro-Akademia's research approach stands out integrating into one tool, many tools of different areas of knowledge for assessing the development potential of enterprises or regions to implement new waste biomass solution into practice. Based on quantitative and qualitative environment-friendly research instruments characteristic for technical, economic and social areas of research, we try to encourage entrepreneurs/local government units to implement the original technologies.

The research methodology used in the RIC Pro-Akademia takes into account the interdisciplinary nature of conducted research in relation to the objectives, tasks and results of scientific challenges. The goal of each research project is to gain new knowledge in the field of technological, economic and environmental issues. An aspect of fundamental importance is to define the impact of technology on the development together of science, the economy, the environment and society. Therefore, the Pro-Akademia's scientific teams consist of engineers, biologists, economists and lawyers, according to the subject of research and planned implementations. To illustrate the achievements of the Pro-Akademia's research team in the field of biofuels solutions, four cases were selected and presented below: "Case 1: Liquid Biofuel Production Based on Slaughterhouse Waste and Animal Fats", "Case 2: Solid Biofuel Production Based on Waste of Onion", "Case 3: Gas Biofuel Production Based on Utilization of CO_2/Methane and Bioethanol" and "Case 4: Integrated Biofuels Production Technology/Methane and Fish Oil Based on Fish Waste".

Case 1: Liquid Biofuel Production Based on Slaughterhouse Waste and Animal Fats

Biofuels derived from materials with a high content of glycerides, such as oils and fats, are one of the most promising sources of replacing conventional diesel with a product with a less negative impact on the environment. However, given the high prices of vegetable oils, such as soybean, rapeseed, sunflower and palm oil, as well as the very negative public perception of using these types of oils and food products for energy purposes (the so-called food vs. fuel dilemma), looking for alternatives which do not compete with food production is extremely important. The Bioprocess Laboratory of RES Pro-Akademia team, in cooperation with the business partners, worked on technology for the production of diesel biocomponent from slaughterhouse waste and expired food products, contributing to maximizing the valorization of this type of waste and changing its characteristics from waste to raw material in a new process as part of biorefinery concept. The process involves the extraction of a fat fraction that also contributes to the reduction of moisture in the residue, which will be recycled to the process. The fat fraction obtained in the process of catalytic hydrogen deoxidation is converted into liquid biocarbons with the composition and properties corresponding to diesel biocomponents.

The use of raw materials such as inedible oils (*Jatropha curcas* and castor), animal fats (tallow, lard, fish oil) and waste (yellow fat) should contribute to increasing the competitiveness of the biodiesel price due to their availability and low cost. The use of such waste opens the way for recycling and better valorization of this type of raw material. Animal fats as potential raw materials for biodiesel production are considered low-quality biodiesel raw materials due to their high content of free fatty acids (FFA) (up to 15%), contrary to soybean oil, which usually contains less than 0.5% FFA (Campana and Airasca 2013).

The high content of FFA makes this kind of waste unsuitable for biodiesel production in the trans-esterification process, which is the most traditional, direct and one-step process. In this process, which is catalysed by a base, soaps are formed, making product separation extremely difficult and the process's mass efficiency being reduced.

In the case of waste of category 2, consisting mainly of post-slaughter waste or outdated food waste, in which animal fats constitute a significant part, there is an additional difficulty, which is the high water content. According to preliminary tests carried out in the Biomass and Waste Valorization Lab, the water content in non-homogeneous slaughter waste is from 22% to 70%. The particularly high water content in this type of waste means that the traditional thermal utilization is energy consuming and extremely expensive. In addition, this type of raw material cannot be subjected to a direct trans-esterification process for the production of biofuel, because this reaction is very sensitive because of the presence of water. The presence of water affects the shift of chemical equilibrium towards the substrates, making the process ineffective. Therefore, taking into account the above, research on obtaining biofuels from slaughterhouse waste was conducted towards obtaining a new product, which is liquid biofuels from waste. An innovative technological approach was applied, which assumes the use of a symbiotic sterilization and extraction of the fat fraction, followed by a hydrogen oxidation process, also called hydrodeoxygenation (HDO). Sterilization combined with extraction allows microbiological protection of the fat fraction, which is then used in hydrogen deoxidation. In this way, the raw material was obtained, which until now has not been used for the production of liquid biohydrocarbons.

The HDO process, where liquid biohydrocarbons are produced, consists of catalytic hydrogenation and cracking of the fat fraction, thanks to the use of hydrogen present in the synthesis gas to obtain anaerobic particles, water and carbon oxides (Campana and Airasca 2013). HDO products can be mixed with fossil fuels, adapting them to existing infrastructure and engines. The main advantages of using HDO for processing fat fraction of waste of category 2 are that this process ensures consistent product quality despite the use of various raw materials and does not generate waste (Pratama and Utomo 2013).

The developed technology allows for a stable and effective valorization of waste, which currently requires energy expenditure to neutralize it. In addition, the solution is characterized by high flexibility against sudden changes in the composition of the fat fraction, which with a large variability in the composition of the raw material is a real and big problem for sensitive alternative systems for the production of biocomponents and biofuels. The solution is important for market success: the technology provides not only a completely alternative method of valorization of the raw material currently considered as highly problematic waste but also the recycling of the residue from the extraction of the fat fraction to the process allowing recovery of the fraction, which can also be used as a fertilizer. Moreover, propane is produced as part of the hydrogen deoxidation process, which due to its energy potential can be considered as an additional source of energy and/or heat. The production of diesel biocomponent from slaughterhouse waste allows the replacement of biofuels obtained, for example, from oil plants characterized by the so-called direct and indirect land use change and higher greenhouse gas emissions.

Case 2: Solid Biofuel Production Based on Waste of Onion

Over the past 30 years, global onion production has increased four times, and Poland is the third largest onion producer in Europe – after Spain and the Netherlands. In 2019, onion harvest in Poland amounted to 5535 thousand (Pratama and Utomo 2013). Onions produced in Poland are intended mainly for food purposes. 30% of the harvest is troublesome post-production waste in the form of scales, root fragments and the first, rejected layer of shell. This means that 1,660,500 tons of onion biomass as post-production waste should end up in a composting plant or a landfill, generating annual costs in the amount of 489 million PLN/115 million EUR (environmental fee: 295.05 PLN per 1 ton × amount of onion waste: 1.645.500 t) (Pratama and Utomo 2019). Therefore, the Polish economy incurs real losses year by year – it loses benefits – in the amount of almost 500 million PLN only due to fees for placing onion waste on landfills.

A team of scientists of the Biomass and Waste Valorization Lab, in cooperation with producers and exporters of frozen onions, developed technology for managing and processing onion waste into biorefinery products and biofuel.

The implementation of results of research will allow onion producers to treat the waste – onion waste biomass as a raw material for biofuels production and biorefinery production. As a result of the implementation of biofuels production and biorefinery technology, it will be possible to produce biofuels and valuable substances: quercetin, bioflavonoids and easily digestible plant fibres as well.

Quercetin is a derivative of flavone and belongs to the line of strong organic antioxidants used in pharmaceutical preparations with anti-inflammatory, anti-allergic and anti-exudative activity, binds heavy metals, removes free radicals and has strong antioxidant properties. Phenolic compounds contained in onion shells are good antioxidants, and plant fibres reduce the risk of developing colorectal cancer and gastrointestinal disorders. They can be used to treat obesity and type 2 diabetes. In this case, bio-refining technologies will provide two valuable products: interested raw materials for the pharmaceutical and cosmetics industry and biofuels, because biorefining residues will be processed into energy fuel, used in onion production processes, especially for the production of cold and freezing of the food onion (Fig. 19.1).

Currently, Polish onion producers have achieved a high place in the onion production rankings in the European Union, have consolidated their position as a reliable supplier of peeled and frozen onion ready for use in gastronomy. However, the high competitiveness of Polish onion suppliers results not only from the low labour costs of growing and initial processing of onions but also from the fact that the production of onions is mostly done by small local companies that buy peeled onions directly from farmers. Onion waste remains with the farmers – husk and onion tips, damaged pieces, etc. finally end up in their fields, which means that the rotting resi-

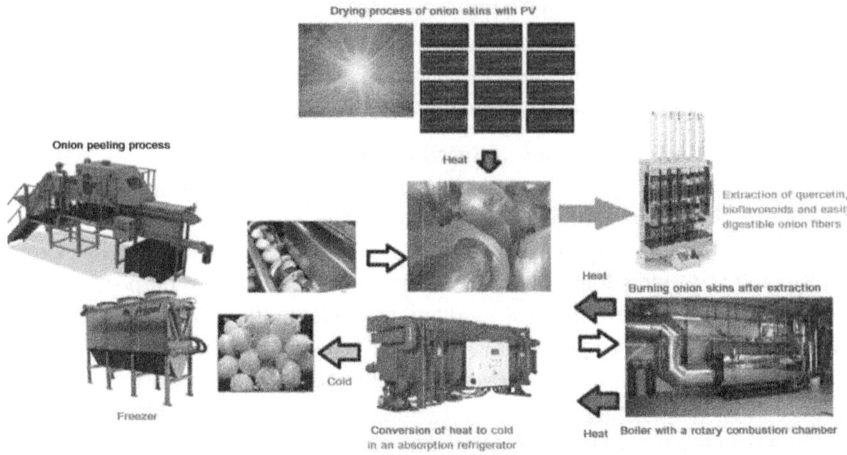

Fig. 19.1 A schematic diagram of the biorefining technology for waste from onion processing

dues release sulphur compounds and pose a growing environmental threat. The scale of onion production and export, which increases from year to year, necessitates onion waste management – utilization or storage in the landfills can dramatically deteriorate the profitability of onion cultivation in Poland.

Vegetable production is relatively simple and low-income production. The implementation of biorefining technology in this case will completely change the situation: it may turn out to be more profitable to extract quercetin, bioflavonoids and vegetable fibres from onion skins than the production of edible onions. An additional benefit of implementing biorefinery technology will be the production of energy: onion skins subject to elemental analysis show a fairly high content of carbon element (C) at 45%, content of hydrogen element (H) at 5.85% and sulphur (S) at 5.3 ppm, at low moisture content and calorific value at the level of approx. 15 MJ/kg (Source: own research, RIC Pro-Akademia). Another, more advanced method of energetic use of onion skins after quercetin extraction is squeezing the juice from onion skins and the production of biogas, rich in methane in an anaerobic process, in a specially dedicated reactor (Campana and Airasca 2013). Methane can power, for example, medium-power fuel cells, and the electricity produced will power a technological line, e.g. a quercetin extractor.[2]

Case 3: Gas Biofuel Production Based on Utilization of CO_2/Methane and Bioethanol

The major problems for currently operated energy systems are energy storage issues, grid stability and CO_2 emissions. The international team of Biomass and Waste Valorization Lab and Cypriot scientists is developing the technology that may contribute to solving all these problems at least partially. The BioElectroCathode project is implemented through close cooperation between the Research and

[2]Acknowledgment: Research was conducted within the project "KSI KSU Consulting for Innovative" co-financed by the European Union from the European Regional Development Fund.

Innovation Centre Pro-Akademia from Konstantynów Łódzki and the Cyprus University of Technology, with the involvement of two companies: ENERES CPM Ltd. from Cyprus and Omni3D Ltd. from Poland. The aim of the research is the application of 3D printing technique for the manufacture of electrochemical bioreactors and developing the innovative methods and materials for the production of cathodes that are used in biosynthesis. This bioprocess allows for conversion of CO_2 present in flue gases or in air into methane and/or ethanol.

Biological or microbiological electrosynthesis (BES or MES) is a way to convert CO_2 into chemical energy carriers like methane (CH_4) or ethanol (C_2H_5OH) that uses electrical energy and microorganisms. BES occurs when a microbial catalyst reduces CO_2 into organics commodities with electrons supplied by an external power source and taken from the cathode of a bioelectrochemical system, designed primarily to perform biological reductive reactions. BES is a novel technology that can (1) convert electrical energy from fluctuating renewable energy sources into organic compound (methane or ethanol) that can be stored, distributed and consumed on demand and (2) utilize CO_2 as a sole carbon source either for methane production that can be then used as a fuel or can be converted into electricity in CHP engines or for ethanol production, i.e. liquid biofuel commonly used in, for example, transport.

Electricity is supplied to microorganisms in the reactor, whereby CO_2 is converted into methane and/or ethanol, depending on the bacterial strain used as a biofilm. Methane-rich biogas (biomethane) or ethanol can be stored, distributed and used, for example, as fuel in vehicles compatible with natural gas. The proposed technology can be successfully applied in biogas plants (over 12,400 in Europe) or in industries generating large amounts of CO_2.

As part of the research and implementation project, the additive technologies (AT), i.e. popular 3D printing, will be tested to build a bioreactor, and modern thermoplastic materials will be used as building materials. 3D printing is applied in many industries. Its importance will grow in the coming years, especially in the field of prototyping and cheap printing. Companies will need energy-saving, very precise printers, for which a wide range of materials will be used. AT can utilize such materials as polypropylene, acrylic, rubber of various hardness, nylon, transparent materials, polystyrene, ABS and metal alloys like tool steel, titanium, stainless steel, aluminium and even glass and many more.

Today, 3D printing is applied not only for prototyping but increasingly also for mass production of ready-made elements, functional and durable products.

In the BioElectroCathode project, Omni3D will use AT techniques to build, i.e. PRINT OUT, the bioreactors. Taking into account the benefits offered by AT technologies, among which eliminating human errors, shortening the costly and time-consuming assembly and multi-part bioreactor production can be mentioned, designing and optimizing of the system will be significantly improved, and the freedom of scientists in design will increase. By changing the design parameters of individual reactor components, such as surface, shape, print resolution, thickness and degree of filling, the structure of the printed device can be easily manipulated and, consequently, the bioreactor can be improved. Several innovative, suitable construction and thermoplastic materials will be tested for bioreactor printing. The chosen materials are relatively inexpensive and easily available and therefore probably will be used in the future for massive, energy-saving and stable production of BES using 3D printing technology.[3]

Case 4: Integrated Biofuels Production Technology/Methane and Fish Oil Based on Fish Waste

The project "FiWastEnergy – EU-Brazilian cooperation for fish waste circular management in PISCIS company" is an example of transferring the European biofuels production technology to Latin America. The project has been elaborated by the scientists of the Biomass and Waste Valorization Lab of RIC Pro-Akademia within the "Low Carbon Business Action (LCBA) in Brazil" programme. The LCBA Brazil effectively contributes to the exchange and uptake of low emission technology, including biofuels production technology through the business and technical cooperation between scientific institution and companies in Brazil and the EU, in a common effort to address the global challenge of climate change.

The scientific challenge of the EU-Brazilian cooperation was implementation of the waste management process, utilizing the problematic fishery waste and conversion of fish waste into biofuels – biogas and fishery oil for electricity and cold production.

The following main outputs of the FiWastEnergy project have been:

- Conceptual and technical design of the waste management process in the company and biogas production technology
- Implementation of the waste management process in the company
- Technical feasibility analysis and profitability analysis
- Project of the biogas production installation in the PISCIS company, active in fishery industry, as well as a biogas micro-power plant, which will be implemented within the same business scenario aiming to use their resources better and to contribute to future circular economy.

Fish residues are potential feedstock for the biogas production. As this is very rare feedstock for biogas, the information about the efficiency of the process of biogas production as well as methane content are very scarce. For example, Pratama et al. (2019) reported the processing of example of fish sample constituted by a mixture of fish waste, fish meal and fish powder with composition as shown in Table 19.1.

The average gas composition obtained could be as follows: 65.7% of methane, 27.0% of carbon dioxide, 2.3 of nitrogen, 0.7% of propane and 0.1% of oxygen. The obtained results demonstrated that the cumulative biogas production achieved steady state after ca. 56 days as demonstrated in Fig. 19.2.

[3] Acknowledgment: Project was co-financed by the National Centre for Research and Development within the M-Era.Net Program.

Table 19.1 The composition of various fish wastes

No.	Parameter	Fish waste (%)	Fish meal (%)	Fish powder (%)
1	Water	7.34	10	12
2	Ash	17.10	18	20.9
3	Fatty	10.37	10	8.4
4	Protein	53.85	60	53.3
5	Crude fibre	0.03	1	–

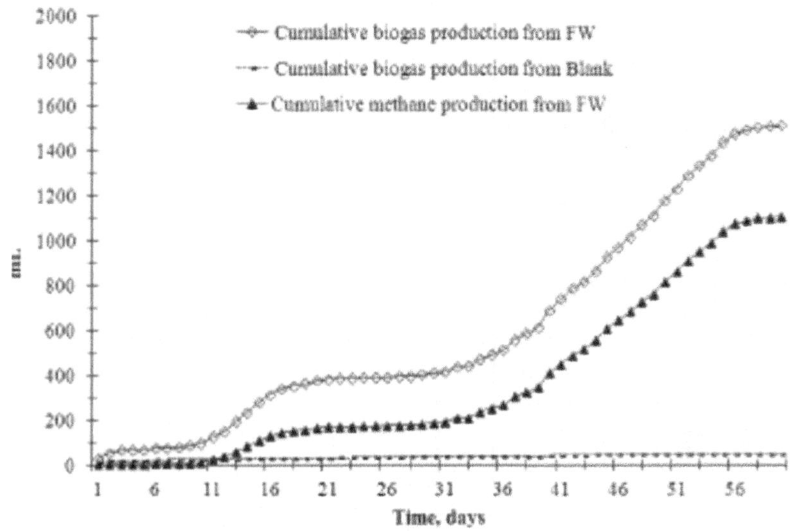

Fig. 19.2 The cumulative biogas production from fish wastes (FW) (open rhombus) and methane production (closed triangles). (Figure taken from Kafle and Kim 2013)

The wastes used for biogas production in this case provide energy equal to 8.120.87 kWh. Focusing on the issue of productivity of waste-to-energy installations, it should be noted that the electricity production results in the production of fertilizers and cold as well.

The total monthly income for the Brazilian SME of the business model presented in Fig. 19.3 is 218.909.29 R$, whereas the total monthly profit considering the investment cost is 107.917.95 R$. Company can gain the potential profit generated by electricity, cold and fertilizer (12.729.10 R$/month vs. 11307.14 R$/month) used for its own purpose and for fish oil as the biofuel sold on the market.

Additionally, considering the savings from the avoidances of waste utilization (474.35 R$), this balance is even much equilibrated within the 5 years of bank loan payback period.

Considering the biofuel production as a specialization of the company, business scenario is very close to be economically feasible and demonstrates the most promising option considering the potential reasonable electricity production.

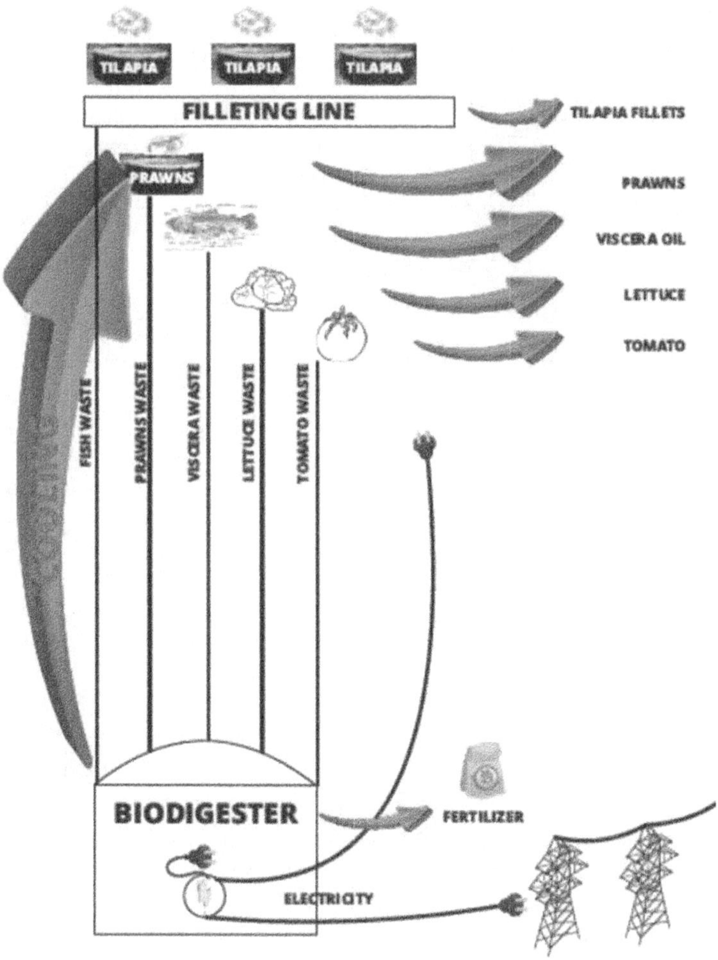

Fig. 19.3 Business model resulting from the implementation of business scenario of biofuel production specialization

An increase in the production scale would help increasing the available streams and consequently the produced electricity, cold, fertilizer and fish oil as well. As the production of fish is connected to the company's production of prawns and vegetables, this option should be considered in the comprehensive manner. In addition, an increase in the installation of biogas plant size would be required, and for the first estimation, the exponential law, i.e. Cost B=Cost A (Size B/Size A)$^{0.85}$ (Matos and Bogel-Łukasik 2013), can be considered.

As it is impossible to directly estimate the cost of the investment related to the increase in the production of fish, prawns and vegetables, it is impossible to provide by how much an increase in the production scale should be made. However, as the proposed technology is still rather nanobiogas plant, it is clear that larger scale of the biogas plant would be more economically and technologically favourable.

Conclusions

The production processes of biofuels and biorefinery processes are characterized by a long payback period of up to 20 years (Institute of Soil Science and Plant Cultivation 2020); simultaneously, the typical service life of the installation is 30–40 years (Kochanska 2013). A major challenge at the implementation stage of biofuels and biorefineries technologies that focus on biomass waste processing is to comply with environmental regulations that perceive them as waste for disposal.

Another limitation is the fact that even if the "know-how" of valorization of different types of waste biomass is available on a laboratory or pilot scale, effective upscaling of this technique is difficult or limited by many difficult technological barriers. First and foremost is the need for standarization the production process to an industrial scale because of difficulty maintaining even relative stability of waste biomass resources. A serious challenge is to provide the heterogeneity of waste biomass, and problems with homogenization of streams subjected to biofuels. The heterogeneity of the substrate is one of the most significant factors that affect the energy consumption of the biofuels production and the overall cost-effectiveness of the biomass valorization process at all.[4]

References

Agarwal S., Kumar A., (2018), *Historical developments of biofuels*, [in:] Biofuels: Greenhouse gas mitigation and global warming. Next generation biofuels and role of biotechnology, ed. Kumar A. Ogita S., Yau Y.Y., Springer: 24–39.

Antizar-Ladislao B. Turrion-Gomez J. L., 2008, Second-generation biofuels and local bioenergy systems, Biofuels, Bioproducts and Biorefining 2, Society of Chemical Industry and John Wiley & Sons: 455–469; https://doi.org/10.1002/bbb.

Baumber A., (2016) Bioenergy crops for ecosystem health and sustainability, Routledge, London.

Biernat K., Perspektywy rozwoju technologii biopaliwowych w świecie do 2050 roku, Chemik 2012, Vol 66 / 11: 1178–1189.

Bio Business, *Detection of business opportunities for entrepreneurs and SMEs in biomass sector in Burgas, Tampere, Pomurje, central Hungary, Nordthueringen and Brandenburg regions.* Study of Detection of Business Opportunities for Entrepreneurs and SMEs in Biomass Sector, Intelligent Energy Europe.

Bioenergy technology – market interest trends, (2019), A report, Curiosity Publishers.

[4] Acknowledgments: The project "FiWastEnergy. EU-Brazilian cooperation for fish waste circular management" was co-financed within the Low Carbon Business Action in Brazil Phase II, EuropeAid/138337/DH/SER/BR.

Bionett, (2007), Najlepsze praktyki w stosowaniu biopaliw w transporcie, Bałtycka Agencja Poszanowania Energii S.A.:6–8, http://bape.com.pl/wp-content/uploads/2014/09/Najlepsze-praktyki-w-stosowaniu-biopaliw-w-transporcie.pdf

B. P. H Campana and M. A. Airasca., "Simpósio Internacional sobre Gerenciamento de Resíduos Agropecuários e Agroindustriais," in *Onion Waste Treatment Preliminary Anaerobic Research Data*, 2013.

Cordis, (2019), https://cordis.europa.eu/article/id/25382-biofuels-for-transport-provide-a-rosy-future-according-to-draft-report/pl (accessible May 8, 2020)

Elder M., Romera J., Bhattacharya A., Sano D., Matsumoto N., Hayashi S., Socioeconomic impact of biofuels in East Asia, (2018), Biofuels and sustainability. Holistic perspectives for policy-making, ed. Takeuchi K., Shiroyama H., Saito O., Matsuura M., SpringerOpen, ebook.

European Commission, "An EU action plan for the circular economy," *Com*, vol. 614, p. 21, 2015.

Fairley P., (2011), Introduction. Next generation biofuels, Nature. 474 (7352) (Supplement): 2–5, June 23,

Furtado A. T., Hekkert M. P., Negro S. O., (2020), Of actors, functions, and fuels: Exploring a second generation ethanoltransition from a technological innovation systems perspective in Brazil, Energy Research & Social Science, Vol. 70, Decmber: 101706. https://doi.org/10.1016/j.erss.2020.101706,

Gold rush for algae, Nature 2009, Vol 461/ 24, 24 September: 460–461

Gheewala S., Damen B., Shi X., (2013), *Biofuels: Economic, environmental and social benefits and costs for developing countries in Asia.* Wiley interdisciplinary reviews: Climate Change. https://doi.org/10.1002/wcc.241.

Ibbara-Yunez A., *The effects of renewable energy policies in electricity markets: global perspectives*, [in:] Energy and Mobility in Smart Cities, ed. W. J. Nuttall, D. V. Gibson, D. Trzmielak, A. Ibarra-Yunez, Institution of Civil Engineering, London 2020: 117–138.

Institute of Soil Science and Plant Cultivation, "http://www.iung.pulawy.pl/eng/," *www.iung.pulawy.pl/eng/*, 2020.

Kochanska E., "Ekonomiczne i technologiczne aspekty rozwoju rozproszonej energetyki opartej o biomasę na przykładzie małej gminy," *Acta Innov.*, vol. 6, pp. 45–67, 2013.

Kafle, G.K.; Kim, S.H., (2013), "Evaluation of the Biogas Productivity Potential of Fish Waste: A Lab Scale Batch Study," *J. Biosyst. Eng.*

Kochanska E., (2012a), Diagnoza potencjału województwa łódzkiego dla produkcji biogazu na bazie odpadów rolno-spożywczych," in *Mikrotechnologie biogazowe jako innowacyjne narzędzie stymulowania rozwoju lokalnego*, Polska Akademia Nauk, Ed. Łódź: 10–20.

Kochanska E., (2012b), Technologie produkcji biogazu na bazie odpadów organicznych jako narzędzie tworzenia przewagi konkurencyjnej przedsiębiorstw rolno-spozywczych," *Mikrotechnologie biogazowe*, Polska Akademia Nauk, Ed. Łódź: CBI Pro-Akademia, 2012, pp. 10–16.

Krieger K., Biofuels heat up. A new generation of industrial plants can make liquid fuels from almost any organic scraps — from corn stalks and wood chips to urban rubbish, Nature 2014, Vol 508, 448–449.

Kupczyk, A., Mączyńska J., Sikora M., Tucki K. Żelaziński (2017). *Stan i perspektywy oraz uwarunkowania prawne funkcjonowania sektorów biopaliw transportowych w Polsce* . Roczniki Ekonomii Rolnictwa i Rozwoju Obszarów Wiejskich. Vol. 104: 39–55. https://doi.org/10.22630/RNR.2017.104.1.3.

Kumar A., (2018), *Global warming, climate change and greenhouse gas mitigation*, [in:] Biofuels: Greenhouse gas mitigation and global warming. Next generation biofuels and role of biotechnology, ed. Kumar A. Ogita S., Yau Y.Y., Springer: 1–16.

Lodzkie Voivodship, "The International Congress of Bioeconomy," (2019) *https://www.lodzkie.pl/strona-glowna/aktualnosci/rozpocz%C4%85%C5%82-si%C4%99-vii-mi%C4%99dzynarodowy-kongres-biogospodarki*, Ministry of Climate, "https://www.gov.pl/web/climate," 2020.

Matos C.T., Bogel-Łukasik R., (2013), Green metrics evaluation of isoprene production by microalgae and bacteria," *Green Chem.*, vol. 15: 2854–2864.

Modeling and Analysis (2016), Biofuels, Bioproducts, Biorefining 11:110–128; DOI: https://doi.org/10.1002/bbb

Orynycz OA. Production of liquid biofuels as the energy source for transport, Enrgetyka i Ekologia, 2012, 8: 471–475.

Perea-Moreno M. P., Samerón-Manzano E., Perea-Moreno A. J., *Biomass as Renewable Energy: Worldwide Research Trends*, Sustainability 2019, *11*, 863; doi:https://doi.org/10.3390/su11030863

Research Highlights. *Positively negative*, Nature 2006, December Vol 444|14: 793.

Panchuk M., Kryshtopa S., Sładkowski A., Panchuk A., (2020), Environmental aspects of the production and use of biofuels in transport, [in:] Ecology in Transport: Problems and Solutions, ed. A. Sładkowski, Springer, Katowice: 115–168.

Pratama A. S., Utomo P.N., (2019), Potential biogas of fish waste as a solution of energy crisis., J. Phys. Conf. Ser., vol. 121: 37–40.

Rogers J. N., Stokes B., Dunn J., Cai H., Wu M., Haq Z., Baumes H., (2016) *An assessment of the potential products and economic and environmental impacts resulting from a billion ton bioeconomy*. Society of Chemical Industry, John Wiley & Sons: 110–127.

Schmidt L. D., Dauenhauer P. J., *Hybrid routes to biofuels*, Nature 2007, Vol 447/21, June: 914–915.

Tsitaa K. G., Kiartzisb S. J., Ntavosa N.K., Pilavachi P. A., (2020), *Next generation biofuels derived from thermal and chemical conversion of the Greek transport sector*, Thermal Science and Engineering Progress, Vol. 17, June: 1–7, https://doi.org/10.1016/j.tsep.2019.100387

Van Geenhuizen M. Schoonman J., *-Renewable energy technologies and potentials for adoption*, [in:] Energy and Innovation: Structural Change and Policy Implications, ed. M. van Geenhuizen, W. J. Nuttall, D. V. Gibson, E. M. Oftedal, Purdue University Press, West Lafayette 2011: 17–36.

Chapter 20
Increasing Flexibility of Biogas Plants Through the Application of Innovative Concepts

Mirko Barz, Hartmut Wesenfeld, Asnakech Laß-Seyoum, Arvid Meibohm, and Sascha Knist

Abstract After years of strong growth, the biogas industry is facing new challenges today. The traditional mode of operation as a baseload combined heat and power plant for heat and power generation is becoming more difficult with an increasing share of fluctuating renewable energy, e.g., from wind and solar in the power grid, and requires innovative concepts for securing an economic operation of the existing plants. In addition to a change in the operating mode towards a demand-oriented power generation by investment in new storage concepts and increased power generation capacities, new concepts for feeding biomethane (and thus a new economic concept) into the gas grids are becoming increasingly important. Feeding biomethane into the gas grids reduces the dependency on natural gas imports and creates opportunities to link the electricity, heating, and transport sectors of the energy economy. The following sections will present and evaluate the solutions currently available on the market and introduce new innovative concepts, which are currently in focus of various research activities.

Keywords Flexibility of biogas utilization · Biomethane production and marketing · Biogas upgrade · Methanation · Power-to-gas

---

M. Barz (✉) · A. Laß-Seyoum · A. Meibohm
HTW Berlin, Faculty of Engineering – Energy and Information, SC Renewable Energies, Berlin, Germany
e-mail: barz@htw-berlin.de; mirko.barz@htw-berlin.de

H. Wesenfeld
Beuth Hochschule für Technik Berlin, Fachbereich II – Mathematik – Physik – Chemie, Berlin, Germany

S. Knist
Graforce GmbH, Berlin, Germany

E. Koukios, A. Sacio-Szymańska (eds.), *Bio#Futures*,
https://doi.org/10.1007/978-3-030-64969-2_20

1 Biogas Technologies and Utilization Pathways: General Overview

Biogas is a mixture of methane (CH_4), carbon dioxide (CO_2), and small amounts of other gas components such as H_2S, NH_3, and water vapor produced by anaerobic digestion of organic compounds. A typical composition of biogas from agricultural biogas plants is shown in Table 20.1.

The composition of the produced biogas varies depending on composition of the used substrates, the conditions in the digester (such as temperature, pH-value, organic loading rate etc.) and as well technological conditions such as the types of the used digester.

The anaerobic digestion process of organic compounds is widely disseminated in nature and takes place, e.g., in swamps, in lake sediments, in landfill sites, and in the digestion systems of ruminants (e.g., cows). The utilization of biogas has a long history. The first utilization of naturally formed biogas (e.g., marsh gases) was reported in China during the times of the Western Han Dynasty, approximately 2000 years ago (Wu, L.B. 1993). The development of simple man-made technologies to produce biogas for cooking and lighting started in the nineteenth century. Such simple technologies (e.g., Chinese-style fixed dome domestic household digester, Indian-style floating drum digester, plug flow digester, or simple plastic tube digester) are still widely disseminated in developing countries around the world, over eight million of them alone in Southeast Asian countries. In Europe, anaerobic digestion has been used to process sewage sludge since the nineteenth century. The first anaerobic wastewater treatment plant in Germany was built by Imhoff in 1910 (Kaltschmitt et al. 2012). During the last decades, a huge increase in biogas production is recognized in the EU, caused by the renewable energy policies in European countries combined with economic, environmental, and climate benefits (Barz & El Bari 2018). The total biogas production reached up to 18 billion m^3 methane (654 PJ) in 2015, representing half of the global biogas production (Scarlat et al. 2018). Between 2009 and 2016, the number of biogas plants in Europe increased to approx. 17,400 units, and most of the plants have been installed in the agricultural sector, using energy plants and/or animal manure as substrates. These agricultural plants (approx. 12,500 units) are followed by biogas plants running on

Table 20.1 Typical biogas composition (Kaltschmitt & Hartmann 2001)

Component	Concentration range in %
Methane	45–75
Carbon dioxide	25–45
Water	2–7 (20–40 °C)
Hydrogen sulfide	20–20,000 ppm
Nitrogen	<2
Oxygen	<2
Hydrogen	<1

sewage sludge (2838 plants), landfill waste (1604 plants), and various other types of waste (688 plants) (EBA 2015). Germany is the leading country in biogas production in Europe and a pioneer country in global biogas production, with approximately 25% of the installed capacity (Achinas et al. 2017). In 2016, around 9200 biogas plants were operated in Germany, and a further increase to 9600 plants was forecast by the end of 2018, 195 of them equipped with technologies for biomethane upgrade (FVB 2018).

Digesters built around the world vary in their design complexity, construction materials, and costs. During most of the digesters in developing countries are low-cost applications, having still a very simple design without heating and mixing devices, the applications in the developed countries like in Europe are more advanced including complex process control and monitoring applications. Typical construction designs for agricultural biogas plants are continuous stirred tank reactors (CSTRs) operated under mesophilic temperature condition. Further design types are plug flow digesters with controlled flow (e.g., for the treatment of the organic fraction of municipal waste), upflow anaerobic sludge blanket (UASB) digester for the treatment of organic materials with a low TS content, and in some cases discontinuously operated dry fermentation applications such as the batch operation/box or garage fermenter (Gromke et al. 2018). Biogas produced in such applications is mostly used for heat and power generation in decentralized combined heat and power (CHP) plants (gas engines reaching 35–40% electrical efficiencies) located at the biogas plants. Caused by new economic and ecological frame conditions, biogas upgrade to produce biomethane is a promising opportunity to increase the flexibility of biogas utilization. Europe is currently the world's leading producer of biomethane for the use as a vehicle fuel or for injection into the natural gas grid which offers a unique low-cost storage opportunity and a multitude of different utilization opportunities such as fuel for CNG vehicles, chemical raw material, or as a combustion fuel for gas engines and gas turbines in areas with high demand on power and heat. According to Scarlat et al. (2018), the capacity of the plants feeding biomethane into the natural gas grid reached 1.459 million m^3 in 2015.

2 Status Quo of Biogas Utilization in Germany

Caused by the favorable support conditions of the Renewable Energy Act (EEG) the development the biogas sector in Germany has developed very quickly. The EEG came into effect in 2000 and defined so-called feed-in tariffs that guaranteed a fixed price to the operators of biogas plants for the electricity they delivered to the grid for 20 years. The first feed-in tariffs and bonuses for biogas paid under the EEG 2004 were as high as 22 ct/kWh$_{el}$ fed into the grid. Starting with a few hundred biogas plants in the early 1990s, about 9600 plants were in operation in 2018 (see Fig. 20.1).

Between 1999 and 2018, the installed capacity increased enormously (from 55 MW in 1999 to 4843 MW in 2018). Biogas currently accounts with 32.2 TWh$_{el}$

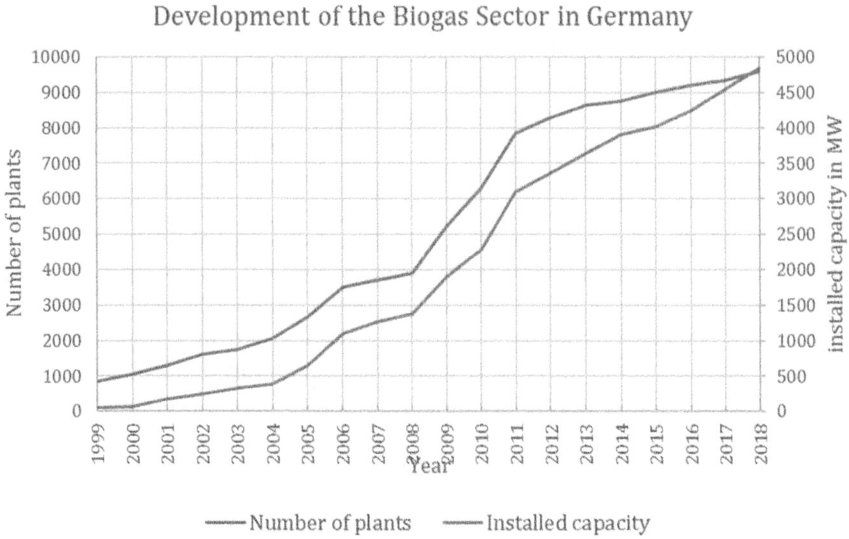

Fig. 20.1 Development of the German biogas sector since 1999. (Data adopted from Fachverband Biogas e.V. 2018)

for 14.2% of electricity generation and with 16.7 TWh_{th} for approx. 10% of the heat supply from renewable energies in Germany (DBFZ 2019). Biogas is mostly produced in small- to medium-scale installations on agricultural farms with an average installed capacity of approx. 500 kW_{el} and in some larger-scale applications (e.g., for waste digestion in urban areas) (Gromke et al. 2018). The most common biogas technology is wet fermentation in continuous stirred tank reactors (CSTRs), operated under mesophilic temperature conditions (mostly in the temperature range between 36 and 38 °C). Most of the plants are used for heat and power generation in CHP plants, receiving the guaranteed feed-in tariffs for the produced electricity according to the EEG regulations. Especially due to the amendments of the EEG in 2004 und 2009, the number and installed electric capacity of biogas plants has increased significantly. Especially between 2006 and 2012 (see Fig. 20.1), many new small- and medium-scale biogas plants in the agricultural sector started operation, creating income and other benefits for the farmers who are usually the plant operator themselves. Typical substrates for the agricultural biogas plants are animal manure from pig and cattle breeding, often used in co-digestion regimes with energy plants like maize or grass silages. According to Gromke et al. (2018), about 78% of the energy supply from biogas plants referred to such energy crops. Caused by reformed regulations of the EEG in 2014 and in 2017 resulting in a cut of the feed-in tariffs for new installations, a drop in the construction of new in biogas plants occurred. The effects will increase after the guarantee period for feed-in tariffs expires after 20 years and a crisis in the sector is already imminent. For this reason, the operator and manufacturer of biogas plants are interested in new and improved opportunities to increase the flexibility of biogas plants and to ensure an economic operation.

3 Technological Options to Increase the Flexibility of Biogas Plant

In Germany, biogas plants were previously designed for the electricity and heat production in baseload operation. Caused by an increasing share of fluctuating renewable energies in the German power sector (currently approx. 40% of renewable power in the German energy mix), new concepts for flexible bioenergy projects are required. This requirement will get an increasing importance since political objectives are targeted to increase the proportion of renewable power (mostly generated from wind parks and PV installations) to at least up to 80% in the German energy mix until 2050. In general, energy from biogas has the potential to generate electricity flexibly on demand (Hahn et al. 2013). Demand-driven biogas plants allow electricity generation specifically at times of peak electricity demand (Persson et al. (2014)). Additional and alternative opportunities to increase the flexibility are the production of biomethane via biogas-upgrade technologies and the participation at sector coupling concepts as, e.g., power-to-gas (PtG) concepts to generate fuels.

3.1 Flexible Power Generation in Demand-Driven Biogas Plants

Since biomass itself is naturally stored energy (e.g., expressed by a caloric value), power produced from biomass and especially from biogas is a good opportunity to flexibly generate electricity on demand. Such a demand-driven production is important for balancing power generation and to ensure the stability of the electric grid system. Studies from the Fraunhofer Institute for Wind Energy and Energy System Technology (IWES) and the German Biomass Research Centre (DBFZ) have shown that a demand-driven biogas supply and a demand-driven power generation from biogas plants are possible and it is a suitable solution also for small- and medium-scale biogas plants.

A regulation defined in the German Renewable Energy Act to promote flexible power generation in biogas plants was the offer of a so-called "flexibility premium" (EEG 2017, § 50b Flexibilitätsprämie für bestehende Anlagen). The flexibility premium itself is a legally secured payment (130 € per additional installed kW_{el}) that enables operators of biogas plants to refinance their investments to increase performance (e.g., for a new internal combustion engine as additional CHP plant and for other components which are required to increase the flexibility of the biogas plant). Therefore, the goal of the flexibility premium was to increase the proportion of a flexible demand-driven electricity production in order to produce as much electricity as possible when the demand for electricity is high without having sufficient other renewable power sources available. With the flexibility premium, plant operators received financial support as soon as they have increased the installed capacity of their plants. A limiting factor in the regulation was the legally declared maximum capacity expansion up to 1350 MW, which was reduced in 2019 in the "Energiesammelgesetz" to a maximum of 1000 MW (DBFZ 2019). As this capacity expansion limit was exceeded in July 2019, the future effectiveness of this financial instrument is currently in question. Nevertheless, the technical requirements and some economic aspects for increasing flexibility by demand-driven operation of biogas plants will be briefly presented.

An easy way to change the operation mode of biogas plants in terms of increasing the flexibility is to install an additional combined heat and power (CHP) unit in order to concentrate the power generation into periods of time when the grid demand is high and the generation of power from wind and solar energy is insufficient. According to Purkus et al. (2018), the technical potential for flexible power generation in Germany would be approx. 40 TWh_{el}. Most of the plants which receive the flexibility premium (40%) are currently small-scale biogas plants in the range of 151–500 kW_{el} followed by approx. 38% in the medium-scale range of 501–1000 kW_{el} generation (Purkus et al. 2018).

Doubling the capacity of the power generation units is the most common solution and easy to implement. In case of an existing 500 kW_{el} CHP unit, an additional 500 kW_{el} unit will be installed. Taking into account the correction factor (1000 kW – 500 kW × 1.1), the additional power results in 450 kW_{el} (DBFZ 2019). With an assumed minimum shift potential of the power generation of 4 hours, most of the existing technological components (existing CHP unit, gas and heat storage installations) can still be used without additional investments for the expansion of heat and gas storage equipment (FNR 2018). For feeding all the generated power into the grid, the operator of the biogas plant can claim a flexibility premium of 58,500 euros/year, or 1.3 cents/kWh (450 kW × 130 euros/kW) (DBFZ 2019). Additional spot-market revenues through efficiency improvements are not included in this calculation. However, the degree of flexibility through doubling of the capacity has to be rated as low. In contrast, a maximum increase in performance is clearly more advantageous. Although the total investment increases, the increase in the flexibility premium for each additional kilowatt is higher than the additional costs. The more consistent improvement of flexibility also leads to an improvement in profitability. According to FNR (2018), the installation of the maximum possible capacity of 2500 kW (up to 5 times the original installed capacity of 500 kW) can generate an additional revenue of approx. 380,000 euros/year. However, it must be taken into account that this fivefold increase of the capacity results in a far more extensive change in the existing system and correspondingly high investment costs. An overview of the required measures is shown in Fig. 20.2.

The first two measures (adjustment of the feeding management and of the process parameter) can control the biogas formation and influence the capacity for power generation in a longer-term perspective. There is no direct investment

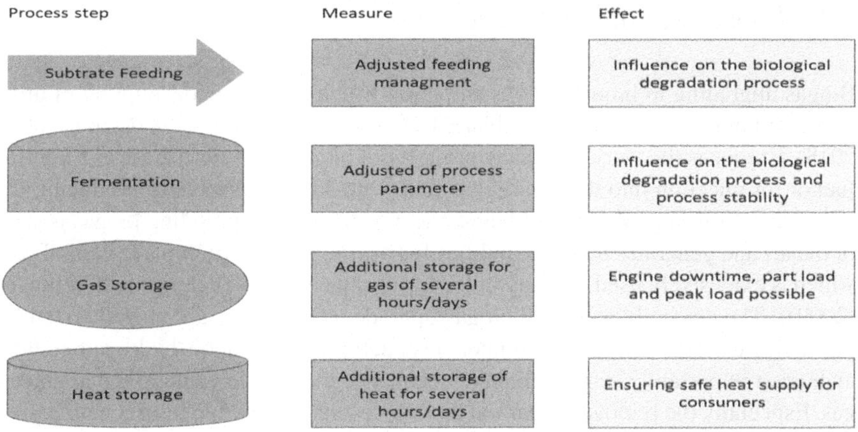

Fig. 20.2 Measures for demand-driven operation of biogas plants. (Adopted from Barchmann et al. 2016)

required. According to Mauky et al. (2017), flexible feeding is not affecting the long-term process stability, and the feeding management also economically benefits from a reduced gas storage demand by a demand-oriented gas production. In contrast, a change in the process parameters (e.g., change of temperature in the fermenter) could damage the microbiology and should only be considered in very careful steps. The measures 3 and 4 (installation of additional gas and heat storage) are directly related to increased investment costs. In particular, large-scale heat storage systems require high investment costs. For a conventional heat storage system with a volume of 1000 m^3, costs of almost 250,000 euros can be expected. The required size of gas storage volumes very much depends on the projected relation between operation times of the CHP units (considering also operation times with part load) and downtime (hours or days), the inclusion of feeding management regimes, and further site-specific features of the biogas plant. Double-membrane gas holders are state of the art which can be installed as low-pressure gas holders on top of the fermenter or added to an existing plant as external gas holders. Specific costs of integrated double-membrane gas storage on top of the fermenter were investigated by the German Biomass Research Center (Barchmann et al. (2016). Resulting in costs between approx. 11 euros/m^3 storage volume for 8000 m^3 large-scale gas storage solutions and approx. 35 euros/m^3 storage volume for small-scale gas storage solutions of 1000 m^3.

In addition to the desired demand-driven power generation, the goal of flexible plant operation of course is to increase the profitability of the biogas plant. Whether this goal can be achieved depends on various technical, economic, and legal issues and has to be checked and optimized for each individual project.

3.2 Biogas Upgrade to Biomethane

Biogas upgrading to biomethane is an option which allows the decoupling of utilization of upgraded biogas (biomethane) from the production process (Gromke et al. 2018). Europe is the world's leading producer of biomethane for the use as a vehicle fuel or for injection into the natural gas grid, with 459 plants in 2015 producing 1.2 billion m^3 (Scarlat et al. 2018). Meanwhile, the process of upgrading biogas is state of the art and generates new possibilities for its use, since it can replace natural gas, which is extensively used in many of the European countries (Petersson &Wellinger (2009)). The currently available biogas-upgrade technologies are characterized by the removal of undesired gas components such as water vapor, hydrogen sulfide, and, most importantly, carbon dioxide to enhance the caloric value of the upgraded gas. Especially the removal of carbon dioxide results in enriched biogas with higher methane content which should be close to the value of natural gas. The removal of carbon dioxide can be achieved by various techniques, such as adsorption, absorption, membrane, or cryogenic upgrading techniques. The most commonly used techniques are:

- *Water scrubbing*: Water scrubbing-based absorption is the most widely and increasingly implemented biogas upgrading technology around the world today (Sahota et al. 2018). The process is based on physical absorption using only water as a solvent for carbon dioxide. The principle of this process is that carbon dioxide has a higher solubility in water than methane.
- *Chemical absorption*: Chemical absorption (often referred to as ammine scrubbing) is characterized by a physical absorption of the gaseous components in a scrubbing liquid followed by a chemical reaction between scrubbing liquid components and absorbed gas components within the liquid phase. Carbon dioxide is not only absorbed in the solvent; it also chemically reacts with the amine present in the solvent. The most commonly used aqueous solutions of amines are monoethanolamine (MEA), diglycolamines (DGA), and diethanolamine (DEA) (Patterson et al. 2011).
- *Pressure swing adsorption (PSA):* Pressure swing adsorption (PSA) is an interesting technology for biogas upgrading, due to compactness of the equipment, low energy requirements, low capital cost, and safety and simplicity of operation (Augeletti et al. 2017). For this reason, PSA is the second most employed technique for biogas upgrading today. Carbon dioxide is separated using physical properties. The biogas is compressed to a pressure between 4 and 10 bars and is fed to a column where it is brought in contact with the adsorbent that will selectively retain carbon dioxide. The adsorbent material is a porous solid usually with a high surface area. Most of the adsorbents employed in the commercial processes are carbon molecular sieves, activated carbons, zeolites, and other materials such as titan silicates.
- *Membrane separation*: The use of membranes for gas cleaning is a well-established technology in chemical industries, and the use for the upgrading of biogas for biomethane production has gained increasing importance especially

during the last decade (Miltner et al. 2016). The technological principle is based on gas dissolution and diffusion into polymer materials (membranes) when a differential pressure is applied on opposing sides of the membrane. The membrane allows carbon dioxide to pass while retaining methane.

- *Cryogenic separation*: Cryogenic separation is based on the principle that different gases liquefy under different temperature-pressure conditions. It is a distillation process operated under very low temperatures (close to −170 °C) and high pressure (around 80 bars). The process consists of cooling and compressing the raw biogas in order to liquefy carbon dioxide, which is then easily separated from the biogas (Chen et al. 2015).

Each of the upgrading technologies mentioned above is suitable for meeting the gas purification requirements for feeding into gas pipelines or vehicle fuel specifications. According to Hoyer et al. (2016), the most common technology applied in the EU in terms of the number of plants is water scrubbing, but it is not possible to generalize which of the techniques is most appropriate. The technologies which are dominating the market besides water scrubbing are pressure swing adsorption, chemical absorption (amine scrubbing) (Bauer et al. 2013), and membrane technologies. This is also reflected in the technology proportions of the biogas-upgrade installations, published by the European Biogas Association in 2015 (see Fig. 20.3). But new developments (e.g., the improvement in the field of membrane technologies) might change the current situation significantly.

Since 2015, the number of the installed biogas upgrading plants steadily increased further. Germany still is the leading country in this development, caused by the high technological standard combined with changing political framework conditions,

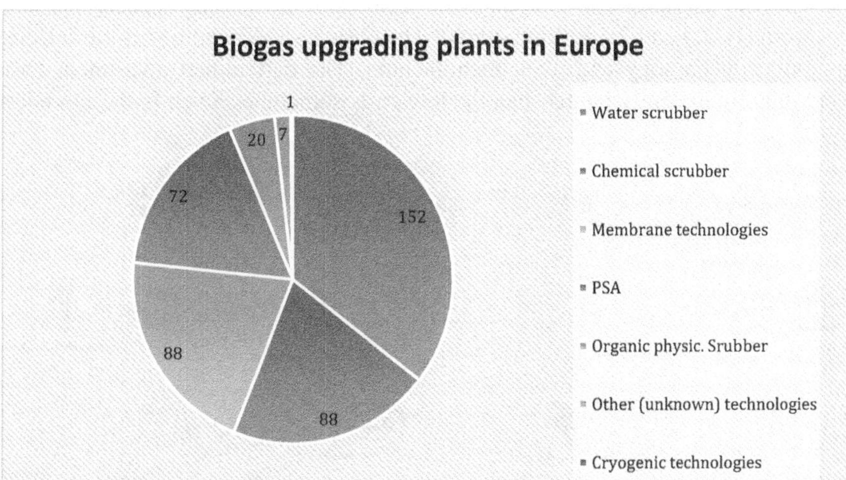

Fig. 20.3 Number of the installed biogas upgrading plants in Europe (data from 2015) according to technologies (Hoyer et al. 2016)

such as reduced feed-in tariffs in the German Renewable Energy Act. In Germany, especially biogas plants with high capacities can gain higher profits by selling biomethane to the gas grid compared to the conventional heat and power production. However, biogas upgrade to biomethane and the use of the natural gas grid as storage opportunity will create flexibility in the sense of stabilizing the electricity grids, only if the gas is used elsewhere (e.g., in urban regions with high heat requirements) for the production of power and heat. For biogas plant operators, biogas upgrade to biomethane is a new opportunity to create flexibility in terms of marketing of the produced biomethane (see Fig. 20.4).

In Germany, 196 biogas upgrading plants were in operation in 2016 with a gas feed-in capacity of 31 PJ (Purkus et al. 2018). In opposite to typical agricultural biogas plants for heat and power generation, the plants with upgrading technologies for biomethane production are using predominately energy crops or organic wastes, collected by public waste management authorities as substrates. According to Gromke et al. (2018), about 77% of the biogas upgrading plants (151 plants) are operated with upgrading capacities between 350 and 700 m^3/h raw biogas, 13% of biogas upgrading plants (26 plants) are smaller than 350 m^3/h raw biogas, and 10% (19 plants) have capacities higher than 700 m^3/h. A feasibility study carried out at HTW Berlin in 2015 concluded that based on the economic frame conditions in Germany at that time (required investment for the biogas-upgrade technologies versus feed-in tariffs for biomethane and electricity), a minimum biogas production capacity of 1000 m^3/h was required to make the biogas-upgrade technologies competitively. The investigated investment costs for a biogas-upgrade plant with a capacity of 250 m^3/h biogas were in the range between 4,000 and 5,600 €/m^3 processing capacity per hour (or respectively 1.0 and 1.4 million euros), while, due to the "economy of scale" costs for a plant with a capacity of 1000 m^3/h raw biogas were only in the range between 2,000 and 2,500 €/m^3 processing capacity per hour (respectively 2.0 and 2.5 million euros). The costs vary depending on the selected upgrading technology and plant manufacturer. The determined investment costs were thus significantly higher than the low costs sometimes given in the literature.

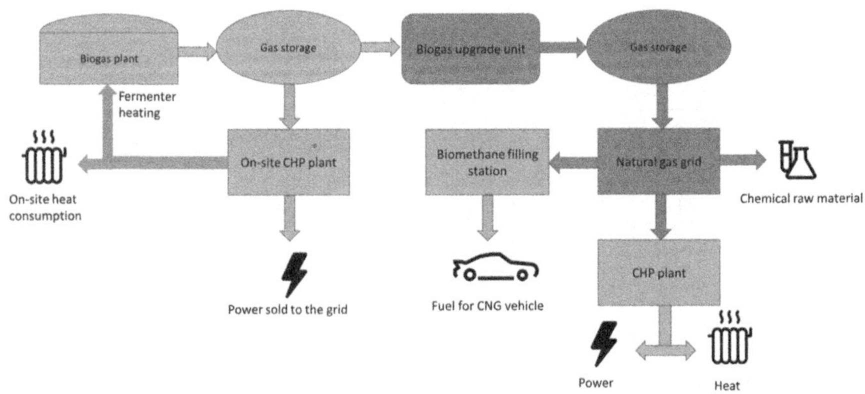

Fig. 20.4 Utilization pathways for biomethane marketing

In addition to that the specific operating costs have to be regarded, which are also higher for small plants compared to large plants. Nonetheless, the authors referred that in the future especially the new opportunities for the use of biomethane in the transport sector will result in a shift from electricity and heat production to biogas upgrading technologies for smaller biogas applications, too. Significant reductions in costs can be expected from new technological developments. Since the biomethane price is in competition with the natural gas price, the future development of biogas-upgrade technologies will strongly depend on governmental policies and subsidies. Currently, the costs of biomethane production are in the range between 40 and 60 €/MWh and so far higher than those of fossil-derived natural gas (which is in the range of 10–2 0€/MWh) (Lambert 2019). For this reason, the use of biomethane is currently only worthwhile with a financial subsidy. In Germany, a strong incentive to invest in biomethane plants was defined by the Renewable Energy Sources Act providing investment security and resulting in the situation that Germany is the leading country in biomethane production in Europe. However, further political support is necessary in the future to continue this development.

3.3 Methanation of Raw Biogas to Biomethane

Direct methanation of biogas is a promising opportunity to increase the flexibility of biogas plants by the creation of new marketing concepts for operators of biogas plants. Biogas consists to approx. 25–45% of carbon dioxide (see Table 20.1) and the additional conversion of this CO_2 to methane can produce 40–60% more methane (depending on the biogas composition) compared to the conventional biogas upgrading technologies. The direct methanation of biogas to biomethane could potentially eliminate the energy-intensive CO_2 separation processes, which are required by the conventional biogas-upgrade technologies and increasing the CH_4 yield significantly (Rusmanis et al. 2019). Above all, the method is gaining in importance because the use of hydrogen generated from renewable electricity creates a good possibility of coupling the energy sectors – electricity, heat, and transport. Since the required hydrogen is usually generated from surplus electricity, the process is also referred to as power-to-methane as one opportunity of the power-to-gas processes. The existing natural gas distribution network and the natural gas storage facilities in Europe offer an enormous storage potential for the methane provided and thus an interim storage facility for excess electricity from fluctuating renewable energy sources such as power from wind turbines and PV installations. The standards and requirements for the injection of biomethane into the gas grids are defined by gas grid operators and national regulations and might slightly differ between different countries. In Germany, the gas network access ordinance requirements define the conditions under which access must be provided by grid operators to their networks to those who want to inject gas into the pipelines. Specifications for biomethane for injection in the natural gas network in Germany are defined in DIN EN 16723-1,2017 and the worksheets DVGW G 260 (Gas quality) and DVGW

Table 20.2 Typical biogas composition (according DIN EN 16723-1, 2017 and DIN EN 16723-2, 2017)

Component	Requirement
Methane	$CH_4 \geq 96$ vol%
Hydrogen Carbon dioxide	$H_2 \leq 2$ vol% $CO_2 \leq 2{,}5$ vol% $CO_2 \leq 4$ vol% (without sensitive components)
Hydrogen sulfide	$H_2S < 5$ mg/m^3
Water	Dew point at −10 °C (pressure level of injection point of the gas grid)

G 262 (Usage of gases from renewable sources in the public gas supply), issued by the German Association for Gas and Water Supply. The main requirements are shown in Table 20.2.

Up to now, carbon dioxide separated from the biogas in conventional biogas-upgrade applications is mostly released into the atmosphere. The aim is to avoid this emission and to convert carbon dioxide with hydrogen into methane. The basic reaction used for the methanation is known as the Sabatier reaction, discovered by the French chemist Paul Sabatier about 100 years ago:

$$CO_2 + 4H_2 \rightleftharpoons CH_4 + 2H_2O \, \Delta H = -165.0\,\text{kJ/mol} \tag{20.1}$$

The reaction is exothermic and takes place in two steps. The first step is the fully reversible water gas shift (Eq. 20.1.1), and the second step is the classical methanation reaction which occurs biologically initiated also in the biogas production process itself (Eq. 20.1.2):

$$1.\text{Step}: \; H_2 + CO_2 \rightleftharpoons CO + H_2O \tag{20.1.1}$$

$$2.\text{Step}: 3H_2 + CO \rightleftharpoons CH_4 + H_2O \tag{20.1.2}$$

Compared to the currently pursued power-to-gas approaches, where carbon dioxide is separated from the raw biogas before the catalytic conversion, e.g., by a biogas-upgrade process takes place, the new methanation concepts are targeted to use the entire raw biogas stream for the methanation process. R&D is currently focusing on biological and thermochemical catalytic methanation concepts.

3.3.1 Biological Methanation

The biological conversion of hydrogen and carbon dioxide to methane is a complex microbiological conversion process to generate methane by means of highly specialized microorganisms (Archaea). Biological methanation uses biological catalysts, i.e., hydrogenotrophic methanogenic microorganisms, to catalyze the methanation reaction. Since microorganisms need special conditions for optimal growth, methanation reactors are usually operated at temperatures between 37 and 65 °C and pressures between 1 and 15 bars (Thema et al. 2019). When implementing biological methanation, a distinction is made between the so-called in situ and ex situ processes. In an in situ methanation system, organic substrate and additional hydrogen are added to the digester where the biogas is produced (Voelklein et al. 2019). In ex situ methanation processes the conversion takes place in a separate external reactor were beside the gases specific nutrients are supplied to the microbial consortium to suit the specific requirements of the hydrogenotrophic methanogens. (Rusmanis et al. 2019).

In Situ Methanation

In situ methanation is the simplest way to perform enhanced hydrogenotrophic methanogenesis. A schema of the biological in situ methanation is shown in Fig. 20.5.

The advantage of the in situ process is the possible integration into the existing biogas infrastructures (biogas or sewage gas plants) and the process-integrated methane enrichment of the biogas. In the literature, achievable methane levels of between 75% and 97% are reported (Kretzschmar 2017). The rate of methane formation depends on the provision of carbon dioxide and the availability of hydrogen in the biogas process. According to Voelklein et al. (2019), the solubilization of hydrogen is the decisive step to make gaseous hydrogen available for microorganisms on a cellular level. With a solubility rate of 0.7 mmol/(l*bar), hydrogen poorly dissolves in water (with solubility rates 24 times less than that of carbon dioxide at 55 °C), and the hydrogen to liquid transfer is the bottleneck of the process. If high methane contents (e.g., >96% according to DVGW worksheets G 262; see Table 20.2) are achieved (it depends on the effectiveness of biological methanation) and feeding of the biomethane into the natural gas grid is the target, the removal of carbon dioxide might not be necessary. But nonetheless, the purification of the gas (removal of unwanted components such as hydrogen sulfide) remains necessary.

Due to the simplicity of the concept and the resulting low investment costs, in situ methanation appears to be a promising option for use in existing small and medium agricultural biogas plants. Such plants are mostly operated under mesophilic temperature conditions (at approx. 37 °C), and an increase of temperature into the thermophile range seems not suitable due to the high energy demand for the fermenter heating. The central question remains whether the high methane contents

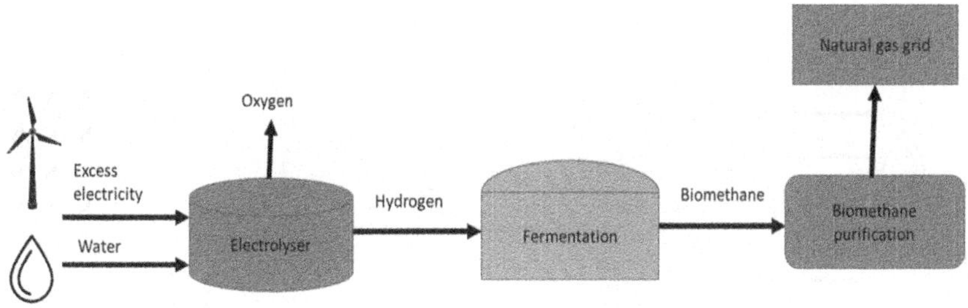

Fig. 20.5 Schema of biological in situ methanation

(greater than 96% according to DVGW) which have been seen in the laboratory tests can also be achieved in technical applications.

Ex Situ Methanation

During ex situ methanation, carbon dioxide and hydrogen are injected into a separate reactor with pure or enriched hydrogenotrophic microorganism and nutrients, where the biogas is upgraded. As shown in Fig. 20.6, it is possible to directly convert carbon dioxide to methane in the raw biogas or to alternatively use other sources of carbon dioxide for the methanation. In terms of designation, carbon dioxide used for biomethane production has to be a product of a biomass conversion process (e.g., carbon dioxide from alcohol fermentation processes in bioethanol plants).

One advantage of ex situ methanation concepts compared to the in situ process is that through splitting the whole process in two reactors (fermentation in AD and the ex situ methanation reactor), the stability of the conventional biogas process is not affected by the upgrading step. In contrast to the in situ concepts, the temperature for the upgrade process can therefore be set independently of the operating temperature of the biogas plant to reach an optimum for thermophilic *Methanothermobacter*. A further advantage is that the ex situ process can handle high volumes of influent gases decreasing the gas retention time even to 1 hour, which minimizes the dimensions of the biogas upgrading chamber (Angelidaki et al. 2018). Possible reactor designs for the ex situ methanation are (a) bubble column reactor, (b) continuous stirred tank reactor (CSTR), (c) membrane reactor, (d) trickle bed reactor, and (e) upflow anaerobic sludge blanket (UASB) reactor. Biogas upgrading efficiencies are reported in several studies in the range between 70 and 98%, depending on the selected reactor design, operation mode (batch operation or continuous operated ex situ methanation), gas retention times, and other process parameters influencing the conversion process. Similar to the in situ concepts, a CO_2 separation after the methanation is not required if the methane concentrations required for feeding into the natural gas grid are reached, but the purification of the gas (e.g., removal of components such as hydrogen sulfide) remains necessary.

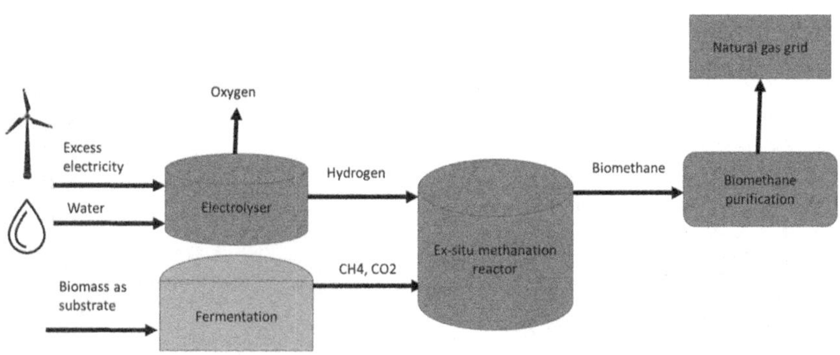

Fig. 20.6 Schema of biological ex situ methanation

Bioelectrochemical Methanation

Bioelectrochemical methanation (also called electromethanogenesis) is a largely unexplored approach to the direct conversion of electricity where methanogenic microorganisms can convert carbon dioxide to methane under electrical stimulation (Kretzschmar 2017). As reported by Schivano et al. (2018) electroactive microbial communities, grown as biofilms on solid electrodes, were able to convert inorganic carbon to methane by using electrons or hydrogen provided electrochemicallyfor the reaction of carbon dioxide to methane. A possible reaction mechanism is shown below:

$$CO_2 + 8H^+ + 8e- \rightarrow CH_4 + 2H_2O \quad (20.2)$$

Electrons can be made available to microorganisms via electrodes integrated directly into the fermenter. A distinction is made between (a) direct electron transfer (microorganisms grow directly on the electrode surface and use the provided electrons) and (b) indirect electron transfer (mediators are reduced on the electrode surface and transmit electrons to microorganisms) from electrodes to microorganisms. An advantage of the bioelectrochemical methanation compared to conventional biological methanation approaches is that hydrogen is not generated external and is fed to the reactor in the gaseous phase or must be dissolved in the liquid phase (Kretzschmar 2017). Until now, the concept of bioelectrochemical methanation is still under research only available in lab scale.

Biological Methanation: Summary

The limiting factor for the technical implementation of all biological methanation concepts is the supply of hydrogen and carbon dioxide to microorganisms. Both are gaseous components and must be in solution in order to be used by microorganisms. Especially due to the poor solubility of hydrogen in aqueous environments, the sufficient supply with hydrogen is the main challenge. Furthermore, it is important to identify potentially suitable efficient microorganisms (e.g., by target-oriented selection and genetic modification) to increase the performance of the process. The current research and demonstration projects produce only small amounts of methane. According to Rusmanis et al. (2019), the first full-scale methanation plant with a full load capacity of 1.502 MW (raw biogas) is operated in Copenhagen, Denmark.

From a techno-economic point of view, biological methanation is rather suitable for small biogas plants (e.g., agricultural biogas plants in the single-digit MW range). Here, integration into the existing biogas plant structure is also possible. The ability to connect such systems to the natural gas grid remains a prerequisite (Kretzschmar 2017).

3.3.2 Thermochemical Catalytic Methanation of Raw Biogas to Biomethane

The thermochemical catalytic conversion of carbon dioxide to methane based on the Sabatier process (see formula 1) is already available in commercial scale and one of the key concepts of the modern power-to-gas technologies. The first commercial large-scale 6 MW_{el} power-to-gas plant was built in Werlte (Germany) within the Audi e-gas project and started operation in 2013. The required CO_2 source is a biogas plant nearby, where carbon dioxide is captured from a biogas-upgrade facility and hydrogen comes from three 2 MW alkaline electrolyzers, powered by renewable electricity. Further applications using alkaline and PEM electrolyzers to produce the required hydrogen are operated in lab scale or pilot plant scale and mostly use pure carbon dioxide, targeted to be supplied by biogas-upgrade facilities (Table 20.3).

Since biogas-upgrade facilities are cost expensive and in the most cases feasible from the economic point of view only at large-scale biogas plants (see Chap. 3.2), further research is required to find solutions for small-scale agricultural biogas plants, too. Growing demands from the biogas industry have motivated the authors to address the specific requirements of smaller biogas plant operators (with a raw biogas production capacity of less than 250 m^3/hour) precisely to generate biomethane for feeding it into the natural gas grid in a new research project. Compared to the currently pursued power-to-gas approaches, where carbon dioxide is separated from the raw biogas before the catalytic conversion, the new project will supply the entire raw biogas stream to the catalytic methanation. The aim of the project is the development of an innovative multicomponent system for direct catalytic methanation of carbon dioxide contained in the raw biogas of biogas plants by reaction with hydrogen produced by using renewable sources. A schema of the project conception is shown in Fig. 20.7.

To achieve this objective, an integrated approach of catalyst zeolite matrix, an optimal integration of gas treatment/gas separation, and a customer-oriented system integration are pursued (Barz & El Bari 2018). To avoid poisoning of the catalyst (zeolite-supported nickel and ruthenium catalysts were used for the catalytic methanation), a comprehensive desulfurization of the biogas is required. A large number of state-of-the-art technologies (physical, chemical, and biological) exist to remove hydrogen sulfide from biogas. To ensure the required low concentrations of <5 mg/m^3 (maximum hydrogen sulfide concentration according to DVGW (2013)), the

Table 20.3 Comparison of Alkaline and Proton Exchange Membrane (PEM) electrolyzer for large-scale PtG projects

Parameter	Alkaline electrolyzer	PEM electrolyzer
Operation temperature [°C]	40–90	50–80
Operation pressure [bars]	1–30	1–50
Efficiency [%]	Up to 80	65–70
Current density [A/cm^2]	0.2–0.6	0.6–1.5
Performance range [MW][*1]	Up to 130	Up to 6
Part-load operational range [%]	20–100	5–100
Load gradients [%/s]	<1	10
Specific energy demand [kWh/m^3 H_2]	4.2–5.8	4.5–6.5
Investment costs (for 6 MW capacity) [€/MW_{el}]	Approx. 750,000	Approx. 1,200,000

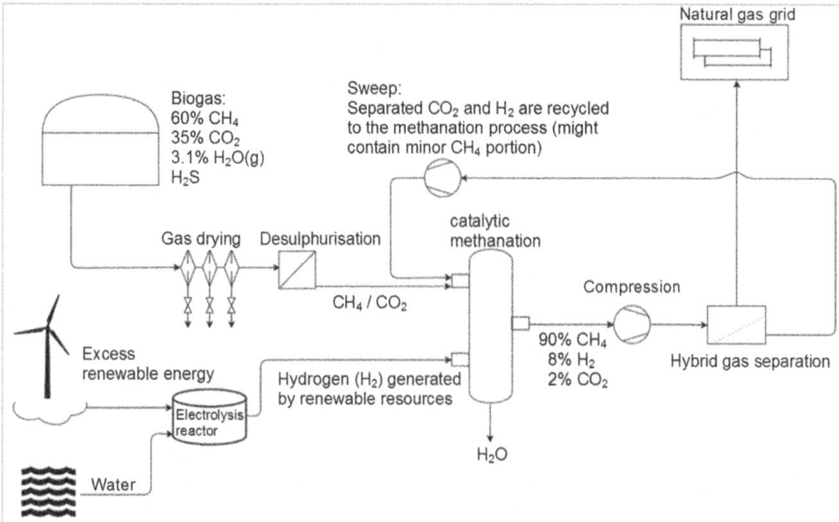

Fig. 20.7 Process schema of the catalytic methanation of biogas

combination of adsorption in iron oxides and activated carbon seems to be a suitable method. The energy released from the following exothermic conversion reaction can be used (and is required) to supply the heat for the biogas digester, which in Germany is mostly operated under mesophilic conditions (at approx. 36 °C). In conventional biogas plants (for heat and power generation) and under German climate conditions, about 20–30% of the thermal energy from the CHP plant (gas engine) is used to provide the thermal energy for digester heating. In case of a complete methanation of biogas, this heat source has to be replaced by other sources (as, e.g., from the exothermic conversion reaction). Approx. 0.72–0.90 kWh thermal energy from the methanation of 1 m^3 (raw) biogas (depending on the initial CO_2 concentration) can be used for digester heating, and a heat surplus is produced which can be used for other applications or heat consumers close by the biogas plant.

In the first phase of the project, the experiments on methanation and gas separation were carried out at different institutions. Methanation experiments at Beuth Hochschule Berlin to increase the methane concentration of 60% in the raw biogas using nickel catalyst resulted in methane yields of >80%, and experiments using ruthenium-based catalyst resulted in methane yields of 70–80% (see Fig. 20.8).

A two-stage membrane separation process is used to achieve the methane concentration of >96%, required for feeding into the gas grid. The membrane itself is the selective barrier that permits the separation of gas components based on diffusion through the membrane material and on the molecule size. The separation is achieved by selectively passing (permeating) of one or more components of the gas stream through the membrane while retarding the passage of other components. The permeate is the portion of gas that diffuses through the membrane, while the retentate is the portion of gas that is retained by the membrane. In the two-stage

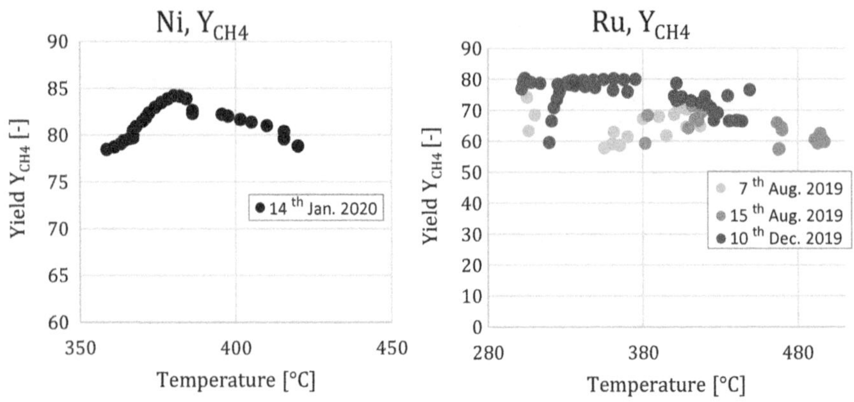

Fig. 20.8 Methane yields after methanation experiments using different catalyst materials

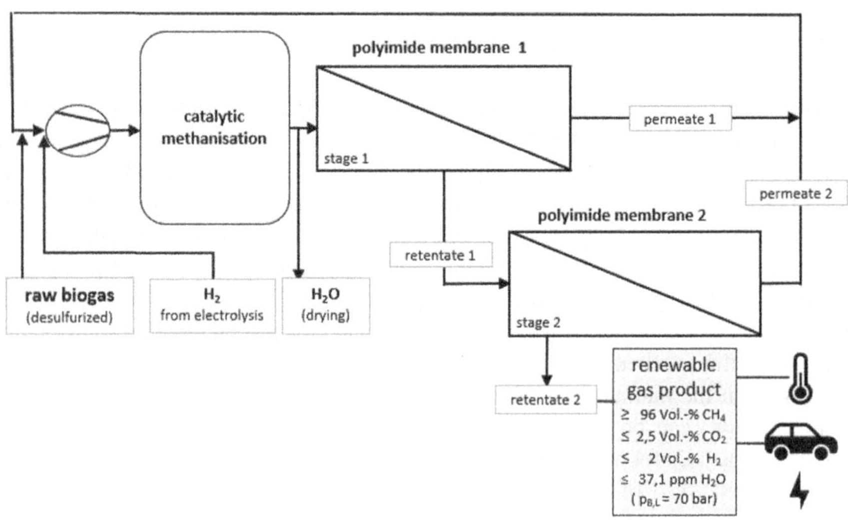

Fig. 20.9 Scheme of the two-stage membrane separation process to concentrate the methane content for gas grid injection

membrane process shown in Fig. 20.9, the retentate 1 is fed as feed 2 into a second membrane. The permeate 2 diffuses through the second membrane, while the retentate 2 is retained by this membrane.

The gas standards defined in the DIN EN 16723-1 and DIN EN 16723-1 for the injection of biomethane into the gas grid can be achieved in a two-stage gas separation with membranes. This has been experimentally determined by the project partner Graforce in laboratory tests, using UBE-hollow fiber polyimide membranes with feed flow rates of 5 l/min at pressure differences of 16 bars at 23 °C. In retentate 2 (see Fig. 20.9), a CH_4 concentration of 98.5 vol% with concentrations of 0.8 vol% H_2 and 0.7 vol% CO_2, respectively, was measured.

Case Study for Small-Scale Biogas Applications The majority of biogas systems in Germany have an electric power output below 500 kW, and they are mostly installed in rural areas where the efficient use of the coproduced heat is hardly possible (Köppel et al. 2009). For this reason, the basic data of a conventional agricultural 500 kW biogas plant with combined heat and power plant will be used to introduce the biogas and energy flows required for the adaption of the new concept into small-scale biogas applications. The considered agricultural biogas plant is using 2200 t of liquid manure from cattle breeding, 6500 t of maize silage, 1100 t of cereal crops, and 1100 t of grass silage per year to produce a total amount of 1.7 Mio m^3 of biogas per year. With an average share of 35% carbon dioxide in the raw biogas, this corresponds to an amount of 590,000 m^3 or respectively 1170 tons of carbon dioxide per year. Based on the reaction equation shown under formula (20.1), 425,500 kg of additional methane (or respectively 590,000 m^3) can be produced from this source caused by the catalytic conversion of carbon dioxide to methane. So compared to the conventional biomethane upgrade technologies, where the maximum annual amount of methane for grid feed-in would be only 1 Mio m^3 (the normal methane yield of the produced biogas), the total amount of biomethane available for the feed-in into the grid will increase to nearly 1.6 Mio m^3/year. For the conversion, an annual amount of about 215,000 kg of hydrogen is required, and about 956,160 kg of water is produced. The water will be recycled and used for the production of hydrogen via electrolysis, covering approx. 50% of the total demand of the water for the electrolysis process.

As a result of experimental investigations in lab scale, a methanization model was created with regard to the application of the concept for such small-scale agricultural biogas plants (see Fig. 20.10).

To balance the quantities of gas streams, the annual raw biogas flow of 1.7 Mio m^3 of biogas per year was converted into an hourly flow of approx. 222 m^3/hour. Assuming a raw biogas composition of 60 vol% CH_4 and 40 vol% CO_2, a CO_2 volume of approx. 89 m^3/h has to be converted hourly into methane. According to the H_2 to CO_2 ratio of 4:1 specified in formula 20.1, 337 m^3/h is required for the methanation. An overview of the gas composition and volume flows is shown in Fig. 20.11.

An energy balance of the catalytic methanation concept compared to the conventional CHP production of agricultural biogas plants is shown in Table 20.4.

Based on the reaction enthalpy of the exothermal process (−165 kJ/mol), the total thermal energy released from the catalytic conversion process is 1216 MWh/year, and since the reactor for the catalytic conversion process has to be cooled to avoid a change in the equilibrium position of the optimal conversion rate and to prevent thermal degradation of the catalysts, this amount of energy can be used to heat the biogas digester. The heat produced corresponds to the average demand of the fermenter heating of a mesophilic biogas plant in Germany (900–1200 MWh/year), and since the conversion occurs in a temperature range of above 300 °C, the temperature level for heating of the biogas reactor is more than sufficient.

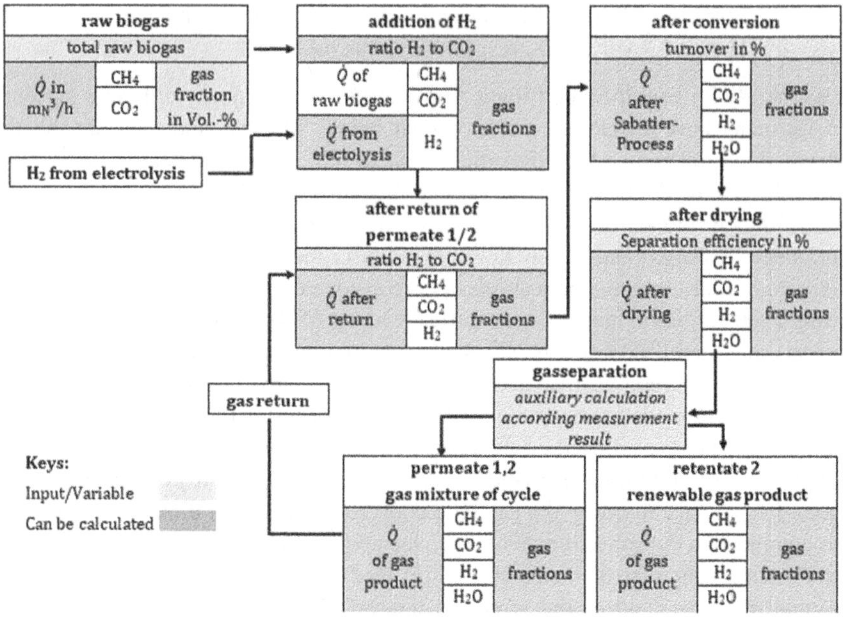

Fig. 20.10 Tabular scheme of the Excel-based methanation model

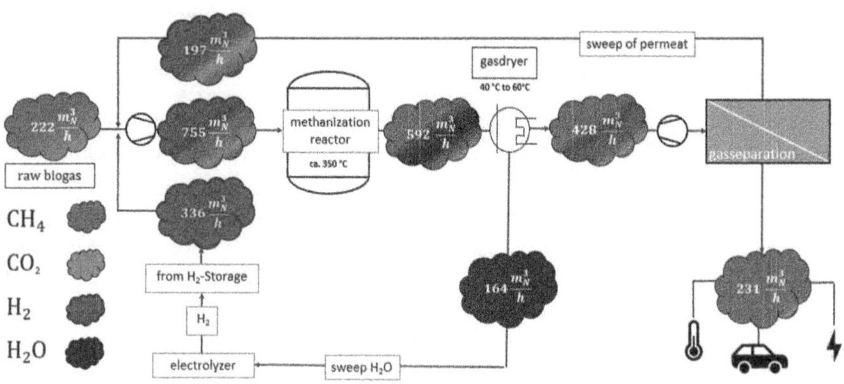

Fig. 20.11 Schematic presentation of the gas volume flows of the methanation model

The standard reaction enthalpy for the hydrogen production via electrolysis shows that 3.54 kWh/m^3 hydrogen (standard conditions) is required. Considering a typical electrical efficiency of up to 80% for commercially available alkaline electrolyzers, a capacity of >1.25 MW is required. To participate on special tariffs in the balancing energy market, even much higher capacities connected with hydrogen storage facilities might be of interest for some applications.

Table 20.4 Comparison of the operation mode of 500 kW biogas plants, plant with CHP unit versus complete methanation

Parameter of typical agricultural biomass plants with CHP units compared to PtG option		
Operation mode	CHP operation	Complete methanation
Electrical capacity [kW]	500	–
Thermal capacity [kW]	520	–
Operation hours [h/year]	8000	8000
Produced (raw) biogas [m^3/year]	1,684,478	1,684,4782
CH_4 concentration (raw) biogas [%]	60	60
CO_2 concentration (raw) biogas [%]	35	35
Produced power [kWh/year]	4000,000	–
Produced thermal energy [kWh/year]	4,400,000	1216,000[a]
Produced biomethane [m^3/year]	–	1,601,728
CH_4 concentration biomethane [%]	–	96
Heat demand fermenter heating [kWh/year]	900,000	900,000
Heat for external use [kWh/a]	3,500,000	316,000

[a]Reaction enthalpy from the exothermal Sabatier reaction

4 Conclusion

Increasing the flexibility of biogas plants will play a significant role in the future energy system. High shares of fluctuating wind and solar power require to compensate fluctuations in power generation by energy storage systems and demand-oriented power generation systems with short ramp-up times. Further flexibility options are the increase of the share of renewable energy in the transportation and heat sectors as the main energy-consuming sectors of our economies.

Flexible power generation in demand-driven biogas plants seems to be the easiest way to increase the flexibility, e.g., simply by the adjustment of the feeding management and/or the increase of the capacity of gas and heat storage facilities and power generation units. The possibility of implementing this flexibility option depends heavily on political and related economic framework conditions. In Germany, e.g., a so-called flexibility premium was defined in the Renewable Energy Act (EEG) 2017 (EEG 2017, § 50b Flexibilitätsprämie für bestehende Anlagen) to motivate operators of biogas plants to invest in measures to increase the performance of their power generation units and to enable them to increase the proportion of a flexible demand-driven electricity production. A limiting factor in the regulation is the legally declared maximum capacity expansion up to a maximum of 1000 MW according to the "Energiesammelgesetz" 2019.

The next promising opportunity for biogas plant operators is the production of biomethane by biogas-upgrade technologies. Various state-of-the-art technologies such as water scrubbing, chemical absorption, pressure swing adsorption, and membrane separation technologies are available on the market. Especially biogas plants with high capacities can gain higher profits by selling biomethane to the gas grid compared to the conventional heat and power production. The limiting factor is that high investments are required and only the methane content of the biogas (in average 60% of the raw biogas) can be fed into the gas grid. A higher methane yield can be achieved by installing additional power-to-gas techniques in the process. The CO_2 contained in the raw biogas is converted into methane by reaction with hydrogen (Sabatier process) so that methane yields are increased by about 50%. The example of Audi e-gas in Werlte shows that the process is marketable. However, it should be noted that the additional methanation step increases the investment costs for the corresponding systems immensely. Especially for the operators of small-scale agricultural biogas plants, these costs are unlikely to be affordable.

Another possible option is the direct methanation of the CO_2 contained in the raw biogas using biological and catalytic methanation processes. The aim is to feed the entire raw biogas stream to methanation. Until now, such methanation concepts for raw biogas are still under research and only available in lab- or pilot-scale applications. The results achieved in the current research projects suggest that marketable solutions will also be available in the short term.

In conclusion, it must be stated that there is currently no generalized best practice solution. It is therefore necessary to check which solution offers the most economic and ecological advantages for each planned project and location based on the general frame conditions.

Acknowledgments The authors acknowledge the financial support from the Institut für angewandte Forschung (IFAF Berlin) for the project "Nachgeschaltete Biogasmethanisierung mit Zeolithmatrix-Katalysatoren (KatMethCon)." The results presented in this chapter are based on a system analysis created as part of this project.

References

Achinas, S.; Achinas, V.; Euverink, G.J. W.: A Technological Overview of Biogas Production from Biowaste, Engineering 3 (2017) 299–307

Angelidaki, I.; Treu, L.; Tsapekos, P; Luo, G.; Campanaro, S.; Wenzel, H.; Kougias, P. (2018): Biogas upgrading and utilization: Current status and perspectives, status and perspectives. Biotechnology Advances, 36(2), 452–466. https://doi.org/10.1016/j.biotechadv.2018.01.011

Augelletti, R.; Conti, M.; Annesini, M. C. (2017): Pressure swing adsorption for biogas upgrading. A new process configuration for the separation of biomethane and carbon dioxide. Journal of Cleaner Production, Volume 140, Part 3, 1390–1398

Barz, M.; El Bari, H.: Ecological Sustainable Waste Management – Energetic Utilization of Organic Waste, S. 1–129, Mensch und Buch Verlag, Berlin, 2018

Barchmann, T.; Mauky, E.; Dotzauer, M.; Stur, M.; Weinrich, S.; Jacobi, H. F.;Liebetrau, J.; Nelles, M.: Expanding the flexibility of biogas plants – substrate management, schedule synthesis and economic assessment, Landtechnik 71(6), 2016, 233–251

Bauer, F.; Hulteberg, C.; Persson, T.; Tamm, D.: Biogas upgrading – Review of commercial technologies, SGC Rapport 2013:270

Chen, Xiao Yuan; Thang, Hoang Vinh; Ramirez, Antonio Avalos; Rodrigue, Denis; Kaliaguineb, Serge: Membrane gas separation technologies for biogas upgrading. RSC Adv., 2015, 5, 24399–24448

DBFZ Deutsches Biomasseforschungszentrum, Schlussbericht zum Vorhaben "Leitfaden Flexibilisierung der Strombereitstellung von Biogasanlagen (LF Flex)," 11/2019

DVGW: Arbeitsblatt G 260 Gasbeschaffenheit, Deutscher Verein des Gas- und Wasserfaches e. V. (2013)

European Biogas Association (EBA), EBA Biomethane & Biogas Report 2015, http://european--biogas.eu/2015/12/16/biogasreport2015/

Hoyer, K.; Hulteberg, C.; Svensson, M.; Jernberg, J.; Nørregard, O.: Biogas Upgrading - Technical Review, REPORT 2016:275, ISBN 978-91-7673-275-5

Renewable Energy Sources Act (EEG) 2017: Act revising the legislation on renewable energy sources in the electricity sector of 22.12.2016. [Original German title: Gesetz zur Einführung von Ausschreibungen für Strom aus erneuerbaren Energien und zu weiteren Änderungen des Rechts der erneuerbaren Energien]. Bundesgesetzblatt Jahrgang 2016 Teil I Nr. 49, ausgegeben zu Bonn am 18. Oktober 2016

Fachverband Biogas (2018). Branchenzahlen 2017und Prognose der Branchenentwicklung 2018, https://www.biogas.org/edcom/webfvb.nsf/id/DE_Branchenzahlen/$file/18-05-25_Biogas_Branchenzahlen-2017_Prognose-2018_end.pdf

Fachagentur Nachwachsende Rohstoffe e. V. (FNR): Leitfaden Flexibilisierung von Biogasanlagen, 1. Auflage, FNR 2018

Gromke, J.F.; Rensberg, N.; Denysenk, V.; Stinner, W.; Schmalfuß, T.; Scheftelowitz M.; Nelles, M.; Liebetrau, J.: Current Developments in Production and Utilization of Biogas and Biomethane in Germany, Chem. Ing. Tech. 2018, 90, No. 1–2, 17–35

Hahnn, H.; Krautkremer, B.; Hartmann, K.; Wachendorf, M.: Review of concepts for a demand-driven biogas supply for flexible power generation, Renewable and Sustainable Energy Reviews 29:383–393, 2013

Petersson, Anneli; Wellinger, Arthur (2009); Biogas upgrading technologies – developments and innovations. IEA Bioenergy, Task 37 - Energy from biogas and landfill gas

Persson, T.; Murphy, J.;Jannasch, A. K.; Ahern, E.; Liebetrau, J.; Trommler, M.; Toyama, J.: A perspective on the potential role of biogas in smart energy grids, IEA Bioenergy 2014, ISBN 978-1-910154-13-7

Kaltschmitt, M.; Hartmann, H.: Energie aus Biomasse – Grundlagen, Techniken und Verfahren; Springer Verlag Berlin, Heidelberg, New York, 2001

Kaltschmitt M., Development of the Biogas Technology, DAAD Advanced Training Course on Renewable Energy Technologies in Thailand, 03.09.2012 – 22.09.2012, King Mongkut's University of Technology, Thonburi (Bangkok)

Köppel, W.; Götz, M.; Graf, F.: Biogas Upgrading for Injection into the Gas Grid, 34 gwf-Gas Erdgas, International Issue, 2009

Kretzschmar, J.: Technologiebericht 4.2b Power-to-gas (Methanisierung biologisch) innerhalb des Forschungsprojekts TF_Energiewende, In: Wuppertal Institut, ISI, IZES (Hrsg.): Technologien für die Energiewende. Teilbericht 2 an das Bundesministerium für Wirtschaft und Energie (BMWi). Wuppertal, Karlsruhe, Saarbrücken, Dezember 2017

Lambert, M.: Tracking progress in scaling up renewable gas production in Europe, Oxford Institute for Energy Studies& Sustainable Gas Institute, Imperial College, London 2019, OIES PAPER: NG 153

Mauky, E.; Weinrich, S.; Jacobi, H.F.; Nägele, H.J.; Liebetrau, J.; Nelles, M.: Demand-driven biogas production by flexible feeding in full-scale - Process stability and flexibility potentials, Anaerobe 46 (2017) 86–95

Miltner, M.; Makaruk, A.; Harasek, M.: Selected Methods of Advanced Biogas Upgrading Chemical Engineering Transactions, 52, 2016, 463 – 468

Patterson, T.; Esteves, S.; Dinsdale, R.; Guwy, A. (2011). An evaluation of the policy and techno-economic factors affecting the potential for biogas upgradation for transport fuel use in the UK. Energy Policy 39, 1806–1816.

Purkus, A.; Gawel, E.; Szarka, N.; Lauer, M.; Lenz, V.; Ortwein, A.; Tafarte, P.; Eichhorn, M.; Thrän, D.: Contributions of flexible power generation from biomass to a secure and cost-effective electricity supply—a review of potentials, incentives and obstacles in Germany, Energy, Sustainability and Society (2018) 8:18

Rusmanis, D.; O'Shea, R.; Wall, D. M.; Murphy, J.D.: Biological hydrogen methanation systems – an overview of design and efficiency, BIOENGINEERED 2019, 10, 1, 604–634 https://doi.org/10.1080/21655979.2019.1684607

Sahota, S.; Shaha, G.; Ghosha, P.; Kapoor, R.; Sengupta, S.; Singh, P. Vijay V.; Sahay, A.; Vijay, V. K.; Thakur, I. S.: Review of trends in biogas upgradation technologies and future perspectives. Bioresource Technology Reports 1 (2018) 79–88

Scarlat, Nicolae; Dallemand, Jean-François; Fahl, Fernando (2018). Biogas: Developments and perspectives in Europe, Renewable Energy 129 (2018) 457 – 472

Schievano, A.; Goglio, A.; Erckert, C.; Marzorati, S.; Rago, L.; Cristiani, P.: Organic waste and bioelectrochemical systems: A future interface between electricity and methane distribution grids, Detritus 01 - 2018 57–63, https://doi.org/10.26403/detritus/2018.6

Thema, M.; Weidlich, T.; Hörl, M.; Bellack, A.; Mörs, F.; Hackl, F.; Kohlmayer, M.; Gleich J; Stabenau C; Trabold T; Neubert M; Ortloff F; Brotsack R; Schmack D; Huber H; Hafenbradl D; Karl J; Sterner M: Biological CO2-Methanation: An Approach to Standardization, Energies 2019, 12, 1670; doi:https://doi.org/10.3390/en12091670

Voelklein, M. A.; Rusmanis, D.; Murphy, J. D.: Biological methanation: Strategies for in-situ and ex-situ upgrading in anaerobic digestion, Applied Energy 235 (2019) 1061–1071

Wu, Libin: 2000 Years' Biogas Technology Development in China, Sustainable Energy News - Europe - No. 2 September 1993

Hoyer, K.; Hulteberg, C.; Svensson, M.; Jernberg, J.; Nørregard, O.: Biogas Upgrading - Technical Review, REPORT 2016:275, ISBN 978-91-7673-275-5

Chapter 21
Case Study for Status and Exploration of Microalgae in Egypt

Guzine El Diwani, N. N. El Ibiari, S. I. Hawash, Sanaa A. Abo El-Enin, Nahed K. Attia, Ola A. Elardy, Elham A. AbdelKader, and Samar A. El-Mekkawi

Abstract This chapter introduces a brief description of the microalgal cell, its presence, its constituents with the benefits of each, the applied technologies of biofuel production and biorefineries through a scope of the physical possibilities in Egypt. Microalgal biomass contains sufficient amounts of carbohydrates and oil for biofuel production especially biodiesel. The technical conversion options for biomass into biofuel can be categorized into three basic methods: chemical, thermochemical, and biochemical conversion. The desired product and the form of energy are the main factors that influence the choice of a specific conversion. Production of algal valuable products such as algal bioactive compounds and algal biofuel products will be presented briefly in this chapter. As a conclusion, the production of the biofuel conforming to ASTM specifications can be achieved by the selection of a techno-economically feasible technology. The feasibility can be performed using microalgal oil after the extraction of valuable products previous to use the remaining oil for biofuel production.

Keywords Biofuels · Nonedible oils · *Jatropha* · Jojoba · Microalgae · Biodiesel · Bio-jet fuel

G. El Diwani (✉) · N. N. El Ibiari · S. I. Hawash · S. A. Abo El-Enin · N. K. Attia · O. A. Elardy · E. A. AbdelKader · S. A. El-Mekkawi
Biofuel Expert Group, Chemical Engineering and Pilot Plant Department, Engineering Research Division, National Research Centre, Cairo, Egypt

E. Koukios, A. Sacio-Szymańska (eds.), *Bio#Futures*,
https://doi.org/10.1007/978-3-030-64969-2_21

1 Introduction

Amazing magic power is stored in a unicellular cell. This magic micrometer cell stores numerous amounts of bioactive compounds and antioxidants to cure immune diseases. Besides the health-care compounds, it contains sufficient amounts of oil to produce bio-aviation fuel introducing green travel. This magic cell is named "microalgae."

Ancient Egyptians were concerned with cosmetics, they used algae as a source of beauty. Egyptian government and investors are interested in this novel industry. Researchers are working hard to discover and modify new approaches in order to optimize microalgal benefits.

In this chapter, the authors introduce a brief description of the microalgal cell, its constituents, and the benefits of each; the applied technologies of biofuel production and biorefineries throw a scoop of the physical possibilities in Egypt.

Egypt has a unique location between the Mediterranean Sea and the Red Sea. Also, it has the Suez Canal that connects the two seas. All the cities located on the Mediterranean Sea, such as Marsa Matruh, Al Behera, Alexandria, Kafr el-Sheikh, Daqahliyah, Dumyat, and Al Sharquiyah, and the Suez Canal, such as Al Ismailia, Port Said, and North Sinai, can be considered suitable for growing algae (Shabaka, 2018). Climate, land availability, nutrients sources, saline and freshwater, and sunny weather are all factors pushing toward the use of algae in biofuel production in Egypt together with the production of valuable coproducts depending on the feasibility studies (Fig. 21.1).

Algae that belong to the plant kingdom are divided into macroalgae such as seaweed of size sometimes reaching meters and microalgae of size in nanometers. The latter is in turn divided into eukaryotic algae and prokaryotic blue-green algae (cyanobacteria) that are nomenclature as photosynthetic bacteria in the recent taxonomy (Dodds & Whiles, 2020).

Algae survive in various environmental conditions. They grow in freshwater, brackish, marine, or hypersaline. It grows ten times more rapidly than terrestrial plants, and less than a tenth of the land is needed to produce an equivalent amount of biomass. Microalgae endure a wide range of temperatures and pH and nutrient variability (Suparmaniam et al., 2019). In Egypt, the five Mediterranean Lakes (Bardawil, Burullus, Edku, Mariut, and Manzala) comprised 867 species of phytoplankton related to 9 algal divisions, 102 families, and 203 genera. Bacillariophyta was the most dominant group, while Cryptophyta, Rhodophyta, and Phaeophyta were rarely existed by only one species. The species diversity of the five lakes can be arranged descendingly as follows: Manzala (383 spp.) > Mariut (376 spp.) > Bardawil (333 spp.) > Burullus (247 spp.) > Edku (183 spp.). The highest number of unique species was recorded in Bardawil (208 spp.), followed by Manzala (128 spp.), then Mariut (85 spp.), Burullus (76 spp.), and Edku (6 spp.). The highest number of unique species (208 spp.) in Lake Bardawil (62.4% of the total species) may be attributed to its hypersaline nature compared with the other oligotrophic lakes (Khairy et al., 2015).

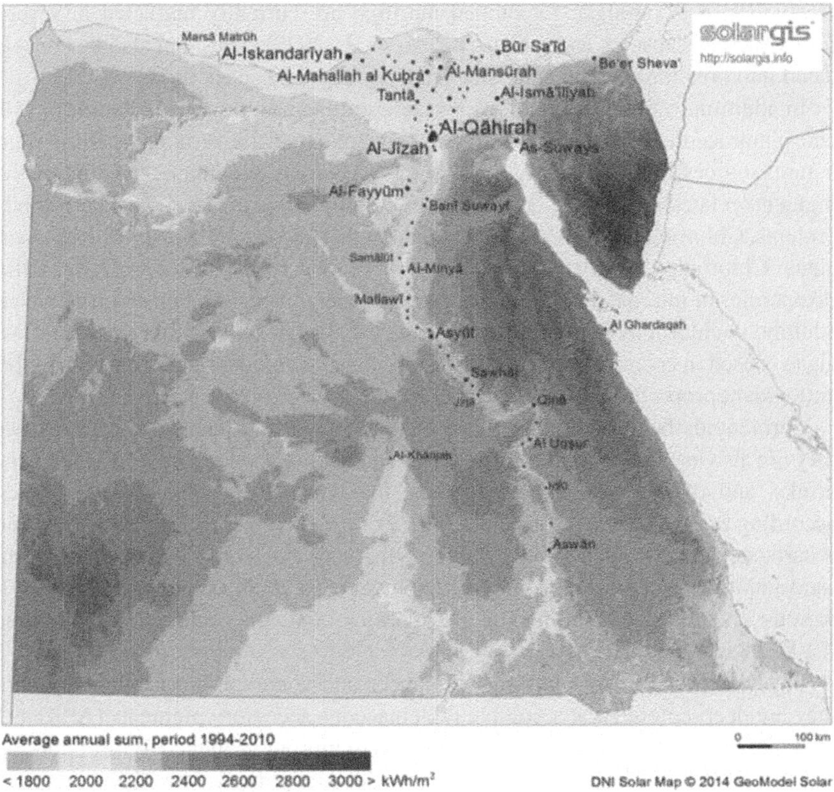

Fig. 21.1 Annual average global horizontal solar radiation in Egypt (Solar gis, 2010)

Microalgae are naturally screening their ecological surrounding for desirable nutrients and energy. For biomass growth, microalgae rely upon adequate quantities of carbon source and light to perform photosynthesis (Elshobary et al., 2019). Depending on the type of cultivation – phototrophic, heterotrophic, mixotrophic, or photo-heterotrophic – the type of metabolism is performed by algae related to the environmental nature and algal adaptation (Suparmaniam et al., 2019).

The chemical compositions of algae strains differ from a strain to another and from a culture technique to another (Rastogi et al., 2018). This depends on physical cultivation factors like temperature, irradiance, pH value, CO_2 supply, aeration rate, and mixing velocity or chemical factors such as nutrient proportions and nutrient composition. For the sake of reaching the desired composition and sufficient biomass of algae, specific cultivation conditions could be applied (Yousuf, 2020). Algae consist mainly of proteins, carbohydrates, lipids, vitamins, and pigments. Much less known to the general public is the assortment of biorefineries that are used in food processing (Molino et al., 2018) – particularly as subsidiary ingredients such as oil, fats, polyunsaturated fatty acids, emulsifiers, thickeners, plasticizers, pigments, natural dyes, sugars, bioactive compounds, and antioxidants (Galasso et al., 2019). Microalgae for human nutrition are currently marketed in several forms such as capsules and tablets (Kumar et al., 2020). They can also be consolidated into breakfast cereals, nutrition bars, cookies, and snack foods.

In addition to food, algae afford a vast assortment of vaccines, vitamins, medicines, nutraceuticals, and other nutrients that are expensive or nonexistent in other natural sources like plants or animal sources. Microalgae contain an abundance of pigments related to light incidence such as chlorophyll, carotenoids, and phycobiliproteins. Chlorophyll, the responsible pigment that absorbs light energy, exists in all algae. Chlorophyll a exists individually in cyanobacteria and Rhodophyta. Other taxonomies of microalgae (Chlorophyta and Euglenophyta) contain chlorophyll b in addition to chlorophyll (Bueno Ariede et al., 2017). The chlorophyll extracted from algae is used in cosmetics due to its ability against the oxidation process since it has antioxidant properties (Bueno Ariede et al., 2017).

Carotenoids from microalgae have a great role as antioxidants and anticancer; they are also used as a vitamin A precursor and natural food colorant for juices, soft drinks, and dairy products. Carotenoids, the liposoluble pigments, are divided according to their chemical structure into two types: carotenes, such as β-carotene, α-carotene, and lycopene, and xanthophylls, such as lutein, canthaxanthin, and astaxanthin. Carotenes and some xanthophylls exist in all taxonomics of algae in various quantities according to the species and the cultivation conditions (El-Mekkawi et al., 2019a).

Phycobiliproteins are used against cancer and some cardiovascular disorders; they are deep-colored water-soluble proteinaceous accessory pigments (Al Khawli et al., 2020). There are two categories of phycobiliproteins: phycoerythrobilins and phycocyanobilins. The phycoerythrins are divided into three different types depending on their spectrum absorption: R-phycoerythrin and B-phycoerythrin that exist in Rhodophyceae and C-phycoerythrin that exists in cyanobacteria and Rhodophyceae (Charoensiddhi et al., 2020).

Microalgae are considered a valuable source of almost all-important vitamins. A wide range of vitamins such as A, B1, B2, B12, C, E, nicotinate, biotin, folic acid, or pantothenic acid exist in several species of microalgae. Vitamins are very sensitive to heat. Thus, vitamin instability should be taken into consideration at harvesting, drying, and extraction steps (Charoensiddhi et al., 2020).

The use of microalgae as animal feed is relatively recent and predominantly aimed at poultry, mainly because it improves growth performance, blood metabolism, and fertility (Abouelezz, 2017). Multiple nutritional and toxicological evaluations demonstrated the suitability of algae biomass as a valuable feed supplement or substitute for conventional protein sources (Abou-Zeid et al., 2015).

Microalgae are an essential food source for aquatic animals; they represent an alternative source of cultured fish. The size range of 1–15 μm algae is most appropriate for filter feeders and 10–100 μm is suitable for grazers (El-Semary, 2018). The diet supplemented with microalgae improves growth and nutrient utilization of shrimps (Sharawy et al., 2020). Microalgae are used as a soil fertilizer, the germination percentage of seeds increased after treating the soil with microalgal powder. After the extraction of oil or carbohydrates from microalgae, most of the nutrients are still present in the leftover biomass. One potential market for this nutrient-rich biomass is biofertilizer (El-Semary, 2018; Dineshkumar et al., 2018). The de-oiled microalgae are used as a biofertilizer.

The use of some microalgal species, especially *Arthrospira* and *Chlorella*, is well established in the skincare market, and some cosmeticians have even invested in their own microalgal production system (LVMH, Paris, France, and Daniel Jouvance, Carnac, France).

Cellulose-containing algae can be used as a renewable feedstock for bioethanol production – although algae are generally known for their low cellulose and hemi-cellulose content, they have enough sugar content to be fermented for bioethanol production (El-Mekkawi et al., 2019b).

Algae have a big role in pollution control, in wastewater treatment facilities, reducing the need for greater amounts of toxic chemicals that are already used, capturing fertilizers in runoff from farms, and reducing CO_2 emissions. The CO_2 can be pumped into a pond, or some kind of tank, on which the algae feed. Alternatively, the bioreactor can be installed directly on top of a smokestack.

Microalgae's main constituents are lipids and fatty acids. Some types of lipids exist for supporting and increasing the elasticity of the membrane, while other sorts of lipids, with specific composition, are interiors and function as storage products.

In recent years, the interest to introduce biofuel to reduce fossil fuel dependence due to increasing its scarcity shortly, and environmentally, biofuel must be introduced to challenge the increased demand for fuel.

Biofuel is expected to have a role in creating a sustainable, economical, and environmentally safe source of energy. The term biofuel relates to liquid or gaseous fuels that are produced from biomass. The main distinction between biofuel and petroleum feedstocks is oxygen content. Biofuel has oxygen constituents from 10% to 45%, while petroleum oil has typically none.

The world over, biofuel is divided into bio-alcohol, biodiesel, bio-jet, and bio-synthetic oils. Microalgae, the unicellular phytoplankton organisms, offer great potential for applications based on biofuel production technology. Microalgal biomass contains sufficient amounts of carbohydrates and oil for biofuel production especially biodiesel.

The technical conversion options for biomass into biofuel can be categorized into three basic methods: chemical, thermochemical, and biochemical conversion (Suparmaniam et al., 2019). The desired product and the form of energy are the main factors that influence the choice of a specific conversion process.

Biochemical processes include fermentation, anaerobic digestion, and bio-photolysis. The fermentation process is simply based on converting starch, sugar, and cellulose of the biomass into alcohol using enzymes, while the solid residue can be used as cattle feed.

The second category of biochemical conversion is anaerobic digestion. The bio-wastes can be digested with bacteria into CH_4 and CO_2 mixture called biogas. The third category is called bio-photolysis. This process is based on the evolution of hydrogen using cyanobacteria and algae. The reaction is identical to electrolysis including splitting of water into oxygen and hydrogen.

Thermochemical conversion describes the thermal decomposition of biomass by different processes such as gasification, pyrolysis, thermochemical liquefaction, and direct combustion. Generally, in the gasification process, the biomass is partially oxidized into a combustible gas mixture at high temperature (873–1073 °C). This mixture is known as syngas and consists of CO, CO_2, H_2, N_2, and CH_4.

Using pyrolysis, biomass can be converted into syngas, bio-oil, and charcoal at a range of temperatures from moderate to high temperature (350–700 °C) in the absence of air. Flash pyrolysis seems to be an applicable technique for the sufficient production of liquid fuels. By thermochemical liquefaction, liquid fuel can be synthesized from wet algal biomass at a temperature ranging from 273 to 350 °C and pressure ranging from 5 to 20 MPa aided by a catalyst in the presence of hydrogen. Via direct combustion, biomass can be utilized as biofuel directly.

The chemical process called transesterification is used to produce biodiesel. Transesterification is the reaction of a fat or oil with an alcohol to form esters and glycerol. A catalyst (acid, alkali, or enzymes) is usually used to improve the reaction rate and yield (Makareviciene & Skorupskaite, 2019). Excess alcohol is used to shift the equilibrium toward the product because of the reversible nature of reaction (Negm et al., 2017). The major influencing factors on the yields of biodiesel are a type of alcohol, reaction temperature, oil-to-alcohol ratio, reaction time, and pressure (Attia et al., 2018). Chemically, most biodiesel consists of alkyl (usually methyl) esters instead of the alkanes and aromatic hydrocarbons of petroleum-derived diesel (Ortiz-Martínez et al., 2019). There are methods for improving this transesterification reaction including ultrasonic methods and microwave methods (Sharma et al., 2016). "In situ transesterification" or "reactive extraction" is another method to produce biodiesel, which combines the steps of lipid (oil) extraction and transesterification to produce biodiesel (Wadood et al., 2019). Integration of these stages could reduce biodiesel production costs. The main concerns in performing in situ reactions are lipid solubilization and prevention of the interference of water or other compounds. Many parameters play important roles in optimizing the process such as reaction temperature, time, and molar ratio of alcohol to oil and catalyst concentration (Chang et al., 2020). Bio-jet fuel can be produced through several technologies such as the flash hydrolysis of algal slurry (Asiedu et al., 2020) and hydro-processing of algal biodiesel through catalytic hydrocracking (Cheng et al., 2019a; Cheng et al., 2019b).

Biofuels derived from microalgae have the potential to (partly) replace petroleum fuels and the first generation of biofuels, but efficacy with which sustainability goals can be achieved is dependent on the life cycle impacts of the processes of microalgae to biofuel (Cheng et al., 2019b). Life cycle assessment (LCA) is a broader methodology that can be used to account for all the environmental impacts of an industrial process. This could include not only energy and greenhouse gases (GHG) but also the consumption of all the materials needed for the production process, water requirements, and emission of many kinds of pollutants (liquid, gaseous, etc.). In other words, the LCA methodology considers in detail the footprint of any given process. As a consequence, wide data sets are required, and calculations tend

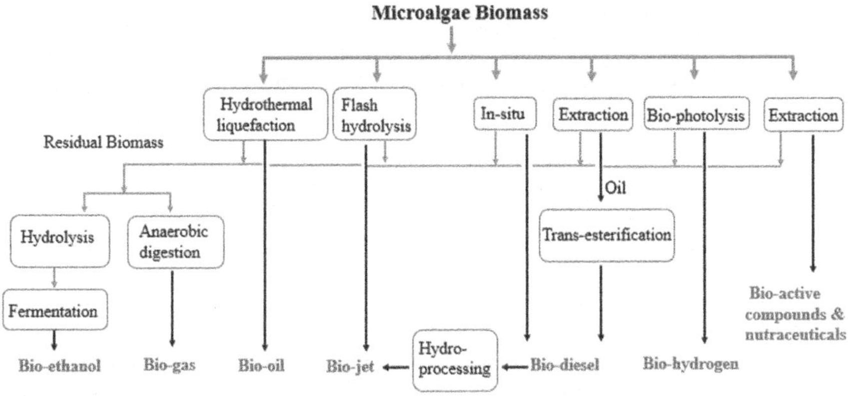

Fig. 21.2 The possibilities of integrated industries based on microalgal biomass

to be more complex and less transparent, and comparability might be more limited (Alves et al., 2017; De Jong et al., 2015; Taher et al., 2020). The integrated industries depend on microalgal biomass are shown in Fig. 21.2 is required to approach LCA.

The Egyptian government has established new sector concerns with renewable energy. Bioenergy represents an essential department that encourages investments in this sector. Biodiesel Miser is one of these recent investments; it is an Egyptian joint-stock company starting, from the second quarter of 2015, producing biodiesel from algae in cooperation with Algasol Company (www.biodieselmisr.com). Alexandria Company for Petroleum Additives (ACPA) is another producer and seller locally and globally of mono-alkyl ester (biodiesel) with advanced facilities to fulfill the growing global demand for standard diesel engines. ACPA production of B 100 meets the European Specifications EN 14214:2003 (www.acpaegypt.com).

After this comprehensive introduction, algal valuable products such as algal bioactive compounds, algal biofuel products, and their production will be presented.

2 Algal Bioactive Compounds

Microalgae play complex multiple roles in the purification process of wastewater. In organically enriched wastewaters, the release of free oxygen is of major significance, promoting aerobic degradation processes by bacteria and other microorganisms. Organic carbon is partly oxidized to free CO_2, serving as a carbon source for the algae. The other important role of microalgae is the accumulation and conversion of wastewater nutrients (mainly dissolved salts of nitrogen and phosphorus) into algal biomass. Two types of wastewater treatment systems are currently available for algae-based treatment, waste stabilization ponds system (WSPs) and high-rate algal ponds (Craggs et al., 2014; Doma et al., 2016).

Biomass of microalgal community (*bloom*) was collected at maximum growth rate from the high-rate algal pond (HRAP) constructed to treat municipal wastewater at Zinin Wastewater Treatment Plant, Giza, Egypt, and dried at 60 °C overnight.

2.1 Extraction of Algal Bioactive Compounds

2.1.1 Carotenoids

Distilled water will be added to the algal cells to be slurry, dry granular citric acid added to the slurry, Jojoba oil was added as (1:1) w/w of algal biomass. The mixture was homogenized at 1000 rpm for approximately 15 minutes. The slurry reaction mixture is stirred for 45 minutes at a temperature of 80 °C, then the mixture is neutralized using potassium hydroxide, and the temperature is reduced to 45 °C. Centrifugation of the mixture was done to obtain two layers: the upper layer was the carotenoid with jojoba oil and the lower layer was biomass and water. Decantation of the upper layer was done, and it was stored in a bottle until HPLC analysis using Agilent Technologies 1100 series liquid chromatograph equipped with an autosampler and a diode-array detector (461 nm) (Nauman et al., 2007; Basily et al., 2018a). The remaining biomass was used for getting lipid, phenolic, flavonoid, protein, and carbohydrate.

2.1.2 Fatty Acids

The remaining algal biomass was mixed with a mixture of n-hexane and isopropanol (3:2 v/v). The extraction process is performed in the ultrasonic bath at room temperature (30–35 °C) under a reflux condenser for 30 minutes. Cell residues were removed by filtration. The filtrate of the solvent mixture was evaporated to enable the gravimetric quantification of total lipid extract. The fatty acid profiles of algal oil are determined using gas chromatography (GC) (Basily et al., 2018a).

2.1.3 Phenolic Compounds

The algal cells were resuspended in 80% ethanol, homogenized, and centrifuged at 2000 rpm for 15 min. The resulting supernatant was centrifuged again at 2000 rpm. The residue was repeatedly extracted with the same solvent until they were colorless. Then, the supernatant was filtered through Millipore filters. The filtrate was evaporated to dryness to give a crude algal ethanolic extract and immediately analyzed. Total phenolic contents were determined with Folin-Ciocalteu reagent using gallic acid as a standard phenol compound. The concentration of total phenol contents was measured as milligram of gallic acid equivalent (GAE in mg/g of the sample) (Hanaa et al., 2009; Sarojini et al., 2013).

2.1.4 Flavonoids

The crushed algal biomass was added to distilled water, centrifuged, and incubated at 4 °C. Then it was transferred to a water bath at 100 °C. The supernatant was filtrated and the filtrate was adjusted to 25 ml with 95% methanol solution and incubated at room temperature for 48 hours for the determination of flavonoid content. Total flavonoid was measured by a colorimetric assay according to Kim et al. (2003) with some modifications (Baviskar & Khandelwal, 2015; Massoumeh et al., 2014).

2.1.5 Protein Content

Total protein content was determined by the Kjeldahl method and then multiplied with a factor of 6.25 to give the total protein content according to Basily et al. (2018a).

2.1.6 Carbohydrate Content

The carbohydrate was measured by a spectrophotometer at 485 nm (Dubois et al., 1956).

2.2 *Uses of Algal Bioactive Compounds*

- The main carotenoids in the microalgal community (*bloom*) are β-carotene and lutein as shown in Table 21.1 and an amount of lycopene (14.7 μg/g). Carotenoid production is highly affected by the growth conditions (Goiris et al., 2012; Mulders et al., 2014; Safafar et al., 2015; D'Alessandro & Filho, 2016). Also, the microalgal community (bloom) synthesizes a significant quantity of carotenoids to protect themselves from damage caused by their culture of environmental stress such as nutrient depletion and high light irradiance (Li et al., 2012) (Table 21.1).

The natural β-carotene is a mixture of *trans* and *cis* isomers (Guedes et al., 2011), has anticancer activity, and is absorbed by the body ten times easily more than synthetic one (Vilchez et al., 2011; Christaki et al., 2013). Each molecule of carotenoids can produce two molecules of vitamin A, so carotenoids possess activity

Table 21.1 Carotenoid contents detected (μg/g)

Algal strains	β-carotene orange-red (μg/g)	Lutein yellow (μg/g)	Lycopene red (μg/g)	Total detected carotenoids (μg/g)
Bloom (B.)	**72.3**	**36.4**	**14.7**	**123.4**

higher than that of vitamin A, which is essential for vision and the correct functioning of the immune system. It is considered as one of the health food colorants because of its strong antioxidant capacity which reduces the harmful effects of free radicals in various disorders. Furthermore, β-carotene could enhance immunity against different infectious diseases (Le & Xiao-Ming, 2010).

Lutein is found in the microalgal community. Several studies have concluded that lutein like zeaxanthin are yellow pigments that are responsible for the maintenance of the natural visual role for eye macula of humans (Friedman et al., 2004), where the other carotenoids are not present or found in few amounts. The eye macula is protected against any adverse photochemical reactions because of the antioxidant activity of carotenoids. The main source for the vision loss in persons over 65 years old has been attributed to the reduction of zeaxanthin and lutein levels (Renju et al., 2014).

Lycopene is present in *bloom* microalgal biomass, it has a red color, and according to different studies, this compound significantly decreased the proliferation of prostate cancer in mice (Christaki et al., 2011). Lycopene reduces cholesterol, decreases the density of lipoprotein levels, and improves rheumatoid arthritis.

The carotenoids cannot be synthesized again by humans or animals and can be obtained through the food or metabolic pathway of the precursor compounds.

The carotenoids that have been extracted from microalgae are considered eco-friendly colorants, so they can be used safely in the food industries instead of the synthesized pigments. Previous studies proved that carotenoids can treat the oxidation damage caused by free radicals (De Jesus et al., 2013).

The advantages of natural carotenoids make it suitable for several pharmaceutical products. The biological features of algal carotenoids such as anti-inflammatory, antioxidant, antitumor, and vitamin A activities contribute to the quality of the product. Also, the natural carotenoids have several applications in cosmetology as anti-aging and sunscreen (Fig. 21.3).

- The microalgae community contains valuable fatty acids for human health with variable quantities. It includes omega-9 and omega-7 fatty acids which belong to the family of monounsaturated fatty acids (MUFAs) and are not similar to essential fatty acids (EFAs) because they can be synthesized by the human body from unsaturated fat. Polyunsaturated FA was present in amounts higher than the monounsaturated FA, and because they cannot be manufactured by the body, they are considered as essential FAs. The essential FAs, ω-3 and ω-6 in particular which are found in large amounts, are very important for the integrity of tissues. γ-Linolenic acid has some cosmetics applications like revitalizing the skin and slowing aging. Linoleic and linolenic acids are essential nutrients for the immune system and tissue regeneration processes. Linoleic acid is also used for the treatment of hyperplasia of the skin. Most studies showed that dietary ω-3 PUFAs from microalgae have a protective effect against heart disease besides reducing hypertension and also important in the development and function of the nervous system (Basily et al., 2018b; Rahul et al., 2016).

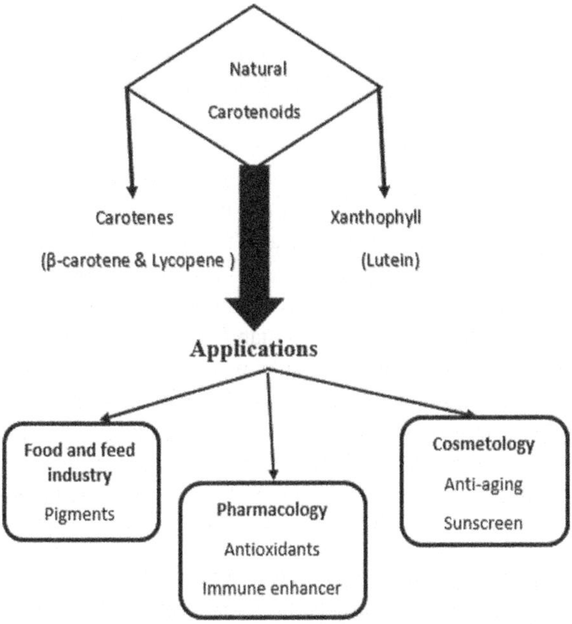

Fig. 21.3 Biological uses and applications of natural carotenoids in microalgae

- The total phenolic content of investigated algal cells (*bloom*) based on the Folin method is 5.36 mg/g. Previous studies demonstrated that the growth conditions are affected in the concentration and composition of phenolic compounds of microalgal biomass. Microalgae contain a variety of phenolic classes. Phenolic compounds are considered to be the most significant and biologically active compounds with various health beneficial properties (Farasata et al., 2014). Many types of research proved that a high diet with a natural phenolic is strongly associated with longer life expectancy, reduced risk of developing some chronic diseases, various types of cancer, diabetes, obesity, improved endothelial function, and reduced blood pressure (Basily et al., 2018a; Shashank & Abhay, 2013).
- The colorimetric determination of flavonoid content in microalgal biomass recorded 1.5 mg/g. Flavonoids are secondary metabolites and can act as strong antioxidants capable to scavenge free radicals that are harmful to a cell of the human body and food products. With increasing the demand of natural products as antiaging agents in pharmaceuticals, nutraceuticals, and cosmetics industries, flavonoids are gaining vital importance with their multiple activities such as antimicrobial, anticancer, and antidiabetic (Basily et al., 2018a; Becker, 2006).
- The determination of protein content in studied microalgal biomass showed 54.3 mg/g.

  The high protein content of microalgal cells is one of the main reasons to consider them as an unconventional source of protein. Proteins are composed of different amino acids. So, the nutritional quality of a protein is determined basically by the proportion content and availability extent of its amino acids (Kraan, 2012). Microalgae can enhance the nutritional value of conventional food preparations and hence have a positive effect on the health of humans and animals.

- The microalgal cells contain 1.5 mg/g of carbohydrates. As in all green plants, one of the most important cell components of microalgae is carbohydrates, which are found in the form of starch, glucose, sugar, and other polysaccharides. Certain polysaccharides have medical effects and functions such as protection against oxidative stress and efficacy on gastric ulcers, wounds, and constipation (Manivannan et al., 2013). From an economical point of view, algal polysaccharides are the most important products manufactured from algae. Carbohydrates are considered active raw materials that play an important role in cosmetics products such as deodorant, hair conditioning, hair waving or straightening, emulsifying, binding agent, viscosity controller, and stabilizing for gel forming and also for skin products such as skin conditioning and emollient. Natural carbohydrates are characterized by antibacterial, antioxidative, anti-inflammatory, antitumor, and antiviral properties (Basily et al., 2018a; Kader et al., 2015).

The microalgae are promising for increasing demand for bioactive compounds like carotenoids, phenolics, flavonoids, proteins, and carbohydrates. Isolation and identification of novel metabolites from microalgae will help to the development of new therapeutic agents, nutraceutical, and food industries. The culture conditions of microalgae are strongly species dependent and significantly affect their metabolic pathways. The communities of microalgae which are collected from municipal wastewater plant contain a relatively high amount of different types of carotenoids. The natural active products are finding an extended range of applications in the cosmetics, pharmaceuticals, food, and feed industries.

3 Algal Biofuel Production

Several biofuels such as bio-oil, biodiesel, and bio-jet can be produced from any vegetable oil. In Egypt, we cannot explore any edible oil for biofuel production, so we look for nonedible oils such as *Jatropha* oil, waste cooking oil, and algal oil.

Egypt's climate is very satisfactory for algae growth due to its sunny weather and consequently the availability of algal oil.

3.1 Bio-oil Production

Spirulina platensis microalgae were supplied from the Microbiology Department, Agriculture Research Center (ARC), Giza, Egypt. Bio-oil can be produced by direct conversion of *Spirulina* microalgae via a hydrothermal liquefaction (HTL) process at a reaction temperature of 300 °C, a pressure of 100 bars, and a reaction time of

30 minutes. The extraction of bio-oil from the reaction mixture was investigated using eight different organic solvents with different polarities. It was found that tetrahydrofuran (THF) gave maximum bio-oil extraction yield with a rather high heating value (HHV ≈ 30 MJ/kg) (Abdel Kader et al., 2015).

Extracted bio-oil can be fractionated into three fractions – heavy oil (48.9%), mid-weight oil (37.8%), and light oil (62.2%) – by using a mixture of THF, ethyl acetate (EAC), and n-hexane (n-HEX). These three oil fractions were characterized using gas chromatography-mass spectrometry (GC-MS) (Abdel Kader et al., 2015).

So it is concluded that *Spirulina* sp. is an ideal feedstock for bio-oil production.

Mid-weight oil is rich in aromatic compounds which can be a promising candidate for partial substitutes for petroleum and asphalt binder, while heavy oil was found to be rich in ester. In Light oil, with higher HHV ≈ 35.83 MJ/kg, the high number of paraffin and low wax content produces a higher percentage of gasoline and diesel fuel when converted into products by an oil refinery (Abdel Kader et al., 2015).

Another attempt for bio-oil production from wet *Spirulina platensis was done* under subcritical water conditions. In this trial, factorial design and response surface methodology were used.

The experimental design was performed to study the effect of the variables (time and temperature) on the process and interaction among variables, while the response was the high % yield by weight of produced biofuel.

Produced bio-crude oil was of higher heating value about 35.77 MJ kg^{-1} and the elemental composition of the liquefied *Spirulina platensis* was comparable to that of petroleum crude oil except for the oxygen and nitrogen content which were is higher in the bio-oil. Also, it was found that the model agreed well with the experimental data, and, within the experimental range considered, the most important factor is the time (Mohammady et al., 2015).

From the obtained results of this investigation, it is confirmed that hydrothermal liquefaction conversion of low lipid content algae into bio-crude oil could be quite promising for achieving effective algae biofuel.

The highest bio-oil yield (60% wt) was achieved at a reaction temperature of 340 °C with a reaction time of 30 minutes via hydrothermal liquefaction of *Spirulina platensis* (Mohammady et al., 2015).

It is recommended new experiments with different microalgae are needed to develop the influence of cell structure and composition on the process yield.

3.2 *Biodiesel Production*

3.2.1 From *Chlorella* Species

Due to its environmental benefits, biodiesel has become an attractive diesel fuel substitute. Technological assessment of biodiesel production from *Chlorella* sp. cultured in open saline water pond was carried out.

Technical and economical assessments were carried out to evaluate biodiesel production from *Chlorella* sp. and their technical benefits and limitations. For processing algal oils, the acid-catalyzed transesterification is the most technically feasible to overcome the complexity of the alkali-catalyzed process (Ashraf et al., 2015). Through continuous transesterification reaction, 97% conversion of triglycerides to methyl esters was achieved at a temperature of 80 °C and a pressure of 400 kPa. The fresh methanol stream, the recycled methanol stream, and the sulfuric acid stream were mixed before being pumped into the transesterification reactor (Dodds & Whiles, 2020). Distillation step is necessary to produce biodiesel (99.6% pure) and glycerol (86% pure), respectively.

Results of process design and preliminary economic evaluation revealed that the total capital investment for 10,000 tons biodiesel per year is $18,270,783 while the total manufacturing cost was $29,386,124 and the return on investment was 46.792% (Ashraf et al., 2015).

An investigation is carried out to develop a model to study the process kinetics for the supercritical transesterification reaction of algae in a well-mixed batch reactor (Ashraf et al., 2015). This kinetic study took into consideration two parameters to consider the effect of methanol-to-algae ratio and reaction time. Excellent fitting between the experimental results and model prediction is observed. The model shows that the optimum methanol-to-algae ratio and reaction time were 26 and 27 minutes, respectively. Complete conversion of triglycerides to biodiesel was achieved at 600 K (Melkert & van Blokland, 2011).

3.2.2 From *Spirulina platensis* Species

The biodiesel production technology using an acid catalyst to the in situ transesterification of microalgae (*Spirulina platensis*), where the main reaction variables that strongly affect the cost of this process were studied. These variables are (1) the catalyst concentration (the larger the catalyst concentration, the more the material costs input), (2) the reacting alcohol volume (also, the more the alcohol volume, the more the material costs input), (3) the temperature (increasing of temperature, and increasing the process of energy requirement), (4) the reaction time (the larger the reaction time, the lower the product amounts yielded and the lower the product profit), and (5) the process stirring (main energy requirement for the reaction agitation). This investigation has been carried out to provide information on the optimum operating conditions that give the best yield while also having the lowest material and energy requirements and, consequently, lowest process costs since the use of the in situ transesterification process as a proper biodiesel production technique is mainly driven by its possible application with relatively low cost.

Microalgae Used

Spirulina platensis microalgae used in biodiesel production were supplied from the Microbiology Department, Soils, Water, and Environment Research Institute, Agriculture Research Center (ARC), Giza, Egypt. This microalgal strain was collected at 3 weeks old. The culture medium used was the same as Zarrouk's medium (Bwapwa et al., n.d.). The cultivation of *Spirulina platensis* was in mini-tanks with dimensions similar to those used in Pelizer (Schlagermann et al., 2012). The cultivation was carried out at 30 °C, 3.5 k lux provided by fluorescent lamps, and a pH of 8.5 ± 0.5. At the end of the culture cycle, algal suspensions were homogenized (Homogenizer Wisetis® HG-15D) for 10 minutes at 1800 rpm, to disrupt the cells and make easier oil extraction, then filtered through centrifuge separator (Beckman CS-6 Centrifuge 3500 rpm, Germany) and finally dried to a constant weight using solar drying beds and stored at 18 °C until use. Sulfuric acid of 98% purity was used as a catalyst in the transesterification process. Methanol (99.9% purity) was used as the reacting alcohol in this study.

Method

Dried microalgae were added carefully to catalyst/alcohol mixture and blended on low setting for several minutes. At this point, the simultaneous extraction and transesterification reaction have been initiated, where the catalyst/alcohol solution attacked the triglyceride (oil) in the microalgal strain and cleaved off a fatty acid chain. The vessels containing the reaction mixtures were then heated and maintained at the temperatures of interest for specified periods.

After the transesterification step, the warm reaction mixture was allowed to cool. The reaction mixture was filtered and the residues were washed three times by resuspension in methanol (45 ml) to recover any traces of FAME product left in the residues. Water was added to the filtrate, to facilitate the separation of the hydrophilic components of the extract, and then poured into a separating funnel. Two layers are observed: hydrophobic layer (hexane, FAME, and glycerides) and hydrophilic layer (water, glycerol, and excess methanol). The top layer in the separation funnel is the produced biodiesel.

Variables Affecting the In Situ Transesterification Process

Effect of Alcohol Volume

One of the most important variables affecting the yield of methyl esters is the molar ratio of alcohol to triglycerides. The stoichiometric ratio for transesterification requires three moles of alcohol and one mole of triglycerides to yield three moles of fatty acid methyl esters and one mole of glycerol. However, transesterification is an equilibrium reaction in which an excess of alcohol is required to drive the reaction to the right (Elmoraghy & Farag, 2012; Fortier et al., 2014).

Hassan et al. (IHI Corporation, 2019) indicated an improvement of the microalgal oil conversion to FAME with increasing alcohol volume, with the lowest FAME equilibrium conversion observed with the reacting molar ratio of the methanol to oil at 1857:1 (a methanol volume of 40 ml for the used 15 gm of microalgae which contains a total lipid content of 10.95% wt. of *Spirulina platensis* biomass). However, with the use of alcohol volumes over 80 ml (i.e., a reacting molar ratio of alcohol to microalgal oil greater than 3714:1) for the in situ transesterification of 15 g microalgal biomass, no significant trends were observed for the FAME yields. Equilibrium conversion was obtained using a fixed reaction time of 8 h, a temperature of 65 °C, and a fixed acidic catalyst molar concentration (0.0154 mol. sulfuric acid) at a constant stirring rate of 650 rpm.

Effect of Catalyst Concentration

One of the most important variables affecting the yield of FAME is the concentration of the acid catalyst. The in situ transesterification with 100% (wt/wt of oil) using sulfuric acid catalyst resulted in the successful conversion of *Chlorella* oil giving the best yields and viscosities of the FAME by E.A. Ehimen et al. (Xu et al., 2018).

Effect of Reaction Time and Temperature

To investigate the influence of reaction time and temperature, a methanol volume of 80 ml was used. Reactions were carried out at different temperatures of 27 °C up to 65 °C using a methanol-to-oil molar ratio of 3714:1, a catalyst concentration of 100% (wt./wt. oil), and a constant stirring rate of 650 rpm. The progress of the microalgal oil to biodiesel conversion process is shown in Fig. 21.4 at different temperature levels. For the samples investigated at room temperature (no process heating), asymptotic FAME conversion value was not reached within the time boundaries of this study. When the in situ transesterification process was carried out at 65 °C, higher equilibrium conversion levels of FAME of 43.1% and 76.22% were attained after a reaction time of 2 hours and 4 hours, respectively. This could be due to the fact that the elevated temperatures improve the initial miscibility of the reacting species, leading to a significant reduction in the reaction time, as observed in Fig. 21.4. Within the investigated experimental conditions, the equilibrium of FAME conversions was observed to reach similar asymptotic values after a reaction time of 8 and 10 hours for temperatures of 50 and 65 °C. Although faster conversion rates could be observed by the use of reaction temperatures greater than the boiling point of the reacting methanol (e.g., 90 °C), the process heating and pressure

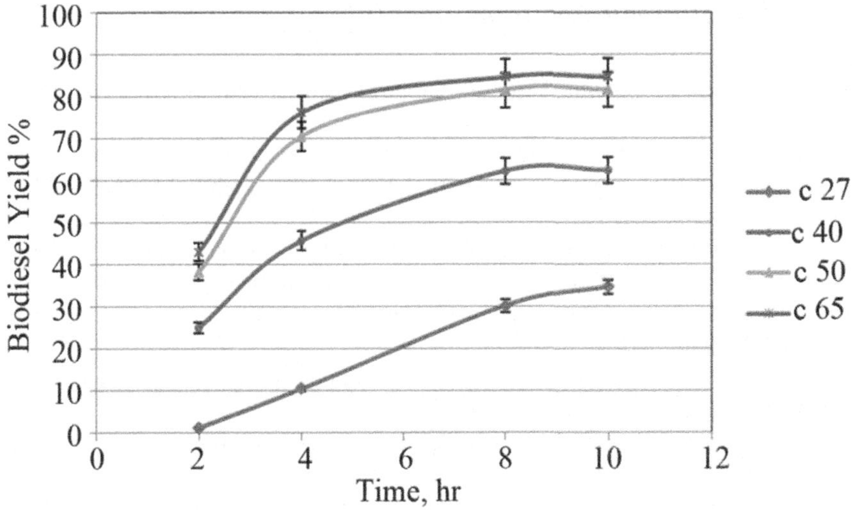

Fig. 21.4 Effect of reaction time at different temperatures on FAME yield % (methanol-to-oil 3714:1 molar ratio, 100% wt. H_2SO_4 with stirring at 650 rpm)

requirements may inhibit the use of such temperature levels. The use of a reaction temperature of 65 °C may, therefore, prove more beneficial, if we consider the total energy consumption and operation cost of the whole biodiesel conversion system.

Effect of Stirring

The effect of stirring on the in situ transesterification process (IHI Corporation, 2019) was performed as a potential process performance strategy. When the in situ transesterification process was conducted without stirring, no reaction would be obtained, and zero conversion of the microalgal oil content to biodiesel is obtained, compared to that for the continuously stirred sample. This indicates that stirring is required to enhance the reaction progress, evidently by aiding the initial miscibility of the reactants.

Quality Assessment of Produced Biodiesel

Once biodiesel is obtained, a series of tests were conducted to establish some properties of the produced biodiesel from microalgae. Viscosity, density, flash point, cold flow properties, and cetane number for produced biodiesel in optimum conditions would be measured by EN methods. The obtained values are comparable with the EN 14214 standards.

3.3 Bio-Jet Production from Algal Oil

Bio-jet fuel can be generated from different feedstocks with different conversion processes.

Bio-jet fuel can be produced from the macro- or microalgal oil (Ashworth, 2006; Cheng et al., 2019c). Algal oil composition indicates its suitability for fuel production (Cheng et al., 2018).

Generally, algal oil quality is suitable for biofuel generation (Cheng et al., 2018) due to its high content of polyunsaturated fatty acids. Biofuels produced from algae show good flow properties even at low temperatures which make it suitable for aviation fuel (Li et al., 2015).

Algae can be cultivated in freshwater that contains nutrients, salty water, wastewater streams, and reverse osmosis water.

A comparison between algae growing in wastewater and reverse osmosis water was carried out; it was found that medium reverse osmosis water was more effective in growing alga especially *Chlorella vulgaris* (Andrade & Costa, 2006).

Microalgae cultivated in wastewater effluent are a suitable feedstock for biofuel production (Pelizer et al., 2001).

The type of feedstocks and production process affect the life cycle assessment of prepared bio-jet (Xu et al., 2018). Oil yield per unit area produced from algae is higher than other plants, so algae are considered as a promising feedstock for biofuel derivation (Refaat, 2009).

Bio-jet fuel can be produced by a hydrothermal liquefaction process followed by catalytic upgrading (Meng et al., 2008).

Life cycle assessment results were highly affected by the source of heating and the heat integration in the hydrothermal liquefaction process and by the solid content in the dewatered algae.

Greenhouse gas emissions from algal bio-jet were reduced by 76% compared to those of conventional jet fuel, (Ashworth, 2006).

Hydrothermal liquefaction was performed at 350 °C for 1 hour. Microalgae produce more bio-oil than macroalgae due to its high lipid content – limacine gave the maximum crude bio-oil yield of 54.42 wt% (Pelizer et al., 2001). Microalgae that contain high nitrogen produce bio-oil with higher nitrogen content, and the contrary is observed in macroalgae (Pelizer et al., 2001). Algal oil generated from both types of algae is characterized by its high viscosity and acidity. So treatment of both oils at 400 °C in the presence of hydrogen for 2 hours with 10% catalyst (Ru/C) produces upgraded bio-oils with higher energy densities and significantly lower N, O, and S contents and viscosities than their corresponding crude bio-oils (Refaat, 2009).

Hydro-processing technologies used a catalyst to get rid of impurities or to reduce the molecular weight of the products. Algal oils could be converted to kerosene very similar to petroleum jet fuel for different purposes uses (Refaat, 2009).

3.3.1 Catalyzed Conversion of Microalgal Biodiesel into Jet Fuel

To produce jet fuel range hydrocarbons long-chain (FAMES) derived from microalgae, the process should accomplish oxygen removal and carbon chain degradation. This process involves the hydro-deoxygenation (HDO) step catalyzed by metallic active sites and hydrocracking step catalyzed by acid active sites.

Most of these catalysts are noble metals such as Pt and Pd. The non-noble transition metals such as Ni and Co are more abundant and cost-effective in comparison to noble metals (Meng et al., 2008).

The Ni-based sulfonated meso-Y zeolite bifunctional catalyst could efficiently catalyze the conversion of microalgae biodiesel to jet fuel range hydrocarbons. The meso-Y zeolite was sulfonated as reported in El-Shimi et al. (2013). Then Ni in the form of 2.97 g Ni $(No_3)_2$ was loaded on the sulfonated meso-Y zeolite by wetness impregnation method. $6H_2O$ and 5.4 g sulfonated meso-Y zeolite were added into 20 ml deionized water, stirred for 5 h, and dried at 80 °C for 8 h. Then the mixture was ground to powder, calcined at 550 °C for 4 h, and reduced at the ambient of hydrogen at 500 °C for 4 h (10% Ni/sulfonated meso-Y zeolite catalyst was eventually prepared) (Ehimen et al., 2010). The catalytic hydro-processing was carried in a fixed continuous flow reaction system as illustrated in Fig. 21.5 (Ehimen et al., 2010).

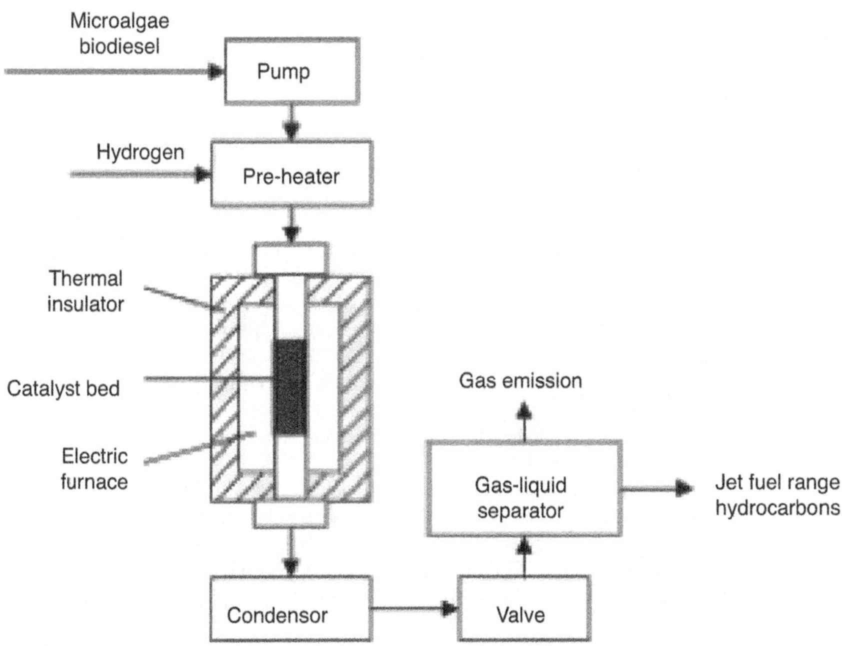

Fig. 21.5 Conversion of microalgae biodiesel to jet fuel range hydrocarbon

4 Conclusion

Biofuels such as biodiesel and bio-jet fuel are important sources of renewable energy, which will, in the near future, replace petroleum fuels. Biofuels, in general, can be produced from several raw agriculture materials and recently from microalgae. Currently, the most important biofuel is bio-jet fuel, as a partial replacement of jet fuel of petroleum origin. The most promising method for bio-jet fuel manufacture is the catalytic hydrocracking of a low-cost nonedible vegetable oil.

The type of product and its quality depend on the process conditions such as reaction time, temperature, and the catalyst type, form, and amount. Heterogeneous catalysts play an important role in this process. The production of the lower biofuel cost conforming to ASTM specifications can be achieved by the selection and preparation of an economic inorganic catalyst and a low oil cost which can be performed by previous extraction of valuable products from micro-algal oil and then use the remaining oil for biofuel production.

References

Abdel Kader E, Hussein Nabila HS, Hussien H, El Diwani G, Hawash SI (2015) "Effect of Extractive Solvents on Bio – oil Production from Microalgae via Hydrothermal Liquefaction," Chemical and Process Engineering Research, Vol. 38, 1–7.

Abdel Kader E, Abo El-Enin SA, El-Ibiari NN, El-Diwani G, Hawash SI (2015) "Hydrothermal Liquefaction of Microalgae (Spirulina Platensis) under Subcritical Water Conditions for Biofuel Production," International Journal of Engineering and Innovative Technology (IJEIT) 4, 12:48–55.

Abou-Zeid AE, El-Damarawy SZ, Mariey YA, El-Mansy MM (2015) Effect of using spirulina platensis and/or *Chlorella vulgaris* algae as feed additives on productive performance of broiler chicks., J. Animal and Poultry Prod., Mansoura Univ 6 (10): 623 – 634.

Abouelezz FMK (2017) Evaluation of spirulina algae (*Spirulina platensis*) as a feed supplement for Japanese quail: nutritional effects on growth performance, egg production, egg quality, blood metabolites, sperm-egg penetration and fertility. Egypt. Poult. Sci. Vol (37)(III): 709–721.

Abu Yousuf (2020) Chapter 1 - Fundamentals of Microalgae Cultivation, Editor(s): Abu Yousuf, Microalgae Cultivation for Biofuels Production, Academic Press, pp: 1–9, ISBN 9780128175361, https://doi.org/10.1016/B978-0-12-817536-1.00001-1.

Al Khawli F, Marti-Quijal FJ, Ferrer E, Ruiz MJ, Berrada H, Gavahian M, Barba FJ, Beatriz de la Fuente (2020) Aquaculture and its by-products as a source of nutrients and bioactive compounds In: Advances in Food and Nutrition Research, Academic Press, ISSN 1043-4526, https://doi.org/10.1016/bs.afnr.2020.01.001.

Alves C.M., Valk M., de Jong S., Bonomi A., Van der Wielen L.A.M, Mussatto S.I. (2017) Techno-economic assessment of biorefinery technologies for aviation biofuels supply chains in Brazil. Biofuels Bioprod. Bioref. 11:67–91, https://doi.org/10.1002/bbb.1711

Ashraf Amin, S. A. AboEl-Enin, G. El Diwani and S. Hawash "Kinetic Model of Algal Biodiesel Production under Supercritical Methanolysis." Journal of Engineering and Applied Sciences 10,1, January (2015).

Asiedu A, Davis R, Kumar S (2020) Catalytic transfer hydrogenation and characterization of flash hydrolyzed microalgae into hydrocarbon fuels production (jet fuel). Fuel. 261:116440.

Andrade MR, Costa JAV (2006) "Mixotrophic Cultivation of Microalga Spirulina platensis Using Molasses as Organic Substrate," Biochemical Engineering Laboratory, Department of Chemistry, Federal University Foundation of Rio Grande (FURG), Rio Grande.

Attia AMA, Nour M, Nada SA (2018) Study of Egyptian castor biodiesel-diesel fuel properties and diesel engine performance for a wide range of blending ratios and operating conditions for the sake of the optimal blending ratio, Energy Conversion and Management, 174: 364–377, ISSN 0196-8904, https://doi.org/10.1016/j.enconman.2018.08.016.

Baviskar JW, Khandelwal SR, Extraction, Detection and Identification of Flavonoids From Microalgae: An Emerging Secondary Metabolite, Int.J.Curr.Microbiol.App.Sci 2015; Special Issue-2: 110–117.

Becker E.W. Micro-algae as a source of protein, Research review paper, online 23 November 2006.

Bueno Ariede M, Marcílio Candido T, Morocho Jacome AL, Robles Velasco MV, de Carvalho JCM, Baby AR (2017) Cosmetic attributes of algae - A review, Algal Research, 25: 483–487

Chang CH, Wei HY, Bo-Yu Chen BY, Tan CS (2020) In situ catalyst-free biodiesel production from partially wet microalgae treated with mixed methanol and castor oil containing pressurized CO_2. The Journal of Supercritical Fluids. 157:104702, https://doi.org/10.1016/j.supflu.2019.104702.

Charoensiddhi S, Abraham RE, Su P, Zhang W (2020) Chapter Four - Seaweed and seaweed derived metabolites as prebiotics In: Advances in Food and Nutrition Research, Editor(s): Toldrá F, Academic Press, Vol. 91, pp. 97–156, ISSN 1043-4526, ISBN 9780128204702, https://doi.org/10.1016/bs.afnr.2019.10.001.

Cheng J, Zhang Z, Zhang X, Fan Z, Liu J, Zhou J, Cen K (2018) Sulfonated mesoporous Y Zeolite with nickel to catalyze hydrocracking of micro algae biodiesel to get fuel range hydrocarbon. 310027, China.

Cheng J, Zhang Z, Zhang X, Fan Z, Liu J, Zhou J (2019a) Continuous hydroprocessing of microalgae biodiesel to jet fuel range hydrocarbons promoted by Ni/hierarchical mesoporous Y zeolite catalyst. International journal of hydrogen energy, 44, 11765–11773.

Cheng J, Zhang Z, Zhang X, Fan Z, Liu J, Zhou J, Cen K (2019b) Sulfonated mesoporous Y zeolite with nickel to catalyze hydrocracking of microalgae biodiesel into jet fuel range hydrocarbons. International journal of hydrogen energy, 44, 1650–1658.

Cheng J, Zhang Z, Zhang X, Fan Z, Liu J, Zhou J, (2019c) Continuous hydroprocessing of microalgae biodiesel to jet fuel range hydrocarbons promoted by Ni/hierarchical mesoporous Y Zeolite catalyst. 272119, China.

Christaki E, Florou-Paneri P and Bonos E. Microalgae: A novel ingredient in nutrition. Int J Food Sci Nutr 2011; 62:794–799.

Christaki E, Bonos E, Giannenas I, Florou-Paneri P (2013) Functional properties of carotenoids originating from algae. J Sci. Food Agric 93:5–11.

Craggs R, Park J, Heubeck S, Sutherland D (2014) High rate algal pond systems for low-energy wastewater treatment, nutrient recovery and energy production, New Zealand Journal of Botany.

De Jesus Raposo M. F., De Morais R. M. S. C., and De Morais A. M. M. B., Health applications of bioactive compounds from marine microalgae, Life Sciences, 2013; vol. 93, no. 15, pp. 479–486.

De Jong S, Hoefnagels A, Faaij R, Slade R, Mawhood R, Junginger M (2015) The feasibility of short-term production strategies for renewable jet fuels – a comprehensive techno-economic comparison, Biofuels Bioprod. Bioref. 9 (6) :778–800

Dineshkumar, R, Subramanian J, Arumugam A, Rasheeq A.A, Sampathkumar, P (2018). Exploring the Microalgae Biofertilizer Effect on Onion Cultivation by Field Experiment. Waste and Biomass Valorization. https://doi.org/10.1007/s12649-018-0466-8.

Dodds WK, Whiles MR (2020) Chapter 9 - Microbes and Plants, In Aquatic Ecology, Freshwater Ecology (Third Edition), Editor(s): Dodds WK, Whiles MR, Academic Press, pp: 211–249, ISBN 9780128132555, https://doi.org/10.1016/B978-0-12-813255-5.00009-0.

D'Alessandro EB, Filho NRA. Concepts and studies on lipid and pigments of microalgae: A review. Renew. Sustain. Energy Rev. 2016; 58:832–841.

Dubois, M., Gilles, K.A., Hamilton, J.K., Rebers PA and Smith, F. Colorimetric method for determination of sugars and related substances. 1956; 28:350–356.
Ehimen EA, Sun ZF, Carrington CG (2010) "Variables Affecting the in Situ Transesterification of Microalgae Lipids," El-Sevier Fuel, Vol. 89, No. 3, pp. 677–684.
El-Mekkawi SA, Abdo SA, Samhan FA, Ali GH (2019a) Optimization of some fermentation conditions for bioethanol production from microalgae using response surface method, Bull Natl Res Cent. 43: 164. https://doi.org/10.1186/s42269-019-0205-8.
El-Mekkawi SA, Hussein HS, El-Enin SA, El-Ibiari NN (2019b) Assessment of stress conditions for carotenoids accumulation in Chlamydomonas reinhardtii as added-value algal products, Bull Natl Res Cent. 43: 130. https://doi.org/10.1186/s42269-019-0169-8
El-Semary NA (2018) Algae and chain aquaculture: An approach towards sustainable agriculture. In: Sustainability of agricultural environment in Egypt: Part II Water-Soil-Plant Nexus. (eds) Negm AM, Abu-hashim M. Springer. pp. 311–316.
El-Shimi HI, Nahed K. Attia, El-Sheltawy ST, El-Diwani GI (2013) " Biodiesel Production from Spirulina-Platensis Microalgae by In-Situ Transesterification Process", Journal of Sustainable Bioenergy Systems, Vol. 3, 224–233.
Elshobary M E, Shady AM, Khairy HM, Essa D, Zabed HM, Qi X, Abomohra A (2019) Influence of nutrient supplementation and starvation conditions on the biomass and lipid productivities of Micractinium reisseri grown in wastewater for biodiesel production. Journal of Environmental Management 250: 109529
Friedman DS, O'Colmain BJ, Mũoz B, Tomany SC, McCarty C, De Jong PT, et al, Prevalence of age-related macular degeneration in the United States. Arch Ophthalmol,2004; 122:564–572.
Galasso C, Gentile A, Orefice I, Ianora A, Bruno A, Noonan DM, Clementina Sansone C, Albini A, Brunet C (2019) Microalgal Derivatives as Potential Nutraceutical and Food Supplements for Human Health: A Focus on Cancer Prevention and Interception, *Nutrients*, *11*(6), 1226; https://doi.org/10.3390/nu11061226
Goiris, K., Muylaert, K., Fraeye, I, Foubert, I., de Brabanter, J., de Cooman, L. Antioxidant potential of microalgae in relation to their phenolic and carotenoid content. J. Appl. Phycol. 2012; 24, 1477–1486.
Guedes AC, Amaro HM and Malcata FX, Microalgae as sources of carotenoids. Mar Drugs 2011; 9:625–644
Haidy S. Basily, Mamdouh M. Nassar, Guzine I. El-Diwani, Sanaa A. Abo El-Enin, Exploration of using the algal bioactive compounds for cosmeceuticals and pharmaceutical applications, Egyptian Pharmaceutical Journal, Vol. 17 No. 2, May–August 2018a
Haidy S. Basily, Mamdouh M. Nassar, Guzine I. El Diwani, Sanaa A. Abo El-Enin, Extraction of algal lipid as a natural cosmetic component, Egyptian Pharmaceutical Journal, Vol. 17 No. 1, January-April 2018b.
Hala S Doma, Sayeda M Abdo, Rehab H Mahmoud, SA El-Enin and G El-Diwani, Production and characterization of biodiesel from microalgae cultivated in municipal wastewater treatment plant, Research Journal of Pharmaceutical, Biological and Chemical Sciences, March–April, 7(2) Page No. 1912, 2016.
Hanaa H. Abd El-Baky, Farouk K. El Baz, and Gamal S. El-Baroty. Phenolics from Spirulina maxima: Over-production and in vitro protective effect of its phenolics on CCl4 induced hepatotoxicity. Journal of Medicinal Plants Research January 2009 ; Vol. 3(1), pp. 024–030.
IHI Corporation "Development of sustainable bio jet fuel derived from Microalgae." NEDO-ADEME Workshop 12. March (2019) Tokyo.
John Ashworth, "Jet Fuel from Micro-algal Lipids." National Renewable Energy Laboratory, NREL (2006).
Joris Melkert, Wouter Beelaerts van Blokland "Bio jet fuel from macro algae." https://www.researchgate.net/publication/275339624, February 2011.
Joseph K. Bwapwa, Akash Anandraj, Cristina Trois "Microalgae processing for jet fuel production." Society of Chemical Industry and John Wiley & Sons, Ltd. 20-18.

Khairy H, Shaltout K, El-Sheekh M, Eassa D (2015). Algal diversity of the Mediterranean Lakes of Egypt. International Conference on Advances in Agricultural, Biological & Environmental Sciences (AABES-2015) July 22-23, London (UK).

Kraan, S., Algal polysaccharides, novel applications and outlook, In: Chang, C.-F. (Ed.), carbohydrates—Comprehensive Studies on Glycobiology and Glycotechnology, InTech, Rijeka, Croatia 2012; pp. 489–532.

Kumar, Singh Gill B, Verma ML, A Verma, Rekha Kushwaha (2020) Chapter 7 - Biotechnological production of high-valued algal astaxanthin and lutein under different growth conditions, Editor(s): Madan L. Verma, Anuj K. Chandel, Biotechnological Production of Bioactive Compounds, Elsevier, pp 191-220, ISBN 9780444643230, https://doi.org/10.1016/B978-0-444-64323-0.00007-2.

Le M and Xiao-Ming L. Effects of lutein and zeaxanthin on aspects of eye health. J Sci Food Agric 2010; 90:2–12

Li T, Cheng J, Huang R, Zhou J, Cen K, Conversion of waste cooking oil to jet fuel with Nickel-based mesoporous Zeolite Y catalyst Bioresour Technol. (2015), 197:289–94.

Makareviciene V, Skorupskaite V (2019) Transesterification of microalgae for biodiesel production. In: Second and Third Generation of Feedstocks, eds: Basile A, Dalena F Elsevier. pp 469–510, ISBN 9780128151624, https://doi.org/10.1016/B978-0-12-815162-4.00017-3.

Manivannan K, Anantharaman P., Balasubramanian T. Evaluation of antioxidant properties Kadam, S. U., Tiwari, B. K., O'Donnell, C. P., Application of novel extraction technologies for bioactives from marine algae. J. Agric. Food Chem. 2013; 61, 4667–4675.

Marian Elmoraghy, Ihab H. Farag "Bio-jet Fuel from Microalgae: Reducing Water and Energy Requirements for Algae Growth." International Journal of Engineering and Science Vol. 1, Issue 2 (Sept 2012), PP. 22–30.

Marie-Odile P. Fortier, Griffin W. Roberts, Susan M. Stagg-Williams, Belinda S.M. Sturm "Life cycle assessment of bio-jet fuel from hydrothermal liquefaction of microalgae." Applied Energy 122 (2014) 73–82.

Massoumeh, F., Ramazan-Ali, K., Seyed, M., Foroogh, N Antioxidant activity, total phenolics and flavonoid contents of some edible green seaweeds from Northern Coasts of the Persian Gulf. Iran. J. Pharm. Res. 2014; 1:163 170.

Massoumeh Farasata, Ramazan-Ali Khavari-Nejada, Seyed Mohammad Bagher Nabavib and Foroogh Namjooyanc Antioxidant Activity, Total Phenolics and Flavonoid Contents of some Edible Green Seaweeds from Northern Coasts of the Persian Gulf, Iranian Journal of Pharmaceutical Research 2014; 13 (1): 163–170.

Meng Z, Chen G, Wang Y (2008) "Biodiesel Production from Waste Cooking Oil via Alkali Catalyst and Its Engine Test," Fuel Processing Technology, Vol. 89, No. 9, pp. 851–857. doi: https://doi.org/10.1016/j.fuproc.2008.02.006.

Mohammady N, Hawash SI, El-Khatib KM, El-Galad MI, El Diwani G (2015) "Biodiesel production from Chlorella Sp: Process Design and Preliminary Economic Evaluation," International Journal of ChemTech Research, ISSN: 0974-4290 Vol. 8, No. 9, pp. 297–304.

Molino A, Iovine A, Casella P, Mehariya S, Chianese S, Cerbone A, Rimauro J, Musmarra D (2018) Microalgae Characterization for Consolidated and New Application in Human Food, Animal Feed and Nutraceuticals Int. J. Environ. Res. Public Health, 15(11): 2436; https://doi.org/10.3390/ijerph15112436

Mulders, K.J.M., Lamers, P.P., Martens, D.E., Wijffels, R.H. Phototrophic pigment production with microalgae: Biological constraints and opportunities. J. Phycol. 2014;50, 229–242.

Nauman Ahamad M., Saleemullah M., Hamid Ullah Shah, Iqtidar A. Khalil and A.U.R. Saljoqi. DETERMINATION OF BETA CAROTENE CONTENT IN FRESH VEGETABLES USING HIGH PERFORMANCE LIQUID CHROMATOGRAPHY, Sarhad J. Agric. 2007; Vol. 23, No. 3.

Negm NA, Sayed GH, Habib OI, Yehia FZ, Mohamed EA (2017) Heterogeneous catalytic transformation of vegetable oils into biodiesel in one-step reaction using super acidic sulfonated modified mica catalyst, Journal of Molecular Liquids, 237: 38–45, ISSN 0167-7322, https://doi.org/10.1016/j.molliq.2017.04.076.

Ortiz-Martínez VM, Andreo-Martínez P, García-Martínez N, Pérez de los Ríos A, Hernández-Fernández FJ, Quesada-Medina J (2019) Approach to biodiesel production from microalgae under supercritical conditions by the PRISMA method. Fuel processing technology 191: 211–222.

Pelizer LH, Danesi EDG, de Rangel CO, Sassano CEN, Carvalho JCM, Sato S, Moraes IO (2001) "Influence of Inoculum Age and Concentration in Spirulina platensis Cultivation," Departamento de Tecnologia Bio- quımico-Farmaceutica, Faculdade de Ciencias Farmaceuticas, Universidade de Sao Paulo, Sao Paulo.

Rahul V, Agrawal P, Sharma M, Shukla S (2016) Total phenolics, flavonoids and antioxidant potential of organic extract of fresh water algal sample collected from a marine lake, Indian Journal of Geo-Marine Sciences, Vol. 45(10), pp. 1320–1326.

Rastogi RP, Pandey A, Larroche C, Madamwar D (2018) Algal Green Energy – R&D and technological perspectives for biodiesel production, Renewable and Sustainable Energy Reviews, 82 (3) 2946–2969, ISSN 1364-0321, https://doi.org/10.1016/j.rser.2017.10.038.

Refaat AA (2009) "Optimization of Biodiesel Production Using Different Techniques," Master Thesis, Cairo University, Cairo.

Renju G. L., Kurup G. M., and Saritha K. C. H. Effect of lycopene from Chlorella marina on high cholesterol-induced oxidative damage and inflammation in rats, Inflammopharmacology, 2014; vol. 22, no. 1, pp. 45–54.

Safafar H, van Wagenen J, Møller P, Jacobsen C (2015) Carotenoids, Phenolic Compounds and Tocopherols Contribute to the Antioxidative Properties of Some Microalgae Species Grown on Industrial Wastewater, Mar. Drugs 13, 7339–7356. https://doi.org/10.3390/md13127069

Sarojini Y, Sujatha B. and Lakshminarayana K. Total phenol content and antioxidant activities of ethanolic extracts of two marine brown macroalgae, INT J CURR SCI 2013; 8: E43–49, ISSN 2250-1770.

Schlagermann P, Göttlicher G, Dillschneider R, Rosello-Sastre R, Posten C "Composition of Algal Oil and Its Potential as Biofuel." Hindawi Publishing Corporation, Journal of Combustion, Volume 2012, Article 14 pages.

Shabaka SH (2018) Checklist of seaweeds and seagrasses of Egypt (Mediterranean Sea): A review. The Egyptian Journal of Aquatic Research, 44(3): 203–212.

Sharawy ZZ, Ashour M, Abbas E, Ashry O, Helal M, Nazmi H, Kelany M, Kamel A Hassaan, Rossi Jr W, El-Haroun E, Goda A (2020) Effects of dietary marine microalgae, Tetraselmis suecica, on production, gene expression, protein markers and bacterial count of Pacific white shrimp Litopenaeus vannamei. Aquaculture Research, 1-13. DOI: https://doi.org/10.1111/are.14566

Sharma AK, Sahoo PK, Singhal S, Joshi G (2016) Exploration of upstream and downstream process for microwave assisted sustainable biodiesel production from microalgae Chlorella vulgaris, Bioresource Technology, 216: 793–800, https://doi.org/10.1016/j.biortech.2016.06.013.

Shashank, K., Abhay, K. Chemistry and biological activities of flavonoids: an overview. Sci. World J., 2013; 1: 1 16.

Solar gis. Free Download of Solar Radiation Maps: Direct Normal Irradiation (DNI). 2010. Available at: http://solargis.info/doc/free-solar-radiation-maps-DNI. Accessed June 4, 2015.

Suparmaniam SH, Lam MK, Uemura Y, Lim JW, Lee KT, Shuit SH (2019) Insights into the microalgae cultivation technology and harvesting process for biofuel production: A review. Renewable and Sustainable Energy Reviews. 115: 109361

Taher H, Giwa A, Abusabiekeh H, Al-Zuhair S (2020) Biodiesel production from Nannochloropsis gaditana using supercritical CO_2 for lipid extraction and immobilized lipase transesterification: Economic and environmental impact assessments, Fuel Processing Technology, 198:106249, https://doi.org/10.1016/j.fuproc.2019.106249.

Vilchez C, Forjan E, Cuaresma M, Bedmar F, Garbayo I and Vega JM. Marine carotenoids: Biological functions and commercial applications. Mar Drugs 2011; 9:319–333.

Wadood A, Rana A, Basheer C, Razzaq SA, Farooq W (2019) In situ Transesterification of Microalgae Parachlorella kessleri Biomass Using Sulfonated Rice Husk Solid Catalyst at Room Temperature. Bioenerg. Res. https://doi.org/10.1007/s12155-019-10060-3

Yu-Ping Xu, Pei-Gao Duan, Feng Wang and Qing Qing Guan " Liquid fuel generation from algal biomass via a two-step process: effect of feedstocks." Biotechnol. Biofuels (2018) 11:83.

Zhen Li, Mingzhe Sun, Qiyu Li, Aifen Li, Chengwu Zhang. Profiling of carotenoids in six Microalgae (Eustigmatophyceae) and assessment of their β-carotene productions in bubble column photobioreactor, Biotechnol Lett. 2012; 34:2049–2053, DOI https://doi.org/10.1007/s10529-012-0996-2.

Part VII
Bioeconomic Catalysts of Socio-Economic Development

Chapter 22
Bioeconomy Education

George Sakellaris

Abstract Bioeconomy tends to become a key technology and a substantial element for development and growth in Europe. It is based on a novel concept including the valorization of natural resources and human manpower in a sustainable way. Multidisciplinary and Cros-Sectorial consideration are the main characteristics of the whole bioeconomy frame, meaning that a wide spectrum of scientific disciplines and technical expertise are required. Adding to this the need for appropriate regulations and the expertise for necessary finances, it is obvious to conclude that a new generation of experts adapted to the current requirements is mandatory. These experts will have also to adapt to the changing dynamics of the current business and market needs. A dedicated bioeconomy education means to satisfy these emerging needs and to prepare this new skilled generation of working force. Many higher education institutions across Europe are facing these challenges, and they are adapting their programmes accordingly. However, a careful analysis of the educational aspect is imperative considering a wide number of lateral parameters such as the regional and sectorial particularities, the concrete detection of the target groups to be educated, the alliance of the educational curricula with socioeconomic priorities, the expected impacts and the potentially raising technical issues.

Keywords Bioeconomy · Circular economy · Education · Training · TVET (Technical and Vocational Education and Training) · Regional · Sectorial

1 An Overview

Since the early 2000s, bioeconomy became a key factor in the EU's development agenda, including three main priorities:

- A new boost for jobs, growth and investment
- A resilient Energy Union with a forward-looking climate change policy
- A deeper and fairer internal market with a strengthened industrial base

G. Sakellaris (✉)
Bioeconomy Bio-Hub of the Czech Republic, Prague, Czech Republic

E. Koukios, A. Sacio-Szymańska (eds.), *Bio#Futures*,
https://doi.org/10.1007/978-3-030-64969-2_22

Fifteen years later, when the 17 Sustainable Development Goals (SDGs) were globally adopted, bioeconomy was the key element in at least 11 of them.

In order to achieve these goals, a concrete strategy needs to be adopted. This strategy not only includes concrete technical steps resulting into economic outcomes but also includes significant societal challenges:

- Sustainably delivering and recycling an increasing amount of biomass to feed a growing world population and substituting chemicals, plastics and fuels for a post-petroleum, low-carbon economy
- Maintaining biodiversity and soil fertility and exploiting new biomass sources, especially aquatic sources and waste streams
- Producing foods that neither harm our health nor that of our planet
- Meeting the CO_2 emission reduction targets set by COP21
- Boosting industrial competitiveness and maintaining jobs in rural and coastal economies

Therefore, the new European economic strength and growth are building based on a new concept including the valorization of natural resources and human manpower in a sustainable way. Those are the concepts of bioeconomy, bio-based economy and circular economy, which they become substantial elements for the future development. In the forthcoming years, more and more processes will be designed on these principles. The current dynamics in business and market requirements are changing. Therefore, it is demanded a new generation of skilled working force, educated according to these needs.

- This education presupposes multidisciplinarity in a wide spectrum of topics, so the competent teams can gradually be complemented by skilled individuals.
- The educational perspective on bioeconomy is aligned with European priorities for growth, offering a good potential for improvement in all levels (technological, economic, social and regulatory).
- Considering the national and regional priorities, the educational programmes are flexible enough and adjustable on a case-by-case basis according to specific demands and conditions.
- The educational curricula consider the development of soft skills in addition to disciplinary knowledge.
- Further education and lifelong learning be considered for the educators according to the new requirements.

These axes are in accordance with the current policies such as (1) the reviewed Bioeconomy Strategy adopted by the EC (October 2018) and (2) the European Stakeholders Manifesto published in November 2017.

In a general frame, besides the main priorities and perspectives, the educational action aims not only to substantially contribute to the specific disciplines of bioeconomy but also will add to the European perspective in institutionalizing the initiative, establishing an educational operation without borders, enhancing networking and mobility and removing obstacles to EU-wide recognition of higher education diplomas and lifelong learning certificates.

Recently, the European Commissioner for jobs, growth, investment and competitiveness, Jyrki Katainen, set out several possible paths for the bloc's education future saying that "By 2025 we should live in a Europe in which learning, studying and doing research is not hampered by borders, but where spending time in another member state to study, learn or work is the norm".

Bioeconomy is a multidisciplinary and Cros-Sectorial concept that requires a concerted cooperation of a large number of scientific disciplines and stakeholders, accompanied by political and economic decision-making processes. In this scenario, the new generation of scientists and managers becomes indispensable, possessing an educational background capable of covering the wide spectrum of this multidisciplinary field. Therefore, the establishment of a specialized education and training concepts become a necessity, and many higher-education institutions are already working towards this effort. Options for greater cooperation on education include creating a new network of European universities, so that they can work seamlessly together across borders. This perspective seems more and more achievable, underlined by several university networks recently established in Europe which contribute into the vocational education on bioeconomy at national or European scales, and helping Europe to become a world pioneer in this domain.

Presently, one of the key concerns of almost all bioeconomy-related policy strategies remains the development of expertise and institutional capacity. Bioeconomy is complex and involves interdisciplinary knowledge. Some countries, like Finland, France or Italy, foster the training of experts, while others, like Austria, Germany and Sweden, also promote stakeholder platforms and Cros-Sectorial alliances as the basis for interdisciplinary exchange. In Europe, only few countries have adopted specific measures related to the improvement of education on bioeconomy in schools and universities. At this point, it is essential to underline the regional character of bioeconomy. It is important to consider the regional context in order to implement potential synergies and coexistence as well as coordinated strategies in a frame of a rather homogeneous ground from economic, social, technological, environmental and even cultural background. In this perspective, educational needs, coordinated actions, information flow, training and required expertise should be tailored according to the principles of the region. (Barrows 1996; Barthel et al. 2013; Bergeret et al. 2018; Bowden & Marton 1998; Burssens et al. 2018; De Besi & McCormick 2015; Egon et al. 2015; European Commission 2005, 2012; Heimann & Svedin 2018; Jaakma 2018; Juvancic & van Leeuwen 2018; Krogh 2013, 2018; Leoussis & Brzezicka 2017; Lindberg 2010; Lundvall et al. 2008; Mastalka & Timonen 2017; McCormick & Kautto 2013; Page 2003; Palsson et al. 2013; Poore 2017; van Vilsteren 2018)

2 A Strategy for Bioeconomy Education

Bioeconomy offers a future opportunity to reconcile economic growth with environmentally responsible action. The first step of the bioeconomy chain is the raw matter production, thus the agro-forestry science topics. Then, innovative processes and technologies from the biosciences are the essential tools. In addition, economy background and new approaches are indispensable in converting the new technologies into beneficial processes. Policy and marketing initiatives are required in opening up of new markets. Other high-tech areas such as information technology, medical technology and new production technologies are required. Social dialogue and understanding the challenges and opportunities of bioeconomy play a decisive role in the demand for new products and services and the innovations and technological developments associated with them; therefore, social and communication skills are becoming mandatory. Finally, the regulatory frames require a basic knowledge of the legal background.

Briefly, the bioeconomy as a concept is a long-term process and requires the concentrated cooperation of a large number of scientific disciplines, as well as political and economic decision-making processes. All the above aspects need to be analysed under the prism of the education in order to be better adapted to specific audiences, regions and processes. In this concept, the objectives should be divided into several groups including:

I. *Research Coordination Objectives*

- *To identify and map* platforms, initiatives and institutions in the BioEast region providing education on bioeconomy at all levels, and to analyse them in terms of objectives, contents, principles, target audiences and impacts.
- *To specify means and methodologies*: Regional analysis with examples for good practice of education curricula, terms of education, practices and priorities, involvement of industry, sources of education, online or virtual courses, tailor-made education, etc. Additionally, analysis of the evaluation system(s) is foreseen.
- *To highlight factors related to the regional characters of bioeconomy* such as (i) homogeneity of structural and operational challenges; (ii) susceptibility, public acceptance and consumers' demand; (iii) presence of stimulating ecosystems at regional level; (iv) importance of smart specialization strategies; (v) existing funding and supporting tools; (vi) presence of key actors at the regional scale; and (vii) existence of regional clusters and national strategies oriented for good regional priorities, and to examine their impact to a specific educational initiative on a case-by-case basis.

II. *Capacity Building Objectives*

- *To enable the potential for pilot courses and specific case studies based on real industry requirements throughout the running of educational courses. To underline the mutual benefits of such approach for private, public sectors and education bodies at all levels.*
- *To define and develop a conceptual basis for a BioEast Platform on Education and Training in Bioeconomy, align the goals of the platform with regional priorities and strategies for bioeconomy, growth and innovation, creating a network for the BioEast Platform on Education and Training in*

Bioeconomy involving the relevant stakeholders from academia, training institutions, industry, policy and government stakeholders.
• *To educate young scientists on how to make better use of information from the value chain and to provide opportunities to find new business models and business opportunities.*
• *To establish a solid contact with other European platforms of this kind and especially with the European Concept of Practice for Bioeconomy Education, for bilateral exchange of information, cooperation and mobility.*

III. *Development and Growth Objectives*

• *To create an infrastructure helping students and young scientists to transfer from relevant theoretical education to the new "bioeconomy concept". The role of the industry can be catalytic. For an effective educational frame, it is necessary the engagement of the industry into the programmes and the increase of student's accessibility. It will therefore guarantee a robust pipeline among industry, faculty staff and students.*
• *To penetrate new markets: The final goal is to enhance the opportunities for penetrating into the market for the newly educated staff. For achieving this goal, issues like technology forecasting, market requirements, international competitiveness, mobility in transnational and trans-sectorial levels, cooperation with non-European markets, etc. should be analysed and incorporated into the educational mechanisms.*

IV. *Societal Objectives*

• By principle, bioeconomy actions have a regional character due to factors such as homogeneity of structural and operational challenges, more susceptible public acceptance and consumers' demand, the presence of stimulating ecosystems at regional level, the importance of smart specialization strategies, the existing funding and supporting tools, the presence of key actors of regional scale, the existence of regional clusters and national strategies oriented for good regional priorities. The expected impact is to highlight these factors and to examine them into a specific educational initiative on a case-by-case basis.

This strategy focuses on its impact to the European competitiveness and growth. The adopted actions to achieve these objectives aim to coordinate activities on bioeconomy education and to fulfil the raising requirements on various levels:

- *On educational level*: The establishment of an active network across Europe aiming to adapt practices, to optimize educational methodologies and to form highly specialized staff.
- *On technology and innovation level*: The formation of a more competitive "bioeconomy-oriented" specialized staff in creation of appropriate bioeconomy-based research, technology and innovation projects.
- *On economic and business level*: The creation of an infrastructure helping students to transfer from relevant theoretical education to the new "bioeconomy

concept". Enhancing the opportunities for penetrating into the market for the newly educated staff.
- *On societal level*: Bioeconomy actions have a regional character. The expected impact is to highlight factors and to examine them into a specific educational initiative.
- *On governance and policy levels*: Integrating the education scheme in a wider strategy or governance platform, highlights to the economic, social, environmental and policy contexts.

3 The Complexity of Bioeconomy Education

Bioeconomy education faces two major challenges. First, it aims to merge very heterogeneous topics and disciplines in a common educational tool and use it into students having very diversified educational background. Second, it has to make the educational procedure, not just a knowledge provider, but also an instrument for problems solving and having a concrete societal impact.

One of the key points is that the principles and the methodologies – including the aspect of interdisciplinarity – generally support both students' acquisition of academic knowledge, skills and competencies for the academic labour market and simultaneously correspond to research methodologies. By applying this approach throughout their education, students actually acquire competencies that are relevant for both university-based careers and non-university academic careers.

Societal changes have had an impact on labour markets with regard to types of jobs, production technologies, company structures and industrial dynamics. Workplaces have become more complex and unpredictable, both technologically and in terms of knowledge, qualifications, competencies, values and attitudes among employers and employees. It has been explained as a shift from the industrial society to the information society, the knowledge society and even to the learning economy and society (Lundvall 2008).

Another important factor is that regardless of discipline, there are also required skills for development, planning, communication, creativity, collaboration, theoretical reflection, problem solving, ethics, action and accountability. These abilities are often referred to as twenty-first-century skills (e.g. Crockett 2016).

4 Forms of Education in Bioeconomy and Specific Target Audiences

The updated European Bioeconomy Strategy addressed the education issue as "Promote education training and skills across the bioeconomy". In fact, this action aims to reduce the shortage of skills by supporting the development of new curricula which respond to the new and evolving needs of stakeholders, by shortening the gap between academia and the industries and by facilitating the design of shared curricula in order to align educational paths with labour market opportunities.

We can distinguish three levels of education in bioeconomy:

- *Education in schools and high schools*: Teaching principles of circularity, acting local and global at the same time and raising interest for bio-based careers.
- *At universities*: New curricula have already been developed, combining life sciences, engineering, economics and marketing, and enabling the dynamics for the development of transversal skills, capable to support the students to become bioeconomy entrepreneurs or managers.
- *Vocational training*: There is a need to match requirements for skills in various sectors involving regional and local actors.
- To the above it has to be also considered a *lifelong learning programme on bioeconomy*.

It is a challenging effort to classify the various groups gaining education on bioeconomy. But it is also indispensable to classify them in order to organize the educational practices and materials according to the needs are specificities of each group.

Sustainability is the aspiration to meet the current needs of education. Further on, the main educational directions are (a) **Economic** (e.g. economic growth, affordability, resilience, energy security), (b) **Social** (e.g. jobs, workforce development, food security, health and safety) and (c) **Environmental** (e.g. energy and water consumption, material intensity, air emissions, ecological impacts).

A number of challenges can be faced in the whole educational frame. These can include:

- Major technical hurdles for development and scale
- A lack of necessary infrastructure
- Uncertainties about understanding environmental, social and economic outcomes
- Lack of access to knowledge, data
- Lack of a formal, collaborative mechanism for sharing knowledge
- Knowledge and technology gaps among the students

In any way, the important role of the bioeconomy education is to provide each group the newest scientific discoveries, technologies and reliable data in easily accessible formats, in a manner that is as comprehensive as possible and consistent with the current status of development and the applicable regulations.

Briefly, we can consider the following groups:

- *Students in all levels of the education*: For the time being, several bioeconomy education programmes have been established in various European universities, either as a part of a wider technology unit or as an autonomous postgraduate course (e.g. master's degree). The aim is the gradual integration of bioeconomy and all levels of education, adapted to the wider current educational requirements and frames. The idea is to demonstrate to students the wider spectrum of possible applications and make them familiar with the exploration of sustainable economic solutions. Knowledge about bioeconomy will become an essential requirement for the new generation of students. The sector not only will provide job possibilities in the future, but also the knowledge will help students to better understand the challenges of tomorrow's society and enhance their skills as responsible citizens. Bioeconomy covers a broad range of sectors, from agriculture and the agri-food industry to fisheries, forestry, biorefineries, chemistry and (bio)energy. Therefore, it is expected that the more advanced is the educational level, the more specific (on regional or sectorial basis) the educational programmes will become.

- *Employees of the related to the bioeconomy industry, sectors and domains*: The related to bioeconomy industry is the key player in implementing the new practices in a perspective to contribute to the development and growth in sustainable conditions. It is therefore obvious that their staff and experts must be updated and specifically trained. This requirement seems to be general independently of the sector or the domain of application. There are two main approaches involving the education of the staff or related industry. In the first case, it is a *bottom-up action* where the industry has detected its needs and specific requirements. They wish their staff to be additionally trained, to be more competent and to adapt to the changing requirements of the market. This approach is more often in countries and regions with more developed economies and a pioneering related industry. In the second case, it is a *top-down action* where the industry has not detected yet the necessity for adaptation and upgrade. In this case, public administration and public consultation with the assistance of the academic community take the initiative to suggest (or even to impose) education to the industry. It is evident that in the two cases the training practices substantially differ and they are tailored to the specific requirements.
- *Public administration and policy making officials in all domains*: It is expected that for most of applications based on bioeconomy, the transition to the new practices is not going to be a smooth or rapid process. For instance, adopting an energy or materials production regime based on renewable resources, it might create many setbacks and obstacles, technically and politically, and it can result to an effort lasting for decades. Similar effects were observed in previous transitions (e.g. from wood to coal or from coal to oil). However, those earlier transitions were not complicated by the so-called grand challenges faced today. In our days, above energy security and food and water security, the impacts of technology and development on climate change are very important. Some events have politically legitimized climate change and its mitigation, and the world finally sworn to action. The bioeconomy holds some of the answers to the economic challenges thrown up by mitigating climate change while maintaining growth and societal well-being. For all this complexed frame, public administrators, public servants, policy makers and regulators are deeply involved. Their actions have a significant societal and economic impact. They should therefore be appropriately trained so they become a substantial group requiring appropriate education.

- *Employees of commercialization chains and business development in the domains related to bioeconomy*: This is a very particular target of trainees, and their training has a dual objective. From one side, they have to increase their level of awareness, knowledge and understanding becoming capable to assimilate the new practices and to comprehend the specific value of their products and services. From the other side, they must reach the capacity to transmit this knowledge to their target markets in order to guarantee a suitable level of communication and mutual understanding. In both cases, they have to deal with significant particularities such as:

  1. The classical perspectives of perceiving and as analysing economy are changing from market approach of static equilibriums into industrial organizations of dynamic networks.
  2. Bioeconomy as a concept gains more and more attention of society, business, politics and the academy.
  3. Bioeconomy can be presented as the complex adaptive system not only in producing high added value but also adapting to the changing environment.
  4. As an adaptive system, bioeconomy sector dynamically changes seeking for new sources of incising productivity and efficiency according to sustainability needs (Mariusz Maciejczak 2017). It is therefore evident that the education has to be specific to a certain sector or a certain region.

- *Employees in economy and finances influenced by the bioeconomy*: A very important domain of the whole bioeconomy business is the domain of finances itself. There is a particular context on the investment and access-to-finance conditions for bio-based industries. Recently, the European Investment Bank published a study analysing in depth the finances dynamics related to bioeconomy. According to this study, bio-based industry projects face issues accessing private capital. Regulation and market and demand framework conditions are perceived as the most important drivers and incentives but also present the biggest risks and challenges for both Bio Based Industries (BBI) project promoters as well as financial market participants to invest in the bioeconomy. The main funding gaps in financing the bioeconomy projects are scaling up from pilot to demonstration projects, and particularly in BBI, they are moving from demonstration to flagship/first-of-a-kind and industrial-scale plants. Existing public financial instruments are utilized, but their catalytic impact could be further enhanced. Policy actions and/or new or modified public financial instruments could de-risk BBI investments and catalyse private capital. The study recommends the following: Establish an effective, stable and supportive regulatory framework for BBI at the EU level, which is essential. Further reinforce awareness about specific funding instruments and the European Fund for Strategic Investments (EFSI), which can match the funding needs of certain BBI projects. Develop a new EU risk-sharing financial instrument dedicated to BBI, potentially taking the form of a thematic investment platform that can meet the needs of BBI projects and mobilize private capital. Explore the creation of an EU-wide contact, information exchange and knowledge sharing platform or other channels to facilitate relationships between BBI project promoters, industry experts, public authorities and financial market participants active or seeking to become active in the bioeconomy (European Investment Bank 2017). It is obvious from the above, but also highlighted in this study, that the experts to be involved in these processes need to reach a comprehensive understanding of bioeconomy principles so they can optimize the outcomes. It is evident that specific educational tools and training methods must be used in this direction.

- *The general public, consumers and stakeholders related to bioeconomy*: There is a common approach versus all those groups of non-specialists or not directly related to the general bioeconomy business. Education of these groups includes two major parts. One is to enhance the general awareness and knowledge not only from the viewpoint of technology applications but also in terms of consequences, influence of the everyday life and its impact to the socioeconomic frame. The second part involves the active participation. Bioeconomy, despite its many applications, has yet to enter the public consciousness as an exciting solution to societal challenges. In order to achieve this goal, the general public and specific groups of stakeholders must develop the certainty that they are part of this process and adopt an attitude of participation, in acting, implementing and decision-making levels. The education in this perspective is rather challenging and requires a societal maturity, devotion and willingness. Many institutions, including the European Commission, have proposed various strategies in order to achieve an effective public education in the domain. They all agree that this action should initiate structured and consistent communication on the bioeconomy across a range of on- and offline platforms with a view to boosting knowledge of the general European public. The expected impact is that bioeconomy will be put more prominent on the agenda at national, local and regional levels.
- *To emphasize the difference between education, awareness and training*: It is important to distinguish the three terms above related to the education. *Awareness* is mostly a pre-educational requirement. It enhances the efficiency of the education, and it is a substantial element in order to achieve homogeneity within the class. Like in other domains, also in bioeconomy awareness can be achieved by many ways and tools and covers a wide horizon of related areas. On the other side, *training* is a post-educational perspective. Usually, it is very specific and it addresses to individuals well aware and educated helping them to increase their capacities in well-defined operations and practices. All three are part of the whole educational framework of bioeconomy.
- *Teaching the teachers*: Teachers can contribute to raising awareness about bioeconomy in future generations. However, given that the bioeconomy concept is relatively new, and also given that it tends to change the educational philosophy in many ways, the range of the existing expertise is not so wide, and it is necessary to consider a "teaching the teachers" stage in order to achieve a critical mass of contributors. There are already some existing methodologies in this perspective, but also funded projects aiming to complete and optimize these methodologies. They aim to bridge the gap in education by giving teachers a fresh perspective into the bioeconomy field and its applications in teaching a wide variety of sub-

jects. These courses contain the necessary information, so, at the end of those, trainees will understand what bioeconomy is, its importance for society and for students and its implications for education. They will also gain the specific capacity and skills to set up their own teaching package according to specific requirements and in the perspective to innovate the classroom practices.

4.1 Raising the Awareness

Part of the education is of course the raising of awareness and dispersion of dedicated information. This process not only is addressing to less awarded targets but also may enhance the lateral knowledge of groups who receive vocational education and dedicated training.

Given that the future development of a smart sustainable and inclusive bioeconomy involves the whole society at large, it is obvious that raising the overall awareness and understanding of the social, economic and environmental impacts of bio-based products and processes is of utmost importance.

There are several ways of increasing the level of awareness. To mention among them: specialized platforms providing such information on a digital or conventional form, dissemination activities, exhibitions and festivals, specific tools, web-based applications, communication initiatives, involvement of the mass media and journalistic initiatives, dedicated actions initiated by various stakeholders and of course the social media.

It is very important to mention that all European-funded projects on bioeconomy require dissemination activities in order to maximize their impact by involving the end users and the society by building community actions in this direction. These activities include knowledge sharing, networking, mutual learning, coordination of joint activities and events, while a proactive collaboration between projects has been promoted. Recently, it has been initiated the European Bioeconomy Network, a proactive alliance of EU-funded projects dealing with bioeconomy promotion, communication and support.

Bioeconomy is too abstract, complex and not easy to understand. Need to address terminology, confusion and misunderstandings and provide scientific basis to communication. Need to address societal and consumer requirements and concerns and provide information on the sustainability and impacts of products and services, to build trust and acceptance. Need to inform not only citizens but also various groups of stakeholders. Also, need to involve young people and multipliers. Important is to consider that the target audiences differ substantially between them; therefore, the communication strategies must be designed accordingly, using innovative channels and tools.

5 The Innovative Character of Bioeconomy Education

To reach a bio-based economy, new knowledge and expertise are required. Today's economy is linear. Materials are taken from nature, turned into something else, used and then disposed of. The process starts all over again by taking virgin materials from nature. This creates two problems: it depletes nature and leaves society with a considerable amount of waste. A transition towards a circular economy, also known as cradle-to-cradle approach, is required. The circular economy features two circles: an inorganic circle that includes metals, nondegradable plastics and building materials and an organic circle that includes composting.

Biomass is all plant and animal material, both raw and processed. Resources are agricultural, food and feed production, forestry and natural and aquatic resources. Aquatic resources include weeds, duckweed and algae. Biomass is a collection of valuable ingredients. It contains sugars, starches, cellulose, lignin, proteins, oils, fats and other specific ingredients. Using mechanical (biotechnical), chemical and thermochemical processes, biomass can be refined and converted into new building blocks for innovative products such as paints, plastics, composites, building materials and fuels.

It will take a generation to transition from a fossil-based to a bio-based society. Today's students will be part of the transition during their working career. If teachers are able to inspire this current generation of students, they can help speed up this transition. The bio-based economy is a perfect stepping stone for inspiration. It inspires young people and universities are seeing growing numbers of students choosing a "greener" education programme.

In recent years, educational institutes have begun to see the relevance of bringing the bio-based economy into classrooms. They have already started developing new courses, minors and programmes on the bio-based economy. Some schools incorporate this topic into existing programmes, while others create new programmes from scratch.

Incorporating a new topic in education also creates opportunities for new innovative (and in most cases digital) formats.

The innovative new skills require a systems approach and Cros-Sectorial mindset. The new educational material on the bio-based economy should include:

- A solid disciplinary knowledge base
- The ability to work in an interdisciplinary team
- A Cros-Sectorial mindset

All this innovation requirement in the educational frame is indispensable. Different disciplines work together through an interdisciplinary, holistic approach, as experts believe that is the key to success. New products and operations will be designed by interdisciplinary teams, considering the socioeconomic, ethical and environmental aspects related to biomass production and carbon capture in an international context. A technological solution alone does not guarantee success on the market.

Moreover, new business opportunities are relevant for students. Chemistry students should realize that plastic can be made from sugar as well as from oil. Agriculture students should realize that their market extends beyond the sugar company and the food business. Sugar can be used to make other products, such as the building blocks for plastics. This gives farmers more outlets for their products. However, if a farmer wants to enter this new market, they will also need to know that it involves a continuous supply instead of a seasonal one as well as more strictly defined specifications. This new Cros-Sectorial mindset is a crucial part of the learning process.

Finally, sharing knowledge and educational resources is also a crucial factor. Cooperation in all levels is a necessity.

6 Regional Aspects of Bioeconomy Education

It has been demonstrated in many occasions that bioeconomy has a very sound regional character. The factors underlying the regional specificity they vary on a case-by-case basis and they can include environmental and climate particularities, specific geographic characteristics, different economic development, diversified general public in terms of perceptions and attitudes, multiple cultural patterns and so on.

It is therefore a highly complexed system, and the transition to a more sustainable context on regional basis requires a variety of knowledge and values in both processing and decision making. Actually, due to its complexity, the decision-making process and the implementation of new forms of governance in general require a very specific knowledge (Swyngedouw 2005, Sassen 2013) which bioeconomy education platforms can provide.

An additional element is that the transformation into a sustainable society requires new challenge-driven innovations and new collaborations between more actors than earlier from different spheres, with a variety of knowledge and practices, including civil society. This underlines the importance of inclusion of different stakeholder groups in all phases of the decision-making processes (Reed 2008). It is therefore obvious that educating these groups is mandatory in order to reach an effective bioeconomy-based development on regional scale.

Since the year 2000, the European Commission decided to enhance its innovation policy by making it applicable in various regions and adopting the so-called regional innovation systems (RIS). The idea of RIS is linked to the raise of regional innovation clusters and aims to support economic growth (Lindenberg 2010, Patterson 2007). In this context, an innovation is the outcome of social processes, occurring in interactions between different actors. In regional science, the outcome of an innovation process is closely related to the regional context and imitational settings (Doloreux and Parto 2005). Of course, the domain of bioeconomy became a part of this frame, and the appropriate education seemed to be an indispensable tool, in order to optimize the outcomes. Currently, innovation policy in Europe is changing.

The goal is now to provide information, methodologies and expertise about smart specialization in the European regions, and this requires large participation and more significant education. A range of actors, from innovation users or other groups representing consumers, non-governmental organizations (NGOs), citizens and workers, are all addressed to be involved in the smart specialization process (Foray et al. 2012), and therefore their needs for related education and training are increasing.

Focusing on the educational matters only, it is important to underline that for the education on regional basis, a number of factors must be considered (Vexler et al. 2014):

- *The training model*: It refers to the specific processes in order to achieve a professionally competent trainee having the expected qualifications, by means knowledge, skills, qualities, experience and individual activity style. Educational quality depends on three main points: aim of education (for which purpose to study), content of education (what to teach) and principles of educational process organization (how to teach).
- *The regional labour market analysis*: Under the labour market is considered the economic relation system, concerning purchases and sells of labour power, where supply and demand and prices for it are shaped. The main labour subjects interacting in the labour market are enterprises, searching for specialist; specialists, searching for job; and unemployed citizens. There are also included educational institutes, training specialists, adapted to the market necessities.
- *Forecast for future*: Education should be "principally prognostic", taking into account development tendencies of one or another brunches, factor dynamics and scientific knowledge in general, creeping scientific and technological discoveries and prospects of their usage in future. Especially in the domain of bioeconomy, this approach is particularly complicated due to its multidisciplinary and the lack of precedents. It has therefore to be established a prognostic professional model based on specialists, and a reliable informational procurement, giving an opportunity of conjoin forecasts of scientific-technical researches with pedagogical and didactic forecasts.
- *Educational institute capabilities*: Educational institute, implementing the programme, should possess material and technical base, corresponding to the acting technical norms adapted to the bioeconomy frames and providing training realization based on a specific bioeconomy curriculum. For educational institute, the following aspects are essential: human resources and their qualification; material and technical potential; educational conditions; educational laboratory equipment; technological potential is a usage of modern educational technologies and means and know-how.
- *External environment factors*: An important attribute of a regional educational system in the domain of bioeconomy is the external environment availability and multidimensional connections with it. An external environment can be an associate of educational system in its searching of development and it may include:
  - Natural factors
  - Social factors

- Cultural factors
- Demographic factors
- Economic factors
- Technological factors
- Political factors

This system complexity is in fact what makes the regional concept of the bioeconomy education so specific and unique on a case-by-case basis. It is vitally important to have a synchronous development, independently from the existing regional differences. In this perspective, it is required a common implementation plan in all regions. This plan should include:

- *Determining criteria and indicators*: The criteria should be economic, environmental and social, and they should be used as a basis for establishing a database with indicators based on geographic levels. The indicators reflect various aspects of national/regional bioeconomy status.
- *Collection of instruments and measures*: A structured database with instruments and measures for regional bioeconomy development, incorporated into a publicly available online search tool.
- *Catalogue of good practices and case studies*: To be created as a result of extensive case study work.
- *Regional bioeconomy profiles*: Developed as a synthesis of the previous steps.
- *Regional bioeconomy network*: Addressing challenges of a stakeholder network at different levels.

7 Restructuring the Educational Platform

Restructuring of public research institutions is an integral part of the co-innovation process. Research capabilities must align with societal needs as a basis for innovation with added societal value. Research institutions should not only adapt to rapidly evolving and new groundbreaking technologies and novel trends in collecting, using and publishing scientific data such as data mining and open access, but they also need to be able to quickly respond to societal changes, new business models and global challenges such as energy, climate, water and food security. Such complex and pressing challenges require a new way of performing research and providing education with an integrated, cross-cutting and multi-actor approach based on trans- and multidisciplinary science.

There have been proposed approaches in achieving this restructure:

- Knowledge should be challenge-oriented rather than driven by scientific curiosity and strike the right balance between basic and applied.
- Knowledge should be transdisciplinary with multiple theoretical challenges and practical methodologies.

- Knowledge should be diverse and socially distributed, with particular attention on social innovation and the inclusion of socially disadvantaged actors and regions.
- Co-creation is the main driver for new knowledge; therefore, it should be reflexive and in dialogue with all stakeholders.
- New rewarding and assessment systems should be installed to ensure high quality control.
- To consider the need for a set of new skills and competencies not only for researchers but also for all actors and stakeholders.

There is growing concern on the impact and economic value of research at public institutions. In general, these organizations – and specifically universities – are not flexible enough to be able to respond quickly to societal needs. Academics can determine their own direction of research without a responsibility or immediate obligation of return to society. However, since universities are often public institutions, supported by taxpayers' money, they should in principle align with societal needs. In that respect, government investments at the national and regional levels are urgently needed to identify new structures and methodologies to effectively and efficiently employ and transfer newly gained knowledge and fully exploit its potential to bring added economic or social value in terms of innovation linked to different sectors.

Based on the above, internal reorganization of institutions should be performed based on a bottom-up approach, looking to societal challenges as a basis for co-creating social innovation to quickly respond to emerging societal needs. Different scientific disciplines from the natural and social sciences should be combined.

8 Conclusions

In order to achieve the goals set, actions must be planned and executed:

Educational systems: New approaches and new curricula and/or changes in existing curricula at several levels will train and retrain academic staff and public researchers to have a mix of skills and new knowledge allowing them to actively take part in the innovation process.

Evaluation criteria: Objective (standardized) quantitative and qualitative evaluation criteria are needed to measure and monitor the social and economic impact of knowledge created at public research institutions and universities as continuous drivers of a flexible and self-correcting restructuring and reorganization process. Such criteria can be developed and targeted at several levels (e.g. institutional, department, research group and on an individual basis). Midterm impact reviews should be carried out to adjust research and action plans accordingly (participatory research) and stimulate a continuously evolving research system.

Outreach activities: A proactive strategic communication road map including outreach activities to engage and enter into dialogue with civil society, different stakeholders and key actors in research and innovation should be at the heart of the public research institution and form the basis of research plans to quickly understand and address societal needs and work towards a dynamic problem-solving and application-driven approach.

Support systems: New support instruments are needed to link basic and applied research and create strong dialogue between all actors and stakeholders involved to ensure knowledge flow and a co-creation and co-innovation approach to maximize the impact and economic value (exploitation of knowledge). Such support systems should be created and funded by government institutions as part of the policies to create an enabling environment for innovation. Existing support systems need to be better coordinated and their interfaces aligned to improve their effectiveness.

References

Barrows H.S. 1996: Problem-based learning in medicine and beyond: a brief overview, In *(Wilkerson L., Gijselaers W.H., eds). Bringing Problem-Based Learning to Higher Education: Theory and Practise. Jossey –Bass Publishers, San Francisco*

Barthel S., Crumley C., Svedin U. 2013: Bio-cultural refugia – Safeguarding diversity of practices for food security and biodiversity. *Global Environmental Change*, 23, 1142–1152.

Bergeret P., Svedin U., Valceschini E. (Editors) 2018: Bioeconomy challenges and implementation: the European research organizations' perspective.

Bowden J., Marton F. 1998: The University of Learning: Beyond Quality and Competence. London:

Burssens S., Jarvenpaa M., Bergeret P. 2018: Restructuring research institutions in view of emerging trends in agricultural research and the bioeconomy: from knowledge to social innovation *in: Bioeconomy challenges and implementation: the European research organizations' perspective* 77–92.

De Besi M., McCormick K. 2015: Towards a Bioeconomy in Europe: National, Regional and Industrial Strategies. *Sustainability*, 7, 10461–10478.

Doloreux, D., Parto, S. 2005: Regional innovation systems: current discourse and unresolved issues. *Technology in Society, 27*(2), 133–153.

Egon N., Alrøe H.F., Thorsøe M.H., Olesen J.E., Sørensen P., Melander B., Fog E. 2015: Knowledge asymmetries between research and practice: A social systems approach to implementation barriers in organic arable farming. *Sociologia Ruralis*, 55 (4), 460–482.

European Commission, 2012: Innovating for Sustainable Growth: A Bioeconomy for Europe Publication Office of the European Office*: Luxembourg*

European Commission, 2005: New Perspectives on the Knowledge-Based Bio-Economy; *European Commission: Brussels, Belgium.*

Foray, D., Goddard, J., Goenaga Beldarrain, X., Landabaso, M., McCann, P., Morgan, K., Nauwelaers, C., Ortega-Argiles, R. 2012: Guide to research and innovation strategies for smart specialisations (RIS3). *Luxembourg: European Union.*

Heimann B., Svedin U. 2018: Innovation and social sciences and humanities in agricultural research *in: Bioeconomy challenges and implementation: the European research organizations' perspective* 93–101.

Jaakma U. 2018: Opportunities and challenges in implementing a Bioeconomy strategy: the university perspective In: *Bioeconomy challenges and implementation: the European research organizations' perspective* 30–40.

Juvancic L., Myrna van Leeuwen 2018: Unleashing regional Bioeconomy potential: the Bioeconomy Regional Strategy Toolkit. *In: Bioeconomy challenges and implementation: the European research organizations' perspective* 60–76.

Krogh L. 2013: The Aalborg PBL model and employability. *In: What did you learn in the real world today. The case of practicum in university Educations (Henriksen L.B., ed.), Aalborg University Press.*

Krogh L. 2018: Education for complex problem solving: Is problem-based learning one answer to today's higher education challenges? in: *Bioeconomy challenges and implementation: the European research organizations' perspective* 19–29.

Leoussis J., Brzezicka P, 2017: Access-to-finance conditions for Investments in Bio-Based Industries and the Blue Economy: *Study InnoFin Advisory for the European Investment Bank & DG Research and Innovation of the EC*

Lindberg, M. 2010: Samverkansnätverk för innovation. En interaktiv och genusvetenskaplig utmaning av innovationspolitik och innovationsforskning. *Luleå: Universitetstryckeriet Luleå.*

Lundvall B.Å., Rasmussen P., Lorenz E. 2008: Education in the Learning Economy: a European Perspective. *Policy Futures in Education*, 6 (6), 681–700.

Maciejczak M. 2017: Bioeconomy as a Complex Adaptive System of Sustainable Development *Journal of International Business Research and Marketing* 2(2) 7–10.

Mastalka M, Timonen L. (Editors) 2017: Perspectives on Bioeconomy - Cases from the ERDI Partner Regions.

McCormick K., Kautto N. 2013: The Bioeconomy in Europe: An Overview. *Sustainability*, 5, 2589–2608.

Page K Commission of the European Communities, 2003: The role of universities in the Europe of knowledge. Brussels: *Communication from the Commission.*

Palsson G., Szerzynski B., Sörlin S., Marks J., Avril B., Crumeley C.,Hackmann H., Holm P., Ingram J., Kirman A., Pardo Buendia M.,Weehuizen R. 2013: Re-conceptualizing the 'Anthropos' in the Anthropocene: Integrating the social sciences and humanities in the global environmental change research. *Envir. Science & Policy*, 28, 3–13.

Pettersson, K. 2007: Men and male as the norm?: a gender perspective on innovation policies in Denmark, Finland and Sweden. *Copenhagen: Nordiska Ministerrådet, NordRegio.*

Poore J.A.C., 2017: Back to the Wild: How Nature is Reclaiming Farmland. *New Scientist*, 3138, 26–29.

Reed, M. S. 2008: Stakeholder participation for environmental management: a literature review. *Biological Conservation*, 141, 2417–2431.

Sassen, S. 2013: When territory deborders territoriality. *Territory, Politics, Governance,* 1(1), 21–45.

Swyngedouw, E. 2005: Governance innovation and the citizen: the Janus face of governance-beyond-the-state. *Urban Studies*, 42(11), 1991–2006.

Vexler, V.A., Bazhenov, R.A., Bazhenova, N.G. 2014: Entity-Relationship Model of Adult Education in Regional Extended Education System *Asian Social Science;* 10(20), 1–14.

van Vilsteren G. E.T. 2018: Innovative education for the Bio-based economy *In: Bioeconomy challenges and implementation: the European research organizations' perspective* 41–47.

Chapter 23
Bioeconomy as Proponent of Digital Meanings Society

Sirkka Heinonen

Abstract This article discusses societal transitions and particularly the one into a new phase after the information society. We will witness the emergence of a digital meanings society, in which the economy is based on the production and consumption of meanings and meaningfulness. Alongside with this techno-economic and sociocultural development or even a philosophical stance, we have seen several alternative characterizations for the new societal phase. Pathways towards digisociety, ecosociety or biosociety and experience society have been anticipated. The focus here will be on biosociety and how it can be intertwined with digital meanings society, propelling it or manifesting it. This is because biosociety may be the very phase that can guide us through the existential crisis that we humans have created by acting like the natural resources were limitless even though our planet does have physical limits. Change, growth, progress, meaningfulness, technology and nature are key ideas in this existential drama. The future will be moulded along the critical lines these ideas are being tackled and rethought.

Biosociety is based on bioeconomy. Bioeconomy instead has biotechnology as its driver. However, it is quintessential on a wider scale, too, how our *relation* to nature and to technology is conceived and practised in biosociety. What are the social constructions, philosophy and culture of life in biosociety? They are of equal importance in creation of a new societal phase, even though economy and technology are key and immediate drivers of change. Consequently, humans' relation to nature on one hand and to technology on the other hand is in the core of achieving sustainable futures. It is interesting to perceive how these notions evolve in time, just like societies evolve through transitions into new phases. Besides reflecting upon the history of such ideas, it is intriguing to dwell on various metaphors of our interaction with nature and technology. Metaphors and myths live in us and affect our thinking, even unconsciously. The myth of Prometheus giving fire – (bio)technology – to us humans is a core myth in this sense to be addressed.

The relation of humans to nature seems to have a widening gap, whereas humans' relation to technology is continuously bridging up. Our relation to nature and our relation to technology need to be rethought and revamped. "Technological

S. Heinonen (✉)
Finland Futures Research Centre, University of Turku, Turku, Finland
e-mail: sirkka.heinonen@utu.fi

E. Koukios, A. Sacio-Szymańska (eds.), *Bio#Futures*,
https://doi.org/10.1007/978-3-030-64969-2_23

somnambulism" describes the dilemma of modern times to willingly sleepwalk through the process of change. It means ignoring the notion that the technologies are not merely aids to human activity but also powerful forces reshaping that activity and its meaning. Biosociety and biotechnology give a meaningful context for humans living on the planet, but their outcomes have to be intentionally positive which is by no means and automatic process.

Keywords Societal transition · Information society · Knowledge society · Digital meanings society · Digisociety · Ecosociety · Biosociety · Prometheus · Myths · Stoicism

1 Introduction to Societal Transformation

Societies evolve in stages or waves, with intermittent transition phases, lasting thousands of years, then hundreds of years and lastly some decades. In the ancient Greece, in Stoic philosophy especially, life was perceived as evolving in cycles. This was not a linear notion of progress but an idea of recurring cyclic developments.[1]

In the modern world, Kondratieff (1925) saw progress to take place through various cycles or waves, changing in some 40–60-year periods. The cycles consist of alternating intervals of high strong sectoral growth and intervals of relatively slow growth.[2] In futures studies, Pentti Malaska (1999, 2001) perceived societies change in recurring stages of stable development, with intermittent transformational periods reconfiguring the system. He saw societal change as a learning process where intentionality is a critical characteristic for the human system as differing from natural systems. In futures studies, there is also a classification for scenarios that can be conceived as different models of development. James (Jim) Dator (2009) proposes four archetypes for scenarios describing societal change, i.e. growth, collapse, discipline and transformation.

The so-called wave theory has a rough equivalent in the school of thinking of transition framework of MLP (multi-level perspective) (Geels 2020). It posits that society evolves in transitions and these transitions come about through interaction processes within and among three analytical levels: niches, sociotechnical regimes and a sociotechnical landscape. In their terms the current regime of information society represents the regime level which will be impacted by innovations and actors arising from the niche level. Whatever term is used – cycle, wave and transition – we as humanity have come a long way with leaps from hunter-gatherer era to agricultural society, and then to industrial age, turned into information society. The

[1] More on the cyclic view in the next section

[2] This theory of waves is much criticized. Joseph Schumpeter named the waves/cycles in honour of Kondratieff. For the next anticipated wave, see Wilenius and Kurki (2017).

current information society (Webster 1995) – knowledge society or network society (Castells 1999) – is again on the brink of transforming into a new societal phase. Although, globally, much of humanity and countries have not yet even reached the stage of the information society, there has been speculation about the content and nature of the next mega-stage of society. Some different estimates of what phase will follow the information society era have been sketched. We have seen several alternative characterizations for the new societal phase. Pathways towards digisociety, ecosociety or biosociety (Rifkin 1998; Ahlqvist 2005) and experience society (Pine and Gilmore 1999; Jensen 2001) have been anticipated (see Fig. 23.1).

The focus in this chapter will be on the notion of biosociety and how it can be intertwined with digital meanings society, propelling it or manifesting it. This is because biosociety may be the very phase that can guide us through the existential crisis that we have created by acting like the natural resources were limitless even though our planet does have physical boundaries. The Club of Rome published its *Limits to Growth* report already in 1972 (Meadows et al. 1972), but only now it has been taken seriously when climate negotiations are proceeding and the UN Sustainable Development Goals (SDGs) have been adopted by many stakeholders. The Club of Rome sees the current situation as Climate Emergency and has accordingly published a Climate Emergency Plan (2020) (https://clubofrome.org/publication/the-climate-emergency-plan/). Change, growth, progress, meaningfulness, technology and nature are key ideas in this existential drama. The future will be moulded along the critical lines these ideas are being tackled and rethought. Rethinking economy is in the forefront. David Korten (2020) has even summed up as the humanity's most important task to be questioning GDP and material growth thinking alongside with those of investing in ensuring the well-being of both people and the planet. Rethinking economy means reinventing prosperity if we want to manage economic growth to reduce unemployment, inequality and climate change (Maxton and Randers 2016).

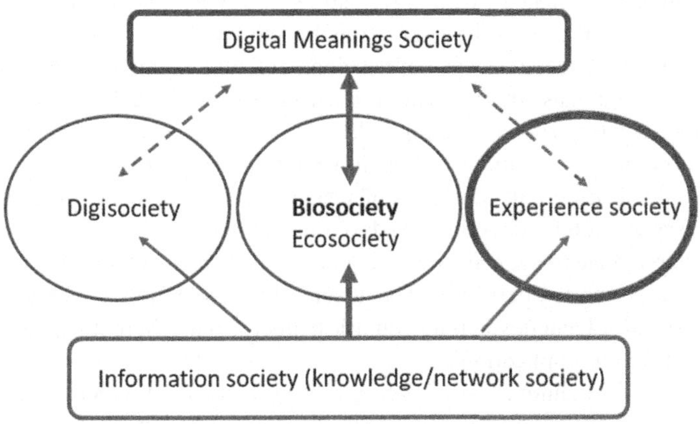

Fig. 23.1 Possible schemes for societal transition into a new phase after the information society

GSDR (2019) emphasizes the urgent need for transformations towards sustainable development. This requires that we strengthen the directionality of science on behalf of a mutually beneficial "moon landing" for humanity and the planet. The 2030 Agenda is suggested as a shared compass for researchers, engineers, science policy makers and funding agencies to increase the relevance and benefits of science and technology for the global community. Besides science, we need the role of companies and citizens to be lifted as well in such transformational efforts for societal development towards sustainability.

In Sect. 2, the concept of biosociety is introduced as conceived in this chapter on "Bioeconomy as proponent of digital meanings society". Then the tentative new concept of digital meanings society is launched and opened up in Sect. 3. Section 4 discusses human relation to nature with a perspective of history of ideas, and accordingly Sect. 5 addresses the role of technology in human interaction and for societal development. Section 6 drills into bioeconomy concretely manifesting digital meanings society. Conclusions are drawn for a learning process where humans' relation to nature and technology is to be reassessed in a fusion of biosociety and digital meanings society.

2 Biosociety

What is biosociety? The prefix "bio" is derived from the Greek word "bios" meaning life. The bio prefix appears in numerous terms denoting a wide range of concepts – from biology, biotechnology, biochemistry and biophilia to biography and biometrics for instance. The origin for the concept of biosociety is thus very telling. Biosociety, i.e. "life society", refers to a society where life in all its forms, its supporting systems and the biosphere are determining conditions, means of subsistence and survival as well as – is claimed in this chapter – source for meaningful existence.

How does the rise of the biosociety happen? The biosociety will be built directly on the foundations created by the information society and utilizes its technologies and ways of working, as well as the arsenal of all previous stages of society. The biosociety is the next major wave of development, with significant and revolutionary effects on humans, their nutrition, environment, industrial processes and the way whole society operates (see Fig. 23.1). In the biosociety, development – both positive opportunities to improve human well-being and threats to face risks – is driven by globalization, the demands of sustainable development, global warming, natural disasters, food production, the development of biofuels and the development of biotechnological and genetic technologies on one hand and by biological warfare and pandemics on the other hand. The advent of the biosociety is global, affecting both industrialized and developing countries, albeit from different perspectives and with very different implications.

Society is always human-based, whereas life is larger than societies. Humans are only one species in the sphere of life. Malaska stressed that in our societal development, the justification for humans' existence is to make life on earth better with humans than without them. A citation from one of his poems called "The Eye of Gaia Is Upon Us" claims that "*The role of humans is to show…that human life is a valuable part of life as a whole, that life with humans is richer and more valuable than without humans*" (Pouru et al. eds. 2018). This reminds us that biosociety can also be a normative stance. Biosociety is something that grows bigger than economy and technology, to become not only a societal but also deeply an existential platform. Wendell Bell (1997, 73) formulated the general purpose of futures studies to be "to maintain or improve the freedom and welfare of humankind". He further pointed out that here could be added "the welfare of all living beings, plants and the Earth's biosphere for their own sakes even beyond what is required for human well-being".

What is bioeconomy? The European Commission defines the bioeconomy as "the production of renewable biological resources and the conversion of these resources and waste streams into value added products, such as food, feed, bio-based products and bioenergy". The bioeconomy aims at reducing our dependence on fossil natural resources, preventing biodiversity loss and creating new economic growth and jobs in line with the principles of sustainable development. Bioeconomy may tackle the major global challenges posed by the growing population and its need for food, water and energy. Many countries have made national strategies for bioeconomy. The French emphasize that bioeconomy is an economy that is moderate, renewable and circular (Roy 2016). Bioeconomy is at its best intertwined with circular economy which minimizes waste by reducing, reusing and recycling. The utilization of waste as raw material to other production processes is not only sustainable, but it can be economically profitable as well. The cradle-to-cradle principle epitomizes this stance in a pragmatic way. Cradle-to-cradle design (Braungart and McDonough 2002) is a biomimetic approach to the design of products and systems that models human industry on nature's processes. In such life-cycle development, materials are conceived as nutrients circulating in healthy, safe metabolisms, and not going from cradle to grave.[3] The blue economy concept by Gunter Pauli (2010) is also a concrete proof of the numerous innovations already available and applicable for bioeconomy. The blue economy goes beyond the globalized and the green economy by presenting a business model that responds to the basic needs of all with what is locally available. It allows producers to offer the best at the lowest prices by introducing innovations that generate multiple benefits for society and local level, not just increased profits.

Bioeconomy is fuelled by biotechnology. The sphere and scope of biotechnology are expanding. Biotechnology is a broad area of biology, involving the use of living systems and organisms to develop or make products (OECD 2011), with many interconnections with related scientific fields. Recently, biotechnology has expanded to cover new scientific fields, such as genomics, recombinant gene techniques, applied immunology and development of pharmaceutical therapies and diagnostic

[3]The cradle-to-cradle concept was originally coined by Walter Stahel, member of the Club of Rome.

tests. The use of biological processes, organisms or systems to produce products that are anticipated to improve human lives is termed biotechnology (Bhatia 2018). By contrast, bioengineering is generally thought of as a related field that more heavily emphasizes higher systems approaches for interfacing with living things. Bioengineering is the application of the principles of engineering and natural sciences to tissues, cells and molecules. This can be considered as the use of knowledge from working with and manipulating biology to achieve a result that can improve functions in plants and animals (Wikipedia, retrieved 22 Aug 2020).

On a parallel level to biosociety, we will witness the emergence of a digital meanings society, in which the economy is based on the production and consumption of meanings and meaningfulness. These two phases of societal developments – biosociety and digital meanings society – could collide, interact and merge. This does not, however, happen automatically. Alongside with the techno-economic and sociocultural development or even a philosophical stance or social mood (Casti 2010), much is based on the political decision-making, business interests and the preferences of citizens. Even though something is technically feasible, and economically even profitable, it may not always be socially acceptable. Promising biotechnical and digital innovations have to go through transparent technology foresight and assessment processes. The next section proposes the concept of digital meanings society, the relation of which to biosociety to be further discussed in Sect. 6.

3 Digital Meanings Society

We are anticipating to witness the emergence of a digital meanings society, in which the economy is based on the production and consumption of meanings and meaningfulness. In Fig. 23.1, there are three options for alternative development of the information society into the next societal phase. Digisociety is a label for depiction of a society where digitalization is the main driving force and a cornucopia for industry and businesses. Information society was based on applications of ICT. Digisociety will be based on applications of the Internet of things (IoT), big data, geographic information systems (GIS) and artificial intelligence (AI) (Heinonen 2016). It is fundamentally technologically driven futures image, lacking attention to wider sociocultural implications. Ecosociety is a futures image of a society that focuses on ecological sustainable development. Here as well, social and cultural aspects may become neglected. Ecosociety can also be described as biosociety when wider approach is adopted beyond ecological sustainability, reaching out to the whole biosphere and including philosophical reflections on life and humans' relation to it (as discussed in Sect. 2). The third option is a futures image of experience society. Experience society is a society where the main driver for economy and lifestyles is search for experiences. Experience economy perceives experiences as tradable commodities. Pine and Gilmore (1999) argue that businesses should orchestrate memorable events for their customers and that memory itself becomes the product, i.e. the experience. Rolf Jensen's (2001) vision of an emerging dream society is closely related to the notion of experience economy. There the main focus has moved to using imagination from using information as a base for making business.

Digital meanings society is a vision of having elements from all the above three societal phase notions. From digisociety, the potential and applications of digitalization are taken to the full, but not as such but after critical technology foresight and assessment. From experience society the idea of immaterial contents for economy and well-being is brought to a more profound level of humans reflecting upon their purpose and role in life. From biosociety this search may find its fulfilment (see Sect. 6).

In the futures projection of digital meanings society, in employment the main goal of workers and corporations would be to produce meanings and meaningfulness, both for themselves and for others. Meaningfulness would derive from the workplace communities instead of corporate culture or predefined professionality. Networks would ensure that communities are not isolated and that cultural meanings circulate freely, strengthening employees' identity construction. The ethos of providing for the common good would accentuate meaningfulness. The same would result from the "rise of humaneness", as highly developed automation would leave to us the tasks through which we can express our humanity. In other words, digitalization fosters the transformation, which reaches far beyond mere technological aspects. Meaningfulness presupposes a "unity", as people feel their life has purpose and meaning when they see themselves as a part of and work for something bigger than themselves (Ruotsalainen et al. 2017).

If the Internet evolves to penetrate all available niches in its environment and ecosystem by following a rhizomatic model of biomimicry, will such growth escort us towards digital meanings society where people using the Internet are empowered by their search for meaning in all their activities? Digitalization is indeed the enabling technology for people, companies, nations and humankind to search for meaning in life – through lifestyles – and to create sustainable solutions (Heinonen 2016). If we perceive the next stage of information society as biosociety – fused with digital meanings society – then the role of the Internet and AI is to help all sectors and actors of society to save energy and move towards renewables, reuse and recycling. The Internet itself and all digital platforms should also function in a way that wastes little energy and natural resources and minimizes the generation of hazardous electronic waste.

4 Human Interaction with Nature: Conceptions and Typologies

In this section first various conceptions of human relation nature and especially to progress or decay of culture and societies are discussed. Then typologies and new metaphoric dynamics of human interaction with nature are proposed. This is intended to provide background to reflections on an ideal human interaction with nature in biosociety.

4.1 Human Relation with Nature Through Progress or Decay or Cycles

Human relation with nature is intertwined with our relation to technology. Consequently, any analysis of the interplay between nature and technology should be embedded in the broader context of the dialogue between humans, nature and culture. Technology can be examined as a domain of artefacts within a culture. Therefore, technology forms part of culture in the counterposition to nature. Human interaction with nature through technology can accordingly be examined from the point of view of cultural evolution. I will attempt this briefly in the following by discussing ideas about the development of civilization – looking first at the shape of the future of society as some ancient philosophers saw it and then moving to address humans' relation to technology in the next section.

Several scholars have reflected upon ancient Greek conceptions of the development of culture, some of them classifying such conceptions according to different models (Sihvola 1989, 2). These classifications can essentially be reduced into two competing models. First, there is the model of historical decay, which was thought to have begun in an original Golden Age when life was easy and nature abundant.[4] The second model is that of cultural progress, considered to have arisen from primitive beginnings and gradually advancing. In addition, there is a third mode, the model of cyclic development (as mentioned previously). Here the development of culture is perceived to consist of successive periods of decay, which after their full cycle start the process anew. The cyclic model may also comprise both of the previous models within a cycle.

The cyclic view is the most important element in ancient descriptions of the world ages. This notion of the endless cycle of human existence and the whole world is of oriental origin. It is based on the belief that analogously to the shifting turns of day and night, and of different seasons within a year, the whole universe is subject to a recurring cycle. The cyclicism of nature is associated with the idea that such alternation moderates the world. The idea of the world periods was probably already known to Anaximandros.[5] Periodic shift of world ages was particularly important for Heraclitus who saw everything in a process of change. Empedocles' world system was also cyclic, though it is difficult to reconstruct its various stages.[6] The Stoics were eager proponents of the cyclic theory for humankind and the whole

---

[4] See, e.g. the Roman Stoic philosopher Seneca who deals with the origin and evolution of culture in great detail within his account of the Golden Age, mainly in Ep. 90. See also Heinonen (2000) for a more general level with Seneca's view of humans as progenitors of civilization and their role in this progress as well as with their approach to nature and technology.

[5] Schwabl (1978), RE Suppl. 15, s.v. Weltalter, 840

[6] Ibid., 841. According to Empedocles' hypothesis, the natural elements (earth, air, fire and water) combine to become One, the primal mover being Love. Then they disintegrate into Many due to Hatred. After the disintegration has reached its maximum level, love starts again its integration process. The cosmic history is an eternal repetition of this cyclic pattern. The ancient doctrine of the world periods was discussed in great detail in Greek philosophy after these thinkers. See Cairns (1962, 206).

universe.[7] The world cycle was seen as a world year (Gr. *megas eniautos*, Lat. *magnus annus* and Sanskr. *mahayuga*) consisting of different world periods (van der Waerden 1979, 254).[8] At the end of a cycle, there was a cosmic upheaval, usually a world fire (Gr. *ekpyrosis*, Lat. *conflagratum*) or a flood.[9]

Plato developed the cyclic theory in his dialogues.[10] In his version, each full cycle was succeeded by another cycle in the opposite direction through a natural catastrophe. This movement, and in particular its change of direction, was directly proportional to the fate of humankind. The forward movement launching a cycle was the age of Kronos and harmony, the poetic Golden Age. A full turn of the cycle was always followed by a cosmic revolution, fire or flood, which destroyed all organic life. After the catastrophe the movement turned backwards in the cycle, creating an era of discord until another catastrophic terminal was reached and the Golden Age returned.[11] A full cycle of the world and humankind consisted of two world years developing in opposite directions, punctuated by a natural catastrophe at regular intervals.[12]

The main principles of the cyclic theory resemble one notion of modern astronomy concerning the birth of the universe and of its evolution, i.e. the "throbbing space" theory.[13] The cyclic theory of the development of the world and humankind

---

[7] In fact, the Epicureans who advocated the idea of linear progress also adopted the cyclic notion of the evolution of nature and culture, though on different grounds than the Stoics. According to Democritus, the world consists of an endlessly large amount of atoms of different size, weight and shape. This creates the variation of phenomena that exists in the world. Since the number of atoms is infinite, while the choice of species being finite, Epicurus concluded that the world had to repeat itself infinitely. All the thinkable combinations of atoms that have formed different worlds must already have existed, and they will appear again. The regeneration of the world is, however, determined by pure chance and not by *logos* as the Stoics claim (Cairns 1962, 222–223).

[8] Heraclitus tells us of the great world year, which consists of 10,800 solar years (*Aet.* 2, 32, 3; Censor. *De die nat.* 18, 10) as pointed out by Schwabl (1978, 841). Moreover, Cairns (1962, 205) refers to another tradition, according to which the length of a world year is 18,000 years.

[9] Cf. the cyclic theory with the Buddhist cosmic evolution theory of the ages of the world (*kalpa*). Besides fire and water, wind varies according to a mathematical series as an element of the destruction of the world. In Buddhism, the theory of the world ages is of old origin. However, it does not represent so much a philosophical, religious or mythic doctrine but rather "secular knowledge" (*lokayatika*). L. de la Vallée Poussin, *Ages of the World* (Buddhist), in Encyclopaedia of Religion and Ethics, vol. 1, ed. By J. Hastings, Edinburgh 1908, 188–189.

[10] For example, Statesman 269c–271c and 272d–273e; Tim. 22a-d and 39df; Republic 546a and 614b–617d, Kritias 109d–110a

[11] Smith (1908, 197–198)

[12] Plato calls the world year the Perfect Year, pointing out that humans know only a little about it. According to Censorinus, Aristotle calls the world year *as maximus pro magnus annus*. Plato Tim. 39de; Censor. De die nat. 18, 11; Schwabl (1978), 843–844.

[13] On the other hand, Hawking (1988, 137) concludes that the "edgeless" model of the universe may abolish the prevailing Big Bang theory. Neither, according to the so-called logarithmic spiral theory of the universe, is the expansion of the universe based on drastic explosion, nor will it end in another such explosion. Instead, all is ultimately considered to expire quietly, and this silence will regenerate a new universe. This view is concerned with the cyclic notion of the development of the world, though without any intervening catastrophes.

was an important doctrine, especially for the Stoics, who also brought distinct ethical elements to the descriptions of the world ages.[14] In earlier accounts in ancient classical literature, ethics did not play a significant role. The Stoics, however, paid main attention to the ethical contents of the theory.[15] According to them, it was essential to examine human relationships with fellow beings, the surrounding world and the state.[16] The Stoics regarded the universe as machinery which is periodically destroyed and regenerated, similar even in detail.[17] To the Stoics, living in accordance with nature was the basic human duty.

The destruction of the world through fire at the end of a cycle is a kind of purification, i.e. *katharsis*, meaning a return to a primitive stage which is the purest part of cosmic evolution. After the fire, all evil has disappeared, and the formation of a new world may begin. Thus, the emerging universe is superior compared to the situation before the fire. This view is ethical. The cyclic development of the world guarantees that the decay following the primitive stage does not proceed eternally. The world fire means a drastic upheaval: it does not only stop the decay process but also implies a real renewal, i.e. return to life in its perfect form.[18] We should note that there is no progress when comparing different cycles with each other. The same beings exist in each cycle in the same roles. All development exists within a single cycle. The Stoics do not see the development of the world as an uninterrupted course towards a goal that is never reached. On the contrary, they hold that the universe reaches its perfection at regular intervals, within each cycle.[19] The famous aphorism concerning the repetitiveness of history *nihil novi sub sole* fully fits the Stoic notion: what is happening now has already happened in the past, and will happen in the future innumerable times.[20]

---

[14] Smith (1908, 196). In explaining the world, the Stoics advocated the cyclic theory and not the continually rising or decaying pattern of development.

[15] One of the most prominent representations of the world ages was provided by Aratus in his *Fainomena* (897–149), where Hesiod's doctrines were moulded in a Stoic view.

[16] Smith (1908, 196)

[17] SVF II 625; Sandbach (1975, 78). Besides the Stoics, the Pythagoreans also probably believed in the cosmic world year and the cyclic return of the world in the same details. Therefore, they have been characterized as absolute astrological fatalists, though on insufficient grounds. Most likely the Pythagoreans, as well as the Stoics, aimed at some kind of compromise between determinism and free will in the explication of fate and the future. H. Thesleff, *The Pythagoreans in the light and shadows of recent research*. Mysticism, Scripta Instituti Donneriani Aboensis 5, ed. by S. S. Hartman and C-M. Edsman, Stockholm 1970.

[18] SVF II 598; Verbeke (1964, 12)

[19] Sihvola (1989, 30) points out that if we consider the myth of the five species as an account of the moral decline of humankind, the cycle of decline may rather be conceived as occurring in the history of each particular species than in the history of humankind as a whole. The golden species or age is the only exception in this chain of declining cycles. It has even been questioned whether the childishness of the silver species was in fact a less valuable characteristic than the strength and warlike nature of the bronze species. von Fritz, K., *Pandora, Prometheus and the Myth of Ages*, RR 10: 238, 1947; Fontenrose, J., *Work, Justice and Hesiod's Five Ages*, CPh 69, 8. 1974.

[20] Verbeke (1964, 10–11)

A primitive society with its traditions based on myths and transmitted from one generation to the next had to give way to a society based on law, reason and rationality. Rationality started to oppose irrationality prevalent in traditional social orders. Greek natural philosophers tried to give grounds to law and justice in an analogy based on nature's own law. They all shared the belief that it is possible to understand nature by means of reason (Gr. *logos*, Lat. *ratio*). They also suggested that nature is "logical" or "rational", i.e. it follows certain laws. The logical understanding of nature meant recognizing the fact that there is a natural explanation for how everything is created. The ancient view of nature included the idea of unorganic nature as well as an organic nature full of life (Sørensen 1984, 18).[21] The semantic scope of the word nature is illustrated by the Greek word *fysis* meaning growth, whereas the Latin word natura referring to giving birth. Lovejoy and Boas (1935) distinguish 66 different meanings for the conception expressed by the term natura.[22]

Accordingly, in the origin of civilization, nature was important to humans **both as a model and a resource**. On the other hand, technology plays a central role in cultural evolution based on human language and the capacity to think, intellect and sagacity. The ancestors of humans have for million years designed, prepared and used tools to facilitate and intensify their interaction with nature and other environment. The first technological revolution can be said to have occurred with the introduction of the first elementary stone axes. It could also be claimed to have taken place with the introduction of fire. This Promethean symbol gave rise to a variety of technologies of fire.

4.2 Typologies and Metaphoric Dynamics of Human Interaction with Nature

After briefly dealing with various dichotomies of nature, and with the human relationship to nature seen as being embedded in cultural evolution (decay, progress or cyclicism), a new typology of human/nature interplay is presented here. Accordingly, the purpose of this section is to present, on the basis of Seneca's texts, a typology of representations of human interaction with nature (Table 23.3).[23] I also intend to give as a background an analysis of some important approaches and previous classifications of such interaction, arranged in illustrative tables according to my interpretation. First, a brief introduction to some of the major classifications of views on nature is given.

[21] For the numerous different meanings of natura in Seneca's philosophical writings and tragedies, see, e.g. Boyle 1987, Seneca's Phaedra, Appendix 1, 213–214.

[22] Most Latin references are based on the texts of Cicero.

[23] The term "interaction" has been chosen to point out the mutual action in the relation between nature and humans. Accordingly, attention is paid both to the attitudes of humans towards nature and the action of nature on humans.

On Previous Classifications

There are several typologies to classify various views on nature. One "watershed" in dividing different approaches to nature is concerned with the dichotomic position of nature as *instrumental* or of *intrinsic value*. Nature is a necessary prerequisite for our existence, which logically justifies the view of nature as of intrinsic value to humans.[24] Such an intrinsic view of nature is associated with the idea of self-sufficiency of nature. Nature could do without humans, having a value in itself, whereas humankind could not survive without nature.[25] However, in modern times, nature is most often approached from the narrow, instrumental viewpoint, and especially from that of economic profit making (Haila 1990, 11–13).

There are, in principle, two main types of ideas concerning the relationship between humans and nature (von Wright 1987a, 59). In the first type of approach, nature is seen as a larger entity to which humans belong, and according to whose overlaying principles they direct their lives. The other view considers nature as an opponent to human beings that must be overcome (Table 23.1). This happens either so that humans defend themselves against its impacts[26] (storms, floods and wild animals) or they domesticate it under their own rule (domestic animals, agriculture and technology).

We could interpret these classes in a way that the unity approach represents the view of nature having intrinsic value (see e.g. Malaska 1997). The opposition approach as divided into the class of struggling for survival is neutral, whereas the class of dominating nature adopts the notion of nature as having instrumental value. Here it must be noted that the distinction between nature as instrumental and having intrinsic value has to be kept apart from the distinction between morally acceptable and non-acceptable attitude towards nature. This becomes understandable when we think that the instrumental view may also be morally acceptable.[27]

Another classification is given by Pietarinen (1994, 290) of four different categories: humanism, utilism, mysticism and naturism (Table 23.2).[28] He emphasizes that the quality of culture and social life in different periods of time has been determined primarily by the dominant idea of the role and rights of humans in nature. The

---

[24] This argument could also be interpreted to justify the instrumental view. Nature could be seen as an instrument for the very existence of humans.

[25] If nature is destroyed, humankind would disappear as well. This point of nature's intrinsic value could, on the other hand, be questioned by asking for whom nature would have value if humans did not exist. Values are usually taken to exist in relation to humans.

[26] Haila (1990, 15) points out that moral responsibility begins only after humans have been liberated from the shackles of nature. Humans cannot be morally demanded something that is not possible for them. The two classes defined through their opposition to nature are also distinguished from each other in the moral dimension. In the group where humans defend themselves against nature, it is a question of survival, whereas in the group where humans conquer nature, it is a question of the moral right to perform such action.

[27] Nature can be conceived as an instrument for some good, morally acceptable cause. Thus, the instrumental view in itself does not deserve moral reprimand.

[28] The term "utilism" does not mean the same as the well-known ethical doctrine of "utilitarianism" (Pietarinen 1991, 581).

Table 23.1 Two main views of the relationship between humans and nature (based on von Wright 1987b)

Approach	Unity	Opposition	
Position	Humans within nature	Humans against nature	
Goal	Accommodate themselves in nature	Defend themselves against nature	Conquer and dominate nature

Table 23.2 Four types of human attitudes towards nature (based on Pietarinen 1994)

Approach	Humanism	Utilism	Mysticism	Naturism
Position	From nature	From nature	With nature	With nature
Goal	To utilize nature's potential for the civilization process	To conquer and dominate nature	To experience unity with nature	To be a species among the others

humanistic approach sees nature as serving humans in perfecting themselves. It is not possible without intervening in nature, without large-scale "humanization" and rational control of nature. This approach sees nature, as such, as an undeveloped, rudimentary and hostile world with great potential for serving the civilization process. Reconciliation between humans and nature is regarded as possible. Technology should be developed in such a way that it serves the goals of humanistic civilization. The humanistic approach is optimistic with respect to the continuing process of perfection, though intervening hardships have to be encountered from time to time (Ibid., 291). It is thought that neither humankind nor nature can perish completely.[29] The basic dilemma of humanism is evident: should an extensive high civilization be developed even at the risk of radical destructive changes in nature – at least in traditional biological nature? Or should we give up traditional ideals of civilization to save nature from damaging effects (Ibid., 291–292)?

The utilistic approach is based on possessive individualism, i.e. desire of individuals to increase their property. Historically, this means a crucial change in the view of nature at the beginning of the modern age. Nature is now seen as a utility in the market economy or as an instrument for producing such utilities. This change of attitude is associated with the demand for harnessing science to serve production. Profit thinking and seeking economic productivity and welfare have been central social forces since antiquity.[30] However, what is new in possessive individualism is the *limitless* search for profit, and it's becoming the main goal in human life (Pietarinen 1991, 582). Utilism sees nature ultimately as nothing but a causally working mechanism and a source for energy and natural resources. Technology is developed and used to increase the efficiency of the exploitation of nature, and this

[29] Compare this to another view called conditional optimism, represented, e.g. by von Wright, which admits the possibility of final destruction. von Wright believes that the direction of development can be changed *if* we wish so. All responsibility rests on our shoulders.

[30] Profit thinking must have been a supporting pillar in the political life in ancient Athens, witnessed by Plato's severe criticism on the search for welfare.

is taken to be self-evidently justified. Science and technology are considered as feasible means for solving theoretical and practical problems of human life. Utilism is the dominating approach today, which directs the planning and development of all Western societies. This dominance is reflected in the techno-optimistic effort which is put in building many information societies. von Wright has often discussed the problematic element in this approach. Utilism cannot clearly justify its goals, which results in a legitimization crisis. Technological progress has become a purpose in itself, which blindly guides humankind towards an unknown goal (von Wright 1987b, 51; 180–182).

The third kind of approach to nature and technology is mysticism. Cultural transitions form a fertile soil for tendencies that abandon scientific and rational thinking in general, and look for a completely different meaning in life. The normless state, typical of a transition, often makes individuals resort to religion and mysticism. Mysticism aims at providing an experience of unity, the feeling of being in unison with something utterly great and powerful. The restrictions of time and space disappear and one becomes merged into the boundless eternity and sanctity. God-centred and nature-orientated mystical movements have been common in the Western culture, and usually they are combined: God is identified with nature; God and nature become one (Pietarinen 1994, 292).

Nature in mysticism has again an appeal to humans probably as a counterbalance to the highly developed utilistic culture. In mysticism, nature is conceived of as spiritual and divine, like humans, too. The attitude to science and technology is negative. These are regarded as violent to the spiritualness of nature and thought to destroy human beings' possibilities of attaining the experience of unity. The spiritualness of nature is believed to represent a force that cannot be destroyed completely (Pietarinen 1994, 292). Mysticism demands natural conservation and protection. And yet, it is faced with the same problem as humanism: how to reconcile humans' subsistence and the respect for nature? This dilemma is even more acute here, since mysticism requires nature that is as intact and original as possible, which would mean abandoning material production. However, return to a primitive natural economy does not necessarily guarantee the growth of spiritual consciousness. Therefore, mysticism remains mainly as a personal solution. It does not provide a general programme for solving satisfactorily the problem of reconciling human life with nature (Ibid., 294).

The fourth kind of approach is naturism. Darwin's evolution theory paved the way for changing the views of living nature, and especially of the role and significance of humans, from anthropocentric to biocentric. It showed how even "homo sapiens sapiens" is part of nature, a biological organism evolved according to nature's own principles.[31] Naturism declares the principle of equity among various

[31] Darwinism could also be used as an argument defending the special position of humans. The intellectual and moral superiority of humans could be argued on the basis of the principle of the survival of the fittest in Darwinism. Such an argument, however, is not valid for justifying the right to act with no respect for other living creatures (Pietarinen 1994, 293). It has been argued, e.g. by Singer, that animals, at least sentient ones, have the same basic moral right to life and freedom as humanity has (Singer, P., *Animal Liberation*. Avon Books, 1978). Further, Paul W. Taylor has

living species and is accordingly based on biocentrical ethics. Because it is obvious that the welfare of humans is developed at the expense of nature, we should give up our privileges and prerogatives (Pietarinen 1994, 293–294). Naturism considers as justified a modest lifestyle which disturbs nature as little as possible. The number of humans living on the earth should be small. Since such a lifestyle cannot be expected to be voluntarily accepted by the majority of humankind, one conclusion is that only two options remain: the decline of nature continues or the industrial system is broken and humans are forced to another lifestyle.

Naturism sees nature as an ecological system based on certain biological principles, and humans are part of it. The development of scientific technology is viewed as detrimental, since it endangers the existence of other species and disturbs respect for nature. An ecosystem can only function well if humans give up their special position and culture, and realize their status as one species among the other species. Otherwise, nature will perish. One of the great problems associated with naturism is the fact that the majority of humankind will not bargain their present welfare for a simple and harsh lifestyle (Pietarinen 1994, 294). But morally sustainable community life cannot be propagated by force. Therefore, the only realistic possibility for widespread naturism would be if the present utilistic civilization were destroyed by some catastrophe to the extent that there would not be any other option for survival.

This classification could be simplified as follows: the first two classes, i.e. humanism and utilism, could represent one main approach in which nature is regarded as being of instrumental value, whereas the other two classes, i.e. mysticism and naturism, clearly correspond to a second main approach, which considers nature as valuable in itself. However, the relationship is more complicated. One could also classify utilism, humanism and mysticism as belonging to anthropocentrism, whereas naturism would represent biocentrism.[32]

4.3 Categories of a New Typology

Bearing the above-presented ways to view the relationship between humans and nature in mind, we can now turn to examine a new kind of typology. It is constructed on the basis of the representations of the interaction between humans and nature in Seneca's writings (Heinonen 2000).[33] This typology consists of three main

developed a system of environmental ethics which is based on the principle of equal respect for the good of all living beings (Taylor, P.W., *Respect for nature: A Theory of Environmental Ethics*. Princeton University Press, 1987).

[32] Both anthropocentrism and biocentrism can regard nature as intrinsically valuable, but they are still different.

[33] After studying the passages in Seneca's texts where the relation of humans with nature is involved, I have come to the conclusion that all such passages could be classified into one of the three categories I propose here. First, all three categories are represented in Seneca's texts. Second, all three categories emphasize a specific cognitive orientation in the history of culture.

categories, which are called "harmony", "exploitation" and "epistemological expansion" (Fig. 23.2). The categories are first characterized and then illustrated in a collective format (Table 23.3). I argue that these three main conceptual categories adequately represent a cognitive map of human-nature-technology relationships. These categories are somewhat overlapping and intertwined, but nevertheless distinguishable as separate entities in dynamic relation to each other. Within each category, the role of nature and the significance of technology are revealed in different positions (see also Malaska 1994). In fact, in various representations of human interaction with nature, the ultimate goal and outcome of corresponding interaction are different. There is also distinct variation in the orientation of activities in social life as expressed in terms of a way of life.

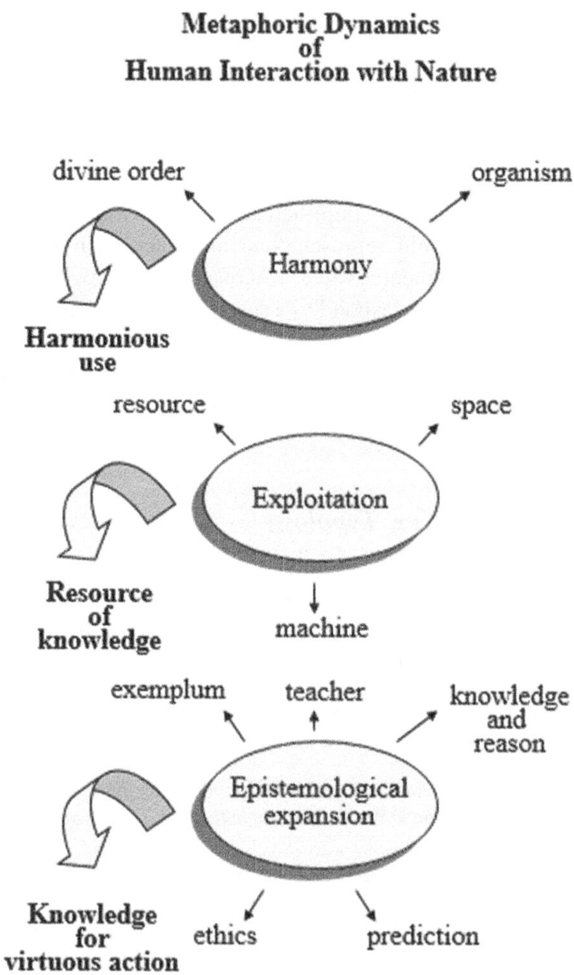

Fig. 23.2 Metaphoric dynamics of human interaction with nature

Table 23.3 A typology of human interaction with nature based on Stoic philosopher Seneca (Heinonen 2000)

Type of representation of interaction	Harmony	Exploitation	Epistemological expansion
Goal of interaction	Happiness, balanced protection, peace, concord	Wealth, prosperity	Knowledge, wisdom
Role of nature	**Partner:** provider of divine order, providence, metaphysical partner	**Slave:** provider of resources slave, instrument	**Teacher:** provider of knowledge model, example, teacher
Role of technology	Provide tools for subsistence; instrument for **cooperation**	Provide tools for utilization of natural resources; instrument for **abuse**	Provide tools for acquiring knowledge; instrument for **learning**
Orientation of activities	Instinct as a way of life	Technology as a way of life	Knowledge as a way of life
Virtues	Tranquility, moderation, modesty, simple lifestyle	Growth thinking, industrious lifestyle	Desire for knowledge mental progress, search for wisdom
Vices	Moral inertia, laziness	Greed, luxurious lifestyle	Intellectual arrogance, hubris, intellectual inertia
Equivalent of societal development	Hunting & gathering society, agrarian society	Industrial society	Late modern society
Legitimization	The right to enjoy	The right to exploit	The right to know
Core	*deus*	*homo*	*scientia*
Mythical-figure	Kronos/Saturnus Tykhe/ Fortuna	Prometheus and Epimetheus	Prometheus revisited → new Prometheus

Each representation also has a conceptual equivalent in the sociological tradition to characterize the development of society as falling into various stages such as hunting and gathering societies, agrarian society, industrial society and late-modern society, be it called the information society, knowledge society, communication society or the biosociety. The basic legitimization, i.e. the logical right related to a certain activity, is different in each category. Therefore, the core or the main focus of the representation of human interaction with nature also varies by category. Furthermore, each representative category encompasses elements of a mythical figure which is portrayed as the metaphor for the character of a specific representation. Different metaphorical contextual settings for each category are also given. The typology is meant to provide a philosophical basis for the information society to be built upon according to the principles of sustainable development – and even further for biosociety.

Harmony

The first category of human interaction with nature is based on the concept of harmony (Table 23.3). In the harmonious representation of human interaction with nature, the goal is a state of balance, concord, peace and happiness.

In Stoicism, the guiding principle is *sequere naturam*, which means living in harmony with nature. It is also one of the most central and repeated precepts of *ars vivendi* in Seneca's *Epistulae Morales*.[34] Following this principle is a necessary prerequisite for human progress towards virtue which is a central doctrine for the Stoics. However, it is not their unique invention. For example, Democritus reflects the idea of imitating nature.[35] Harmony with nature requires the concordance of humans with their own nature (Ep. 4, 8). This kind of *concordia* was realized in the mythic Golden Age, when humans lived concretely in a natural state.

The role of nature in the category of harmony is primarily that of the divine order. The Stoic Seneca claims nature to be equivalent of *deus*. Accordingly, the role of nature is that of a divine order, God, or, in a sense, metaphysical partner. In this respect, the category of harmony resembles mysticism (see Table 23.2). Harmony refers to the balanced state of companionship between humans and nature as embedded between two extremes: absolute subordination to nature and control over nature.

The role of technology in this representation of human interaction with nature is to provide an instrument for cooperation and synergy between humans and nature. Technology, here, is a tool for subsistence in nature. The application of technology is not supposed to exceed the limits of necessity. On the other hand, there is no need for a wide array of technological applications, since nature lavishly provides humans with nourishment and other commodities. In the Golden Age, which is the "purest" representative of this category, technology was not even needed. The orientation of activities is based on instinct as a way of life. Representatives of this category actively seek neither technology nor knowledge. They are primarily guided by natural instincts, and their needs are directly satisfied by nature.

In the category of harmony, the ideal virtues are tranquillity and moderation, modesty and simple lifestyles, which are recommended both in Stoicism and in old Roman ethical codes. The corresponding vices are moral inertia or laziness. If life is easy, there is a risk of ignoring moral efforts. Laziness can be considered as a vice also originating from the ease of harmonious life with nature. However, laziness is a vice rather from the point of view of Lutheran ethics of hard work. It is also reprimandable from the point of view of old Roman thinking, which considered *industria* as virtuous. Nevertheless, it should be borne in mind that in the antiquity, *otium* or leisure free from obligations can be regarded as a most desirable state of affairs.[36]

---

[34] See, e.g. Ep. 5, 4; 41, 9; 98, 14; 118, 12.

[35] Reinhardt 1953, 808. Citations in *Epistulae Morales* show that Epicurus advocated the same view (Ep. 16, 7; 118, 12). *Sequere naturam* was a major principle in Stoic ethics already since Zeno and Cleanthes. The same admonition is present in Seneca's tragedies as well (e.g. Phaedra 481): *proinde vitae sequere naturam ducem*. Boyle 1987, Seneca's Phaedra, 166.

[36] The topic of *otium* was widely discussed in classical literature. See, e.g. Seneca's treatise *De*

This representation of human interaction with nature can be interpreted to pose as a close equivalent of the agrarian or – better still – preceding hunting and gathering societies. As a whole, this category is by and large ideal and mythical. There is no complete concretization of this category on earth, not even among the so-called primitive people. A typical example of this category is in the mythical states of the Golden Age and the Paradise.[37]

The rationality or legitimization of this category as expressed in terms of rights could be described as the right to enjoy. The core of the interaction is *deus*, since the prevailing idea of nature is intrinsic and manifested by the divine order. In the Golden Age, gods even lived among the humans.

The mythical figure representing this category is Kronos or its Latin equivalent Saturnus as the divine ruler of the Golden Age. Its mythical emblem could also be Fortuna, i.e. the goddess of fate, the Roman equivalent of the Greek Tykhe. Fortuna lavishly provides humans with good fortune, but as a fickle figure, she can suddenly impose misfortune as well. Such an approach also fits in the concept of the divine order as the ultimate agent in this category.[38]

Exploitation

The second category representing human interaction with nature in this new typology centres around the idea of exploitation (Table 23.3). In explicit contrast to the previous category is the view that humans can dominate nature. Seneca clearly saw the human attempt to dominate nature. Here, the metaphor of nature as something to be followed (leader) has changed into the metaphor of nature as something to be conquered or exploited (prey).

In the representation built on exploitation, the ultimate goal of action is a state of wealth, prosperity and even luxury. The goal is to acquire, specifically, material wealth. Exploitation is a derivative of conquest. The dream of conquering nature is one of the oldest that has, in Mumford's words (1963, 37), "flowed and ebbed" in the mind of human beings. When analysed at a deeper level, the modern crisis of sustainability can be traced to a drive to dominate nature that is evident in Western science and technology (Orr 1992).[39] The idea of domination over nature rests upon the idea of separation of human beings from nature. However, the human enterprise

---

Otio.

[37] The humans of the Golden Age cannot, strictly speaking, be compared to agrarian, or hunting and gathering societies. This is because agriculture was not practiced and animals were not hunted. Neither did they have to put much effort into gathering food: all that nature lavishly produced could be picked at arm's length.

[38] Distinction must here be made, though, between the Stoic view of divine order as Providentia, which always knows what is best for humans, and Fortuna, a fickle divinity imposing things on humans at her own will.

[39] Exploitation and domination of nature is built upon a mistaken coupling of the particular mode of mechanical progress with an unjustifiable sense of increasing moral superiority of humans (Mumford 1966, 4).

cannot really be separated from the natural world even in our minds, because no such separation exists in nature (Wackernagel and Rees 1996, 4).[40]

The category of exploitation is intertwined with the power of knowledge. Bacon thought that if humans can systematically pose questions to nature and modify the answers, they can also control reality, i.e. utilize natural resources and direct natural forces according to their plans and desires. The questions made to nature are tests, while the answers obtained are natural laws. Science as a means of power is called technology (von Wright 1987a, 27).[41] Bacon was convinced that human mastery over nature would greatly benefit individuals and society. Similar, rationalist views on happiness were shared, for example, by Thomas More in his utopia and Campanella in his *Civitas Solis*.

Exploitation leads to the idea of domination. von Wright (1987b, 72) claims that Christian transcendentalism and the modern technological way of life have the same origin.[42] The idea that human beings *can* dominate nature germinated from the same seed as the idea that they *must* turn away from nature towards the supernatural. However, in the Middle Ages, the idea of the technological way of life is beyond the reach of humans. This is because it is not the same as the idea that humans can dominate nature. Technology means a *systematic use of science to achieve such domination over nature*. Science in the Middle Ages sought to serve humans' search for God, not their mundane ambition. The orientation of this category resembles partly humanism and partly utilism (Table 23.2).

In this category of human interaction with nature, the main focus of the relationship manifests itself in exploitation. Such exploitation is further divided, on one hand, into moderate utilization and, on the other hand, into excessive exploitation or abuse. It can be seen as a process, the speed of which can be slow or accelerated.

The role of nature in this representation is the provider of resources. Metaphorically, nature becomes slave and instrumental machine to cater to humans' various needs. The role of technology is crystallized into providing a tool for utilization of natural resources. In the extreme form of this category, the role of technology becomes that of an instrument for abuse. The orientation of activities is based on technology as a way life.

In this "exploitation" category, the predominant virtue is growth thinking. Growth thinking as an idea of continuously increasing one's material wealth, and

---

[40] In terms of energy and material flow, the human economy is a fully dependent subsystem of the ecosphere. Wackernagel and Rees (1996, 4) suggest that humanity's role in nature should be studied in much the same way as that of any other large consumer organism. Humankind has become a major (and in many cases the dominant) species in practically every significant ecosystem on the planet.

[41] Science could be thus conceived as comprising both the power to know the laws and the power to use the laws.

[42] As von Wright (1987b, 185) points out, the ultimate legitimizating grounds for our modern technological way of life has been the notion that God has given humans the right to dominate nature. Later, however, as the power of humans in relation to nature grew, their self-sufficiency increased as well. Accordingly, the right of humans to dominate nature became self-made and needed no other justification.

thereby general welfare, is rather a modern concept. As a virtue in this category, industriousness or industrious lifestyle could better still represent both a Stoic and old Roman attitude. The corresponding vices are greed and luxurious way of life. *Avaritia* and *luxuria* are vices that Seneca as a Stoic frequently disapproves of. Luxurious way of life as a vice must be understood here as a forum for lacking virtues of modesty and simple way of life, which were highly esteemed in old Roman ethical codes.

This type of representation of human interaction with nature finds an elaborated equivalent in modern industrial society or consumer society, where material wealth is often seen as the pervading overall goal in society. In Seneca's texts, this type of human interaction with nature is seen as the end of the Golden Age. In other words, through the introduction of the above-mentioned vices, the attitude of humans towards nature was no longer to follow nature, but to extract material good from nature.

The rationality or legitimization of this category, as expressed in terms of rights, could be described as the right to use or the right to exploit. It is a dynamically evolved and accentuated form of the right to enjoy, the rationality of the previous category of harmony. The core of the interaction in this category of exploitation is *homo*, since the prevailing idea of nature is instrumental.

The mythical figure for this category is Prometheus, symbol of technological progress with its subsequent drawbacks and implications. Another appropriate mythical figure here is Prometheus' brother, Epimetheus, or "he who reflects the past". He represents the neglect of long-term futures thinking, which is prevalent in this category. Short-term profit thinking is the preferred approach. In addition, Epimetheus is stupid and curious enough to accept Pandora's box with dire consequences for humankind. The metaphorical setting is a "mine" for exploring the riches of the earth or a "test field" for exploring the secrets of nature. After the enjoyable Golden Age, nature becomes an exploitable "gold mine".

Epistemological Expansion

The third category representing human interaction with nature in the proposed typology focuses on the idea of "epistemological expansion" (Table 23.3). Human interaction with nature can be represented here as epistemological expansion, where the role of nature is that of a model, example or teacher. Such interaction is based on a certain apprenticeship of humans to the teachings of nature. In the representation focusing on epistemological expansion, the goal of interaction is a state of accumulated knowledge and wisdom. In other words, through nature, humans can raise their level of knowledge, not only quantitatively but also qualitatively. Thus, the goal is to acquire, specifically, immaterial wealth. Indirectly, this can, of course, lead to the growth of material wealth as well. The point is that the growth of immaterial wealth is considered a superior goal.

The role of nature as a provider of knowledge is twofold. First, nature can provide humans with knowledge of herself, of all that relates to nature. Second, nature can provide humans with knowledge, or indirectly, at least, vehicles for knowledge of themselves and precepts for behaviour. Both these levels are present in Seneca's thinking.

Epistemological expansion deserves to be a category on its own, since it deals exclusively with the immaterial level of human interference with nature, whereas the other two dwell upon the material level as well. Besides, in the exploitation category, the power of knowledge (and technology) was applied for the purposes of material exploitation, whereas, here, the power of knowledge is used for learning and moral progress. Seneca reminds us that the growth of understanding is itself the greatest reward from studying nature. Instead of financial remuneration, cognitive satisfaction is being sought. Furthermore, in the category of exploitation, nature is a resource to be exploited, whereas in the category of epistemological expansion, nature is a resource to be respected (i.e. to be used with respect). In this category, nature is a resource for knowledge, though not intended merely for knowledge exploitation (as in the category of exploitation) but for knowledge creation. In the category of exploitation, technology dominates the orientation of activities, and technology means a systematic use of science to achieve such domination over nature. Instead, in the category of epistemological expansion, the search for knowledge is the dominating orientation. This means a systematic use of science and knowledge to achieve domination over technology and allegiance to nature.

von Wright (1987a, 56) reminds us that knowledge can be a way of life. Then the value of knowledge manifests itself in the human thirst for knowledge. Knowledge as a way of life means a desire for learning and understanding for the sake of learning and understanding, without any other goals. Therefore, this category could be characterized as focusing on knowledge as a way of life, whereas in the previous category of exploitation, technology symbolizes the dominant way of life.

Aristotle called the human being a rational animal. Rationality is undoubtedly the characteristic that most profoundly distinguishes humans from (other) animals in the biosphere. Accordingly, one can say that the human drive for rational perfection means striving for the realization of "self" or "real nature". Knowledge as a way of life is, to humans, the most typical life pattern where their human nature is most clearly manifest. Therefore, we can share Aristotle's view that "theoretical life", i.e. the increasing knowledge of the world and realization of basic issues of life, provides the highest and purest happiness.[43] Humans should strive for rational perfection, but they should beware of the illusion of perfect complacency. This category of epistemological expansion can be seen as a struggle for rational perfection.

The answer to the question about the value of rationality is twofold. Knowledge is as much an instrument of happiness as of misery, of good deeds as of malice.[44] Knowledge as a way of life can turn out to be the highest and the lowest alternative for an individual. A presentiment of this ambiguity of human rational capacities contributes to the depth and beauty of the myth of Prometheus, which is tragic.[45] It

[43] See von Wright (1987, 57).

[44] von Wright (1987, 59). The same ambivalence and potential for good and evil dwell in the concept of technology (see more on this in Chap. VII). The underlying idea and the analogy are of course that technology is based on some knowledge.

[45] The same is valid for the paradise myth and the legend of Faust, in a sense that they present us humans as torn between two dominating powers, light and darkness.

presents a being who has overestimated himself but basically fights for a rightful cause. The smothering of Promethean drive in humans would mean crippling their humanity, depriving them of their freedom and, from the religious point of view, cutting their closest relations to divinity.[46] The Promethean endeavour for knowledge and rationality is seen in this category of epistemological expansion as a refined effort to support the moral progress of humans.

There are two rational relationships between humans and nature: theoretical and practical. The rational theoretical relation to nature, or objective reality, is called a scientific approach. Its core idea is the realization that we cannot force nature to adapt herself to our notions about her structure and functioning, but that we have to adapt our opinions to fit the reality. It is questionable whether the ancient thinkers fully succeeded in rationalizing their theoretical approach to nature. Their science remained, to a large degree, speculations about the structure of nature. Consequently, it had a subjective tone which makes it look unscientific to us (von Wright 1987b, 85–86).

The rational practical approach to nature could be called judgement. Its core idea is to set our goals with respect to the impact of present wishes on a longer term. The ancient culture was more aware than ours of the necessity to rationalize the practical approach to reality. This is understandable, since it was a strange idea to the Greeks that the rational theoretical approach to reality could be used to dominate nature or intervene and direct the causal relationships of nature for the benefit of human needs. von Wright (1987b, 85) claims that they easily adopted the view that saw the *obedience to the order of nature as the highest norm of a rational way of life*. Instead, we often forget that the rationality of our goals is determined by nature, i.e. by causal requirements and consequences. In order to overcome the difficulties in adapting to technological progress, a counterpart of rational practical life would be needed to the rationalization of the theoretical approach which took place in Europe three centuries ago within the scientific revolution.[47]

Practical impacts of the scientific revolution form a new stage in the history of humankind. So far, there are a few experiences in the applicability of the technological way of life, which is the rationality of the previous category of exploitation. The problems of the human species that arise from ongoing global integration through technology baffle us. The new technological way of life has, from the point of view of the history of ideas, roots that reach throughout the entire development of culture from antiquity through the Christian Middle Ages to this day.[48]

[46] Antirationalism means self-deceit of humanity. Antirationalism has to be opposed to in our modern age when some ideologies inhibit the free growth of human rationality. On the other hand, rational optimism has to be opposed to in the ages when life is threatened by the flattening impact of self-admiration. Both of these struggles continue to be actual throughout world history.

[47] Making humans as rational actors is partly a question of education and partly, or perhaps primarily, a question of the organizing and planning of society. Mere formulation of the problem may be a first step towards the solution. Problems that are not clearly presented cannot be tackled with any success.

[48] von Wright (1987b, 86) points out that the knowledge of such links and roots may not make us

The role of technology in this category is to provide tools for acquiring knowledge. Technology is primarily an instrument for learning. The orientation of activities is based on knowledge as a way life.

In the category of epistemological expansion, the characteristic virtues are a desire for knowledge, mental progress and search for wisdom, whereas the vices are intellectual arrogance or intellectual inertia. Of the virtues, especially the search for wisdom reminds us of Stoicism, since it is associated with the Stoic virtue of search for moral progress. In fact, the Stoics equate virtue with wisdom (Long and Sedley 1987, 41H, 61B, 65T). The most virtuous human being is the Stoic *sapiens*.

In the modern discussion on societal developments, this category has its equivalent in the concept of late-modern society, information society or biosociety. This is not yet realized but represents a potential. The rationality or legitimization of this category, as expressed in terms of rights, could be described as the right to know. In its extreme form, this means the limitless right to know – to study nature. The difference between this and the category of exploitation is that the right to know and explore nature, in this case, is filtered through a sense of moral responsibility. The core of the interaction in this category of epistemological expansion is *scientia*.

The mythical figure of this category is forward-looking Prometheus, but this time a new Prometheus. The Prometheus thus revised is a figure representing search for knowledge for the purpose of learning from nature and not for the purpose of merely enjoying or exploiting nature in a limitless manner. The metaphorical setting in this category is nature as a school – a place to learn.

In comparison with the previously presented typologies, the unity approach can be said to correspond to the categories of harmony and epistemological exploitation in the new typology, while the opposition approach resembles the category of exploitation (cf. Table 23.2). In another comparison, the approach of utilism corresponds to the category of exploitation, while the approaches of mysticism and naturism resemble the category of harmony in the new typology. The approach of humanism partly covers the category of exploitation and partly the category of epistemological expansion (cf. Table 23.3).

The categories could be regarded as a chronological and dynamic structure of thesis-antithesis-synthesis. The thesis of enjoying the fruits of nature (harmony) was transformed into the antithesis of abusing nature (exploitation), which requires the synthesis of learning from nature (epistemological expansion). The categories do not have to be dealt with as a chronological sequence. They can also be regarded to exist simultaneously, giving a choice of different approaches to nature for humankind to choose from.

wiser in solving practical problems. However, it may alleviate the sense of helplessness which easily besets modern thinking humans when they view the changes in the ways of life enabled by science and realized by its applications.

5 Human Relation to Technology

Elements of humans' approach to technology were partly present already in the previous section discussing humans' relation to nature. This is because the interaction of humans with both nature and technology is an integrated triangle with mutual interconnections and implications. This section moves on to further decipher humans' relation to technology and knowledge.

5.1 The Promethean Thirst for Knowledge Through Technology

Technological knowledge is a genre of human knowledge. This section contemplates the Promethean thirst for knowledge as an element of the Promethean complex.

The ambiguity of technology on one hand related to work and on the other hand associated with a desire for immortality is worth closer inspection. We can do this by way of myths and metaphors. The theft of fire from Zeus by Prometheus is an example of the usurpation of divine prerogative. The myth of Prometheus has parallels with Hindu mythology. The Greek word *ambrotos* for gods' food ambrosia resembles the Sanskrit word *amrta*. The literal meaning of both words is immortality. The name Prometheus may be related to the Hindu hero Pramantha or Prithu who steals the cow of immortality from the gods to help mortals (van de Braak 1995, 15).

The French philosopher of science Gaston Bachelard (1949, 21–31) proposes the idea of the Prometheus complex for analysing the relation of humans with technology. He elaborates to include in the Prometheus complex all the tendencies that impel us to know as much as our fathers and even more. In a broader context, there is a desire to know as much as our teachers and even more than our teachers (van de Braak 1995, 21). The Prometheus complex means that human beings are obsessed with superior technology. *They believe that through technology, they themselves become superior.* The Prometheus complex can be characterized as the Oedipus complex of the intellectual life.[49] According to Bachelard only this Promethean complex permits us to comprehend the interest that is always aroused by the myth of the "father" of fire.[50] To know facts and to make things are needs characteristic of human beings. There is a veritable will of intellect embedded in humans. This thirst for knowledge should not be underestimated, since it is a prime motivator in societal development. However, it should not be immediately associated with utility

[49] However, Bachelard (1949) warns that it should not be confused with the Freudian Oedipus complex.

[50] Gaston Bachelard, *La psychanalyse du feu*, Gallimard, Paris, 1949, p. 31

considerations. The modern Western culture of technology is often symbolized by the figure of Prometheus unbound, fascinated by science, technology and machines and pursuing technological achievements obsessively.

However, there is duality in Western attitudes and dialectic between opposite points of view. Therefore, Prometheus must be revisited, i.e. another approach to technology than obsession is to be searched for (Heinonen 2000). As half of humankind is fascinated by high technology, the other half is already partially in tune with more a contemplative attitude. Besides Bacon's comments on science as dominion over nature, there is also his more insistent view that knowledge should be applied in works of compassion. In fact, Bacon contends that men should seek knowledge and practical skills, not for pleasure of mind, invention, superiority over others, profit, fame, power or any of these inferior things but for the benefit and use of life (Pacey 1992, 114;172). Here Bacon strikes a Stoic chord, since the Stoics regard as *adiafora* – not exactly inferior but indifferent – all other things except virtue. Applying knowledge for benefiting life may well be interpreted as representing virtue. van de Braak (1995, 22) points out that the phenomenon of rivalry is significant in the Prometheus complex. It manifests itself as a rivalling thirst for knowledge.[51] It has led to the modern technology race among industrial nations. Such technology race is illustrated by several nations' urge and investments for space exploration even though huge global challenges still remain unsolved on our planet. The Millennium Project has a framework of 15 global challenges and updates foresight knowledge that would help alleviate these challenges. The UN Agenda propagates rather similarly 17 sustainable development goals. They sum up humankind's most urgent objectives. What is needed is the stakeholders will to act upon them.

Prometheus is a god of primitive apprenticeship. He is a mediator of technology between gods and human beings. van de Braak (1995, 23) hypothesizes that for early human beings, technology was the object of social prohibition. According to this view, substantial progress of technology is therefore not natural but has to break through taboo. Each new "species" of technology would depend on transgressing social interdictions. Simultaneously, the natural phenomenon of any new, major technology in the history of the human species is rapidly mixed in with complex and confused items of social experience.[52] As related to this breaking up of taboos, a sense of guilt is generated. The myth of Prometheus can be interpreted to reflect the guilty conscience of prehistoric humans who ventured to deal with fire. Fire was taboo for mortals in prehistoric times, and in the same way, it is prohibited for today's children. However, the more forbidden something is, the more attractive it

[51] van de Braak (1995, 22) draws an analogy to scientific communities having "founding fathers." A scholar may wish to beat the "king" of the scientific community in order to win the favour of the "queen", i.e. the reader or the colleagues.

[52] van de Braak (1995, 21–22) compares the approach to technology with a child having to obtain personal knowledge of fire only by disobedience. A child is tempted to become a little Prometheus and play with matches, i.e. fire. And yet, the first thing to be learnt from fire as a child is that we must not touch it. Therefore, the social interdiction is our first general knowledge of fire.

becomes (van de Braak 1995, 19–20). The irresistible temptation to break new grounds in technology may be linked with the human will to break taboos in spite of subsequent feelings of remorse.

What the Prometheus myth teaches us is not only the ambivalence in relationship between human beings and technology but above all the representation of technology as transgression in relation to gods. Salomon (1981, 4) draws our attention to the fact that there is a connection between the Greek myth of Prometheus and the Jewish myth of Adam and Eve: in both cases it is the same secular heresy which emancipates human beings with divine powers and condemns them to live in the insecurity of their own work. They experienced the same exclusion and the same uncertainty of their fate. Regardless of the obvious differences between these two myths, in both cases there is still transgression and a price to be paid. Adam and Eve lost the Garden of Eden after having tasted the fruit of knowledge, while Prometheus was chained to a rock for having stolen the fire, i.e. for having given technology and knowledge to humans. Eve yielded to temptation and it is curiosity that resulted in her destruction, expelling her, Adam and her descendants from the garden where God would have rather had them rejoicing under his protection. Prometheus in his turn threw himself into conflict with Zeus who hid the fire from the human race and thus menaced their survival. It was in an act of revolt that he turned to Zeus. This act had a double outcome: Prometheus substituted the natural fire with the technology of fire, and it was a clever trick that caught Zeus by surprise (Salomon 1981, 6).

Transgression did not have the same sense and significance, while the sanction is the same in both these myths: expulsion from the blessed place of God, i.e. Eden, or expulsion from the magnificent era, i.e. the Golden Age. Adam and Eve, like Prometheus, find out that the condition of labour is attached to life. It is a couple leaving paradise. Or it is the man with two faces, Prometheus and Epimetheus, the Forethought and the Afterthought, who leave the Golden Age. At the same time Prometheus was condemned to labour, the riches that had previously arisen from the ground disappeared as a counteraction to Prometheus' theft. Pandora lifted the lid of the box and irrecoverably distributed bad and good to the human race (Salomon 1981, 6).[53]

A parallel figure for Prometheus is Tantalus who was punished for an effort to prove himself as equal or superior to the gods. He wanted to prove he knew more than the gods. He was punished with eternal hunger and thirst in Tartarus. Do human beings have eternal hunger and thirst for new technology? Did this originate from ancient times?[54]

[53] Pandora's box is actually an incorrect expression. It was a great vase that Pandora brought in her arms. By raising the lid, terrible afflictions escaped and spread over the earth. Hope alone was left in the vase (and for humankind), since it did not fly away (NLEM 1985, 93).

[54] According to van de Braak (1995, 20), technical and scientific curiosity may have a model in sexual curiosity. Because sexuality is mysterious, all mysterious things evoke sexuality (Bachelard 1949, 185). All knowledge is originally intimate, personal knowledge, i.e. bodily experiences by the senses. This is being referred to by the expression "carnal knowledge" (from lat. *carnus*). An illustration in point is given by the non-metaphorical use of the term "know" in the Bible: "And Abraham knew his wife Sarah."

Is the hunger for knowledge at the same time hunger for power? Knowledge is power, but is it good for human beings to know? Our perspective will deepen when we recognize these issues in myths. von Wright (1987a, 28) emphasizes that the myth of Prometheus is an expression of the problem associated with the human right to develop rational talents.

5.2 Hubris Followed by Nemesis

The Prometheus complex of human beings can lead to an everlasting technological race and invasion of ever new galaxies of technological ideas and applications. According to the Stoics, human beings have a divine sparkle that can kindle their development towards god. However, this progress is supposed to take place in *virtus* only. What are the limits of science, and what are the limits of technological advances? Maybe *hubris* could be redefined as progress that is void of *virtus*?

Along with advances in biotechnology and modern gene technology, human beings have obtained the arsenal for manipulating life itself, thus becoming god. The control of our own evolution is within reach. Prometheus is revisited, since the acquisition of gene technology is paramount to the acquisition of the technology of fire. It marks the beginning of a new phase of technology and its social implications. The purposes stated for gene technology are often cited to provide plants and species of better quality or to produce new medicines, etc. However, genetic manipulation offers opportunities that evoke emotional responses and feelings of awe. This reminds us of the original taboo attached to a new technique. Mining and melting were tabooed in ancient times. Metallurgy did not rise until about 3500 B.C., simultaneously with writing. On the whole, the taboo on intervening both in physical nature and human nature has loosened its boundaries. Yet, the speed in which technical opportunities arise to create human beings in their own image is so high that it arouses increasing resistance against manipulation of human nature. By that token, a revival of the original taboo on technology cannot be ignored (van de Braak 1995, 27).

The possibilities for expressing *hubris* are broadened along with the evolution of technology. Wagensberg (1997, 95) classifies the development of technology into three phases according to what is the relation of human beings' action to matter. First, he sees the era of **forming matter**. This age started when the first hominoids discovered that certain motions of the hand could be strengthened with the aid of a new concept – the tool. This was the beginning of the Stone Age. The core of technological activities was to find matter and force it into some form.

The second era was concerned with **transforming matter**. Metal was what the stone axe lacked: human beings needed a good cutting edge. This marked the beginning of the Bronze Age 4500 years ago. However, bronze did not exist as such in nature. Therefore, human beings had to transform matter before they could form it. By combining various materials in a talented way, they discovered what they were looking for. The world of Seneca was already living in the era of transforming matter. Wagensberg claims that the majority of innovations serve a purpose. We were on the same level in the world that was invisible, since it was either infinitely small or infinitely large. Technology such as the computer started to help us find ourselves anew in this world.

The third era that we are now living consists of **inventing matter**. There is the possibility of molecular conception. We can consult a list of different properties that allow the invention of desired materials. However, as Wagensberg (1997, 95) asks, is it possible to invent all kinds of material? For example, is it possible to invent living matter? This has so far been reserved to gods. At the moment we already know how to copy or how to correct the work of gods.

The first stone tool, the first bronze tool and Dolly the sheep are symbols of the three ages of humankind. We are living the creation of this third age – the age of making matter becoming organic part of biosociety. Living matter transmits information in two ways: genetically (by genes via heredity) or culturally (in all other ways). Dolly was an example of a radical novelty emerged from the shortcut between genetic information and cultural information. The spark took four billion years to ignite, but finally it did it. Dolly the sheep (generated in 1996) is a practical recognition of a theoretical possibility. It is naturally pregnant with all kinds of bifurcations, of which some will be beneficial while the others perhaps criminal. The world of ideas does not recognize any limits; everything is permitted. How could you even regulate someone's imagination? However, we are all living in the material world, which already represents the third type. The possibilities for succumbing to *hubris* are now in this third age manifold.

Mazlish's (1993, 161) thesis is that biogenetics is the latest step in the attempt to form animals into machines, this time in reality better than in imagination, as was the case with the seventeenth-century animal-machine. Mazlish goes on to elaborate on biogenetic revolution and claims that the modification of the environment and of genes must interact in evolution. He sees that the evolutionary nature of humans – differing profoundly in degree from that of any other animal – has for a long time led them to shape external nature in the image of their own imaging and imagining mind. In this context synthetic biology is opening up even more avenues for making matter.

Human beings have become proud, since they control nature. As von Wright (1987b, 86) claims, the idea that human beings can intervene and direct the causality of nature has become "innate" in us. Therefore, it is easy for us to forget that the rationality of our goals is determined by nature, i.e. causal requirements and consequences.

6 Bioeconomy Manifesting Digital Meanings Society

Bioeconomy at its best could manifest digital meanings society. Then, economy would use the full potential of digitalization, ICT, big data and AI to build a biosociety where the social mood encompasses humans as part of nature and responsible actors for sustainable futures. A challenge for biosociety is to combat the rising metaphor of a human being as a slave, dominated not by nature but by technology. Humans think to have risen beyond nature by technology. They have forgotten that they are part of nature, and paradoxically they have become a part of technology. Bioeconomy and biotechnology could clear the vision of humans having meaningful life as conceived part of nature. Then, humans' relation nature would represent the category of epistemological expansion (Fig. 23.2). The creation of knowledge would be directed to developing bioeconomy in ways that are meaningful to governments, companies and citizens alike.

As stated in Sect. 3, the role of digitalization, the Internet and AI is to enable all sectors and actors of society to save energy and move towards renewables (Heinonen and Karjalainen 2019), reuse and recycling. The Internet itself and all digital platforms should also function in a way that wastes little energy and natural resources and minimizes the generation of hazardous electronic waste. This would mean that digitalization will provide the platform and empowerment for biosociety to function in a meaningful way.

Kevin Kelly (2016) reminds us of the huge potential technology has already in the existing resources. Real sustainable economic growth does not stem from new resources but from existing ones that are rearranged to make them more valuable. According to him "growth comes from **remixing**". In biosociety such remixing possibilities even increase when biotechnology is combined with other scientific fields as described in Sect. 2. Accordingly, biosociety will be a period of productive remixing.

In order to avoid nemesis of ecological or genetic impacts, measures have to be taken. Salomon (1998, 61–62) proposes the categorical imperative of precaution that requires consideration of the unexpected, unforeseen and negative consequences resulting from technological change on future generations. Accordingly, the principle of **responsibility** should be imbued into the process of technological change. Such a conscious ethic on a global scale also means that instead of the "ethics of proximity", the "ethics of distance" must be adopted. In the case of technology, most impacts are concerned with the future perspective, often undeterminably long. We should ensure that the prerequisites for the Aristotelian maxim of good life are not diminished for future generations. At the moment, we are hardly prepared to act like we were face to face with the heirs of the present civilization in the sixth or eighth millennium.[55] The Western logic of rationality has traditionally had a shorter time perspective.

Ethical considerations concerning new technology soon become evident. As related to the application of technology in general, some have proposed that there should be a Hippocratic oath in engineering such as that in medicine (Pacey 1992, 112). Such an oath would, for example, commit high value to technologies that enhance possibilities for the fulfilment of human potential and low value to weapons development. Some rules of conduct and recommendations would be needed

[55] With certain nuclear waste, the time scale of impacts is even longer.

especially for biotechnology and gene technology. Are human beings permitted to change living creatures genetically? Are human beings becoming the creators of themselves? Is *homo autofabricus*, or literally "self-made man", being created (van de Braak 1995, 24)? Furthermore, the scope of possible applications, for example, in gene technology and synthetic biology, is enormous. Will all the unforeseen possible negative impacts from such technological activity represent biological, environmental and physiological effects only? Or could it be that technological *hubris* propelled by the Prometheus complex might end in a fundamental *nemesis*, sowing seeds for changing the very nature of human beings, besides the possible destruction of nature and the earth's biosphere? What are the limits of human beings' urge to build a second nature within nature, so proudly presented by Cicero (De nat. deor. 2,15)? This question is still valid in the biosociety.

The relationship between human beings and nature has for centuries been ruled by the philosophical tradition characterized by the ideas of struggle and domination based on knowledge. This tradition has denied the real roots of humankind that lie in nature. In the face of looming threat of the destruction of human beings, nature and the earth, the counterpoint leads into a lethal duel. This situation requires a new contract. Serres (1994, 11) states that the social contract, which is supposed to prevent human beings from destroying each other, is no longer sufficient. He proposes a natural contract that would procure peace between humans and nature and which would help us to sustain life and all that used to be regarded as created, such as seas, oceans, mountains and rivers – the whole body of Mother Earth. *Nemesis*, due to ideological and technological domination over nature, could be avoided if humans changed their relationship with nature from parasitic to symbiotic. A clear model for such a symbiotic contract with nature would be provided by the Stoic doctrine *sequere naturam*.

von Wright (1987a, 44) questions of what value is freedom if it leads human beings to slavery again? In voicing this question, we begin to understand that the tragedy of the Prometheus myth is deeper than the victory of violence over freedom. The concepts of hubris and nemesis or the philosophy of balance originating in Greek tragedies can be interpreted in various aspects. Is the transition of human beings to the slavery of machines – i.e. of their own inventions made for mastering natural forces – nemesis followed by their hubris? What is then hubris or the raising of ourselves? Is it our takeover of nature? The nemeses are then environmental impacts.

von Wright (1987a, 55) states that today, as human beings we have not fought our way to freedom from nature, since we have not learnt self-control in the utilization of the magic word of science and technology for satisfying our whims and lusts.

According to Jeremy Rifkin (1998), we are living a historic transition into the age of biotechnology. He calls the process where information and life sciences are fusing into a single powerful technological and economic force the foundation for "biotech century". He sees it as a promise of a cornucopia of genetically engineered plants and animals to feed a hungry world and genetically derived sources of energy and fibre to propel commerce to build a renewable world. Indeed, biosociety would go beyond ecological transition; it would require restructuring all core systems of our societies – those of food, energy, cities, economy and infrastructures. Production and consumption patterns should become sustainable, adopting the potential of circular economy.

In biosociety the energy can be generated with renewables, and for citizens it will become meaningful to be able to produce not only renewable energy but also be able to self-produce it on their rooftops. The energy system will become not only renewable-based but also more efficient when artificial intelligence and digital data from multiple sources are utilized and optimized. The whole system of agriculture should be transformed to follow the principles of regeneration to qualify for real biosociety. In biosociety, food production will also move to cover new modes – such as that of producing food without soil. With the help of electricity produced using renewable energy, carbon dioxide extracted from the air and microorganisms, we can produce protein-rich food without the need for land or raising of cattle (Heinonen and Karjalainen 2019). This environmentally friendly method may in the future surpass traditional agricultural food production methods. The agricultural sector is globally the second largest producer of greenhouse gas emissions, after the energy sector. On earth, new farmland is established by clearing out forests while world population and demand for food increase. According to the researchers' calculations, the efficiency of this new process of creating protein-rich food from air is 10 times greater than agriculture, and 100 times greater than meat production when only environmental consequences are considered.[56] Accordingly, biosociety will expand the conventional notion of agriculture, farming and forestry. Forests may emerge as commodities and services for tourism and health sector (see, e.g. Beresford-Kroeger 2019), rather than raw material for paper and pulp.

Biosociety is regenerative and dynamic. In cybernetics a living organism is no longer seen as a permanent form but rather a network of activity. With this new definition of life, the philosophy of becoming supersedes that of being. Rifkin (1998) conceives thus life and mind becoming intricately bound to the notion of "processing" change. Biosociety itself is in the continuous process of change. There are seven core characteristics (all starting with the letter R) that can be considered as prerequisites for meaningful biosociety:

1. Recognition (of humans' role as part of nature)
2. Respect (life as intrinsic value)
3. Rethinking (conventional concepts and relations critically revisited)
4. Remixing (using the potential of technology convergence)
5. Regeneration (changing systems as regenerative, esp agriculture, forestry and communities)

[56] Just 2 years after the Neo-Carbon Energy study was published, the spin-off company Solar Foods is in full swing to commercialize the process, and has started production. The European Space Agency (ESA) has already expressed an interest in using the technology for space travel. The technology could also have applications in developing countries where land is precious. This is how short the innovation distance can be between science fiction and the everyday realization.

6. Responsibility (according to Rifkin 1998, constant anticipation and response are the central dynamics of life; responsibility is built on technology foresight and long termism)
7. Resilience (futures resilience is a result from systematic foresight and responsible action towards renewable energy transformation and peer-to-peer activities; Heinonen and Karjalainen 2019)

7 Conclusions

There is short-sightedness behind many problems, especially environmental problems. Systems thinking, futures thinking and long-term reflections are needed (Masini 1993; Bell 1997; Glenn et al. 2009). If the time horizon of planning were extended, many problems related to governance, society and ethics would be more easily resolved. A new nature treaty must therefore be concluded, based on a new kind of economic, technological and ethical union. "Nature always wins - either with or without humans", Pentti Malaska emphatically stated on many occasions (Pouru et al. eds. 2018). Such a new nature partnership could be taken to environmental education in schools to strengthen environmental awareness and futures thinking. It is often the case that children and young people – hopes and actors for the future – educate their parents about recycling practices, for example, thus forming a forefront to biosociety. More researched information on the state of the environment and the means by which everyone can influence their own activities and choices could be produced in the teaching materials of schools. Biosociety in a meaningful context goes beyond economy and technology, and requires an educational reform. Warning of the future is not a doomsday forecast but a recommendation to follow another path if the current highway leads to environmental degradation. Climate change is the last straw on humankind's back unless appropriate action is taken. Bardi (2017) reminds us how growth is slow, but collapse is rapid as already Seneca realized. Linear growth thinking that relies on depleting natural resources should be urgently replaced by neo-growth thinking where economic growth minimizes the waste of natural resources and builds on immaterial growth (Malaska 2010; Heinonen 2013). Incremental or sectoral transformation is not enough; we need new vital blueprints for ensuring our future on earth (Botkin 1990; Martin 2007). Martin (2007) calls this century consequently a make-or-break century.

We can simulate two major development paths for the human interaction with nature through technology, and as reflected for biosociety context. Using myths and metaphors may help us in this search for a meaningful biosociety. If the challenge of Prometheus revisited succeeds in readjusting our views of nature and of technology, we will have better possibilities in building society A – an information society based on sustainable development and furthermore biosociety as based on digital meanings society.

However, if Prometheus revisited does not succeed in detaching the current way of technological thinking from environmentally destructive patterns, there will be society B as an outcome. It represents a society of denaturalization and technohumanization. In other words, it signifies the end of nature as independent entity. Human beings have transformed nature into artefacts (*natura secunda*), not by alienating themselves from it but by penetrating into nature ever deeper and by manipulating it to the interest of human beings. Human beings are not seen as part of nature, and therefore the idea of sustainable development cannot be conceived as an inward motivation. Human beings are regarded merely as resources attached to computers and have value only as part of such production processes. Society B also denotes technohumanization, which allows technology, in its turn, to penetrate ever deeper into human beings physically, mentally and socially. In a dystopic form, i.e. technohumanized information society or biosociety, the computers, mobile phones, other sophisticated equipment of ICTs and AI evolve into artificial limbs, ultimately replacing the human brain and soul.

It is a learning process where human beings' relations to nature and technology are to be reassessed. The information society should produce not only technology but, above all, such knowledge that will direct us to a post-modern society based on sustainable development. The role of technology should be seen as that of a mediator between human beings and nature, not as an instrument for exploiting or devastating nature.

Nature has been able to create a conscious being with knowledge – a human being. Now it is the turn of humans to regenerate themselves as wise beings. This requires a map of knowledge for navigating in the landscape of eco-consciousness. All the necessary knowledge for adopting wise attitude to nature and technology is already at hand. What is needed is the will for change. Biosociety based on sustainable development could also be an example of the wisdom society. The wisdom that is of relevance here would be formed of knowledge provided by the information society and digitalization (concerning us humans, nature and technology) and of moral values and responsibility (concerning human relation with fellow beings, nature and technology). The information society based on sustainable development would reach to grow into biosociety, built on life-sustaining knowledge with a moral dimension.

This provides us with a preferred vision of biosociety. It does not happen automatically, though. Urgent action and new kind of thinking are needed, alongside with leadership and pioneers – both companies and individuals. Pioneers can show us a way forward to biosociety (Heinonen and Karjalainen 2019b; Heinonen 2017).

A new notion of the relationship between human beings, nature and technology was attempted to provide fertile soil for building the modern biosociety on the basis of the principles of sustainable society and utilizing digitalization. This means recommending the approach of epistemological expansion as described in the typology presented in Sect. 4, based on the ancient Stoic view of nature and replenished with the requirements in the context of modern information and biosociety. The Stoic approach can be used to break up the myth of separation between humans and nature. Such breaking up would help us to see that human society is a subsystem of the ecosphere, and further of the biosphere. We have to learn to live as part of nature. Humanity is dependent on nature, but not vice versa. Instead of managing natural resources, we have to manage knowledge, technology and ourselves.

The myths live in us, even if unconsciously. We can also learn from them. Prometheus revised proposes a union with nature based on partnership, and a learning contract instead of domination. In this union, technology should not, however, dictate our commitments but assume the metaphor of instrument and "revelator" of its potential impacts on society. After all, human beings have not yet truly turned out to be rational animals. The metaphor of nature as a teacher would better lead us towards rationalism in the Stoic sense where following nature means following reason. Thus, there are both rational and existential reasons for aiming at digital meanings orientated biosociety as a preferred future.

References

Ahlqvist, Toni (2005). From information society to biosociety? On societal waves, developing key technologies and new professions. Technological Forecasting & Social Change 72 (2005) 501–519.

Bachelard, Gaston (1949). La psychoanalyse du feu. Paris, Gallimard.

Bardi, Ugo (2017). The Seneca Effect. Why Growth is Slow, but Collapse is Rapid. Springer.

Bell, Wendell (1997). Foundations for Futures Studies. Vol 1 Human Science for a New Era. Transaction Publishers, Brunswick.

Beresford-Kroeger, Diana (2019). To Speak for the Trees. My Life's Journey from Ancient Celtic Wisdom to a Healing Vision of the Forest. Random House Canada.

Bhatia, Saurabh (2018). History, scope and development of biotechnology. In: Introduction to Pharmaceutical Biotechnology Vol I. https://iopscience.iop.org/book/978-0-7503-1299-8/chapter/bk978-0-7503-1299-8ch1

Botkin (1990): Botkin, Daniel B. Discordant Harmonies. A New Ecology for the Twenty-First Century. New York. 241 p.

van de Braak (1995): van de Braak, Hans. The Prometheus Complex. Man's Obsession with Superior Technology. Amersfoort, the Netherlands. 209 p.

Braungart, Michael & McDonough, William (2002). Cradle-to-Cradle: Remaking the Way We Make Things. New York: North Point Press.

Cairns (1962): Cairns, G.E. Philosophies of History. Meeting of East and West in Cycle-Pattern Theories of History. New York

Castells, Manuel (1999). The Rise of the Network Society.

Casti, John L. (2010). Mood Matters: From Rising Skirt Lengths to the Collapse of World Powers. Springer.

Club of Rome: Climate Emergency Plan (2020). https://clubofrome.org/publication/the-climate-emergency-plan/

Dator, James (2009). Alternative Futures at the Manoa School. Journal of Futures Studies 14(2):1-18. http//www.jfs.tku.edu.tw/14-2/A01.pdf

GSDR = Global Sustainable Development Report 2019 (2019) by Independent Group of Scientists appointed by the Secretary-General: Future is Now - Science for Achieving Sustainable Development. United Nations, New York, 252 p. https://sustainabledevelopment.un.org/content/documents/24797GSDR_report_2019.pdf

Glenn et al. (2009): Glenn, Jerome, Gordon, Theodore & Florescu, Elizabeth. Futures Research Methodology V 3.0. cd, Millennium Project. Washington D.C..

Geels, F. W. (2020). Micro-foundations of the multi-level perspective on socio-technical transitions: Developing a multi-dimensional model of agency through crossovers between social constructivism, evolutionary economics and neo-institutional theory. Technological Forecasting and Social Change, vol 152, March 2020.

Haila (1990): Haila, Yrjö. Vihreään aikaan (Green Age). Tutkijaliitto, Jyväskylä. (in Finnish).

Hawking, Stephen (1988). A Brief History of Time. From the Big Bang to Black Holes. Bantam Books.

Heinonen, Sirkka & Karjalainen, Joni (2019). Electrification in Peer-to-Peer Society – a New Narrative for Sustainable Futures. FFRC eBook 1/2019. https://utushop.utu.fi/p/2986-electrification-in-peer-to-peer-society-a-new-narrative-for-sustainable-futures/

Heinonen, S. & Karjalainen, J. (2019b). Pioneer Analysis as a Futures Research Method for Analysing Transformations. In: Poli R. & Valerio M. (eds) Anticipation, Agency and Complexity. Anticipation Science, vol 4. Springer, Cham. https://doi.org/10.1007/978-3-030-03623-2_5

Heinonen, S. (2000). Prometheus Revisited – Human Interaction with Nature through Technology in Seneca. Doctoral dissertation. Helsinki University. Commentationes Humanarum Litterarum Vol. 115, the Finnish Society of Sciences and Letters, Helsinki 2000, 232 p.

Heinonen, S. (2013). Neo-Growth in Future Post-Carbon Cities. Journal of Futures Studies. Tamkwang University.

Heinonen, S. (2017). Pioneer Analysis and International Cultural Changes Positrend and Negatrend Analysis in the Identification of Cultural Change In: Osmo Kuusi, Sirkka Heinonen and Hazel Salminen (eds.). How Do We Explore Our Futures? Acta Futura Fennica no 10, Helsinki: The Finnish Society for Futures Studies.

Heinonen, Sirkka (2016). The Future of the Internet as a Rhizomatic Revolution Towards A Digital Meanings Society. In: Winter, Jenifer & Ono, Ryota (eds.) (2016). The Future Internet. Alternative Visions. Public Administration and Information Technology 17, 75–90. http://www.springer.com/gp/book/9783319229935?wt_mc=Internal.Event.1.SEM.ChapterAuthorCongrat

Jensen, Rolf (2001). Dream Society: How the Coming Shift from Information to Imagination Will Transform Your Business.

Kelly, Kevin (2016). The Inevitable. Understanding the 12 Technological Forces that Will Shape Our Future. New York.

Kondratieff, N. (1925). The Major Economic Cycles.

Korten, David (2020). Wellbeing versus GDP: The Challenge and Opportunity of Human Development in the 21st Century.

Long, A A & Sedley, D N (1987). Hellenistic Philosophers 2 vols. Cambridge.

Lovejoy, A. O. & Boas, G. A. (1935). A Documentary History of Primitivism and Related Ideas. I: Primitivism and Related Ideas in Antiquity. Baltimore. Ch. 2.

Malaska, Pentti (2010) A more innovative direction has been ignored. In Grönroos, Riitta (ed.) Understanding Neo-Growth: An Invitation to Sustainable Productivity. TeliaSonera Finland Plc, p. 200–210. http://www.sonera.fi/media/13069ab55806de22e8955bc2a3f1afeab17b28bd/Understanding_Neogrowth.pdf

Malaska, Pentti (1999a). A Conceptual Framework for the Autopoietic Transformation of Societies. FUTU Publication 5(99). Turku School of Economics and Public Administration. Finland Futures Research Centre (FFRC).

Malaska, Pentti, (2001) A Futures Research Outline of a Post-Modern idea of Progress. Futures 33, 225–243.

Malaska (1994): Malaska, Pentti. Progress, Nature and Technology in Late-Modern Transition. Futures: Vol 26 (5). pp. 529–541.

Malaska (1997): Malaska, Pentti. Sustainable Development as Post-Modern Culture. TURKU School of Economics and Business Administration. Finland Futures Research Centre. FUTU publication 1. 24 p.

Martin, James (2007). The Meaning of the 21st Century. A vital blueprint for ensuring our future. London.

Mazlish (1993): Mazlish, Bruce. The Fourth Discontinuity. The Co-evolution of Humans and Machines. New Haven and London: Yale University Press.

Masini (1993): Masini, Eleonora. Why Futures Studies? London. 144 p.
Maxton, Graeme & Randers, Joergen (2016). Reinventing Prosperity. Managing Economic Growth to Reduce Unemployment, Inequality, and Climate Change. A report to the Club of Rome. Vancouver.
Meadows, Donella, Meadows, Dennis, Randers, Joergen & Behrens III, William (1972). The Limits to Growth. Report to the Club of Rome. Universal Books.
Mumford (1963): Mumford, Lewis. Technics and Civilization. Orlando, Florida. 495 p.
Mumford, Lewis (1966). Technics and Human Development. The Myth of the Machine. USA. 342 p.
NLEM (1985). New Larousse Encyclopedia of Mythology. Introduction by Robert Graves. Middlesex: Hamlyn Publishing (twentieth impression).
OECD (2011). Key biotechnological Indicators. OECD, December 2011. http://www.oecd.org/science/inno/49303992.pdf
Orr, D. (1992). Ecological Literacy: Education and the Transition to a Postmodern World. Albany, NY: State University of New York Press.
Pacey, Arnold (1992). The Culture of Technology. MIT. Massachusetts.
Pauli, Günther (2010). The Blue Economy. 10 years, 100 innovations, 100 million jobs. Report to the Club of Rome. Paradigm Publications. https://www.theblueeconomy.org/
Pietarinen (1994): Pietarinen, Juhani. The Principal Attitudes of Humanity towards Nature. In: Philosophy, Humanity and Ecology: Vol I. Philosophy of Nature and environmental ethics. H. Odera Oruka, ed. Nairobi, Kenya: ACTS Press. African Centre for technology Studies and the African Academy of Sciences. pp. 290–294.
Pietarinen (1991): Pietarinen, Juhani, Principal attitudes towards nature, 581–587. In: Pekka Oja & Risto Telema (eds), Sport for All. Proceedings of the World Congress on Sport for All, Tampere, Finland 3-7 June 1990. Amsterdam: Elsevier Svience Publishers.
Pine, Joseph & Gilmore, James (1999). Experience Economy. Harvard Business School Press. Boston.
Pouru, Laura, Wilenius, Markku, Holstius, Karin & Heinonen, Sirkka (2018). Collection of poems written by Pentti Malaska. In: Pentti Malaska. A Visionary and Forerunner. Futures Series 9. The Finnish Society for Futures Studies, Helsinki, p. 215. http://www.tutuseura.fi/wp-content/uploads/2018/04/Futures-Series_9_Pentti-Malaska.pdf
Rifkin, Jeremy (1998). The Biotech Century. Harnessing the Gene and Remaking the World. New York.
Roy, Claude (2016). Les potentiels de la bioéconomie. Futuribles Janvier-Février 2016, nro 410, 69–80.
Ruotsalainen, Juho; Heinonen, Sirkka; Karjalainen, Joni & Parkkinen, Marjukka (2017). Peer-to-peer work in the digital meaning society 2050. European Journal of Futures Research 4(10). http://link.springer.com/article/10.1007%2Fs40309-016-0092-2
Salomon (1981): Salomon, Jean-Jacques. Prométhée empetré. La résistance au changement technique. Preface by Etienne Davignon. Collection futuribles. Paris 1981.
Salomon (1998): Salomon, Jean-Jacques. Sur les dérives de la raison technicienne. Futuribles. No 232. June. pp. 57–66.
Sandbach, F H (1975). The Stoics. Cambridge.
Schwabl (1978): Schwabl, H.. Weltalter. *RE Suppl. Band 15. pp. 783–850.*[translate non-title words]
Serres, Michel (1994). Le contrat naturel. Paris: Editions Francois Bourin. 1990. 180 p.
Sihvola (1989): Sihvola, Juha. Decay, Progress, the Good Life? Hesiod and Protagoras on the Development of Culture. Commentationes Humanarum Litterarum 89. Societas Scientiarum fennica. The Finnish Society of Sciences and Letters. Helsinki. 174 p.
Smith (1908): Smith, K. F. Ages of the World. Greek and Roman. Encyclopaedia of Religion and Ethics: Vol I. ed. J. Hastings. Edinburgh. pp. 192–200.

Sørensen (1984): Sørensen, V. Seneca. The Philosopher at the Court of Nero. Chicago.Serres, Michel (1990). Le contrat naturel. Editions Francois Bourin, Paris.
Verbeke, Gérard (1964). Les Stoïciens et le progrès de l'histoire. Revue Philosophique de Louvain. tome 62, 5–38.
Wackernagel, Mathis & Rees, William E. (1996). Our Ecological footprint. Reducing Human Impact on the Earth. Bioregional Series. British Colombia. New Catalyst Books.
Wagensberg, Jorge (1997). Le court-circuit de Prométhée. La Recherche 299, juin, 95.
Wilenius, M. & Kurki, S. (2017). K-Waves, Reflexive Foresight, and the Future of Anticipation in the Next Socioeconomic Cycle. Roberto Poli (Ed.) Handbook of Anticipation. Theoretical and Applied Aspects of the Use of Future in Decision Making. Cham: Springer Nature.
Webster, Frank (1995). Theories of the Information Society. Routledge, Oxford.
van der Waerden (1979): van der Waerden, B. L., Religiöse Bruderschaft und Schule der Wissenschaft, Zürich.
von Wright, Georg Henrik (1987a). Vetenskapen och förnuftet (in Swedish). Borgå. Söderström & C:o Förlags AB.
von Wright, Georg Henrik (1987b). Humanismen som livshållning. (In Swedish, Humanism as a way of life) Helsinki

Chapter 24
New Humanism: A Vital Component of Sustainable Socio-technical Change

Silke Van Cleuvenbergen and Gaston Meskens

Abstract The adagio that we cannot solve our societal problems with the same methods that create them is well known. The vision that inspiration and motivation for 'new methods' need to come from deeper thinking about who we are as individuals and groups and about how to deliberate these problems with each other is less popular. This vision is the point of departure of the New Humanism Project. Tackling complex social problems such as climate change, poverty and the various forms of social depression and oppression comes down to a fair dealing with their complexity. This requires ethical competence and the preparedness to engage in joint public reflexivity 'in face of that complexity', taking into account our interests, hopes, hypotheses, beliefs and concerns.

However, our current methods of democracy, scientific research and education stimulate conflict, polarisation and competition rather than public reflexivity and the development of ethical competence. Therefore, we need to rethink and reform those methods into interaction methods that are inclusive, pluralistic, transdisciplinary and deliberative. We believe these interactive methods will not only enable more effective sustainable development towards a domain such as bioeconomy. They also have a chance to be perceived – in a form of inner ownership – as fair by anyone in our society.

Keywords Education · Art · Humanism · Reflexivity · Ethics · Complexity · Co-existence · Philosophy · Connectedness

1 Introduction to Chapter

The adagio that we cannot solve our societal problems with the same methods that create them is well known. The vision that inspiration and motivation for 'new methods' need to come from deeper thinking about who we are as individuals and groups and about how to deliberate these problems with each other is less popular.

S. Van Cleuvenbergen (✉) · G. Meskens
The New Humanism Project, Arts Institute, Antwerp, Belgium

E. Koukios, A. Sacio-Szymańska (eds.), *Bio#Futures*,
https://doi.org/10.1007/978-3-030-64969-2_24

This vision is the point of departure of the New Humanism Project. Tackling complex social problems such as climate change, poverty and the various forms of social depression and oppression comes down to a fair dealing with their complexity. This requires ethical competence and the preparedness to engage in joint public reflexivity 'in face of that complexity', taking into account our interests, hopes, hypotheses, beliefs and concerns.

However, our current methods of democracy, scientific research and education stimulate conflict, polarisation and competition rather than public reflexivity and the development of ethical competence. Therefore, we need to rethink and reform those methods into interaction methods that are inclusive, pluralistic, transdisciplinary and deliberative. We believe these interactive methods will not only enable more effective sustainable development towards a domain such as bioeconomy. They also have a chance to be perceived – in a form of inner ownership – as fair by anyone in our society.

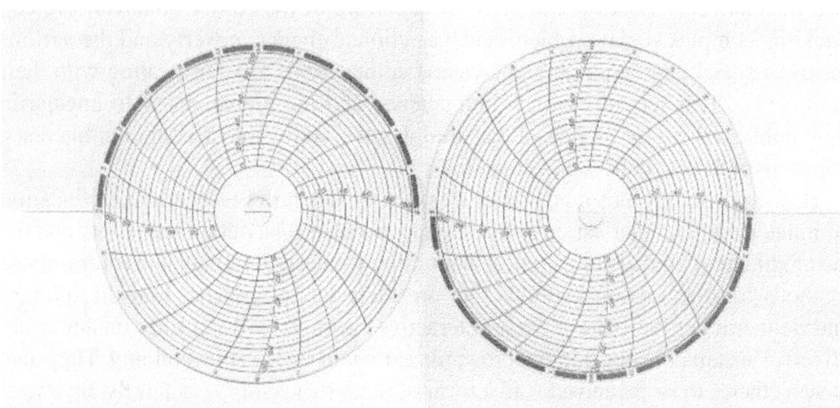

2 Dear Reader

The New Humanism Project provides a statement but does not preach 'the truth'. It is a dialectical process in which we invite everyone to think about the path towards a new humanism.

Because of our belief in the power of multilayered language and perspective, we approached this chapter in a visual way. We included a series of images that we found fit for the content of this chapter. In addition, a number of concepts are supported in the form of creative schemes. Every text paragraph has its own 'visual translation'. In this way we hope not only to go beyond language as a form of written words but to also trigger those with a rather limited time while scouting through this book.

Yet the most important is this: you, and only you have the choice to approach the text as you see fit. Make a movement or dance with it, if you like. However you do it. Experience, feel, and process it your way. Above all, let it seep in. This chapter, and its interpretation, is all yours. Enjoy!

Any Ideas? Comments? Thoughts?

Share them with us or join our happenings, network and reflections (www.newhumanism.org).

Write us at silke@newhumanism.org and/or gaston@newhumanism.org.

3 The Need for a New Humanism?

3.1 The Stakes Are High

How can we make this world a better place for all? Sketching what goes wrong in our world today, the picture does not look very bright… structural poverty; expanding industrialisation and urbanisation and consequent environmental degradation; waste of precious resources, water, food and products; adverse manifestations of technological risk; hardly controllable virus pandemics; economic exploitation; anticipated overpopulation; derailed financial markets, etc., all of this adding up to old and new forms of social, political and religious oppression and conflict. Last but not least, regardless of whether or not they are directly affected by the global problems sketched above, more and more people feel lost in their personal life. In search for meaning, recognition and self-confirmation, they feel overburdened and exhausted by the rage of life and stuck in the labyrinth of often conflicting social norms, codes and expectations. As a result, psychological distress and depression are becoming the fundamental personal disorders of our modern times. So how can we make this world a better place for all? The stakes are high and the need to take more action is manifest.

3.2 *However, the Adagio Goes That the Recipes Are Known…*

To tackle the socio-political challenges, we have international politics run by nation states and democracy organised through party politics and elections. We have globalised markets steered by competition and profit and education programmes that prepare workers and specialists to function in any socio-technical role the big system of our society requires. And relying on those modern methods, it is said that the only thing we need is leaders showing commitment and political will to do what reason and science tell us to do, and entrepreneurs and consumers showing a sense of responsibility and the will to cooperate in executing the plans.

In our personal spheres, we have the checks and overcodes that bring order in our life and that help us to realise our goals: a proper education, a partner, kids, a career, money, a yearly holiday, a house and a car. We have religions with gods that prescribe what to do and what not, and we have religions without gods that prescribe what to do and what not, and in case we are insecure or mess up, professional life coaches and spiritual leaders are there to help us to get back on track. And our mental deviations and disorders are now categorised and analysed down to the finest detail, and the market has medication and tailor-made therapies to cure any form of them.

The rationality and necessity of these methods are key, it is said, because the world is complex, societal problems have no easy solutions and the outlook on our personal paths of life is troubled by multiple uncertainties and ambiguities. And, at the same time, we are told by political and corporate elites and self-declared spiritual gurus that we should not be naïve but simply accept that our society lives by the grace of competition and self-protection and that altruism and spontaneous solidarity of haves with have nots are nothing but nice philosophical ideas. People are selfish, it is said, and will always put the individual before the common good, regardless whether they are rich or poor. Politics and markets cannot be but competitive and conflictual power games, the theory goes, where the most popular opinion or product wins. And are the multiple acts of terror and aggression of the last years not to prove in themselves that security and defence are the only working remedies against the so-called unavoidable human evil?

Also our personal lives are said to be driven by competition and self-protection. The fact that the rich mate with the rich and the poor with the poor is simply a law of nature, we are told. And of course you are not necessarily doomed, as you can always work yourself a class up, that is, you are always free to try. And what about love and sex then? Well there is the love and sex from the movies and the video games, and there is the love and sex in 'real life'. And in that real life, you better pragmatically conform to the norms and find your better half, as any alternative life form can only lead to chaos, pain and misery….

3.3 And Here Is the Thing

Our modern methods and overcodes of social and political life may be seen as signs of social, political and even moral progress, given that they are the results of historical emancipatory struggles away from the often brutal oppressions by the pre-modern elites of emperors and priests, but in essence they are *not* designed to cope with the complexity, uncertainty and ambiguity of that life. Although each of them has its own history, one could almost say their common feature is that they were rather designed to *escape* confrontation with that complexity, uncertainty and ambiguity. Today,…

Modern education

prepares you for a job,

not for (the complexity of) life.

Modern religion

is (still) designed to 'relieve' you (and your innocent children) from doubt with regard to your origin and destiny and from choice stress with regard to the Path of Life,

but it (still) relies on (competing) dogmatic power structures strategically promoting collective beliefs in fictional 'truths' that cannot be proven.

The modern love relationship

is as much a construction meant to help you to resist lust and curiosity and to streamline doubt about your feelings as it is the materialisation of a 'bond' of love,

but breakups, cheating, disappointment and pain seem to be basic consequences of that construction as much as feelings of belonging, security and joy.

Modern science

is organised as a quest for measurable and usable truth at the service of politics and the market,

but it is not designed to advise on issues open to value-based interpretations and troubled by uncertainties that cannot be resolved (yet).

Modern markets

are organised as systems of competition that reward strategic insight and profit,

but they are unable to demarcate their own ethical boundaries.

Modern politics

is (still) organised as a conflict of opinion relying on political self-promotion and simple ideologies (including that of the nation state),

but it is unable to deal with thoughtful nuance as well as with populist misuse.

…

3.4 In Sum

We will not save our planet and humanity in a society that remains blind for the fact that our current 'modern' methods to make sense of and organise our coexistence are too primitive to grasp the complexity of that coexistence and are actually *denying* instead of recognising who we really are as human individuals. In other words, our traditional methods of making sense of our coexistence (politics, science and education) are no longer able to grasp the complexity of these social problems. In addition, it is important to realise that these methods and overcodes are not 'errors' of the motor of modernity but rather strategic tools. They prepack, streamline and exploit our human quest for belonging and recognition (as lover, as spiritual mind, as consumer, as citizen, etc.) at the service of the contemporary elites of emperors, entrepreneurs and priests who need these methods and overcodes to legitimise and safeguard their own power and privileges.

3.5 The New Humanism Project

And here we are. The adagio that we cannot solve our societal problems with the same methods that (facilitate to) create them is well known. The vision that inspiration and motivation for 'new methods' need to come from deeper thinking about who we are as individuals and groups and about how to deliberate these problems with each other is less popular. This vision is the point of departure of the New Humanism Project. The New Humanism Project aims to facilitate dialogue about that vision and, consequently, about what these 'new methods' should be and can be. The idea is that, in order to tackle societal problems such as climate change, pandemics, poverty and the various forms of social depression and oppression, we first need to rethink and reform the formal methods we use to make sense of our coexistence, namely, the methods of education, scientific research and democracy.

3.6 What Was Wrong with the 'Old' Humanism?

Why would we need a 'new humanism'? We aim to present here a vision on our individual and collective being and capacity transcending the humanist one that emerged as a reaction against oppression by the pre-modern elites of emperors and priests. While liberating ourselves from this oppression was of course a good thing as such, throughout the following ages of social, scientific and technological progress, humanity has built up a self-confidence leading to the current 'hyper-rationality' driving education, science, economics and politics today. In that sense, the New Humanism Project explores a new way of looking at the problems the world is facing. It rejects cynical post-whatever defaitism as well as 'back to the good old and simple times' nostalgia. Alternatively, we want to present a 'pragmatic ethics' view on who we are, what we can know and should know and how we can deliberate the issues, and we believe this view is essential for how we organise our coexistence in general, and education, science and politics in particular.

4 The Art of a New Humanism

The New Humanism philosophy is a living and fluid philosophy, in constant development through dialogue with others. But here are our key ideas.

4.1 The 'Fact of Complexity': New Characteristics of Our Modern Coexistence

Problems such as climate change, environmental pollution, pandemics, unsustainable food production and consumption and loss of biodiversity are complex social problems troubled by multiple uncertainties and often incommensurable value judgements. In addition, typical for these complex social problems is that they are all interconnected, which means they all need to be tackled together in a holistic perspective.

Dealing fairly with these complex problems comes down to dealing fairly with their complexity, and that requires the joint preparedness of all of us to become 'reflexive in face of that complexity', trying to understand 'the bigger picture and yourself in it', each of us with our specific interests, hopes, hypotheses, beliefs and concerns. That kind of reflexivity can thus be seen as an ethical attitude in the face of that complexity, and as a motivation to seek rapprochement with each other and to engage in 'public reflexivity' to deliberate the problems. In the New Humanism Project, we argue that this kind of deliberation, as a form of public reflexivity, is marked by two fundamental principles:

When it comes to give meaning to and decide on what is a personal meaningful life and on how to live together, we are all equal and we have no reference other than each other. In our care for personal and general well-being, we can only make use of one absolute reference value: the possibility of a continuous engagement in deliberation as equal human beings. All other possible value references (specific ideas, facts, values, statements, roles, responsibilities, objects, systems, etc.) are relative and need to be incorporated as subject of that deliberation.

However, 'being reflexive in face of complexity' is not an intellectual exercise we can choose to do or not, detached from reality. In whatever position, situation or role in our daily life, we are all impacted by complexity, and we have impact on complexity itself. In addition, with globalisation and the interconnectedness of our current socio-economic practices, it is clear that complex social problems now have global dimensions. Today, we have to understand that, as individuals enjoying an acceptable standard of living, all of our choices with respect to the food we eat, the clothes we wear, the consumer products we buy, the energy we consume, the means of transport we use and so on have some effect somewhere else on earth. As a consequence, ethical reasoning with respect to those choices requires us to look beyond our familiar local 'comfort zones' and to think as 'citizens of the world' or cosmopolitans who try to evaluate the consequences of their choices, and who are motivated to understand their specific place, role, responsibilities and rights in the bigger picture of it all.

In the New Humanism Project, we translate this vision by saying that the 'fact of complexity' brings along three new characteristics of our modern coexistence (see scheme):

(Moral) connectedness

We are connected with each other 'in complexity'. We cannot any longer escape or avoid it.

Fair dealing with each other implies a fair dealing with the complexity that binds us.

(Social) vulnerability

In complexity, we became dependent on each other (we can only know, feel, understand and act 'together'). At the same time, we should care for the vulnerability of ignorance and confusion, but also of that of 'mandated authority' and the next generations.

(Sense for) engagement

Our experiences now extend from the local to the global. As intelligent reflective beings, to become involved in deliberating issues of general societal concern became a new source of meaning and moral motivation.

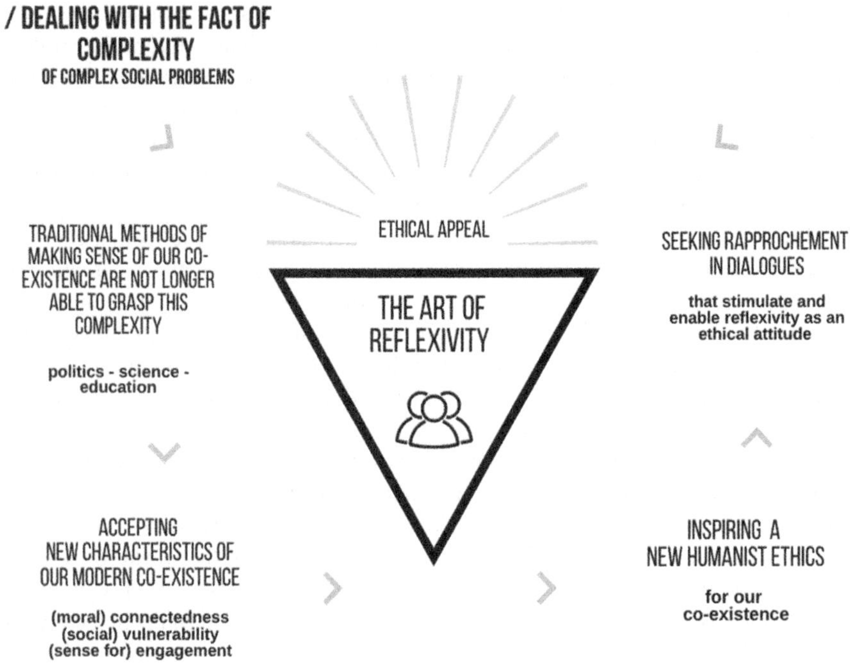

4.2 The Need for and Power of an Ethical Competence

Our responsibility to adopt reflexivity as an ethical attitude and to reason and act as a cosmopolitan essentially leans on our capacity to do so. Understanding the bigger picture, the complexity of social problems and the consequences of our acts, roles, rights and responsibilities in relation to them therefore requires developing reflexivity as a skill, or thus the ability to see that bigger picture and yourself in it, with your interests, hopes, hypotheses, beliefs and concerns.

In the New Humanism Project, we see reflexivity as both skill and ethical attitude as the two essential elements – interdependent and mutually influential – of an ethical competence needed to fairly deal with each other 'bound in complexity' (see scheme). We argue that ethical competence for reflexivity can be stimulated and fostered in dialogues that 'work' emancipatory and (compassionately) confrontational at the same time. From this perspective, it becomes clear that there is a need to reform the old modes of politics, research and education into interaction methods that are inclusive, pluralistic, transdisciplinary and deliberative. We believe these interactive methods, as forms of inner ownership, will not only enable more effective governance of complex social problems, they may also be perceived as fair by anyone concerned.

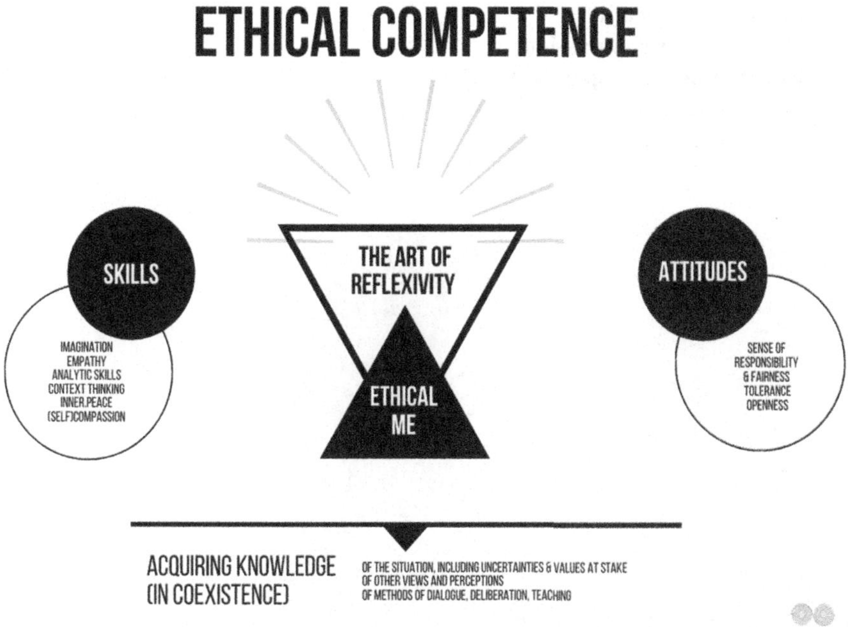

4.3 The Art and Right of Reflexivity

From on the concept that a person with the right skills and attitudes can develop into an ethically competent person ('ethical me') arises the idea of a mastery in the art of reflexivity. An artist of reflexivity is in constant interaction with the complexity that appeals to all.

Accompanied by this form of artistry is the 'right' to practise this art. After all, 'ethical me' can only 'express' this mastery in constant coexistence. In this case, a dialectical process emerges between the art of reflexivity and the right to reflect with others. The other must give me that right; otherwise I can never master the art. In a New Humanist society, everyone aspires to reason, feel and communicate with the art of reflexivity, and everyone has the right to do so…'.

From a holistic perspective, we see (in an ever-growing list) these crucial components:

…
The art of vulnerability
The art of responsibility
The art of fluidity
The art of acceptation
The art of perplexity
The art of knowing/not knowing
The art of intuition

The art of physicality
The art of (self)compassion
The art of multiperspectivity
The art of empathy
The art of self-wisdom
...

5 On Track to a New Humanism (Strategy of the New Humanism Project)

The path to a New Humanism is a continuous path. We do not intend to preach a new truth but rather co-reflect towards a philosophical utopia, one that will fill and shape and reform itself.

But what are the ways to get there? In search of new methods that rely on and stimulate ethical competence and reflexivity, we have started different tracks including workshops, happenings and networking.

Our first track focuses on education, why?

#We believe that there are reasons to conclude that, still today, most approaches to education now mainly prepare our children to function as one-dimensional uncritical subjects in the systems of our society.

#We believe everyone has the right to an education that would aim to make them more resilient to all kinds of capitalist, conformist and fundamentalist manipulations of our coexistence and to help them to live better with each other in our complex society.

#We believe that for children, youngsters and adults, becoming self-critical world citizens is not their duty; it is their human right.

#We believe that reforming our formal and non-formal education systems with a focus on ethical competence and reflexivity can inspire and motivate (future) changemakers to make us shift away from our current methods and overcodes of social and political life driving on competition, conformism, polarisation and conflict. We need to find new structures and new forms not to, evolve towards new fixed systems but towards a constantly reinventing fluid society.

However, to incubate this change, we need innovations in social sciences and humanities that can help achieve and manage the necessary changes and reform towards building joint ethical competence and reflexivity. Therefore we want to build a network of exchanging forces that analyse in which forms this can be implemented, 'glocally'.

Together we ask ourselves and others these following questions:

Can you learn how to make ethical choices?

What do we need to make ethical choices?

How can we initiate systemic change from a human core and vice versa?

Which systems can we work on? And how? What is feasible, in which contexts?

What kind of education do we need to enable us to live better with each other in our complex society?

What are the implications for form, content and curriculum?

What does this mean for the relation teacher–pupil?

What might be the role of art, philosophy and mindfulness?

What is a world citizen? How to become one?

How can education contribute to becoming a world citizen?

Do we need to become world citizens?

How do our views relate to the other work in this book?

Join us for further reflection!

www.newhumanism.

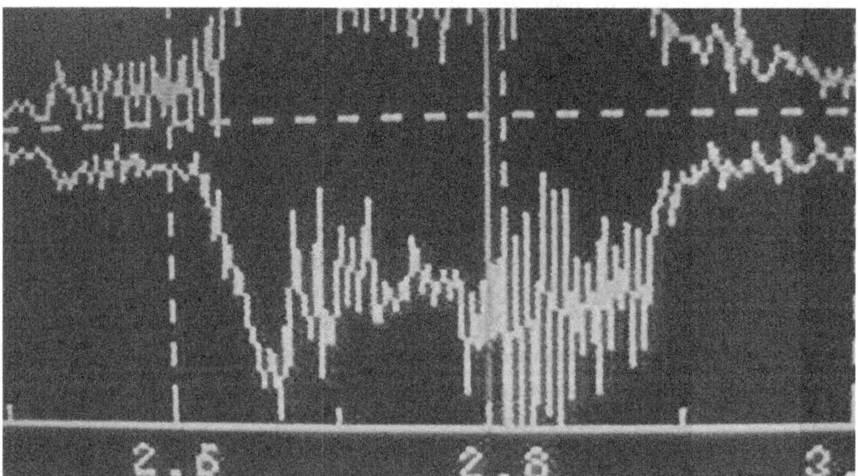

Part VIII
The Way Ahead, Key Trends and Lessons

Chapter 25
Bioeconomy in the Twenty-First Century: Global Trends Analysis Perspective

Aleksandr Chulok

Abstract Bioeconomy as an ecosystem faces several fundamental changes, triggered by at least two wild cards – events with low probability, but huge effects – pandemic COVID-19 and low oil prices. They influence the whole landscape of global trends that constitute bioeconomy in the beginning of the year 2020. The new bio-reality will develop by five blocks: demand; supply; infrastructure; science, technology and innovation (STI); and regulation. Major changes will reveal in new consumer patterns, which shift from eco, responsible and green principles to efficient, rational and pragmatic behaviour, dramatic transformation of global value, production and logistic chains and new target KPI for world's governments from economic growth and prosperity to survivance and national security. Will bioeconomy become a new general standard for business and society depends on proper identification and further management of global trends, presented in this chapter.

Keywords Blended foresight · Data-driven analytics · Strategic analytics · Technological road maps · Global trends scanning · New competitiveness factors · Demand-driven innovations · Responsible behaviour

1 Introduction

By the beginning of 2020, the term "bioeconomy" was firmly among the current global trends. For example, the top ten most important and rapidly growing areas identified using the iFORA big data mining system comprised a cluster of bioeconomic trends. According to various estimates, the size of bioeconomy (and most experts include almost all human activities into it) is expected to reach over $8 trillion by 2025. Bioeconomy as a subject of research did not leave indifferent neither

A. Chulok (✉)
S&T Foresight Centre, Institute for Statistical Studies and Economics of Knowledge, Higher School of Economics, Moscow, Russia
e-mail: achulok@hse.ru

E. Koukios, A. Sacio-Szymańska (eds.), *Bio#Futures*,
https://doi.org/10.1007/978-3-030-64969-2_25

the disciples of Vernadsky's noosphere theory (Vernadsky 1926) nor the followers of another Soviet scientist, Kondratiev, and his wave theory (Grinin et al. 2012).

Researchers' approaches to classifying bioeconomy are so many and varied, ranging from seeing it as a multilevel economic system to production chain with a large number of participating sectors (EC 2019a, b, c, d, e; OECD 2019a). Increased attention of the state, society and businesses to ethical and environmental issues noted in 2019 (IRENA 2019a, b; Ronzon and Sanjuánc 2019; Perea-Moreno et al. 2019; OECD 2019a, b, c) implied a cloudless future for the bioeconomy, along with an unconditional victory over the dominant economic, science and technology paradigms of the twentieth century, including those related to carbon economy (GBS 2018a, b, c; EB 2018; EC 2018a, b, c; Keswani 2020).

However, in just the first few months of 2020, two events unexpected by any of the key stakeholders challenged not just bioeconomy but all ongoing debates and the customary junctions of the present day. The pandemic COVID-19 and the collapse of oil prices are the two wild cards (events with a low probability of occurring, but producing major interindustry effects) which, if not bring all previously made economic forecasts and assumptions to an end, then definitely call them into question.

This chapter presents a multidisciplinary and cross-factor analysis of the future of bioeconomy in the twenty-first century, conducted through the prism of global trends, which are, by definition, major, steady tendencies in the science and technology, socio-economic, political, value, and environmental spheres with significant and sometimes destructive effect on the economy, industry and science.

2 Methodology

Identifying key trends that affect the structure of the economy and society is not a novel objective; it has been pursued by a number of prominent macroeconomists such as Kondratiev (Grinin et al. 2012), Schumpeter (Schumpeter 2011), Ansoff (Ansoff 1979; Martinet 2010), Aghion (Aghion and Howitt 1992; Aghion 2001), Solow (Solow 1970), Perez (Perez 2002), Wallerstein (Wallerstein 1979) and, finally, Paul Romer (Romer 1990; Krugman et al. 2005) who in 2018 received the Nobel Prize in Economics for assessing the impact of technological innovations on economic growth, along with many others in the context of identifying, classifying and evaluating factors affecting economic growth; developing industrial organisation theory; and analysing life cycles of products and sectors.

For several decades global trends remained the focus of decision makers' attention: heads of government in developed and developing countries and managers of large transnational corporations, academic organisations and leading universities. In Japan, studying global trends and the challenges they create (grand challenges) was the basis of a series of national foresight exercises since the late 1970s, conducted using the Delphi method – a special expert polling technique covering between 25,000 and 500 respondents.

Germany, the UK, the Czech Republic, France and Austria made adapting to (and, if possible, managing) global trends the central element of their science, technology and innovation policies. Achieving competitiveness at the national and corporate levels wouldn't be possible without identifying a target set of global trends (Table 25.1).

Initial list of global trends for bioeconomy was prepared by iFORA system and count for more than 100 different trending areas. Then the trends bioeconomy is likely to face in the future were assessed through the prism of seeing bioeconomy as an ecosystem comprising the following five key blocks:

1. Households' and companies' demand for bioeconomic technologies, products and services
2. Supply from industries and sectors supplying bioeconomy companies in value chains
3. Infrastructure representing both physical and natural environments
4. Science, technology and innovation which support bioeconomy's development as a system, and its specific elements
5. Regulation by national and international institutions which set the "rules of the game" for bioeconomy

3 Exploring the Reference Point

At the beginning of 2020, the "reference point" for bioeconomy, at least for the next 10 years, was hardly questioned by the majority of experts: these were the total digitisation and robotisation, reduced share of manual labour, wide adoption of new business models based on sharing economy, growth of creative industries, development of alternative energy, active biotechnological experiments, increased average life expectancy, "silver economy" of older people, increased mobility and globalisation (Babu and Debnath 2019; FFRC 2019; FAO 2016, 2018, 2019a, b; EC 2020; IEA 2017, 2018a, b, 2019; OECD 2020).

Over the next decade, a majority of the developed countries were expected to complete the transition to a new production paradigm associated with accelerated adoption of advanced information, communication and production technologies (including robotics), medical and biotechnologies and new materials and application of the so-called "cross-cutting" or "horizontal" technologies in production and social practices, with a significant multiplicative potential: they can be applied in various sectors of the economy, changing the cost structure and allowing producers direct access to end users. Certain international experts integrate the above technologies into the "fourth industrial revolution" concept,[1] bright examples of which include "Industry 4.0", the industrial Internet and the Internet of things. The

[1] https://www.weforum.org/about/the-fourth-industrial-revolution-by-klaus-schwab

Table 25.1 Selected organisations involved in systemic research and monitoring of global trends

Organisation	Project/report title	Methodological approach
International organisations		
OECD	The Future of Education and Skills Education 2030[a] Science, Technology and Innovation Outlook[b] Embracing Innovation in Government: Global Trends[c] Meeting Policy Challenges for a Sustainable Bioeconomy: Designing a Policy Agenda[d]	A large number of international reviews describing the impact of various global trends on the economy, science and society
European Commission	Global Trends to 2030 Challenges and Choices for Europe[e] The Future of Government 2030[f] European Commission Foresight Fiches: "Global Trends to 2030[g]" Shaping the Future in a Fast-Changing World Global Trends to 2030[h] A Sustainable Bioeconomy for Europe: Strengthening the Connection Between Economy, Society and the Environment. Updated Bioeconomy Strategy[i]	Global trends in public administration, science, environment and technology, affecting competitiveness of the EU member states
UNESCO	Futures Literacy,[j] The Blue Dot[k] Education 2030: Incheon Declaration and Framework for Action for the Implementation of Sustainable Development Goal 4[l] EDCRUNCH Global Conference[m]	Studies of competencies required for future development and education prospects
World Economic Forum	Interactive database of studies of global trends affecting the future[n] The fourth industrial revolution could mean the end of traditional manufacturing in Africa[o] These will be the world's megacities in 2030[p]	Numerous studies of global trends in areas such as competencies for the future, robotisation, artificial intelligence, Industry 4.0 and urbanisation
Global companies		
McKinsey Global Institute	Navigating a world of disruption[q] Trends & Global Forces[r] The consumer sector in 2030: Trends and questions to consider[s]	Assessing the impact of global trends on companies' competitiveness and business strategies
Gartner	Top 10 global technology trends[t] Interactive database of global IT development trends[u]	Analysis of global trends' dynamics, primarily of technology-related ones, among other things in the scope of the "hype cycle" concept, and their impact on the economy and society

(continued)

Table 25.1 (continued)

Organisation	Project/report title	Methodological approach
PWC	Five Megatrends and Their Implications for Global Defence & Security[v]	Monitoring five megatrends and assessing their impact on the economy and society
Frost & Sullivan	World's Top Global Mega Trends to 2025 and Implications to Business, Society and Cultures[w] TechVision Interactive Global Technology Trends Database[x]	Describing more important social, technological and economic trends using a single format
International research centres		
NISTEP	Digest of Japanese Science and Technology Indicators[y] Science and Technology Foresight and Science and Technology Trends[z]	National Japanese Foresight study based on Delphi polling Digests of main S&T indicators' analytics
KISTEP STEPI	KISTEP 10 Emerging Technologies[aa] STEPI Annual Research Topics[ab]	Reviews of major global technology trends and of markets they create
RAND	Science and Technology Global Topics[ac] The Global Technology Revolution 2020, In-Depth Analyses[ad] An Analysis of Global Societal Trends to 2030 and Their Impact on the EU[ae]	Reviewing global trends affecting international politics, society, environment, healthcare and security
HSE Institute for Statistical Studies and Economics of Knowledge	Russian S&T Foresight 2030[af] Global Technology Trends Monitoring[ag]	Over 500 global trends described in various open-access studies Global trends identified through big data mining using the iFORA system

Source: author's compilation

[a]https://www.oecd.org/education/2030/E2030%20Position%20Paper%20(05.04.2018).pdf
[b]https://www.oecd.org/sti/
[c]https://www.oecd.org/gov/innovative-government/innovation2018.htm
[d]https://www.oecd-ilibrary.org/science-and-technology/policy-challenges-facing-a-sustainable-bioeconomy_9789264292345-en
[e]https://ec.europa.eu/assets/epsc/pages/espas/index.html
[f]https://ec.europa.eu/jrc/en/events/future-of-government-2030-plus
[g]https://ec.europa.eu/digital-single-market/en/news/european-commission-foresight-fiches-global-trends-2030
[h]https://ec.europa.eu/epsc/publications/other-publications/shaping-future-fast-changing-world_en
[i]https://ec.europa.eu/research/bioeconomy/pdf/ec_bioeconomy_strategy_2018.pdf#view=fit&pagemode=none
[j]https://en.unesco.org/themes/futures-literacy
[k]https://unesdoc.unesco.org/ark:/48223/pf0000261487/PDF/261487eng.pdf.multi
[l]https://en.unesco.org/themes/education2030-sdg4
[m]https://iite.unesco.org/events/edcrunch-2019-global-conference-on-technology-in-education/
[n]https://www.weforum.org/system-initiatives
[o]https://www.weforum.org/agenda/2019/08/how-africa-can-adapt-to-the-digital-revolution/

(continued)

Table 25.1 (continued)

[p]https://www.weforum.org/agenda/2018/10/mapping-the-world-s-new-megacities-in-2030/
[q]https://www.mckinsey.com/featured-insights/innovation-and-growth/navigating-a-world-of-disruption
[r]https://www.mckinsey.com/business-functions/strategy-and-corporate-finance/how-we-help-clients/trends-and-global-forces
[s]https://www.mckinsey.com/industries/consumer-packaged-goods/our-insights/the-consumer-sector-in-2030-trends-and-questions-to-consider
[t]https://www.gartner.com/smarterwithgartner/gartner-top-10-strategic-technology-trends-for-2018/
[u]https://www.gartner.com/smarterwithgartner/category/it/trends/
[v]https://www.pwc.com/gx/en/government-public-services/assets/five-megatrends-implications.pdf, https://www.pwc.co.uk/issues/megatrends.html
[w]https://www.thegeniusworks.com/wp-content/uploads/2016/01/Megatrends-2025-Frost-and-Sullivan.pdf
[x]https://ww2.frost.com/research/industry/techvision/
[y]https://www.nistep.go.jp/en/?cat=3
[z]https://www.nistep.go.jp/en/?page_id=56
[aa]http://www.kistep.re.kr/en/c3/sub4.jsp#none
[ab]http://www.stepi.re.kr/eng/app/eTopics/list.jsp?cmsCd=CM0237
[ac]https://www.rand.org/topics/science-and-technology.html
[ad]https://www.rand.org/pubs/technical_reports/TR303.html
[ae]https://www.rand.org/randeurope/research/projects/global-societal-trends.html
[af]https://prognoz2030.hse.ru/
[ag]https://issek.hse.ru/trendletter/

"Industry 4.0" concept was suggested in Germany as a national strategic initiative[2] as early as in 2011,[3] and then taken up by many developed and developing countries as a major S&T development goal.

The radical changes in the technological environment, innovation and economic ecosystems, cultural values and social patterns that have been accumulating since the last global crisis have led to a number of significant shifts in the global agenda.[4] Trying to retain the local, temporarily monopolistic "uniqueness premium" has led to a shift towards the need to achieve global, universally recognised competitiveness of products and services, companies and countries. For example, in August 2019 more than 180 US companies announced that increasing investors' profits should not be their only goal; meeting the interests and needs of the public was also important.[5]

There's evidence of moving on from specific priority research areas or narrow research topics to S&T packages which comprise a platform component in the form

[2] https://ec.europa.eu/growth/tools-databases/dem/monitor/sites/default/files/DTM_Industrie%204.0.pdf

[3] https://www.bmbf.de/de/zukunftsprojekt-industrie-4-0-848.html

[4] https://publications.hse.ru/mirror/pubs/share/direct/263131918

[5] https://www.businessroundtable.org/business-roundtable-redefines-the-purpose-of-a-corporation-to-promote-an-economy-that-serves-all-americans

of digital solutions.[6,7] According to optimistic estimates, the first notable effects, primarily in the field of nanoelectronics, photonics, nanobiotechnology, medical products and equipment and neuro-electronic interfaces, could be expected as early as in the next 5 years. Most likely, the particularly important advances of the next decade will include molecular production of macroscopic objects ("desktop nano-factories") and progress in the atomic design domain. In turn, new energy sources, in all their numerous forms and manifestations, can fundamentally alter the existing balance of power in economic and political spheres alike, at least beyond the 2020 horizon when the cumulative effects of their introduction bring about corresponding changes.

Thus the narrow, focused effects of the new industrial revolution such as increased labour productivity, improved logistics or reduced material or energy intensity of production processes "build up" to cover the entire chain comprising creation, development, sales, and disposal of products or services.[8,9] Value chains are radically changing, moving on from "knowledge economy" to "action economy" which implies an increased role of system integrator companies offering "turnkey solutions" based on quickly "packaged" best available technologies adapted to match specific demand formats. Such "task managers", guided by a long-term vision of the economy and using cutting-edge technologies, are likely to make up the bulk of business models of the future.

As for the beginning of 2020, the Product-as-a-Service (PaaS) concept and business models based on it are rapidly gaining popularity in almost all sectors of the global economy.[10] New links in production chains are emerging (product- and geography-wise alike), while certain long-established ones disappear (e.g. intermediaries such as retail companies), with profits redistributed between the remaining participants. The development of platform technologies which allow to change the structure of production costs and provide direct access to end users will lead to growth of markets based on network solutions: the all-pervasive communication technologies and services make "c2c" (customer-to-customer) interaction many times more efficient (the Uber platform being a bright example). Meanwhile the very structure of global markets will be radically transformed: sales of many "non-primary" products and services such as entertainment content (e.g. network games, relevant information, etc.) already exceed many conventional markets considered "unshakable" in terms of sales.

Numerous economic models which provided the basis for modern science, technology and innovation policy tools are dying out, and replaced by completely

[6] Measuring the Digital Transformation. A Roadmap for the Future. OECD 2019 (https://doi.org/10.1787/9789264311992-en)

[7] Gokhberg L., Sokolov A., Chulok A. Russian S&T Foresight 2030: identifying new drivers of growth // Foresight. 2017. Vol. 19. No. 5. P. 441–456. doi

[8] http://www3.weforum.org/docs/WEF_Innovate_Europe_Report_2019.pdf

[9] http://www3.weforum.org/docs/WEF_Our_Shared_Digital_Future_Report_2018.pdf

[10] https://www2.deloitte.com/insights/us/en/topics/strategy/as-a-service-business-model-flexible-consumption.html

different ones. The basic prerequisites applied to develop them are fundamentally changing. By 2020 we could no longer talk about rational consumer behaviour since now it's often irrational; the expectation economy is emerging, largely due to radically new needs and new approaches to using information; the nature and structure of transaction costs are changing, which in turn significantly affects the structure and organisation of modern businesses.

New global markets were emerging in medicine and healthcare, whose growth could be closely linked to advances in diagnostic and treatment methods based on personalised medicine principles, application of monitoring technologies at home and remote provision of medical services in line with the safety and efficiency requirements. According to various expert estimates, the market for functional food products will recently reach $305 billion; the global neurocomputer interface market will reach $1.7 billion by 2022, and the diagnostic system market can come close to about $9 billion in 2019 already, with the average annual growth rate of up to 30%. In December 2018 China announced the birth of genetically modified twins (though it hasn't been confirmed by the scientific community yet). In turn, active application of biotechnology in production of biopharmaceutical and medical products will allow to make highly promising products such as biodegradable materials, diagnosticums, implants, vital medicines, cell lines, etc.

The above factors affect demand for S&T development legal regimes: from individual cases and specific practices of applying new regulations in an experimental format to realising the need for a radical transformation of almost all institutions responsible for efficient organisation of social and economic processes,[11] for example, regulating e-commerce, preventing cybercrime, etc.

Significant changes were awaited taking place in the employment structure. They are driven by the transition to a new production paradigm, including information technology (artificial intelligence), biotechnology and robotics. Despite numerous estimates made by future scientists as early as 50 years ago, replacement of routine human labour with machines only began in the last few years. This includes, among other things, "smart factories" managed by 1–2 specialists, "smart agriculture" centred around the next-generation farmer and "smart cities" with intelligent transport and energy systems. For a majority of developed countries, the issue of social security will come to the fore (e.g. in single-industry towns). Many professions currently seen as essential will obviously become irrelevant.[12] The competency profile allowing one to be competitive on the labour market will change drastically. A new educational system will be required to meet these challenges:

*We see how sophisticated the whole landscape of global bioeconomy trends was be the beginning of 2020. Now we move to identification drivers and forks for bioeconomy after wild cards. (*IEA 2020a, b*)*

---

[11] https://openknowledge.worldbank.org/bitstream/handle/10986/31334/9781464813726.pdf?sequence=2&isAllowed=y

[12] For example, many companies are already replacing human workers with telephone answering software; warehouse logistics are completely automated; initial medical diagnostics are performed by ongoing monitoring devices, etc.

4 13 Bioeconomy 2.0 Trends Till 2030

By overlaying the wild card events of the last few months over the global trends that set the key directions for bioeconomy's development at the beginning of 2020, 13 trends were identified which will determine the future of bioeconomy for the next 10–15 years (Fig. 25.1).

1. *New consumer patterns:* the economic, environmental and epidemiological events of 2020 will result in consumers' making more careful choices of food products or when making significant for the family budget purchases (such as, e.g. motor vehicles or foreign trips). This will inevitably lead, at least in the next couple of years, to manufacturers' getting the "sovereignty" back: they will decide what to make and how to make it, while customers will find it difficult to impose their conditions like they were able to during the era of customised products, domineering consumerism and personalised consumption.

   On the other hand, the share of household expenditures on healthy food and pharmaceuticals, which help strengthen immunity and increase resistance to diseases, is likely to increase. A vivid example was the temporary spike of lemon and ginger prices (in some regions they grew tenfold compared with the usual level). The "food as fuel" concept may also become popular, as opposed to the hedonistic "food as an art" attitude of the previous years.

   The more selfish behaviour of the population primarily concerned with surviving and recovering from the planetary-scale crisis, combined with persistently low oil prices, can significantly affect the demand for electric cars which in the current situation may turn out to be not so attractive compared to

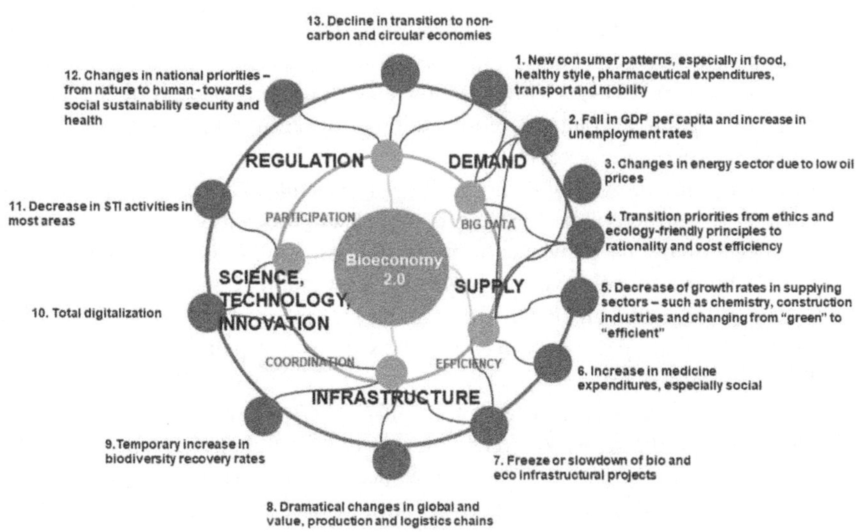

Fig. 25.1 The future of a bioeconomy as an ecosystem: a view from 2020 forks and drivers

conventional ones, providing not only complete freedom of movement but also an element of prestige.

The fate of the sharing economy looks ambiguous: for many of its participants, especially in the food sharing, car sharing and car-pooling segments (such as BlaBlaCar), consumption can be literally toxic. Fear of crowds can persist as a psychological trauma for several years to come. Whether bioeconomy players will be able to adapt to these new consumer behaviour patterns is a big question.

2. *Decline in per capita GDP and increase in unemployment rates* will remain in place for the next 2–3 years even according to the most optimistic estimates. Recovery of the global economy will occur unevenly, both country-wise and in specific sectors and regions. The demand for bioeconomy as an ecosystem and its individual elements will obviously depend on the rate and scale of the recovery processes. However, the trends may be multidirectional here too. On the one hand, general impoverishment of the population and companies is unlikely to lead to increased demand for bioeconomic products and services, which for the most part have not yet managed to offer more competitive prices compared to conventional alternatives. On the other hand, large armies of the unemployed may play the role of an absorber during the emerging shift towards bioeconomy, substituting various missing elements with their labour. At least such opportunities remain in agriculture, forestry, fishing industries and urban development.
3. *Changes in the energy sector*, one of the main battlegrounds between bioeconomy and conventional economy, will be inevitable due to low oil prices. The world encountered these processes at different stages of the energy sectors' transformation. A number of countries, primarily European ones, have almost completed the transition to a non-carbon economy, while in others, e.g. in Russia, conventional energy sources still dominate. Whether biofuel will become a key source of energy (as the world's biofuel associations predicted at the beginning of 2019) is a big question, at least for the period until 2025, especially taking into account the reorientation of the agro-industrial sector to meet people's basic need for food.
4. *Changing priorities, from ethics and green principles to practicality and cost-efficiency*, consumers' and producers' alike, will inevitably create a challenge for bioeconomy one of whose priorities is responsible behaviour. However, in the medium term (5–7 years), humanity may embrace the "integrating with nature" trend again, albeit at a different level. In the longer term (10–15 years), the events taking place now may even be seen as a trigger for development of Bioeconomy 2.0.
5. *Reduced growth in supplying sectors, such as chemical production and construction, and a shift from "green" to "efficient"* will largely set the vector for development of bioeconomy for the next decade. The confident growth of "bio-related" segments in certain subsectors such as varnishes, paints, glues, packaging, materials and many others will have to remain in place in the new competitive conditions. And if as late as at the end of 2019 many companies

declared that profit was not their primary goal, giving priority to transparency, ethics and responsible behaviour, now finding themselves on the verge of extinction businesses may return to the roots of wild capitalism, when price and product quality were the key competitiveness factors.

6. *Increased medical expenditures, especially public ones*, will create important niches for certain bioeconomy segments such as life science, agriculture and green chemistry. The global "human enhancement" trend which for years acted as a powerful driver for development of biotechnologies to support people's improvement and enhancement is likely to make a pit stop for "human survival"; however, the latter trend can also support growth of bioeconomy in the future. A separate group of trends related to combating diseases, including those caused by malnutrition and reduced biodiversity, will receive a boost in the near future as well.
7. *Freeze or slowdown of "bio" and "eco" infrastructure development projects* due to reduced growth rate of the global economy and problems with finding "long" money at low interest rates can lead to accumulated negative infrastructural effects for bioeconomy. Frozen construction projects such as plants and factories for processing biomass or waste and universal transition to digital technologies, creating countrywide networks of electric vehicle charging stations, may "fall out of time": they were started on the basis of a certain image of the future, specific production chains and demand patterns, but will be completed (if they are ever completed) in a totally different world. Meanwhile without an advanced and efficient infrastructure, growth of bioeconomy wouldn't be possible.
8. *Dramatic changes in global value, production and logistics chains* can produce mixed effects on bioeconomy as an ecosystem. On the one hand, many stakeholders of bioeconomic transformation for a long time have been trying to integrate into the existing chains, often acting as disruptive innovators or relying on the "pull" of demand by environmentally responsible consumers. However, having so far failed to pass the "breakeven point", without additional incentives, bioeconomy may roll back to the very beginning of its development. On the other hand, the uniqueness of the current world situation is in the fact that practically all economic ties, some of which took decades to be established, are now if not severed, than put on an indefinitely long pause, both as a result of the pandemic and the collapse of energy markets. The world's economy has taken a step from being global and efficient towards becoming isolated and natural, locked within national borders. It is in just such circumstances, as the institutional economy teaches us, new industry standards emerge. If one manages to integrate into the process of these standards' formation essentially becoming such a (next generation) standard, subsequent development becomes an integral part of the general evolution. If bioeconomy turns into an integral part of, and a necessary condition for, the world's new value chains, it will be able to make a significant progress in dominating over other economic concepts.
9. *A temporary increase in the recovery rates of biodiversity* may be a more positive consequence of the global crisis for bioeconomy. The shutdown of many

industries that significantly pollute the environment and the self-isolation of millions of people have already produced evidence that the nature is beginning to recover. However, in a number of areas, the damage caused by humanity is so great that without targeted efforts and significant investment, recovery wouldn't be possible. Whether humanity will be able to achieve its own recovery along with a harmonious recovery of the nature remains to be seen; at least many of the unemployed now have much more time to take part in bioprogrammes such as garbage collection, planting of trees and organic farming.

10. *Total digitisation*, the rate of which exceeded the rate of COVID-19 proliferation, connected all countries and regions into a single ecosystem, essentially making Professor Vernadsky's noosphere theory come true at a new, digital level. Whether the bioeconomy will be able to take advantage of the new infrastructure remains to be seen, but in a number of sectors, digitisation and digital ecosystems already have prominent positions. For example, according to HSE and Rosselkhozbank estimates, implementing the precision and smart agriculture concept allows to reduce production costs by at least 10–15%. Meanwhile the Industry 4.0 concept which implies total digitisation and robotisation, on the whole, will undergo significant changes. On the one hand, countries and economic sectors that have already invested in technological modernisation will try to complete it at a faster rate. On the other hand, those caught by the pandemic at "crossroads" will need to make the difficult choice between increasing economic efficiency and maintaining social stability. It would be quite problematic to explain to ten unemployed people, at least in the near future, that their jobs are now being done by a single robot (according to various estimates, that's going to be the human-machine replacement rate).

    Another area where digitisation may turn out to be a major development driver is research and education. In the near future, academic mobility in the physical form will be seriously reduced, but willingness of scientists, students and teachers to collaborate may open up new opportunities for bioeconomic research. It will be easier to meet the fully planetary in scope COVID-19 challenge by joining countries' efforts and competencies. However, the overall level of investment in innovation and S&T development will be moderate.

11. *The decrease in most areas of STI activities* caused by the global crisis may affect bioeconomy to a lesser extent. For biotechnological experiments which allowed to significantly advance the life science segment of the bioeconomy, the coming years may see increased control by the society and the state after they have actually seen the potential consequences of biological warfare or hacker attacks. On the other hand, demand for human enhancement-related research, at least in traditional areas such as vaccines, will obviously significantly increase. The world's inability to promptly come up with an S&T response to the pandemic will probably lead governments and companies to realise the need to increase investments in basic research, and step up the opportunities for rapid application of its results to solve applied problems, thus promoting emergence of new formats for national innovation systems. For example, if it would have already been possible to use digital doubles (high-fidelity

mathematical models) for medical purposes, developing a COVID-19 vaccine and testing it on volunteers could take not years but months.

In general, research areas related to development of bioeconomy will undergo a transformation adapting to the new priorities of countries, society and businesses.

12. *Changes in national priorities from the environment to people, towards social sustainability, security and healthcare*, will lead not just to provision of "helicopter" basic income and application of other financial mechanisms to support the public but also to more radical transformation of the functions of the state as an institution. "The economy must be economical – this is the demand of the time", said Leonid Brezhnev in 1980 at the communist party congress. Paradoxically, the Soviet leader's statement of 40 years ago may be very relevant in the current realities. It's echoed by the words of the Nobel laureate in economics J. Stiglitz who just a few years ago urged heads of state to recognise the fact that per capita, GDP was no longer the best indicator for comparing countries' success. The Nobel laureate suggested including in the number of relevant indicators the KPIs related to social stability, responsible and environmentally friendly behaviour. Now countries seem to be competing only in terms of the share of those cured from COVID-19 and the rate of lifting lockdown. For bioeconomy, these trends will mean the need to develop cost-effective technologies and products with socially significant properties.
13. *Less active transition to non-carbon and circular economy* due to both low oil prices which make carbon economy look more attractive than before and the pandemic which negatively affects many of the fundamental activities of circular economy, e.g. separate collection of waste which became almost impossible because of uncontrolled proliferation of toxic waste from COVID-19 patients and carriers.

On the other hand, the changes in value chains mentioned earlier and many countries' orientation towards self-sufficiency (including food security) may push people and businesses if not to accept non-carbon and circular economies as complete "turnkey" concepts, then at least to sensible adoption of certain technologies and products they offer (such as, e.g. home bioreactors).

5 Bioeconomy Beyond 2030

Which transformations, trends and wild cards for bioeconomy could be beyond 2030?

5.1 Changing Competitiveness Factors

Network models for reaching consumers in the framework of sharing economy, environmentally and ethically friendly behaviour, creating a comfortable environment for work and creativity, transparency and social responsibility, will determine companies' success in the market for at least the next 10–15 years. Zero or even negative profits no longer hinder growth, as most of today's companies clearly demonstrate. Being a technology leader, including in the area of digitisation, will remain important until 2035–2045 – the horizon when many experts expect singularity will be achieved. What factors are going to determine businesses' competitiveness after 2050? Key factors are: ability to fully identify, and make full use of each person's potential, on the basis of big data analysis, reliable adaptive infrastructure and maximum diversification. The ability to deliver emotions and literally make dreams come true will become the key to achieving market success, provided this economic term survives to 2100.

5.2 Changing Production Factors

Labour, land and capital – the basic production factors according to the classic twentieth-century economists – will be digitised as much as possible (all the way up to digital factories), and robotised. In the next 10–15 years, the focus will be on optimising production and organisational processes (among other things on the basis of AI and big data), diversifying funding sources through crowdfunding and a fierce war for personnel and advanced technologies. Human capital with new competencies will become a key growth factor by 2040–2050. Afterwards it will all be about time: the time it takes to bring a product or service to the market, meet demand and respond to customer requests. Perhaps by 2100 time will replace conventional money, including cryptocurrencies whose popularity will come to an end after proliferation of quantum computers.

5.3 Changing Image of Business

Large or small, local or transnational, conventional or digital, business will pass all these forks by 2035, and enter the era of IT giants who'd be providing shares in planetary business as a cloud service. After 2050 the very concept of a corporation is likely to radically transform into a social function, putting full stop in the dispute between the two economists we have begun the story with. Only about 30–40 years, separate the mobile application economy which has already started to grow from the full-fledged neuroeconomy of impressions, with consumers connected to digital platforms being able to create anything at all using the power of thought. Perhaps after 2050 the phrase "I am the state" will become literally true, and by 2100 a single network will be in place connecting individual entrepreneurs armed with technological gadgets from all over the world into one large corporation.

5.4 Changing Role of Managers

The ability to find the right solutions in a situation of constant uncertainty is the skill efficient leaders would need in the next 5–10 years, while having a vision of the future and being willing to rally a team around them will allow them to remain so until 2035. Complete digitisation of business processes will challenge human leaders closer to 2045. Afterwards it would be more about creators, architects or engineers in a broad sense rather than company managers and executives. Whether people will remain in charge after 2050 or all management roles will be taken by artificial intelligence remains to be seen, but the working week should definitely be reduced, by 2100 at the latest.

5.5 Changing External Factors: Wild Cards

Events with a low probability of happening but producing major effects if they do will accompany businesses throughout the century, ranging from environmental shocks affecting the conventional resource sectors (flooding of oil wells, changing agroclimatic zones) and total ethical certification in the horizon until 2035 to the prospects of colonising outer space and new planets after 2050. Cyberhacking, escalation of military conflicts, a sharp decline in energy and other resources' costs will compete for the title of companies' worst nightmare with total nationalisation and planned economy 2.0. However, human immortality may turn out to be the truly wild card, which certain future scientists expect to be achieved if not by 2045, then by 2100 for sure.

6 Conclusion

Thus, the future of bioeconomy in the twenty-first century will depend on the rate, scale and success of passing the following key forks in 2020:

The nature and scale of economic support (provided to the population, economic sectors and business areas)

The re-establishment of previously existing or creation of new logistics and production chains

The role of innovation and technology: infrastructural support or "disruptive" breakthroughs
The factors affecting companies' competitiveness: cost and value structure and ethics
The struggle for "sovereignty" between producers and consumers
"Smart" technologies replacing not just routine operations but also highly skilled professionals
Education: format wars – "old school" versus "zoom pandemic"

References

Aghion P., Howitt P. (1992). *A Model of Growth Through Creative Destruction.* Econometrica. 60 (2): 323–351, https://doi.org/10.2307/2951599.

Aghion P. (2001) *Schumpeterian growth theory and the dynamics of income inequality.* Econometrica. 70 (3): 855–882, https://doi.org/10.1111/1468-0262.00312.

Ansoff I. (1979) 2007 *Strategic Management.* New York: Palgrave Macmillan.

Babu S. C., Debnath D. (2019) *Biofuels, bioenergy and food security: Technology, institutions and policies.* Academic Press. DOI: https://doi.org/10.1016/C2015-0-00851-6

BMU (2019) *Waste Management in Germany 2018: Facts, data, diagrams.* Berlin: Federal Ministry for the Environment, Nature Conservation and Nuclear Safety (BMU), Germany. Available at: https://www.bmu.de/fileadmin/Daten_BMU/Pools/Broschueren/abfallwirtschaft_2018_en_bf.pdf, accessed 22.02.2020.

EB (2018) *European Bioplastics.* Berlin: «European Bioplastics» Association. Available at: https://docs.european-bioplastics.org/publications/EUBP_Facts_and_figures.pdf, accessed 22.02.2020.

European Commission (2018a) *A sustainable Bioeconomy for Europe: strengthening the connection between economy, society and the environment. Updated Bioeconomy Strategy,* Brussels: European Commission. Available at: https://ec.europa.eu/research/bioeconomy/pdf/ec_bioeconomy_strategy_2018.pdf#view=fit&pagemode=none, accessed 21.02.2020.

European Commission (2018b) *Biomass production, supply, uses and flows in the European Union: First results from an integrated assessment,* Brussels: European Commission. Available at: http://publications.jrc.ec.europa.eu/repository/bitstream/JRC109869/jrc109869_biomass_report_final2pdf2.pdf, accessed 21.02.2020.

European Commission (2018c) *Joint survey on bioeconomy policy developments in different countries,* Brussels: European Commission. Available at: https://ec.europa.eu/knowledge4policy/sites/know4pol/files/jrc112081_joint_survey_report_final.pdf, accessed 21.02.2020.

European Commission (2019a) *Report on the implementation of the Circular Economy Action Plan,* Brussels: European Commission. Available at: https://eur-lex.europa.eu/legal-content/EN/TXT/?qid=1551871195772&uri=CELEX:52019DC0190#footnote75, accessed 21.02.2020.

European Commission (2019b) *Assessing the impacts of the EU bioeconomy on third countries,* Brussels: European Commission. Available at: http://publications.jrc.ec.europa.eu/repository/bitstream/JRC117364/scienceforpolicy-follador-final14102019.pdf, accessed 21.02.2020.

European Commission (2019c) *Waste statistics,* Brussels: European Commission. Available at: https://ec.europa.eu/eurostat/statistics-explained/index.php?title=Waste_statistics#Waste_treatment, accessed 21.02.2020.

European Commission (2019d) *Brief on algae biomass production,* Brussels: European Commission. Available at: http://publications.jrc.ec.europa.eu/repository/bitstream/JRC118214/kcb_brief_algae_biomass_production_online_version.pdf, accessed 21.02.2020.

European Commission (2019e) *Alternative global transition pathways to 2050: Prospects for the bioeconomy,* Brussels: European Commission. Available at: https://publications.jrc.ec.europa.

eu/repository/bitstream/JRC118064/jrc_techreport_bioeconomy2050_online.pdf, accessed 21.02.2020.

European Commission (2020) *Building a monitoring system for the EU bioeconomy. Technical report*, Brussels: European Commission. Available at: https://publications.jrc.ec.europa.eu/repository/bitstream/JRC119056/bemonitoring_finaltopubsy_withbackcover_gm.pdf, accessed 21.02.2020.

FAO (2016) *How sustainability is addressed in official bioeconomy strategies at international, national and regional levels. An overview.* Rome: The Food and Agriculture Organization of the United Nations. Available at: http://www.fao.org/3/a-i5998e.pdf, accessed 15.05.2020.

FAO (2018) *Assessing the contribution of bioeconomy to countries' economy. A brief review of national frameworks.* Rome: The Food and Agriculture Organization of the United Nations. Available at: http://www.fao.org/3/I9580EN/i9580en.pdf, accessed 15.05.2020.

FAO (2019a) *Towards sustainable bioeconomy - Lessons learned from case studies.* Rome: The Food and Agriculture Organization of the United Nations. Available at: http://www.fao.org/3/ca4352en/ca4352en.pdf, accessed 15.05.2020.

FAO (2019b) *Indicators to monitor and evaluate the sustainability of bioeconomy. Overview and a proposed way forward.* Rome: The Food and Agriculture Organization of the United Nations. Available at: http://www.fao.org/3/ca6048en/ca6048en.pdf, accessed 15.05.2020.

FFRC (2019) *Precursors to a "Good" Bioeconomy in 2125: Making Sense of Bioeconomy & Justice Horizons. First Foresight Report of the BioEcoJust Project.* Turku: Finland Futures Research Centre. Available at: https://www.utupub.fi/bitstream/handle/10024/148181/eBook_2-2019.pdf?sequence=1&isAllowed=y, accessed 15.05.2020.

GBS (2018a): *Conference Report*, Berlin: Global Bioeconomy Summit. Available at: https://gbs2018.com/fileadmin/gbs2018/GBS_2018_Report_web.pdf, accessed 22.02.2020.

GBS (2018b): *Future Opportunities and Developments in the Bioeconomy – a Global Expert Survey*, Berlin: Global Bioeconomy Summit. Available at: https://gbs2018.com/fileadmin/gbs2018/Downloads/Bioeconomy_Global_Expert_Survey.pdf, accessed 22.02.2020.

GBS (2018c): *Update Report of National Strategies around the World*, Berlin: Global Bioeconomy Summit. Available at: https://gbs2018.com/fileadmin/gbs2018/Downloads/GBS_2018_Bioeconomy-Strategies-around-the_World_Part-III.pdf, accessed 22.02.2020.

Grinin L., Devezas T., Korotayev A. (2012) *Kondratieff Waves: Dimensions and Prospects at the Dawn of the 21st Century.* Volgograd: Uchitel.

IEA (2017) Technology Roadmap. Delivering Sustainable Bioenergy. Paris: International Energy Agency. Available at: https://www.ieabioenergy.com/wp-content/uploads/2017/11/Technology_Roadmap_Delivering_Sustainable_Bioenergy.pdf, accessed 15.05.2020.

IEA (2018a) *The IEA Bioenergy Annual Report 2018.* Paris: International Energy Agency. Available at: https://www.ieabioenergy.com/wp-content/uploads/2019/04/IEA-Bioenergy-Annual-Report-2018.pdf, accessed 22.02.2020.

IEA (2018b) *Gasification of waste for energy carriers. A review.* Paris: International Energy Agency. Available at: https://www.ieabioenergy.com/wp-content/uploads/2019/01/IEA-Bioenergy-Task-33-Gasification-of-waste-for-energy-carriers-20181205-1.pdf, accessed 22.02.2020.

IEA (2019) Will energy from waste become the key form of bioenergy in Asia? Paris: International Energy Agency. Available at: https://www.iea.org/articles/will-energy-from-waste-become-the-key-form-of-bioenergy-in-asia, accessed 15.05.2020.

IEA (2020a) Biofuels in the time of Covid-19: Staying the course on clean transport fuels in Latin America. Paris: International Energy Agency. Available at: https://www.iea.org/commentaries/biofuels-in-the-time-of-covid-19-staying-the-course-on-clean-transport-fuels-in-latin-america, accessed 15.05.2020.

IEA (2020b) Outlook for biogas and biomethane: Prospects for organic growth. Paris: International Energy Agency. Available at: https://www.iea.org/reports/outlook-for-biogas-and-biomethane-prospects-for-organic-growth, accessed 15.05.2020.

IRENA (2019a) *Solid Biomass Supply for Heat and Power: Technology Brief* Abu Dhabi: International Renewable Energy Agency. Available at: https://www.irena.org/-/media/Files/

IRENA/Agency/Publication/2019/Jan/IRENA_Solid_biomass_supply_2019.pdf, accessed 22.02.2020.

IRENA (2019b) *Advanced biofuels: What holds them back?* Abu Dhabi: International Renewable Energy Agency. Available at: https://www.irena.org/-/media/Files/IRENA/Agency/Publication/2019/Nov/IRENA_Advanced-biofuels_2019.pdf, accessed 22.02.2020.

Keswani C (2020) *Bioeconomy for Sustainable Development.* Springer. DOI: https://doi.org/10.1007/978-981-13-9431-7

Krugman P, Wells R, Romer P (2005) *Macroeconomics*. London: Macmillan Higher Education

Martinet A-C (2010) *Strategic planning, strategic management, strategic foresight: The seminal work of H. Igor Ansoff.* Technological Forecasting and Social Change. 99(7):1485–1487. https://doi.org/10.1016/j.techfore.2010.06.024

OECD (2019a) *Meeting Policy Challenges for a Sustainable Bioeconomy. Designing a Policy Agenda*, Paris: OECD. Available at: https://www.oecd-ilibrary.org/science-and-technology/policy-challenges-facing-a-sustainable-bioeconomy_9789264292345-en, accessed 13.05.2020.

OECD (2019b) *Bio-economy and the sustainability of the agriculture and food system*, Paris: OECD. Available at: https://www.oecd-ilibrary.org/agriculture-and-food/bio-economy-and-the-sustainability-of-the-agriculture-and-food-system_d0ad045d-en, accessed 13.05.2020.

OECD (2019c) *Innovation ecosystems in the bioeconomy*, Paris: OECD. Available at: https://www.oecd-ilibrary.org/science-and-technology/innovation-ecosystems-in-the-bioeconomy_e2e3d8a1-en, accessed 13.05.2020.

OECD (2020) *Digitalisation in the bioeconomy: Convergence for the bio-based industries*, Paris: OECD. Available at: https://www.oecd-ilibrary.org/science-and-technology/innovation-ecosystems-in-the-bioeconomy_e2e3d8a1-en, accessed 13.05.2020.

Perea-Moreno M-A, Samerón-Manzano E, Perea-Moreno A-J (2019) *Biomass as Renewable Energy: Worldwide Research Trends*. Sustainability, vol. 863 , no 11. https://doi.org/10.3390/su11030863

Perez C (2002) *Technological Revolutions and Financial Capital: The Dynamics of Bubbles and Golden Ages*. London: Elgar.

Romer P (1990) *Endogenous Technological Change*. Journal of Political Economy, 98(5):71–102.

Ronzon T, Sanjuánc AI (2019) *Friends or foes? A compatibility assessment of bioeconomy-related Sustainable Development Goals for European policy coherence.* Journal of Cleaner Production, vol. 254. https://doi.org/10.1016/j.jclepro.2019.119832

Schumpeter J (2011) *The Theory of Economic Development: An Inquiry into Profits, Capital, Credit, Interest, and the Business Cycle*. New Brunswick, London: Transaction Publishers.

Solow R (1970) 2006 *Growth Theory - An Exposition*. Oxford: Oxford University Press.

Vernadsky V (1926) 1997. *The Biosphere.* McMenamin (ed.) New York: Copernicus.

Wallerstein I (1979) *The Capitalist World-Economy (Studies in Modern Capitalism).* Cambridge: Cambridge University Press.

Wu W, Hasegawa T, Ohashi H, Hanasaki N, Liu J, Matsui T, Fujimori S, Masui T, Takahashi K (2018) *Global advanced bioenergy potential under environmental protection policies and societal transformation measures*. GCB Bioenergy, no 10, pp. 1041–1055. https://doi.org/10.1111/gcbb.12614

Chapter 26
Responsible Innovation in Industry: The Role of Firm's Multi-Stakeholder Network

Jolita Ceicyte, Monika Petraite, Vincent Blok, and Emad Yaghmaei

Abstract Responsible innovation (RI) receives a high interest from the society; however, its practical implementation remains complicated because of the lack of operationalisation that would allow integrating RI into existing organisational processes, especially, in industrial contexts. Most prior research on RI in the industry seek to move forward firm's innovation policies and strategies to include RI principles. They are, however, limited to include firm' behavioural choices and institutional requirements in their analysis. In an integrative view, a multi-stakeholder network approach to reveal how firm's multi-stakeholder network shapes responsible innovation practices throughout the different stages of the innovation process of the firm is applied. Data from ten case studies are used to validate different levels of responsibility, through engagement with the firm's multi-stakeholder network. Based on that, a new conceptualisation of the RI is proposed, consisting of two levels: first level defined by legal and contractual responsibilities and the second level defined by moral responsibilities.

Keywords Responsible innovation · Multi-stakeholder network · R&D firm · RRI

J. Ceicyte (✉) · M. Petraite
Innovation and Entrepreneurship Research Group, School of Economics and Business, Kaunas University of Technology, Kaunas, Lithuania
e-mail: jolita.ceicyte@ktu.lt; monika.petraite@ktu.lt

V. Blok
Philosophy Group, Wageningen University, Wageningen, The Netherlands
e-mail: vincent.blok@wur.nl

E. Yaghmaei
Department of Values, Technology & Innovation, Faculty of Technology, Policy and Management, TU Delft, Delft, Netherlands
e-mail: e.yaghmaei@tudelft.nl

E. Koukios, A. Sacio-Szymańska (eds.), *Bio#Futures*,
https://doi.org/10.1007/978-3-030-64969-2_26

1 Introduction

Responsible innovation (RI) emerged in R&D intensive contexts due to the uncertainty of the impact and long-term consequences of innovation for societal and environmental development (Zwart et al. 2014; Pavie et al. 2014). In defining responsible innovation, Von Schomberg (2012, p. 9) argues that "a transparent, interactive process by which societal actors and innovators become mutually responsive to each other with a view to the (ethical) acceptability, sustainability and societal desirability of the innovation process and its marketable products (in order to allow a proper embedding of scientific and technological advances in our society)". Despite rising interest in using the RI construct, systematic studies of RI in the industry remain cumbersome because of conceptual ambiguity. The extant literature presents the concept of RI in industry, particularly in R&D contexts, which are limited (Van de Poel et al. 2017; Ceicyte and Petraite 2018; Yaghmaei 2018). The approach of this study is to provide an analytical frame of RI in the industry to show the effects of firm's multi-stakeholder network on the RI initiatives.

It is theoretically preferred that all the related stakeholders should be enrolled from the very ideation phase of innovation development (Swiestra and Rip 2007; Sutcliffe 2011; Geoghegan-Quinn 2012). Yet this imperative is challenging to apply in a real practice as firms act upon economic logic and tend to include economic stakeholders, and mostly evaluate innovation impact ex post, or at the closing stages of innovation development (Blok, Hoffmans and Wubben 2015). Although R&D-driven innovations emerge in collaboration between complementary actors of the network, based on collaboration, knowledge and competence exchange (Pittaway et al. 2004), the problem of R&D intensive firm's innovation network is that they still follow the typical business network logic and tend to involve supply chain actors, based on their value added in innovation processes (both public and private), while noneconomic stakeholders remain to be of lower importance due to additional required resources (Blok, Lemmens 2015). A similar tendency can be seen in the innovation process, where mainly technological and commercial issues are considered in order to gain competitiveness, and socio-ethical aspects are given less importance. Whereas, in RI literature, mostly normative aspects concerning RI are examined, economic motives regarding stakeholder inclusion are mostly neglected (Porcari et al. 2015a, b; Garst et al. 2017).

Thus, in order to develop a theory, the specific research question of this paper focuses on an R&D-intensive firm's multi-stakeholder influence on responsible innovation practices implementation along the different innovation stages. An exploratory study design is applied in order to find out how RI activities are implemented along the innovation process in R&D-intensive firms. This study followed an inductive research approach, consisting of four stages. At first, existing literature on the integration of multi-stakeholder networks and RI processes in the industry was consulted. As a second step, secondary data analysis of R&D-intensive firms selected for interviews was collected. Third, interviews with R&D-intensive firms were conducted. Finally, in order to explore the existing reality, an analysis of the primary interview data was performed, and linkages to the existing theory in RI in the industry were discussed. A contribution to research literature is based on demonstration of a firm's multi-stakeholder network impact on the application of RI practices along innovation stages. The paper is structured as follows. The next section presents a literature review in the field of RI and multi-stakeholder network. Section 2 provides a theoretical framework explored during interviews. In Sect. 3, a qualitative research methodology is developed based on the multiple-case study method. Section 4 provides findings based on comparative analysis of cross-industrial RI behaviour towards multi-stakeholder network integration along with different innovation phases. In Sect. 5, the implications of the findings for theory and practice are discussed. The paper ends with concluding remarks.

2 The Role of Multi-Stakeholder Networks in Firm's Responsible Innovation Management: A Literature Review

2.1 Responsible Innovation in the Industry

RI studies are still evolving in different directions, but so far, much research on the concept and practice of RI has focused on the development of different frameworks incorporating political and ethical considerations into the innovation process (Long and Blok 2017). Stakeholder inclusion into innovation processes as a common feature of open innovation is a product of coordinated multiple networks and applies in RI practices (Blok et al. 2015; Yaghmaei 2016). RI in the open innovation vein (Long and Blok 2017), which is oriented to mitigate the Grand Challenges by collaborative initiatives (Von Schomberg 2013), requires specific organisational competencies and practices to reach responsibly considered innovative outcomes serving the societal and environmental benefit.

These specific organisational competencies and practices are aligned with the RI process dimensions coined by Stilgoe et al. (2013), consisting of anticipation, inclusion, reflexivity and responsiveness (AIRR). Anticipation is defined as involving "systematic thinking aimed at increasing resilience while revealing new opportunities for innovation and the shaping of agendas for socially-robust risk research" (Stilgoe et al. 2013: 1570). Inclusion of various stakeholders is essential in offering more opportunities to solutions in dealing with the Grand Challenges (Cagnin et al. 2012) while sharing different, maybe even opposing, opinions, combining different interests, etc. Reflexivity, the third dimension of RI, is determined as an institutional and scientific practice, which is "holding a mirror up to one's activities, commitments and assumptions, being aware of the limits of knowledge and being mindful that a particular framing of an issue may not be universally held" (Stilgoe et al. 2013, 1753). The last dimension, responsiveness, is about being capable of changing or shaping the direction of innovation in response to values of stakeholders and broader society (Lubberink et al. 2017). In practice, these process dimensions should be integrated into the firm's innovative activity and applied iteratively (Genus and Iskandarova 2017), so the more responsible outcomes of innovation would be achieved. However, as empirical research shows, neither firms nor stakeholders are willing to share the responsibility for the innovation risks (Blok et al. 2015).

Conceptual links between RI and other disciplinary theories such as network theory help to better answer fundamental questions on RI antecedents and operationalisation. Stahl et al. (2017), for example, contributed to the RI theory by developing RI as a meta-responsibility, which can govern existing networks of responsibilities. In their study, mainly formal and task responsibilities were analysed in the alignment between different actors in firms to meet the overall societal goals.

2.2 The Role of the Multi-Stakeholder Network in the Responsible Innovation Process

Innovation, being a complex phenomenon, is, therefore, best viewed at a systemic level (Smith 2000). The previously introduced definition by von Schomberg's (2013) suggests that RI is not only output but also a "transparent and interactive" process, where the members ("societal actors and innovators") are a part of a broader system. Societal actors and innovators seek to bring value like "(ethical) acceptability, sustainability and societal desirability" within innovation process and "its marketable products" by collaboration ("become mutually responsive to each other"). The consequence of such a conceptualisation of RI is to become one interdependent system in shaping both innovation processes and outcomes for the benefit of environment and society as innovation with and for society.

This resonates with the network approach, which emphasises interdependencies between related actors and stimulates new ways of collaboration (Frels et al. 2003). From the network theory perspective, different actors are interconnected and are analysed in a systemic manner. As Ritter and Gemünden (2003) note, cutting-edge innovations are developed in collaboration between complementary actors of the network. This is also relevant in responsible innovation context because complex problems can only be mitigated by different actors working together by gathering information, resources, activities and capabilities; it is then possible to achieve a solution that could not be performed by organisations alone (Bryson and Crosby 2006). Still, RI requires extending the network beyond the value chain actors and embraces a multi-stakeholder network representing various societal groups and public interest.

Application of network theory to multi-stakeholders is elaborated in Rowley's (1997) study. Multi-stakeholder networks are defined as "networks in which actors from civil society, business and governmental institutions come together to find a common approach to an issue that affects them all" (Roloff 2008: 238). Complex problems that are difficult to pin down by one actor are attempted to be solved in such networks. RI, by its definition, is a complex challenge for the firm, as it demands the integration of various stakeholders' interests and ex ante assessment of innovation impacts along the development process.

Roloff (2008) argues that firms in multi-stakeholder networks should be understood as equal to other members in multi-stakeholder networks rather than initiators or leaders. According to Roloff (2008), multi-stakeholders are "tripartite", meaning that firms, civil society and governmental organisations should be equally participating in innovation processes. The multi-stakeholder network can also be analysed as a representation of formal and informal institutions (i.e. legal and societal norms), and actors manifesting them, from the perspective of impact on firm's decision-making, external and internal stakeholders given their positioning to the firm and the level of network transaction formalisation. Since the analysis of multi-stakeholders' role in RI processes is based on their position towards the firm, thus, Freeman's (1984) view of internal and external stakeholders provides a valuable perspective.

External Stakeholders Firm's external stakeholders shape the RI framework and form a network of multiply interactions in the innovation decision-making process. External stakeholders can be affected or can affect the firm. Usually, firm's external stakeholders are value chain-based actors, i.e. customers, distributors, suppliers, creditors, user communities, financial institutions, governmental and non-governmental organisations, local communities or broader society, etc. In the case of high-technology industries, external stakeholders, especially innovation and supply chain partners, are highly important, since there are many collaborations and alliance formations for joint innovation projects (Stuart 2000). However, interactive learning for innovation requires intensive trust-based interaction (Lund Vinding 2006); therefore, in order to maintain collaboration with external stakeholders, it is crucial to building trust and cognitive understanding. This force firms to behave responsibly towards their external stakeholders to ensure long-term relationships and, at the same time, attract new potential external stakeholders. External stakeholders usually are also the end users or customers, so they play an essential role in the long-term success of a firm's innovative activity and have a direct impact on innovation market return.

Internal Stakeholders Firm's internal stakeholders are employees or teams who work within an organisation. Internal stakeholders are working in various areas within an organisation, like management, marketing, manufacturing, sales, etc. (Blyler and Coff 2003). In order to achieve better results in a firm, it is essential to motivate and engage with internal stakeholders.

These two positions of stakeholder against a firm highlight the interdependencies between the industrial organisations and their environment. As such, multi-stakeholder networks should open up the innovation process to engage both external and internal stakeholders. Firms vary in the extent to which they can engage stakeholders, as research has shown that firms need organisational competencies and practices to involve stakeholders positively in corporate activities (Greenwood 2007).

To summarise, multi-stakeholder networks demand the coordination of innovation activities from firms to manage collaboration constructively and thus hold the coordinating agency to bring stakeholders together to commonly explore new alleys to innovate by composing various interests and expectations.

2.3 Innovation Process for Responsible Innovation

There are several possibilities to conceptualise the innovation process, but mostly the terminology depends upon the industry (Dreyer et al. 2017). Considering the new product development process, innovation funnel (Hayes et al. 1988; Tidd et al. 2001) is an adopted practice. In specific sectors like pharmaceuticals and medical engineering, the innovation process is differentiated into preclinical research, clinical development, regulatory review and post-market safety monitoring (FDA n.d.).

Innovation stage-gate model (Cooper 2008) is also being used to conceptualise the RI process. RI studies in public (Macnaghten and Owen 2011) and private (Blok et al. 2015) contexts applied the stage-gate model because of the "gates" that highlight the specifics of decision-making in every stage including innovation launch with regard to the progress made and actual work on innovation to be done. Blok et al. (2015) in the study of RI in the private sector applied three-stage framework: first phase (discovery and scoping stages), middle phase (business case and development stages) and third phase (testing and validation and launch stages). Similarly, in this paper, exploration, development and implementation are used for RI analysis in the business context (Dreyer et al. 2017). This approach allows minimising possible deviations caused by the individual innovation process design within a single firm.

Exploration Also, it is called idea generation. During this stage, the identification of the issue and exploration of potential solutions are made. The uncertainty and various unknown aspects regarding the project idea are dominating in this stage. Therefore, this stage can also lead both to the project idea development and killing (Dreyer et al. 2017). During this stage, all related stakeholders should be integrated into a firm's ideation stage to anticipate possible risks of innovation. In reality, however, the wider society is usually integrated into the very last stage before the innovation commercialisation (Blok et al. 2015); thus, stakeholders do not have possibilities to express their opinion towards innovation by approving or contesting the innovation idea.

Development stage includes research, development and engineering (Tushman 1977). During this phase risk management plays a key role; thus at the end of this process, benefits, risks and values both for the firm and society have to be clearly stated (Dreyer et al. 2017). In parallel with RI, the need for the wider society's inclusion into the development phase is required in order to align innovation with the society's expectations.

Implementation phase consists of R&D, manufacturing, market tooling up, coordination and administration of the new product (Tushman 1977). During this stage, it is essential to focus on the value for the society. Compliance with regulations and standards of the product must be ensured during the implementation stage. At the end of this stage, product is usually launched (Dreyer et al. 2017). Regarding RI, the feedback from the stakeholders is important in the way that if the innovation violates socio-ethical, environmental aspects, it should be considered for a modification or eliminated from the market.

Responsibilities During the Innovation Process

So far, a few scholars have investigated on the different levels and dimensions of RI (Wickson and Carew 2014; Iatridis and Schroeder 2016; Dreyer et al. 2017; Stahl et al. 2017; Yaghmaei 2018). RI dimensions are usually classified into three different types of responsibilities: legal, contractual and moral.

Legal responsibilities are "based on laws and a jurisprudence providing a framework of obligations, but which is dependent on the laws applicable within a specific jurisdiction, e.g., a particular country" (Dreyer et al. 2017, 11). Regarding the industrial context, legal responsibilities are concerning product liability aspects such as negligence, various defects, warranty, etc., as well as being transparent regarding the product and its possible negative impacts (Dreyer et al. 2017; Iatridis and Schroeder 2016).

Contractual responsibilities in industrial contexts concern mutual obligations that depend on the agreement of two or more parties. There is more room for interpretations regarding the contractual responsibilities, but still, the responsibilities of the parties are quite clear and can end up with penalties where a breach occurs (Dreyer et al. 2017).

Moral responsibilities are based on the value system and mostly depend on a cultural context. Moral responsibilities are open to interpretations since they are embedded in different societal contexts and norms, industries, etc. (Dreyer et al. 2017).

RI scholars seem to focus on the importance and elaboration on moral responsibility, thus excluding other types of responsibilities, i.e. legal and contractual. According to Garst et al. (2017), this could be due to the non-commercial research and innovation as a central field of study by RI scholars. Wickson and Carew (2014), for example, made a typology of RI by classifying it into the four levels: (1) exemplary, (2) great, (3) good and (4) routine. Every level consists of processes and practices that indicate a certain level of RI. Although the criteria mapping for RI is oriented towards projects' evaluation and funding in the context of public policy, the classification itself is combining different types of responsibilities, like legal, contractual and moral, in order to achieve an exemplary level regarding RI.

The need to integrate both legally prescribed procedures and informal interactions for stakeholder inclusion in the public policy context is also emphasised in Cuppen's et al. (2015) research. This confirms the need to integrate different types of responsibilities when understanding RI. Inspiration also comes from CSR studies that tend to incorporate various responsibilities in an industrial context (Garst et al.

2017), where "being socially responsible means operating in a way that meets or exceeds society's ethical, legal, and public expectations" (Gomez-Bezares et al. 2017, 4).

Another reason for integrating legal and contractual responsibilities together with moral responsibilities into the whole RI concept is that sometimes legal and contractual responsibilities can be of a lower level in emerging markets compared to developed countries (Hadengue et al. 2017). In this case, firms have different legal and contractual responsibilities in different societal contexts. *Based on that, in this article, the first level of RI is understood as a firm's compliance with legal and contractual responsibilities. Second level of RI is regarded as being more complicated and requiring to integrate moral responsibility aspects into the innovation process of the firm.*

3 Research Methodology

A case study research approach to find out how a firm's multi-stakeholder network influences the application of RI practices along innovation stages in R&D-intensive firm was adopted. In fact, the nature of this research question is the most important rationale behind choosing our research method (Yin 2009). For research areas where there is little knowledge about the topic as well as adequate literature with practical experience, case study research is an appropriate way to create a theoretical basis using empirical evidence (Eisenhardt and Graebner 2007; Gioia et al. 2013). Since there is a limited understanding of the main factors for the implementation of RI activities in business practices, therefore, the case study method is a suitable method for this study. The unique position of the case study method is held as it can be dealt with a variety of evidence, such as interviews, documents, artefacts and observations. The interpretive perspective in case study research method provides inductive research by connecting theory and data, depends on multiple sources of evidence and guides data collection and analysis strategies. To this end, a case study method was undertaken following replication logic in case selection (Ireland and Hine 2007).

3.1 Research Setting

A purposeful sampling approach (Patton 2002) was used, so the firms were preselected due to their (1) R&D intensity and (2) expressed approach to responsibly developing their innovations. Once a firm fulfilled the preselection criteria, six main selection criteria were deployed to meet the focus of this study. Ten R&D-intensive innovative firms were selected as based on the following criteria:

1. Firms that are R&D intensive

2. Firms that address Grand Challenges in their innovative activity
3. Firms that are awarded as being successful in innovative activities by national institutions or (and) being exemplary in sustainability and CSR
4. Firms that state their responsible orientation towards innovative activities in firms' strategy, mission or vision
5. Firms that their innovative activities are based on standards, regulations or codes of conducts
6. Firms that are open to society or cooperating with civil society

All the selected firms (Table 26.1) operate in medical engineering and healthcare sectors. Firms in these sectors tend to operate in various types of networks within their innovation development. Further, firms in medical engineering and pharmacy sectors tend to deliver high value-added concerning innovation and must comply with international standards to grow and operate in (international) multi-stakeholder networks that include both formal and informal institutions depending on the export market. Pilot projects within selected firms are developed within a single home country. Details of the ten case firms are provided in Table 26.1. In order to ensure anonymity, codes are given instead of the real names (Gioia et al. 2013).

Data Collection Two main sources of evidence to triangulate the data were used: (1) desk research and archival records and (2) in-depth semi-structured interviews in order to capture firms' RI activities within innovation processes, and to find out how RI is enforced in the firms. To get acquainted with the firms, at first, secondary data was collected from all the possible access of information about the firms, like press releases, newsletters, videos from conference participated, financial statements, official reports and membership in various associations and clusters. One member of our team has participated in seminars led by selected firms, e.g. an innovation management seminar in Firm 2 regarding the stage-gate method during their innovation development.

The in-depth semi-structured interviews were conducted with the CEOs or innovation project managers of the case firms. The interviewees came from two use case groups, namely, "the high-level strategy manager, CEO in our case study to raise awareness and convince to engage", "the CTO or innovation project manager for assessing RI and mapping of the responsibility framework to firm and implement RI".

Interview guideline was prepared in advance in coherence with the research question raised in this study. To gain new insights with regard to RI in industry, the interviews were open-ended, so the interviewees could freely express their point of view. This allowed to gather comparable results and unfold the research question by giving the opportunity to firms to move the conversation in any direction in the field of RI. The interviews were conducted between August 2015 and October 2017 at the workplace of the case firms. All the interviews were recorded upon the agreement of interviewees and lasted about 70 minutes on average. Follow-up emails and phone calls with some of the interviewees helped to clarify some missing data about the inclusion of multi-stakeholder into exact innovation phase.

Table 26.1 List of firms and interviews

Case firm	Manufacturing sector	Size[a]	Interviews
Firm 1	Manufacturing, 26.60 Irradiation, electromedical and electrotherapeutic apparatus manufacturing	Small	1 – CEO 2 – Quality manager 3 – Innovation manager
Firm 2	Manufacturing, 20.13 Other basic organic chemical manufacturing	Large	1 – Project manager 2 – Innovation manager
Firm 3	Manufacturing, 21.20 Pharmaceutical preparation manufacturing	Medium	1 – Technology manager
Firm 4	Manufacturing, 20.42 Perfume and Toilet Preparation Manufacturing	Medium	1 – CEO
Firm 5	Manufacturing, 32.50 Medical and odontology equipment, instrument and maintenance manufacturing	Medium	1 – CEO
Firm 6	Professional, scientific and technical services, 72.19 Other scientific research and development in life sciences and engineering	Micro	1 – CEO
Firm 7	Professional, scientific and technical services, 71.12 Engineering and related technical consulting service	Small	1 – CEO
Firm 8	Professional, scientific and technical services, 72.11 Research and development in biotechnology	Micro	1 – CEO 2 – Technology manager
Firm 9	Information and relations, 63.99 All other information services	Micro	1 – CEO
Firm 10	Information and relations, 62.01 Computer software programming services	Small	1 – Quality manager 2 – Innovation manager

Note: Case studies took place between September 2015 and October 2017
[a]Firm's size is defined under Eurostat (n.d.), where *micro* enterprises have less than 10 persons employed; *small* enterprises have 10–49 persons employed; *medium*-sized enterprises have 50–249 persons employed; *large* enterprises have 250 or more persons employed

A total of 13 interviews with 15 interviewees from 10 different firms were made. All interviews were held in the Lithuanian language in a face-to-face manner. The interviews were conducted between September 2015 and October 2017 at the workplace of the case firms. All interviews were recorded upon agreement with the interviewees and lasted from 42 to 75 minutes. Follow-up emails and phone calls with some of the interviewees helped to clarify some missing data.

Case Study Analysis The interview transcripts were transferred to, coded and analysed with qualitative and mixed methods data analysis software Maxqda12. The data were then thematically analysed. This entailed condensing the data set by assigning codes to the text of varying size such as words, sentences and paragraphs (Miles and Huberman 1994). Considering a thematic coding analysis approach, initially coding from the archival records was started, right after used relevant codes for interview analysis. A peer-review approach is applied within the duration of the work to fulfil triangulation purposes.

The within-case study analysis was used to deeply understand and recognise the unique patterns of each case (Eisenhardt 1989). As the following step, cross-analysis was used to compare these patterns within the cases, and differences and similarities across the cases were checked.

Case Firms All the case firms are R&D intensive and operate in the healthcare industry in Lithuania. To ensure anonymity, codes are given instead of the real names (Gioia, Corley, and Hamilton 2013). The evidence of RI in these firms was found concerning a business philosophy promoting ethical and social awareness; some of the Grand Challenges are reflected and integrated into the innovation orientation of the firm; firms are awarded as being successful in innovative activities and being exemplary in sustainability and/or corporate social responsibility by national institutions; firms state their responsible orientation towards innovative activities in firms' strategy, mission or vision; firms' innovative activities are based on standards, regulations or codes of conducts; firms operate in various types of networks during their innovation implementation; firms collaborate with a variety of external stakeholders during their innovation implementation.

4 Findings

Several common themes emerged from the case studies, which further confirmed the need to develop the levels of RI for innovation phases. There is a common recognition by the interviewees (Table 26.1) for the need for embedding responsibility into the innovation processes as well as engaging stakeholders along the whole value chain. They identified that new stakeholder engagement techniques help the firm to extend its impact beyond the existing network. Further, all interviewees agreed on engaging relevant stakeholders in the early stages of their new product, because stakeholder feedbacks are useful for creating and finalising the basic specification of a product or service.

The literature review and case study research identified the need to distinguish different levels of RI (first level, legal and contractual responsibilities; second, moral responsibilities) within three innovation development phases (exploration, development and implementation) embedded in multi-stakeholder networks.

4.1 Exploration Phase

First Level of RI: Legal and Contractual Responsibilities Standards and regulations constrain case firms in manufacturing due to various levels of responsibility (Table 26.2). Empirical data suggests that legal requirements are necessary to ensure at least the minimum level of responsibility in a healthcare firm, so these firms are concentrating on compliance with the law at first.

Despite the strict rules in the medical sector, it actually can foster creativity and new solutions; thus, institutional regulations could lead to positive outcomes. In the case of RI context, there are various social, environmental and ethical issues that have to be mitigated by various actors worldwide; thus new regulations and standards towards sensitive issues foster creativity and responsibility embedded in new innovative solutions.

To wrap up, the first level of RI (legal and contractual responsibilities) during the exploration phase in R&D firms that are constrained by various regulations is oriented towards making sure that any innovative idea is, first of all, compatible with existing law. Additionally, these innovative ideas usually come from such key stakeholders as researchers.

Second Level RI: Moral Responsibilities Moral responsibilities are induced by actors of a firm's multi-stakeholder network, such as lead users, professional colleagues and scientists during the exploration phase (Table 26.3). A prominent example of the induced moral responsibility was the case of Firm 5, where their most popular medical innovation was enforced by the professional colleague of the firm's CEO. He came with an idea of how the existing problem could be solved for the benefit of society. Instead of having surgery, the professional colleague suggested to a firm's CEO (Firm 5) a technology that would allow more people to get cured with lower costs. The openness of a firm's CEO towards its multi-stakeholder network and his willingness to try out new solutions was a success for a firm.

The importance of a purposeful medical innovation that would specifically solve or mitigate social issues was supported by Firm 8 and Firm 9. Firm 8 relies on their end users into the innovation idea development and tries to find out whether the innovative idea makes sense and would solve existing issues of the potential customer. The moral aspects during the exploration stage in Firm 9, university start-up,

Table 26.2 First level of RI during the innovation exploration phase

Type of firm	Characteristic	Exemplary quotes of exploration
Manufacturing	Restrictions and regulations	"The American government has banned research with babies' steam cells, and ethical research suddenly is beginning. Researchers do not do any research with babies' steam cells, and it turns out that it is possible to genetically modify any kind of cell and turn it into the steam cell. Therefore, only such a big limitation has made it possible to discover new ways" (Firm 1)
	Compliance with law	"Firms do not search for ideas inside the firm in the medical sector. Ideas are coming from scientists. Pharmacy is such an industry that it is necessary to comply with the law and requirements to commercialise innovation. Therefore, innovative ideas have to come from outside because it is just impossible to finance such expensive research" (Firm 3)

are enforced by scientists with whom they work together in the same university building. Although scientists are not official members of the start-up, nevertheless Firm 9 established excellent relationships with them and continuously seek the feedback of scientists. Due to scientists, Firm 9 is able to anticipate possible negative consequences of medical innovation and gets valuable insights on the new solutions that are scientifically available. Then Firm 9 is able to adjust their medical innovation upon scientists' feedback.

In the case of Firm 1, the informal knowledge cluster played an important role to foster moral responsibilities during the innovation process. The CEO of the Firm 1 belongs to the cluster, in which they share and check their potential innovative ideas. They encourage each other to implement those ideas by helping with the knowledge and competences. As it was expressed by Firm 1, they have the same moral values to solve the health issues of the society; therefore, they trust and support each other. High relevance of trust among the cluster members gives the possibility to share insights and issues about the potential idea in the early phase innovation. Also, Firm 10 emphasised the importance of the coherence of moral values – they work closely with the best hospital in Lithuania. The hospital applies the highest standards of medical activities and has a willingness to improve its medical technologies for the benefit of society. Firm 10 has emphasised that if they create a medical innovation that satisfies the hospital, they are sure that the developed medical innovation has a lot of potential. As Firm 10 revealed, not all hospitals are willing to change for the benefit of society as it requires extra time, learning and cost resources.

In contrast, Firm 2 and Firm 4 mostly rely on their internal stakeholders during the innovation process. Firm 2 has its committee of internal experts, which anticipates the potential of the innovation idea and deliberates possible risks regarding society and the environment. Firm 4, for instance, was the only firm that identified its employees as a source for innovative ideas. Firm 4 organises monthly workshops, where they gather and brainstorm potential innovative ideas and anticipate its potential risks and benefits for society.

To summarise, the second level of RI referring to moral responsibilities and related actions is described by additional efforts to create a more responsible innovation which is induced by the multi-stakeholder network members like professional colleagues, clusters, associations, etc. in healthcare firms. During the exploration phase, healthcare firms tend to invest more of time in finding an idea that would solve an existing problem and thus create an added value for society, at the same time leading to a success of a firm. In this process, the role of the professional community is significant.

To sum up, in the exploration phase, RI is enforced by institutional regulations, because it ensures the minimum basis of responsibility during the innovation process and its outcomes, and, at the same time, it can lead to creativity and search for new solutions. In most cases, moral responsibilities – the second level of RI – are induced by case firms' multi-stakeholder network, consisting of such stakeholders like members of clusters, end users, scientists and professional colleagues.

Table 26.3 Second level of RI during the innovation exploration phase

Type of firm	Characteristic	Exemplary quotes of exploration
Manufacturing	Informal network to share ideas	"I belong to one informal network, where there is no jealousy; instead, we treat each other as family members. The very crucial aspect which unites us in this informal network is the comfort of sharing scientific knowledge, ideas and reflections from our innovative activities. All the members of this network are scientists and various professionals, and that gives a lot of valuable insights" (Firm 1)
	Existing societal problem	"My colleague doctor was raising a question of how we could heal the patient without a need to make surgery and make it more available for most of society. Our most popular innovation for knee joint started when the doctor came to me with his idea, which was patented and could easily save many peoples' health in an affordable way" (Firm 5)
	Anticipation	"You have to anticipate your possible innovation idea for a few years ahead. Of course, it is quite difficult to do that in the scientific field, but you cannot act against nature" (Firm 2)
	Inner workshops and committees	"We organise monthly workshops with our employees. We remind them of our strategic directions, and they have to come up with new innovative ideas" (Firm 4) "Inside the firm, we have our committee of experts, who decide whether this innovative idea is worth developing" (Firm 2)
Professional, scientific and technical services	Collaboration with professional community and scientists	"Ideas for innovation come from informal chats in a professional community, and then you suddenly realise that this idea has potential and can possibly solve the existing problem" (Firm 7) "If we had been working separately with scientists, we would lose the edge and our competitive advantage in a long-term perspective, because we would be stuck with old knowledge and technologies, so we could not progress and develop technologies that would help better solve the problems with DNA" (Firm 8)
Information and relations	Direct communication with the users and clients	"We ask our end-users, for example: "Would this kind of product help you out in your everyday life? Would it make your illness a bit easier to live with?"" (Firm 9) "Our client is the best hospital in the country. This is a fantastic place to discover ideas and realise it together with them. Because if we would create innovation that would fulfil the needs of this hospital, we can be sure that we developed the innovation with the highest potential and best features" (Firm 10)

4.2 Development Phase

First Level RI: Legal and Contractual Responsibilities In the development phase, the variety and the role of multi-stakeholder networks differ across the sectors (Table 26.4). Case firms in manufacturing experience a dominating role of institutional standards and regulations. Legal requirements help to develop innovation more responsibly and come up with more responsible innovative solutions. For these case firms, it is critically important to develop innovation at its best. At the same time, the prototype has to comply with existing regulations and standards to pass accreditation and ethics committees. Nevertheless, external stakeholders are still integrated during the innovation development phase as they offer opportunities to reflect on the prototype that it would meet the expectations of the initial idea. Thus, in the development phase, it is clear that the first level of RI consisting of legal responsibilities is ensured by the external stakeholders such as ethics and accreditation committees.

Second Level RI: Moral Responsibilities All the case firms in manufacturing exceed standards and regulations as the first level of RI by pursuing to fulfil the expectations of their external stakeholders and to prove their value and capabilities for external stakeholders (Table 26.5). These firms strive to take additional actions towards responsibility that can be described as moving into the second level of RI – moral responsibilities. Case firms remain embedded in their multi-stakeholder networks to increase the value of innovation towards society and environment because their multi-stakeholder network is rich in knowledge, experience and competences that could be used for RI development. Exceeding requirements and adding some additional operations or actions related to a more RI process distinguish firms from other firms in the same field or sector.

Meanwhile, firms in professional, scientific and technical services and information and relations sectors are much less restricted by law and standards; therefore, they take actions related to the second level of RI – moral responsibilities. By taking actions related to moral responsibilities, these firms try to prove their credibility by

Table 26.4 First level RI (legal and contractual responsibilities) during the development phase

Type of firm	Characteristic	Exemplary quotes of the development phase
Manufacturing	Meeting the standards and regulations	"In this phase, standards and regulations are inevitable, because it is the basis for the continuous development of the whole system towards the right way" (Firm 1)
	External evaluation	"The new product has to be approved by the committee of ethics, too. It is not important to them if I am a Nobel prize winner, or not, because the procedure is very strict" (Firm 1)
		"Accreditation committee approves if my product satisfies all the requirements, and only after this approval, it is possible for us to move further with our innovation" (Firm 5)

Table 26.5 Second level of RI during the development phase

Type of firm	Characteristic	Exemplary quotes of development
Manufacturing	Organisational culture	"Our firm is just too big that it could act irresponsibly. We, the employees, are all acting responsibly in every stage is the most important thing" (Firm 2)
	Building trust via exceeding the standards	"We feel responsible for our customers. Standards and regulations are mandatory, but you have to do it not for the regulations. It has to be done for the benefit of the human. We have a long-term strategic plan to stay in the market, so we cannot rely just on the standards, we have to exceed it. We can lose the trust very quick, but to regain it again is almost impossible. My boss once taught me one lesson that I will never forget – better lose money, not confidence. You can't buy confidence" (Firm 4)
	Life cycle	"Basically, in every stage of innovation development, we try to think about the life cycle costs, we integrate sustainability aspects in our innovation development" (Firm 2)
	Reflexivity using stakeholders' feedback	"If we have positive research results, then we include many people like potential clients in our further product development. Clients' opinion is important for us because they will use our product eventually. Maybe it is our sector, but it is essential to stop after each innovation phase and reflect on how the processes are going" (Firm 3)
	Integrating the multidisciplinary team	"We ask for the help of scientists when we feel that cannot make it ourselves. We don't risk developing it ourselves" (Firm 4)
Professional, scientific and technical services	Searching for trustworthy partners	"When we decide to develop an innovative idea, we search for partners outside the firm; usually, we purchase scientific research, so, we collaborate with scientists because we seek for the highest quality" (Firm 7) "It took about half a year until we found trustworthy partners for our innovation development. We had situations that some partners would disappoint us by disappearing or being not capable of delivering the quality we ask for. I feel quite certain about our current network of partners who I trust to, and this is a really nice feeling because they can take of certain things and we can be sure about their quality, and, eventually, about our quality" (Firm 8)
	Integrating feedback of the stakeholders	"We integrate coaches because they know best how our innovation works with the people, like how they breathe, feel, how to observe the physical exertion. During the development process, their reflections regarding our innovation are the most important" (Firm 6)

(continued)

Table 26.5 (continued)

Type of firm	Characteristic	Exemplary quotes of development
Information and relations	Multidisciplinary team	"To proceed with the prototype that we had, we needed certain knowledge, we could not risk by finishing it ourselves. <...> We have a team of professionals in medicine, who consult and help us with our innovation development, so we can be sure that we will achieve the best quality possible" (Firm 9)

institutionalising their innovative activities on their own, for example, installing quality management. Upon receipt of legal credibility, firms strive for external stakeholders even more actively than firms in manufacturing. External stakeholders, like lead and end users, clusters, professional colleagues and universities, play a crucial role in RI development in the healthcare sector.

Reflexivity practices are actively applied in manufacturing firms in order to meet the expectations of future clients. In comparison, case firms in professional, scientific and technical services and information and relations tend to apply reflexivity practices together with their end users.

In conclusion, in the development phase, institutional standards and regulations form the basis for further innovation development in manufacturing firms. Compliance with the law ensures a firm's credibility and trust towards the multi-stakeholder network for collaboration. The firms in professional, scientific and technical services and information and relations sectors are not necessary to comply with the law, but they still enforce themselves to have legal credibility by accrediting themselves voluntarily, because this helps to ensure the basis for responsibility in innovation development phases. Lead users, end users and scientists are the most important during innovation development phase, because of their knowledge based on their everyday experience in the field where innovation is created. Clusters and professional colleagues are valuable for their complex knowledge and competences.

4.3 Implementation Phase

First Level of RI: Legal and Contractual Responsibilities After the product launch, firms expressed the importance of their clients (hospitals or individuals using the innovation) (Table 26.6). On the one hand, the negative feedback from the clients, which is institutionalised, can lead the firm to the so-called blacklist, which means that the firm would have to suspend its activity. On the other hand, the excellent reputation of the firm makes it more reliable and trustworthy towards their multi-stakeholder network. Thus, this first level of RI during the implementation phase is described by close connections with multi-stakeholders who are internal to

Table 26.6 First level of RI (legal and contractual responsibilities) during implementation phase

Type of firm	Characteristic	Exemplary quotes of implementation
Manufacturing	Final approval	"Finally, our New product committee finishes their job by ensuring that this new product is really safe, that the features didn't changed during the development process and it functions as it was planned, they evaluate the risks for the environment" (Firm 2)
	Institutional feedback system	"After the launch of the new product, the feedback system is very important. Usually, medical institutions are giving these institutional feedbacks depending on patients' wellbeing due to our products. If there are some negative consequences for patients, we try to find out what and why it happened. However, in this case, if some aspects where unanticipated throughout the innovation process, after some negative consequences for patients, our innovation could be eliminated, and we must check our innovation again, including clinical studies" (Firm 1)

firms like new product committee and external like hospitals. This collaboration is based on legal and contractual responsibilities during the innovation implementation phase.

Second Level of RI: Moral Responsibilities At the final stage of the innovation process, firms tend to integrate noneconomic or other than value chain-based stakeholders (Table 26.7). Firm 3, for instance, tends to integrate external noneconomic stakeholders to get feedback about the prototype of the innovation. However, the feedback from the potential customers cannot fundamentally change the innovation, since product design questions are the main focus of the inquiry, rather than conceptual concerns. As the interviewees expressed, the inclusion of noneconomic stakeholders in the late phase of innovation development is due to the protection of intellectual property. This corresponds with existing literature (Blok et al. 2015) that firms tend to integrate noneconomic stakeholders only in the last stage of the innovation development phase due to information asymmetry.

For the case firms in professional, scientific and technical services, external value chain-based stakeholders like lead and end users are dominating in the last phase of innovation. This could be due to the specificity of the innovation, which is aimed at specific groups of clients, like coaches, laboratories, etc.

The rest of the case firms depend on personal feedback with their clients. Firm 4, for instance, expressed that they put extra efforts for the feedbacks by dealing with it individually. For them, any kind of feedback is a possibility to make their innovation more responsible regarding quality and features. Firms try to be responsive to their clients; therefore, the feedback system is taken responsibly.

To sum up, during the implementation phase, the second level of RI is mostly related to responsiveness to different stakeholders. The communication with external stakeholders is based on the different kinds of feedback mechanisms that help to improve healthcare innovations of the case firms continually.

Table 26.7 Second level of RI during the implementation phase

Industry	Indicator	Exemplary quotes of implementation
Manufacturing	Organisational feedback system	"Clients are included when we need their opinion about the usage of the product, its package" (Firm 3) "Our administrator works responsibly on the clients' feedbacks. Every feedback is analysed carefully, and then we check the negative aspects of our products all together in a firm, and we try to come up with best solutions" (Firm 4) "We have the so-called Quality journal, where we register all the complaints, defects, and other types of feedback, which help us to improve our products. Our clients' feedbacks are very important for our professional development" (Firm 5)
Professional, scientific and technical services	Reflection upon the feedback of end-users	"The most important feedback comes from our coaches, who work directly with our final product" (Firm 6) "Scientist are giving us the most valuable feedback towards our developed technology" (Firm 8)
Information and relations	Responsiveness towards stakeholders' feedback	"Usually, when we launch a new product, we meet with doctors, and we discuss the product, i.e. does it meet the initial idea regarding features and quality, what could be improved, what new functions could be added, that are crucial to the doctors" (Firm 9) "What is important is that we communicate with our end-user after the product launch. Because they work with our product, they know best what is essential to healing people better. Sometimes there are situations where we are almost done with the product launch, and then, suddenly, some doctor suggests us a new feature that could be integrated into the product. Sometimes it is feasible, and sometimes we postpone it to the next version of our product. But we are responsive to our end-users, and they are aware of what and how we develop our innovation" (Firm 10)

5 Discussion and Conclusions

The purpose of this paper was to examine the role of a firm's multi-stakeholder network role on RI practices in R&D-intensive firms. For this reason, the case study was set out. The adopted multiple-case study method proved to be suitable for extracting the impact of firm's multi-stakeholder network on firm's RI.

The findings provide evidence that firms prioritise their reputation and trustworthy relationships towards multi-stakeholder networks. Without external stakeholders who usually bring potential ideas for innovations, R&D-intensive firms would not be capable of developing and commercialising the innovation because of the high costs in the exploration stage. As empirical results show, being embedded in multi-stakeholder networks brings R&D-intensive firms' opportunities to be aware of what society needs and what innovative ideas could solve existing problems, so at the same time it gives value for society. It thus is having more potential to be successful innovations. R&D-intensive firms coordinate multi-stakeholder network to maintain access and benefit from network knowledge, reputation and other intangible but also tangible resources. Thus, a firm's initiative to organise innovation network behaviour under the principles of RI within the multi-stakeholder network is a useful practice for achieving RI. Therefore, this study contributes to the theory on the influencing role of firm's multi-stakeholder network on firm's RI practices.

Furthermore, this study sheds light on the fragmented responsibility issue and how it could be managed via the firm's multi-stakeholder network. Empirical evidence, as based on our case studies, suggests that institutional standards and regulations in some cases form the basis for RI and could be interpreted as the first level of responsibility in innovation development. This confirms the conceptual model of Pavie et al. (2014) and previous research in RI in the industry (Petraite et al. 2017), where the first stage of compliance with the standards and law is only a precondition to enter the network. However, in some cases, standards and regulations are also necessary in order to mitigate some specific social, ethical or ecological issues. When the firm goes beyond the existing standards and regulations, this could be understood as the second level responsibility in innovation management.

In order to achieve RI, different types of responsibility are required, like contractual, legal and moral responsibility (Dreyer et al. 2017). Although there is a tendency to focus more on moral responsibilities due to the context of emerging technologies, however, in the case of bringing RI into commercial innovation context, contractual and legal responsibilities are important dimensions in implementing RI. The systemic view is needed to understand how firms are embedded in different dimensions of responsibility.

Also, as our empirical results show, R&D firms tend to engage with the stakeholders during the whole innovation development process. This is in contrast with the findings in the food industry (Blok et al. 2015), where firms did not tend to engage with stakeholders during the innovation development stage. A possible reason for that could be the differences across the industries because, in the health care sectors, there is a long-standing tradition to engage end users into the innovation process. Accordingly, managerial and policy implementations could be suggested to foster RI in industry.

There are several areas for future research that emerge from our paper. First, while our study is limited by the number of firms involved and the absence of firm's stakeholders and network members in interviews, there may be future research focusing on a broader inclusion of a firm's network members. Second, while there are some insights into how multi-stakeholder networks influence performance, our study was limited to focus on the underlying decision process, which is important as firms face difficulties in maintaining the multi-stakeholder network. Third, further research would appear to be necessary to include other industrial sectors in future studies to validate the results of our study and increase the generalizability of our findings, which is currently limited only to the healthcare industry. A few scholars have looked at different RI modes in specific industries such as food industry (Blok et al. 2015) and robotics (Stahl et al. 2014). Still, there is a lack of evidence about how firms can apply different levels of RI along different innovation stages.

This paper highlights the levels of RI in multi-stakeholder networks, and this could be seen as a natural next step to investigate conditions to harmonise RI implementation and multi-stakeholder network's agendas given the industrial differences.

To conclude, the implementation of RI in R&D firms is dependent on their overall multi-stakeholder network, which forces firms to comply with the minimum requirements of the legal environment and enforces additional actions towards moral responsibility. The impact of the whole stakeholder networks of R&D firms allowed to come up with the two levels of RI – first level, defined by legal and contractual responsibilities, and second level – defined by moral responsibilities. So far, the moral responsibility was the main focus in RI field, which created a distinction with the general industry and decreased the possibilities for more conventional R&D firms to implement RI in their innovative activities. The paper revealed the different impacts of various actors from the firms' stakeholder network, thus contributing to a better understanding of the influencing factors for R&D firms to establish RI practices. Based on that, policy implications should be formed accordingly to foster the integration of different stakeholders into the industrial innovation for positive social change.

Acknowledgements This paper is partially based on a PhD thesis entitled "Implementing Responsible Innovation at the Firm Level" by Jolita Ceicyte (defended on 28 August 2019).

References

Blok, V., Lemmens, P. (2015). The emerging concept of responsible innovation. Three reasons why it is questionable and calls for a radical transformation of the concept of innovation. In Responsible Innovation 2 (19–35). Springer International Publishing.

Blok, V., Hoffmans, L., Wubben, E. F. M. (2015). Stakeholder engagement for responsible innovation in the private sector: Critical issues and management practices. *Journal on Chain and Network Science*, 15(2), 147–164.

Blyler, M., Coff, R. W. (2003). Dynamic capabilities, social capital, and rent appropriation: Ties that split pies. *Strategic Management Journal*, 24(7), 677–686.

Bryson, J. M., Crosby, B. C. (2006). Leadership for the common good. Creating a culture of collaboration, 367–396.

Cagnin, C., Amanatidou, E., Keenan, M. (2012). Orienting European innovation systems towards grand challenges and the roles that FTA can play. *Science and Public Policy*, 39(2), 140–152.

Ceicyte, J., Petraite, M. (2018). Networked responsibility approach for responsible innovation: Perspective of the firm. *Sustainability*, 10(6), 1720.

Cooper, R. G. (2008). Perspective: The stage-gate® idea-to-launch process—update, what's new, and nexgen systems. *Journal of Product Innovation Management*, 25(3), 213–232.

Cuppen, E., Brunsting, S., Pesch, U., Feenstra, Y. (2015). How stakeholder interactions can reduce space for moral considerations in decision making: A contested CCS project in the Netherlands. *Environment and Planning A*, *47*(9), 1963–1978.

Dreyer, M., Chefneux, L., Goldberg, A., von Heimburg, J., Patrignani, N., Schofield, M., Shilling, C. (2017). Responsible innovation: A complementary view from industry with proposals for bridging different perspectives. *Sustainability*, 9(10), 1719.

Eisenhardt, K. M. (1989). Building theories from case study research. *Academy of Management Review*, 14(4), 532–550.

Eisenhardt, K. M., Graebner, M. E. (2007). Theory building from cases: Opportunities and challenges. *Academy of Management Journal*, 50(1), 25–32.
Eurostat (n.d.). Small and medium-sized companies (SMEs). Available online at: http://ec.europa.eu/eurostat/web/structural-business-statistics/structural-businessstatistics/sme.
FDA (n.d.). The Drug Development Process. Available online at: https://www.fda.gov/forpatients/approvals/drugs/
Freeman, R.E., (1984). Strategic management: a stakeholder approach. ([Reprint]) Pitman Series in Business and Public Policy Pitman, Boston.
Frels, J. K., Shervani, T., Srivastava, R. K. (2003). The integrated networks model: Explaining resource allocations in network markets. *Journal of Marketing*, 67(1), 29–45.
Garst, J., Blok, V., Jansen, L., Omta, O.S.W.F. (2017). Responsibility versus Profit: The Motives of Food Firms for Healthy Product Innovation. *Sustainability*, *9*, 2286.
Genus, A., Iskandarova, M. (2017). Responsible innovation: its institutionalisation and a critique. *Technological Forecasting and Social Change*, 128, 1–9.
Geoghegan-Quinn, M. (2012). Responsible Research and Innovation. Europe's ability to respond to societal challenges. Proceedings of the Science in Dialogue–Towards a European Model for Responsible Research and Innovation, Odense, Denmark, 23–25.
Gioia, D. A., Corley, K. G., & Hamilton, A. L. (2013). Seeking qualitative rigor in inductive research: Notes on the Gioia methodology. *Organizational Research Methods*, 16(1), 15–31.
Gomez-Bezares, F., Przychodzen, W., & Przychodzen, J. (2017). Bridging the gap: How sustainable development can help companies create shareholder value and improve financial performance. *Business Ethics: A European Review*, 26(1), 1–17.
Greenwood, M. (2007). Stakeholder Engagement: Beyond the Myth of Corporate Responsibility. *Journal of Business Ethics* 74 (4), 315–327. https://doi.org/10.1007/s10551-007-9509-y.
Hadengue, M., de Marcellis-Warin, N., Von Zedtwitz, M., Warin, T. (2017). Avoiding the Pitfalls of Reverse Innovation: Lessons Learned from Essilor One company's experiences suggest how the specific challenges of reverse innovation may be anticipated and overcome. *Research-Technology Management*, 60(3), 40–47.
Hayes, R. H., Wheelwright, S. C., Clarke, K. B. (1988). Dynamic Manufacturing: Creating the Learning Organization. New York, NY: The Free Press.
Iatridis, K., Schroeder, D. (2016). Responsible Research and Innovation in Industry: The Case for Corporate Responsibility Tools. Springer, Cham.
Ireland, D.C., Hine, D. (2007). Harmonising science and business agendas for growth in new biotechnology firms: Case comparisons from five countries. *Technovation*, 27, 676–692.
Long, T., Blok, V. (2017). Integrating the management of socio-ethical factors into industry innovation: towards a concept of Open Innovation 2.0. International Food and Agribusiness Management Review, 1–24, Wageningen Academic Publishers.
Lubberink, R., Blok, V., van Ophem, J., and Omta, O. (2017). Lessons for Responsible Innovation in the Business Context: A Systematic Literature Review of Responsible, Social and Sustainable Innovation Practices. *Sustainability*, 9(5), 721.
Lund Vinding, A. (2006). Absorptive capacity and innovative performance: A human capital approach. *Economics of Innovation and New Technology*, 15(4–5), 507–517.
Macnaghten, P., Owen, R. (2011). Environmental science: good governance for geoengineering. *Nature*, 479(7373), 293–293.
Miles, M. B., Huberman, A. M. (1994). Qualitative data analysis: An expanded sourcebook. Sage.
Patton, M. Q. (2002). Two decades of developments in qualitative inquiry: A personal, experiential perspective. *Qualitative Social Work*, 1(3), 261–283.
Pavie, X., Scholten, V., Carthy, D. (2014). Responsible Innovation: From Concept to Practice. World Scientific Publishing Co. Pte. Ltd.: Singapore.
Petraite, M., Pavie, X., Ceicyte, J., Janiunaite; B., & Carthy, D. (2017). Managing innovation in multicultural environments: An imperative of responsibility within interorganizational networks. In Rozkwitalska, M., Sułkowski, Ł., Magala, S. (Eds.) Intercultural Interactions in the Multicultural Workplace: Traditional and Positive Organizational Scholarship, (pp. 137–154). Springer International Publishing: Cham, Switzerland.

Pittaway, L., Robertson, M., Munir, K., Denyer, D., & Neely, A. (2004). Networking and innovation: a systematic review of the evidence. *International Journal of Management Reviews*, 5(3–4), 137–168.
Porcari, A., Borsella, E., Mantovani, E. Stahl, B., Flick, C., Ladikas, M., Hahn, J., Brem, A., Yaghmaei, E., Søraker, J.H., Barnett, S.J., Chatfield, K., Ikonen, V., Leikas, J., Obach M. (2015a). Responsible-Industry executive brief: implementing responsible research and innovation for ICT for an ageing society, Italian Association for Industrial Research, Rome.
Porcari, A., Borsella, E., & Mantovani, E. (2015b). A Framework for Implementing Responsible Research and Innovation in ICT for an Ageing Society. Rome: Agra Editrice.
Ritter, T., Gemünden, H. G. (2003). Network competence: Its impact on innovation success and its antecedents. *Journal of Business Research*, 56(9), 745–755.
Roloff, J. (2008). Learning from multi-stakeholder networks: Issue-focused stakeholder management. *Journal of Business Ethics*, 82(1), 233–250.
Rowley, T. J. (1997). Moving beyond dyadic ties: A network theory of stakeholder influences. *Academy of management Review*, *22*(4), 887–910.
Smith, K. (2000). Innovation as a systemic phenomenon: rethinking the role of policy. Enterprise and innovation management studies, 1(1), 73–102.
Stahl, B. C., Obach, M., Yaghmaei, E., Ikonen, V., Chatfield, K., & Brem, A. (2017). The Responsible Research and Innovation (RRI) maturity model: Linking theory and practice. *Sustainability*, *9*(6), 1036.
Stahl, B., N. McBride, K. Wakunuma and C. Flick (2014). The empathic care robot: a prototype of responsible research and innovation. *Technological Forecasting and Social Change*, 84: 74–85.
Stilgoe, J., Owen, R., and Macnaghten, P. (2013). Developing a framework for responsible innovation. *Research Policy*, 42(9), 1568–1580.
Stuart, T. E. (2000). Interorganizational alliances and the performance of firms: A study of growth and innovation rates in a high-technology industry. *Strategic Management Journal*, 791–811.
Sutcliffe, H. (2011). A Report on Responsible Research and Innovation. Available online at: http://ec.europa.eu/research/sciencesociety/document_library/pdf_06/rri-report-hilary-sutcliffe_en.pdf.
Swiestra, T., Rip, A. (2007). Nano-ethics as NEST-ethics: Patterns of Moral Argumentation About New and Emerging Science and Technology. *Nanoethics* 1, 3–20, https://doi.org/10.1007/s11569-007-0005-8.
Tidd, J., Bessant, J., and Pavitt, K. (2001). Managing Innovation: Integrating Technological, Market and Organisational Change. Chichester: John Wiley & Sons
Tushman, M. L. (1977). Special boundary roles in the innovation process. *Administrative Science Quarterly*, 587–605.
Van de Poel, I., Asveld, L., Flipse, S., Klaassen, P., Scholten, V., Yaghmaei E. (2017) Company strategies for responsible research and innovation (RRI): a conceptual model, *Sustainability*, 9, 1–18.
Von Schomberg, R. (2012). Prospects for technology assessment in a framework of responsible research and innovation. Technikfolgen abschätzen lehren: Bildungspotenziale transdisziplinärer Methoden, 39–61.
Von Schomberg, R. 2013. A vision of responsible research and innovation. In Responsible innovation, ed. R. Owen, M. Heintz, and J. Bessant. London: Wiley.
Wickson, F., Carew, A. L. (2014). Quality criteria and indicators for responsible research and innovation: Learning from transdisciplinarity. *Journal of Responsible Innovation*, 1(3), 254–273.
Yaghmaei, E. (2016). Addressing responsible research and innovation to industry: introduction of a conceptual framework, ACM SIGCAS Computers and Society, 45, 294–300
Yaghmaei, E. (2018). Responsible research and innovation key performance indicators in industry: A case study in the ICT domain. *Journal of Information, Communication and Ethics in Society*, 16, 214–234.
Yin, R. K. (2009). Case study research: Design and Methods. SAGE publications. Thousand oaks.
Zwart, H., Landeweerd, L., Van Rooij, A. (2014). Adapt or perish? Assessing the recent shift in the European research funding arena from 'ELSA' to 'RRI'. *Life Sciences, Society and Policy*, 10(1), 11.

Chapter 27
Inclusion and Resilience in the Bioeconomy

Lotte Asveld

Abstract New technological developments such as CRISPR-Cas, advanced genetic sequencing and the digitalization of agriculture offer promising prospects to realize the potential of a sustainable bioeconomy. At the same time, enormous challenges abound such as the pressure on biodiversity and the associated risk of pandemics, climate change and the ever-increasing global economic inequality. The bioeconomy can play a beneficial role in this; however, this will only be possible if the bioeconomy is developed on the basis of inclusion. In this chapter I will explain the relevance of inclusion for the bioeconomy and describe some of the sociotechnical developments where inclusion should be realized in order to build a resilient and sustainable bioeconomy. These developments include biosphere capacity, global biobased value chains, digital genetic resources and the digitalization of agriculture. I will conclude with the question of who bears responsibility for an inclusive bioeconomy.

Keywords Inclusion · Resilience · Agency · Global value chains · Digital sequence information · Digitalization of agriculture

1 Introduction

The bioeconomy is an appealing concept that integrates the promise of economic prosperity with that of ecological stability, by replacing fossil resources with biomass. Recent advances in bioengineering such as CRISPR-Cas and other synthetic biology approaches offer novel pathways to modify micro-organisms into high-performing production platforms for a wide range of products (Straathof et al. 2019). The use of digital sequence information (DSI) promises unlimited access to a wide range of promising biological production pathways. At the same time, digi-

L. Asveld (✉)
Biotechnology and Society, Technical University of Delft, Delft, The Netherlands
e-mail: L.Asveld@tudelft.nl

E. Koukios, A. Sacio-Szymańska (eds.), *Bio#Futures*,
https://doi.org/10.1007/978-3-030-64969-2_27

talization of agriculture allows for highly efficient and sustainable production of biomass to feed these novel micro-organisms.

Such increasingly advanced technologies give rise to questions about economic justice, resembling issues that emerged when genetic modification first arrived on the scene. Who will benefit from these sophisticated technologies? Who will own and distribute them? How will they affect global economic inequalities, especially considering the fact that a lot of the biomass will be sourced from the global south and processed in the global north? And how can they contribute to sustainability and the protection of biodiversity?

In this chapter I want to stress that social and economic inclusion in global biobased value chains can help to develop a sustainable as well as resilient bioeconomy. I consider this issue to complement the question of the desirability of a particular technology, policy or agricultural production system per se. This implies that the desirability of biobased applications should be determined based not only on the specific (social, economic, environmental) impact that it has on a specific region but also on how it has come about and whether the associated value chains have been developed in line with the needs, values and knowledge of local biomass producers and (potential) local consumers of biobased products. I claim that when a biobased application is developed in an inclusive manner, it is more likely to be sustainable in a broad sense and is also likely to lead to more resilient biobased value chains. In this chapter I will discuss specific avenues to realize such inclusion.

I will first discuss the notion of inclusion and its general relevance to a resilient bioeconomy. I will elaborate my claim about the central importance of inclusion with reference to four themes: biosphere capacity, reliability in value chains, control over genetic resources and digitalization of farming.

2 Inclusion

The concept of inclusion has gained attention as an element of the approach of responsible research and innovation where it refers to the engagement of a wide range of voices in the development of new technologies in order to increase both the legitimacy and the acceptance of innovations (Stilgoe et al. 2013), thereby moving away from an innovation system dominated by technological experts. In this approach, inclusion is mainly considered to be a process where several stakeholders can provide input on the desirability of the design of a specific innovation. Such inclusion is generally seen as a prerequisite to achieve societally desirable outcomes (Sonck et al. 2017).

Other authors focus more specifically on inclusion as an outcome of innovation, rather than as a prerequisite. In the context of the bioeconomy, inclusive innovation has been defined as a ‘new way of doing things (that) may improve the lives of the most needy’ (Bryden et al. 2017). In this approach, the actual benefits of an innovation take central stage. This focus on improving the livelihoods of the most vulnerable is also prominent in the approach of inclusive agricultural value chains (Devaux et al. 2016) and inclusive innovation for development in general (Heeks et al. 2014). These latter approaches show a development from solely offering products tailored to the most needy (e.g. frugal innovation) towards more comprehensive tactics in which also the living conditions and general well-being of vulnerable groups are taken into account (Ros-Tonen et al. 2019).

Since the transition to a sustainable bioeconomy comprises complex changes in the global south as well as in the global north, I think that inclusion should be as wide-ranging as possible. So inclusion should benefit the most needy, but it should also take into account the perspectives of those in the global north. Therefore I understand inclusion as the fair distribution of risks and benefits associated with a specific socio-technological development (such as the biobased economy), implying that any relevant technological or economical is designed while taking into account the values, knowledge and interests of all actors involved.

However there is ample evidence that inclusion as a process is not always feasible. Relevant stakeholders may be unavailable, reluctant to participate (Sonck et al. 2017), not capable of standing up for their own needs and values due to cultural, economic or biophysical hurdles (Ostrom 2005) or because of socio-economic or institutional incentives that hinder openness, for instance for researchers who find it hard to discuss uncertainties in their work publicly (Wickson and Carew 2014) or commercial actors who need to protect private interests (Blok and Lemmens 2015). These are serious barriers that should be addressed to achieve true inclusion. These barriers point out that inclusion is more than a participation exercise, but also requires capacity building and institutional support (Postal et al, 2020b). Without the right conditions, the laudable aim of inclusion could instead lead to an empty legitimation stunt that serves only the interests of the most powerful. In the remainder of the chapter, I aim to point to some socio-technological avenues within the bioeconomy that offer openings for true inclusion.

3 Biosphere Capacity

A major issue that permeates all political and scientific evaluations of the bioeconomy is the carrying capacity of the biosphere. With biosphere I refer to all living parts of the earth. How much biobased resources for human consumption are this biosphere able to sustain? The Covid-19 pandemic has put a spotlight on the ecological risks of the current agro-industrial complex. By pushing the frontiers of our agricultural system further into previously undisturbed ecosystems, we risk allowing pathogens to escape from their ecological niche (Wallace 2016). Can we realistically feed, clothe, warm, transport, cool, etc. ourselves to a considerable part based on sustainably produced biomass?

3.1 How to Assess Sustainability?

Two major issues stand out here: how to assess the sustainability of particular products and how to assess the sustainability of the bioeconomy as a whole? First is the question about how to reliably ascertain that biobased applications and the associated value chains are sustainable. Different models rely on different indicators and assumptions and hence will lead to varying output (Matthews et al. 2019). Allocation of CO_2 emissions for woodchips has for instance been a major issue in the debate concerning the sustainability of burning these woodchips. Should these CO_2 emissions be allocated to the place of origin of the woods, or the place where the woodchips are burned? A similar discussion arose around the inclusion of ILUC effects of biomass for biofuels. Should indirect land use change be taken into account or should it not? (Asveld 2016) This variance in assumptions and indicators is problematic for a sustainable bioeconomy. Various actors can shape sustainability assessments according to their needs. This undermines the credibility of a sustainable bioeconomy.

Second, such models to account for sustainability usually rely on specific indicators that are easy to quantify such as CO_2 emissions or land use. Therefore they necessarily leave out many aspects that are hard to quantify or that are related to a specific biobased application in very complex ways, such as local socio-economic effects (Flipse 2014; Parada et al. 2017), or aspects of uncertainties relating to future developments (Matthews et al. 2019). Models that focus on quantifiable indicators are not suitable to answer what *kind* of bioeconomy is most sustainable in a broad, holistic sense. Should a sustainable bioeconomy be able to compete with the fossil economy in terms of productivity? In other words, does sustainable equate efficiency? Or should sustainability instead focus on ecological stability, social impact and technological appropriateness, meaning that we do not opt for high-tech solutions if they do not fit the local cultural and economic context, even if that means biomass will not be utilized in the most efficient way?

3.2 An Inclusive Understanding of Sustainability

What is needed is a shared understanding of the ecological and economic underpinnings of the bioeconomy (Veraart and Blok 2020). Various conflicting perspectives on the bioeconomy exist, ranging from a kind of business as usual approach in which the economy continues to grow but based on biomass instead of fossil resources to an economy that seeks to minimize the use of bioresources in order to assure a healthy economy within the ecological, planetary boundaries (Richardson 2012; Vivien et al. 2019). A possible way to deal with this is the application of a constructive form of sustainability assessment that allows for the integration of a wide range of perspectives to achieve an interdisciplinary, anticipatory type of sustainability assessment (Matthews et al. 2019). Such an approach can be considered more inclusive than the current, prevalent methods of sustainability assessment.

A shared understanding of sustainability should encompass the perspectives of both stakeholders in the global north and the global south. The bioeconomy presents a new frontier in our relationship with natural resources that requires a solid philosophical underpinning with regard to environmental values. What does natural mean to us? Does it have a specific value in itself that should be cherished or is it simple resources that should be exploited for all its magnificent bounty? What does sustainability imply? Can it imply that sometimes we prioritize the interests of non-human animals and nature over human interests? How can we build a fair and prosperous world within the ecological limits that we are facing?

Moreover, we can only expect those that manage natural resources to do that sustainably, if their interests and values are reflected in the set-up of those value chains. It is of vital importance to the entire planet to protect biodiversity, and so far we haven't done such a great job (WWF 2020). Further loss of biodiversity increases the risks of a new pandemic (Quammen 2012). This presents an urgent need to present those living in biodiverse-rich countries with an economic, social and cultural incentive to protect biodiversity (Berkes et al. 2009). I believe that inclusive biobased value chains can present such an incentive because they can connect the sustainable management of natural resources with financial gains.

4 Resilient Value Chains

The above observation brings us to the first pressing issue facing the bioeconomy at present. There is a need to include local producers of biomass in order to achieve reliable and resilient biobased value chains, built on economic fairness (Asveld 2019). Global catastrophic events such as pandemics and climate change can have huge impacts on global biobased value chains. Inclusive value chains that build on local knowledge, values and interests, and that give agency to local actors, can be expected to be more resilient compared to value chains that rely on remote technical expert knowledge and control (Sumane et al. 2018). This will have beneficial effects in three ways: more resilience, more sustainability due to less uncertainty and commercially more viable value chains.

4.1 Resilience in Value Chains

Resilience refers to the ability of a system to respond to a threat or hazard (Doorn 2017). Resilience can occur in ecological systems but also in more broader socio-ecological systems (Walker et al. 2004). Resilience can come about through diversity (Rammel and Van de Bergh 2003), innovation, adaptive management and learning (Doorn 2017). Crucial to any system being able to respond to a threat or hazard is the agency of their social constituents (Brown and Westaway 2011). Therefore the resilience of global biobased value chains depends on the agency of the various actors involved.

Agency can be understood as the capability to shape one's own life. Amartya Sen has defined an agent as 'someone who acts and brings about change, and whose achievements can be judged in terms of her own values and objectives, whether or not we assess them in terms of some external criteria as well' (Sen 1999, p. 19). Only if all actors in a global biobased value chain can exercise their agency can such a value chain be sustainable and resilient. A high degree of autonomy amongst relevant participants helps to build institutions that can sustainably manage resources (Becker and Ostrom 1995).

4.2 *Managing Uncertainties to Achieve Sustainability*

Many uncertainties shroud how to develop sustainable and economically fair biobased value chains (Asveld and Stemerding 2017; Kamali et al. 2018). This is often the case because the chains usually span global networks yet biomass production takes place in distinct contexts (Meckenstock et al. 2016). There may be uncertainties about which crop is best suited to local conditions, what are local sustainable soil management practices, what processing technologies are feasible given the local circumstances, what are reliable local means of transportation and what is needed to convince producers of biomass to commit to a new biobased value chain (Robaey et al. forthcoming).

Including local stakeholders in the set-up of biobased value chains – taking into account their needs, values, wishes and knowledge – can help mitigate these uncertainties (Pretty 1995); local producers often have valuable knowledge about land management, their natural environment and the associated biomass (Sumane et al. 2018). This knowledge is indispensable for achieving a system that sustainably manages local resources (Folke et al. 2011). Numerous examples have shown that institutions based on intimate local knowledge and with the input of participants closely connected to the specific environment achieve better results compared to an imposed central authority or an orientation only on global market (Becker and Ostrom 1995). Many biobased value chains will be connected to the global market. The challenge lies in also connecting to the local realities of sustainable resource management. The interests of local producers can only be adequately recognized and taken into account when they have an actual chance to speak up (Postal et al. 2020). Including local biomass producers is thus expected to have both epistemic and moral benefits (Wals 2007). Epistemic benefits refer to the reduction of uncertainties, while moral benefits refer to the fair distribution of risks and benefits.

4.3 *Commercially Successful Value Chains*

Another advantage of inclusion is its potential contribution to commercial success. Currently, approaches to sustainable agriculture that focus mainly on technological aspects while neglecting local stakeholders often fail from a business perspective

(Hounkonnou et al. 2012). Such lack of inclusion has already negatively affected many biobased value chains, either because they failed economically (Hounkonnou et al. 2012) or because the biomass producers were not committed to deliver their produce to the biobased value chain (De Hoop et al. 2016; Balkema and Pols 2015). Inclusion, on the other hand, can lead to robustness and commercial success, for instance because a realistic expectation exists towards the capabilities of local stakeholders to invest and shoulder the associated risks (Devaux et al. 2016) or because they build on the prevalent skills and knowledge of local producers (Harper et al. 2015).

4.4 Challenges to Inclusive Value Chains

Several challenges exist to realize inclusion in global biobased value chains. Inclusion can only be meaningful if those who are to be included have a real choice. If producers of biomass have a choice, they have a better negotiation position and will be more able to withstand monopolistic tendencies (Harper et al. 2015). Inclusion without other options is not inclusion – it is coercion (Kleine et al. 2012).

Inclusion thus requires the capability to be included to begin with (Simpson and Basta 2018). This capability depends on actors having specific skills and access to resources, reliable infrastructure and education (Frediani 2010). For example, the capability of handling sophisticated technologies may be a prerequisite for the capability to be included in an advanced biorefinery. Or the capability to access relevant information at low costs (Becker and Ostrom 1995). It is difficult to truly include biomass producers and stakeholders who lack basic skills and resources. Possibly companies need to take actions beyond the private sphere and invest in public goods such as infrastructure and education.

Another challenge lies in the cultural differences between actors in a global biobased value chain. Many companies developing biobased applications come from Western countries, while many producers of biomass live in the global south. But also between partners from different countries in the global north or between partners from the global south. To ensure that a sufficient amount of trust emerges between various participants usually requires extensive time and effort (Lundy et al. 2005).

These challenges suggest that building inclusive biobased value chains asks a lot of commercial partners compared to using fossil resources. Such actors may need to build alliances with NGOs and governments to shoulder the burden collectively. However, once inclusive biobased value chains are up and running, they can be expected to be resilient and viable. They should be able to withstand catastrophic global events such as a pandemic, because the actors involved are autonomous and can proceed even when they are disconnected from their partners in the value chain.

5 Control over Genetic Resources

5.1 An Ongoing International Conflict

Another pressing issue for the bioeconomy is the use of digital sequence information (DSI). The development of ever faster genetic sequence technologies in combination with DNA synthesis and gene editing techniques such as CRISPR-Cas has rapidly increased the use of DSI. Digital sequence information consists of DNA information that can be spread via digital channels. A researcher can download genetic information from anywhere in the world and use it to modify or construct any organism. The use of DSI can greatly enhance the search for sustainable sources of energy, materials and medicine.

The use of DSI brings forth questions on the status and economic value of genetic information. How relevant is the origin of genetic information for determining how to share its economic value? Who can have access to the information and who reaps the benefits of this information? This question in turn ties in with diverging views on nature and how we should treat natural resources. As such, the issue of how to deal with DSI opens up fundamental questions that should be answered in order to reach a shared understanding of a desirable and sustainable bioeconomy.

The Nagoya Protocol on Access and Benefit Sharing issued by the United Nations and ratified by 114 parties stipulates that for any genetic resource used either commercially or academically, its origin has to be documented, and it may only be used when the country of origin has given its explicit consent through a material transfer agreement (MTA). Such a MTA contains conditions about the access and benefit sharing relating to the specific genetic resource. The supporting principle is that genetic information is a resource that belongs to the country of origin and that other actors cannot take this resource without due compensation (Bagley 2016).

The issue currently being discussed within the UN Convention in biological diversity is whether DSI should also fall within the scope of the Nagoya Protocol. Developing countries say it should because they want to protect their genetic wealth and demand an equal share in the possible benefits deriving from that wealth. Most developed countries instead claim that DSI should not fall under the Nagoya Protocol because it would be practically impossible to determine the origin of the many bits of genetic digital resources available (Rabitz et al. 2020). Imposing an access and benefit requirement to the use of DSI would set very high barriers for scientific development, while the use of DSI offers a great potential for the bioeconomy.

5.2 The Relevance of the Origin of Genetic Information

Three contentious issues stand out here. First is the question whether the origin of genetic information holds any specific relevance to the actual genetic code. The perspective of many bioengineers is that genetic code is comparable to computer code and that a cell is a programmable entity (Calvert 2012). In this view, the origin of the genetic code is completely irrelevant; all that matters is its function (Roosth 2017). However, others may argue that genetic information is actually only relevant within a specific biological context. That is where it has an impact and that is also how it was shaped in the first place. Such a perspective likely aligns more with the holistic vision of nature and environment as often found with indigenous people (Right to food and nutrition Watch 2018). One's view on this matter is connected to the second question about the value of the origin of genetic information.

5.3 The Value of the Origin of Genetic Information

This second questions refers to how the value of the origin of genetic information can be determined and how it should be rewarded. In other words, what is a fair distribution of benefits in the context of DSI? Is a given party entitled to receive access to benefits simply because some genetic resource happened to originate in their backyard, also when there is no physical impact on their territory from extracting that resources? Does it matter that some actors have more means to make a profit from the genetic resources than others? Do they owe other actors something because of that? It may be fair to say that the knowledge of indigenous people should be rewarded if that helped others to identify valuable genetic resources, but does that extend to their simply living next to some genetic resources they had not previously recognized as valuable?

However many developing countries see themselves as stewards of genetic wealth. Without their stewardship, genetic resources wouldn't have been available to begin with. As such it would be unfair to allow the developed countries, who have the capacity to exploit this wealth, reap all the benefits, particularly because DSI could lead to patented applications which would not be freely accessible to researchers in developed countries. This would prevent actors with less available technological means to exploit genetic resources, while actors in developed countries could reap all the benefits. Access to genetic resources should therefore be safeguarded (Rabitz et al. 2020). Additionally the sharing of both economic and non-economic benefits, such as profits, joint ownership of intellectual property rights, or funds for conservation, can provide an incentive for developing countries to safeguard local biodiversity, although the effectiveness of this remains up for dispute (Rosendal 2006).

5.4 Fair Sharing of Benefits

The third question again revolves around fairness. If countries or communities of origin should be rewarded for genetic information, what would be a fair reward? It has been notoriously hard to share benefits for genetic resources with indigenous communities (Schroeder and Lucas 2013). The existing international treaties such as the CBD and the Nagoya Protocol leave it up to countries to decide amongst themselves what a fair sharing of benefits amounts to (Morgera 2017). The continuing international disagreement over this issue has led developing countries to walk out on negotiations on the future global governance of biodiversity (Rabitz et al. 2020). This disagreement jeopardizes the effort to establish a common ground for the sustainable and fair use of natural resources. For the bioeconomy to move forward and realize its full potential, the issue of what fair sharing of genetic resources implies needs a robust and widely supported answer.

6 Digitalization of Agriculture

Precision agriculture and related technologies known as digital or smart agriculture hold the promise of making agriculture more productive and sustainable (OECD 2019). Digital agriculture comprises technologies such as artificial intelligence, robots and Internet of things and involves huge data processing (Wolfert et al. 2017). Think for instance of sensors that link to a 'smart' tractors that can respond to real-time input. Some authors see a possible link between such real-life data and more sophisticated genetic modification of plants, i.e. the design of crops might be better adapted to specific local circumstances (Clapp and Ruder 2020). This set of technologies is expected to improve efficiency within food production systems as well as post-farm monitoring and hence improve food security, safety and sustainability (Wolfert et al. 2017) as well as improving animal health (Rotz et al. 2019).

Questions have been raised about the impact of such technologies on the existing agricultural system. Critics fear that such large integrated systems will come at the expense of the autonomy of farmers, who are reduced more and more to small elements in a big digital agricultural production machine (Clapp and Ruder 2020). Other concerns involve the alienation between farmers and their animals (Blok and Gremmen 2018) and an increased industrialization of agriculture at the expense of agroecological practices (Rotz et al. 2019).

These developments raise many questions for agriculture in general, such as a need for new modes of governance (Wolfert et al. 2017) and the effects on skills of farmers, economics, knowledge and innovation systems, privacy and power relations (Klerkx et al. 2019). I want to point out here that digitalization of agriculture raises specific questions for the bioeconomy.

On the one hand, it might prove to be a vital element in the exploitation of agricultural residues. A main barrier to effectively using such residues often lies in conflicting interpretations about how much residue can be taken off the land. Another barrier lies in ensuring the quality of the biomass on offer (Asveld et al. 2015). Agricultural waste typically shows a lot of divergence in quality. It might be rotten or it might contain stones. Digital farming technologies can greatly enhance both the reliability and the efficiency of managing waste streams as feedstock for biobased production platforms. Soil quality can be monitored. If needed the amount of waste left on the field can be adapted. Residues can be scanned for cleanliness and overall quality.

However promising such technologies might be, their actual contribution to a sustainable and resilient agriculture will depend on many factors, such as many other authors have already pointed out. They might help solve some of the barriers to an efficient bioeconomy, but they might also take away agency from the producers of biomass, and as argued above, such agency is indispensable for a resilient and sustainable bioeconomy (which does not necessarily equal an efficient bioeconomy).

7 Responsibilities

One last theme requires mentioning here, which is responsibility. Who is morally responsible for ensuring a resilient, sustainable and fair bioeconomy? In the case of building a sustainable, inclusive and resilient bioeconomy, the main type of responsibility is forward-looking responsibility, i.e. the responsibility to see to it that a certain end comes about (Van de Poel 2011). To assign a responsibility towards a specific end (such as a sustainable and inclusive bioeconomy) to an agent, implies that the specific agent has at least the capability to influence that end and that there is a causal relationship between the agent's actions and the envisioned ends (ibid.).

Governments and global governance bodies such as the UN have a large role to play as they provide the regulations and institutional backdrop against which the bioeconomy takes shape, as is also evidenced by the current debate on DSI and subsidies for bioenergy. They can influence global trading systems and local demand for biobased products. They should take this responsibility with a clear eye on public values such as sustainability, economic stability and health while at the same time taking into account the many uncertainties present in the developing bioeconomy.

They can do this by adopting a strategy of governance by experimentation, where they ensure learning trajectories for new biobased innovations that stimulate learning on institutional, moral and impact aspects (Asveld 2016). For instance, the variance in bilateral agreements around the use of material genetic resources can be seen as experiments of how a fair sharing of access and benefits can come about. From this variance, general lessons could be drawn to set the scene for experiments with agreements on the use of DSI.

A substantial responsibility also falls on the shoulders of biobased companies. Other than governments they do not carry the responsibility for the societal goal of achieving a sustainable, resilient and inclusive bioeconomy. However, through their innovation trajectories, they can exercise considerable influence on the composition of biobased products, extending to the design of associated feedstocks. It is a moral responsibility of companies to contribute societal goals, and they can do so through the choices they make in their innovation trajectories (Van de Poel et al. 2017).

Although the influence of companies is limited in some respects, as are their resources, there are still many instances in innovation trajectories that bring forth opportunities to respond to societal needs and concerns around biobased products, for instance in the choice of feedstock, partners or product portfolio (Sonck et al. 2020). Supporting broad societal goals can also be in the self-interest of companies, for instance when considering to invest in public goods in countries where they derive biomass from. If a company invests in local infrastructure or local educational facilities in a specific country, this can help build local capabilities for producing reliable and sustainable biomass. However, also companies themselves need specific capabilities to support an inclusive bioeconomy, such as the right tools and incentives to reflect on their own goals and stakeholder engagement (ibid.)

Additionally, producers of biomass also need to take on new responsibilities if the biobased economy is to be successful (Asveld et al. 2015). Farmers do not carry the moral responsibility to ensure the future of a sustainable biobased economy. However they can be held morally accountable for a sustainable farming system that stimulates biodiversity, if they have the capability to influence this system. Whether they indeed have this capability will differ per context and is also a much-debated issue politically. What I hope has become clear from the above overview that for a sustainable bioeconomy, farmers *should* have the capability to build a sustainable system for producing biomass for all kinds of purposes and that actors such as governments and companies have both a moral imperative and an interest in supporting this capability.

8 Conclusion

Inclusion is a central element in building a sustainable and resilient bioeconomy, especially in the light of catastrophic events such as a global pandemic. Inclusion can help to give agency to local stakeholders, which in turn will enable them to create sustainable, resilient and commercially viable value chains. Inclusion should also be a main consideration in the governance of genetic resources and in the employment of digital agriculture. This not only a moral imperative that may create economic justice, it is also instrumentally important. Without inclusion, the implementation of new technologies might create barriers to a sustainable bioeconomy by reducing agency of local participants in the bioeconomy.

Inclusion can only be meaningful when individuals have the capability to be included to begin with. If individuals do not understand what is asked of them, for instance, their answers are rather meaningless. Responsibility for inclusion implies that given actors such as governments invest in capabilities of relevant actors such as farmers to be included, such as proper education, infrastructure and platforms for participation. Such capabilities are essential to the success of a resilient, sustainable bioeconomy.

Literature

Asveld, L., Ganzevles, J., & Osseweijer, P. (2015). Trustworthiness and responsible research and innovation: the case of the bio-economy. *Journal of Agricultural and Environmental Ethics, 28*(3), 571–588.

Asveld, L. (2016). The need for governance by experimentation: The case of biofuels. *Science and engineering ethics, 22*(3), 815–830.

Asveld, L., & Stemerding, D. (2017). *Social learning in the bioeconomy: The Ecover case*. In Van de Poel, I., Asveld, L., & Mehos, D. C. (Eds.). New Perspectives on Technology in Society: Experimentation Beyond the Laboratory. London: Routledge.

Asveld, L. (2019). Towards including social sustainability in green and sustainable chemistry. *Current Opinion in Green and Sustainable Chemistry, 19*, 61–65.

Bagley, M. A. (2016). Digital DNA: The nagoya protocol, intellectual property treaties, and synthetic biology. *Intellectual Property Treaties, and Synthetic Biology (February 2016). Virginia Public Law and Legal Theory Research Paper*, (11).

Balkema, A., & Pols, A. (2015). Biofuels: sustainable innovation or gold rush? Identifying responsibilities for biofuel innovations. In *Responsible Innovation 2* (pp. 283–303). Cham: Springer.

Becker, C. D., & Ostrom, E. (1995). Human ecology and resource sustainability: the importance of institutional diversity. *Annual review of ecology and systematics, 26*(1), 113–133.

Berkes, F., Kofinas, G. P., & Chapin, F. S. (2009). Conservation, community, and livelihoods: sustaining, renewing, and adapting cultural connections to the land. In *Principles of ecosystem stewardship* (pp. 129–147). Springer, New York, NY.

Blok, V., & Lemmens, P. (2015). The emerging concept of responsible innovation. Three reasons why it is questionable and calls for a radical transformation of the concept of innovation. In *Responsible innovation 2* (pp. 19–35). Springer, Cham.

Blok, V., & Gremmen, B. (2018). Agricultural technologies as living machines: toward a biomimetic conceptualization of smart farming technologies. *Ethics, Policy & Environment, 21*(2), 246–263.

Brown, K., & Westaway, E. (2011). Agency, capacity, and resilience to environmental change: lessons from human development, well-being, and disasters. *Annual review of environment and resources, 36*, 321–342.

Bryden, J., Gezelius, S. S., Refsgaard, K., & Sutz, J. (2017). Inclusive innovation in the bioeconomy: concepts and directions for research.

Calvert, J. (2012). Ownership and sharing in synthetic biology: A 'diverse ecology' of the open and the proprietary?. *BioSocieties, 7*(2), 169–187.

Clapp, J., & Ruder, S. L. (2020). Precision technologies for agriculture: Digital farming, gene-edited crops, and the politics of sustainability. *Global Environmental Politics, 20*(3), 49–69.

De Hoop, E., Pols, A., & Romijn, H. (2016). Limits to responsible innovation. *Journal of Responsible Innovation, 3*(2), 110–134.

Devaux, A., Torero, M., Donovan, J., & Horton, D. E. (Eds.). (2016). *Innovation for inclusive value-chain development: successes and challenges*. Washington, D.C.: International Food Policy Research Institute.

Doorn, N. (2017). Resilience indicators: Opportunities for including distributive justice concerns in disaster management. *Journal of Risk Research, 20*(6), 711–731.

Flipse, S. (2014). Environmental Life Cycle Assessments as Decision Support Systems within Research and Development Processes: Solutions or Confusions for Responsible Innovation?. *International Journal of Business and Management, 9*(12), 210.

Folke, C., Jansson, Å., Rockström, J., Olsson, P., Carpenter, S. R., Chapin, F. S., ... & Elmqvist, T. (2011). Reconnecting to the biosphere. *Ambio, 40*(7), 719.

Frediani, A. A. (2010). Sen's Capability Approach as a framework to the practice of development. *Development in practice, 20*(2), 173–187.

Heeks, R., Foster, C., & Nugroho, Y. (2014). New models of inclusive innovation for development. *Innovation and Development, 4*(2), 175–185. https://doi.org/10.1080/2157930X.2014.928982

Hounkonnou, D., Kossou, D., Kuyper, T. W., Leeuwis, C., Nederlof, E. S., Röling, N., & Van Huis, A. (2012). An innovation systems approach to institutional change: smallholder development in WestAfrica. *Agricultural systems*, *108*, 74–83.

Kamali, F. P., Borges, J. A. R., Osseweijer, P., & Posada, J. A. (2018). Towards social sustainability: Screening potential social and governance issues for biojet fuel supply chains in Brazil. *Renewable and Sustainable Energy Reviews*, *92*, 50–61.

Klerkx, L., Jakku, E., & Labarthe, P. (2019). A review of social science on digital agriculture, smart farming and agriculture 4.0: New contributions and a future research agenda. *NJAS-Wageningen Journal of Life Sciences*, *90*, 100315.

Lundy, M.; Gottret, M.V.; Ashby, J. (2005) Learning alliances: An approach for building multistakeholder innovation systems. ILAC Brief 8 p. 4.

Matthews, N. E., Stamford, L., & Shapira, P. (2019). Aligning sustainability assessment with responsible research and innovation: Towards a framework for Constructive Sustainability Assessment. *Sustainable Production and Consumption*, *20*, 58–73.

Meckenstock, J., Barbosa-Póvoa, A. P., & Carvalho, A. (2016). The wicked character of sustainable supply chain management: evidence from sustainability reports. *Business Strategy and the Environment*, *25*(7), 449–477.

Morgera, E. (2017). Fair and Equitable Benefit-Sharing: History, Normative Content and Status in International Law. *Normative Content and Status in International Law (June 22, 2017).'Benefit-sharing' in E Orlando and L Krämer (eds), Encyclopedia of Environmental Law: Principles of Environmental Law (EE, 2017).*

Ostrom E. (2005) Doing Institutional Analysis Digging Deeper Than Markets and Hierarchies. In: Menard C., Shirley M.M. (eds) Handbook of New Institutional Economics. Springer, Boston, MA. https://doi.org/10.1007/0-387-25092-1_31

Parada, M. P., Osseweijer, P., & Duque, J. A. P. (2017). Sustainable biorefineries, an analysis of practices for incorporating sustainability in biorefinery design. *Industrial Crops and Products*, *106*, 105–123.

Pretty, J. N. (1995). Participatory learning for sustainable agriculture. *World development*, *23*(8), 1247–1263.

Postal, A. M., Kamali, F. P., Asveld, L., Osseweijer, P., & da Silveira, J. M. F. (2020). The impact of sugarcane expansion in Brazil: Local stakeholders' perceptions. *Journal of Rural Studies*, *73*, 147–162.

Quammen, D. (2012). *Spillover: animal infections and the next human pandemic*. WW Norton & Company.

Rabitz, F., Reynolds, J. L., & Tsioumani, E. (2020). Emerging Technologies in Biodiversity Governance: Gaps and Opportunities for Action. *Available at SSRN 3653021*.

Rammel, C., & van den Bergh, J. C. (2003). Evolutionary policies for sustainable development: adaptive flexibility and risk minimising. *Ecological economics*, *47*(2–3), 121–133.

Richardson, B. (2012). From a fossil-fuel to a biobased economy: The politics of industrial biotechnology. *Environment and Planning C: Government and Policy*, *30*(2), 282–296.

Robaey, Z., Asveld, L. & Sinha, K. (forthcoming) Inclusive biobased value chains: a means/goals approach to practice\

Roosth, S. (2017) *How life got made*. University of Chicago Press: Chicago.

Rosendal, G. K. (2006). Balancing access and benefit sharing and legal protection of innovations from bioprospecting: Impacts on conservation of biodiversity. *The journal of environment & development*, *15*(4), 428–447.

Rotz, S., Duncan, E., Small, M., Botschner, J., Dara, R., Mosby, I., ... & Fraser, E. D. (2019). The politics of digital agricultural technologies: a preliminary review. *Sociologia Ruralis*, *59*(2), 203–229.

Schroeder, D., & Lucas, J. C. (2013). Benefit Sharing: From Biodiversity to Human Genetics—An Introduction. In *Benefit Sharing* (pp. 1–7). Springer, Dordrecht.

Sen, A. (2001). *Development as freedom*. Oxford Paperbacks.

Simpson, N. P., & Basta, C. (2018). Sufficiently capable for effective participation in environmental impact assessment?. *Environmental impact assessment review*, *70*, 57–70.

Sonck, M., Asveld, L., Landeweerd, L., & Osseweijer, P. (2017). Creative tensions: mutual responsiveness adapted to private sector research and development. *Life Sciences, society and policy*, *13*(1), 14.

Sonck, M., Asveld, L., & Osseweijer, P. (2020). Meta-responsibility in corporate research and innovation: A bioeconomic case study. *Sustainability*, *12*(1), 38.

Stilgoe, J., Owen, R., & Macnaghten, P. (2013). Developing a framework for responsible innovation. *Research policy*, *42*(9), 1568–1580.

Straathof, A. J., Wahl, S. A., Benjamin, K. R., Takors, R., Wierckx, N., & Noorman, H. J. (2019). Grand research challenges for sustainable industrial biotechnology. *Trends in biotechnology*, *37*(10), 1042–1050.

Šūmane, S., Kunda, I., Knickel, K., Strauss, A., Tisenkopfs, T., des Ios Rios, I., ... & Ashkenazy, A. (2018). Local and farmers' knowledge matters! How integrating informal and formal knowledge enhances sustainable and resilient agriculture. *Journal of Rural Studies*, *59*, 232–241.

The Right to Food and Nutrition Watch (2018) *When food becomes immaterial, confronting the digital age*. Issue 10.

Van de Poel I. (2011) The Relation Between Forward-Looking and Backward-Looking Responsibility. In: Vincent N., van de Poel I., van den Hoven J. (eds) Moral Responsibility. Library of Ethics and Applied Philosophy, vol 27. Springer, Dordrecht. https://doi.org/10.1007/978-94-007-1878-4_3

Van de Poel, I., Asveld, L., Flipse, S., Klaassen, P., Scholten, V., & Yaghmaei, E. (2017). Company strategies for responsible research and innovation (RRI): A conceptual model. *Sustainability*, *9*(11), 2045.

Vivien, F. D., Nieddu, M., Befort, N., Debref, R., & Giampietro, M. (2019). The hijacking of the bioeconomy. *Ecological economics*, *159*, 189–197.

Walker, B., Holling, C. S., Carpenter, S. R., & Kinzig, A. (2004). Resilience, adaptability and transformability in social–ecological systems. *Ecology and society*, *9*(2).

Wals, A. E. (Ed.). (2007). *Social learning towards a sustainable world: Principles, perspectives, and praxis*. Wageningen Academic Publishers.

Wickson, F., & Carew, A. L. (2014). Quality criteria and indicators for responsible research and innovation: Learning from transdisciplinarity. *Journal of Responsible Innovation*, *1*(3), 254–273.

Wolfert, S., Ge, L., Verdouw, C., & Bogaardt, M. J. (2017). Big data in smart farming–a review. *Agricultural Systems*, *153*, 69–80.

WWF (2020) Living Planet Report 2020 - Bending the curve of biodiversity loss. Almond, R.E.A., Grooten M. and Petersen, T. (Eds). WWF, Gland, Switzerland.

Index

E. Koukios, A. Sacio-Szymańska (eds.), *Bio#Futures*,
https://doi.org/10.1007/978-3-030-64969-2

D

F

Q

R

S

'ufacturer's authorised representative in the EU is Springer
'stomer Service Centre GmbH, Europaplatz 3, 69115 Heidelberg,
vou have any concerns regarding our products, please
'tSafety@springernature.com

by CPI Group (UK) Ltd, Croydon, CR0 4YY
28/09/2026
02224061-0012